AF615998

NIBSC
-2 JUL 2008
LIBRARY

# Proteoglycan Protocols

# METHODS IN MOLECULAR BIOLOGY™

## *John M. Walker, Series Editor*

183. **Green Fluorescent Protein:** *Applications and Protocols,* edited by *Barry W. Hicks, 2002*
182. **In Vitro Mutagenesis Protocols, 2nd ed.**, edited by *Jeff Braman, 2002*
181. **Genomic Imprinting:** *Methods and Protocols,* edited by *Andrew Ward, 2002*
180. **Transgenesis Techniques, 2nd ed.:** *Principles and Protocols,* edited by *Alan R. Clarke, 2002*
179. **Gene Probes:** *Principles and Protocols,* edited by *Marilena Aquino de Muro and Ralph Rapley, 2002*
178. **Antibody Phage Display:** *Methods and Protocols,* edited by *Philippa M. O'Brien and Robert Aitken, 2001*
177. **Two-Hybrid Systems:** *Methods and Protocols,* edited by *Paul N. MacDonald, 2001*
176. **Steroid Receptor Methods:** *Protocols and Assays,* edited by *Benjamin A. Lieberman, 2001*
175. **Genomics Protocols**, edited by *Michael P. Starkey and Ramnath Elaswarapu, 2001*
174. **Epstein-Barr Virus Protocols,** edited by *Joanna B. Wilson and Gerhard H. W. May, 2001*
173. **Calcium-Binding Protein Protocols, Volume 2:** *Methods and Techniques,* edited by *Hans J. Vogel, 2001*
172. **Calcium-Binding Protein Protocols, Volume 1:** *Reviews and Case Histories,* edited by *Hans J. Vogel, 2001*
171. **Proteoglycan Protocols,** edited by *Renato V. Iozzo, 2001*
170. **DNA Arrays:** *Methods and Protocols,* edited by *Jang B. Rampal, 2001*
169. **Neurotrophin Protocols,** edited by *Robert A. Rush, 2001*
168. **Protein Structure, Stability, and Folding,** edited by *Kenneth P. Murphy, 2001*
167. **DNA Sequencing Protocols,** *Second Edition,* edited by *Colin A. Graham and Alison J. M. Hill, 2001*
166. **Immunotoxin Methods and Protocols**, edited by *Walter A. Hall, 2001*
165. **SV40 Protocols**, edited by *Leda Raptis, 2001*
164. **Kinesin Protocols,** edited by *Isabelle Vernos, 2001*
163. **Capillary Electrophoresis of Nucleic Acids, Volume 2:** *Practical Applications of Capillary Electrophoresis,* edited by *Keith R. Mitchelson and Jing Cheng, 2001*
162. **Capillary Electrophoresis of Nucleic Acids, Volume 1:** *Introduction to the Capillary Electrophoresis of Nucleic Acids,* edited by *Keith R. Mitchelson and Jing Cheng, 2001*
161. **Cytoskeleton Methods and Protocols**, edited by *Ray H. Gavin, 2001*
160. **Nuclease Methods and Protocols,** edited by *Catherine H. Schein, 2001*
159. **Amino Acid Analysis Protocols,** edited by *Catherine Cooper, Nicole Packer, and Keith Williams, 2001*
158. **Gene Knockoout Protocols,** edited by *Martin J. Tymms and Ismail Kola, 2001*
157. **Mycotoxin Protocols,** edited by *Mary W. Trucksess and Albert E. Pohland, 2001*
156. **Antigen Processing and Presentation Protocols,** edited by *Joyce C. Solheim, 2001*
155. **Adipose Tissue Protocols,** edited by *Gérard Ailhaud, 2000*
154. **Connexin Methods and Protocols,** edited by *Roberto Bruzzone and Christian Giaume, 2001*
153. **Neuropeptide Y Protocols**, edited by *Ambikaipakan Balasubramaniam, 2000*
152. **DNA Repair Protocols:** *Prokaryotic Systems*, edited by *Patrick Vaughan, 2000*
151. **Matrix Metalloproteinase Protocols,** edited by *Ian M. Clark, 2001*
150. **Complement Methods and Protocols,** edited by *B. Paul Morgan, 2000*
149. **The ELISA Guidebook**, edited by *John R. Crowther, 2000*
148. **DNA–Protein Interactions:** *Principles and Protocols* **(2nd ed.)**, edited by *Tom Moss, 2001*
147. **Affinity Chromatography:** *Methods and Protocols*, edited by *Pascal Bailon, George K. Ehrlich, Wen-Jian Fung, and Wolfgang Berthold, 2000*
146. **Mass Spectrometry of Proteins and Peptides,** edited by *John R. Chapman, 2000*
145. **Bacterial Toxins:** ***Methods and Protocols***, edited by *Otto Holst, 2000*
144. **Calpain Methods and Protocols**, edited by *John S. Elce, 2000*
143. **Protein Structure Prediction:** *Methods and Protocols*, edited by *David Webster, 2000*
142. **Transforming Growth Factor-Beta Protocols**, edited by *Philip H. Howe, 2000*
141. **Plant Hormone Protocols**, edited by *Gregory A. Tucker and Jeremy A. Roberts, 2000*
140. **Chaperonin Protocols**, edited by *Christine Schneider, 2000*
139. **Extracellular Matrix Protocols**, edited by *Charles Streuli and Michael Grant, 2000*
138. **Chemokine Protocols**, edited by *Amanda E. I. Proudfoot, Timothy N. C. Wells, and Christine Power, 2000*
137. **Developmental Biology Protocols, Volume III**, edited by *Rocky S. Tuan and Cecilia W. Lo, 2000*
136. **Developmental Biology Protocols, Volume II,** edited by *Rocky S. Tuan and Cecilia W. Lo, 2000*
135. **Developmental Biology Protocols, Volume I,** edited by *Rocky S. Tuan and Cecilia W. Lo, 2000*
134. **T Cell Protocols:** *Development and Activation*, edited by *Kelly P. Kearse, 2000*
133. **Gene Targeting Protocols,** edited by *Eric B. Kmiec, 2000*
132. **Bioinformatics Methods and Protocols,** edited by *Stephen Misener and Stephen A. Krawetz, 2000*
131. **Flavoprotein Protocols,** edited by *S. K. Chapman and G. A. Reid, 1999*
130. **Transcription Factor Protocols,** edited by *Martin J. Tymms, 2000*
129. **Integrin Protocols,** edited by *Anthony Howlett, 1999*
128. **NMDA Protocols**, edited by *Min Li, 1999*
127. **Molecular Methods in Developmental Biology**: Xenopus and Zebrafish, edited by Matthew Guille, 1999
126. **Adrenergic Receptor Protocols,** edited by *Curtis A. Machida, 2000*
125. **Glycoprotein Methods and Protocols:** *The Mucins*, edited by *Anthony P. Corfield, 2000*
124. **Protein Kinase Protocols,** edited by *Alastair D. Reith, 2001*
123. ***In Situ* Hybridization Protocols (2nd ed.),** edited by *Ian A. Darby, 2000*
122. **Confocal Microscopy Methods and Protocols**, edited by *Stephen W. Paddock, 1999*

METHODS IN MOLECULAR BIOLOGY™

# Proteoglycan Protocols

Edited by

**Renato V. Iozzo, MD**

*Department of Pathology, Anatomy, and Cell Biology*
*Thomas Jefferson University*
*Philadelphia, PA*

**Humana Press** **Totowa, New Jersey**

999 Riverview Drive, Suite 208
Totowa, New Jersey 07512



This publication is printed on acid-free paper. ∞

ANSI Z39.48-1984 (American Standards Institute) Permanence of Paper for Printed Library Materials.

Cover design by Patricia F. Cleary.

Cover illustration: Space-filling model of the small leucine-rich proteoglycan decorin. The sugars in the background are repeating units of glucuronic acid and N-acetylglucosamine (Artwork kindly provided by Charles C. Reed).

For additional copies, pricing for bulk purchases, and/or information about other Humana titles, contact Humana at the above address or at any of the following numbers: Tel: 973-256-1699; Fax: 973-256-8341; E-mail: humana@humanapr.com, or visit our Website at www.humanapress.com



Printed in the United States of America. 10 9 8 7 6 5 4 3 2 1

Library of Congress Cataloging-in-Publication Data

Proteoglycan protocols / edited by Renato V. Iozzo.
p. cm. -- (Methods in molecular biology ; 171)
Includes bibliographical references and index.
ISBN 0-89603-759-2 (alk. paper)
1. Proteoglycans--Laboratory manuals. I. Iozzo, Renato V., 1950- II. Methods in molecular biology (Totowa, N.J.) ; v. 171.
QP552.P73 P757 2001
572'.68--dc21

00-061317

# Preface

Proteoglycans are some of the most elaborate macromolecules of mammalian and lower organisms. The covalent attachment of at least five types of glycosaminoglycan side chains to more than forty individual protein cores makes these molecules quite complex and endows them with a multitude of biological functions. *Proteoglycan Protocols* offers a comprehensive and up-to-date collection of preparative and analytical methods for the in-depth analysis of proteoglycans. Featuring step-by-step detailed protocols, this book will enable both novice and experienced researchers to isolate intact proteoglycans from tissues and cultured cells, to establish the composition of their carbohydrate moieties, to generate strategies for prokaryotic and eukaryotic expression, to utilize methods for the suppression of specific proteoglycan gene expression and for the detection of mutant cells and degradation products, and to study specific interactions between proteoglycans and extracellular matrix proteins as well as growth factors and their receptors. The readers will find concise, yet comprehensive techniques carefully drafted by leading experts in the field.

Each chapter commences with a general Introduction, followed by a detailed Materials section, and an easy-to-follow Methods section. An asset of each chapter is the extensive notation that includes troubleshooting tips and practical considerations that are often lacking in formal methodology papers. The reader will find this section most valuable because it is clearly provided by experienced scientists who have first-hand knowledge of the techniques they outline. In addition, most of the chapters are well illustrated with examples of typical data generated with each method. I have made a special effort to ensure a detailed, step-by-step rendition of the methods that would help in reproducing each experimental protocol.

*Proteoglycan Protocols* is divided into three sections. Part I focuses on the general protocols dedicated to the isolation and purification of proteoglycans from tissues and cells of vertebrates and invertebrates. In nineteen chapters, the most commonly used protocols are reviewed by expert scientists. The reader will be able to isolate proteoglycans from specialized tissues, cell culture, body fluids, analyze the glycosaminoglycan side chains and the protein core, and investigate intracellular biosynthetic events using both qualitative and quantitative approaches.

Part II focuses on the expression, detection, and degradation of proteoglycans. Twenty chapters cover most of the current knowledge regarding these issues of proteoglycan biology. The reader will find protocols describing various approaches in recombinant gene expression systems, together with theoretical and practical gene targeting approaches, using both antisense technology and somatic cell targeting. A variety of unique approaches will enable the reader to inhibit glycosaminoglycan synthesis, to identify cell mutants in proteoglycan biosynthesis, and to study the degradation of various glycosaminoglycans by using chemical and enzymatic methods. In addition, strategies are described for the identifications in adult human tissues and fluids of proteoglycan degradation products.

Part III comprises twelve chapters focused on understanding the complex interactions between proteoglycans—either the protein or the glycosaminoglycan moiety—and various extracellular matrix proteins, growth factors, and receptors. This is a blooming area that will undoubtedly expand in the near future as we enter the era of proteomics. The reader will find protocols devoted to the theoretical and practical understanding of specific proteoglycan interactions using affinity-based approaches, including affinity coelectrophoresis, affinity chromatography, and optical biosensor and phage display technologies. In addition, proteoglycan interactions with receptor tyrosine kinases and lipoproteins are also covered.

*Proteoglycan Protocols* has been developed through the efforts of ninety-seven scientists representing ten countries and three continents. I wish to thank all of the authors for their contributions and their outstanding job in summarizing complex protocols and for providing extensive notations. The latter, I believe, will make a crucial difference in the success rate of each experimental strategy.

I would like to thank the Series Editor, John Walker, and the production team at Humana Press for their contributions to the successful realization of this multiauthor book. I am particularly indebted to my assistant Kit Foster for her dedication and continuous effort in this endeavor. She played a key role in the assembly, editing, formatting, general organization, and index preparation of this volume.

From my personal vantage point, I learned immensely during the process of assembling and editing the chapters. I am thankful for that and amply rewarded.

***Renato V. Iozzo, MD***

# Contents

Preface ........ v

Contributors ........ xi

I Isolation and Purification

1 Isolation of Proteoglycans from Cell Cultures and Tissues
***Masaki Yanagishita*** ........ *3*

2 Isolation of Proteoglycans from Tendon
***Kathryn G. Vogel and Julie A. Peters*** ........ *9*

3 Purification of Proteoglycans from Mineralized Tissues
***Neal S. Fedarko*** ........ *19*

4 Purification of Perlecan from Endothelial Cells
***John Whitelock*** ........ *27*

5 Isolation and Characterization of Nervous Tissue Proteoglycans
***Yu Yamaguchi*** ........ *35*

6 Analysis of Proteoglycans and Glycosaminoglycans from *Drosophila*
***William D. Staatz, Hidenao Toyoda, Akiko Kinoshita-Toyoda, Kimberlly Chhor, and Scott B. Selleck*** ........ *41*

7 Cartilage and Smooth Muscle Cell Proteoglycans Detected by Affinity Blotting Using Biotinylated Hyaluronan
***James Melrose*** ........ *53*

8 Preparation of Proteoglycans for N-Terminal and Internal Amino Acid Sequence Analysis
***Peter J. Neame*** ........ *67*

9 High–Specific-Activity $^{35}$S-Labeled Heparan Sulfate Prepared from Cultured Cells
***Nicholas W. Shworak*** ........ *79*

10 Selective Detection of Sulfotransferase Isoforms by the Ligand Affinity-Conversion Approach
***Nicholas W. Shworak*** ........ *91*

11 Particulate and Soluble Glycosaminoglycan-Synthesizing Enzymes: *Preparation Assay and Use*
***Geetha Sugumaran and Jeremiah E. Silbert*** ........ *103*

12 Disaccharide Composition of Hyaluronan and Chondroitin/Dermatan Sulfate: *Analysis with Fluorophore-Assisted Carbohydrate Electrophoresis*
**Anna H. K. Plaas, L. West, Ronald J. Midura, and Vincent C. Hascall** ..... **117**

13 Integral Glycan Sequencing of Heparan Sulfate and Heparin Saccharides
**Jeremy E. Turnbull** ..... **129**

14 Analytical and Preparative Strong Anion-Exchange HPLC of Heparan Sulfate and Heparin Saccharides
**Jeremy E. Turnbull** ..... **141**

15 Proteoglycans Analyzed by Composite Gel Electrophoresis and Immunoblotting
**George R. Dodge and Ralph Heimer** ..... **149**

16 Quantitation of Proteoglycans in Biological Fluids Using Alcian Blue
**Madeleine Karlsson and Sven Björnsson** ..... **159**

17 Cellulose Acetate Electrophoresis of Glycosaminoglycans
**Yanusz Wegrowski and Francois-Xavier Maquart** ..... **175**

18 Disaccharide Composition in Glycosaminoglycans/Proteoglycans Analyzed by Capillary Zone Electrophoresis
**Nikos K. Karamanos and Anders Hjerpe** ..... **181**

19 Intact and Oligomeric Glycosaminoglycans Analyzed with Capillary Electrophoresis
**Nikos K. Karamanos and Anders Hjerpe** ..... **193**

II Expression, Detection, and Degradation

20 Recombinant Expression of Proteoglycans in Mammalian Cells: *Utility and Advantages of the Vaccinia Virus/T7 Bacteriophage Hybrid Expression System*
**David J. McQuillan, Neung-Seon Seo, Anne M. Hocking, and Camille I. McQuillan** ..... **201**

21 Purification of Recombinant Human Decorin and Its Subdomains
**Anna L. McBain and David M. Mann** ..... **221**

22 Prokaryotic Expression of Proteoglycans
**Alan D. Murdoch and Renato V. Iozzo** ..... **231**

23 Proteoglycan Gene Targeting in Somatic Cells
**Giancarlo Ghiselli and Renato V. Iozzo** ..... **239**

24 Constitutive and Inducible Antisense Expression
***Melissa Handler and Renato V. Iozzo*** ........................................ ***251***

25 Cell-Mediated Transfer of Proteoglycan Genes
***Jens W. Fischer, Michael G. Kinsella, David Hasenstab, Alexander W. Clowes, and Thomas N. Wight*** ........................ ***261***

26 Morphological Evaluation of Proteoglycans in Cells and Tissues
***Stephanie L. Lara, Stephen P. Evanko, and Thomas N. Wight*** ... ***271***

27 Expression and Characterization of Engineered Proteoglycans
***Wei Luo, Chunxia Guo, Jing Zheng, and Marvin L. Tanzer*** ......... ***291***

28 Intracellular Localization of Engineered Proteoglycans
***Tung-Ling L. Chen, Peiyin Wang, Seung S. Gwon, Nina W. Flay, and Barbara M. Vertel*** ........................................ ***301***

29 Selection of Glycosaminoglycan-Deficient Mutants
***Xiaomei Bai, Brett Crawford, and Jeffrey D. Esko*** ...................... ***309***

30 Xyloside Priming of Glycosaminoglycan Biosynthesis and Inhibition of Proteoglycan Assembly
***Timothy A. Fritz and Jeffrey D. Esko*** ........................................ ***317***

31 Inhibition of Heparan Sulfate Synthesis by Chlorate
***H. Edward Conrad*** ........................................................................ ***325***

32 Detection of Proteoglycan Core Proteins with Glycosaminoglycan Lyases and Antibodies
***John R. Couchman and Pairath Tapanadechopone*** .................. ***329***

33 Proteoglycan Core Proteins and Catabolic Fragments Present in Tissues and Fluids
***John D. Sandy*** .............................................................................. ***335***

34 Degradation of Heparan Sulfate by Nitrous Acid
***H. Edward Conrad*** ........................................................................ ***347***

35 Degradation of Heparan Sulfate with Heparin Lyases
***Laurie A. LeBrun and Robert J. Linhardt*** .................................. ***353***

36 Degradation of Chondroitin Sulfate and Dermatan Sulfate with Chondroitin Lyases
***María José Hernáiz and Robert J. Linhardt*** ................................ ***363***

37 In Vitro Assays for Hyaluronan Synthase
***Andrew P. Spicer*** .......................................................................... ***373***

38 Hyaluronidase Activity and Hyaluronidase Inhibitors: *Assay Using a Microtiter-Based System*
***Susan Stair Nawy, Antonei B. Csóka, Kazuhiro Mio, and Robert Stern*** .......................................................................... ***383***

39 Detecting Hyaluronidase and Hyaluronidase Inhibitors: *Hyaluronan-Substrate Gel and -Inverse Substrate Gel Techniques*
***Kazuhiro Mio, Antonei B. Csóka, Susan Stair Nawy, and Robert Stern* ..... 391**

III INTERACTIONS

40 Affinity Coelectrophoresis of Proteoglycan–Protein Complexes
***James D. San Antonio and Arthur D. Lander* ..... 401**

41 Binding Constant Measurements for Inhibitors of Growth Factor Binding to Heparan Sulfate Proteoglycans
***Kimberly E. Forsten and Matthew A. Nugent* ..... 415**

42 Interaction of Proteoglycans with Receptor Tyrosine Kinases
***David K. Moscatello and Renato V. Iozzo* ..... 427**

43 Measuring Single-Cell Cytosolic $Ca^{2+}$ Concentration in Response to Proteoglycans
***Sandip Patel and Renato V. Iozzo* ..... 435**

44 Serum Amyloid A Peptide Interactions with Glycosaminoglycans: *Evaluation by Affinity Chromatography*
***John B. Ancsin and Robert Kisilevsky* ..... 449**

45 Interactions of Lipoproteins with Proteoglycans
***Kevin Jon Williams* ..... 457**

46 Hyaluronan and Hyaluronan-Binding Proteins: *Probes for Specific Detection*
***Bryan P. Toole, Qin Yu, and Charles B. Underhill* ..... 479**

47 Novel Confocal-FRAP Analysis of Carbohydrate–Protein Interactions Within the Extracellular Matrix
***Philip Gribbon, Boon C. Heng, and Timothy E. Hardingham* ..... 487**

48 Regulatory Roles of Syndecans in Cell Adhesion and Invasion
***J. Kevin Langford and Ralph D. Sanderson* ..... 495**

49 Optical Biosensor Techniques to Analyze Protein–Polysaccharide Interactions
***David G. Fernig* ..... 505**

50 Phage Display Technology to Obtain Antiheparan Sulfate Antibodies
***T. H. van Kuppevelt, G. J. Jenniskens, J. H. Veerkamp, G. B. ten Dam, and M. A. B. A. Dennissen* ..... 519**

51 Tissue-Specific Binding by FGF and FGF Receptors to Endogenous Heparan Sulfates
***Andreas Friedl, Mark Filla, and Alan C. Rapraeger* ..... 535**

Index ..... ***547***

# Contributors

JOHN B. ANCSIN • *Department of Pathology, Queen's University; the Syl and Molly Apps Research Center, Kingston General Hospital, Kingston, Ontario, Canada*
XIAOMEI BAI • *Department of Cellular and Molecular Medicine, Glycobiology Research and Training Center, University of California, San Diego, La Jolla, CA*
SVEN BJÖRNSSON • *Department of Clinical Chemistry, Central Hospital, Kristianstad, Sweden*
TUNG-LING L. CHEN • *Department of Cell Biology and Anatomy, Finch University of Health Sciences/The Chicago Medical School, North Chicago, IL*
KIMBERLLY CHHOR • *Department of Molecular and Cellular Biology, The University of Arizona, Tucson, AZ*
ALEXANDER W. CLOWES • *Department of Pathology, University of Washington School of Medicine, Seattle, WA*
H. EDWARD CONRAD • *Department of Biochemistry, University of Illinois, Urbana, IL*
JOHN R. COUCHMAN • *Department of Cell Biology and The Cell Adhesion and Matrix Research Center, University of Alabama at Birmingham, Birmingham, AL*
BRETT CRAWFORD • *Department of Cellular and Molecular Medicine, Glycobiology Research and Training Center, University of California, San Diego, La Jolla, CA*
ANTONEI B. CSÓKA • *Department of Pathology, School of Medicine, University of California at San Francisco, San Francisco, CA*
M. A. B. A. DENNISSEN • *Department of Biochemistry, Faculty of Medical Sciences, University of Nijmegen, Nijmegen, The Netherlands*
GEORGE R. DODGE • *Bone and Cartilage Research Laboratory, Department of Research, A. I. duPont Hospital for Children, Wilmington, DE*
JEFFREY D. ESKO • *Department of Cellular and Molecular Medicine, Glycobiology Research and Training Center, University of California, San Diego, La Jolla, CA*

STEPHEN P. EVANKO • *Department of Pathology, University of Washington School of Medicine, Seattle, WA*
NEAL S. FEDARKO • *Division of Geriatrics, Department of Medicine, Johns Hopkins University, Baltimore, MD*
DAVID G. FERNIG • *School of Biological Sciences, University of Liverpool, Liverpool, UK*
MARK FILLA • *Department of Pathology and Laboratory Medicine, University of Wisconsin-Madison, Madison, WI*
JENS W. FISCHER • *Department of Pathology, University of Washington School of Medicine, Seattle, WA*
NINA W. FLAY • *Department of Cell Biology and Anatomy, Finch University of Health Sciences/The Chicago Medical School, North Chicago, IL*
KIMBERLY E. FORSTEN • *Department of Chemical Engineering, Virginia Polytechnic Institute and State University, Blacksburg, VA*
ANDREAS FRIEDL • *Department of Pathology and Laboratory Medicine, University of Wisconsin-Madison, Madison, WI*
TIMOTHY A. FRITZ • *Department of Biochemistry and Biophysics, University of California, San Francisco, CA*
GIANCARLO GHISELLI • *Department of Pathology, Anatomy, and Cell Biology, Thomas Jefferson University, Philadelphia, PA*
PHILIP GRIBBON • *Wellcome Trust Center for Cell Matrix Research, School of Biological Sciences, University of Manchester, Manchester, UK*
CHUNXIA GUO • *Department of Biostructure and Function, University of Connecticut Health Center, Farmington, CT*
SEUNG S. GWON • *Department of Cell Biology and Anatomy, Finch University of Health Sciences/The Chicago Medical School, North Chicago, IL*
MELISSA HANDLER • *Center for Neurologic Diseases, Harvard Medical School, Boston, MA*
TIMOTHY E. HARDINGHAM • *Wellcome Trust Center for Cell Matrix Research, School of Biological Sciences, University of Manchester, Manchester, UK*
VINCENT C. HASCALL • *Department of Biomedical Engineering, The Lerner Research Institute, The Cleveland Clinic Foundation, Cleveland, OH*
DAVID HASENSTAB • *Department of Pathology, University of Washington School of Medicine, Seattle, WA*
RALPH HEIMER • *Department of Biochemistry and Molecular Pharmacology, Thomas Jefferson University, Philadelphia, PA*
BOON C. HENG • *Wellcome Trust Centre for Cell Matrix Research, School of Biological Sciences, University of Manchester, Manchester, UK*

MARÍA JOSÉ HERNÁIZ • *Division of Medicinal and Natural Products Chemistry, Department of Chemistry; Department of Chemical and Biochemical Engineering, University of Iowa, Iowa City, IA*

ANDERS HJERPE • *Department of Immunology, Microbiology, Pathology, and Infectious Diseases, Division of Pathology, Karolinska Institute, Huddinge University Hospital, Huddinge, Sweden*

ANNE M. HOCKING • *Center for Extracellular Matrix Biology, Institute of Biosciences and Technology, Texas A & M University System Health Science Center, Houston, TX*

RENATO V. IOZZO • *Departments of Pathology, Anatomy, and Cell Biology, Thomas Jefferson University, Philadelphia, PA*

G. J. JENNISKENS • *Department of Biochemistry, Faculty of Medical Sciences, University of Nijmegen, Nijmegen, The Netherlands*

NIKOS K. KARAMANOS • *Department of Chemistry, University of Patras, Patras, Greece*

MADELEINE KARLSSON • *Department of Engineering and Natural Science, Växjö University, Växjö, Sweden*

AKIKO KINOSHITA-TOYODA • *Department of Moleular and Cellular Biology, The University of Arizona, Tucson, Arizona*

MICHAEL G. KINSELLA • *Department of Pathology, University of Washington School of Medicine, Seattle, WA*

ROBERT KISILEVSKY • *Department of Pathology, Queen's University; the Syl and Molly Apps Research Center, Kingston General Hospital, Kingston, Ontario, Canada*

ARTHUR D. LANDER • *Department of Developmental and Cell Biology, University of California, Irvine, Irvine, CA,*

J. KEVIN LANGFORD • *Arkansas Cancer Research Center, Department of Pathology, University of Arkansas for Medical Sciences, Little Rock, AR*

STEPHANIE L. LARA • *Department of Pathology, University of Washington School of Medicine, Seattle, WA*

LAURIE A. LEBRUN • *Division of Medicinal and Natural Products Chemistry, Department of Chemistry; Department of Chemical and Biochemical Engineering, University of Iowa, Iowa City, IA*

ROBERT J. LINHARDT • *Division of Medicinal and Natural Products Chemistry, Department of Chemistry; Department of Chemical and Biochemical Engineering, University of Iowa, Iowa City, IA*

WEI LUO • *Department of Biostructure and Function, University of Connecticut Health Center, Farmington, CT*

DAVID M. MANN • *Clearant, Inc., Rockville, MD*
FRANCOIS-XAVIER MAQUART • *Laboratory of Biochemistry, CNRS, Medical Faculty, University of Reims, France*
ANNA L. MCBAIN • *Clearant, Inc., Rockville, MD*
CAMILLE I. MCQUILLAN • *Center for Extracellular Matrix Biology, Institute of Biosciences and Technology, Texas A & M University System Health Science Center, Houston, TX*
DAVID J. MCQUILLAN • *LifeCell Corporation, Branchburg, NJ; Center for Extracellular Matrix Biology, Institute of Biosciences and Technology, Texas A & M University System Health Science Center, Houston, TX*
JAMES MELROSE • *Institute of Bone and Joint Research, Royal North Shore Hospital, St. Leonards, Sydney, Australia*
RONALD J. MIDURA • *Department of Biomedical Engineering, The Lerner Research Institute, The Cleveland Clinic Foundation, Cleveland, OH*
KAZUHIRO MIO • *Department of Pathology, School of Medicine, University of California at San Francisco, San Francisco, CA; Lion Corporation, Kanagawa, Japan*
DAVID K. MOSCATELLO • *Coriell Institute for Medical Research, Camden, NJ*
ALAN D. MURDOCH • *Wellcome Trust Centre for Cell Matrix Research, Biochemistry Division, School of Biological Sciences, University of Manchester, Manchester, UK*
SUSAN STAIR NAWY • *Department of Pathology, School of Medicine, University of California at San Francisco, San Francisco, CA*
PETER J. NEAME • *Department of Biochemistry and Molecular Biology, University of South Florida, College of Medicine, Shriners Hospital for Children, Tampa, FL*
MATTHEW A. NUGENT • *Department of Biochemistry, Boston University School of Medicine, Boston, MA*
SANDIP PATEL • *Department of Pharmacology, University of Oxford, Oxford, UK*
JULIE A. PETERS • *Department of Biology, The University of New Mexico, Albuquerque, NM*
ANNA H. K. PLAAS • *Center for Research in Skeletal Development and Pediatric Orthopedics, Shriners Hospital for Children, Tampa, FL*
ALAN C. RAPRAEGER • *Department of Pathology and Laboratory Medicine, University of Wisconsin-Madison, Madison, WI*
JAMES D. SAN ANTONIO • *Department of Medicine, Thomas Jefferson University, Philadelphia, PA*
RALPH D. SANDERSON • *Arkansas Cancer Research Center, Department of Pathology, University of Arkansas for Medical Sciences, Little Rock, AR*

JOHN D. SANDY • *Section Chief in Biochemistry, Shriners Hospital for Children, Tampa, FL*

SCOTT B. SELLECK • *Department of Molecular and Cellular Biology and Arizona Cancer Center, University of Arizona, Tucson, AZ*

NEUNG-SEON SEO • *Center for Extracellular Matrix Biology, Institute of Biosciences and Technology, Texas A & M University System Health Science Center, Houston, TX*

NICHOLAS W. SHWORAK • *Department of Medicine, Harvard Medical School; Angiogenesis Research Center, Beth Israel Deaconess Medical Center, Boston, MA*

JEREMIAH E. SILBERT • *Department of Medicine, Harvard Medical School; Division of Rheumatology/Immunology/Allergy, Brigham and Women's Hospital; Edith Nourse Rogers Memorial Veterans Administration Medical Center, Bedford, MA*

ANDREW P. SPICER • *Rowe Program in Genetics, Department of Biological Chemistry, School of Medicine, University of California, Davis, Davis, CA*

WILLIAM D. STAATZ • *Department of Molecular and Cellular Biology and Arizona Cancer Center, University of Arizona, Tucson, AZ*

ROBERT STERN • *Department of Pathology, School of Medicine, University of California at San Francisco, San Francisco, CA*

GEETHA SUGUMARAN • *Department of Medicine, Harvard Medical School; Division of Rheumatology/Immunology/Allergy, Brigham and Women's Hospital; Edith Nourse Rogers Memorial Veterans Administration Medical Center, Bedford, MA*

MARVIN L. TANZER • *Department of Biostructure and Function, University of Connecticut Health Center, Farmington, CT*

PAIRATH TAPANADECHOPONE • *Department of Cell Biology and The Cell Adhesion and Matrix Research Center, University of Alabama at Birmingham, Birmingham, AL*

G. B. TEN DAM • *Department of Biochemistry, Faculty of Medical Sciences, University of Nijmegen, Nijmegen, The Netherlands*

BRYAN P. TOOLE • *Department of Anatomy and Cellular Biology, Tufts University School of Medicine, Boston, MA*

HIDENAO TOYODA • *Department of Molecular and Cellular Biology, The University School of Arizona, Tucson, AZ*

JEREMY E. TURNBULL • *School of Biosciences, University of Birmingham, Edgbaston, Birmingham, UK*

CHARLES B. UNDERHILL • *Lombardi Cancer Center, Georgetown University Medical Center, Washington, DC*

T. H. van Kuppevelt • *Department of Biochemistry, Faculty of Medical Sciences, University of Nijmegen, Nijmegen, The Netherlands*

J. H. Veerkamp • *Department of Biochemistry, Faculty of Medical Sciences, University of Nijmegen, Nijmegen, The Netherlands*

Barbara M. Vertel • *Department of Cell Biology and Anatomy, Finch University of Health Sciences/Chicago Medical School, North Chicago, IL*

Kathryn G. Vogel • *Department of Biology, The University of New Mexico, Albuquerque, NM*

Peiyin Wang • *Department of Cell Biology and Anatomy, Finch University of Health Sciences/The Chicago Medical School, North Chicago, IL*

Yanusz Wegrowski • *Laboratory of Biochemistry, CNRS, Medical Faculty, University of Reims, France*

L. West • *Center for Research in Skeletal Development and Pediatric Orthopedics, Shriners Hospital for Children, Research Labs, Tampa, FL*

John Whitelock • *CSIRO, Molecular Science, North Ryde, Sydney, NSW, Australia*

Thomas N. Wight • *Department of Pathology, University of Washington, Seattle, WA*

Kevin Jon Williams • *The Dorrance H. Hamilton Research Laboratories, Division of Endocrinology, Diabetes, and Metabolic Diseases, Department of Medicine, Jefferson Medical College of Thomas Jefferson University, Philadelphia, PA*

Yu Yamaguchi • *Neurobiology Program, The Burnham Institute, La Jolla, CA*

Masaki Yanagishita • *Department of Hard Tissue Engineering, Graduate School, School of Dentistry, Tokyo Medical and Dental University, Tokyo, Japan*

Qin Yu • *Molecular Pathology Unit, Massachusetts General Hospital East, Charlestown, MA*

Jing Zheng • *Department of Biostructure and Function, University of Connecticut Health Center, Farmington, CT*

# I

# Isolation and Purification

# 1

# Isolation of Proteoglycans from Cell Cultures and Tissues

**Masaki Yanagishita**

## 1. Introduction

Proteoglycans are a class of glycosylated proteins characterized by the presence of glycosaminoglycans as a carbohydrate component, which endow them with unique biological as well as biochemical properties. Therefore, isolation of proteoglycans from various biological sources such as cell cultures and tissues could be achieved by ordinary molecular purification procedures utilizing their general molecular properties and by those taking advantages of the presence of glycosaminoglycan moiety. This chapter focuses on the latter experimental procedures, which are particularly useful for obtaining total proteoglycan and glycosaminoglycan species from various biological sources. These protocols could be followed by purification procedures specific to individual proteoglycan species (i.e., by using antibodies to core proteins, or binding proteins to specific sequences of glycosaminoglycans) to select specific molecules.

Two major classes of well-established purification procedures aiming at the presence of glycosaminoglycans in the molecule have been used extensively. They include (1) density gradient ultracentrifugation, and (2) anion-exchange chromatography. The former procedure makes use of the fact that the glycosaminoglycan moiety of proteoglycans has high specific gravity, so, proteoglycans with a large number of glycosaminoglycans have high molecular densities (often as high as those of nucleic acids). Therefore, such a procedure is particularly suited for the purification of proteoglycans with many glycosaminoglycan chains (e.g., aggrecan, the major proteoglycan in cartilage tissues). On the other hand, the procedure is not suitable for isolating proteoglycans with only few glycosaminoglycans and high protein contents (e.g., "small, leucine-rich proteoglycans" and cell surface heparan sulfate proteoglycans). Purification procedures belonging to the latter class take advantage of high negative charges contributed by sulfate groups and carboxyl groups universally present in glycosaminoglycans. Thus, they can be widely used

From: *Methods in Molecular Biology, Vol. 171: Proteoglycan Protocols*
Edited by: R. V. Iozzo © Humana Press Inc., Totowa, NJ

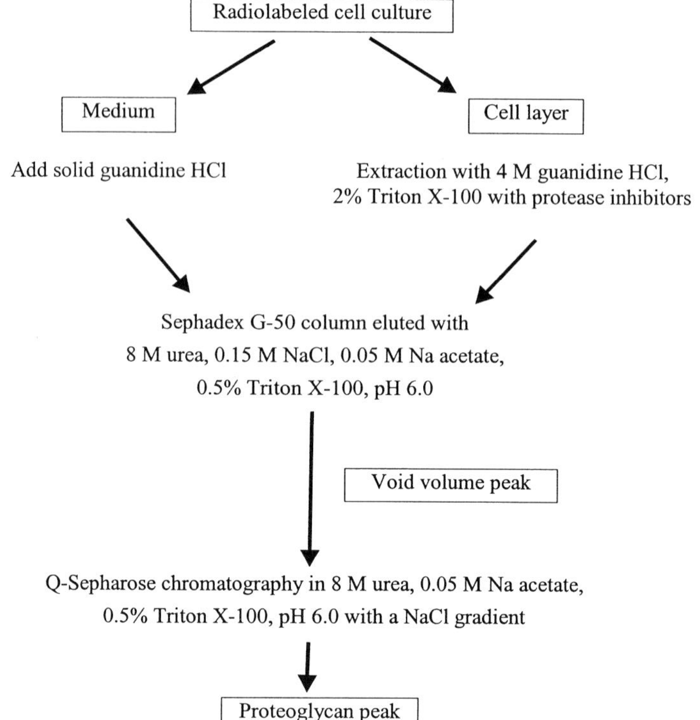


Fig. 1. A flow diagram of proteoglycan isolation from a radiolabeled cell culture.

for isolating any species of proteoglycans and glycosaminoglycans, and are the main subject of following discussion.

This chapter outlines a proteoglycan-isolating protocol from a metabolically radiolabeled cell culture (*see* **Fig. 1**) as an example case, which addresses technical problems often encountered when working with small quantities of proteoglycans. Other reviews on similar subjects can be consulted for more detailed discussions ***(1–3)***.

## 2. Materials

### 2.1. Extraction

1. Buffer: 4 *M* guanidine HCl, 0.05 *M* Na acetate, pH 6.0, containing 2% (w/v) Triton X-100 and protease inhibitors (1 m*M* phenylmethylsulfonyl fluoride, 10 m*M* N-ethylmaleimide, 10 m*M* disodium ethylenediaminetetraacetic acid ***(4)***.
2. Stock solutions (100 × concentrated) of phenylmethylsulfonyl fluoride and N-ethylmaleimide are prepared in ethanol and added to the guanidine HCl buffer just prior to use (they are unstable in aqueous solutions).

### 2.2. Solvent Exchange

1. Sephadex G-50, fine (Amersham Pharmacia Biotech). Preswelling of Sephadex can be done in hot water (off the heater), which achieves sterilization, degassing, and shortening of swelling time. Extreme caution should be exercised when adding Sephadex powder to

boiling water, to avoid flushing. A convenient concentration of gel (50% slurry) can be made by mixing 5 g of Sephadex G-50 with 100 mL of water. Bacteriostatic agents (e.g., 0.02% Na azide) should be added for long-term storage.
2. 10 mL plastic disposable pipet (Falcon).
3. Glass wool, #3950 (Corning).
4. Buffer: 8 *M* urea, 0.20 *M* NaCl, 0.05 *M* Na acetate, 0.5% Triton X-100, pH 6.0.

### 2.3. Anion-Exchange Chromatography

1. Q-Sepharose, fast flow (Amersham Pharmacia Biotech). Q-Sepharose has to be preequilibrated with the low salt buffer as described below.
2. Low salt buffer: 8 *M* urea, 0.20 *M* NaCl, 0.05 *M* Na acetate, 0.5% Triton X-100, pH 6.0.
3. High salt buffer: 8 *M* urea, 1.5 *M* NaCl, 0.05 *M* Na acetate, 0.5% Triton X-100, pH 6.0.
4. Gradient former (a simple configuration can be made with two beakers).
5. Peristaltic pump.

### 2.4. Gel Filtration Chromatography

1. Superose 6, HR 10/30 (Amersham Pharmacia Biotech).
2. Buffer: 4 *M* guanidine HCl, 0.05 *M* Na acetate, 0.5% (w/v) Triton X-100, pH 6.0.

## 3. Methods

General experimental procedures introducing radioactive precursors into proteoglycans using cell cultures and tissue cultures are reviewed elsewhere *(1)*.

### 3.1. Extraction

1. After removing media from the cell layer, approximately 2 mL of extraction buffer (per 35-mm-diameter cell culture dish) is added to a culture plate (*see* **Note 1**).
2. Proteoglycans are extracted within 2–3 h of constant shaking at 4°C.
3. Passing the extract through a 1-mL pipet tip up and down approximately 10 times reduces the viscosity of the solution caused by DNA.
4. Extraction of proteoglycans from tissues: As used originally in the extraction of proteoglycan from cartilage tissue *(3)*, 4 *M* guanidine generally provides excellent solubilization of proteoglycans.
5. When tissues are to be extracted, ordinarily approximately 10 times volume of 4 *M* guanidine HCl buffer successfully solubilizes proteoglycans from finely minced tissues in 12 h at 4°C.
6. Additional consideration should be made when extraction of cell-associated proteoglycans is attempted; i.e., inclusion of sufficient amounts of detergent may be necessary (such as 2% Triton X-100) for the solubilization of proteoglycans *(5)*.
7. Secreted proteoglycans in cell culture media are generally already soluble, but, in order to minimize interactions between highly charged proteoglycans and other molecules, direct addition of solid guanidine HCl (0.53 g of solid guanidine HCl per milliliter of media makes 4 *M* guanidine HCl solution) is frequently used.

### 3.2. Solvent Exchange

1. In order to prepare the extracted proteoglycans in 4 *M* guanidine HCl for the anion-exchange chromatography procedure in the next step, guanidine HCl has to be replaced with a solvent compatible with the procedure.
2. A preferred solvent is a urea buffer, since it disrupts molecular interactions by interfering with the formation of hydrogen bonds.
3. A convenient buffer-exchange procedure can be done by gel filtration (such as Sephadex G-50 chromatography) using a small disposable pipet. This process is also very conve-

nient to remove unincorporated radioactive precursors, when radiolabeled cell cultures were extracted with a 4 *M* guanidine HCl solvent.

4. Preparation of Sephadex G-50 column: Pour preswollen Sephadex G-50 (50% slurry) into a 10-mL plastic disposable column (which has been cut at the top with a file and plugged with glass wool at the bottom) to make 8 mL bed volume.
5. Remove excess water and equilibrate the column with 8 *M* urea buffer (a total of 9 mL is sufficient to equilibrate the column).
6. Carefully prepare a flat gel surface with a glass Pasteur pipet and remove excess urea buffer. Apply a 2-mL sample and discard the eluent.
7. After the entire sample is in the column, carefully overlay 3 mL of buffer and collect eluent until the entire buffer is in the column (3-mL fraction collected). This fraction contains proteoglycans and other macromolecules in 8 *M* urea buffer, while leaving small molecules in the original sample (guanidine HCl, isotope precursors, etc.) behind in the column.
8. At this point, the column can be safely disposed of as a radioactive waste.

## 3.3. Anion-Exchange Chromatography

### 3.3.1. Preparation of Q-Sepharose Column (see **Note 2**) and Sample Application

1. 2 mL of preequilibrated Q-Sepharose (1 mL of Q-Sepharose can bind up to 3–5 mg of proteoglycans) is packed into a small column (10-mL plastic pipet cut by a file and plugged with glass wool at the bottom).
2. Alternatively, 2 mL of Q-Sepharose is mixed with the sample in 8 *M* urea buffer and gently shaken for 1 h, then packed into the column; this latter method gives uniform binding of proteoglycans to Q-Sepharose, resulting in a better flow property, especially when a large quantity of materials is used.
3. After sample application, the column is washed with 10 mL of the low-salt buffer.
4. Then the column is connected to a gradient former and eluted with a total of approximately 40 mL of buffer with a flow rate of 10–15 mL/h.
5. Every 1- to 2-mL fraction is collected and monitored for NaCl concentration by conductivity measurement (*see* **Fig. 2**). Eluent fractions are monitored for proteoglycans (by radioactivity detection or colorimetric procedures; a convenient and sensitive colorimetric procedure using Safranin O is described in *ref.* ***6***).
6. Typically, heparan sulfate proteoglycans are eluted in a peak at approximately 0.5 *M* NaCl and chondroitin sulfate proteoglycans at 0.65 *M* NaCl.

## 3.4. Gel Filtration Chromatography

1. Proteoglycans with differing molecular weights purified by Q-Sepharose chromatography are further separated by gel filtration chromatography. Superose 6 (or equivalent gel) is suitable for resolving small proteoglycans ("small, leucine-rich proteoglycans", cell surface heparan sulfate proteoglycans, etc.) from large proteoglycans (e.g., aggrecan, versican, and perlecan) and partially degraded proteoglycans.
2. Superose 6 is equilibrated in 4 *M* guanidine HCl buffer containing 0.5% Triton X-100 (*see* **note 2**), and up to 0.5 mL of sample can be loaded to the column.
3. The column is eluted at a flow rate of 0.4 mL/min and each 1-min fraction is collected (*see* **Fig. 3**).
4. Eluent fractions are analyzed for proteoglycan content (by radioactivity detection or colorimetric procedures).

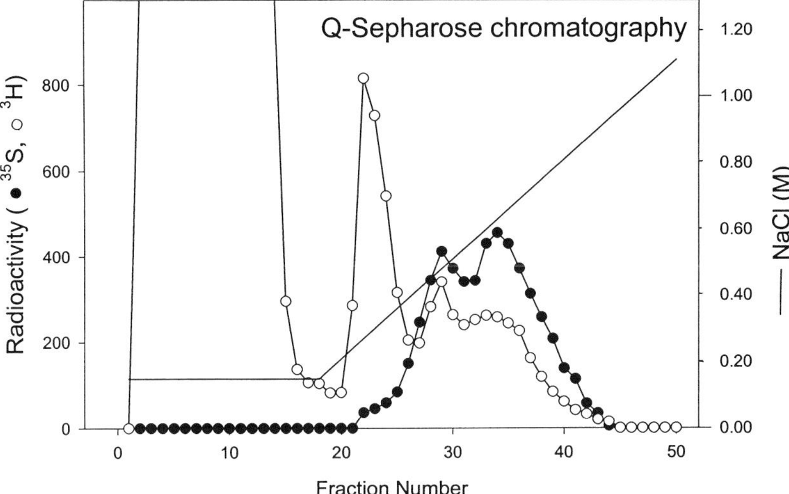


Fig. 2. Q-Sepharose anion-exchange chromatography. A sample from a cell culture that was metabolically radiolabeled with [$^{3}$H]leucine and [$^{35}$S]sulfate was analyzed. A large peak containing $^{3}$H-labeled proteins was efficiently separated from $^{35}$S-labeled proteoglycan peaks eluting later in high-salt fractions.

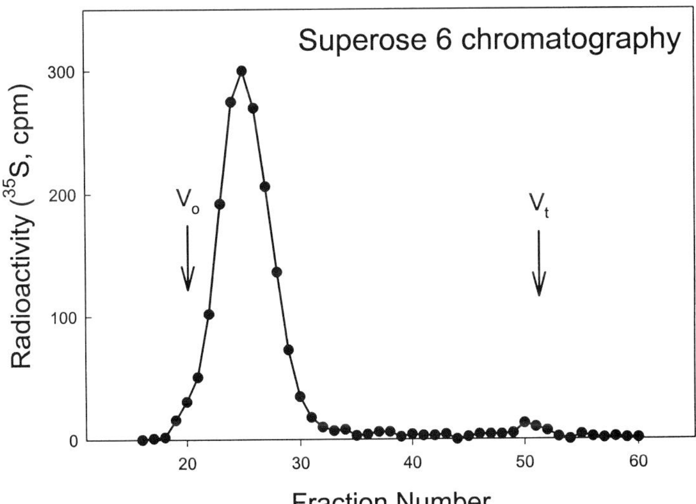


Fig. 3. Superose 6 chromatography of $^{35}$S-labeled cell surface heparan sulfate proteoglycan.

## 4. Notes

1. Extraction of cell-associated proteoglycans from cultured cell requires the use of detergent *(5)*. Usefulness of 4 *M* guanidine HCl in the presence of 2% Triton X-100 has been rigorously demonstrated.
2. One of the major technical problems associated with anion-exchange chromatography of proteoglycans, especially when purifying molecules present in small quantities (e.g.,

isolation of proteoglycans from cell cultures), is poor recovery of materials. This can be, in most cases, overcome by the use of detergents (either nonionic or zwitterionic). Routinely, the use 0.5% (w/v) Triton X-100 (or NP-40) dramatically improves recovery of proteoglycans (even glycosaminoglycans) from ion-exchange columns. Most nonionic detergents (such as Triton X-100, NP-40) possess strong absorbance in the ultraviolet (UV) range, thus making UV tracing for protein detection difficult. If this causes problems in analyses, non-UV-absorbing, nonionic detergents such as Genapol X-100™ (Calbiochem) can be used. Also, when the removal of detergents is required in later experimental steps, the use of those with high critical micellar concentrations (such as CHAPS™, Calbiochem) in place of Triton X-100 is beneficial.

## References

1. Hascall, V. C., Calabro, A., Midura, R. J., and Yanagishita, M. (1994) Isolation and characterization of proteoglycans. *Meth. Enzymol.* **230,** 390–417.
2. Yanagishita, M., Midura, R. J., and Hascall, V. C. (1987) Proteoglycans: Isolation and purification from tissue cultures. *Meth. Enzymol.* **138,** 279–289.
3. Hascall, V. C. and Kimura, J. H. (1982) Proteoglycans: Isolation and characterization. *Meth. Enzymol.* **82,** 769–800.
4. Oike, Y., Kimata, K., Shinomura, T., Nakazawa, K. and Suzuki, S. (1980) Structural analysis of chick-embryo cartilage proteoglycan by selective degradation with chondroitin lyases (chondroitinases) and endo-beta-D-galactosidase (keratanase). *Biochem. J.* **191,** 193–207.
5. Yanagishita, M. and Hascall, V. C. (1984) Proteoglycans synthesized by rat ovarian granulose cells in culture; isolation, fractionation, and characterization of proteoglycans associated with the cell layer. *J. Biol. Chem.* **259,** 10,260–10,269.
6. Lammi, M. and Tammi, M. (1988) Densitometric assay of nanogram quantities of proteoglycans precipitated on nitrocellulose membrane with Safranin O. *Anal. Biochem.* **168,** 352–357.

# 2

# Isolation of Proteoglycans from Tendon

**Kathryn G. Vogel and Julie A. Peters**

## 1. Introduction

Proteoglycans make up less than 1% of the dry weight of a dense connective tissue such as tendon *(1)*. Most of this proteoglycan is a small molecule called decorin. Because decorin has only one glycosaminoglycan chain, it cannot be separated from other proteins by the CsCl density-gradient centrifugation method that was originally used to purify aggrecan from cartilage.

The basic approach described in this chapter is to extract proteoglycans from the tissue with 4 *M* guanidine, a solution that will denature collagen and disrupt most kinds of noncovalent molecular interactions. The cross-linked collagen of adult tendon remains insoluble during this extraction, allowing the proteoglycans and other soluble proteins to be separated from the bulk of the tissue by centrifugation. Proteoglycans are then separated from other extracted proteins by ion-exchange chromatography, taking advantage of the anionic nature of the glycosaminoglycan chain.

This method has been used to quantitate and isolate proteoglycans from the tensile (proximal) and compressed (distal) regions of bovine deep flexor tendon. These mechanically distinct regions of flexor tendon are characterized by differences in proteoglycan amount and type *(2)*. The method is equally applicable to isolation of proteoglycans from human tendon or from other dense connective tissues *(3)*. Once isolated, the large proteoglycans can be separated from smaller ones by sieve chromatography. These isolated proteoglycans and their unique core proteins and glycosaminoglycan chains are of sufficient purity to then be examined by specific analytical techniques or used in functional assays.

## 2. Materials

### 2.1. Quantitation of Glycosaminoglycans

1. Single-edge razor blades.
2. Heated water bath.

From: *Methods in Molecular Biology, Vol. 171: Proteoglycan Protocols*
Edited by: R. V. Iozzo © Humana Press Inc., Totowa, NJ

3. Microcentrifuge.
4. Lyophilizer.
5. Papain digestion solution: 0.05 *M* sodium acetate, 5 m*M* L-cysteine HCl, 10 m*M* EDTA, pH 5.5. Add 50 μg/mL papain just before starting digestion.
6. DEAE cellulose (Whatman, DE52).
7. Disposable columns (Bio-Rad Poly Prep, 0.8 × 4 cm).
8. Column elution solutions: 0.02 *M* HCl, 1 *M* HCl.
9. Uronic acid detection reagent (orcinol reagent): 2 g of orcinol (5-methylresorcinol; 3,5-dihydroxytoluene monohydrate) in 50 mL of 1.5% $FeCl_3$, 850 mL of conc. HCl, and 150 mL of water ***(4)***. Orcinol is sensitive to light and is irritating to eyes, the respiratory system, and skin. Dissolve orcinol in $FeCl_3$ solution, then add HCl and water. Use caution and work in a fume hood when adding water to concentrated HCl.
10. Uronic acid standards: stock solution of 100 μg/mL of glucuronolactone diluted to standards of 5 μg/mL, 10 μg/mL, 20 μg/mL, and 30 μg/mL.
11. Pyrex test tubes.
12. Heat block.
13. Spectrophotometer.

## 2.2. Isolation of Proteoglycans

1. Single-edge razor blades.
2. Tissue-grinding device (we have used a food mill made by Arthur Thomas, Incorporated, Philadelphia, PA).
3. Lyophilizer.
4. 16-mL centrifuge tubes with sealing screw cap (Nalge Nunc).
5. 4 *M* guanidine buffer: 4 *M* guanidine, 0.05 *M* sodium acetate, pH 6.0. Add protease inhibitors just before starting extraction: 5 m*M* benzamindine, 5 m*M* phenylmethylsulfonyl fluoride, and 1 m*M* N-ethylmaleimide.
6. Nutator (Clay-Adams) or other rocking device.
7. Centrifuge (Beckman J2, JA-17 rotor).
8. 7 *M* urea buffer: 7 *M* urea, 0.05 *M* sodium acetate, pH 6.0.
9. 7 *M* urea buffer plus protease inhibitors: shortly before use add 5 m*M* benzamindine, 5 m*M* phenylmethylsulfonyl fluoride, 1 m*M* N-ethylmaleimide.
10. Dialysis tubing (MWCO: 12,000–14,000).
11. DEAE cellulose (Whatman, DE52).
12. Disposable columns, Bio-Rad Poly-Prep, 0.8 × 4 cm.
13. Column elution solutions: 7 *M* urea buffer + 0.1 *M* NaCl; 7 *M* urea buffer + 0.2 *M* NaCl; 7 *M* urea buffer + 0.8 *M* NaCl.

## 2.3. SDS Polyacrylamide Gel Electrophoresis of Proteoglycans

1. 2-mL screw-capped centrifuge tubes.
2. 95% ethanol.
3. Microcentrifuge.
4. Enzyme digestion buffer: 0.1 *M* Tris base, brought to pH 8.0 with acetic acid.
5. Chondroitinase ABC (Seikagaku, 5 U/vial) resuspended in 0.5 mL of 0.1 *M* Tris-HCl, pH 8.0 mL (10 U/mL).
6. Gel sample buffer: for 100-mL, combine 50 mL stacking gel buffer, 5 g sodium dodecyl sulfate, 20 g glycerol, 4 mg bromophenol blue, and water to volume.
7. Electrophoresis apparatus and power source.

**Table 1**
**Water Content of Adult Flexor Tendons**

| Tissue | 1 | 2 | 3 | |
|---|---|---|---|---|
| Tens. | 65.6 | 61.5 | 62.7 | |
| | 65.2 | 62.1 | 63.4 | $\bar{X}$ = 63.7 + 1.7 % |
| | 66.6 | 62.1 | 63.7 | |
| Comp. | 70.2 | 70.6 | 73.1 | |
| | 70.8 | 67.3 | 71.1 | $\bar{X}$ = 71.0 + 1.7 % |
| | 70.9 | 70.9 | 73.7 | |

Samples of fresh tendon from animals between 1 and 2 years of age were cut into small pieces with a razor blade, weighed, lyophilized, and weighed again to determine the percent of tendon wet weight that was water. Triplicate samples of about 130 mg each were assessed. *Tens.*, tissue from the tensile/proximal region of tendon; *Comp.*, tissue from the compressed/distal region of tendon. Samples are from the front (1) and rear (2) leg of one animal and the rear (3) leg of another. Note that the water content is lower in the tensile tendon.

8. Polyacrylamide: 30% acrylamide, 0.8% bis-acrylamide.
9. Separating gel buffer: 0.75 *M* Tris, 4 m*M* EDTA, pH 8.8.
10. Stacking gel buffer: 0.25 *M* Tris, 4 m*M* EDTA, pH 6.8.
11. 0.5% ammonium persulfate.
12. TEMED.
13. Electrode buffer stock: 0.1 *M* Tris, 0.768 *M* glycine, 8 m*M* EDTA, pH 8.7. After adjusting pH (if necessary), add 0.4% SDS. The upper buffer is diluted 1/4; the lower buffer is diluted 1/8.
14. Gel fixative: 50% methanol, 7% acetic acid, 43% $dH_2O$.
15. Coomassie blue staining solution: 0.1% Coomassie Brilliant Blue R-250, 40% methanol, 10% acetic acid, 50% $dH_2O$.
16. Gel destain: 25% methanol, 7% acetic acid, 68% $dH_2O$.
17. Alcian blue staining solution: 0.5% Alcian blue (8GX, Sigma Chemical Co), 7% acetic acid, 93% $dH_2O$.
18. Alcian blue destain solution: 7% acetic acid, 93% $dH_2O$.

## 3. Methods

### *3.1. Quantitation of Glycosaminoglycans*

1. Dissect tendon tissue free of muscle and fat, cut into chunks about 3 mm on a side, weigh, lyophilize, weigh again (*see* **Note 1** and bovine tendon water content, **Table 1**).
2. Add papain digestion solution to tissue at 50 mg dry tissue/1 mL enzyme in 2-mL screw cap tubes. Incubate at 65°C until tissue pieces have disappeared (6–18 h, *see* **Note 2**).
3. Centrifuge and remove the supernatant. Dilute supernatant with equal volume of $dH_2O$ to assure low salt concentration.
4. Pour columns containing 0.3 mL DE-52 in water (*see* **Note 3**).
5. Apply 1 mL of diluted sample to DEAE cellulose in column; collect the flow-through (*see* **Note 4**).
6. Rinse column with 20 mL of 0.02 *M* HCl to remove loosely adherent molecules. Elute the glycosaminoglycans into a clean tube with 2 mL of 1 *M* HCl.

**Table 2**
**Uronic Acid Content of Adult Flexor Tendons (μg uronic acid/mg dry weight)**

| Tissue | 1 | 2 | 3 | |
|---|---|---|---|---|
| Tens. | 3.1 | 2.6 | 2.1 | |
| | 2.8 | 2.2 | 2.3 | $\bar{X} = 2.6 \pm 0.4$ |
| | 3.3 | 2.2 | 2.4 | |
| Comp. | 7.1 | 7.1 | 6.0 | |
| | 5.8 | 6.8 | 6.3 | $\bar{X} = 6.4 \pm 0.8$ |
| | 5.2 | 7.9 | 5.7 | |

Samples of dry tendon were digested with papain, passed over a small column of DEAE cellulose, and the isolated glycosaminoglycans assessed for uronic acid content. Note that tissue from the compressed region of tendon contains a higher amount of uronic acid.

7. Determine the amount of GAG uronic acid spectrophotometrically using the orcinol reagent and uronic acid standards *(4)*. Combine 0.5 mL of glycosaminoglycan sample or standard with 1.5 mL of orcinol reagent in Pyrex test tube, place in heat block at 100°C for 20 min, cool to room temperature, read OD at 670 nm within 30 min, and calculate micrograms of uronic acid per milligram of tissue dry weight (*see* **Notes 5–7**, and tendon uronic acid content, **Table 2**)

## 3.2. Isolation of Proteoglycans

1. Chop the tissue into chunks ~3 mm on a side and freeze immediately with liquid nitrogen (*see* **Note 8**).
2. Powder the frozen chunks of tissue (*see* **Note 9**).
3. Extract tissue in 4 *M* guanidine buffer + protease inhibitors for 24 h at 4°C with rocking. A good ratio of tissue wet weight to extraction fluid is about 1/20 (*see* **Note 10**). The efficiency of proteoglycan extraction from powdered tendon was higher than extraction from tissue chunks (*see* **Tables 3** and **4**).
4. Centrifuge in Beckman JA-17 fixed-angle rotor at 3440*g* (5000 rpm) for 10 min; carefully remove the supernatant.
5. Repeat the extraction with fresh buffer, centrifuge, and combine supernatants (*see* **Note 11**).
6. Dialyze the extract into 7 *M* urea buffer + 0.1 *M* NaCl (*see* **Notes 12** and **13**).
7. Apply 0.5 mL of dialyzed extract to a 1-mL column of DEAE cellulose equilibrated in 7 *M* urea buffer (*see* **Note 14**).
8. Rinse the column with 8 mL of 7 *M* urea buffer + 0.1 *M* NaCl followed by 3 mL of 7 *M* urea buffer + 0.2 *M* NaCl.
9. Elute proteoglycans with 1.5 mL of 7 *M* urea buffer + 0.8 *M* NaCl (*see* **Note 15**).

## 3.3. SDS Polyacrylamide Gel Electrophoresis of Proteoglycans

1. Precipitate the proteoglycans in ethanol to prepare samples for gel electrophoresis. For example, mix 50 μL of sample from the ion-exchange column with 400 μL of ethanol in a small centrifuge tube. Let stand at –20°C for several hours or in –70°C freezer for at least 1h.

**Table 3**
**Extraction of Proteoglycan from Adult Flexor Tendons**

| | Tensile | | Compressed | |
|---|---|---|---|---|
| | Chunks | Powder | Chunks | Powder |
| 7 M Urea | | | | |
| | 1. 47 | 41 | 63 | 131 |
| | 2. 17 | 35 | 28 | 41 |
| Total | 64 | 76 | 91 | 172 |
| 7 M Urea + *0.1 M NaCl* | | | | |
| | 1. 141 | 195 | 285 | 624 |
| | 2. 79 | 110 | 131 | 122 |
| Total | 220 | 305 | 416 | 746 |
| 4 M Guanidine | | | | |
| | 1. 137 | 176[a] | 748 | 1335 |
| | 2. 83 | 330 | 239 | 112 |
| Total | 220 | 506 | 987 | 1447 |

[a]Due to tissue swelling, 15 mL of solution was added.

Chunks or powdered tendon was extracted with 7 *M* urea, 7 *M* urea + 0.1 *M* NaCl, or 4 *M* guanidine. In each case the extraction started with 200 mg dry weight of tissue and 8 mL of extraction solution. After 24 h the extracts were centrifuged, the supernatants removed, and an additional 8 mL of extraction solution added to the pellet for another 24 h. The amount of uronic acid in the supernatant of each extraction is shown, in micrograms. 1, first extraction; 2, second extraction of same tissue.

2. Spin samples in the microcentrifuge in cold for 10 min at 10,000 rpm, remove supernatant by suction, rinse pellet and tube with 1 mL cold ethanol, let stand at –20°C for at least 1h, and spin again. Dry pellet with brief lyophilization (*see* **Note 16**).
3. Solubilize pellet in 25 µL of gel sample buffer and heat to 100°C for 5 min in a heat block. If the samples are to be reduced, add dithiothreitol or β-mercaptoethanol to the gel sample buffer before heating.
4. The core protein of proteoglycans having chondroitin sulfate or dermatan sulfate glycosaminoglycan chains can be seen after removal of the glycosaminoglycans by digestion with chondroitinase ABC (*see* **Note 17**). Resuspend the dry proteoglycan pellet in 25 µL of digestion buffer, add 1 µL (0.01 unit) of enzyme, and incubate at 37°C for 1 h. Add 25 µL of 2X gel sample buffer and heat to 100°C for 5 min in the heat block.
5. Pour 4–16% gradient gel, 1.5 mm thick (*5, see* **Notes 18** and **19**). For one gel, make solutions containing 4% and 16% acrylamide. For 4%: 6.75 mL separating buffer, 1.8 mL acrylamide + bis, 4.14 mL water, 135 µL 10% SDS, 13.5 µL TEMED, and 675 µL 0.5% ammonium persulfate. For *16%:* 3.38 µL 2X separating buffer, 7.2 mL acrylamide + bis, 2.12 mL water, 135 µL 10% SDS, 13.5 µL TEMED, and 675 µL 0.5% ammonium persulfate. The 16% solution is stirred while pumping it between glass plates, as it is diluted by the 4% solution.
6. Pour 4% stacking gel using a 15-lane comb.

**Table 4**
**Efficiency of Proteoglycan Extraction[a]**

| | Tensile | Compressed |
|---|---|---|
| 7 *M* urea | 16% | 12% |
| 7 *M* urea + 0.1 *M* NaCl | 65% | 51% |
| 4 *M* Guanidine | 108% | 100% |

[a] Uronic acid in the papain digest of each tissue was set to equal 100% tens. = 467 μg, comp. = 1454 μg.

The amount of uronic acid solublized by two sequential 24-h extractions in each extraction solution (**Table 3**) is compared to the amount of uronic acid measured after papain digestion of tissue from the same animal (sample 2, **Table 2**). Note that adding 0.1 *M* NaCl to 7 *M* urea greatly increased the amount of proteoglycan that was solublized from the tissue. Virtually all proteoglycan was solublized by 4 *M* guanidine.

7. Load 10– 50 μL of sample into each lane, as appropriate. Put molecular weight standards in one lane. If samples were digested with chondroitinase ABC, it is useful to prepare one sample containing only 1 μL of enzyme and digestion buffer.
8. Run electrophoresis at 8 mA for approximately 18 h (*see* **Note 20**). Cooling to 16°C may help uniformity but is not necessary. If desired, carry out Western blot procedures immediately after electrophoresis.
9. Put the gel into gel fixative solution for 1 h.
10. Cover the gel with Coomassie blue staining solution for 1h with gentle rocking. Pour off the stain. Cover the gel with several changes of destain solution until the gel background is clear.
11. To stain for proteoglycans, cover the gel in Alcian blue staining solution for 1 h. Pour off the stain and destain with 7% acetic acid until gel background is clear (*see* **Fig. 1** and **Notes 21** and **22**).

## 4. Notes

1. Many plastic tubes will lose weight during lyophilization. It is best to remove dry tissue from the tube to obtain dry weight.
2. Tissue digestion can be encouraged by shaking the tube and by adding additional papain. The digest may appear cloudy.
3. Precycle DE52 through washes in 0.5 *N* HCl, $dH_2O$, 0.5 *N* NaOH, and then rinse in $dH_2O$ until filtrate is near pH 7. Used resin can be recovered by the same treatment and used again.
4. The DE52 resin is kind; the columns will not run dry even when no fluid remains above the resin. Columns can be placed in a tube to collect eluent and simply moved to the next tube for the next elution fluid.
5. Place a glass marble on each tube while in the heat block to diminish evaporation of acid. Cool tubes quickly and uniformly by transferring the tubes to a rack and placing the rack in a pan of water.
6. Dilute the sample with $dH_20$ as needed (1/2 or 1/5) to obtain readings that fall within the range of standards. Authentic chondroitin sulfate was recovered from the columns with efficiency >95%. Multiply by 3.2 to convert amount of uronic acid to chondroitin sulfate.
7. Orcinol reagent is stable for 6 wk when kept in the refrigerator in a dark glass container.

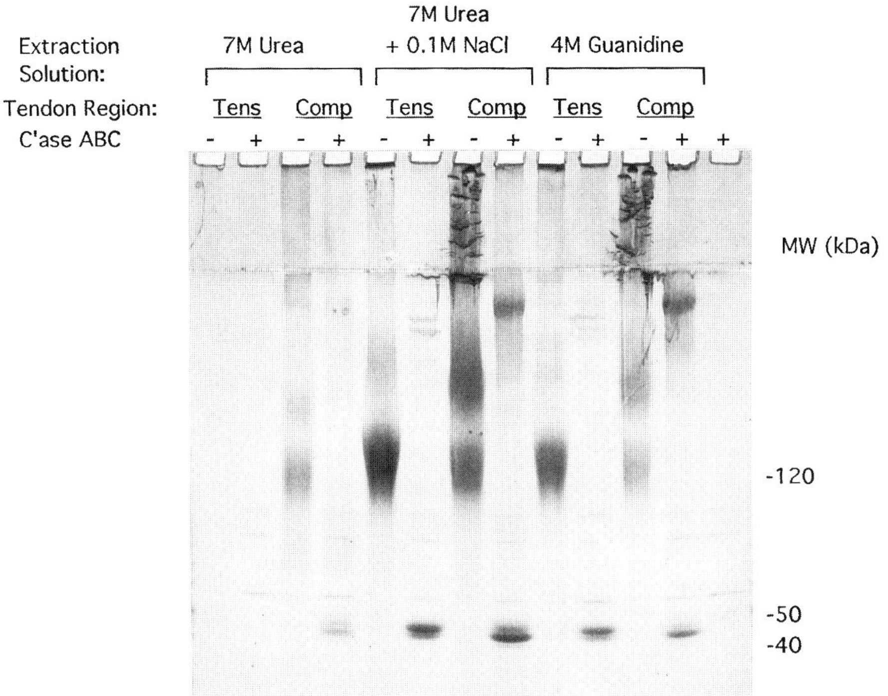


Fig. 1. SDS/Polyacrylamide gel electrophoresis of proteoglycans. Proteoglycans were extracted, isolated by ion-exchange chromatography, precipitated with ethanol, incubated without or with added chondroitinase ABC, and loaded onto a 4–16% gradient gel. After electrophoresis the gel was stained with Coomassie blue and Alcian blue. An approximately equal amount of uronic acid (12.5 μg) from samples extracted with 7 *M* urea + 0.1 *M* NaCl and 4 *M* guanidine was loaded onto each lane; it was not possible to precipitate this much material from the 7 *M* urea extracts. *Tens,* from tensile region of tendon; *Comp,* from compressed region of tendon. Note that a proteoglycan migrating just above 120 kDa (decorin) is the predominant molecule in tensile samples. A proteoglycan that enters the running gel but migrates more slowly (biglycan) is present only in the compressed samples. The core proteins of these small proteoglycans migrate close together at about 45 kDa. Large proteoglycan (aggrecan) is present in the stacking gel of compressed samples but does not form a discrete band. The high-molecular-weight band appearing after chondroitinase ABC digestion of samples from compressed tissue is the aggrecan core protein with some keratan sulfate chains.

8. To freeze tissue chunks for storage, put a few at a time into a half-liter plastic container, add a small amount of liquid nitrogen, cover loosely, and shake the container. This makes it possible subsequently to remove individual chunks of frozen tissue. Be sure to maintain a vent to allow gas to escape during freezing.

9. Tendon is difficult to powder. To avoid gumming up the mill, it is necessary to keep the grinding surfaces cold by adding liquid nitrogen continually. Some loss of tissue is inevitable. For smaller amounts of tissue one can use a tissue macerator (a shaking steel ball and chamber) cooled with liquid nitrogen.
10. The ratio of extraction fluid to tissue can be varied, depending on the goal of the extraction. Tensile tendon swells a great deal during extraction, whereas compressed tendon and cartilage do not. Powdered tendon swells more than chunks of tissue. For highest extraction efficiency it is important to have a large supernatant volume and a small pellet after centrifugation. However, a large supernatant volume is not desirable during the subsequent dialysis.
11. With sufficient fluid volume, virtually 100% of the proteoglycan can be removed from powdered tendon with two sequential extractions in 4 *M* guanidine buffer (*see* **Tables 3** and **4**). In contrast, extraction with 7 *M* urea solublized less that 20% of the proteoglycan. Addition of 0.1 *M* NaCl to 7 *M* urea increased extraction efficiency fourfold compared to extraction in 7 *M* urea alone, but still solublized only 50–65% of the total proteoglycan (*see* **Table 4**). **Figure 1** indicates that these two extraction solutions solublize the same proteoglycans.
12. It is necessary to remove 4 *M* guanidine in order to carry out ion-exchange chromatography. Efficient dialysis is accomplished during three sequential 24-h dialysis steps using 5 volumes of 7 *M* urea buffer for each step. This will reduce guanidine concentration to less than 0.04 *M*, a level that does not impede glycosaminoglycan binding to the anion-exchange resin.
13. Although extraction in 7 *M* urea + 0.1 *M* NaCl is less efficient than extraction in 4 *M* guanidine, it eliminates the need to dialyze samples before ion-exchange chromatography.
14. With a larger extract volume, one can use a larger DEAE cellulose column. The extract can be pumped onto the column and eluted with a continuous gradient of NaCl from 0.1 to 0.8 *M*. Proteoglycans will elute at about 0.25 *M* NaCl ***(6)***.
15. The yield of decorin after ion-exchange chromatography and sieve chromatography was about 150 µg/g wet weight of adult tensile bovine tendon ***(6)***.
16. Precipitate proteoglycans in 8 volumes of ethanol. Precipitation from 7 *M* urea + NaCl does not present a problem. However, it is sometimes useful to precipitate the 4 *M* guanidine extract. In this case it is important to remove all supernatant and rinse the tube carefully. If any guanidine remains in the sample, it will form a nasty precipitate with SDS in the gel sample buffer and make electrophoresis impossible.
17. The resuspended chondroitinase ABC enzyme from Seikagaku can be kept in the refrigerator for a year.
18. All electrophoresis buffers should be made with Tris base and brought to proper pH with HCl.
19. It is not necessary to run gradient gels to see proteoglycan. However, the 4–16% gel is useful for visualizing intact biglycan and decorin and their core proteins on the same gel.
20. Large proteoglycans such as aggrecan will not enter the separating gel. As the gel is running, it is possible to see "crinkly" diffraction lines in the stacking gel of lanes containing the large proteoglycan.
21. The various concentrations of methanol suggested for destaining solutions are designed to reduce methanol consumption. After destaining in 7% acetic acid, the gel will be somewhat swollen; it will return to size when stored in the solution containing 25% methanol. If the protein bands have faded, just add a few drops of Coomassie blue to this final solution.
22. Gel electrophoresis can be used to visualize intact proteoglycans in a tissue extract, without ion-exchange purification, by staining only with Alcian blue. The gel should be washed

with at least three changes of destain solution over a 24-h period to assure complete removal of SDS; then stain the gel with Alcian blue and destain in 7% acetic acid. If SDS remains in the gel, it will bind Alcian blue and make the gel impossible to destain. Do not stain samples of the total extract with Coomassie blue, because this produces a blue smear that will obscure the proteoglycans.

## Acknowledgements

Kathryn G. Vogel wishes to thank Dick Heinegard, Lund, Sweden, and Larry Fisher, Bethesda, Maryland, for teaching her many of the techniques and tips reported here. Supported by the National Institutes of Health, AR36110.

## References

1. Vogel, K. G. and Heinegard, D. (1985) Characterization of proteoglycans from adult bovine tendon. *J. Biol. Chem.* **260,** 9298–9306.
2. Vogel, K. G. (1995) Fibrocartilage in tendon: a response to compressive load, in *Repetitive Motion Disorders of the Upper Extremity* (Gordon, S. L., Blair, S. J., and Fine, L. J., eds.) American Academy Orthopedic Surgery, Rosemont, IL, pp. 205–215.
3. Berenson, M. C., Blevins, F. T., Plaas, A. H. K., and Vogel, K. G. (1996) Proteoglycans of human rotator cuff tendons. *J. Orthoped. Res.* **14,** 518–525.
4. Brown, A. H. (1946) Determination of pentose in the presence of large quantities of glucose. *Arch. Biochem.* **11,** 269–278.
5. Laemmli, U. K. (1970) Cleavage of structural proteins during the assembly of the head of bacteriophage T4. *Nature* **227,** 680–685.
6. Vogel, K. G., Sandy, J. D., Pogany, G., and Robbins, J. R. (1994) Aggrecan in bovine tendon. *Matrix Biol.* **14,** 171–179.

# 3

## Purification of Proteoglycans from Mineralized Tissues

**Neal S. Fedarko**

### 1. Introduction

The isolation and purification of proteoglycans derived from mineralized tissues requires sequential dissociative extraction and fractionation techniques that were first employed on developing dental enamel *(1)* and then on bone *(2)*. The basic method involves preliminary extraction in nondemineralizing denaturing buffers, followed by extraction with demineralizing and denaturing solutions combined. The denaturing buffer extract is considered to be the osteoid-associated proteoglycan pool, while the demineralizing buffer extract is taken as the mineral-associated pool of proteoglycans. Following extraction procedures, the samples are subjected to anion-exchange chromatography and size-exclusion chromatography for purification.

The protocols listed below, while specifically detailing the purification of proteoglycans from bone, are applicable directly to other mineralized tissue such as teeth. The methodology is slightly modified from previously described chromatographic purification schemes *(3–6)*. Proteoglycans are profiled during isolation by a special acrylamide gradient gel electrophoresis system (3–15% acrylamide and 0.06–0.08 % *bis*-acrylamide) and staining with Stains-All. Chromatographic fractions are profiled for proteoglycan content using a modified carabazole assay for uronic acid content *(7)*.

### 2. Materials

#### *2.1. Tissue Preparation and Extraction*

1. Impact grinder cooled by liquid nitrogen (e.g., Spex Freezer/Mill tissue pulverizer).
2. Liquid nitrogen.
3. Denaturing buffer (0.5 L):
   a. 191.08 g guanidine HCl (4 *M*).
   b. 3.028 g Tris, pH 7.4. (0.05 *M*).
   c. 6.56 g 6-aminocaproic acid (0.1 *M*).
   d. 0.382 g benzamidine HCl (0.005 *M*).

From: *Methods in Molecular Biology, Vol. 171: Proteoglycan Protocols*
Edited by: R. V. Iozzo © Humana Press Inc., Totowa, NJ

e. 0.625 g N-ethylmaleimide (0.01 *M*).
f. 0.087 g phenylmethylsulfonyl fluoride (0.001 *M*).
g. Make up to a total volume of 0.5 L with distilled, deionized $H_2O$.
4. Demineralizing buffer (0.5 L):
a. Denaturing buffer above plus
b. 95.05 g EDTA tetrasodium salt (0.5 *M*).
5. Nutator or orbital shaker at 4°C.

## 2.2. Extract Concentration and Desalting

1. 40% (w/v) polyethylene glycol 6000 in distilled, deionized $H_2O$.
2. Formamide buffer (0.5 L):
a. 200 mL formamide (40% v/v, 10 *M*) (*see* **Note 1**).
b. Sodium phosphate buffer, pH 6.0 (40 m*M*).
i. 39 mL of a 0.2 *M* stock of monobasic sodium phosphate (27.8 g $NaH_2PO_4$ in 1 L $H_2O$).
ii. 61 mL of a 0.2 *M* stock of dibasic sodium phosphate (53.85 g $Na_2HPO_4 \cdot 7H_2O$ or 71.7 g $Na_2HPO_4 \cdot 12H_2O$ in 1 L $H_2O$).
c. 2.92 g NaCl (0.1 *M*).
d. 0.5 mL Tween 20 buffer.
e. Make up to a total volume of 0.5 L with distilled, deionized $H_2O$.
3. ToyoPearl TSK-GEL HW 40(S) resin packed in a 1.0 × 25 cm Omnifit column (*see* **Note 2**).

## 2.3. Anion Exchange Chromatography

1. ToyoPearl Super Q-650S resin packed in an Omnifit column, 1.0 × 5 cm (*see* **Note 2**).
2. Equilibration buffer (0.5 L):
a. 200 mL formamide (40% v/v, 10 *M*).
b. Sodium phosphate buffer, pH 6.0 (40 m*M*).
i. 39 mL of a 0.2 *M* stock of monobasic sodium phosphate (27.8 g $NaH_2PO_4$ in 1 L $H_2O$).
ii. 61 mL of a 0.2 *M* stock of dibasic sodium phosphate (53.85 g $Na_2HPO_4 \cdot 7H_2O$ or 71.7 g $Na_2HPO_4 \cdot 12H_2O$ in 1 L $H_2O$).
c. 2.92 g NaCl (0.1 *M*).
d. 0.5 mL Tween 20 buffer.
e. Make up to a total volume of 0.5 L with distilled, deionized $H_2O$.
3. Elution buffer (0.5 L):
a. Equilibration buffer plus
b. 58.44 g NaCl (2 *M*).

## 2.4. Size-Exclusion Chromatography

1. Formamide buffer: same as equilibration buffer for anion exchange.
2. Guanidine SEC buffer (0.5 L):
a. 286.6 g guanidine HCl (6 *M*).
b. Sodium phosphate buffer, pH 6.0 (0.1 *M*).
i. 97.5 mL of the 0.2 *M* stock of monobasic sodium phosphate.
ii. 152.5 mL of the 0.2 *M* stock of dibasic sodium phosphate.
c. 0.5 mL Tween 20.
d. Make up to a total volume of 0.5 L with distilled, deionized water.
3. Pharmacia Superose 6 HR 10/30 and/or Phenomenex BioSep SEC S-4000 column.

## 2.5. Electrophoretic Analysis

### 2.5.1. Stock Solutions

1. Solution 1. 30% acrylamide stock: 30 g acrylamide to a total volume of 100 mL $H_2O$.
2. Solution 2. 2 % *bis*-acrylamide stock: 2 g *N*, *N*-methylene-*bis*-acrylamide to a total volume of 100 mL $H_2O$.
3. Solution 3. Tris-$PO_4$ buffer (stacking buffer):
   a. 6 g Tris base (0.5 *M*).
   b. 25 mL 1 N $H_3PO_4$.
   c. pH 6.8 with concentrated $H_3PO_4$.
   d. Make up to a total volume of 100 mL $H_2O$.
4. Solution 4. Tris-HCl buffer (separating buffer):
   a. 18.16 g Tris base (1.5 *M*).
   b. 15 mL 2 *N* HCl.
   c. pH 8.8 with concentrated HCl.
   d. Make up to a total volume of 100 mL $H_2O$.
5. 10 % sodium dodecyl sulfate (SDS) 100 g SDS in 100 mL $H_2O$.
6. 10 % ammonium persulfate (APS) 100 mg ammonium persulfate in 1 mL $dH_2O$.
7. *N*, *N*, *N'*, *N'*-tetramethylethylenediamine (TEMED).
8. Stains-All stock solution:
   a. 0.1 % (w/v) Stains-All (1-ethyl-2-|3-(1-ethylnaphtho|1,2d|-thiazolin-2-ylidene)
   b. Ethylpropenyl|naphtho|1,2d|thiazolium bromide) in formamide (*see* **Note 3**).

### 3.5.2. Working Gel Solutions (*see* **Note 4**).

1. Special 3% stacking gel:
   a. 10 mL solution 1 (30% acrylamide).
   b. 3 mL solution 2 (2% Bis).
   c. 25 mL solution 3 (separating buffer),
   d. 1 mL 10% SDS solution.
   e. 61 mL $dH_2O$.
2. 3% separating gel
   a. 10 mL solution 1 (30% acrylamide),
   b. 3 mL solution 2 (2% Bis),
   c. 25 mL solution 4 (separating buffer),
   d. 1 mL 10% SDS solution,
   e. 61 mL $dH_2O$.
3. 15% separating gel:
   a. 50 mL solution 1 (30% acrylamide).
   b. 4 mL solution 2 (2% Bis).
   c. 25 mL solution 4 (separating buffer),
   d. 1 mL 10% SDS solution.
   e. 20 mL $dH_2O$.
4. Gel fixative solution (4 L):
   a. 1.0 L isopropanol (25% v/v).
   b. Make up to 4.0 L with $H_2O$.
5. Stains-All working solution:
   a. 192 mL distilled, deionized $H_2O$.
   b. 3 mL Solution 4.
   c. 75 mL isopropanol.

d. 15 mL formamide.
e. 15 mL 0.1 % (v/v) Stains-All stock.

### 2.6. Modified Carbazole Assay

1. Stock reagents.
   a. 0.025*M* sodium tetraborate: $10H_2O$ in sulfuric acid. Dissolve 0.95 g of sodium tetraborate decahydrate in 2.0 mL of hot $H_2O$ and add 98 mL of ice-cold concentrated $H_2SO_4$ carefully with stirring. Stable indefinitely if refrigerated.
   b. 0.125% carbazole in absolute ethanol or methanol (analytical grade). Dissolve 125 mg of carbazole in 100 mL of absolute ethanol.
   c. Standards: 200 ng–20 mg D-glucuronolactone in 250 mL $H_2O$.

## 3. Method

### 3.1. Tissue Preparation and Extraction

1. Immediately after excision, trim bone free of sutures and soft adherent tissue.
2. To facilitate processing, cut tissue into ≤1-cm pieces using sterile side cutters or bone biters.
3. Process bone into a fine powder under liquid nitrogen using a Spex impact mill.
4. For every gram of powdered bone, add 0.1 L denaturing extraction buffer and mix the suspension using a Nutator or orbital shaker for 48–72 h at 4°C.
5. Centrifuge at 800*g* for 10 min, aspirate, and save supernatant.
6. Rinse remaining pellet twice with 100 mL of fresh denaturing extraction buffer and combine rinses.
7. Add 0.1 L of demineralizing buffer for every gram of bone and extract for 72 h at 4°C.
8. Centrifuge at 5000*g* to remove any insoluble material (*see Note 5*).

### 3.2. Extract Concentration and Desalting

1. Denaturing buffer extracts and demineralizing buffer extracts are made 20% in polyethylene glycol (PEG) by adding either an equal volume of the 40% PEG stock or by the appropriate dry weight of PEG (*see* **Note 6**).
2. Chill samples at 4°C for 1 h to overnight with mixing on an orbital shaker or a nutator.
3. Centrifuge samples at 5000*g* for 15 min.
4. Resuspend pellet in 2–5 mL of chilled formamide buffer and subject to HPLC desalting.
5. Preparative desalting and buffer exchanging utilizes a $1.0 \times 25$ cm Omnifit column packed with ToyoPearl TSK-GEL HW 40(S) resin equilibrated in the formamide equilibration buffer and a flow rate of 0.5 mL/min. Up to 5.0 mL are injected onto the column and 1 min fractions collected. The void volume (fractions 8–16) are collected and pooled, an aliquot taken for SDS-PAGE analysis (*see* below), and the remainder of the pool subjected to anion-exchange chromatography.

### 3.3. Anion-Exchange Chromatography

1. Equilibriate a ToyoPearl Super Q-650S column in low-salt (0.1 *M* NaCl) formamide buffer at a flow rate of 2.0 mL/min.
2. Inject desalted extracts directly onto the string anion-exchange column and elute material by an initial isocratic segment of 10 min at 0.1 *M* NaCl, followed by an 80-min linear gradient to 1.0 *M* NaCl, and finally, a 30-min linear gradient to 2.0 *M* salt. The flow rate throughout is 2.0 mL/min, absorbance is monitored at 280 nm, and 1-min fractions are collected.

3. Profile aliquots from fractions with significant A 280-nm absorbance by SDS PAGE (*see* below). Profile proteoglycan elution positions by a modified carbazole (uronic acid) assay (*see* **Note 7**).

### 3.4. Size-Exclusion Chromatography

1. Molecular size separations are performed using two different systems. The columns and flow rates used are a Superose 6 HR 10/30 FPLC column at a flow rate of 0.25 mL/min and a Phenomenex BioSep SEC S-4000 column at a flow rate of 0.5 mL/min. The Superose 6 column is used to further isolate individual proteoglycans, while the BioSep SEC S-4000 column is used for final purification and estimation of size.
2. Pooled fractions from anion exchange are concentrated by PEG precipitation (*see* above).
3. Pellets are resuspended in the formamide equilibration buffer and injected onto the Superose 6 column.
4. 1-min fractions are collected as absorbance at 280 nm is monitored and 50-µL aliquots of each fraction are diluted to 250 µL with $H_2O$ and taken for the carbazole assay (*see* below) to profile uronic acid-containing proteoglycans.
5. Pooled fractions that have been subjected to anion exchange and subsequent size-exclusion chromatography on Superose 6 are concentrated by PEG precipitation.
6. The pellet is resuspended in the 6 *M* guanidine-containing buffer and resolved on the BioSep SEC S-4000 column. Again, absorbance is followed at 280 nm and aliquots are analyzed for uronic acid (proteoglycan) content.
7. Both columns are calibrated using protein standards *(8)*.

### 3.5. Electrophoretic Analysis

#### 3.5.1. Pouring Gradient Gels (see Note 8).

1. 3–15% gradient large gels (16 × 14 × 0.15 cm)
   a. 14.6 mL 3% in left chamber.
   b. 14.5 mL 15% in right chamber with mixing flea stir bar.
   c. Add in order: 35 µL 10% APS to both chambers (mixing with pipet in chamber without stir bar); 2.75 µL TEMED to 3% side; 2.5 µL TEMED to 15% side.
   d. Open stopcocks and start pumps (flow rate ≤ 1 mL/min).
   e. It should take about 40–45 min to pour the separating gel and another 40 min to polymerize.
   f. Top off gel with saturated butanol until ready to pour stacking gel.
2. 3% stacking gel for large gels (16 × 14 × 0.15 cm)
   a. To 20 mL of "special 3% stacking gel solution" add: 300 µL APS; and 30 µL TEMED.
3. Small gels (10 × 10.5 × 0.15 cm)
   a For small gels, 5 mL of working gel solution is used per chamber of the gradient former.
   b. Each chamber receives 25 µL APS.
   c. The 3 % acrylamide working solution receives 2.6 µL TEMED, while the 15% working solution receives 2.5 µL of TEMED.
   d. The 5 mL of stacking gel working solution (special 3%) receives 25 µL APS and 5 µl TEMED.

#### 3.5.2. Running Gradient Gels

1. Samples are reduced with 2 m*M* dithiothreitol and heated at 100°C for 10 min before loading. Samples in the formamide buffer do not require the addition of gel sample buffer and can be loaded directly onto the gel.

2. Electrophoresis is carried out on SDS slab gels (16 ×14 × 0.15 c m) at a constant current of 9 mA per gel for 14 h (30 mA/gel in 6 h with water a buffer cooling system) or until the prestained carbonic anhydrase molecular weight marker (30,000 daltons) reaches the bottom of the gel.
3. For small gels (10×10.5×0.15 cm), electrophoresis is carried out at 25 mA/gel in 2 h.
4. At the end of electrophoresis, gels are rinsed in the 25% isopropanol through 4 changes over a 12-h period to remove SDS. Residual SDS will interfere with Stains-All solubility.
5. Gels are reacted with the Stains-All working solution in a light-protected container.
6. Once bands are readily apparent, the gels are destained in $H_2O$. Stains-All reacts with proteins and proteoglycans to generate different colored bands based on band component acidity. An advantage of Stains-All over other proteoglycan–glycosaminoglycan staining procedures is that it also enables contaminating proteins and glycoproteins to be visualized.

### 3.6. Modified Carbazole Assay

1. Cool samples and standards (250 µL) in an ice bath.
2. Add 1.5 mL of ice-cold sulfuric acid-sodium tetraborate reagent with mixing and cooling on ice.
3. Heat the mixtures at 100°C for 10 min.
4. Cool rapidly in the ice bath.
5. Add 50 µL of carbazole reagent and mix well.
6. Reheat at 100° C for 15 min.
7. Cool rapidly to room temperature and determine the absorbance at 525 nm.

The combination of anion-exchange chromatography and size-exclusion chromatography enables separation of the three major proteoglycans present in bone—a versican-like proteoglycan, biglycan, and decorin. The small leucine-rich proteoglycans, biglycan and decorin, elute early on the salt gradient in anion exchange and are included and elute later on the size-exclusion columns.

## 4. Notes

1. Fresh formamide should be used and, once opened, the bottle stored tightly capped at –20°C until further use.
2. Omnifit glass columns were adapted for HPLC with Tefzel tubing (1/16-in. outer diameter × 0.01 in. internal diameter) by boring out the ferrules with a 1/16-in drill bit ***(6)***.
3. Stains-All, or 1-ethyl-2-l3-(1-ethylnaphthol1,2dl-thiazolin-2-ylidene)-2-, ethylpropenyllnaphthol1, 2dlthiazolium bromide, is available from Eastman Kodak Co. with a catalog number of 2718. An alternative to Stains-All profiling of proteoglycans is Alcian blue staining ***(9)***.
4. Working solutions are stable for 3–4 mo when stored at 4°C in flasks wrapped with aluminum foil (to eliminate light exposure).
5. Following incubation with demineralizing buffer, it is common in bone samples derived from adult animals to have a significant amount of insoluble material that pellets after centrifugation. This majority of this material consists of highly cross-linked type I collagen, though minor amounts of matrix glycoproteins and proteoglycans may also be present. In general, the older the sample donor, the higher are the levels of insoluble cross-linked material.

6. Incubation of the 40% PEG in water suspension at 50°C facilitates dissolution.
7. Alternatives to the carbazole assay to follow the uronic acid content as a marker of the glycosaminoglycans in proteoglycans would be one of the numerous stoichiometric dye binding assays such as Safranin O ***(10)*** or 1,9-dimethylmethylene blue ***(11)***.

## References


1. Termine, J. D., Belcourt, A. B., Miyamoto, M. S., and Conn, K. M. (1980) Properties of dissociatively extracted fetal tooth matrix proteins. II. Separation and purification of fetal bovine dentin phosphoprotein. *J. Biol. Chem.* **255,** 9769–9772.
2. Termine, J. D., Belcourt, A. B., Conn, K. M., and Kleinman, H. K. (1981) Mineral and collagen-binding proteins of fetal calf bone. *J. Biol. Chem.* **256,** 10,403–10,408.
3. Fisher, L. W., Termine, J. D., Dejter, S. W., Jr., Whitson, S. W., Yanagishita, M., Kimura, J. H.; Hascall, V. C.; Kleinman, H. K.; Hassell, J. R.; and Nilsson, B. (1983) Proteoglycans of developing bone. *J. Biol. Chem.* **258,** 6588–6594.
4. Fisher, L. W., Hawkins, G. R., Tuross, N., and Termine, J. D. (1987) Purification and partial characterization of small proteoglycans I and II, bone sialoproteins I and II, and osteonectin from the mineral compartment of developing human bone. *J. Biol. Chem.* **262,** 9702–9708.
5. Beresford, J. N., Fedarko, N. S., Fisher, L. W., Midura, R. J., Yanagishita, M., Termine, J. D. and Robey, P. G. (1987) Analysis of the proteoglycans synthesized by human bone cells in vitro. *J. Biol. Chem.* **262,** 17,164–17,172.
6. Fedarko, N. S., Termine, J. D., and Robey, P. G. (1990) High-performance liquid chromatographic separation of hyaluronan and four proteoglycans produced by human bone cell cultures. *Anal. Biochem.* **188,** 398–407.
7. Mayes, E. L.V. (1984) Protein, in *Methods in Molecular Biology* (Walker, J. M., ed.), Humana, Totowa NJ, pp. 5–12.
8. Chaplin, M. F. (1986) Monosaccharides, in *Carbohydrate Analysis: A Practical Approach.* (Chaplin M.F. and Kennedy, J. F. eds.), IRL Press, Oxford, pp. 1–36.
9. Wall, R. S.and Gyi, T. J. (1988) Alcian blue staining of proteoglycans in polyacrylamide gels using the "critical electrolyte concentration" approach. *Anal. Biochem.* **175,** 298–299.
10. Carrino, D. A., Arias, J. L., and Caplan, A. I. (1991) A spectrophotometric modification of a sensitive densitometric Safranin O assay for glycosaminoglycans. *Biochem. Int.* **24,** 485–495.
11. Muller, G. and Hanschke, M. (1996) Quantitative and qualitative analyses of proteoglycans in cartilage extracts by precipitation with 1,9-dimethylmethylene blue. *Connect. Tissue Res.* **33,** 243–248.

# 4

# Purification of Perlecan from Endothelial Cells

**John Whitelock**

## 1. Introduction

Perlecan, the major heparan sulfate (HS) proteoglycan of basement membranes and other connective tissues, is a modular molecule with five structural domains. It has been isolated from various sources including kidney and placenta, but most commonly from the mouse Englebroth Holm Swarm (EHS) tumor *(1,2)*. It has also been purified from cultured cell lines such as endothelial cells *(3)* or fibroblasts *(4)*. When it has been extracted from tissue or from the extracellular matrix of cultured cells, it has been first solubilized using chaotropic agents such as urea or guanidine. The effect of these dissociating agents on the structure and function of isolated proteoglycans is not known, but it is thought that some of the denaturation may be irreversible. Human perlecan is found predominantly as a heparan sulfate proteoglycan, but has been isolated from carcinoma cell lines as an undecorated protein core and from placental tissue decorated with chondroitin sulfate/dermatan sulfate. The molecular weight of the protein core is approximately 470 kDa, while the size of the mouse homolog is around 400 kDa. Domain I is the N-terminal domain, and it contains a cluster of three potential glycosaminoglycan attachment sites. Domain II has homology to the LDL receptor, with strict conservation of the cysteine residue positions. Domain III shares homology with the short arm of the laminin $\alpha$-chain. Domain IV is the largest domain with a molecular weight in excess of 200 kDa and is made up of numerous immunoglobulin-like repeats similar to those found in the neural cell adhesion molecule. Domain V is the C-terminal domain, and is made up of three regions with homology to the globular domains of the laminin $\alpha$-chain, interspersed with four epidermal growth factor-like repeats.

From: *Methods in Molecular Biology, Vol. 171: Proteoglycan Protocols*
Edited by: R. V. Iozzo © Humana Press Inc., Totowa, NJ

## 2. Materials

1. Human endothelial cells were isolated from either umbilical veins or arteries as described previously *(5)*. Dr. J. Gamble provided C11-STH, the endothelial cell line at the Hanson Cancer Center in Adelaide, Australia.
2. Endothelial cell growth medium: For 1 L of Medium 199 (Earles salts), open 1 packet of Medium 199 powder (Gibco) and dissolve powder in 900 mL of pyrogen-free $H_2O$ (Baxter), add 2.2 g of NaHCO3, 100 U of penicillin, and 100 µg of streptomycin. Adjust the pH to 7.2, make the total volume up to 1 L, sterilize by filtration using a 0.22-µm filter and store at 4°C. Before using the medium in cell cultures, add 0.3g/L of bovine brain extract (BBE), prepared as described *(6)* (*see* **Note 1**), 1 mg/L of heparin (Sigma cat. no. H3149) (*see* **Note 2**), 15% v/v fetal bovine serum (*see* **Note 3**).
3. Flasks that were used to grow endothelial cells were first coated with 10 µg/mL of fibronectin for 2 h at 37°C. A 2% gelatin solution can be used in place of fibronectin.
4. DEAE Sepharose column (Pharmacia).
5. Anti-perlecan immunoaffinity column. Weigh out 2 g of dry CnBr-activated Sepharose 4B matrix (Pharmacia). This makes a column of bed volume 7 mL. Swell and wash the powder in 1 m*M* HCl and then equilibrate the gel with approx. 50 mL of 0.1 *M* $NaHCO_3$, 0.5 *M* NaCl, pH 8.3 (coupling buffer). Dialyze the anti-perlecan monoclonal antibody (A71, A74, or A76) against the coupling buffer (allow approx. 5 mg of antibody per milliliter of swelled Sepharose). Mix the antibody solution with the prepared gel and mix overnight at 4°C on a rocking platform. The ratio of buffer to swelled gel should be at least 2/1 to achieve efficient mixing. This time can be shortened to 4 h at room temperature. Centrifuge at 1000*g*, aspirate the supernatant, and wash the gel on a sintered glass funnel under vacuum. Block the remaining active binding sites by incubating the gel with 1 *M* ethanolamine, pH 8.0 (this should be made fresh) for 2 h at room temperature on a rocker. Centrifuge at 1000*g*, aspirate the supernatant, and wash the gel with coupling buffer. Perform alternate washes of the elution buffer to be used and the coupling buffer. Equilibrate the column in phosphate-buffered saline and store the column at 4°C
6. Dialysis tubing—10,000-Mr cutoff membrane (Spectropor).
7. Sucrose.
8. Superose™ 6 prepackaged HR 10/30 column and Pharmacia FPLC™ system.
9. Coomassie Plus Protein Assay Kit (Pierce).
10. SDS-PAGE gradient gels, reagents, and electrophoretic equipment (Bio-Rad).
11. Composite agarose-acrylamide gels (*see* **Note 4**).
12. Anti-perlecan protein core antibodies; A71 (specific for domain I), A74 (specific for domain V) and A76 (not characterized) *(7)*.
13. Anti-heparan sulfate antibody, 10E4 (Seikagaku).
14. Anti-chondroitin sulfate antibody, CS-56 (Sigma).
15. ELISA plates, reagents, and microtiterplate reader

## 3. Methods

### *3.1. Human Endothelial Cell Culture*

1. Cultures of human arterial and venous cells are maintained by replacing the conditioned medium three times every week.
2. The cells can also be grown in roller culture bottles (900 $cm^2$). These will need to be coated with fibronectin (10 µg/mL; 10 mL per roller bottle) for 2 h at 37°C before seeding the cells.
3. Roller cultures are fed twice weekly with 120 mL of endothelial cell growth medium.

4. Once the cells condition the medium, it is removed, centrifuged at 1500*g* for 10 min to remove cell debris and stored at –20°C. (*see* **Notes 5** and **6**).

### 3.2. DEAE Chromatography

1. Equilibrate a 100 mL bed volume DEAE Sepharose column with approximately 4 volumes of 20 m*M* Tris, 250 m*M* NaCl, 10 m*M* EDTA, 1 m*M* benzamidine, pH 7.5 (DEAE running buffer) at 4°C.
2. Thaw and filter the conditioned medium (approx. 2-L batches) using either glass wool or a sintered funnel over a vacuum.
3. Apply the medium to the column at a flow rate of 1–2 mL/min.
4. Wash the medium through the column with the DEAE running buffer and continue to wash the column with this buffer until a baseline is achieved (measuring the absorbency at a wavelength of 280 nm) (*see* **Note 7**).
5. Elute bound molecules (including the perlecan) from the column with 20 m*M* Tris, 1 *M* NaCl, 10 m*M* EDTA, 1 m*M* benzamidine, pH 7.5 into a collection vessel which is on ice.
6. Pool the perlecan containing fractions (approx. 60–80 mL) and store at 4°C.
7. Remove a small sample (~200 µL) for testing in the ELISA based screening assay.
8. Regenerate the DEAE column by washing it with 2 volumes of 20 m*M* Tris, 2 *M* NaCl, 10 m*M* EDTA, 1 m*M* benzamidine, pH 7.5 and reequilibrating it with the DEAE running buffer.

### 3.3. Anti-Perlecan Immunoaffinity Chromatography

1. Equilibrate the column with 20 m*M* phosphate buffer with 1 *M* NaCl, pH 7.5 (immunoaffinity running buffer).
2. Apply the pooled perlecan-containing fractions directly to the immunoaffinity column and recirculate over the column for approx. 6 h or overnight at 4°C, at a flow rate of 1 mL/min.
3. Keep the flow through for testing in the ELISA-based screening assay.
4. Wash the column with approx. 50 mL of the immuno-affinity running buffer or until a baseline is achieved.
5. Elute the bound perlecan from the immunoaffinity column with either 0.1 *M* Tris, pH 7.5 containing 3 *M* $MgCl_2$ or PBS containing 6 *M* urea. (*see* **Note 8**).
6. Pool the perlecan peak (approx. 15 mL).
7. Wash the column with the immunoaffinity running buffer. (Another perlecan peak may elute at this stage; the two peaks should be pooled.)
8. The column is washed with 50 volumes of immunoaffinity running buffer containing 0.08% $NaN_3$ and stored at 4°C.
9. Dialyze the perlecan containing peak against three changes of PBS at 4°C overnight and then concentrate to approx. 1–2 mL by removing the PBS through the dialysis tubing by laying it on a bed of sucrose.
10. Once the volume is reduced to approx. 1–2 mL, the clips on the dialysis tubing can be readjusted (to prevent reswelling) and redialyzed against PBS overnight at 4°C to remove any sucrose.
11. The immuno-purified perlecan can be assayed for purity and quantity, and stored in small aliquots at –70°C in siliconized tubes (*see* **Note 9**).

### 3.4. Gel Filtration Chromatography

1. Equilibrate a Superose 6 prepackaged HR 10/30 (Pharmacia) column with 0.5 *M* $CH_3COONa$, 0.05% Tween-20, pH 7.5 at 0.4 mL/min.

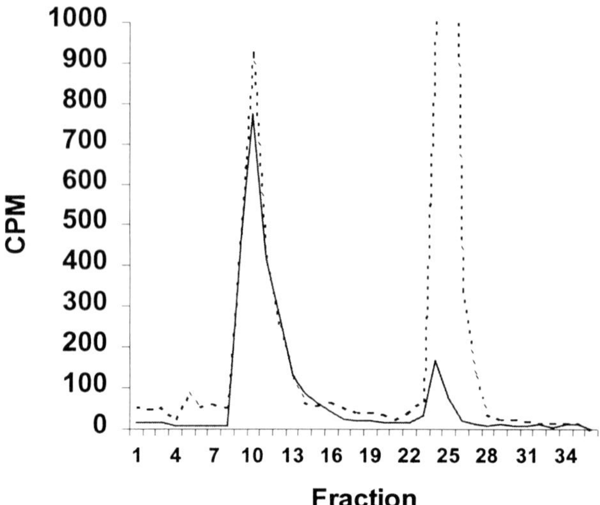


Fig. 1. Elution profile of immunopurifed human perlecan using a Superose 6 10/30 prepacked column. Samples of $^3$H-labeled perlecan were run through a Superose 6 10/30 column as described. HUAEC-derived perlecan is shown by the dashed line, whereas HUVEC-derived perlecan is shown by the unbroken line. The $V_0$ of the column was 8 mL (fraction 10), whereas the $V_t$ was 20 mL (fraction 25). Notice the large amount of free $^3$H-glucosamine in the HUAEC sample.

2. Set the fraction collector to collect fractions every 2 min (0.8-mL fraction volume).
3. Apply sample via the loop while maintaining the flow rate at 0.4 mL/min.
4. Fractions can be monitored for either the presence of proteins by measuring the absorbency at a wavelength of 280 nm, or for radioactivity by taking a sample and mixing it with scintillant and counting.
5. This method is essentially as described by Melrose and Ghosh *(8)*. It is particularly suitable for separating biosynthetically labeled perlecan from free label (*see* **Fig. 1**), as well as separating $^3$H-labeled or $^{35}SO_4$-labeled HS chains that are liberated from the protein core of perlecan by alkaline borohydride elimination (*see* **Fig. 2**).
6. Perlecan-enriched fractions can be prepared using this methodology. This is particularly suitable if immuno-affinity chromatography is not possible. It is important to remember that these fractions will also contain other large molecules such as the chondroitin sulfate proteoglycan, versican. This can be monitored by the screening of fractions with the anti-chondroitin sulfate monoclonal antibody, CS-56 (*see* **Note 10**).

### 3.5. Preparation of Perlecan HS (Alkaline Borohydride Elimination)

1. Prepare 2 *M* $NaBH_4$.
2. Prepare 0.1 *M* KOH.
3. Mix an equal volume of the above solutions and add this to a volume of labeled perlecan solution in the ratio 2/1 v/v ($NaBH_4$/KOH:perlecan).
4. Cover with parafilm, and incubate overnight at 45°C.
5. Neutralize with an equal volume of 3.6 *M* $CH_3COOH$ (*see* **Note 11**).

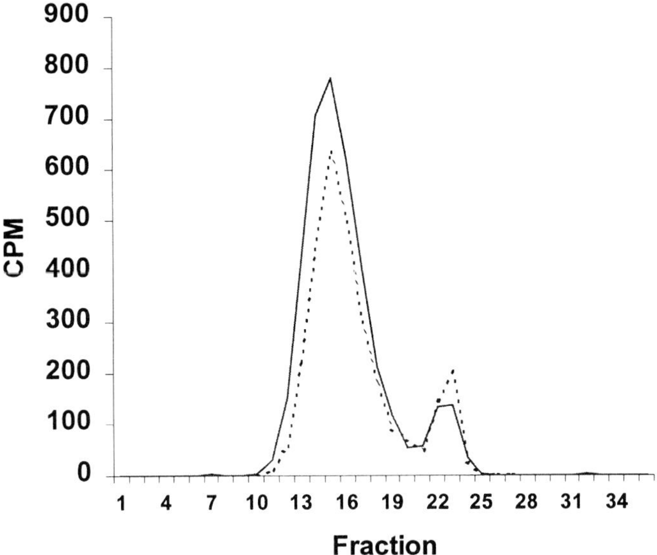


Fig. 2. Elution profile of HS derived from immunopurified perlecan using a Superose 6 10/30 prepacked column. Samples of $^3$H-labeled HS were run through a Superose 6 10/30 column as described. HUAEC-derived HS is shown by the dashed line, whereas HUVEC-derived HS is shown by the unbroken line. The *V0* of the column was 8 mL (fraction 10), whereas the *Vt* was 20 mL (fraction 25). The molecular weight of the two types of HS was estimated to be 40 kDa using previously published standards for chondroitin sulfate *(8)*.

### 3.6. Ethanol Precipitation for Concentrating HS

1. To the sample to be concentrated, add 4 volumes of 1.3% $CH_3COOK$ in 95% ethanol.
2. Incubate at –20°C for 3–16 h.
3. Isolate the precipitated labeled perlecan by centrifugation in a microfuge at room temperature for 5 min.
4. Aspirate the supernatant, wash twice (1–5 mL) with salt-free 95% ethanol.
5. The perlecan can be either lyophilized or dissolved in buffer.

### 3.7. Detection and Quantitation of Perlecan

1. The presence of perlecan in purified and semipurified fractions is monitored using an ELISA-based screening assay using one antibody that reacts with the protein core (A76) and another that reacts with the heparan sulfate (10E4). This assay can also be used to screen for the presence of chondroitin sulfate on the protein core or other co-purifying extracellular matrix proteins (*see* **Note 12**).
2. We have found that the heparan sulfate attached to the protein core of perlecan interfered with the BCA reaction, resulting in an overestimation of the amount of perlecan present. Therefore, to avoid these errors, use the dye-binding protein assay, Coomassie Plus Protein Assay (Pierce), as it is more reliable on the amount of protein core present.

### 3.8. Electrophoresis

1. Standard SDS-PAGE methodology can be employed to analyze perlecan in gels and Western blots. It is recommended to run 4–15% gradient gels, as the intact perlecan runs at the

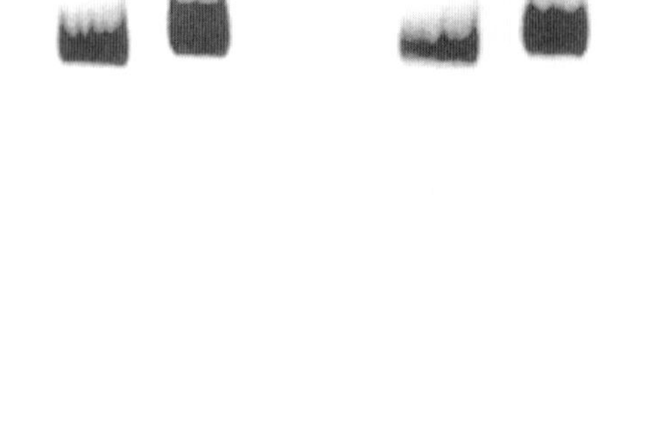


Fig. 3. COA-PAGE immunoblot of perlecan immunopurifed from a HUVEC cell line and primary HUAEC cultures. Lane 1 represents 3 μg of perlecan immunopurified from a HUVEC cell line (C11-STH). Lane 2 represents 3 μg of perlecan immunopurified from primary cultures of HUAECs. Both samples were run through CO-A/PAGE gels and blotted to nitrocellulose. The nitrocellulose was cut into strips and probed with the anti-protein core antibodies as indicated.

bottom of the application wells. The use of 3–8% gradient gels may facilitate the migration of perlecan into the running gel (*see* **Note 13**).

2. SDS-PAGE Gels can be stained with either (*see* **Note 14**):
   a. 0.25% Coomasie blue in 40% methanol, 50% $H_2O$, 10% $CH_3COOH$.
   b. 0.08% Azure A in $H_2O$.
   c. 0.1% Alcian blue in 50% ethanol.
   d. Or a combination of the above with a $AgNO_3$ stain.
3. Composite agarose-polyacrylamide gel electrophoresis (COA-PAGE) can also be employed to analyze the purified perlecans (*see* **Fig. 3**). These gels have the advantage that the perlecan migrates through the pores of the gel and the distance the molecules travel depends on the ratio of their mass to their charge. This has the advantage that perlecans from different sources may be separated on the basis of charge (*see* **Fig. 3**). (*see* **Note 15**). They can be blotted to nitrocellulose on a flat bed dry blotter and probed with antibodies (*see* **Fig. 3**).

## 4. Notes

1. Commercially prepared endothelial cell growth supplement or purified growth factors such as recombinant human fibroblast growth factors (FGF) or vascular endothelial cell growth factors (VEGF) may be used in place of the BBE. They are used at a final concentration of 10 ng/mL.
2. Heparin is not required when using either FGF-2 or VEGF. Cells may also be grown heparin-free by using 15% v/v human serum.

3. After 2 weeks of storage at 4°C, the glutamine will have to be re-added at a final concentration of 2 m*M* (4 m*M* for smooth muscle cells). The medium should be discarded after 4 wk.
4. These are prepared essentially as described by McDevitt and Muir ***(9)*** (*also see* Chapter15).
5. Either $^3$H-glucosamine, $^{35}$S-$Na_2SO_4$, or $^{35}$S-methionine can be added to the cultures (5 µCi/mL) to produce biosynthetically labeled perlecan. The choice of labeling reagent depends on whether you want to label the protein core or the glycosaminoglycan chains.
6. Endothelial cells derived from primary tissue can be maintained at confluence for 1 wk. The endothelial cell line could be maintained at confluence for longer periods.
7. Biosynthetically labeled perlecan can be prepared using the same methodology as described for the unlabeled molecule. Its purification can be monitored by sampling 100 µL of each fraction and mixing it with scintillant (Insta-gel; Packard) and counting in a scintillation counter (LKB 1217 Rackbeta). Radiolabeled perlecan can also be concentrated using ethanol precipitation.
8. When using the 3 *M* $MgCl_2$ to elute the column, make the solution fresh before use and keep the $MgCl_2$ stored in a desiccator.
9. Collect the perlecan fractions in siliconized Eppendorf tubes to minimize loss due to adsorption.
10. The buffers used for the gel filtration step, which may be used instead of the immunoaffinity step and results in the isolation of perlecan-enriched fractions, may contain 6 *M* urea as a dissociative agent. This is particularly pertinent to the isolation of perlecan from tissue sources.
11. When performing the $NaBH_4$/KOH step (overnight at 45°C), choose a screw-capped tube or cover the tube with parafilm. This prevents the cap popping due to the pressure build up caused by the release of $H_2$ gas. Also, choose a 15-mL tube (for solutions up to 2 mL) so that during the neutralization step the $H_2$ gas that is liberated is allowed to escape without the loss of the sample.
12. 0.1% casein in PBS is used as both the blocking agent and diluent, due to the fact that it gives superior backgrounds with the anti-HS antibody, 10E4.
13. If you plan to run gels with a separate stacker, do not remove the stacker before staining or blotting to nitrocellulose.
14. The SDS must be washed out of the gels with 3 changes in either $H_2O$ or 50% ethanol prior to staining with Azure A, Alcian blue, or silver.
15. These gels can be stained with 0.08% Azure A in $H_2O$ by placing them on a sheet of plastic to act as a support.

## Acknowledgments

The author would like to acknowledge Dr. Anne Underwood for discussion and critical review of the chapter; Penny Bean, Sue Mitchell, and Debbie Lock for excellent technical assistance; and the Australian Government through the Co-Operative Research Center scheme for funding.

## References

1. Hassell, J. R., Robey, P. G., Barrach, H. J., Wilczek, J., Rennard, S. I., and Martin,G. R. (1980). Isolation of a heparan sulfate-containing proteoglycan from basement membrane. *Proc. Natl. Acad. Sci. (USA)* **77,** 4494–4498.
2. Paulsson, M., Yurchenco, P. D., Ruben, G. C., Engel, J., and Timpl, R. (1987). Structure of low density heparan sulfate proteoglycan isolated from a mouse tumor basement membrane. *J. Mol. Biol.* **197,** 297–313.

3. Saku, T. and Furthmayr, H. (1989). Characterization of the major heparan sulfate proteoglycan secreted by bovine aortic endothelial cells in culture. *J. Biol. Chem.* **264,** 3514–3523.
4. Heremans, A., Cassiman, J.-J., Van Den Berghe, H., and David, G. (1988). Heparan sulfate proteoglycan from the extracellular matrix of human lung fibroblasts. *J. Biol. Chem.* **263,** 4731–4739.
5. Weis, J. R., Sun, B., and Rodgers, G. M. (1991). Improved method of human umbilical arterial endothelial cell culture. *Thromb. Res.* **61,** 171–173.
6. Maciag, T., Cerundolo, J., Ilsley, S., Kelley, P. R., and Forand, R. (1979). An endothelial cell growth factor from bovine hypothalamus: identification and partial characterization. *Proc. Natl. Acad. Sci. (USA)* **76(11),** 5674–5678.
7. Whitelock, J. M., Murdoch, A. D., Iozzo, R. V., and Underwood, P. A. (1996). The degradation of human endothelial cell-derived perlecan and release of bound basic fibroblast growth factor by stromelysin, collagenase, plasmin, and heparanases. *J. Biol. Chem.* **271(17),** 10,079–10,086.
8. Melrose, J. and Ghosh, P. (1993). Determination of the average molecular size of glycosaminoglycans by fast protein liquid chromatography. *J. Chromatogr.* **637,** 91–95.
9. McDevitt, C. A. and Muir, H. (1971). Gel electrophoresis of proteoglycans and glycosaminoglycans on large pore composite polyacrylamide-agarose gels. *Anal. Biochem.* **44,** 612–622.

# 5

## Isolation and Characterization of Nervous Tissue Proteoglycans

Yu Yamaguchi

### 1. Introduction

Nervous tissues contain a variety of proteoglycans *(1–3)*. In tissues, proteoglycans are present predominantly in extracellular matrices and on cell surfaces. Proteoglycans in extracellular matrices are generally extracted in soluble fractions by physiological buffers without detergents, whereas cell surface proteoglycans that are anchored to the plasma membrane either by transmembrane domains or glycosylphosphatidylinositol (GPI) linkages are fractionated into membrane fractions. Some proteoglycans in extracellular matrices requires chaotropic agents such as urea and guanidine for solubilization *(4)*. It has been shown that a total of 20 µg of proteoglycans (as protein) can be isolated from 1 g (wet weight) of embryonic rat brain tissues, in which 8 µg are from the soluble fraction and 12 µg are from the membrane fraction *(1)*. In the case of adult rat brain, a total of 35 µg of proteoglycans are isolated, 20 µg in the soluble fraction and 15 µg in the membrane fraction. In general, the soluble fraction contains predominantly chondroitin sulfate proteoglycans (CSPGs), whereas heparan sulfate proteoglycans (HSPG) are enriched in the membrane fraction.

Herndon and Lander *(1)* reported that ~25 distinct proteoglycan core proteins can be identified in embryonic and adult rat brain. Among these bands for putative proteoglycan core proteins, 16 were identified as chondroitin sulfate proteoglycans and 9 were heparan sulfate proteoglycans. Not all of these bands represent distinct proteoglycan core proteins, since some represent proteolytic fragments. Thus far, more than 20 proteoglycans that are molecularly defined have been shown to be expressed in embryonic and adult brain. It should be noted that there is a great deal of difference in the amount of these proteoglycans in the brain at the protein level, and not all of these proteoglycans have been isolated from the brain in biochemical quantities. Among CSPGs, brevican, versican, and phosphacan are abundant in adult brain, whereas neurocan is abundant in embryonic brain *(5,6)*: All these CSPGs have been isolated in biochemical quantities from the brain. Less is known about the relative

From: *Methods in Molecular Biology, Vol. 171: Proteoglycan Protocols*
Edited by: R. V. Iozzo © Humana Press Inc., Totowa, NJ

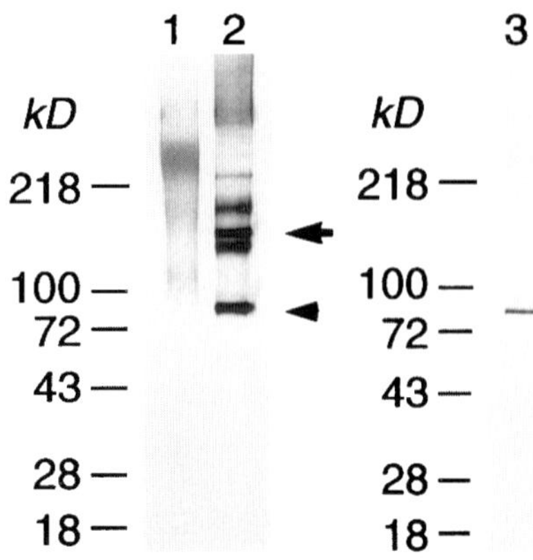


Fig. 1. SDS-PAGE analysis of the total proteoglycan fraction isolated from adult rat brain (modified from **ref. 9**). The total proteoglycan fraction was isolated from soluble extracts of adult rat brain by the methods described here. (*Lanes 1, 2*) The total proteoglycan fraction was digested with (*lane 2*) or without (*lane 1*) protease-free chondroitinase ABC (Seikagaku America) and analyzed on 8–16% gradient gel (Novex) under nonreducing condition. Proteins were visualized by silver staining. *Arrow* and *arrowhead* indicate the 145-kDa (full-length) and 80-kDa (N-terminally truncated) forms of brevican core protein, respectively. Peptide sequencing also identified the ~180-kDa band above the 145-kDa brevican core protein is phosphacan and the ~125-kDa band just below the 145-kDa brevican core protein is neurocan. (*Lane 3*) The 80-kDa brevican core protein was purified by reverse-phase HPLC on a Vydac C4 column.

abundance of individual HSPGs, but at least glypican (glypican-1) and cerebroglycan (glypican-2) have been isolated for the purpose of N-terminal sequencing *(7)*.

The procedure described here was originally reported by Herndon and Lander *(1)* and modified by us for large scale purification *(5,8,9)*. This procedure provides a straightforward way to isolate fractions that are highly enriched for various proteoglycans, by taking advantage of negative charges of glycosaminoglycan chains. Purification of individual proteoglycans in native forms, however, requires methods specific for each proteoglycan, such as affinity chromatography and immunoaffinity chromatography. For instance, neurocan and phosphacan were purified using immunoaffinity chromatography on monoclonal antibodies *(10)*. Brevican can be isolated by using affinity chromatography on a column of tenascin-R fragment *(11)*, while versican can be isolated by affinity chromatography on hyaluronan *(6)*. Although it is possible to purify individual proteoglycans by the combination of conventional chromatographic procedures, such approaches are usually not so easy. This is because proteoglycans behave similarly in most biochemical separation methods due to the large, negatively charged glycosaminoglycan chains shared by all these molecules. Purification of individual core proteins is less challenging. For this purpose, after digesting glycosaminoglycan chains with chondroitinase, heparitinase, or both, mixtures of core proteins can be fractionated by HPLC (*9*; *see also* **Fig. 1,** lane 3), preparative PAGE, or other protein separation methods. However, since some core proteins have similar molecular weights (e.g., all glypican core proteins migrate around ~60 kDa), such core proteins are more difficult to purify.

## 2. Materials

1. *Brain tissues.* This procedure has been used for rat, mouse, bovine, and human brain tissues. For large scale purification of proteoglycan-enriched fractions as described below, we usually start with 100 g of brain tissue. For most applications, we use brain tissues frozen on Dry Ice and stored at –80°C. It is preferable to strip blood vessels and meninges from brains, if contamination of proteoglycans derived from these tissues is to be minimized.
2. Sorvall GSA rotor or equivalent.
3. 250-mL centrifuge bottles for GSA rotor.
4. Sorvall T865 ultracentrifuge rotor or equivalent.
5. 30 mL ultracentrifuge tubes (e.g., Sorvall Ultrabottles, polycarbonate).
6. Polytron homogenizer.
7. Gradient former.
8. DEAE-Sepharose Fast Flow.
9. Buffers: Buffer A: 4 m*M* Hepes, pH 7.5, 0.3 *M* sucrose, 0.15 *M* NaCl, and protease inhibitors*(*see* **item 17** below).
10. Buffer B: 50 m*M* Tris-HCl, pH 8.0, 0.15 *M* NaCl, 1% CHAPS, and protease inhibitors (*see* **Note 1**).
11. Buffer C: 50 m*M* Tris-HCl, pH 8.0, 0.15 *M* NaCl, and 0.5% CHAPS.
12. Buffer D: 50 m*M* Tris-HCl, pH 8.0, 0.25 *M* NaCl, and 0.1% CHAPS.
13. Buffer E: 50 m*M* Tris-HCl, pH 8.0, 0.25 *M* NaCl, 0.1% CHAPS, and 6 *M* urea.
14. Buffer F: 50 m*M* sodium formate, pH 3.5, 0.2 *M* NaCl, 0.1% CHAPS, and 6 *M* urea.
15. Buffer G (preelution buffer): 100 m*M* Tris-HCl, pH 8.0, 0.2 *M* NaCl, and 0.5% CHAPS.
16. Buffer H (elution buffer): 50 m*M* Tris-HCl, pH 8.0, 1 *M* NaCl, and 0.5% CHAPS.
17. 1 m*M* EDTA, 1 µg/ml pepstatin A, 0.5 µg/ml leupeptin, 0.4 m*M* PMSF (final concentrations); add these inhibitors to buffers just before use.

## 3. Method

### 3.1. Homogenization and centrifugation

One round of ultracentrifugation processes 25–30g of brain tissue. Four times this amount (100 g) can be processed in a day by doing four rounds of ultracentrifugation. The following is our standard procedure for the isolation of soluble proteoglycans from 100–120 g of rat brain.

1. Chill four 250-ml centrifuge bottles in a freezer. Set up a Polytron homogenizer in a cold room (*see* **Note 2**).
2. Weigh pieces of frozen brain tissues and distribute 25–30 g of tissues per centrifuge bottle. Keep the bottles on ice.
3. Take one bottle and add buffer A up to ~1 cm below the neck of the bottle.
4. Insert the probe of a Polytron homogenizer into the centrifuge bottle and homogenize at speed 4–5 for 30 s. Keep the bottle on ice during homogenization. Put the bottle back on ice. Repeat step 4 for the other three bottles.
5. Homogenize each bottle again for 30 s.
6. Adjust the balance of bottles with buffer A. Centrifuge in a GSA rotor at 12,400*g* for 30 min at 4°C.

7. Transfer supernatants into a beaker on ice.
8. Fill eight 30-ml ultracentrifuge tubes with the supernatant. Balance precisely. Every tube must be filled up to the neck, otherwise the tube will break during ultracentrifugation and become almost impossible to remove from the rotor. Use buffer A to balance and fill the tubes.
9. Ultracentrifuge in the Sorvall T865 rotor at 60,000 rpm (~380,000*g* ) for 60 min at 4°C.
10. Collect supernatants (= soluble fraction) in a beaker on ice. Collect pellets from the bottom of the ultracentrifuge tubes with a spatula and keep them on ice until solubilization (see step 13).
11. Repeat steps 8–10 for the rest of the homogenates.
12. Combine supernatants from four rounds of ultracentrifugation (~1000 mL). Clarify by filtration through a 0.2 μm pore filter, if necessary. If the supernatants are not applied to DEAE column on the same day, add CHAPS at 0.5% to prevent aggregation during storage.
13. For isolation of the membrane fraction, combine pellets from step 10 in a Dounce or a Teflon-on-glass homogenizer, add ~50 mL of buffer B, and homogenize. Incubate homogenates for 1–4 h at 4°C, if necessary. After solubilization, remove insoluble aggregates by centrifugation at 380,000*g* for 1 h (= membrane fraction).

### 3.2. Isolation of Proteoglycan-Enriched Fractions by Anion-Exchange Chromatography

The following procedure is for the isolation of a fraction enriched for soluble proteoglycans. For the membrane fraction, the size of the column and the volume of buffers should be reduced based on the volume of the sample. We perform the entire process of chromatography in a cold room.

1. Apply the combined soluble fraction onto a column of DEAE-Sepharose Fast Flow preequilibrated with buffer C. For 1000 mL extract, a column with 25 to 30-mL bed volume is sufficient. It takes several hours to load this volume of extracts onto the column. We usually let the extract pass through the column overnight by gravity flow.
2. Wash the column with 10 column volumes of buffer C. A peristaltic pump may be used during the washing and elution steps. For a 25 to 30-mL column of DEAE Fast Flow, set the pump at around 3 mL/min.
3. Wash the column with 10 column volumes of buffer D.
4. Wash the column with 10 column volumes of buffer E.
5. Wash the column with 5 column volumes of buffer F.
6. Wash the column with buffer G until pH returns to ~8.
7. Elute with 5–6 column volumes of 0.2–1 *M* NaCl linear gradient in (e.g., 75 mL of buffer G and 75 mL of buffer H for a column of 25-mL bed volume). Collect fractions of 2 mL.
8. Determine protein concentration in each fraction.
9. Analyze fractions by SDS-PAGE. Proteoglycans appear as diffuse smears. To identify what types of proteoglycans are present in the fractions, treat them with carrier-free chondroitinase ABC (protease-free chondroitinase ABC; Seikagaku America, Rockville, MD) and analyze by SDS-PAGE. Bands appear after chondroitinase digestion represent CSPG core proteins (*see* **Note 3**). *See* **Fig. 1** for a representative result.

## 4. Notes

1. Unlike Triton X-100, CHAPS has a small micelle size and is rather easy to remove from the sample if necessary. However, CHAPS is more expensive than Triton X-100. If this poses a problem, the CHAPS in buffers C, D, E, and F can be replaced with Triton X-100. In this

case, the large portion of Triton X-100 will be washed out and replaced with CHAPS during the preelution with buffer G. Note that Triton X-100 does not efficiently solubilize GPI-anchored proteins, including glypicans. Also Triton X-100 interferes with Absorbance Measurements.

2. Homogenization with a Dounce or a Teflon-on-glass homogenizer is preferable when smaller amounts of brain tissues are processed. For a large-scale isolation, as described here, the use of these types of homogenizers is not very practical.
3. For the analysis of proteoglycan core proteins, the use of carrier-free chondroitinase ABC is important. Most of the commercial glycosaminoglycan lyases contain BSA as a stabilizer, which migrate as a thick, 67-kDa band, rendering the identification of core proteins in this region impossible. No carrier-free heparitinase or heparinase is available commercially at present. For the analysis of heparan sulfate proteoglycan core proteins, nitrous acid digestion may be useful. The presence of carrier BSA in heparitinase will not be a problem when the isolated fractions are analyzed by immunoblotting or after iodination of core proteins *(1)*.

## References


1. Herndon, M. E. and Lander, A. D. (1990) A diverse set of developmentally regulated proteoglycans is expressed in the rat central nervous system. *Neuron* **4,** 949–961.
2. Yamaguchi, Y. (2000) Chondroitin sulfate proteoglycans in the nervous system, in: *Proteoglycans: Structure, Biology and Molecular Interactions*, Iozzo, R. ed., Marcel Dekker, New York, NY, in press.
3. Yamaguchi, Y. (2000) Lecticans: organizers of the brain extracellular matrix. *Cell. Mol. Life Sci.,* **57,** 276–289.
4. Iwata, M. and Carlson, S. S. (1993) A large chondroitin sulfate proteoglycan has the characteristics of a general extracellular matrix component of adult brain. *J. Neurosci.* **13,** 195–207.
5. Yamada, H., Fredette, B., Shitara, K., Hagihara, K., Miura, R., Ranscht, B., Stallcup, W. B., and Yamaguchi, Y. (1997) The brain chondroitin sulfate proteoglycan brevican associates with astrocytes ensheathing cerebellar glomeruli and inhibits neurites outgrowth from granule neurons. *J. Neurosci.* **17,** 7784–7795.
6. Schmalfeldt M., Dours-Zimmermann M. T., Winterhalter K. H., and Zimmermann D. R. (1998) Versican V2 is a major extracellular matrix component of the mature bovine brain. *J. Biol. Chem.* **273,** 15,758–15,764
7. Stipp, C. S., Litwack, E. D., and Lander, A. D. (1993) Cerebroglycan: an integral membrane heparan sulfate proteoglycan that is unique to the developing nervous system and expressed specifically during neuronal differentiation. *J. Cell. Biol.* **124,** 149–160.
8. Yamada, H., Watanabe, K., Shimonaka, M., and Yamaguchi, Y. (1994) Molecular cloning of brevican, a novel brain proteoglycan of the aggrecan/versican family. *J. Biol. Chem.* **269,** 10,119–10,126.
9. Yamada, H., Watanabe, K., Shimonaka, M., Yamasaki, M., and Yamaguchi, Y. (1995) cDNA cloning and the identification of an aggrecanase-like cleavage site in rat brevican. *Biochem. Biophys. Res. Commun.* **216,** 957–963.
10. Rauch, U., Gao, P., Janetzko, A., Flaccus, A., Hilgenberg, L., Tekotte, H., Margolis, R. K., and Margolis, R. U. (1991) Isolation and characterization of developmentally regulated chondroitin sulfate and chondroitin/keratan sulfate proteoglycans of brain identified with monoclonal antibodies. *J. Biol. Chem.* **266,** 14,785–14,801.
11. Miura, R., Aspberg, A., Ethell, I. M., Hagihara, K., Schnaar, R. L., Ruoslahti, E., and Yamaguchi, Y. (1999) The proteoglycan lectin domain binds sulfated cell surface glycolipids and promotes cell adhesion. *J. Biol. Chem.* **274,** 11,431–11,438.

6

# Analysis of Proteoglycans and Glycosaminoglycans from *Drosophila*

**William D. Staatz, Hidenao Toyoda, Akiko Kinoshita-Toyoda, Kimberlly Chhor, and Scott B. Selleck**

## 1. Introduction

The fruitfly, *Drosophila melanogaster,* has provided a powerful model organism for the study of development. Analysis of patterning in a variety of species from the nematode, *C. elegans*, to frogs, chicks, and mice have demonstrated that the fundamental mechanisms of morphogenesis are conserved among diverse species. In recent years genetic studies in *Drosophila* have shown that proteoglycans (PGs) and their associated glycosaminoglycans (GAGs) are critical for normal development. While there exists a sophisticated set of genetic and molecular tools for the study of *Drosophila*, methods for the analysis of PGs and GAGs are not nearly as well developed. We have begun to divise such methods in order that genetic and biochemical studies can be merged to understand better the function of PGs during development at the molecular level.

We describe here methods for the analysis of PGs from *Drosophila* larvae and tissue culture cells, as well as quantitative methods for characterizing GAGs from different *Drosophila* developmental stages. The PG methods are scaled-down modifications of those used for vertebrate tissues and cultured cells *(1)* and are sufficiently sensitive to detect PGs immunochemically in DEAE column fractions from 25–50 third-instar larvae (*see* **Fig. 1**). The most practical and perhaps the only quantitative approach to studying GAG structure is the analysis of the unsaturated disaccharides derived enzymatically from GAGs. The most common detection method used for this type of analysis on material from vertebrate sources is ultraviolet absorption at 232 nm. This method is not sensitive enough for the microdetermination of samples derived from 7–10 third-instar *Drosophila* larvae or comparable amounts of other *Drosophila* tissue. The postcolumn fluorometric detection method using 2–cyanoacetamide, which we describe here, is especially well suited for analyzing unsaturated disaccharides from *Drosophila* GAGs *(2)* (*see* **Fig. 2**).

From: *Methods in Molecular Biology, Vol. 171: Proteoglycan Protocols*
Edited by: R. V. Iozzo © Humana Press Inc., Totowa, NJ

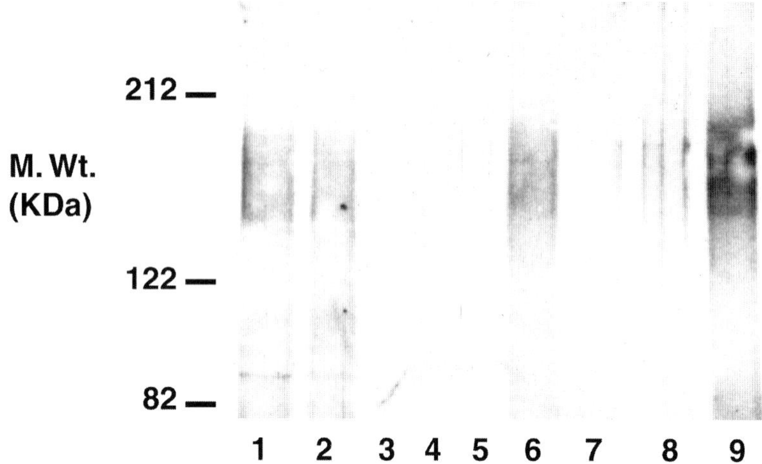


Fig. 1. Isolation of the proteoglycan Dally from third-instar *Drosophila* larvae by chromatography on DEAE Sepharose. Forty third-instar larvae were homogenized and fractionated on a DEAE column as described under **Subheading 3.1.** Thirty micrograms of protein from each fraction were separated by SDS-PAGE, blotted onto PVDF membrane (Millipore), and immunostained using a polyclonal anti-*Dally* antiserum at 1:15,000 dilution followed by goat anti-rabbit-HRP conjugate at 1:5000 and ECL (Amersham Pharmacia Biotech) according to the manufacturer's instructions. The lanes contain (1) flow-through, (2) column buffer (first wash), (3) column buffer (second wash), (4) low salt buffer, (5) 6 *M* urea, (6) pH 3.5 wash. (7) pH 8.0 wash, (8) 2 *M* NaCl (first wash), (9) 2 *M* NaCl (second wash). Dally elutes in the 2 *M* NaCl fractions as a smear between 90 and 200 kDa.

## 2. Materials

### 2.1. Purification of Proteoglycans from Drosophila

1. Homogenization buffer: 50 m*M* Tris HCl, 0.15 *M* NaCl, pH 8.0 1% CHAPS 2 × Complete Protease Inhibitors (Roche Molecular Biochemicals) (*see* **Note 1**). (6 *M* urea may be added to all buffers except the pH 8.0 wash and the high-salt buffer, or it may be added as a separate urea wash buffer [*see* **Note 2**]).
2. Column buffer: 50 m*M* Tris-HCl, 0.15 *M* NaCl, pH 8.0, 0.5% CHAPS, 1 × Complete Protease Inhibitors. (6 *M* urea: *see* **Note 2**.)
3. Low-salt buffer: 50 m*M* Tris-HCl, 0.25 *M* NaCl, pH 8.0, 0.1% Triton X-100, 1 × Complete Protease Inhibitors. (6 *M* urea: *see* **Note 2**.)
4. Urea wash buffer (used only if no urea is in the previous buffers [*see* **Note 2**]). 50 m*M* Tris-HCl, 0.25 *M* NaCl, pH 8.0, 6 *M* urea, 0.1% Triton X-100, 1 × Complete Protease Inhibitors.
5. pH 3.5 buffer: 50 m*M* Sodium formate, 0.25 *M* NaCl, pH 3.5, 0.1% Triton X-100, 2 m*M* PMSF, 10 m*M* EDTA, 6 *M* urea. Urea is always included in this buffer (*see* **Note 2**).
6. pH 8.0 wash buffer : 50 m*M* Tris-HCl, pH 8.0, 0.5% CHAPS, 2 m*M* PMSF, 10 m*M* EDTA.
7. High-salt buffer: 50 m*M* Tris-HCl, 2 *M* NaCl, pH 8.0, 0.5% CHAPS, 2 m*M* PMSF, 10 m*M* EDTA.
8. A 0.5 × 3 cm chromatography column.

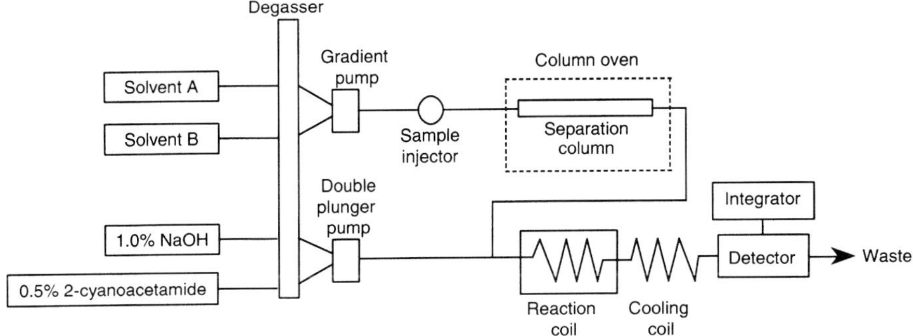


Fig. 2. Flow diagram of the postcolumn HPLC system used for analyzing the unsaturated disaccharides from *Drosophila* glycosaminoglycans. The sample is injected into the solvent pathway and separated on the column at 55° C with a NaCl gradient formed by the gradient pump. The column eluate is then mixed with 2-cyanoacetamide and NaOH, supplied by the double-plunger pump, and reacted at 125°C to form fluorescent products that, after cooling, are detected by the fluorescence detector. All solvents are degassed before they enter the separation and detection pathways. Solvent A is 1.2 m*M* *tert-n*-butylammonium hydrogen sulfate in 8.5% acetonitrile. Solvent B is 0.2 *M* NaCl in solvent A.

9. 0.3 mL of DEAE Sepharose Fast Flow (Amersham Pharmacia Biotech). (*see* **Note 3**).
10. A supply of *Drosophila*. For the volumes used in this chapter, initial experiments should be done with 40 pupae or third-instar larvae, 60 adults, 80 second-instar larvae, or 400 embryos or first-instar larvae. These numbers will have to be optimized for the needs of each individual experiment. (*see* **Note 3**).

### 2.1.1. For the Alternative Stepped NaCl Gradient Elution

1. Aliquots of column buffer (above) with total NaCl concentrations of 0.25, 0.5, 0.75, 1.0, 1.5, and 2.0 *M*. Because it is significantly less expensive, 0.1% Triton X-100 may be substituted for CHAPS in these buffers.

## 2.2. Purification of Proteoglycans from Drosophila S-2 Cells

1. CellFECTIN Reagent® (Life Technologies) (*see* **Note 4**).
2. Insect tissue culture medium. HQ-CCM3 serum-free insect cell culture medium (Hyclone) or Shields and Sang M-3 insect cell culture medium (Life Technologies, Sigma, or Hyclone) are both suitable. The latter must be supplemented with 12.5% fetal calf serum (Life Technologies, Sigma, or Hyclone) (*see* **Note 5**).
3. Transfection vector F449 *(3)* with the gene of interest under the control of the Hsp70 promoter. *See* Note 6.
4. To generate stable cell lines, a selection plasmid, such as pH8C0 (4) or pHGC0 (5), which confers methotrexate resistance (*see* **Note 6**).
5. Log-phase cultures of S-2 cells. S-2 cells are routinely maintained in dishes or flasks as nonadherent cells with a mean generation time of 24 h, and are split 1/5 every 5 days. To ensure optimum transfection rates, the cells should be split 1/1 in fresh medium the day before transfection. A thorough review of *Drosophila* cell culture can be found in Cherbas et al. *(6)*.

**Table 1**
**Disaccharide Standards from Chondroitin Sulfate**



| | | $R^1$ | $R^2$ | $R^3$ |
|---|---|---|---|---|
| 1 | ΔDi-0S | H | H | H |
| 2 | ΔDi-4S | H | $SO_3^-$ | H |
| 3 | ΔDi-6S | $SO_3^-$ | H | H |
| 4 | ΔDi-UA2S | H | H | $SO_3^-$ |
| 5 | ΔDi-diSE | $SO_3^-$ | $SO_3^-$ | H |
| 6 | ΔDi-diSB | H | $SO_3^-$ | $SO_3^-$ |
| 7 | ΔDi-diSD | $SO_3^-$ | H | $SO_3^-$ |
| 8 | ΔDi-triS | $SO_3^-$ | $SO_3^-$ | $SO_3^-$ |

6. Growth medium containing $10^{-7}$ *M* methotrexate, if stable cell lines are being generated.
7. TBS: 50 m*M* Tris-HCl, 0.15 *M* NaCl, pH 7.4, containing Complete Protease Inhibitor coctail (Roche Molecular Biochemicals).
8. 2 mL of DEAE Sepharose Fast Flow (Amersham Pharmacia Biotech).
9. A 1.4 × 4 cm glass chromatography column (Bio-Rad). A disposable polyproplyene column with a 3-mL capacity would also work.
10. The column and buffer materials under **Subheading 2.2.**

## 2.3. Extraction of Glycosaminoglycans from Drosophila

1. GAG extraction solution: 0.5% SDS, 0.1 *M* NaOH, 0.8% $NaBH_4$.
2. 1.0 *M* sodium acetate.
3. 1.0 *M* HCl.
4. 80% Ethanol.
5. Ethanol.
6. 200- to 300-µm pore disposable filter column (Fisher Scientific).
7. *Drosophila* or *Drosophila* tissue sample.

## 2.4. Microdetermination of Chondroitin Sulfate in Drosophila

1. Crude GAG solution (*see* **Subheading 3.3.**).
2. Digestion buffer: 0.2 *M* Tris–acetate buffer (pH 8.0).
3. Chondroitinase ABC (E.C. 4.2.2.4) at 10 IU/mL (Seikagaku America).
4. Chondroitinase ACII (E.C. 4.2.2.5) at 10 IU/mL (Seikagaku America).
5. Unsaturated disaccharide standards from chondroitin sulfate (Seikagaku America). *See* **Table 1**.

## 2.5. Microdetermination of Heparan Sulfate in Drosophila

1. Crude GAG solution (*see* **Subheading 3.3.**).
2. Ultrafree MC Durapore microcentrifugal filtration units with 0.45 cm pores (Millipore).
3. Ultrafree MC Biomax–5 microcentrifugal filtration units with a nominal molecular-weight exclusion of 5000 daltons (Millipore).

**Table 2**
**Disaccharide Standards from Heparin/Heparan Sulfate**

| | | $R^1$ | $R^2$ | $R^3$ |
|---|---|---|---|---|
| 1 | ΔUA-GlcNAc | H | Ac | H |
| 2 | ΔUA-GlcNS | H | $SO_3^-$ | H |
| 3 | ΔUA-GlcNAc6S | $SO_3^-$ | Ac | H |
| 4 | ΔUA2S-GlcNAc | H | Ac | $SO_3^-$ |
| 5 | ΔUA-GlcNS6S | $SO_3^-$ | $SO_3^-$ | H |
| 6 | ΔUA2S-GlcNS | H | $SO_3^-$ | $SO_3^-$ |
| 7 | ΔUA2S-GlcNAc6S | $SO_3^-$ | Ac | $SO_3^-$ |
| 8 | ΔUA2S-GlcNS6S | $SO_3^-$ | $SO_3^-$ | $SO_3^-$ |

4. Ultrafree–MC DEAE microcentrifugal filtration units with DEAE-derivatized membranes (Millipore).
5. 0.3 *M* sodium phosphate buffer (pH 6.0).
6. Loading buffer: 50 m*M* sodium phosphate buffer (pH 6.0) containing 0.15 *M* NaCl.
7. Elution buffer: loading buffer containing 1.0 *M* NaCl.
8. Heparin lyase mixture: 200 mIU/mL each of heparin lyase I (heparinase, E.C. 4.2.2.7), heparin lyase II (heparitinase II), heparin lyase III (heparitinase I, E.C. 4.2.2.8) (all from Seikagaku America).
9. Digestion buffer: 0.1 *M* sodium acetate buffer (pH 7.0) containing 10 m*M* calcium acetate.
10. Unsaturated disaccharide standards from heparin/heparan sulfate (Seikagaku America; *see* **Table 2**).
11. A Speed-Vac centrifugal evaporator (Savant).

### 2.6. HPLC Analysis

1. 1.2 m*M* tetra-*n*-butylammonium hydrogen sulfate in 8.5 % acetonitrile.
2. 0.2 *M* NaCl in 1.2 m*M* tetra-*n*-butylammonium hydrogen sulfate and 8.5% acetonitrile.
3. 0.5% 2-cyanoacetamide.
4. 1.0% NaOH.
5. HPLC system shown in **Fig, 2**. For a detailed description of equiptment used by the authors, *see* **Note 7**.

## 3. Methods

### 3.1. Purification of Proteoglycans from Drosophila

#### 3.1.1. Carry out all steps on ice or at 4°C.

1. Homogenize 40 third-instar larvae thoroughly in 3 mL of lysis buffer on ice, using 25–30 strokes with a Dounce homogenizer with the B pestle. One can also use a motorized pestle

in a glass homogenizer. Pieces of adult fly cuticle and larval mouth parts will remain as black fragments in the homogenate, but the soft tissues should be completely dissociated. (*see* **Note 1**).

2. Centrifuge the lysate to clarify and remove excess lipid. To remove the major debris, centrifuge the lysate at 12,000*g* for 5 min at 4°C. Then recentrifuged supernatant at 30,000*g* for 30 min so that any lipid is removed from the surface of the clarified supernatant. (*see* **Note 8**).
3. Place 1 mL of DEAE Sephaeose suspension in a 15-mL conical centrifge tube and let the resin settle. Remove the overlying buffer and resuspend the resin in 10 mL of high-salt buffer. Equilibrate the resin with occasional mixing for 30 min, then let the resin settle and remove the buffer. Wash the resin twice with 10 mL column buffer by resuspension and settling.
4. The clarified supernatant is mixed with 0.3 mL of DEAE Sepharose (Amersham Pharmacia) and rocked at 4°C for 1 h (*see* **Note 8**).
5. The Sepharose resin is then allowed to settle and the overlying liquid is replaced with 1 mL of column buffer. Save the removed liquid (and all further fractions) on ice for analysis at the end of the column run.
6. The resin suspension is rocked for 5 min at 4°C, then allowed to settle and the overlying buffer is removed and saved as above. The resin is then resuspended in 1 mL of column buffer and packed into a 0.5 × 3 cm column. If the resin still contains dark flakes from the tanning reaction, repeat the washing step until the flakes are no longer visible to the naked eye. Then pack the column.
7. The column is washed successively with the following buffers. Collect fractions equivalent to 1 column volume:
   a. 3 column volumes of column buffer.
   b. 3 column volumes of low-salt buffer.
   c. 3 column volumes of urea wash buffer (*see* **Note 2**).
   d. 3 column volumes of pH 3.5 wash buffer. At this pH the carboxy side chains of Asp and Glu residues are protonated, thus releasing them from the column. Sulfates on GAG chains, however, remain negatively charged and bound to the column.
   e. 4–5 column volumes of pH 8.0 wash buffer to return the pH of the column to 8.0. Check the pH of the 10 µL effluent with 1 µL of 0.1% phenol red. If it is still yellow, continue washing until the effluent is red-orange.
8. To elute the sulfated PGs from the column, add 1 column volume of high-salt buffer and let it drain to the surface of the resin. Close the column, add 1 column volume of buffer, and let it equilibrate for 20–30 min. Then reopen the column and collect the eluate. Close the column, and repeat the elution step with 1 column volume of high-salt buffer. Then wash with the remaining high-salt buffer without closing the column (*see* **Note 9**).
9. Store fractions on ice, and assay 10 µL of each fraction for protein content using the BCA or other suitable protein assay.
10. Samples can then be dialyzed, precipitated, and assayed by PAGE or other means. (*see* **Fig. 2**).

### 3.1.2. Alternative Stepped NaCl Gradient Elution

1. Follow the standard procedure through **step 7b**.
2. Wash the column successively with 2-mL aliquots of column buffer containing 0.25, 0.5, 0.75, 1.0, and 1.5 *M* NaCl.
3. Elute the most tightly bound material from the column with 3 mL of buffer with 2.0 *M* NaCl.

4. Dialyze and analyze fractions as for the standard protocol.

## 3.2. DEAE Isolation of Sulfated Proteoglycans from Drosophila S-2 cells

### 3.2.1. Transfection of Drosophila S-2 Cells

1. Add 2 µg each of expression vector and selection plasmid DNA to be transfected and 10 µg CellFECTIN reagent to 0.5 mL of serum-free medium, vortex briefly, and incubate at room temperature for 20–40 min. (*see* **Note 6**).
2. Resuspend $2 \times 10^6$ S-2 cells in mid-log phase in 0.5 mL of serum-free medium. (*see* **Note 5**).
3. Gently mix the DNA/CellFECTIN solution with the cell suspension and plate in a 35 mm-diameter culture dish or in a well of a 6-well culture plate and incubate at 23°C for 4–6 h.
4. Add 1 mL of fresh growth medium to the cells and return to the incubator overnight.
5. In the morning, add 1 mL of fresh growth medium to the cells and return to the incubator for 7–8 h.
6. Gently resuspend the cell pellets by centrifugation at 300*g* for 5 min. Then wash the cells once by resuspending and repelleting in serum-free medium to remove excess CellFECTIN reagent and DNA.
7. Replate the cells in 3 mL of fresh growth medium. If stable cell lines are to be created, add the selection medium at this time. Use $2 \times 10^{-7}$ *M* methotrexate for cells co-transfected with the selection plasmid pH8 C0.
8. Transient transfectants can be grown for 12–48 h before use. Stable transfectants are grown for 1 wk. Aliquots are then frozen in liquid nitrogen.

### 3.2.2. Harvesting Cells and Fractionation PGs

1. Heat-shock $3 \times 10^9$ log-phase S-2 cells transfected with the desired gene cloned into F449 or another suitable vector (*see* **Note 6**): 30 min at 37°C, 5 min on ice, 2–4 h recovery at 23°C.
2. Resuspend nonattached cells and put in a centrifuge tube on ice.
3. Remove attached cells by scraping in 2 × 5 mL of ice-cold TBS, pH 7.4, containing protease inhibitors, and add to centrifuge tube.
   **Carry out the following work at 4°C.**
4. Spin cells for 5 min at 3000*g*, remove the supernatant, and wash the cells twice with 10 mL of TBS.
5. Homogenize the cells in 10 mL of homogenization buffer, on ice, 3 min, with a motorized Teflon pestle in a glass homogenizer.
6. Centrifuge homogenate at 12,000*g*.
7. Place 6 mL of DEAE Sephaeose suspension in a 50-mL conical centrifge tube and let the resin settle. Remove the overlying buffer and resuspend the resin in 40 mL of high-salt buffer. Equilibrate the resin with occasional mixing for 30 min, then let the resin settle and remove the buffer. Wash the resin twice with 40 mL of column buffer by resuspension and settling.
8. Apply supernatant to a 2-mL DEAE Sepharose column and elute with the protocol described for whole animal lysates (*see* **Subheading 3.1.**).
9. Dialyze aliquots of each fraction and analyze by PAGE/Western blotting for the presence of the desired core proteins.
10. Store the remaining fractions at –20°C.

## 3.3. Extraction of Glycosaminoglycans from Drosophila

1. To extract GAGs from *Drosophila*, lyophilize 100 adult flies, 100 larvae or 250 µL of dechorionated embryos, or wash with distilled water and then lyophilize.

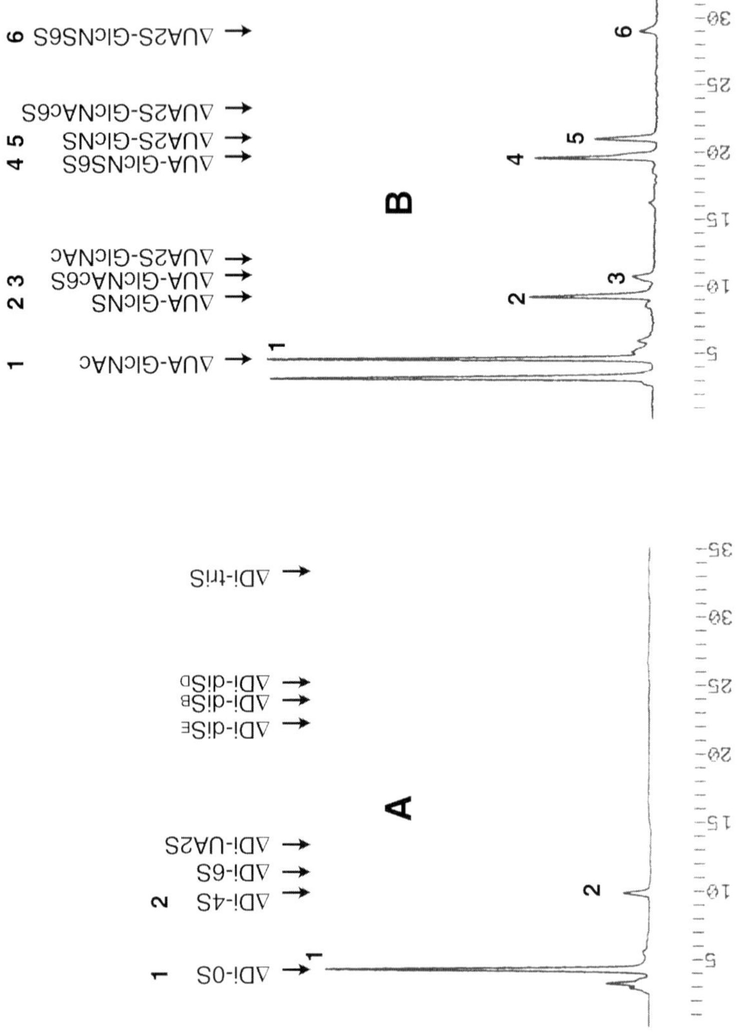
Fluorescence intensity
Retention time (min)
A
1 ΔDi-0S
2 ΔDi-4S
ΔDi-6S
ΔDi-UA2S
ΔDi-diSE
ΔDi-diSB
ΔDi-diSD
ΔDi-triS
B
1 ΔUA-GlcNAc
2 ΔUA-GlcNS
3 ΔUA-GlcNAc6S
ΔUA2S-GlcNAc
4 ΔUA-GlcNS6S
5 ΔUA2S-GlcNS
ΔUA2S-GlcNAc6S
6 ΔUA2S-GlcNS6S

2. Homogenize samples (up to 20 mg of lyophilized samples) in 1.0 mL of acetone cooled on ice. Extract for 30 min on ice, then centrifuge 15 min at 2500*g* at 4°C. Wash the precipitate 3 times with 1.0 mL of ice-cold acetone and dry under vacuum.
3. Extract samples in 1.0 mL of GAG extraction solution for 16 h at room temperature with constant stirring.
4. Neutralize samples with 200 µL of 1.0 *M* sodium acetate and 300 µL of 1.0 *M* HCl and centrifuge 10 min at 2500*g*, 4°C. Remove particulate matter from the supernatant with a 300-µm-pore disposable filter column.
5. Add 200 µL of 1.0 *M* HCl to the filtrate and remove insoluble materials by centrifugation for 10 min at 2500*g*, 4°C, then add 7 mL of ethanol to the supernatant and allow the GAGs to precipitate for 2 h at 0°C.
6. Centrifuge 10 min at 2500*g* at 4°C and remove supernatant. Wash the precipitates once with 1 mL of 80% ethanol and once with 1 mL of ethanol. These washes should be done at 4° C. Dry the precipitate under vacuum. At this stage the samples can be stored at –20°C until needed.
7. For use in determining the chondroitin sulfate and heparan sulfate disaccharide ratios (following protocols), dissolve the crude GAG pellets in 250 µL of distilled water.

### 3.4. Microdetermination of Chondroitin Sulfate in Drosophila

1. Dilute 20 µL of crude GAG solution to 100 µL with distilled water.
2. Add 10 µL of digestion buffer, 5 µL of chondroitinase ABC solution, and 5 µL of chondroitinase ACII solution to 20 µL of the diluted crude GAG solution in a 500-µL polypropylene microcentrifuge tube, then incubate at 37°C for 3 h.
3. Spin down the tube in an Eppendorff centrifuge for 30 s, then heat the tube for 2 min at 100°C and spin down again.
4. Submit 8 µL of the digest to HPLC analysis using the method under **Subheading 3.6.** Use 2 ppm of each unsaturated disaccharide mixture as standard. A typical chromatogram is shown in **Fig. 3A**.

### 3.5. Microdetermination of Heparan Sulfate in Drosophila

1. Add 400 µL of elution buffer to the Ultrafree–MC DEAE insert and spin at 5000*g* for 1 min.
2. Empty the microcentrifuge tube. Then add 400 µL of loading buffer to the insert and spin at 5000*g* for 1 min. Transfer the insert to a new microcentrifuge tube.
3. Add 50 µL of 0.3 *M* sodium phosphate buffer (pH 6.0) to 230 µL of crude GAG solution.
4. Filter the mixture through a Ultrafree MC (Durapore®, 0.45 µm).
5. Add the filtrate to the Ultrafree–MC DEAE insert and spin at 5000*g* for 1 min. Pass the sample over the membrane twice.
6. Transfer the insert to a new microcentrifuge tube. Add 400 µL of loading buffer to the insert and spin at 5000*g* for 1 min.
7. Transfer the insert to a new microcentrifuge tube. Add 100 µL of elution buffer to the insert and spin at 5000*g* for 1 min. Repeat 3 times. This fraction is the eluate.

---

Fig. 3. HPLC profiles of CS and HS disaccharides. Typical chromatograms of unsaturated disaccharides from chondroitin sulfate (A) and heparan sulfate (B) from adult *Drosophila*. The positions at which disaccharide standards migrate are indicated by the arrows. The numbers above the standards correspond to the numbers of their respective elution peaks. It is interesting to note that the only sulfated chondroitin disaccharide detected in *Drosophila* is ΔDi-4S.

8. Add the eluate to a Ultrafree MC (Biomax–5) and spin until the retentate is 30 μL.
9. Add 50 μL of distilled water to the retentate and spin. Repeat four times.
10. Remove the retentate to a new tube with a pipette, and rinse the membrane four times with 20 μL of distilled water. Add the washes to the retentate.
11. Dry the sample in a centrifugal evaporator, then dissolve it in 12 μL of distilled water.
12. Add 5 μL of digestion buffer and 5 μL of heparin lyase mixture to 5 μL of the partial purified heparan sulfate from **step 11** in a 500-μL polypropylene microcentrifuge tube, then incubate at 37°C for 16 h.
13. Centrifuge for 30 s at 12,000*g*, then heat the tube for 2 min at 100°C and recentrifuge.
14. Dry the digest in a centrifugal evaporator.
15. Add 10 μL of water, then inject 8 μL of the sample into the HPLC for analysis using the method under **Subheading 3.6.** Use 5 ppm of each unsaturated disaccharide mixture as standard. A typical chromatogram is shown in **Fig. 3B**.

### 3.6. HPLC Analysis

A flow diagram of the HPLC system is shown in **Fig. 2**. For a detailed description of the HPLC system used in this method, (*see* **Note 9**).

1. Before injecting the sample, heat the column to 55°C and equilibrate with 1.2 m*M* tetra–*n*–butylammonium hydrogen sulfate and 2 m*M* NaCl in 8.5% acetonitrile. Warm the postcolumn reaction coil to 125°C.
2. Inject the disaccharides and elute at a flow rate of 1.1 mL/min with an NaCl gradient consisting of the following segments:
   a. 0–10 min : 1–4% 0.2 *M* sodium chloride.
   b. 10–11 min: 4–15% 0.2 *M* sodium chloride.
   c. 11–20 min: 15–25% 0.2 *M* sodium chloride.
   d. 20–22 min: 25–53% 0.2 *M* sodium chloride.
   e. 22–29 min: 53% 0.2 *M* sodium chloride.

   All gradient components contain 1.2 m*M* tetra–*n*–butylammonium hydrogen sulfate and 8.5% acetonitrile. The column is then reequilibrated with 1% 0.2 *M* sodium chloride for 20 min.
3. To detect the eluted disaccharides, aqueous 0.5% (w/v) 2–cyanoacetamide solution and 1% (w/v) sodium hydroxide are added to the column effluent at the flow rate of 0.35 mL/min. The mixture is then passed through the reaction coil at 125°C, and following cooling, the effluent is monitored fluorometrically at 410 nm with an excitation wavelength of 346 nm.

## 4. Notes

1. *Drosophila* larvae and pupae can have high levels of protease activity. It is therefore imperative that all work be done in the cold and in the presence of a complete cocktail of protease inhibitors. A mixture of protease inhibitors that inhibit serine-, cysteine-, and metalloproteases must be included in the lysis buffer and the first 2–3 column wash buffers. The remaining buffers should contain at least 2 m*M* PMSF and 10 m*M* EDTA. The inclusion of these inhibitors will help prevent the degradation of the core proteins of the proteoglycans being isolated. Several manufacturers supply suitable protease mixtures in liquid or tablet form.
2. To help solubilize proteoglycan, 6 *M* urea may be added to the homogenization buffer. In that case it should be included in all buffers except the pH 8.0 wash buffer and the high salt buffer. Alternately, 6 *M* urea may be used as a separate wash step (urea buffer) to help dissociate protein aggregates from the column. Urea should always be included in the pH 3.5 buffer. It should be noted that urea in solution readily oxidizes to uric acid. Thus, all

urea-containing solutions should be prepared from the highest-quality urea available and just prior to use. Including urea in the homogenization buffer may increase the yield of proteoglycans, but it denatures the core protein and must be removed if the samples are to be subjected to digestion by GAG lyases.

3. DEAE Sepharose Fast Flow has a maximum binding capacity of 0.11–0.16 mmol/mL resin, which is equivalent to 3–4 mg of a 90-kDa protein or proteoglycan. Thus, a 0.3 mL column has a theoretical capacity for PGs from several hundred third-instar larvae.
4. Calcium phosphate has also been used extensively to transfect S-2 cells, and electroporation has been used with *Drosophila* KC167 cells. These methods are well reviewed by Cherbas et al. ***(6)***.
5. *Drosophila* cells can be maintained in a number of media. Most commonly used for S-2 cells are M3 *(7)*, D-22 ***(8)***, and Schneider's medium ***(9)***, all of which are commercially available (Sigma, Life Technologies) and must be supplemented with 5–12.5% fetal bovine serum. Purchase fetal bovine serum that has been tested for use with tissue culture cells, because not all lots of serum support insect cell growth. Hyclone's HQ-CCM3 is a serum-free medium that works well with S-2 cells and, lacking the need for serum supplementation, is less expensive than the other media. Cherbas et al. ***(6)*** report that HQ-CCM3 is also preferable to M3 for methotrexate and G418 selection.
6. The most commonly used expression vectors used in *Drosophila* cells put the gene of interest under the control of promoters for HSP 70, metallothionein, or ecdysone (reviewed by Cherbas et al. ***[6]***). Since S-2 cells take up several hundred to several thousand copies of the transfected DNA, it is not necessary to have the resistance gene used to generate stable cell lines on the same plasmid as the gene of interest. In all but a few exceptional cases, co-transfecting a selection plasmid which carries a selectable resistance gene with the gene of interest in a separate expression vector provides stable cell lines expressing the desired gene product. Methods using selection plasmids carrying resistance to methotrexate, G418, and hygromycin can be found in Ashburner ***(10)***.
7. The chromatographic equipment used by the authors includes an L-7000 gradient pump and D-7500 chromato–integrator (Hitachi Instruments) and a model 7125 sample injector with a 20-µL loop (Rheodyne). Samples were separated on a 4.6 mm × 150 mm Senshu Pak DOCOSIL A column (Senshu Scientific, Tokyo, Japan).The postcolumn reaction system consists of a double-plunger pump, AA–100–S (Eldex Laboratories), a CH–30 column heater, a FH–40 dry reaction bath, and a TC–55 thermocontroller (Brinkmann Instruments). Samples are detected with a RF–10AXL fluorescence spectrophotometer (SHIMADZU SCIENTIFIC).
8. The wound-healing response of *Drosophila* and other insects involves a rapid tanning reaction that converts components in the hemolymph into dark, cuticle-like material. When *Drosophila* are homogenized, this reaction can produce a fine suspension of particles that are not always removed by centrifugation at modest speeds. These particles can plug columns and disrupt biochemical fractionation. Two rounds of centrifugation, followed by loading and washing the ion-exchange resin in batches, allows the particles to be washed away before the resin is packed into a column.
9. By washing 1 column volume of elution buffer through the column, then closing the column and letting the resin equilibrate for 20–30 min with 1 column volume of fresh buffer, most of the bound ligand will be released into the second elution buffer fraction.
10. The most practical quantitative approach to the analysis of the unsaturated disaccharides derived enzymatically from GAGs is detection of their ultarviolet absorption at 232 nm. However, detection systems are not sensitive enough for the microdetermination of biologi-

cal samples. To improve the detection limits and specificity, pre- and postcolumn detection techniques have been investigated. The postcolumn method using 2-cyanoacetamide ***(11)*** was specifically developed to detect optimally unsaturated disaccharides from small amounts of *Drosophila* GAGs.

## References


1. Herndon, M. E. and Lander, A. D. (1990). Diverse set of developmentally regulated proteoglycans is expressed in the rat central nervous system. *Neuron* **4,** 949–961.
2. Toyoda, H., Kinoshita-Toyoda, A., and Selleck, S. B. (2000). Structural analysis of glycosoaminoglycans in *Drosophila* and *C. elegans* and demonstration that *tout-velu*, a *Drosophila* gene related to EXT tumor suppressors, affects heparan sulfate in vivo. *J. Biol. Chem.* **275,** 2269–2275.
3. Struhl, G. (1985). Near-reciprocal phenotypes caused by inactivation or indiscriminate expression of the *Drosophila* segmentation gene *ftz. Nature* **318,** 677–80.
4. Rebay, I., Fleming, R. J., Fehon, R. G., Cherbas, L., Cherbas, P., and Artavanis-Tsakonas, S. (1991). Specific EGF repeats of Notch mediate interactions with Delta and Serrate: implications for Notch as a multifunctional receptor. *Cell* **67,** 687–699.
5. Bunch, T. A. and Goldstein, L. S. (1989). The conditional inhibition of gene expression in cultured *Drosophila* cells by antisense RNA. N*ucleic Acids Res.* **17**(23), 9761–9782.
6. Cherbas, L., Moss, R., and Cherbas, P. (1994). Transformation techniques for *Drosophila* cell lines. *Meth. Cell Biol* **44**, 161–179.
7. Shields, G. and Sang, J. H. (1977). Improved medium for the culture of *Drosophila* embryonic cells. *Drosophila Information Service* **52,** 161.
8. Echalier, G. and Ohanessian, A. (1970). In vitro culture of *Drosophila melanogaster* embryonic cells. *In Vitro* **6,** 162–172.
9. Schneider, I. (1964). Differentiation of larval Drosophila eye-antennal discs in vitro. *J. Exp. Zool.* **156,** 91–103.
10. Ashburner, M. (1989). *Drosophila: A Laboratory Manual.* Cold Spring Harbor Press, Cold Spring Harbor, NY.
11. Toyoda, H., Yamamoto, H., Ogino, N., Tioda, T.,and Imanari, T. (1999). Rapid and sensitive analysis of disaccharide composition in heparin and heparan sulfate by reversed-phase ion pair chromatography on 2 µm porous silica gel column. *J. Chromatogr. A* **830,** 197–201.

# 7

# Cartilage and Smooth Muscle Cell Proteoglycans Detected by Affinity Blotting Using Biotinylated Hyaluronan

**James Melrose**

## 1. Introduction

Chondrocytes and smooth muscle cells synthesise the large CS-rich proteoglycans aggrecan *(1)* and versican *(2)* respectively. Both proteoglycans are capable of interacting with hyaluronan to form molecular aggregates that have important tissue specific functional roles to play. Aggrecan is a major matrix component of cartilage, the aggrecan aggregates are physically entrapped within the collagenous extracellular matrix of this tissue, and it is the collective interplay between this collagenous network and the aggrecan aggregates that equips this tissue with its unique viscoelastic and hydrodynamic properties and the ability to provide an almost frictionless weight-bearing surface to articulating joints *(3)*. Smooth muscle cell versican, in comparison, is a quantitatively minor component of blood vessels but nevertheless it may influence the physicochemical properties of the vessel wall *(4)*. Although the exact functional role of versican within blood vessels has yet to be fully elucidated, it is known to accumulate in intimal lesions during atherosclerosis and is implicated in the entrapment of low-density lipoprotein in the arterial wall during atherogenesis. Such interactions are likely to influence both the viscoelasticity and permeability of the vessel wall.

The interaction of aggrecan and versican with hyaluronan, which forms the basis of the assembly of massive macromolecular proteoglycan arrays within connective tissues, particularly in cartilage, is mediated via an amino-terminal globular domain that extends from the core protein, the so-called G1 domain or hyaluronan-binding region (HABR) *(3,6)*. This interaction is further stabilized via a ternary complex formed with a small glycoprotein link protein that displays an affinity both for hyaluronan and for the G1 domain *(7)*. A further amino-terminal globular domain in aggrecan, termed the G2 domain, has also been identified. This domain is absent in versican, and despite

From: *Methods in Molecular Biology, Vol. 171: Proteoglycan Protocols*
Edited by: R. V. Iozzo © Humana Press Inc., Totowa, NJ

considerable sequence homology with the G1 domain, it does not bind hyaluronan and its exact functional role has yet to be defined. Since both the HABR and the link protein display high affinities for hyaluronan, these hyaluronan-binding proteins (HABPs) have been utilized as probes for the detection and quantitation of hyaluronan in biological fluids by enzyme-linked immunosorbent assays (ELISAs) ***(8–10)***, for the localization of hyaluronan in tissue sections by immunohistological techniques ***(11–17)*** or for demonstration of hyaluronan species on Western blots ***(18)*** (*see also* Chapter 46).

The recent development of biotinylated hyaluronan ***(19)*** and its application as an affinity probe ***(20–22)*** now provides a means of monitoring not only electrophoretically resolved intact aggrecan and versican monomer, but also proteolytically derived fragments of these proteoglycans containing functional G1 domains that are likely to be generated in certain disease states. The technique employed, affinity blotting, uses similar methodology to Western blotting but does not utilize specific antibodies. Affinity blotting is so named not only to differentiate it from the related Western technique but also to emphasise the functional nature of the interaction between the G1 domain and a nonsubstituted stretch of the biotinylated hyaluronan affinity probe that forms the basis of the technique. This chapter provides a simplified protocol for the preparation of G1-containing aggrecan fragments and also demonstrates the utility of the affinity blotting technique for the visualization of G1-containing proteoglycan species resolved by sodium dodecyl sulfate polyacrylamide gel electrophoresis (SDS PAGE) or composite agarose polyacrylamide gel electrophoresis (CAPAGE).

## 2. Materials

### 2.1. Equipment and Chemicals

1. Electrophoresis system for SDS PAGE (Novex Xcell II mini electrophoresis system).
2. Blotting system for transfer of SDS PAGE gels (Novex Xcell II blot module).
3. Novex model 3540 programmable power supply.
4. Prepoured 4–20% polyacrylamide gradient Tris-glycine gels (Novex).
5. SilverXpress silver staining kit (Novex).
6. Electroelution cell (ISCO little blue tank) ***(23)***.
7. Vertical electrophoresis system for CAPAGE (Hoefer SE 600).
8. Blotting system for semidry transfer of CAPAGE gels (Bio-Rad).
9. Platform rocker or orbital shaker.
10. Agarose gel-bond support film (FMC Bioproducts, Rockland, ME, USA).
11. Acrylamide: *bis*-acrylamide 40% w/v stabilized liquid concentrate (19/1, $C = 5\%$)(Bio-Rad).
12. Agarose, low-electroendosmosis grade (Bio-Rad) .
13. Low-molecular-weight hyaluronan, 170 kDa (Fidia, Abano Terme, Italy) (*see* **Note 1**).
14. Avidin conjugated to alkaline phosphatase, 750 U/mg protein (Sigma).
15. EDC (1-ethyl-3-[3-dimethylamino)-propyl]-carbodi-imide) (Sigma).
16. NHS-LC-biotin (sulfosuccinimidyl-6-(biotinamido)-hexanoate) (Pierce).
17. Nitrocellulose (0.22 μm) (Schliecher and Schuell or Novex).
18. Chromaphor® green dye (Promega).
19. Prestained and protein molecular-weight standards for SDS PAGE (Novex).

### 2.2. Antibodies and Enzymes

1. Recombinant proMMP-3 (DuPont Pharmaceuticals) ***(24)***.
2. Rabbit anti-VDIPES (DuPont Pharmaceuticals) (*see* **Note 2**).
3. Mouse anti-VDIPEN (MAb BC-4) ***(25)*** (*see* **Note 2**).
4. Mouse anti-G1 domain of aggrecan (MAb 1-C-6, $IgG_1$ iso-type) (*see* **Note 3**) ***(26–27)***.
5. Mouse anti-KS (MAb 5-D-4, Ig M iso-type, ICN) (*see* **Note 3**) ***(28)***.
6. Mouse anti-link protein (MAb 8-A-4, $IgG_{2b}$ iso-type) (*see* **Note 3**) ***(26–27)***.
7. Rabbit anti-bovine versican ***(29)***.
8. Alkaline phosphatase conjugated goat anti-rabbit IgG (Promega).
9. Alkaline phosphatase-conjugated goat anti-mouse IgG (Promega).
10. Alkaline phosphatase-conjugated goat anti-mouse IgM (Kirkegaard and Perry).

### 2.3. Buffers

1. Buffer A (extraction buffer): 4 *M* GuHCl buffered with 0.5 *M* sodium acetate, pH 5.8, 10m*M* EDTA, PMSF (2 m*M*), 10m*M* benzamidine, and 10m*M* 6-amino hexanoic acid.
2. Buffer B (trypsin digestion buffer): 50 m*M* Tris-HCl, pH 8.2, 0.15*M* NaCl, 10m*M* $CaCl_2$, and 0.02% w/v $NaN_3$.
3. Buffer C (preparative SDS PAGE buffer): 25 m*M* Tris, 192m*M* glycine, pH 8.3, containing 0.035% w/v SDS, for use with in-situ band staining with Chromaphor® green (*see* **Note 4**).
4. Buffers D1 and D2 (electroelution cell buffers): ***(24)***, main tank buffer D1: 25 m*M* Tris, 192 m*M* glycine buffer, pH 8.3, 0.035% w/v SDS (90 mL); sample trap buffer D2 (where gel pieces are placed), 2.5 m*M* Tris, 19.2 m*M* glycine buffer, pH 8.3, containing 0.0035% w/v SDS.
5. Buffer E (MMP-3 digestion buffer): 50 m*M* Tris-HCl, pH 7.5, 10 m*M* $CaCl_2$, 0.2 *M* NaCl, 0.05% Brij 35, 0.02% $NaN_3$.
6. Buffer F (anion-exchange running buffer): 50 m*M* Tris-HCl, pH 7.2, 0.15 *M* NaCl, 7 *M* urea.
7. Buffer G (SDS PAGE running buffer): 25 m*M* Tris, 192 m*M* glycine, 0.1% w/v SDS, pH 8.3.
8. Buffer H (CAPAGE running buffer): 10 m*M* sodium acetate, pH 6.3, 1 m*M* sodium sulfate.
9. Buffer I (Western transfer buffer): 12 m*M* Tris, 98 m*M* glycine, 20% v/v methanol, pH 8.3.
10. Buffer J (transfer buffer for semidry transfer of CAPAGE gels to nitrocellulose): (Bio-Rad Trans-blot) 25 m*M* Tris, 192 m*M* glycine, pH 8.3.
11. Buffer K (affinity-blotting wash buffer): 50 m*M* Tris, 500 m*M* NaCl, 0.05% w/v Tween 20, 0.02 % w/v $NaN_3$, pH 7.2.
12. Buffer L (affinity-blotting blocking buffer): buffer K but containing 0.1% w/v Tween 20.
13. Buffer M (Western-blotting wash buffer): 50 m*M* Tris, 200 m*M* NaCl, 1% w/v BSA, pH 7.2, 0.02% w/v $NaN_3$.
14. Buffer N (Western blocking buffer): buffer M containing 5% w/v BSA.

## 3. Methods

### 3.1. Purification of Cartilage Aggrecan by CsCl Density Gradient Centrifugation

1. Freeze-shattered cartilage powder (5 g) is extracted with 50 mL of buffer A for 48 h at 4°C with constant end-over-end stirring.
2. The extract is recovered by centrifugation (10,000g × 10 min at 4°C) and brought to a starting density of 1.42 g/mL with solid CsCl. Ultracentrifugation is undertaken using a

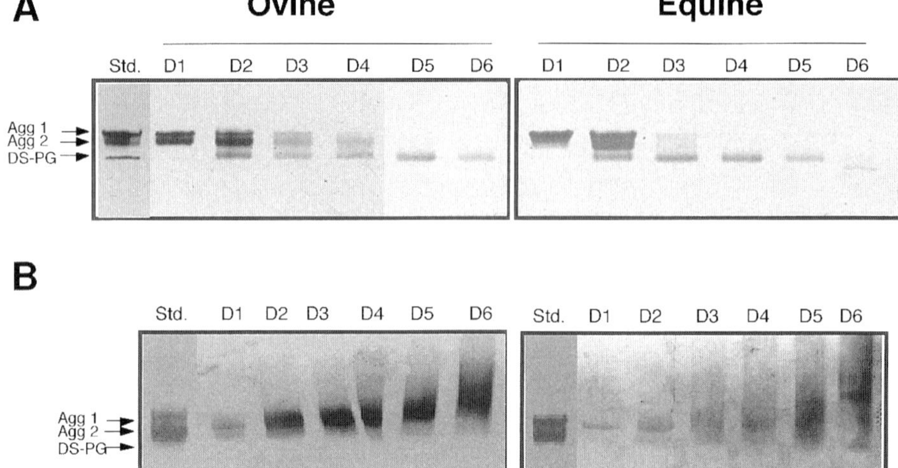


**Fig. 1.** (A) CAPAGE (toluidine blue) and (B) biotinylated hyaluronan affinity blots of ultracentrifuge fractions of the 4 *M* GuHCl ovine and equine articular cartilage extracts (**Subheading 3.1**). A crude ovine articular cartilage proteoglycan sample (1 μg hexuronic acid) that contained the Aggrecan-1, -2 (Agg-1, -2) and dermatan sulfate proteoglycan (DS-PG) populations was run in the standard (Std) lanes. Each of the other lanes had a loading equivalent to 1.4 μL of the original ultracentrifuge fraction except the D1 lanes, which, due to their higher proteoglycan contents, were diluted 1 in 10 and 0.14 μL of the original ultracentrifuge fraction was run to avoid overloading.

TV865B 8 × 17 mL titanium vertical rotor at 205,000g for 20 h at 10°C employing a Sorvall OTD 65 ultracentrifuge.

3. The tubes are manually fractionated into six equal fractions (D1–D6). The density of fractions is determined by weighing a known volume. The hexuronic acid *(30)* and protein contents *(31)* of fractions are also monitored by established methods.
4. Individual fractions are exhaustively dialyzed against distilled water and freeze-dried. The D1 fraction of density > 1.55 g/mL serves as purified aggrecan monomer, while the D2–D6 fractions contain G1-containing proteoglycan fragments of intermediate to low buoyant density and link proteins. *See* **Figs. 1** and **2.**

### 3.2. Preparation, Isolation, and Demonstration of G1-Containing Aggrecan Proteoglycan Fragments

1. Freeze-shattered bovine nasal cartilage powder (10 g) is digested with trypsin (0.1 mg/mL) in buffer B (100 mL)for 24 h at 37°C.
2. The tissue residue is collected and washed three times with 50 mL 1*M* NaCl.
3. The washed trypsinised tissue residue is then extracted for 48 h with 100 mL buffer A.
4. The clarified extract (100 mL) and HA Sepharose 4B *(32)* (100 mL gel mixture) are placed inside dialysis tubing and dialyzed overnight at 4°C against 2L of distilled water to reduce the [GuHCl] to ~0.4*M* (associative conditions).
5. The gel/sample mixture is poured into a column and eluted at room temperature with 0.4 *M* GuHCl to remove nonbound material. Bound material is fractionated by eluting with a

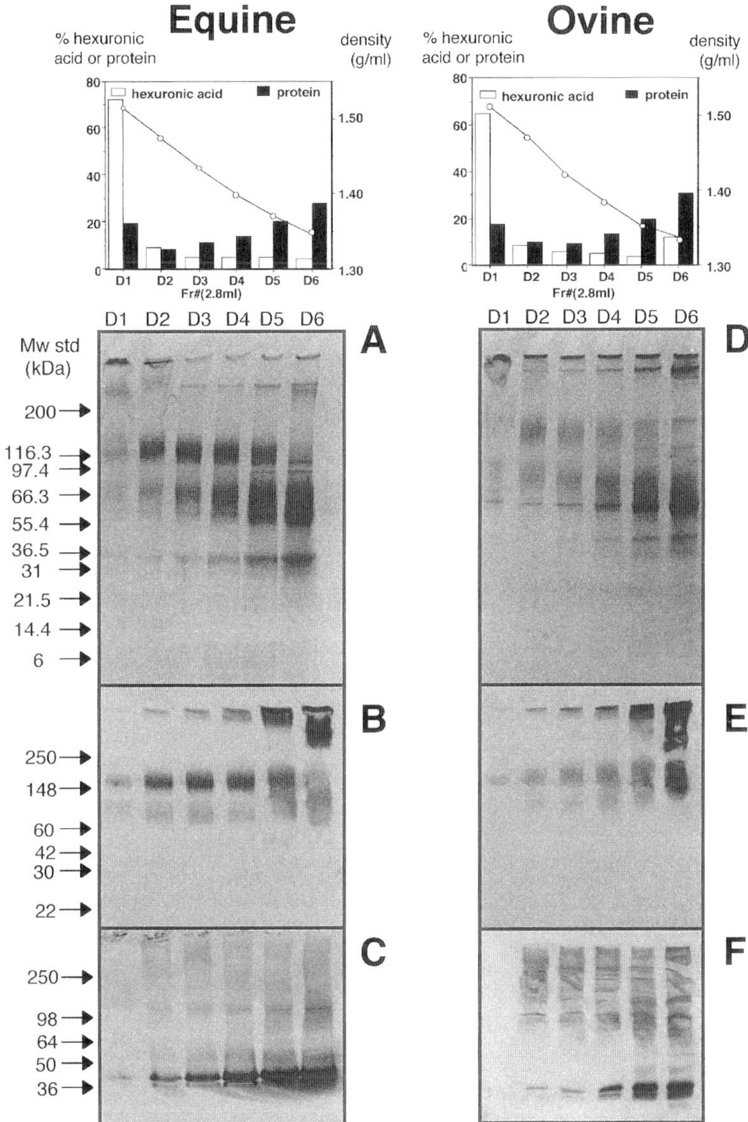


**Fig. 2.** Analytical 4–20% SDS PAGE of the same ultracentrifuge samples as examined in **Fig. 1**. Gels in segments A and D were stained with colloidal Coomassie to visualize the distribution of proteins; proteoglycan species were visualized in B and E by staining with toluidine blue. Replicate gels were also transferred to nitrocellulose (C and F), and G1 domain-containing proteoglycan species were visualized by affinity blotting. The top of the figure illustrates the distribution of protein *(31)*, proteoglycan (as hexuronic acid) ***(30),*** and density of the ultracentrifuge samples. The equivalent of 4.1 μL of original ultracentrifuge fraction was loaded per lane. The migration position of Novex broad-range and See Blue prestained standard proteins are also indicated on the left of the figure.

linear gradient of 0.4–4 *M* GuHCl in 50 m*M* Tris-HCl, pH 7, over ~6 bed volumes. Fractions of 10 mL are collected at a flow rate of 10 mL/h. Aliquots of individual fractions are assayed for protein and proteoglycan (as sulfated glycosaminoglycan) ***(33)***. The gradient is measured manually using a conductivity monitor.

6. Ice-cold ethanol (3 mL) is added to 1 mL aliquots of individual fractions and the samples shaken at 4°C. The precipitates are collected by centrifugation, washed with 75% ethanol, air-dried and redissolved in SDS PAGE application buffer (0.5 mL).
7. Aliquots (10 µL/lane) of the ethanol-precipitated samples are examined by 4–20% SDS PAGE and blotted to nitrocellulose, then probed with anti-link protein (Mab 8-A-4). HABPs are identified by affinity blotting. *See* **Fig. 3.**

### 3.3. Isolation of Aggrecan G1-Proteoglycan Species by Preparative SDS PAGE

1. Freeze-shattered ovine articular cartilage (10 g) is digested with 0.1 mg/mL trypsin in 100mL buffer B for 24 h at 37°C.
2. The tissue residue is collected and washed three times with 50 mL 1*M* NaCl.
3. The washed trypsinized tissue residue is extracted for 48 h with 100mL buffer A, the extract recovered and dialyzed exhaustively against distilled water and freeze-dried.
4. The freeze-dried material is redissolved in 0.4 M GuHCl and the HABPs isolated by HA affinity chromatography as indicated under **Subheading 3.2** but utilizing isocratic elution with 4 *M* GuHCl to remove the bound HABPs; these are dialyzed against distilled water and freeze-dried (*see* **Note 5**).
5. Purified HABP samples are redissolved in 125 m*M* Tris-HCl buffer, pH 6.8, containing 10% v/v glycerol and 1% w/v SDS application buffer, and heated at 95°C for 5 min.
6. HABP samples from HA affinity (5 mg dry weight) are electrophoresed at 125 V for 90 min in 1.5 mm 4–20% gradient gels using buffer C as running buffer. Chromaphor green stain is added to the upper electrode chamber shortly after commencement of the run in order to visualize protein band separation within the gel during the run (*see* **Note 4**).
7. After electrophoresis, the gel is placed in 5% v/v acetic acid for 10 min to improve the visualization of the protein bands. Appropriate regions of the gel are then excised and cut into 1 $mm^3$ pieces.
8. Proteins are electroeluted from the gel pieces using a "little blue-tank electroelution device" (Isco) at 2.5 mA/trap for 2 h at 4°C ***(23)***. Each of the electroelution buffer tanks of this device is filled with 90 mL buffer D1. The sample traps are filled with 10 mL buffer D2.
9. The electroeluted samples are examined by analytical SDS PAGE and Western and affinity blotting to confirm the identities of the purified HABPs (*see* **Notes 6 and 7**). *See* **Fig. 4.**

### 3.4. Aggrecan MMP-3 Digestions: Detection of G1 Domain Containing Proteoglycan Species by Western and Affinity Blotting

1. rProMMP-3 samples ***(24)*** are activated in 1 m*M* APMA in buffer E for 3 h at 37°C, then applied to HiTrap desalting columns (Pharmacia) to remove the APMA and diluted to a concentration of 40 µg/mL by protein assay ***(31)***.
2. D1 aggrecan samples (0.8 mg/mL) are dissolved overnight at 4°C in buffer E.
3. Proteoglycan samples ([0.4 mg/mL] final) are digested at 37°C for up to 18 h with 20 µg/mL final MMP-3.
4. Aliquots (0.2 mL) of the digests are removed after 0, 4, and 18 h and added to 50 µL of 20 m*M* EDTA to stop enzymatic activity and stored at –20°C until required.

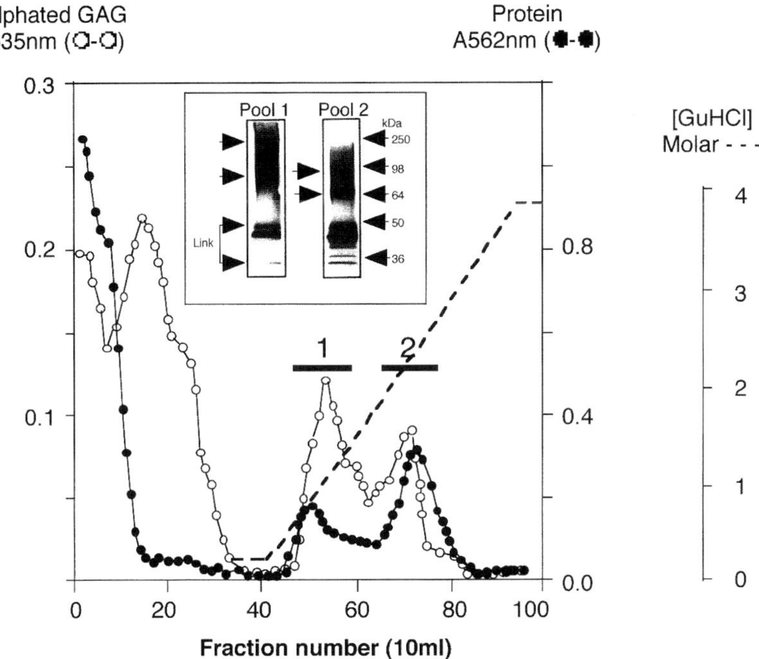


B

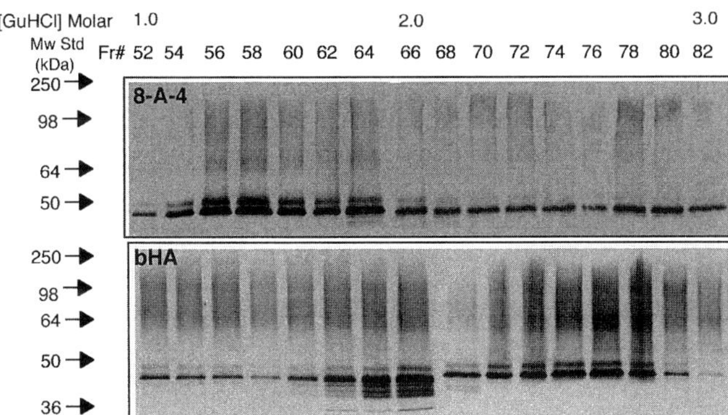


**Fig. 3.** (A) Fractionation by hyaluronan (HA) affinity chromatography *(32)* of hyaluronan-binding proteins (HABPs) in an extract of trypsinized bovine nasal cartilage, showing the distribution of proteoglycan species (as sulfated GAG) *(33)* and protein *(31)* and visualization of HABP pools 1 and 2 by affinity blotting (inset). (B) Aliquots of individual fractions were also examined by 4–20% gradient SDS PAGE, blotted to nitrocellulose, and link protein species identified using Mab 8-A-4 and HABPs identified by affinity blotting. The migration positions of Novex See Blue prestained standards are also indicated on the left-hand side of the figure.

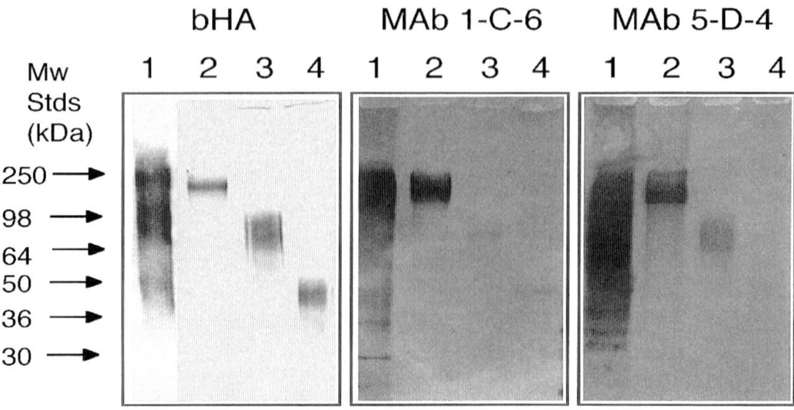


**Fig. 4.** Analysis of G1 domain-containing proteoglycan species isolated by preparative SDS PAGE of a trypsinized ovine articular cartilage sample by 4–20% SDS gradient PAGE and Western (Mab 1-C-6; anti-G1 domain; MAb 5-D-4 anti-KS) and affinity blotting (biotinylated hyaluronan). Lane 1 represents the crude trypsinized cartilage extract, which had been purified by HA affinity chromatography and lanes 2–4 show aggrecan fragments isolated from this by preparative SDS PAGE (*see* **Subheadings 3.2** and **3.3**). Lane 2 contained a high-molecular-weight core protein fragment containing the G1 and G2 domains and also some of the KS-rich region of aggrecan. Lane 3 contained a G1–G2 aggrecan core protein fragment with very little KS substitution, while lane 4 contained free G1 domain devoid of KS. Lanes 1–4 were loaded at 10 μg dry weight HABP per lane. (*see also* **Notes 4, 5** and **7**).

5. Aliquots of proteoglycan digests are examined by 4–20% Tris-glycine gradient PAGE and either silver stained or transferred to nitrocellulose membranes prior to examination by Western and affinity blotting to identify G1-containing proteoglycan species. *See* **Fig. 5.**

### 3.5. Detection of Versican in Smooth Muscle Cell Media Samples by CAPAGE and Western and Affinity Blotting

1. Monolayer cultures of smooth muscle cells are established from explant outgrowth primary human aorta explant cultures *(22,34)*.
2. 48 h media changes (100 mL) from fifth-passage subconfluent cultures are collected and stored at –20°C.
3. The media proteoglycans are ethanol precipitated with 3 volumes of ice-cold ethanol with shaking overnight at 4°C. The precipitate is collected by centrifugation, washed with 75% w/v ice-cold ethanol, and air dried.
4. The proteoglycan pellet is redissolved in 20 mL buffer F and applied to a column of 5 mL DEAE Sepharose CL6B preequilibrated in the same starting buffer. Nonbound material is washed from the column with 6 bed volumes of starting buffer. Bound material is then eluted with 4 bed volumes of 4 *M* GuHCl and the sample is dialyzed exhaustively against distilled water and freeze-dried.
5. The smooth muscle cell media proteoglycans are examined by CAPAGE, blotted to nitrocellulose, and versican identified by Western and affinity blotting. *See* **Fig. 6.**

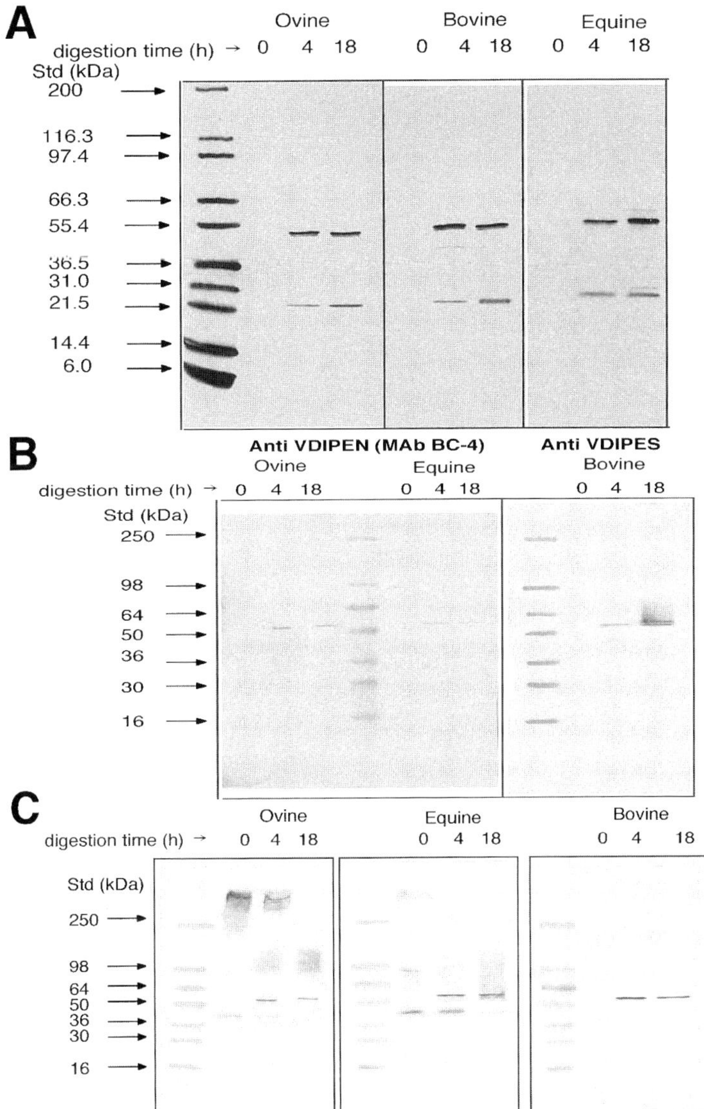


**Fig. 5.** Assessment of the generation of free G1 domain of ovine articular, bovine nasal, and equine articular cartilage aggrecan monomer by digestion with stromelysin (rMMP-3). (A) Silver-stained gel segment. (B) Western blots identifying the terminal peptide sequences VDIPEN and VDIPES on the G1 fragments generated by cleavage within the interglobular domain by MMP-3 using neo-epitope antibodies BC-4 and anti-VDIPES, respectively. (C) Demonstration of HABPs in the aggrecan digests by affinity blotting. The migration position of Novex broad-range and see blue prestained standard proteins are also indicated on the left of the figure. The aliquots of digestion mixtures electrophoresed each contained 5 μg PG monomer in the starting material prior to digestion/electrophoresis.

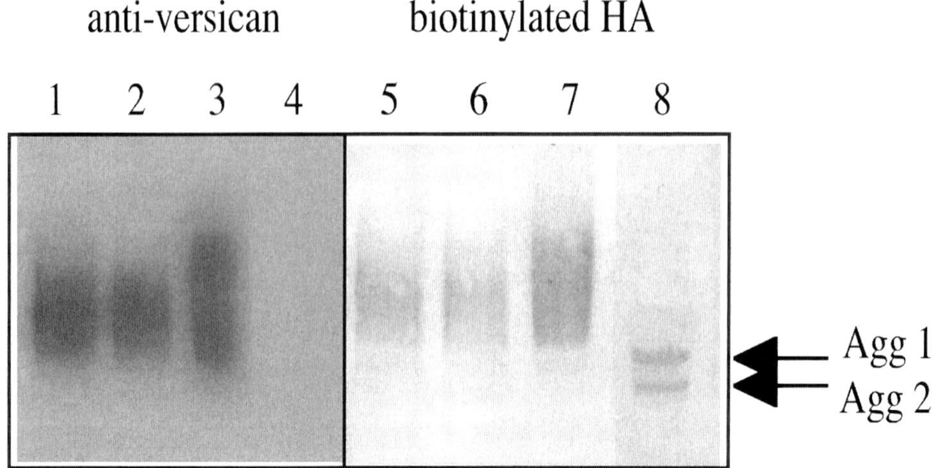


**Fig. 6.** Detection of smooth muscle cell (SMC) versican in media samples of three human SMC cell line monolayer cultures by immunoblotting using an antibody to bovine versican ***(29)*** *(lanes 1–4)* and by affinity blotting ***(20,21)*** *(lanes 5–8)*. The samples (10 μg hexuronic acid ***(30)*** *(lanes 1–3 and 5–7)* were separated by CAPAGE ***(21)*** and transferred to nitrocellulose by semi-dry blotting. Purified ovine cartilage aggrecan (1 μg hexuronic acid) was run in lanes 4 and 8 as an internal standard.

### 3.6. Western and Affinity Blotting (see Note 6)

1. SDS PAGE gels are transferred to nitrocellulose membranes at 200 mA constant current for 2 h using buffer I ***(35)***.
2. Affinity blots are blocked for 3 h in buffer L.
3. Western blots are blocked in buffer N for 3 h.
4. Affinity blots are incubated for 2 h with 2 μg/mL biotinylated hyaluronan diluted in buffer K (*see* **Note 6**).
5. MAb 1-C-6 (1/1000 dilution), 8-A-4 (1/1000 dilution), 5-D-4 (1/5000 dilution), or BC-4 (1/1000 dilution) are diluted in buffer M and allowed to bind for 2 h with constant shaking (*see* **Note 3**).
6. Affinity blots are then washed in buffer K (4 × 5 min). Western blots are washed in buffer M (4 × 5 min).
7. Appropriate secondary detection reagents are then added. For Affinity blots, avidin alkaline phosphatase, 2 μg/mL in buffer K; for Western blots, either goat anti-mouse IgG or IgM, or goat anti-rabbit IgG alkaline phosphatase conjugates 1/5000 dilution in buffer M. After a further 1 h, the membranes are washed again as in **step 6**.
8. A solution of NBT/BCIP in 0.1 *M* Tris-HCl, pH 9.5, is then added for the development step, which is allowed to proceed for up to 20 min at room temperature.
9. The blots are then washed in several changes of distilled water and dried.

### 3.7. Composite Agarose Polyacrylamide Gel Electrophoresis (CAPAGE) (21)

1. Proteoglycan samples are dissolved in 8 *M* urea, 10 m*M* sodium acetate, pH 6.3, and 1 m*M* sodium sulfate (CAPAGE dissociation buffer) overnight at 37°C then diluted 1/1 with 10 m*M* Tris-acetate/1 m*M* $Na_2SO_4$, pH 6.3, 60% w/v sucrose, and 0.01% w/v bromophenol blue (CAPAGE application buffer).
2. CAPAGE slab gels (0.15 × 14 × 14 cm) of 0.6% w/v agarose and 1.2% w/v acrylamide are cast in buffer H and left to polymerize for at least 2 h at 4°C. The gel is then preequilibrated overnight in 4 *M* urea in buffer H.
3. Proteoglycan samples are electrophoresed in a Hoefer SE 600 vertical slab gel electrophoresis system in running buffer (buffer H) at 50 V for ~5 min, then the voltage is increased to 150 V and the sample electrophoresed until the bromophenol blue marker dye has migrated ~2–3 cm (~45 min).
4. The electrophoresis running buffer is maintained at a temperature of 10 ± 2.0°C during the run, using an external cooler.
5. The gels are stained 30 min in toluidine blue (0.02% w/v) in 0.1 *M* acetic acid, destained in 0.5 *M* acetic acid 30 min, then distilled water 5 min and dried onto the hydrophilic side of agarose gel-bond support films. Final destaining is then completed on this dried film in a few minutes in distilled water.

### 3.8. Semidry Transfer of Proteoglycans from CAPAGE Gels to Nitrocellulose

1. CAPAGE gel segments (14 × 5.5 × 1.5 mm) are electroblotted to nitrocellulose membranes using buffer J in a Bio-Rad semidry blotter at 5.5 mA/cm$^2$ for 30 min *(21)*.
2. Immuno and affinity blots are undertaken as outlined under **Subheading 3.6**.

### 3.9. Preparation of HA-Sepharose 4B (22)

1. A suspension of 100 mL of hydrated gel ε-aminohexyl-Sepharose 4B is gently mixed end-over-end for 12 h at room temp with 100 mL of a 170 kDa HA solution, 1mg/mL, in milli Q distilled water.
2. The pH of the solution is adjusted to 4.75 with 0.1N HCl and 0.3g solid EDC is gradually added, the pH is maintained at 4.75 with small additions of 0.1 *N* HCl.
3. After 2 h the sample is brought to pH 7 with 0.1 *N* NaOH and the mixture dialyzed against several changes of distilled water.
4. The HA-Sepharose 4B gel is stored in 0.02% w/v sodium azide at 4°C. One mL of HA Sepharose 4B gel can bind at least 300 µg of purified bovine nasal cartilage G1 domain under the conditions outlined above in **Subheading 3.3**.

### 3.10. Preparation of Biotinylated Hyaluronan

1. The method is a modification of that of Pouyani and Prestwich (1994) ***(19)***. Low-molecular weight (170-kDa) hyaluronan (500 mg) is dissolved with overnight end-over-end stirring at 4°C in 100 mL distilled water.
2. Adipic dihydrazide (4 g) is added to this solution and the pH is adjusted to 4.75 using 0.1 *N* HCl.
3. Solid EDC (1-ethyl-3-[3-dimethylamino)-propyl])-carbodi-imide), (0.5 g) is gradually added to this solution and the pH of the reaction mixture maintained at 4.75 by the addition of small aliquots of 0.1 *N* HCl.

4. After 2 h at room temperature there is no further rise in pH; the reaction mixture is therefore adjusted to pH 7 with 1*N* NaOH and exhaustively dialyzed against distilled water (4 × 5l × 18h). Then the hydrazido-hyaluronan is freeze-dried.
5. Hydrazido-hyaluronan (100 mg) is redissolved overnight in 20 mL 0.2 M $NaHCO_3$ with end-over-end stirring at 4°C.
6. Sulfosuccinimidyl-6-(biotinamido)-hexanoate (NHS-LC-biotin, 100 mg) is added to the reaction mixture and end-over-end stirring continued for 18 h at room temperature.
7. Five milliliters of a 1*M* Tris free base solution is then added and the sample dialyzed exhaustively against distilled water to remove free biotin, then freeze-dried.
8. The level of substitution of *D*-glucuronic acid residues of the biotinylated hyaluronan with biotin typically is 35.7%, which represents 0.92 μmol biotin/mg biotinylated hyaluronan ***(20)***.

## 4. Notes

1. Tengblad (1979) ***(32)*** provides details on how to depolymerize high-molecular-weight HA to an appropriate size range if a small-molecular-weight HA is not available.
2. The Mab anti-VDIPES* recognizes the major MMP cleavage site between the G1 and G2 globular domains of bovine cartilage aggrecan. This antibody is equivalent to Mab BC-4 (anti-VDIPEN*), which recognizes the MMP cleavage site in the interglobular domain in most other mammalian species investigated. *Single-letter code for amino acid nomenclature.
3. The 1-C-6, 8-A-4, and 5-D-4 Mabs are available from the Developmental Studies Hybridoma Bank, Department of Biological Sciences, The University of Iowa, Ames, IA, USA. Mab 5-D-4 is also available commercially from ICN Biomedicals, and Seikegaku Corporation, Tokyo, Japan.
4. A lower concentration of SDS to that conventionally used in SDS PAGE ***(36)*** must be used when staining proteins insitu with Chromaphor/green dye, since the levels of SDS normally used in the Laemmli method prevent staining with this dye.
5. HABP pools 1 and 2 (*see* **Fig. 3**) may be further fractionated by dissociative gel permeation chromatography (6 *M* GuHCl) and/or anion-exchange chromatography in 8 M urea to separate link protein and free G1 domain from other G1-containing proteoglycan species.
6. Although reducing conditions are frequently used in Western blotting to facilitate protein unfolding and thus allow appropriate antigen presentations for effective interaction with Mabs ***(25–28)*** reducing conditions must be avoided in affinity blotting since binding of HA to the G1 domain requires a native (disulfide bond stabilized) conformation in this region of the aggrecan core protein.
7. Extractigel D affinity columns (Pierce) may be utilized to remove residual SDS from the HABP pools. Free G1 domain generated as under **Subheading 3.3** has been used for the histochemical visualization of HA in rat lung tissues ***(17)***. The G1–G2 and G1–G2–KS pools prepared by preparative SDS PAGE (*see* **Subheading 3.3**) have also been used as substrates for MMPs using similar methodology to that presented in **Fig. 5**.

## Acknowledgments

Professor Bruce Caterson, School of Molecular and Medical Biosciences, University of Cardiff, Wales, UK, is acknowledged for his gift of the neo-epitope antibody BC-4. Dr. Michael Pratta, Du Pont Pharmaceuticals, Wilmington, Delaware, USA kindly provided the anti-VDIPES Mab and recombinant pro MMP-3 ***(24)*** used in these studies.

## References

1. Doege, K. J. (1995) Aggrecan, in *Guidebook to the Extracellular Matrix and Adhesion Proteins* (Kreis, T. and Vale, R., eds.) Oxford University Press, Oxford, New York, Tokyo, pp. 17–18.
2 Zimmerman D. R. (1995) Versican, in *Guidebook to the Extracellular Matrix and Adhesion Proteins*. (Kreis, T. and Vale, R., eds.), Oxford University Press, Oxford, New York, Tokyo, pp. 100–101.
3 Mow, V. C. and Mak, A. F. (1986) Lubrication of diarthrodial joints, in *Handbook of Bioengineering* (Skalak, R. and Chien, S., eds.), McGraw-Hill, New York, NY, pp. 5.1–5.34.
4. Berenson, G. S., Radhakrishnamurthy, B., Srinivasan, S. R., Vijayagopal, P., Dalferes, E. R., and Sharma, C. (1984) Recent advances in molecular pathology: carbohydrate-protein macromolecules and arterial integrity—a role in atherogenesis. *Exp. Mol. Pathol.* **41,** 267–287.
5. Hardingham, T. E. and Muir, H. (1972) The specific interaction of hyaluronic acid with cartilage proteoglycans. *Biochim. Biophys. Acta* **279,** 401–405.
6. Hascall, V. C. and Heinegård, D. (1974) Aggregation of cartilage proteoglycans. The role of hyaluronic acid. *J. Biol. Chem.* **249,** 4232–4241.
7. Caterson, B. and Baker, J. (1978) The interaction of link proteins with proteoglycan monomers in the absence of hyaluronic acid. *Biochem. Biophys. Res. Commun.* **80,** 496–503.
8. Kongtawelert, P. and Ghosh, P. (1990) A method for the quantitation of hyaluronan (hyaluronic acid) in biological fluids using a labeled avidin-biotin technique. *Anal. Biochem.* **185,** 313–318.
9. Fosang, A. J., Hey, N. J., Carney, S. L., and Hardingham, T. E. (1990) An Elisa based assay for hyaluronan using biotinylated proteoglycan G1 domain (HA binding region). *Matrix* **10**, 306–313.
10. Bray, B. A., Hsu, W., and Turino, G. M. (1994) Lung hyaluronan as assayed with a biotinylated hyaluronan-binding protein. *Exp. Lung Res.* **20,** 317–330.
11. Asari, A., Miyauchi, S., Kuriyama, S., Machida, A., Kohno, K., and Uchiyama, Y. (1994) Localization of hyaluronic acid in human articular cartilage. *J. Histochem. Cytochem.* **42,** 513–522.
12. Asari, A., Miyauchi, S., Takahashi, T., Kohno, K., and Uchiyama, Y. (1992) Localization of hyaluronic acid, chondroitin sulfate and CD44 in rabbit cornea. *Arch. Histol. Cytol.* **55,** 503.
13. Edelstam, A. A. B., Lundkvist, O. E., Wells, A. F., and Laurent, T. C. (1991) Localization of hyaluronan in regions of the human reproductive tract. J. *Histochem. Cytochem.* **39,**1131–1135.
14. Hellström, S., Laurent, C., and Yoon, Y.-J. (1994) Distribution of hyaluronan in the middle and inner ear. A light microscopical study in the rat using a hyaluronan binding protein as a specific probe. *J. OtorhinoParyngal. Relat. Spec.ORL* **56,** 253–256.
15. Ripellino, J. A., Bailo, M., Margolis, R. U., and Margolis, R. K. (1988) Light and electron microscopic studies on the localization of hyaluronic acid in developing rat cerrebellum. *J. Cell. Biol.* **106,** 845.
16. Ripellino, J. A., Klinger, M. M., Margolis, R. U., and Margolis, R. K. (1985) The hyaluronic acid binding region as a specific probe for the localization of hyaluronic acid in tissue sections. Application on chick embryo and rat brain. *J. Histochem. Cytochem.* **33,** 1060.
17. Rosenbaum, D., Peric, S., Holecek, M., and Ward, H. E. (1997) Hyaluronan in radiation induced lung disease in the rat. *Radiat. Res.* **147,** 585–591.
18. Lee, H. G. and Cowman, M. K. (1994) An agarose gel electrophoretic method for analysis of hyaluronan molecular weight distribution. *Anal. Biochem.* **219,** 278–287.
19. Pouyani T. and Prestwich, G. D. (1994) Biotinylated hyaluronic acid: a new tool for probing hyaluronate-receptor interactions. *Bioconjugate Chem.* **5,** 370–372.

20. Melrose, J., Numata, Y., and Ghosh, P. (1996) Biotinylated hyaluronan: a versatile and highly sensitive probe capable of detecting nanogram levels of hyaluronan binding proteins (hyaladherins) by a novel affinity detection procedure. *Electrophoresis* **17,** 205–212.
21. Melrose, J., Little, B., and Ghosh, P. (1998) Detection of aggregatable proteoglycan populations by affinity blotting using biotinylated hyaluronan. *Anal. Biochem.* **256,** 149–157.
22. Melrose, J., Whitelock, J., Xu, Q., and Ghosh, P. (1998) Pathogenesis of abdominal aortic aneurysms: possible role of differential production of proteoglycans by smooth muscle cells. *J. Vasc. Surg.* **28,** 676–686.
23. Allington, W. B., Cordry, A. L., McCullough, G. A., Mitchell, D. E., and Nelson, J. W. (1978) Electrophoretic concentration of macromolecules. *Anal. Biochem.* **85,** 188–196.
24. Rosenfeld, S. A., Ross, O. H., Corman, J. I., Pratta, M. A., Blessington, D. L., Feeser, W. S., and Freimark, B. D. (1994). Production of human matrix metalloproteinase-3 (stromelysin) in *Escherichia coli. Gene* **139,** 281–286.
25. Hughes, C. E., Caterson, B., Fosang, A. J., Roughley, P. J., and Mort, J. S. (1995) Monoclonal antibodies that specifically recognise neoepitope sequences generated by aggrecanase and matrix metalloproteinase cleavage of aggrecan: application to catabolism in-situ and in-vivo. *Biochem. J.* **305,** 799–804.
26. Caterson, B., Christner, J. E., Baker, J. R., and Couchman, J. R. (1985) Production and characterization of monoclonal antibodies directed against connective tissue proteoglycans. *Fed Proc. Fed. Am. Soc. Exp. Biol.* **44,** 386–393.
27. Caterson, B., Calabro, T., and Hampton, A. Monoclonal antibodies as probes for elucidating proteoglycan structure and function in *Biology of the Extracellular Matrix* (Wright, T. and Mecham eds, R., eds.),Academic Press, New York, NY,1987.
28. Caterson, B., Christner, J. E., and Baker, J. R. (1983) Identification of a monoclonal antibody that specifically recognises corneal and skeletal keratan sulfate. *J. Biol. Chem.* **258,** 8848–8854.
29. Heinegård D., Bjorne-Persson A., Cöster L., Franzén S., Gardell A., Malmström M., Paulsson, M., Sandfalk, R., and Vogel, K. (1985) The core proteins of large and small interstitial proteoglycans from various connective tissues form distinct subgroups. *Biochem J.* **230**:181–194.
30. Bitter, T. and Muir, H. M. (1962) A modified uronic acid carbazole reaction. *Anal. Biochem.* **4,** 330–334.
31. Smith, P. K., Krohn, R. I., Hermanson, G. T., Mallia, A. K., Gartner, F. H., Provenzano, M. D., Fujimoto, E. K., Goeke, N. M., Olson, B. J., and Klenk, D. C. (1985) Measurement of protein using bicinchoninic acid. *Anal. Biochem.* **150,** 76–85.
32. Tengblad, A. (1979) Affinity chromatography on immobilised hyaluronate and its applications to the isolation of hyaluronate binding proteins from cartilage. *Biochim. Biophys. Acta* **578,** 281–289.
33. Farndale R. W., Buttle, D. J., and Barrett, A. J. (1986) Improved quantitation and discrimination of sulfated glycosaminoglycans by use of dimethylmethylene blue. *Biochim. Biophys Acta* **883,** 173–177.
34. Whitelock, J., Mitchell, S., and Underwood, P. A. (1997) The effect of human endothelial cell derived proteoglycans on human smooth muscle cell growth. *Cell Biol. Int.* **21,** 181–189.
35. Towbin, H., Staehelin, T, and Gordon, J. (1979) Electrophoretic transfer of proteins from polyacrylamide gels to nitrocellulose sheets. Procedure and some applications. *Proc. Natl. Acad. Sci. (USA )* **76,** 4350–4354.
36. Laemmli, U. K. (1970) Cleavage of structural proteins during the assembly of the head of bacteriophage T4. *Nature* (Lon.) **227,** 680–685.

# 8

# Preparation of Proteoglycans for N-Terminal and Internal Amino Acid Sequence Analysis

**Peter J. Neame**

## 1. Introduction

Amino acid sequence analysis of proteoglycans is performed using many of the same methods that are used for conventional proteins and glycoproteins, with some specific modifications that result from the glycosaminoglycans that are attached to the protein core. Amino acid sequence analysis of proteoglycans is more challenging than for conventional proteins for two reasons. First, because proteoglycans are large molecules, they have the same problems that are inherent in sequence analysis of larger proteins. The higher molecular weight provides for relatively high levels of nonspecific cleavage of the protein during Edman chemistry, resulting in a rapidly increasing background. Second, the presence of glycosaminoglycans, which tend to bind water, can exacerbate the problem of nonspecific hydrolysis of peptide bonds to such an extent that it may be impossible to obtain an N-terminal on a proteoglycan that is over 70% glycosaminoglycan. In an ideal situation, it is difficult to determine more than 10 amino acids from the N-terminal of an intact proteoglycan.

On the other hand, the glycosaminoglycan chains provide a unique handle that enables proteoglycans or glycosaminoglycan-containing peptides to be separated from other proteins or peptides by a combination of size-exclusion chromatography and anion-exchange chromatography. By judicious use of glycosaminoglycan-specific endoglycosidases, it is possible to separate different families of proteoglycans from each other.

Classically, proteins have been best identified or characterized by amino acid sequencing (Edman degradation). The chemistry has changed little since its original conception *(1)*, although the analytical tools available have improved enormously. In general, proteins and proteoglycans are most easily identified or characterized by obtaining internal peptide sequence, rather than an N-terminal sequence. This eliminates any problems caused by blocked N-termini and allows the investigator to make

From: *Methods in Molecular Biology, Vol. 171: Proteoglycan Protocols*
Edited by: R. V. Iozzo © Humana Press Inc., Totowa, NJ

multiple attempts to obtain unequivocal data. Direct analysis of the N-terminal is best achieved from a sample that is Western blotted onto polyvinyl difluoridene (PVDF) to reduce the risk that the N-terminal may derive from a lower-molecular-weight contaminant that, while minor on a mass basis, represents a significant contaminant on a molar basis.

Recently, it has become possible to reliably analyze peptides derived from proteolytic digestion by mass spectrometry. The resulting list of fragment masses can then be compared to a database of fragment masses and proteins identified with a high degree of confidence. This is both cheaper and quicker than Edman degradation. However, a major limitation at present is that the protein must be present in the databases to be identified, or it must be so similar in sequence to a protein that is in the databases that most peptides have identical masses. In practice, this means >98% sequence identity and a restriction to human and murine molecules, as the number of known proteins in these species begins to approach the complete repertoire found in the genome. Proteins that have extensive posttranslational modification cannot readily be identified in this way.

Mass spectrometry has also made inroads into direct amino acid sequence analysis; individual peptides can be separated and partially degraded on the mass spectrometer, and the fragments that are released analyzed to determine changes of mass. This methodology does not allow differentiation of leucine or isoleucine, but in the hands of a skilled operator can provide considerable useful data. Extensive novel amino acid sequence is still difficult to determine by mass spectrometry, but it is a valuable adjunct to classical methods *(2)*.

It is unlikely that individual investigators will have direct access to mass spectrometers or automated sequencers, so the methodologies described here are designed to enable the investigator to prepare samples that are suitable for mass spectrometry or Edman degradation performed by a core facility. As this chapter is designed to help individuals working with proteoglycans, it will be assumed that glycosaminoglycan chains will be left intact. The majority of core laboratories are not familiar with extensively glycosylated proteins and may need some education on the unique properties of these molecules.

To obtain good yields of protein, transfers from one container to another should be avoided as much as possible, particularly in the later stages of purification. Losses are generally a result of surface adsorption; it is easy to lose as much as 5 µg protein/container. If 20 µg of protein is present at the start, there will not be much left after three transfers from one container to another.

The methods described here have proved to be generally useful for structural analysis of proteoglycans (and several other proteins) in the author's lab. There are a number of more detailed descriptions of some of the techniques mentioned; the reader is referred to publications from the Protein/DNA Technology Center at The Rockefeller University *(3)* and the Howard Hughes Biopolymer/Keck Foundation Biotechnology Resource Laboratory at Yale *(4,5)*.

This overview consists of three parts: (1) preparation of proteoglycans for N-terminal analysis, (2) proteolytic digestion of proteoglycans, and (3) separation of peptides by column-based two-dimensional chromatography. The example of two-dimensional chromatography described here is size-exclusion chromatography followed by reversed-phase chromatography. However, the first dimension could be ion-exchange chromatography or affinity chromatography *(6)*. Reversed-phase high-performance liquid chromatography (HPLC) is the final preparation method of choice for Edman degradation, as the solvents are completely volatile.

## 2. Materials

### 2.1. Sample Preparation

1. Reduction and alkylation buffer 10× concentrate: 1 *M* Tris-HCl, 0.1 *M* EDTA pH 8.5.
2. Saturated guanidine HCl (this can be purchased as an 8 *M* solution in 100-mL bottles from Sigma or Pierce).
3. Dithiothreitol; keep refrigerated and preferably desiccated and under nitrogen.
4. Chromatography columns (anion-exchange, Mono-Q; large-pore size-exclusion, Superose 6, Sephacryl 400-HR, or Sepharose 4B).
5. It may be useful to perform size-exclusion chromatography in chaotropic conditions (4 *M* guanidine HCl or 6 *M* urea). Use high-quality reagents for this—Sigma Ultra grade, for example—and filter through a 0.2-μm filter.
6. Dialysis membrane with a molecular-weight cutoff of 12,000 or desalting columns (PD-10, Amersham Pharmacia Biotech).
7. Iodoacetic acid or iodoacetamide should be used fresh and kept in the dark. Both reagents should be essentially colorless crystalline salts. Free iodine released by these reagents will derivatize tyrosine and can be detected by smell and color.
8. Savant Speedvac.
9. Waterbath/heating block capable of reaching 54° C.
10. Eppendorf tubes (avoid glass and polystyrene containers for dilute solutions of proteins).
11. For Western blotting, a high-capacity form of Immobilon, such as Immobilon-P (Millipore) should be used.

### 2.2. Enzymes

#### 2.2.1. Glycosidases

Depending on the particular problem being addressed, it may be useful to remove glycosaminoglycan chains and/or N-linked and O-linked oligosaccharides. Note that carrier protein (usually bovine serum albumin) is often added to these enzymes, so removing glycosaminoglycan chains before fragmentation by proteases may cost more and be more trouble than it is worth. However, if N-terminal analysis from a blotted sample, or proteolytic mapping using in-gel digestion are desired, then removal of glycosaminoglycan chains may be useful (*see* **Note 1**). Detailed descriptions are inappropriate here, as they will depend on the particular problem being addressed.

1. Heparatinase, chondroitinase, keratanase (protease free), available from Seikagaku America. Note that all of these enzymes will leave "stubs," the undigested remnant of the glycosaminoglycan chains, attached to the protein core, and so, upon analysis, gaps will appear in the amino acid sequence.

2. N-glycosidase F (removes N-linked oligosaccharides to leave aspartate in the protein), O-glycosidase (cleaves between N-acetyl galactosamine and serine/threonine), both available from Roche Molecular Biochemicals.

### 2.2.2. Proteases (see **Note 2**)

1. Sequencing-grade trypsin (Boehringer Mannheim or Calbiochem): dissolve to obtain a concentration of 1 mg/mL in 0.1% trifluoroacetic acid. This can be stored for up to a week at 4°C and may be stored longer if frozen. Trypsin cleaves at the C-terminal of lysine and arginine. If the following amino acid is proline, cleavage does not usually occur.
2. Endoprotease Lys-C (Boehringer Mannheim): dissolve in water to obtain a concentration of 1 mg/ml. Use immediately. Endoprotease Lys-C has the same characteristics as trypsin, but does not cleave at arginine.
3. Proteolytic digestion buffer for both enzymes: 50 m*M* Tris-HCl, pH 8.0.

## 2.3. Postdigestion Size-Exclusion Separation

This is the first stage of a two-dimensional separation of digestion products. The objective is to separate peptides by size into those that are easy to resolve by reversed-phase chromatography (< 2.5 kDa, <20 amino acids) and those that are difficult to separate (typically glycopeptides or peptides with attached glycosaminoglycan chains).

1. A high-resolution, low-column-volume chromatography system is strongly recommended to avoid losses. The Amersham Pharmacia Biotech FPLC system is particularly suitable, but other HPLC systems can also be used. A UV monitor set to 280 nm is required.
2. Columns: Superose 6 or 12 (Amersham Pharmacia Biotech) are ideal, but other columns with similar porosity could be used.
3. Running buffer: 4 *M* guanidine HCl in 20 m*M* phosphate, pH 6.5 (*see* **Note 3**).

## 2.4. HPLC

### 2.4.1. Hardware

It is assumed that an HPLC is available (note that while a fast protein liquid chromatography (FPLC) can separate simple mixtures of peptides by reversed-phase HPLC, it usually has too high a dead volume for high-resolution applications); if not, most protein analysis core facilities will perform peptide separations. HPLC eluant should be monitored at 220 nm; a lower wavelength will give greater sensitivity but at the cost of increased baseline drift. A low-volume flow cell (< 5 µl internal volume) is necessary for use of 2.1-mm-internal diameter-columns. While most HPLCs have integrators attached, a chart recorder is quite adequate. It is important that there be some way of monitoring the absorbance in real time, either via a display on the monitor, a chart recorder, or an integrator display that updates rapidly.

### 2.4.2. Column

Highest sensitivity will be obtained with a column with an internal diameter of 2.1 mm or less. Flow rates in the range of 200–250 µL/min are suitable. Column length is not particularly important, although longer columns (25 cm) will give better resolution. Optimal column life will be obtained with C18 (octadecylsilane, or ODS columns). A pore size of 300 Å is necessary to separate molecules of the size of peptides.

As a general rule, the smaller the particle size, the better is the separation. Vydac columns with 5-μm particles give good separations, but other column manufacturers also make suitable columns. The column should be stored in 50% methanol in water.

#### 2.4.3. Solvents

Solvent A: 1.54 g of trifluoroacetic acid (this *must* be UV-transparent) in 1 L of HPLC-grade water (0.1% TFA v/v). It is not possible to measure TFA accurately by pipetting. Caution: trifluoroacetic acid is corrosive and causes severe burns on contact with skin.

Solvent B: 1.31 g of trifluoroacetic acid in 300 mL of HPLC-grade water and 700 mL of HPLC-grade UV-transparent acetonitrile (0.85% TFA v/v in 70% aqueous acetonitrile v/v). Caution: acetonitrile may cause liver damage with chronic exposure.

Solvents will remain usable for 2–3 wk.

## 3. Methods

### 3.1. Sample Preparation

#### 3.1.1. Preparation for Direct N-Terminal Analysis

Direct N-terminal analysis is not a particularly helpful method for identification of a proteoglycan. The hydrated glycosaminoglycan chain(s) generally result in high backgrounds, due to nonspecific peptide bond cleavage. However, it is the only method suitable for defining the N-terminus and has proved to be valuable for identifying, for example, the N-termini of the mature forms of the leucine-rich proteoglycans decorin and biglycan *(7)* (not obvious from their cDNA sequences) and for identifying processing sites in aggrecan *(8)*.

Methods for proteoglycan purification that are described in other chapters of this volume should be followed. To get clear sequence data, it is essential to pay attention to the following.

1. **Water quality.** Very high quality water is essential for all of the later stages of purification.
2. **Urea.** Some grades of urea can be contaminated with cyanates. If present, cyanates will block the N-terminal by forming a carbamylate. Urea can be purified by passing it down an ion-exchange column.
3. **Sample concentration.** The sample must be in a small volume, and therefore at a high concentration. Typical amounts that are appropriate for application to a sequencer are in the region of 10 pmol protein in a volume not greater than 100 μl. For a proteoglycan with a molecular weight of 250,000 (250 μg = 1 nmol), this corresponds to a concentration of 25 μg/mL. The highest concentration of proteoglycan will be found at the apex of peaks when they are eluted from chromatography columns. This will also be likely to be the region where they are at their highest purity. As a general rule, pooling fractions will reduce concentration and add contaminants to the sample and should be avoided.
4. **Sample quantitation.** By far the best method for defining the amount of a protein is amino acid analysis. Proteoglycans may also be quantitated by the dimethylmethylene blue (DMMB) assay *(9)* if the percentage of glycosaminoglycan is known. Quantitation of proteoglycans is somewhat empirical if dry weight cannot be obtained or the size of the core protein is unknown.

5. **Sample contaminants.** Because Edman degradation relies on repetitive transitions from basic conditions (during coupling of phenylisothiocyanate to primary amines on the protein) to highly acidic condition (during cleavage of the derivatized N-terminal amino acid), it will not work in the presence of buffers. For this reason, reversed-phase HPLC is the method of choice for final isolation. Unfortunately, it is inappropriate for intact proteoglycans. Methods that are suitable for intact proteoglycans are Western blotting onto a polyvinyldifluoridene membrane (such as Immobilon, sold by Millipore) or extensive dialysis against water. Buffer exchange into water on a Sephadex G25 column (for example, Amersham Pharmacia Biotech PD 10 columns) is an option, but there is the risk of the proteoglycan precipitating in the column. Quantitate the proteoglycan after desalting.

A method of identifying the location of proteoglycan on Western blots that we have found useful for N-terminal analysis is to perform a blot onto Immobilon P using conventional methods. A thin ribbon of membrane is then removed from the center of the lane(s) on the membrane and is blocked with casein and developed using a proteoglycan-specific antibody. This will then precisely identify the location of the proteoglycan on the remainder of the membrane. The unstained proteoglycan-containing region can then be cut out and used directly for sequencing.

### *3.1.2. Reduction and Alkylation*

The objective is to convert cystine (and cysteine) into a stable, nonoxidizable derivative. Reduction and alkylation also enables proteins to be digested more completely by opening globular domains and removes the possibility that peptides held together by disulfide bonds might be sequenced (which will confusingly give two or more N-termini).

If the proteoglycan does not have globular domains, this stage can be eliminated. Minimal losses will occur if the proteoglycan is exchanged into digestion buffer using a desalting column as described above. It is possible to digest samples on blots or in gels. *See* **Subheading 3.1.1.** and **Note 1**.

1. Dissolve proteoglycan in the smallest volume possible of water or low-concentration (<20 m*M*) buffer.
2. Add 1 volume of 8 *M* guanidine-HCl (if the proteoglycan has been collected from a size-exclusion column run in chaotrope, then it may already be in guanidine HCl).
3. Add 0.2 volume reduction and alkylation buffer to give a final concentration of 100 m*M* Tris-HCl, pH 8.5.
4. Make up a fresh 100-m*M* solution of dithiothreitol in water (15.3 mg/mL) and add enough to the proteoglycan solution to bring the concentration to 2 m*M* (1/50th volume from **step 3**).
5. Briefly degas on a Savant Speedvac, if available. If not, minimize the amount of headspace by choosing a small container. To degas, start the Speedvac spinning, and once it has come up to speed, apply a vacuum for 5 min. Release the vacuum and, when air ceases to bleed into the chamber, stop the Speedvac.
6. Gently blow nitrogen into the headspace of the tube to remove the majority of the air. Incubate under nitrogen for 45 min at 54°C.
7. Cool to room temperature and add freshly dissolved iodoacetic acid or iodoacetamide to bring the final concentration to 4.4 m*M*. It is most convenient to do this by making a 50 × concentrate (40.9 mg/mL)

8. Incubate for 20 min in the dark.
9. Desalt into digestion buffer using a G25 column or a PD10 column. Desalting by dialysis is also possible (although not recommended), and should be performed in the dark. When using Amersham Pharmacia Biotech's desalting columns, use the following protocol:
   a. Equilibrate with 5 column volumes in the buffer of choice (for example, digestion buffer).
   b. Add sample in <0.2 column volumes. Allow to run into column. For a PD-10 column, up to 1.5 mL can be effectively desalted.
   c. Wash sample into column with equilibration buffer until (sample volume + buffer volume) = 0.4 column volumes. For a PD-10 column, this will be 2.5 mL.
   d. Elute protein with 2 X sample volumes of buffer.

   It is useful to collect 0.5-mL fractions and measure the absorbance of the eluate at 280 nm; a very rough approximation is that a 1-mg/mL solution will have an absorbance of 1 at 280 nm. However, this value depends on the numbers of tyrosine and tryptophan residues in the protein and varies considerably.

### 3.2. Digestion

It is important to maintain an enzyme:substrate ratio high enough to ensure that complete digestion is obtained but low enough to ensure that background from the enzyme is immaterial. For high levels of substrate, a ratio of approximately 1:40 is appropriate. For low levels of substrate, 1:25 will be more useful. It is useful to perform a blank digest without substrate that can be used to obtain a baseline.

1. Prepare sample (your sample will either be in solution, will be on a PVDF membrane, or will be in a gel slice). For proteoglycans, a sample in solution will be assumed. Methods for digestion of blotted proteins or samples in gels have been published elsewhere in this series *(3,4)*.
2. Dry weight is the best method for substrate quantitation but is not recommended, as it can be difficult to redissolve the sample. Absorbance at 280 nm will give a very rough approximation; assume that 1 mg/mL solution will have an absorbance of 1.
3. Add enzyme at a molar ratio between 1:25 and 1:40. For example, assume that the molecular weight of the enzyme is 25,000 (approximately correct for both endoprotease Lys-C and trypsin); 1 nmol would be 25 µg. If the proteoglycan substrate has a molecular weight of 100,000, then 1 nmol would be 100 µg of core protein and 1 µg of enzyme would be appropriate. Avoid digesting less than 10 µg; losses will likely become excessive. A recommended minimum is 50 µg.

### 3.3. Postdigestion Size-Exclusion Separation

For most proteins, peptides that derive from proteolytic digestion can be separated by direct application of the digest to a reversed-phase HPLC column (*see* **Subheading 3.4.**). Proteoglycans, by virtue of the fact that they have large glycosaminoglycan chains attached to them, require an additional step. Glycosaminoglycan-containing peptides and proteins can cause deterioration of reversed-phase columns, but their most annoying characteristic is that they elute as extremely broad peaks that will give a background to many of the other peaks. It is therefore worthwhile to insert a size-exclusion chromatography step into the protocol. This has the added advantage that the position of the glycosaminoglycan chain can readily be deduced by identifying the glycosaminoglycan-containing peptide(s) by Edman degradation. A further advantage is that peptides are separated into pools that have similar molecular weights and are more easily characterized. A good example of this can be found in the characterization of a glycosaminoglycan-containing peptide from epiphycan, in which the peptide elutes as several broad peaks spread out along the entire acetonitrile gradient ***(10)***.

The following protocol is appropriate for a Superose 12 or Superose 6 column on an Amersham Pharmacia Biotech FPLC system. It is useful to keep 10% of the digest for analysis on a reversed phase column (*see* **Subheading 3.4.**).

1. Equilibrate column in 4 *M* guanidine-containing buffer at a flow rate of 0.5 ml/min.
2. Set UV monitor to 280 nm.
3. The chart recorder or integrator should be set up so that full-scale deflection corresponds to an absorbance of 0.02.
3. Load sample (typically 0.5 mL for a Superose 6 column) into loop.
4. Inject.
5. Collect fractions (0.5 mL). The void volume will appear at approximately 7 mL and the *VT* will be at approximately 18 mL.
6. The UV trace is only a guide; do *not* pool peaks based on what you can see, as you will only be detecting peptides that contain tryptophan and tyrosine.
7. Fractions can be concentrated up to twofold on a Speedvac and injected directly onto a reversed-phase column (*see* **Note 4**).
8. Fractions containing glycosaminoglycan-substituted peptides can be detected using the DMMB assay ***(9)*** (*see* **Note 5**).

Because the column is running in guanidine-HCl, proteins and peptides will have approximately twice the hydrodynamic volume that they would have under "native" conditions. The exclusion molecular weight for a Superose 6 column is thus approximately 1,000,000. It will separate peptides as small as 1 kDa from those of 2 kDa.

### 3.4. HPLC

It will be necessary to calculate an approximate lag time between the flow cell and the exit point of tubing where you will be collecting peaks. This will be (60 × (internal volume of the tube coming out of the flow cell + half the volume of the flow cell) / flow rate) seconds (*see* **Subheading 2.4.**).

1. Prepare solvents (If helium sparging is used to remove dissolved gases, perform this before addition of trifluoroacetic acid. In practice, helium sparging is not generally necessary.
2. Purge pumps.
3. If sample pH is > 7, acidify sample(s) by addition of glacial acetic acid (0.05 volume). This is necessary to avoid damaging silica-based columns.
4. Equilibrate column in solvent A and run a blank chromatogram. This would also be a useful time to run approximately 10% of the original digest to determine where peaks are likely to be.
5. Load digested proteoglycan into sample loop (100–200 μL).
6. Inject and start a gradient of 0–100% solvent B in 45–90 min.
7. Collect peaks into Eppendorf tubes by hand. Note that there is likely to be a very large void volume peak. There may occasionally be peptides here, but this is rare. Most peptides will not elute before 8 min.

   When collecting peaks, start collecting after the UV absorbence has left the baseline and add the lag time that you have previously calculated. Stop collecting at the point where the UV trace is approximately halfway back to the baseline, plus the lag time. There are fraction collectors that can do some of this process automatically, but it is much more reliable to do it by hand and label the tubes based on the peak elution time (*see* **Fig. 1**).

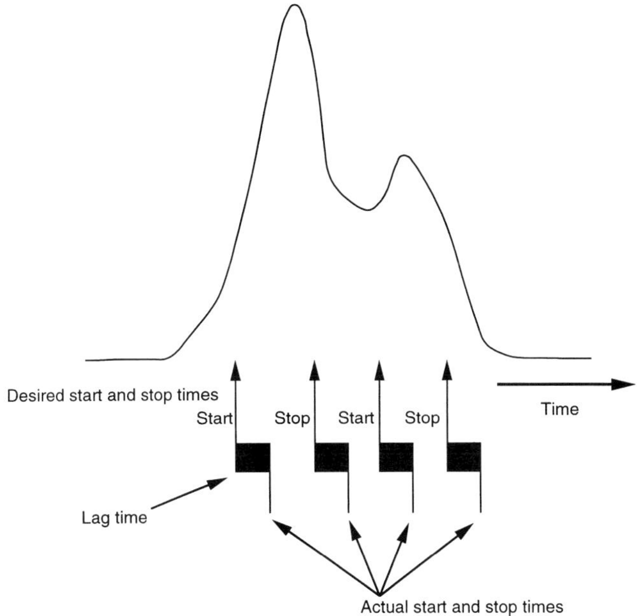


Fig. 1. The use of lag time to ensure that peaks are collected into a single tube. To ensure that peaks from reversed-phase HPLC analyses are collected into single tubes, it is necessary to know the time between a peptide arriving in the flow cell (which translates into an absorbance change) and the peptide exiting the tube downstream from the flow cell. In a typical scenario, the tube after the flow cell will be 20–30 cm long and will have an internal diameter of 0.1 mm, resulting in a volume of 20 µL. If the flow cell has a volume of 5 µL, then this means that at 250 µL / min, the lag time between the peak entering the flow cell and it appearing at the outlet will be 6 s. In practice, the point that is most important is when the apex of the peak is in the flow cell. For this, a volume of half the flow cell + the outlet tube will give a lag of 5.4 sub at 250 µL/min and 6.75 sub at 200 µL/min. It is also possible to determine this empirically by injecting a colored dye into the system. Note that for maximum peptide purity, it is important to collect the peak, but not the leading and trailing edges of the peak.

8. Tubes can either be dried in a Savant Speedvac or frozen for storage. If peptides are being sent to a core facility, there is no need to ship them cold if they have been dried. Transferring peptides from one container to another may result in complete loss.

At the end of the run, or series of runs, rinse the injection loop with deionized water in both the load and inject positions to remove salt.

The relative size of peaks will be largely related to the amino acids that are present in the peptides. The largest peaks will likely have tryptophan as one of the constituent amino acids. The next largest peaks will likely contain tyrosine. Smaller peaks may be from contaminants or will contain no aromatic amino acids. Peaks that contain single peptides will be symetrical or may tail slightly. If the HPLC separation has been preceded by gel filtration, it is possible to estimate purity by determining how many fractions a given peak is in. Typically, a single peptide should not elute from the size-exclusion volume in a volume greater than 1 ml for a 0.5 ml sample size (two or three 0.5-ml fractions).

An estimate of the likely amount of a peptide can be obtained from the UV trace. With the UV monitor set at 220 nm and using a 2.1-mm-internal-diameter column, a "typical" peak of 15 pmol of a 12-amino acid peptide should have a peak height of 0.01–0.02 AU.

## 4. Notes

1. It is possible to digest samples on blots or in gels. This is described in more detail in another volume of this series ***(3,4)***. An earlier version of the blot digestion procedure has also been published ***(11)***. These methods work well for globular proteins, but in the case of proteoglycans it can be difficult to obtain a sample density that provides an adequate signal-to-noise ratio. Proteoglycans often do not stain well with Coomassie blue or Ponceau S, so it can be difficult to identify the location of the substrate. It is unusual to be able to load a sufficient quantity of proteoglycan onto an SDS-PAGE gel to obtain the necessary amount and concentration of material. As little as 25 pmol of material can be digested (2.5 μg of the small proteoglycan, decorin), although 2–4 times this much is recommended. For reasonable results, this should be in a total gel weight of 50 mg. A blank area of gel should be used as a control.
2. Chymotrypsin tends to cut proteins in too many places to be useful; it typically cuts at the C-terminal of tyrosine, tryptophan, and phenylalanine, but will also cut at leucine. Other enzymes (endoprotease Glu-C, for example) can be a little unpredictable. Endoprotease Glu-C (V8 protease) will generally cut on the C-terminal of glutamate, but can also cut at aspartate.
3. Guanidine HCl (4 *M*) as a chaotrope for size-exclusion chromatography has the advantage that it gives good yields and prevents peptides precipitating. However, it is corrosive, and crystals of salt can cause scoring of moving parts in pumps. To avoid damage to hardware, chromatography systems should be washed thoroughly after use. Guanidine also precludes use of direct SDS-PAGE analysis, which would make it difficult to analyze GAG-containing peptides using Western blotting. Urea is an alternative chaotrope, but it should be free of cyanates, which will carbamylate amino groups (the N-terminal and lysine side chains). Urea can be cleaned by passing it down an ion-exchange column. All buffers should be filtered through a 0.2 μm filter. This is particularly important for the high-concentration chaotropes.

   If high-salt buffers are used, syringe-type pumps, such as those found on the Pharmacia FPLC, are easier to use than the piston-based pumps found on most HPLC systems. Some HPLCs have a pump design that enables the rear of the piston to be irrigated with water; this eliminates the problem of salt buildup on the rear of the piston and subsequent scoring of the piston and damage to the pump seals.
4. Some HPLC users prefer not to inject guanidine onto their columns. It is good courtesy to use your own HPLC column, but in fact the guanidine will not damage anything as long as it is at a pH below 7 and it is not left in contact with moving parts.
5. The fractions containing glycosaminoglycan can be concentrated on a Speedvac, desalted, and applied directly to the sequencer. While this may give more than one sequence, if a cDNA-derived protein sequence is available, it should be possible to deconvolute the amino acids at each cycle to determine where in the protein the glycosaminoglycan is attached. On occasion, it is possible to separate glycosaminoglycan-substituted peptides by reversed-phase HPLC, although the peaks are extremely broad ***(6,10)***.

## References


1. Edman, P. (1956) Mechanism of the phenylisothiocyanate degradation of proteins. *Nature* **177,** 667–668.
2. Williams, K. R., Samander, S. M., Stone, K. L., Saylor, M., and Rush, J. (1996) Matrix-assisted laser desorption ionization mass spectrometry as a complement to internal protein sequencing, in *The Protein Protocols Handbook* (Walker, J. M., ed.), Humana Press, Totowa, NJ, pp. 541–555
3. Fernandez, J., and Mische, S. M. (1996) Enzymatic digestion of membrane-bound proteins for peptide mapping and internal sequence analysis, in *The Protein Protocols Handbook* (Walker, J. M., ed.), Humana Press, Totowa, NJ, pp. 405–414.
4. Stone, K. L. and Williams, K. R. (1996) Enzymatic digestion of proteins in solution and in SDS-polyacrylamide gels, in *The Protein Protocols Handbook* (Walker, J. M., ed), Humana Press, Totowa, NJ, ), pp. 415–425.
5. Stone, K. L. and Williams, K. R. (1996) Reverse-phase HPLC separation of enzymatic digests of proteins, in *The Protein Protocols Handbook* (Walker, J. M., ed.), Humana Press, Totowa, NJ, pp. 427–434.
6. Barry, F. P., Rosenberg, L. C., Gaw, J. U., Gaw, J. U., Koob, T. J., and Neame, P. J. (1995) N- and O-linked keratan sulfate on the hyaluronan binding region of aggrecan from mature and immature bovine cartilage. *J. Biol. Chem.* **270,** 20516–20524.
7. Fisher, L. W., Hawkins, G. R., Tuross, N., and Termine, J. D. (1987) Purification and partial characterization of small proteoglycans I and II, bone sialoproteins I and II, and osteonectin from the mineral compartment of developing human bone. *J. Biol. Chem.* **262,** 9702–9708.
8. Sandy, J. D., Neame, P. J., Boynton, R. E., and Flannery, C. R. (1991) Catabolism of aggrecan in cartilage explants. Identification of a major cleavage within the interglobular domain. *J. Biol. Chem.* **266,** 8683–8685.
9. Farndale, R. W., Buttle, D. J., and Barrett, A. J. (1986) Improved quantitation and discrimination of sulphated glycosaminoglycans by use of dimethylmethylene blue. *Biochim. Biophys. Acta.* **883,** 173–177.
10. Johnson, H. J., Rosenberg, L., Choi, H., Garza, S., Höök, M., and Neame, P. J. (1997) Characterization of epiphycan, a small proteoglycan with a leucine-rich core protein. *J. Biol. Chem.* **272,** 18709–18717.
11. Fernandez, J., DeMott, M., Atherton, D., and Mische, S. M. (1992) Internal protein sequence analysis: enzymatic digestion for less than 10 μg of protein bound to polyvinyldifluoride or nitrocellulose membranes. *Anal. Biochem.* **201,** 255–264.

# 9

## High–Specific-Activity $^{35}$S-Labeled Heparan Sulfate Prepared from Cultured Cells

**Nicholas W. Shworak**

### 1. Introduction

Cultured cells are a facile reagent for elucidating the molecular mechanisms that regulate the biosynthesis of heparan sulfate (HS) ***(1–3)***. However, a typical confluent flask (~20 million cells) produces only a small amount of HS (1–2 μg), which is at or below the detection limit of many nonradioisotopic techniques. Fortunately, this limitation can be circumvented by the metabolic labeling of cells with $Na_2{}^{35}SO_4$. Sulfate from the culture medium is transported into the cytoplasm, where it is incorporated into the biosynthetic sulfate donor adenosine 3'-phosphate 5'-phosphosulfate (PAPS), which is transported into the Golgi apparatus ***(4)***. Specific biosynthetic enzymes transfer a sulfonyl group from PAPS onto maturing glycosaminoglycan chains.

The high sulfate content of glycosaminoglycans ensures that metabolic labeling with $Na_2{}^{35}SO_4$ is relatively selective and allows for extreme sensitivity. Although PAPS is the obligate sulfate source of all cytosolic and membrane-bound sulfotransferases, glycosaminoglycan biosynthesis typically accounts for ~70% of PAPS consumption ***(5–7)***. Consequently, $^{35}SO_4{}^{-2}$ from the culture medium is preferentially incorporated into glycosaminoglycan chains. Cells can even be incubated with virtually carrier-free $Na_2{}^{35}SO_4$ (~$1.5 \times 10^6$ Ci/mol), which allows for the production of $^{35}$S-labeled HS with an exquisitely high-specific-activity (theoretically ~$1.2 \times 10^8$ Ci/mol, given that there are ~80 sulfates per chain) ***(3,8)***. Such extreme measures are beneficial for applications that require large amounts of radioactive HS. One example is structural analysis for the identification of rare modifications, such as 3-*O*-sulfated disaccharides. Such species can comprise less than 1% of HS sulfate, and their isolation can require extensive manipulations that result in substantial loss of starting material ***(3,9,10)***. Consequently, large amounts of input radioactivity are necessary for detection of the final product. A second application is detection of sulfotransferase enzymatic activity by the

From: *Methods in Molecular Biology, Vol. 171: Proteoglycan Protocols*
Edited by: R. V. Iozzo © Humana Press Inc., Totowa, NJ

ligand affinity conversion approach (Chapter 10). Although each assay requires a small amount of HS as substrate (100,000 cpm), the analysis of a large number of samples mandates a substantial supply of $^{35}S$-labeled HS.

This chapter describes protocols for generating ~100 µCi of high-specific-activity $^{35}S$-labeled HS from cultured L cells. Suggestions are included for adapting this method to other cell types and for obtaining low-specific-activity material.

## 2. Materials

### 2.1. General

1. Water should be deionized, then either distilled or processed through a Milli-Q (Millipore)-type filtration system. Analytic-grade chemicals are used throughout.
2. *Hand-held radioactivity detectors.* A Geiger-Müller probe provides the most sensitive direct detection of unshielded β-emissions. A sodium iodide scintillation probe detects the secondary γ-radiation emitted when shielding is bombarded by β-rays (Bremsstrahlung effect).
3. *Materials for radioactive work.* Disposable bench coat and latex gloves, plastic shielding (Owl Scientific) wrapped with sufficient lead foil to capture secondary Bremsstrahlung photons. Collection of radioactive waste is best performed with a vacuum aspiration setup comprised of two serially coupled side-arm flasks. The first is the fluid reservoir and the second collects accidental overflow.
4. *Plastic ware.* All plastic tubes/flasks for containment of radioactivity should have lids that provide a complete seal. Required items include 25- and 75- $cm^2$ flasks (Nunc), 15-mL polypropylene conical tubes (Fisher # 05-539-5), and 2-mL screw-cap polypropylene tubes with a cap that contains a sealing Neoprene O-ring (Sarstedt # 500 65 716).
5. *Pipetters.* A full complement of micropipetters (Rainin) and aerosol-resistant disposable tips. An Eppendorf Repeater (VWR # 53512-500).
6. *Bottle-top filters.* For sterilization use 0.22-µm, and for removal of particles use 0.45 µm.
7. *Liquid scintillation.* Vials, 7-mL capacity (Fischer Scientific), aqueous compatible scintillant (e.g., Ecolume, New England Nuclear), and a liquid scintillation counter.
8. *Centrifuges.* A microfuge and a low-speed tabletop centrifuge.
9. A liquid-nitrogen source.

### 2.2. Labeling Cell Surface Proteoglycans

1. *Tissue culture setup.* Apart from standard equipment, a specialty gas mix of 5% $CO_2$ balanced with air should be placed near the laminar-flow hood. A plastic tube from the low-pressure regulator is connected onto a plugged 5- or 10-mL pipet, which acts as sterile filter.
2. *Appropriate sterile medium.* L cells use Dulbecco's modified Eagle's medium containing 10% fetal bovine serum, 100 units/mL of penicillin G, and 100 µg/mL of streptomycin sulfate (Life Technologies, Inc.).
3. *PBS (phosphate-buffered saline) and trypsin/EDTA.* PBS is 0.22-µm filtered and stored at room temperature (Sigma # P4417). Sterile trypsin/EDTA (# 25300-062, Life Technologies) is distributed in 50-mL aliquots. Reserve stocks are stored at –20°C and working stock is stored at 4°C.
4. *PBS/$Na_2SO_4$ and Trypsin/EDTA/$Na_2SO_4$.* Supplement the above solutions to 1 m*M* $Na_2SO_4$.
5. *Labeling medium base.* A sulfate-free base mixture is generated by combining the ingredients of **Table 1** in the ascribed order (the 100 × sulfate-free salt mixture is 4% KCl, 2% $MgCl_2 \cdot 7H_2O$ and 1.4% $NaH_2PO_4 \cdot H_2O$, autoclaved and stored at room temperature). The mixture is 0.22-µm filtered into an autoclaved 500-mL bottle. The gassing pipet (**item 1**), is fully inserted into the bottle and flow is set to initiate light bubbling. The medium is

**Table 1**
**Base Mixture for Cell Labeling**

| Stock solutions | | Final | |
|---|---|---|---|
| Composition | Source, sterilization, storage | concentration | Volume |
| $H_2O$ | | | 211 mL |
| 1.5 *M* D-glucose | 0.22-µm filtered, 4 or –20°C | 5.54 m*M* | 0.92 mL |
| 5 *M* NaCl | Autoclaved, room temperature | 100 m*M* | 5.0 mL |
| 4.4% $NaHCO_3$ | 0.22-µm filtered, 4°C | 0.22% | 12.5 mL |
| 1 *M* HEPES, pH 6.5 | 0.22-µm filtered, 4°C | 12.5 m*M* | 3.13 mL |
| 1 *M* HEPES, pH 7.9 | 0.22-µm filtered, 4°C | 12.5 m*M* | 3.13 mL |
| 11% Sodium pyruvate | 0.22-µm filtered, –20°C | 0.011% | 0.25 mL |
| 0.75% Phenol red | Room temperature | 0.00038% | 0.13 mL |
| 100 × Sulfate free salts | 0.22-µm filtered, room temperature | 1× | 2.5 mL |
| 50 × MEM amino acids | Life Technologies, #11130-051, 4°C | 1 × | 5.0 mL |
| 100 × BME vitamins | Life Technologies, #21040-050, –20°C | 1 × | 2.5 mL |
| 200 m*M* L-glutamine | 0.22-µm filtered, –20°C | 1 m*M* | 1.25 mL |
| 2% $CaCl_2$ | Autoclaved, room temperature | 0.02% | 2.5 mL |
| | | Total volume: | 250 mL |

fully $CO_2$ equilibrated when the indicator dye turns a stable orange color. The bottle is tightly capped and stored at 4°C (*see* **Note 1**).

6. *Carrier-free* $Na_2{}^{35}SO_4$ (1,495,000 Ci/mol, ~1 mCi/µL) (ICN). Centrifuge the vial before opening to pool radioactivity away from the lid. (*see* **Note 2**).
7. *0.1 M* $Na_2SO_4$ (10 mL). Dilution of stock from **Table 2**; sterilize by 0.22-µm filtration.
8. *Ice-chilled grate.* To generate a stable, level, chilled work surface, place a metal grate (VWR # 54848-316) on a level bed of ice in a large, flat Styrofoam tray.

### 2.3. Purification of Proteoglycans by Preparative DEAE Microchromatography

1. The stocks listed in **Table 2** are required to generate the DEAE column buffers listed in **Table 3**. The stocks and DEAE column buffers are stored at room temperature unless stated otherwise. All column buffers are 0.45-µm filtered.
2. *Preparative wash buffer.* A 50-mL solution of 150 m*M* NaCl, 0.03% Triton X-100, and 30 µg/mL BSA is made from **Table 2** stocks, then stored at 4°C.
3. *Preparative elution buffer.* A 50-mL solution of 1 *M* NaCl, 0.03% Triton X-100, 30 µg/mL BSA, and 2.5 µg/mL glycogen is made from **Table 2** stocks, then stored at 4°C.
4. *GUTE* (100 mL). 4 *M* guanidine·HCl (Fluka # 50935), 50 m*M* Tris pH 7.4, 1 m*M* EDTA. (*see* **Note 4**).
5. *Chondroitin sulfate* (from shark cartilage, Sigma C-4384). Make 2 mL of a 20-mg/mL solution in water, store at –20°C.
6. *Column blocking mix.* To 1.8 mL of fetal bovine serum add 0.2 mL of the above chondroitin sulfate solution. Spin in a microfuge at maximum speed for 10 min and transfer the supernatant to a clean tube.
7. *DEAE microcolumn assembly.* A gravity-flow microcolumn is assembled in the barrel of a plastic 1-mL syringe. Onto the tip, fit a Luer adapter with ~4 cm of tubing (trim a "butterfly" catheter; VWR # VT7251). This outflow tract can be plugged by inserting the point of a "yellow" tip (seal the wide end by flame-melting), whereas the top is sealed

**Table 2**
**Stock Solutions for Generating DEAE Column Buffers**

| Stock | Comments |
|---|---|
| 1.0 *M* Tris (pH 7.4), 500 mL | Autoclave sterilize. |
| 1.0 *M* $NaC_2H_3O_2$ (pH 5.0), 500 mL | |
| 5.0 *M* NaCl, 500 mL | Autoclave sterilize. |
| 0.5 *M* EDTA, pH 8.0, 500 mL | Insoluble at low pH, add NaOH pellets while stirring and monitor pH. |
| 18% (w/v) CHAPS, 100 mL | Sigma (# C3023) To 80 mL of stirring water, gradually add solid. Cover with Parafilm, stir overnight, adjust volume, 0.45-μm filter. Lot variability *see* **Note 3**. |
| 10 *M* Urea, 500 mL | A saturated solution. Complete dissolution requires heating at 37°C. If crystals form upon storage, dissolve by reheating. |
| 10% (w/v) Sodium azide, 100 mL | Extremely toxic. Wear mask, gloves, and coat when weighing out powder. |
| 1.0 *M* $Na_2SO_4$, 100 mL | |
| 20% (v/v) Triton X-100, 100 mL | Extensive stirring required. Store at 4°C. |
| 10 mg/mL BSA, 10 mL | RIA grade bovine serum albumin (BSA) (Sigma # A7888). Reserve aliquots (2 mL) stored at –80°C, working aliquots stored at –20°C. |
| 20 mg/mL Glycogen, 1 mL | Boehringer Mannheim (# 901 393) |

with a microstopper (Thomas Scientific, # 8751-K10). A bottom frit is punched out of porous polyethylene (Labconco # 4330115) with a #2 cork bore, then inserted into the barrel. Flush the syringe with water, then add 320 μL of 50/50 slurry of DEAE Sepharose Fast Flow (Pharmacia #17-0709-01). Adjust to generate a 160-μL matrix bed.

8. *Blocking the DEAE column* (*see* **Note 5**). Drain the column by gravity, then apply 200 μL of column blocking mix. After 10 min, wash the column by the sequential application of 1 mL of equilibration buffer, three applications of 1 mL of column wash, 1 mL of equilibration buffer, 1 mL of GUTE, and 1 mL of equilibration buffer. Store as plugged columns containing ~1 mL of equilibration buffer.
9. *Final column preparation.* Before sample application drain the column. Resuspend the bed by forcefully squirting, in rapid succession, two 0.5-mL volumes of equilibration buffer directly at the matrix. Allow the bed to repack by gravity flow, then repeat. Pass through 1 mL of column wash, followed by 1 mL of urea/acid wash. The column is ready for loading (*see* **Subsection 3.2, step 1**).

## 2.4. Isolation of $^{35}S$-labeled HS.

1. 10 *M* NaOH.
2. *10 M $NaBH_4$ suspension.* Wear gloves and a mask when weighing out this very toxic powder. $NaBH_4$ is extremely volatile, and open solutions should be manipulated in a fume hood. Resuspend 3.78 g in 10 mL (final volume) of 0.1 *M* NaOH, then distribute into 0.5-mL aliquots, snap-freeze on liquid nitrogen and store at –80°C. When aliquoting, mix the stock frequently to ensure even suspension.
3. *NaOH/$NaBH_4$.* Thaw an aliquot of 10 *M* $NaBH_4$ on ice, mix briefly and then transfer 9

**Table 3**
**Recipes for Most DEAE Column Buffers (100 mL each)**

| | Equilibration buffer | | Urea/acid wash | | Column wash | | Preparative loading buffer (PLB) | |
|---|---|---|---|---|---|---|---|---|
| | **Table 2** Stock | | | | | | | |
| | Final concentration | Stock volume | Final concentration | Stock volume | Final concentration | Stock volume | Final concentration | Stock volume |
| Tris | 50 m*M* | 5 mL | | | | | | |
| Sodium acetate | | | 50 m*M* | 5 mL | 50 m*M* | 5 mL | 69.4 m*M* | 6.9 mL |
| NaCl | 150 m*M* | 3 mL | 150 m*M* | 3 mL | 1558 m*M* | 31 mL | 222 m*M* | 4.4 mL |
| EDTA | 1 m*M* | 200 µL | 1 m*M* | 200 µL | 1 m*M* | 200 µL | 1.39 m*M* | 278 µL |
| CHAPS | 0.6% | 3.3 mL | 0.6% | 3.3 mL | 0.6% | 3.3 mL | 0.833% | 4.63 mL |
| Urea | | | 6 *M* | 60 mL | 6 *M* | 60 mL | 8.33 *M* | 83 mL |
| Sodium azide | 0.02% | 200 µL | 0.02% | 200 µL | 0.02% | 200 µL | 0.028% | 280 µL |
| $Na_2SO_4$ | | | 1 m*M* | 100 µL | 1 m*M* | 100 µL | 1.39 m*M* | 139 µL |
| $H_2O$ | | 88.3 mL | | 28.2 mL | | | | |

μL of suspension to 89.5 μL of 10 *M* NaOH, mix, and use immediately. The 10 *M* $NaBH_4$ stock is snap-frozen and returned to –80°C.

4. 8.54 *M* ammonium formate/1.7 *M* HCl. To 8.75 mL of 10 *M* ammonium formate, add 1.5 mL of concentrated HCl (11.6 *M*). Store at room temperature.
5. *PCI*-phenol /choroform/isoamyl alcohol, 25/24/1 (v/v/v) (100 mL). Phenol is extremely corrosive and toxic. PCI preparation and precautions are described extensively by Sambrook et al. (*11*).
6. Absolute ethanol (500 mL). Store at –20°C.
7. 1 *M* ammonium acetate, pH 7.0 (50 mL). Store at room temperature.
8. 5 *M* NaCl. *see* **Table 2**.
9. 10 mg/mL BSA. *see* **Table 2**.
10. 0.0005% Triton X-100 (15 mL). Diluted from **Table 2** stock.
11. 2-mg/mL glycogen (2 mL). Diluted from **Table 2** stock.
12. 5-U/mL chondroitinase ABC (Sigma # C-2905) (1 mL). Dissolve in PBS; aliquots of 100 μL are generated, snap-frozen on liquid nitrogen, and stored at –20°C (*see* **Note 6**).
13. Set incubator or oven at 46°C.

## 3. Methods

### 3.1. Labeling Cell Surface Proteoglycans

1. *Cell maintenance.* L cell cultures are maintained in exponential growth at 37°C, under 5% $CO_2$ in DMEM with 10% FBS.
2. *Prepare an exponentially growing culture for $^{35}$S-labeling.* A 75-cm$^2$ flasks is inoculated at 38,000 cells/cm$^2$. Two days later the flask should be ~50% confluent.
3. *Prepare the laminar-flow hood. Turn off* the laminar-flow hood so as to prevent aerosolized radioactivity from entering the hood's inner machinery. Cover the work surface with disposable bench coat. Assemble and organize all required hardware, including appropriate shielding, within the hood. A stable rack is required to hold the test tube containing the labeling medium. Double-glove, *see* **Note 7**.
4. *Prepare labeling medium.* In a 15-mL conical tube containing 4 mL of labeling base mixture (*see* **Table 1**), add 0.4 mL of fetal bovine serum, 4.6 μL of sterile 0.1 *M* $Na_2SO_4$ and then 200 mCi of carrier-free $Na_2{}^{35}SO_4$ (~200 μL), but do not mix yet. Gently blow 5% $CO_2$ gas over the surface and then cap the tube shut. (*see* **Notes 8–10**).
5. *Remove culture medium.* Hold the flask at an angle and aspirate off as much medium as possible. Gently pipet 5 mL of PBS down the side of the flask so as not to disrupt the monolayer, swirl to wash the monolayer surface, and aspirate as above. Repeat with a second PBS rinse and aspirate.
6. *Add labeling medium.* Stand the flask upright. Using a 5-mL pipet, carefully mix the labeling medium by pumping up and down once, then transfer to the bottom surface of the culture flask. To prevent accidental contamination, *avoid generating bubbles* — do not blow out the final drop from the pipet tip. Carefully place the pipet back into the empty labeling medium tube, then remove the pipet from the pipet gun. Monitor your gloves for radioactivity; if they are contaminated, replace the outer glove.
7. *Gas and incubate the flask.* Place the 5% $CO_2$ pipet just into the neck of the flask and blow a gentle gas stream of over the medium for about 10 s. Do not generate bubbles. Tightly cap the flask and tilt to spread the medium gently over the cell monolayer. Monitor gloves and flask with a β-probe, and if clean place flask in incubator for ~16 h, (*see* **Note 11**). Hang a "Radioactivity in Use" sign on the incubator. Monitor and clean the culture hood.

8. *Wash monolayer.* Chill the flask on an ice-chilled grate (*see* **Subheading 2.2., item 9**) for 10 min. The resulting negative pressure prevents expulsion of radioactive vapors. Remove the lid. Aspirate off the radioactive medium and, using a repeat pipetter (*see* **Subheading 2.1., item 5**), rinse the flask with 5 mL of ice-cold PBS containing 1 m*M* $Na_2SO_4$ (as described in step 5). Repeat for a total of 10 rinses, (*see* **Note 12**).
9. *Release extracellular proteoglycans by trypsinization.* Add 2 mL of typsin/EDTA/$Na_2SO_4$, cap flask, and distribute solution over the monolayer. Incubate at room temperature for 30 min, redistribute the solution, and incubate at 37°C for 10 min. Transfer the cell suspension to a polypropylene 15-mL conical tube. Rinse the flask with 0.5 mL of PBS and combine with the cell suspension. Cells are pelleted by centrifugation at 1500*g* and the supernatant (~2.5 mL), which contains cell surface proteoglycans, is collected.

### 3.2. Purification of Proteoglycans by Preparative DEAE Microchromatography

1. *Column loading.* Proteoglycans are partially purified and concentrated with a DEAE microcolumn, prepared by **Subheading 2.3., steps 7–9**. All effluents are to be considered as radioactive and disposed of accordingly. The proteoglycan-containing supernatant is combined with 6.5 mL of PLB (*see* **Table 3**). A pipetman is used to load 1-mL portions. Apply slowly so that the sample runs down the wall and only minimally disturbs the gel bed. *see* **Note 13**.
2. *Column washing.* Wash with 4 mL of urea/acid wash, 2 mL of equilibration buffer, and 2 mL of preparative wash buffer.
3. *Elution of labeled proteoglycans with high salt.* A 2-mL polypropylene screw-cap tube is positioned under the column. Let 200-µL of elution buffer run through the column, then repeat 4 more times to generate 1.0 mL of eluate, (*see* **Note 14**).
4. *Determine radioactivity.* Close the tube with an O-ring-containing lid (*see* **Note 15**). Briefly vortex and centrifuge. Remove 0.5 µL for liquid scintillation counting.

### 3.3. Isolation of $^{35}$S-labeled HS

1. *Cleave GAGs from proteoglycans.* Add 10 µL of NaOH/NaBH4 mix to the eluate, vortex, and centrifuge briefly. Incubate in a 46°C oven for ~16 h (*see* **Note 16**) .
2. *Reaction neutralization.* Breakdown of borohydride liberates hydrogen gas, which can lead to excessive foaming and radioactive contamination. To prevent this outcome, chill the reaction on a bed of ice for 10 min. Uncap slowly to gently release pent up gases. Carefully add 60 µL of 8.54 *M* ammonium formate/1.7 *M* HCl to the sample meniscus. Bubbling will occur as the residual borohydride degrades. Cap, vortex, and spin briefly, then return to ice and slowly uncap. Transfer 0.5 mL to a second screw-cap tube.
3. *Phenol extraction.* Add 0.75 mL of cold (taken directly from 4°C) PCI to each tube, then cap. Vortex the tubes at maximum for 30 s each. Separate the phases by centrifugation at 10,000*g*, 4°C, for 10 min. Remove as much top (aqueous) phase as possible without taking any interphase material (denatured proteins) and transfer into a single 1.5-mL microfuge tube, (*see* **Note 17**).
4. *Ethanol precipitation of GAG.* Distribute the aqueous phase among four 2-mL screw-cap tubes and add water so that each tube contains ~260 µL total volume. Add 1.25 mL of absolute ethanol per tube, and vortex briefly, (*see* **Note 18**). Mark a line down the side of each tube. Place samples in a microcentrifuge with the line toward the outside, so as to indicate the anticipated site of pellet formation. Spin at 10,000*g*, room temperature, for 30 min (*see* **Note 19**). Gently remove the tubes from the centrifuge and place in a rack so

**Table 4**
**Chondroitinase ABC Reaction**

| Stock solution | Final concentration | Required volume |
|---|---|---|
| Entire $^{35}$S-labeled GAG sample | | 139 µL |
| 5 *M* NaCl | 30 m*M* | 0.9 µL |
| 1 *M* ammonium acetate (pH 7.0) | 25 m*M* | 3.75 µL |
| 10 mg/mL BSA | 67 µg/mL | 1 µL |
| 2 mg/mL Glycogen | 13 µg/mL | 1 µL |
| 5 U/mL Chondroitinase ABC | 20 mU/reaction | 4 µL |
| | Total volume: | 150 µL |

that the pellet formation site is oriented to the front. The GAG pellets are extremely small and very difficult to see. At best, they might only be detected as a slightly cream-colored hazy line when viewed from a tangential aspect. If extremely high incorporation is achieved, the $^{35}$S generates sufficient secondary Bremsstrahlung photons to detect the pellet with a hand-held sodium iodide tube. Carefully aspirate off ethanol to generate a nearly "dry" pellet, (*see* **Note 20**).

5. *Dissolve GAG.* Place 10 µL of water onto the pellet of each tube and sit for 10 min. Add 50 µL of water to the first tube, vortex briefly and centrifuge, then transfer the contents to the second tube, vortex, and spin. Repeat until the forth tube contains the entire 90 µL. Repeat this process with a 49-µL water wash. Remove 0.5 µL and determine radioactivity by liquid-scintillation counting. Typically, ~70% of the DEAE $^{35}$S is recovered.
6. *Chondroitinase treatment.* The chondroitin component is eliminated by GAG lyase digestion in the reaction shown in **Table 4.** Incubate the reaction at 37°C for 1 h, add an additional 4 µL of chondroitinase ABC and incubate an additional hour.
7. *Purify HS.* The reaction proteins are next removed by phenol extraction followed by ethanol precipitation to purify the $^{35}$S-labeled HS away from chondroitin degradation products: Add 1.2 mL of PCI to the reaction, cap, and vortex at maximum for 30 s. Separate the phases by centrifugation at 10,000*g*, room temperature, for 2 min. Transfer the aqueous phase to a 2-mL screw-cap tube containing 4 µL of 5 *M* NaCl. To enhance recovery, add 50 µL of water to the initial reaction tube, vortex, spin, then recover and pool with the first aqueous phase (~200 µL total volume). Add 1 mL of absolute ethanol, centrifugation at 10,000*g*, room temperature, for 30 min and aspirate, as described in **Note 20**. Add 200 µL of 0.0005% Triton X-100, let sit at room temperature for 10 min, then resuspend by two quick vortex/spin cycles. Remove 0.5 µL and determine radioactivity by liquid-scintillation counting. Snap-freeze sample on liquid nitrogen and store at –80C°.

## 4. Notes

1. The described recipe maximizes L cell production of HS. Others might find it more convenient to use a preformulated medium such as sulfate-free BME (Life Technologies, # 31300) or MEM (Life Technologies, # 22300). The medium is $CO_2$ equilibrated to maximize reproducibility. Equilibrate the medium prior to the addition of $^{35}$S so as to minimize the possibility of contamination.
2. High levels of $^{35}$S can be safely manipulated when diligent respect is paid to the three basic principles governing radioactive lab work: (1) minimize personal exposure time and (2) use appropriate shielding. In particular, always consider the secondary γ-radia-

tion emissions, as this is an underappreciated point for high-level $^{35}S$ work. High-level stocks of $^{35}S$ should always be stored in lead containers. (3) Monitor for contamination. Individuals who use levels greater than 5 mCi at once should wear a radiation badge and should undergo a urine test shortly after use, to rule out personal contamination. The institution's radiation protection office should be consulted in advance regarding the policies and procedures for use and disposal of such materials.

3. Certain lots of CHAPS (3-[(3-chloamidopropl)-dimethylammonio]-1-propanesulfonic acid) contain an impurity that causes a gradual precipitant formation, which clogs columns. To test for the contaminant, store the 18% CHAPS stock at 4°C for 1 wk. If a precipitate has formed, then the supplier must send a replacement from a different lot.
4. Solubility of guanidine·HCl is highly dependent on purity. High quality material is readily soluble at 4 *M*, whereas less expensive grades are insoluble.
5. This treatment of the column matrix blocks "nonspecific" sites that will irreversibly bind GAG. Consequently, quantitative recover of extremely small amounts of radiolabeled proteoglycans (<1 fmol or <200 pg of a 200-kDa molecule) can be obtained.
6. Under these conditions the enzyme can withstand multiple freeze–thaw cycles. PBS must be used; if water is used the activity is destroyed by the freezing process.
7. Wear two sets of latex gloves to minimize risk of radioactive contamination. It is advisable to have a separate pipet gun dedicated for radioactive work, as this equipment is easily contaminated.
8. This recipe makes sufficient medium to cover a monolayer in a flask that has an actual growth surface area of <75 $cm^2$. Some flasks have larger surfaces, which can lead to drying out of the monolayer. In this case, the use 5 mL base medium, 0.5 mL fetal bovine serum, 5.75 μL of 0.1 *M* $Na_2SO_4$, and 200 mCi of $Na_2{}^{35}SO_4$.
9. The labeling mixture has a total sulfate concentration of ~200 μ*M*. This is the minimum concentration to prevent undersulfation of GAGs in L cells (***12***); thus, $^{35}S$-labeled HS is of normal structure and maximal specific activity. The formulation has only three sulfate sources, fetal bovine serum, cold sodium sulfate, and carrier-free $Na_2{}^{35}SO_4$ which respectively provide about 70, 30, and 100 μ*M*. Antibiotics are purposefully omitted, as these can contribute substantial amounts of sulfate, which will reduce net incorporation. Expect ~$2 \times 10^8$ cpm of $^{35}S$-labeled HS. If reduced radioactivity is required, decrease the input $Na_2{}^{35}SO_4$ and compensate with cold carrier to maintain total sulfate at 200 μ*M*.
10. Certain cell types (e.g., CHO) efficiently salvage sulfate from amino acid degradation and do not generate undersulfated GAG, even when grown in sulfate-free media (***13,14***). For such cells, efficient incorporation can be obtained with lower external sulfate concentration. CHO cells can be labeled using 4 mL of Ham's F12 medium (contains 6 μ*M* sulfate) supplemented with either 10% fetal bovine serum and 100 mCi $Na_2{}^{35}SO_4$, or *dialyzed* fetal bovine serum and 8 mCi $Na_2{}^{35}SO_4$ (both conditions employ a specific activity of ~270,000 Ci/mol, so comparable incorporations are obtained).
11. It is important that the incubator shelf be level so that the cell monolayer does not dry out. The flask is sealed to prevent off-gassing of volatile $^{35}S$-labeled metabolites, which certain cell types can generate. If sealing is insufficient, then such compounds may be captured by placing the flask on a bed of activated charcoal during the incubation period. After the incubation, monitor the incubator to ensure that contamination has not occurred.
12. Multiple rinses are necessary to remove "free" $^{35}S$, which can copurify with GAG and confound later steps. The majority of the input radioactivity is contained within the culture medium and the first two washes. Certain institutions require that high-level radioactive waste be kept in a minimal volume. For this reason, it may be necessary to collect this material separate from other contaminated waste fluids.

13. This procedure was originally optimized to recover quantitatively small levels of labeled proteoglycans from L cells, but can be modified for proteoglycans from other sources. The loading and initial washing pH facilitate removal of acidic proteins, whereas the urea disrupts protein complexes and prevents hydrophobic interactions with the column matrix. These very stringent conditions may not allow for efficient binding of free GAG chains, or proteoglycans with GAG chains that are short, undersulfated, or derive from GAG biosynthetic mutants. To optimize recoveries for problem samples, reduce the NaCl concentration in both UAW and the preparative load buffer (equivalent proportional reductions), then monitor total $^{35}S$ recovery and its corresponding susceptibility to digestion with GAG lyases. An optimal NaCl level maximizes the recovery of lyase-sensitive material while minimizing non-lyase-sensitive material.
14. Elution is performed by the application of multiple small aliquots so as to reduce the flow rate of the eluate. This maximizes sample recovery and minimizes sample volume.
15. Watch the cap to ensure that the O-ring becomes compressed during closing. A perfect seal is required to prevent leakage of sample during vortexing.
16. Classically, β-elimination is performed under basic conditions with high concentrations of borohydride. The base cleaves the glycosidic bond to a substituent that is in a β-orientation relative to an electron-withdrawing group. For proteoglycans, cleavage occurs between the linkage region xylose and the corresponding serine residue of the core protein. This works best with intact proteins rather than proteolytically degraded material, since cleavage cannot occur if the carboxyl group of the serine is free. High borohydride is included in general polysaccharide analysis to prevent the "peeling reaction" — a sequential cyclic process whereby the terminal sugar undergoes oxidative modification followed by β-elimination. High borohydride is not required for proteoglycans, since the linkage xylose does not undergo peeling. Do not use high levels of borohydride, as violent bubbling occurs upon neutralization, which will cause extensive radioactive contamination.
17. To maximize recovery, hold the tube vertical and remove the aqueous phase by drawing with a P-1000 pipetter from the gradually descending meniscus. When the meniscus approaches the interphase, stop and transfer the harvest to a clean tube. Next, tip the tube on a ~45° angle to break the surface tension between the aqueous and organic phases (slight tapping of the tube may be required). The aqueous phase will detach from the lower tube wall and form a ball on the uppermost surface. Place a P-200 tip inside the ball and withdraw solution until the tip gets close to the protein interphase. Pool the recovered material.
18. Efficient (i.e., > 90% recovery) precipitation of GAG requires a minimum of 4.5 volumes of ethanol, as well as carrier glycogen and a minimum of 100 m*M* NaCl. In high ethanol, sodium has low solubility. The described protocol dilutes the sample so that the salt remains in solution during the precipitation of GAG.
19. To prevent pellet loss, it is best not to let the tubes sit for any length of time. It is preferable to turn the centrifuge off just short of 30 min, so that the samples can be removed the instant the centrifuge stops.
20. Slow aspiration is necessary to prevent pellet loss. This is accomplished with a 25-gage needle connected to the vacuum line. For right handed people, tubes are held in the left hand with the pellet positioned to the right side (providing a side-on view of the pellet). Initially, hold tubes vertically and maintain the needle bezel at the level of the meniscus and parallel to the tube wall. Always point the needle opening away from the pellet. Aspiration speed is controlled by slowly lowering the needle, and by slightly pressing the bezel against the tube wall. A slow speed ensures complete fluid removal without leaving behind residual drops. As the meniscus approaches the pellet region, start tilting the tube to the right so

that the pellet is under fluid for as long as possible. Ultimately, the tube will be nearly horizontal with the pellet at the lowest point and the bezel will be a pressed against the opposite tube wall directly above the pellet.

## References

1. Bame, K. J. and Esko, J. D. (1989) Undersulfated heparan sulfate in a chinese hamster ovary cell mutant defective in heparan sulfate *N*-sulfotransferase. *J. Biol. Chem.* **264,** 8059–8065.
2. Bame, K. J., Lidholt, K., Lindahl, U., and Esko, J. D. (1991) Biosynthesis of heparan sulfate. Coordination of polymer-modification reactions in a chinese hamster ovary cell mutant defective in *N*-sulfotransferase. *J. Biol. Chem.* **266,** 10,287–10,293.
3. Shworak, N. W., Shirakawa, M., Colliec-Jouault, S., Liu, J., Mulligan, R. C., Birinyi, L. K., and Rosenberg, R. D. (1994) Pathway-specific regulation of the synthesis of anticoagulantly active heparan sulfate. *J. Biol. Chem.* **269,** 24,941–24,952.
4. Ozeran, J. D., Westley, J., and Schwartz, N. B. (1996) Kinetics of PAPS translocase: evidence for an antiport mechanism. *Biochemistry* **35,** 3685–3694.
5. Klaassen, C. D. and Boles, J. W. (1997) Sulfation and sulfotransferases 5: the importance of 3'- phosphoadenosine 5'-phosphosulfate (PAPS) in the regulation of sulfation. *Faseb J.* **11,** 404–418.
6. Sjoberg, I. and Malmstrom, A. (1982) Biosynthesis of dermatan sulphate in cultured fibroblasts. Characterization of newly synthesized glycans from cells and microsomes. *Eur. J. Biochem.* **128,** 29–34
7. Kimura, J. H., Caputo, C. B., and Hascall, V. C. (1981) The effect of cycloheximide on synthesis of proteoglycans by cultured chondrocytes from the Swarm rat chondrosarcoma. *J. Biol. Chem.* **256,** 4368–4376
8. Shworak, N. W., Fritze, L. M. S., Liu, J., Butler, L. D., and Rosenberg, R. D. (1996) Cell-free synthesis of anticoagulant heparan sulfate reveals a limiting activity which modifies a nonlimiting precursor pool. *J. Biol. Chem.* **271,** 27,063–27,071.
9. Liu, J., Shworak, N. W., Fritze, L. M. S., Edelberg, J. M., and Rosenberg, R. D. (1996) Purification of heparan sulfate D-glucosaminyl 3-*O*-sulfotransferase. *J. Biol. Chem.* **271,** 27,072–27,082.
10. Liu, J., Shworak, N. W., Sinay, P., Schwartz, J. J., Zhang, L., Fritze, L. M., and Rosenberg, R. D. (1999) Expression of heparan sulfate D-glucosaminyl 3-*O*-sulfotransferase isoforms reveals novel substrate specificities. *J. Biol. Chem.* **274,** 5185–5192.
11. Sambrook, J., Fritsch, E. F., and Maniatis, T. (1989) *Molecular Cloning: A Laboratory Manual*, 2nd Ed., Cold Spring Harbor Laboratory Press, Cold Spring Harbor, MA.
12. Humphries, D. E., Silbert, C. K., and Silbert, J. E. (1986) Glycosaminoglycan production by bovine aortic endothelial cells cultured in sulfate-depleted medium. *J. Biol. Chem.* **261,** 9122–9127.
13. Keller, J. M. and Keller, K. M. (1987) Amino acid sulfur as a source of sulfate for sulfated proteoglycans produced by Swiss mouse 3T3 cells. *Biochim. Biophys. Acta* **926,** 139–144.
14. Esko, J. D., Elgavish, A., Prasthofer, T., Taylor, W. H., and Weinke, J. L. (1986) Sulfate transport-deficient mutants of Chinese hamster ovary cells. Sulfation of glycosaminoglycans dependent on cysteine. *J. Biol. Chem.* **261,** 15,725–15,733.

# 10

# Selective Detection of Sulfotransferase Isoforms by the Ligand Affinity-Conversion Approach

**Nicholas W. Shworak**

## 1. Introduction

### *1.1. Perspective*

The profound functional diversity of heparan sulfate (HS) proteoglycans is largely engendered by the regulated structural complexity of the glycosaminoglycan (GAG) component (reviewed in ***ref. 1***). Ultimately, the status of the heparan sulfate biosynthetic machinery dictates fine structure that encodes binding sites for protein ligands. In part, regulation of fine structure is made possible through the existence of multiple isoforms for most of the HS sulfotransferases *(2,3)*. Examination of the heparan sulfate 3-*O*-sultotransferase (3OST) family shows that each isotype exhibits a unique sequence specificity *(4)*. All 3OST enzymes place a sulfate group in the 3-*O*-position of glucosamine residues within heparan sulfate; however, each isoform recognizes and modifies a distinct sequence context. Consequently, only 3OST1 generates the antithrombin-binding site, whereas 3OST3 makes binding sites for gD, an envelope protein of herpes virus *(4,5)*. The existence of a multitude of sulfotransferases necessitates a facile approach for the specific detection of individual isoforms present in crude extracts.

This chapter describes establishing a ligand affinity-conversion assay, based on a procedure that has been used to measure selectively the activity of 3OST1 *(6)*. Extensive suggestions are included to assist others in the application of this approach to other glycosaminoglycan biosynthetic enzymes. Thus, the methods should be viewed as modifiable elements of a general approach to measure the sequence specific activity of sulfotransferases that generate binding sites for protein ligands.

### *1.2. Principle, Advantages, and Limitations of the Ligand Affinity-Conversion Approach*

The ligand affinity-conversion approach is based on two tenets that permit a two-step assay. (1) An appropriately selected HS source will contain biosynthetic pre-

From: *Methods in Molecular Biology, Vol. 171: Proteoglycan Protocols*
Edited by: R. V. Iozzo © Humana Press Inc., Totowa, NJ

cursors for the binding site of a protein ligand. Precursors are "partially" synthesized structures that require modification by a single sulfotransferase to generate a functional binding site. Fulfillment of this criterion enables the cell-free synthesis of ligand-binding sites. Thus, cell extracts containing the sulfotransferase of interest are incubated with $^{35}$S-labeled HS and the sulfate donor adenosine 3'-phosphate 5'-phosphosulfate (PAPS). The sequence specific action of the enzyme *converts* precursors into functional binding sites. (2) The *affinity* of the ligand for its completed binding site is substantially higher than for the corresponding precursor structure(s). Such a ligand is used in the second assay step to capture the newly generated, high-affinity portion of $^{35}$S-labeled HS. Thus, sulfotransferase activity is indirectly quantified through the production of HS that contains high-affinity ligand-binding sites. The measured enzyme activity is expressed in arbitrary units of affinity conversion.

The pros and cons of this technique are determined by the reaction conditions, which differ from typical HS sulfotransferase assays. Classically, activity is determined directly by quantifying the transfer of radioactivity from the labeled sulfate donor (PAP$^{35}$S) onto various GAG acceptors ***(7)***. This approach is quite facile for measuring the abundant *N*-sulfotransferases ***(8)*** but detection of rare enzymes in crude cell extracts requires extensive modifications ***(9)***. Subcellular fractionation is frequently required to overcome the high levels of background incorporation, which result from endogenous sulfotransferases and sulfate acceptors contained within whole cell extracts. The affinity-conversion approach eliminates high backgrounds by using $^{35}$S-labeled HS in conjunction with unlabeled PAPS. This arrangement allows for the selective detection of a specific sulfotransferase activity within even crude cell extracts ***(6)***.

The $^{35}$S-labeled HS substrate is generated by metabolically labeling cells with $Na_2{}^{35}SO_4$, diluted to any desired specific activity. Indeed, cells can even be labeled with undiluted $Na_2{}^{35}SO_4$ to generate a substrate of ~$1.2 \times 10^8$ Ci/mol ***(6)***. Consequently, cell-free reactions can contain as little as 350 attomol ($10^{-18}$ mol) of substrate and can detect as little as 0.35 attomol of product ***(6)***. In contrast, sensitivity of the classic assay is restricted by the specific activity of PAP$^{35}$S. Micromolar concentrations are require for enzyme activity, which limits practical specific activity to ~400 Ci/mol ***(8,10)*** and minimum detection to ~50 pmol of product. Thus, the affinity-conversion approach also provides for tremendous sensitivity.

It is important to note that assay specificity can be limited by the structural heterogeneity of HS, which allows HS to function as a substrate for multiple sulfotransferases. (The affinity-conversion assay minimizes this problem by exploiting a protein ligand to focus detection on only the reaction(s) that generate the ligand's binding site.) Nevertheless, it is possible that a ligand-binding site can be generated by distinct enzymes that modify different precursor structures. Thus, a single HS sample may contain multiple precursor structures and so could detect the multiple activities. Clearly, care must be taken in characterizing extracts from novel sources and in identifying the particular HS sample that is most selective for monitoring the enzyme of interest. In this regard, cell mutants that are defective in the production of specific ligand binding sites ***(11,12)*** may serve as idea sources for an exquisitely selective HS substrate.

## 2. Materials

Refer to Chapter 9, Subheading 2.1., for required general materials.

### 2.1. Generation of Crude Cell Extract

1. A cell line that contains the enzymatic activity of interest.
2. PBS and trypsin/EDTA. *See* Chapter 9, **Subheading 2.2., item 3**.
3. Lysis/sample buffer (100 mL): 0.25 *M* sucrose, 1% Triton X-100, make from stocks of 1 *M* sucrose (autoclaved) and 20% Triton X-100. Store at 4°C.
4. Bradford reagent (Bio-Rad).

### 2.2. Cell-Free Synthesis of Ligand-Binding Sites

1. 1 *M* MES, pH 7.0 (100 mL). The sodium salt of 2-[*N*-morpholino]ethanesulfonic acid (MES) is obtained from Sigma. Solutions are stored in a dark bottle at 4°C.
2. 20% Triton X-100; 10 mg/mL BSA and 20-mg/mL chondroitin sulfate. *See* Chapter 9, Table 2 and **Subheading 2.3., item 5**.
3. 2 *M* $MnCl_2$ (15 mL). Filter sterilize and store at room temperature.
4. 1 *M* $MgCl_2$ (15 mL). Filter sterilize and store at room temperature.
5. 10-mg/mL protamine chloride (Sigma) (10 mL). Aliquots (1 mL) are stored at –20°C.
6. 25 m*M* PAPS (0.5 mL). 3'Phosphoadenosine 5' phosphosulfate (PAPS) (Sigma). Dissolve in ethanol/water (50/50 v/v), make 25-µL aliquots, snap-freeze on liquid nitrogen, and store at –80°C, *see* **Note 1**.
7. $^{35}$S-labeled heparan sulfate. *See* **Note 2**.
8. Cell extract. From the method of **Subheading 3.1.**
9. Quantitation standard of sulfotransferase of interest, *see* **Note 3**.
10. Standard buffer. The buffer for quantitation standards, *see* **Note 4**.
11. Sample buffer. The cell extract lysis buffer, *see* **Subheading 2.1., item 3**.
12. 96-well PCR tube plates with sealing tape (Stratagene #410088 and #410152) are ideal for analysis of large numbers of samples.
13. A PCR cycler (optional). Convenient for incubations of large sample sets. Machines with heated lids are preferred to prevent sample evaporation and lid condensation.
14. 167 m*M* NaCl, 13.3-µg/mL glycogen (10 mL). Store at room temperature.
15. A centrifuge capable of holding 96-well plates.
16. A multichannel pipetter (for analysis of large sample sets).

### 2.3. Quantitation of Protein-Heparan Sulfate Complexes by Solid-Phase Capture

1. 96-Well dot-blot apparatus (Schleicher and Schuell).
2. Whatman 3 mm paper.
3. Nitrocellulose membrane (BA85, Schleicher and Schuell).
4. 6 × binding buffer (15 mL). 60 m*M* Tris-HCl, pH 8.0, 3 *M* NaCl, 1.2-mg/mL BSA, store at 4°C. (*see* **Note 5**).
5. Wash buffer (100 mL). 10 m*M* Tris-HCl, pH 8.0, 0.5 *M* NaCl. (*see* **Note 5**).
6. A one-hole punch. Obtain from any stationary store.

## 3. Methods

Upon verification of two criteria, this assay can be tailored to detect the activity of any HS sulfotransferase that creates a high-affinity binding site for a protein ligand. First, the protein ligand of interest must preferentially bind either to a subpopulation of HS or to HS from only a subset of cell types. Second, two specific cell lines must be identified—an expressive line that produces high levels of the ligand-binding site and a nonexpressive line that produces little or no ligand-binding site. If necessary, one might wish to mutagenize the former to create the latter ***(11,12)***. The nonexpressive line is the source of the $^{35}$S-labeled HS substrate, whereas the expressive cell line supplies the crude extract with the sulfotransferase activity of interest (*see* **Subheading 3.1.**). These reagents are required for establishing conditions appropriate for sulfotransferase activity (*see* **Subheading 3.2.**). The ligand of interest is used to fractionate in-vitro-synthesized reaction products and thereby determine the percentage of $^{35}$S-HS chains containing newly generated binding sites (*see* **Subheading 3.3.**). This last step requires the greatest degree of customization, and the reader is advised to consider alternate approaches to measure protein·GAG interactions that are described extensively in Part III. Once established, the assay can be used to monitor sulfotransferase activity of crude cell or tissue extracts, or of derived purified fractions.

### 3.1. Generation of Crude Cell Extract

1. Grow tissue culture cells in a 25-cm$^2$ flask to the desired growth state.
2. Wash monolayers twice with 2 mL of PBS (at room temperature), as described in Chapter 9, **Subheading 3.1.,** step 5.
3. *Remove cells by trypsinization.* Add 1 mL trypsin and incubate at 37°C until cells round up, (*see* **Note 6**). Add 3 mL of tissue culture medium containing 10% fetal bovine serum and dislodge cells from the flask by tituration.
4. *Wash cells.* Transfer the suspension to a 15-mL conical tube containing 10 mL of PBS (room temperature). Mix, then pellet cells by centrifugation at 500*g* for 5 min. Remove as much supernatant as possible and resuspend the pellet in 1 mL of PBS. Transfer the suspension to a 1.5-mL microfuge tube and pellet cells at 1000*g* for 5 min. Aspirate off medium. Optional. If desired, pellets can be snap-frozen on liquid nitrogen and stored at –80°C.
5. *Lyse cells.* Place pellets on ice and add 0.5 mL of 0.25 *M* sucrose with 1% Triton X-100 (ice chilled). Cells are disrupted by vortexing on high for at least 30 s. Sit on ice for 10 min, then vortex for an additional 30 s (*see* **Note 7**).
6. *Pellet nuclei.* Centrifuge 10,000*g* for 10 min, then transfer the crude lysate (supernatant) to a clean tube.
7. Determine protein content by Bradford procedure ***(13)***, (*see* **Note 8**).
8. Snap-freeze samples on liquid nitrogen and store at –80°C.

### 3.2. Cell-Free Synthesis of Ligand-Binding Site

1. *Reaction design.* Reactions for experimental samples and for quantitation standards should all contain the same components (*see* **Table 1**). The standard curve should have at least six different enzyme concentrations as well as two zero enzyme (background control) points. The curve range should span the anticipated sample activities. All sample reactions should contain equal amounts of cell extract protein, (*see* **Note 9**).
2. *Reaction setup.* Reactions are assembled on ice using individual 0.6-mL tubes for a small number of samples, or a 96-well plate (with adhesive plastic lid) for a large sample set. First pipet the required amounts of standard buffer and sample buffer (*see* **Subheading 2.2 items 10** and **11**). Then add the standards and individual samples.

**Table 1**
**Design of Assay Reactions**

| | | Standards | Samples |
|---|---|---|---|
| Standard | | 0–10 µL | 0 µL |
| Standard Buffer | | 10–0 µL | 10 µL |
| | Subtotal: | 10 µL | 10 µL |
| Sample | | 0 µL | X µL |
| Sample Buffer | | 10 µL | 10–X µL |
| | Subtotal: | 10 µL | 10 µL |
| Reaction Mix | | 30 µL | 30 µL |
| | Total: | 50 µL | 50 µL |

3. *Reaction mix.* A 10% excess of reaction mix is generated, on ice, by combining the **Table 2** ingredients in the listed order. (*see* **Notes 10** and **11** regarding optimization for certain sulfotransferases and sources of crude extract). After adding the mix to each reaction, tubes are sealed, mixed by inversion (so as not to make bubbles), briefly centrifuged, then placed in a 37°C incubator for 3 h to overnight, (*see* **Note 12**). If 96-well plates are employed, it is convenient to incubate the reaction in a PCR cycler.
4. *Removal of protein.*
   a. *Small sample sets.* To each tube, add 150 µL of 167 m*M* NaCl, 13.3 µg/mL glycogen and 300 µL of PCI. Vortex each tube at maximum for 20 s, then separate the phases by centrifugation at 10,000*g*, at room temperature for 2 min (*see* Chapter 9, **Subheading 4, Notes 17** and **19**). Remove as much aqueous (top) phase as possible (but do not take interphase material) and transfer to a 1.5-mL screw-cap tube containing 1 mL of absolute ethanol. Centrifugation at 10,000*g*, at room temperature, for 30 min and aspirate, as described in Chapter 9, **Subheading 4, Note 20**. The "wet" pellet is directly resuspended in ligand binding buffer (**Subheading 3.3.**, step 1a). (*see* **Note 13**).
   b. *Large sample sets.* Heat the plate in a PCR cycler at 95°C for 15 min to denature proteins, then cool the plate to room temperature. Mix the samples by inverting the plate and then centrifuge at maximum speed for 10 min to pellet the protein. Using a multichannel pipetter, transfer 40 µL of supernatant to a clean tube plate.

### 3.3. Quantitation of Protein-Heparan Sulfate Complexes by Solid Phase Capture

This procedure exploits the observation that nitrocellulose does not bind free HS but does bind the protein component of a protein·HS complex. Consequently, filtration through nitrocellulose can be used to specifically capture protein·HS complexes for quantitation.

1. Form protein·HS complexes.
   a. *Small sample set.* Make a 1 X binding buffer (33 mL per sample provides a 10% excess) by combining appropriate amounts of 6 X binding buffer and water with pro-

**Table 2**
**Mix Composition for a Single Reaction**

| | Concentration | | |
|---|---|---|---|
| | Stock | Working | Volume per reaction |
| $H_2O$ | | | ~15.5 µL |
| MES, pH 7 | 1 *M* | 50 m*M* | 2.5 µL |
| Triton X-100 | 20% | 1% | 2.5 µL |
| $MnCl_2$ | 2 *M* | 10 m*M* | 0.25 µL |
| $MgCl_2$ | 1 *M* | 5 m*M* | 0.25 µL |
| Chondroitin sulfate | 20 mg/mL | 0.4 mg/mL | 1 µL |
| BSA | 10 mg/mL | 1.2 mg/mL | 6 µL |
| PAPS | 25 m*M* | 0.5 m*M* | 1 µL |
| $^{35}$S-HS | ~$10^5$ cpm/µL | $10^5$ cpm | ~1 µL |
| | | | Total: 30 µL |

tein ligand (for determining appropriate ligand concentration *see* **Note 14**). Add 30 µL per tube, let sit 10 min, then lightly vortex samples (do not make bubbles) and incubate at room temperature for at least 20 min.

b. *Large sample set.* A 5 X binding buffer is generated and 10 µL are added per sample. Plates are capped with fresh sealing tape, mixed by inversion, briefly spun to pellet liquid, and incubated at room temperature for at least 20 min.

2. *Assemble dot blot apparatus.* Cut two pieces of Whatman paper and one piece of nitrocellulose membrane, sized for the apparatus (~11 cm × 8 cm), (*see* **Note 15**). On the filter support plate, position both pieces of Whatman paper, on top of this place the nitrocellulose membrane, and finally add and clamp down the sample manifold. Put 100 µL of wash buffer in all of the wells, and let sit for 10 min to saturate the membrane. Apply a gentle vacuum so as to just remove residual liquid from wells. Add 100 µL of wash buffer to all wells that are to receive sample, and to all adjacent wells. Proceed to **step 3**. In the meantime, capillary action will absorb much of the applied fluid.
3. *Determine input radioactivity.* Samples are centrifuged at maximum for 5 min to pellet insoluble material. For each sample, remove 2.5 µL (small sample set) or 4 µL (large set), mix with 3 mL of scintillant, and place in a liquid scintillation counter.
4. *Capture complexes.* All sample wells should still be wet (*see* **Note 16**). If any are dry, add 50 µL of wash buffer. Transfer 25 µL (small set) or 40 µL (large set) of each sample to it's respective manifold well. When applying, pump the pipet up and down once to mix the sample with the preexisting well buffer. Apply a gentle vacuum. Once all wells are dry, maintain the vacuum and rinse each sample well with 200 µL of wash buffer. Once dry, remove vacuum, disassemble the apparatus, and place the membrane in a fume hood to expedite air drying.
5. *Quantitation of bound radioactivity.* The manifold's O-rings should leave an impression in the membrane that reveals the location of each samples "dot." A one hole punch is used to excise each dot, which is placed in a 7 mL scintillation vial containing 3 mL of scintillant. Radioactivity levels are then determined by liquid scintillation counting.
6. *Calculate background-corrected percentage bound.* For each reaction, first calculate the percentage of $^{35}$S-HS that is converted to ligand-bound material:

% bound = cpm bound (**step 5**) ÷ input cpm (**step 3** cpm × 10) × 100%

Average the percentage bound for the two zero enzyme controls and then subtract this background value from all standards and experimental samples, (*see* **Note 17**).

7. *Calculate raw units of activity*. Plot the data from the standard curve with units of standard on the ordinate (Y axis) and the background corrected percentage bound on the abscissa (X axis), (*see* **Note18**). Use computerized curve fitting (e.g., Microsoft Excel) to evaluate whether a line, second-order polynomial, or power function is the best fit (has the highest correlation coefficient, $r$), (*see* **Note 19**). The equation for the curve of best fit is then used to calculate the units of activity for each experimental reaction (*see* **Note 20**); that is, replace $x$ in the equation with the background-corrected percentage bound.
8. *Correct raw activity for extract effects (optional)*. If one wishes to compare activities between extracts from different tissue or cell sources, then a correction factor is determined for each extract source (*see* **Note 21**). Raw activities are multiplied by the appropriate extract specific correction factor.

## 4. Notes

1. PAPS is considered to be extremely unstable. Immediately before use, thaw required aliquots on ice. Tubes containing residual solution should be appropriately marked, snap-frozen on liquid nitrogen, and returned to –80°C.
2. The procedures in Chapter 9 are used to prepare high-specific-activity $^{35}$S-labeled HS from a cell line that produces HS precursors for the sulfotransferase of interest (the nonexpressive cell line, *see* Methods). Labeling conditions are described in Chapter 9, **Note 9**. For certain sulfotransferases the desired precursor may be a biosynthetic intermediate; thus, it might be desirable to produce partially undersulfated HS, so as to potentially increase the relative abundance of a particular precursor structure. Undersulfation can be accomplished by omitting the cold sulfate and by replacing the fetal bovine serum with Nutridoma-ND (Boehringer-Mannheim, 1% final concentration) or dialyzed fetal bovine serum (100 mL against 4 L of PBS overnight at 4°C). With L cells, high levels of undersulfation are not achieved until the sulfate concentration is reduced below 20 µM.
3. The most reliable determination of sulfotransferase activity requires that each sample set include a standard curve. Ideally, the standard should be a pure sample of the sulfotransferase of interest (either native or recombinantly expressed). In lieu of pure enzyme, partially purified material or well-characterized cell extract will suffice and are suitable for establishing initial assay conditions.
4. The standard should be stored under conditions that stabilize activity. For example, the storage buffer for 3OST1 standards is 50 m*M* MOPS, pH 7.0, 0.6% CHAPS, 1% glycerol, 20 m*M* NaCl, and 100 µg/mL BSA. Stocks, at varying concentrations (for 3OST1, log dilutions from 1 U/µL down to 0.001 U/µL) are snap-frozen on liquid nitrogen and stored at –80°C. Under such conditions, 3OST1 enzyme activity is stable over several cycles of freeze-thaw. If the enzyme of interest loses activity upon freeze-thaw, then standards should be distributed into single-use aliquots. If the standard is stored in a high salt that inhibits activity, then generate aliquots of a single high concentration. For each assay run, this stock must be sufficiently diluted with low-salt or salt-free buffer so that the final reaction salt concentration has little or no inhibitory effect.
5. The described binding and wash buffers were designed to quantitate FGF·HS complexes. Other ligands are likely to require buffer optimization. This is accomplished with pilot runs using equal counts of $^{35}$S-labeled HS isolated from the nonexpressive and the expressive cell line (defined under Methods). The objective is to identify conditions that maximize the expressive/nonexpressive cpm ratio of filter-bound HS. Initially, each chain

type is incubated with the ligand of interest in NaCl concentrations ranging from 50 m*M* to 1 *M*. A comparable series of wash buffers is generated so that washing is performed at the same salinity as binding. This process reveals the optimal salinity for binding, which is next held constant while varying the sodium concentration of the wash buffer. The following points should be considered if altering other constituents. Each nitrocellulose well maximally binds ~30 μg of protein. Each reaction contains 5–10 μg of BSA, which is included as carrier to prevent loss of ligand to plastic surfaces. Higher concentrations of BSA will compete with the ligand for nitrocellulose and prevent efficient capture of complexes. The buffer pH of 8.0 was chosen to reduce the positive charge on histidine residues and thereby reduce nonspecific ionic interactions; however, optimization of pH may be required.

6. Trypsinization times vary considerably (from 0.5 to 5 min) between cell types and the age of trypsin solution, which self-degrades over time. It is best to monitor the reaction with an inverted microscope. Initially the cell edges will brighten and eventually cell processes will retract to create a round ball.
7. To monitor lysis, transfer 2 μL of suspension to a microscope slide containing 20 μL of 0.25 *M* sucrose with 1% Triton X-100, add a cover slip, and observe microscopically. Cells should be predominantly converted into nuclei that are naked, or contain some membrane tags. Efficiency can be monitored with a Coulter counter, since the released nuclei are smaller than the starting cells. Some cell types may require additional vortexing.
8. Protein yields can range from 0.5 to 2 μg/μL. All samples and protein quantitation standards (e.g. BSA), should contain the same final volume of lysis buffer (0.25 *M* sucrose, 1% Triton X-100), since Triton X-100 can interfere with the Bradford assay.
9. Crude extracts can contain a complex mixture of activators and inhibitors, and so may not always show a linear dose-response. Indeed, excess extract can completely inhibit sulfotransferase activity. When analyzing a novel extract, one should first run a dose-response curve to identify the optimal level of extract. Maximal assay precision is obtained if one is working in the linear portion of the curve; however, sensitivity limitations may preclude this luxury for some extracts.
10. The described reaction mixture was optimized for 3OST1 activity. Other enzymes may require optimization for pH and concentrations of divalent/monovalent cations, PAPS, and other constituents. The chondroitin sulfate is included, at noninhibitory concentrations for 3OST1, as a carrier for high-specific-activity $^{35}$S-HS. For other enzymes, one might find glycogen or heparan sulfate to be preferable. Certain investigators use protamine chloride (*see* **Subheading 2.2., item 5**) to enhance activity. Care should be exercised if combining protamine and chondroitin sulfate, as an insoluble precipitate can result. HS sulfotransferases are typically inhibited at sodium concentrations above 50 m*M*. Consequently, reaction conditions should be adjusted if one wishes to analyze plasma or other high-salt samples.
11. If one wishes to compare activities between different cell or tissue types, then additional establishment steps are necessary. Extracts from selected sources may require unique conditions to detect enzymatic activity (this is particularly true for certain tissue extracts). Subsequently, one must determine a correction factor for each extract type, to account for extract specific inhibition or activation (*see* **Note 21**).
12. The reaction is self-quenching due to chemical and enzymatic degradation of PAPS. A protein precipitate may form during the incubation, depending on extract source and concentration. This is not usually a problem and is removed in subsequent steps.
13. PCI extraction is labor intensive but advantageous. Protein is most effectively removed by the extraction, whereas the subsequent ethanol precipitation both removes salts and

allows for sample concentration. Overall, a much cleaner sample is obtained, which imposes much fewer technical limitations on subsequent approaches to form and quantitate ligand·HS complexes. When setting up an assay, PCI extraction is initially preferred. One can shift to heat deproteinization once the assay is established, appropriate controls are available, and high-throughput analysis is required.

14. In general, the concentration of the ligand in the binding buffer should be 10- to 100-fold greater than dissociation constant ($k_{diss}$) of the protein·HS complex. Given that

$$\frac{[\text{ligand}\cdot\text{HS}_{\text{binding site}}]}{[\text{free HS}_{\text{binding site}}]} = \frac{[\text{ligand}_{\text{total}}]}{K_{\text{diss}}}$$

It follows that 90–99% of the binding sites will be complexed at the above ligand levels, respectively. Overly high ligand concentrations should be avoided. Excessive ligand will increase assay background by binding to low-affinity sites. Moreover, sensitivity will be reduced if the protein-binding capacity of the membrane is exceeded, (*see* **Note 5**).

15. Always cut membranes for the entire sample surface of the apparatus; the vacuum will not draw sample if the filter support is only partially covered. Before running samples, the clamp tension should be adjusted to ensure that all wells are independently sealed. Perform a test run by passing 100 µL of dye (e.g., bromophenol blue) through each well and inspecting the resulting stained membrane. Sufficient tension should be applied to prevent the tracking of dye between wells.
16. Quantitation is most accurate if the sample loads uniformly across the well surface. Uneven loading can cause saturation of protein-binding sites with resultant loss of complexes. Initially, it is advisable to generate an autoradiograph to verify that samples loaded evenly, without leaking. Autoradiographs readily shows uneven loaded wells as blotchy doughnuts rather than uniform spots. Doughnuts can occur when the membrane and filters are too dry, which results in lateral capillary draw at the well periphery. Doughnuts also appear with low well volumes, where most of the sample occurs in the meniscus edge. In either case, just before loading add an extra 100 µL of wash buffer, wait 10 min for capillary draw and then load samples.
17. This background-corrected value will underestimate the true amount of complex formed because the solid support quenches radioactivity and because some complexes escape capture. However, these factors do not affect the final activity determination, as all experimental samples and standards will be affected equivalently. Since quantitation of spot radioactivity is relative, one may alternatively use a betascope or phosphorimager to analyze the entire filter digitally. This approach is preferable for the analysis of large sample sets. Such digital approaches are good for relative comparisons between samples but can profoundly underestimate the absolute c.p.m. of captured complex. If desired, the calculation of absolute cpm requires correction for the counting efficiencies of the digitalization equipment and the liquid scintillation counter employed.
18. The plotting of the independent variable (enzyme activity) on the ordinate defies mathematical convention but simplifies data analysis, because the equation of the resulting curve is used directly to calculate sample activity. If the independent variable is plotted on the abscissa, then the resulting curve equation must be manipulated algebraically. This becomes potentially problematic for second-order polynomials, which are solved by the quadratic equation, which yields two possible solutions. Fortunately, this confusion is avoided by nonconventional data plotting.
19. The standard curve will initially be linear but gradually approaches a plateau when the bulk of the $^{35}$S-labeled HS precursor is converted into high-affinity material. If the entire

HS population were to function as precursor, then linearity would only be approximated until ~30% of substrate was expended. In reality, a precursor pool may comprise only a small fraction of total HS *(6)*, consequently linearity may occur over only a narrow range of enzyme activities. Fortunately, the deviation from linearity is mathematically predictable, which allows for an expansion of the useful range of the standard curve. Nonlinear data are best fit by a second-order polynomial if the departure from linearity is small. With higher enzyme activities the plateau is approached, which best fits a power function. In this latter case, assay precision deteriorates near the plateau. Consequently, one should discard any high-activity points that lie within an obvious plateau.

20. Given the above complexities of curve fitting, extrapolation should not be performed. Any samples that lie outside the curve should be rerun at adjusted concentrations to obtain conversion values that lie within the reliable portion of the standard curve.
21. To determine the correction factor, *pure* standard is preferred. For each extract type, a standard curve is run in the absence and presence of cell extract and raw sample activities are calculated (*see* **Subheading 3.3., Steps 6** and **7**). The extract specific correction factor is the slope obtained by plotting observed activity (abscissa) versus the activity of added standard (ordinate).

## References


1. Rosenberg, R. D., Shworak, N. W., Liu, J., Schwartz, J. J., and Zhang, L. (1997) Heparan sulfate proteoglycans of the cardiovascular system. Specific structures emerge but how is synthesis regulated? *J. Clin. Invest.* **99,** 2062–2070.
2. Shworak, N. W., Liu, J., Petros, L. M., Zhang, L., Kobayashi, M., Copeland, N. G., Jenkins, N. A., and Rosenberg, R. D. (1999) Multiple isoforms of heparan sulfate D-glucosaminyl 3-O-sulfotransferase. Isolation, characterization, and expression of human cDNAs and identification of distinct genomic loci. *J. Biol. Chem.* **274,** 5170–5184.
3. Aikawa, J. and Esko, J. D. (1999) Molecular cloning and expression of a third member of the heparan sulfate/heparin GlcNAc N-deacetylase/ N-sulfotransferase family. *J. Biol. Chem.* **274,** 2690–2695.
4. Liu, J., Shworak, N. W., Sinay, P., Schwartz, J. J., Zhang, L., Fritze, L. M., and Rosenberg, R. D. (1999) Expression of heparan sulfate D-glucosaminyl 3-O-sulfotransferase isoforms reveals novel substrate specificities. *J. Biol. Chem.* **274**(8)**,** 5185–5192.
5. Shukla, D., Liu, J., Blaiklock, P., Shworak, N. W., Bai, X., Esko, J. D., Cohen, G. H., Eisenberg, R. J., Rosenberg, R. D., and Spear, P. G. (1999) A novel role for 3-*O*-sulfated heparan sulfate in herpes simplex virus 1 entry. *Cell* **99,** 13–22.
6. Shworak, N. W., Fritze, L. M. S., Liu, J., Butler, L. D., and Rosenberg, R. D. (1996) Cell-free synthesis of anticoagulant heparan sulfate reveals a limiting activity which modifies a nonlimiting precursor pool. *J. Biol. Chem.* **271,** 27,063–27,071.
7. Silbert, J. E. (1967) Biosynthesis of heparin. *N*-Deacetylation of a precursor glycosaminoglycan. *J. Biol. Chem.* **242,** 5153–5157.
8. Brandan, E. and Hirschberg, C. B. (1988) Purification of rat liver N-Heparan-sulfate sulfotransferase. *J. Biol. Chem.* **263,** 2417–2422.
9. Kusche, M., Bäckström, G., Riesenfeld, J., Petitou, M., Choay, J., and Lindahl, U. (1988) Biosynthesis of heparin. O-sulfation of the anti-thrombin binding region. *J. Biol. Chem.* **263,** 15,474–15,484.
10. Bame, K. J. and Esko, J. D. (1989) Undersulfated heparan sulfate in a Chinese hamster ovary cell mutant defective in heparan sulfate *N*-sulfotransferase. *J. Biol. Chem.* **264,** 8059–8065.

11. Bai, X. and Esko, J. D. (1996) Mutant defective in heparan sulfate hexuronic acid 2-*O*-sulfation. *J. Biol. Chem.* **271,** 17,711–17,717.
12. Shworak, N. W., Shirakawa, M., Colliec-Jouault, S., Liu, J., Mulligan, R. C., Birinyi, L. K., and Rosenberg, R. D. (1994) Pathway-specific regulation of the synthesis of anticoagulantly active heparan sulfate. *J. Biol. Chem.* **269,** 24,941–24,952.
13. Bradford, M. M. (1976) A rapid and sensitive method for the quantitation of microgram quantities of protein utilizing the principle of protein-dye binding. *Anal. Biochem.* **72,** 248–254.

# 11

# Particulate and Soluble Glycosaminoglycan-Synthesizing Enzymes

## Preparation Assay and Use

**Geetha Sugumaran and Jeremiah E. Silbert**

## 1. Introduction

The general features of the biosynthetic assembly of all proteoglycans (*see* **refs. *1–9*** for reviews), except the keratan sulfate portions of cartilage and cornea (*see* **Note 1**), consist of sequential: (1) synthesis of the core protein; (2) xylosylation of specific Ser moieties of the core protein; (3) addition of two galactose (Gal) residues to the xylose (Xyl); (4) completion of a common tetrasaccharide linkage region by addition of a glucuronic acid (GlcA) residue; (5) addition of an N-acetylgalactosamine (GalNAc) or N-acetylglucosamine (GlcNAc) residue to initiate the chondroitin/dermatan or heparan glycosaminoglycan, respectively; (6) repeat addition of hexosamine residues alternating with GlcA residues to form the large heteropolymer glycosaminoglycan chains; and (7) modification of these glycosaminoglycan chains by variable N-deacetylation/N-sulfation, and/or O-sulfation, and variable epimerization of GlcA to iduronic acid (IdceA). The Xyl may occasionally be 2-phosphorylated in some chondroitin sulfate ***(10)*** and heparan sulfate ***(11)***, and one or both of the Gal residues of the chondroitin sulfate linkage region may be 4-O- ***(12)*** or 6-O-sulfated ***(13)***. However, Gal sulfation has not been found in the identical oligosaccharide linkage region of heparin/heparan sulfate ***(14)***. In addition, glycoprotein-like N-linked glycosylation and/or O-linked glycosylation takes place before or while the synthesis of the oligosaccharide linkage region and glycosaminoglycans are being formed.

With the single exception of the formation of UDP-Xyl from UDP-GlcA in microsomes ***(15)***, the formation and modifications of the precursor sugar nucleotides take place in the cytosol, which are then transported through the membrane vesicles into the cisternae by means of an antiport mechanism ***(16)***. The assembly of the link-

From: *Methods in Molecular Biology, Vol. 171: Proteoglycan Protocols*
Edited by: R. V. Iozzo © Humana Press Inc., Totowa, NJ

age region on the core protein followed by glycosaminoglycan polymerization, sulfation, and epimerization occurs completely isolated from the cytoplasm within the intracellular membrane system composed of the endoplasmic reticulum (ER), transfer vesicles, Golgi apparatus, and secretory vesicles. Thus, the precursor core protein moves from rough ER to smooth ER to Golgi for subsequent O-linked glycosylation, completion of the N-linked oligosaccharides, and glycosaminoglycan formation.

There is considerable evidence that during the course of biosynthesis, all proteoglycan core proteins are continuously attached to the inner membrane surfaces of the ER or Golgi apparatus *(3)*. However, in contrast to cell surface, where proteoglycans may become integral parts of the membrane, proteoglycans do not appear at any time to become integral structural components of the membranes where the biosynthesis is taking place.

The location, type, and amount of glycosaminoglycan produced is dependent on the core protein that is synthesized, but other than the presence of a Ser-Gly-containing peptide, it is not clear what proteins become proteoglycans or what determines which will have chondroitin/dermatan sulfate substituents and which will have heparin/heparan sulfate.

The enzymes responsible for the synthesis of the linkage region common to chondroitin/dermatan sulfate (GalNAc-containing) and heparin/heparan sulfate (GlcNAc-containing) proteoglycans are Xyl transferase, Gal transferase I, Gal transferase II, and GlcA transferase I. These enzymes act sequentially to transfer Xyl, Gal, and GlcA from their respective sugar nucleotide substrates to the nascent core protein. Animal cell mutants that lack any one of these glycosyltransferases are incapable of synthesizing both chondroitin/dermatan sulfate and heparin/heparan sulfate chains, indicating that the same glycosyltransferases are involved in the synthesis of the linkage region for both types of glycosaminoglycan substituents *(17)*.

The transfer of the first GlcNAc or GalNAc from UDP-GlcNAc or UDP-GalNAc to the linkage oligosaccharide is the first step that provides specificity for formation of chondroitin/dermatan sulfate or heparin/heparan sulfate, respectively. Accordingly, it has been reported that this first GlcNAc and the first GalNAc may be added by enzymes different from the GlcNAc transferase *(18)* and the GalNAc transferase *(19)* involved in polymerization (*see* **Note 2**).

Chondroitin/dermatan sulfate chain elongation requires alternating transfer of GlcA and GalNAc residues, which appear to be catalyzed by two separate *(20)* glycosyltransferase proteins while heparin/heparan sulfate chain elongation requires alternating transfer of GlcA and GlcNAc residues, shown to be catalyzed by a single protein co-polymerase *(21)*.

The glycosaminoglycan chains of chondroitin are modified by 4-O- and/or 6-O-sulfation of most of the GalNAc residues and occasionally 2-O-sulfation of some GlcA residues. Dermatan sulfate is formed by epimerization at the polymer level *(22)* of some or most of the GlcA residues to IdceA adjacent to 4-O-sulfated GalNAc, followed by 2-O-sulfation of many of the IdceA residues *(23)*. The glycosaminoglycan chains of heparin/heparan are modified by N-deacetylation of some (heparan) or essentially all (heparin) GlcNAc residues with concurrent N-sulfation catalyzed by a bifunctional N-deacetylase/N-sulfotransferase enzyme *(24)* to form GlcNS residues. This is then followed by 3-O- and/or 6-O-sulfation of GlcNS and epimerization of adjacent GlcA residues to IdceA accompanied by 2-O-sulfation catalyzed by specific enzymes *(1,7)*. Sulfation of glycosaminoglycans takes place with adenosine 3'-phosphate 5'-phosphosulfate (PAPS) as the sulfate donor apparently in the same Golgi location as polymerization *(4)* and apparently while the nascent glycosaminoglycan chains are actively growing rather than after completion of the chains *(25)*. Details regarding the heparin/heparan sulfate sulfotransferases are discussed in Chapter 9.

With the exception of the polysaccharide chain-initiating Xyl transferase, which is found partially in the ER *(26)*, the other enzymes are firmly attached to the Golgi membranes *(3)*, but with some being found in serum or culture medium of cells. Although this membrane-bound nature has hampered the purification of these enzymes, the advent of new techniques and the presence of already soluble forms has allowed progress in the purification and cloning of many of these enzymes.

Microsomal fractions were found to contain all the enzymes responsible for the synthesis of the glycosaminoglycans, as well as the endogenous nascent proteoglycan acceptors or primers at different stages of growth on which chondroitin or heparan sulfate chains could be built. Usually extensive polymerization in cell-free systems can be achieved only with the presence of endogenous membrane-attached nascent material capable of being the primer for glycosaminoglycan formation, while oligosaccharides serve only as substrates for the addition of one or a few sugars *(27)*. Thus, the membrane organization of the polymerizing enzymes and proper positioning of the nascent proteoglycan attached to the membrane in juxtaposition to the enzymes is necessary for polymerization. This is consistent with subsequent findings that endogenous acceptors are attached to Golgi membrane during glycosaminoglycan synthesis. However, one cannot determine the exact concentrations of the different membrane-attached endogenous acceptors and thus cannot determine whether the enzyme concentrations or acceptor concentrations are rate limiting. Therefore, it is best to use exogenously added, defined substrates for assaying the different enzyme activities.

## 2. Materials

### 2.1. Tissue Sources

1. Chick embryo epiphyseal cartilage, human skin fibroblasts, and bovine endothelial cell cultures have been the best sources for obtaining microsomal preparations enriched for chondroitin 6-sulfate synthesis (*see* **Note 3**).
2. Swarm rat chondrosarcomas, mouse mastocytomas or cell cultures, human skin fibroblasts, and bovine endothelial cell cultures have been the best sources to obtain microsomal preparations for chondroitin 4-sulfate synthesis (*see* **Note 4**).
3. Skin fibroblast cultures and bovine endothelial cell cultures have been the best sources to obtain microsomal preparations for dermatan sulfate synthesis.
4. Mouse mastocytomas or cell cultures have been the only source for microsomal preparations for heparin synthesis.
5. EHS (Englebreth-Holm-Swarm) mouse sarcoma (*see* **Note 5**), liver, skin fibroblast cultures, and endothelial cell cultures have been the best sources for obtaining microsomal preparations for heparan sulfate synthesis.

## 2.2. Donor Substrates

1. Sugar nucleotides: Radiolabeled ($^{14}C$ or $^{3}H$) or nonlabeled UDP-Xyl, UDP-Gal, UDP-GlcA, UDP-GalNAc, UDP-GlcNAc are obtained from commercial sources.
2. PAPS: Radiolabeled ($^{35}S$) or nonlabeled PAPS are obtained from commercial sources. Alternatively, PAP$^{35}$S can be synthesized by published techniques using yeast enzyme preparations incubated with ATP and [$^{35}$S]sulfate *(28)*.

## 2.3. Exogenous Acceptor Substrates

1. Deglycosylated core proteins are obtained by treating the proteoglycans with chondroitin ABC lyase or heparan lyase to remove the chondroitin/dermatan sulfate chains or heparan sulfate chains and then treating with polyhydrogen fluoride in pyridine with anisole as scavenger *(29)* in order to remove the GlcA-Gal-Gal-Xyl linkage region (*see* **Note 6**).
2. Partially deglycosylated chondroitin sulfate core protein is prepared by digestion with testicular hyaluronidase to remove variable amounts of glycosaminoglycan, followed by mixtures of either β-N-acetylgalactosaminidase, β-glucuronidase, and/or β-galactosidase to remove nonreducing terminal GalNAc, GlcA, and Gal as substrates for the GalNAc I, GlcA I, and Gal I and II transferases, respectively (*see* **Note 7**).
3. Alternatively, commercially available 4-methyl umbelliferyl-β-D-xyloside (Xyl-methyl umbelliferyl) is used as a substrate for addition of the first Gal.
4. The Gal-Xyl-methyl umbelliferyl product from **item 3**, or commercially available β-methyl galactoside (Gal-methyl), is used as a substrate for addition of the second Gal.
5. The Gal-Gal-Xyl-methyl umbelliferyl product from **item 4**, or commercial Gal-Gal-β-methyl, is used as a substrate for GlcA I transferase.
6. GlcA-Gal-Gal-β-methyl can also be formed by incubation of Gal-Gal-β-methyl together with phenyl β-D-GlcA in the presence of 0.5 U of β-glucuronidase (a transglycosidase) prepared from rat liver *(30)*.
7. Chondroitin 4-sulfate, chondroitin 6-sulfate, and dermatan sulfate are obtained from commercial sources (*see* **Note 8**).
8. Nonsulfated chondroitin and dermatan are obtained commercially or by chemical desulfation based on solvolysis of the pyridinium salts of chondroitin sulfate or dermatan sulfate in dimethyl sulfoxide in the presence of small amounts of water or methanol *(6)*.
9. Deacetylation of GalNAc or GlcNAc units is accomplished with 1% hydrazine sulfate heated to 100°C for various time intervals to obtain partially or completely deacetylated products *(6)*.
10. Nonsulfated labeled or nonlabeled glycosaminoglycans can be prepared by synthesis using appropriate microsomal preparations and sugar nucleotides in the absence of PAPS. Alternatively, nonsulfated labeled glycosaminoglycans can be isolated from media of cells grown with [$^{14}$C]- or [$^{3}$H]glucosamine in the presence of 10 m*M* or 50 m*M* chlorate for production of nonsulfated chondroitin or heparan, respectively *(31)*.
11. Even-numbered oligosaccharides terminating in uronic acid (radioactively labeled or unlabeled, sulfated or nonsulfated) are derived from chondroitin sulfate or chondroitin by variable degrees of degradation with short-term incubations using commercially available glycosaminoglycan hydrolases (hyaluronidase). Testicular hyaluronidase will not work on the IdceA residues of dermatan sulfate.
12. Odd-numbered oligosaccharides are obtained from even-numbered oligosaccharides after digestion with the appropriate β-glucuronidase.
13. Alternatively, dermatan and dermatan sulfate oligosaccharides as well as chondroitin and chondroitin sulfate oligosaccharides are obtained by partial degradation with chondroitin

ABC, AC, or B lyases, resulting in even-numbered oligosaccharides containing a Δ–uronyl residue at the nonreducing terminal *(32)*. This nonreducing terminal Δ–uronyl residue is removed with mercuric acetate *(33)*, resulting in odd-numbered oligosaccharides.

14. The resulting even-numbered or odd-numbered oligosaccharides are fractionated by gel filtration (usually Sephadex G-50) chromatography to obtain the desired size oligosaccharides.
15. The [C-5-$^3$H]chondroitin for assay of epimerization of GlcA to IdceA information of dermatan is prepared by incubating chondroitin-synthesizing systems with [C-5-$^3$H]glucose *(34)*. If 10 m*M* chlorate is present in the incubation, there will be no sulfation.
16. Heparin and heparan sulfate are obtained from commercial sources (*see* **Note 9**).
17. Desulfated heparin and completely desulfated, re-N-sulfated heparin is obtained commercially.
18. Desulfated heparan is prepared chemically as in **item 8**, yielding a polymer consisting of desulfated GlcN or GlcNAc alternating with IdceA or GlcA.
19. Preparation of heparin/heparan sulfate oligosaccharides is more difficult, in part because there is no readily available source for any heparin/heparan sulfate hydrolases. Commercially available lyases (heparinase, heparanases) *(6)* similar to those that degrade chondroitin/dermatan sulfate are used in short-term incubations with heparin/heparan sulfate glycosaminoglycans and oligosaccharides to provide even-numbered oligosaccharides containing a Δ–uronyl residue at the nonreducing terminal. As with chondroitin/dermatan oligosaccharides, this nonreducing terminal Δ–uronyl residue can be removed with mercuric acetate to obtain odd-numbered oligosaccharides.
20. Alternatively, even-numbered [GlcA-GlcNAc]*n*-GlcA-α-Man heparan oligosaccharide units are obtained by nitrous acid degradation for deaminative cleavage of N-sulfated GlcN units while the acetylated GlcN units are not affected *(6)*. The nonreducing terminal GlcA can then be removed by digestion with β-glucuronidase to obtain odd-numbered oligosaccharides. Oligosaccharides are separated and characterized for size by gel chromatography in the same fashion as that used for chondroitin/dermatan oligosaccharides.
21. An alternative method to provide nonsulfated, fully N-acetylated heparan oligosaccharides is the use of capsular polysaccharide from Escherichia coli K5, which has the same [GlcA-GlcNAc]*n* structure *(35)* as the nonsulfated precursor glycosaminoglycan that is generated as the first step of heparin/heparan sulfate biosynthesis. This capsular polysaccharide can be modified chemically by partial N-deacetylation as in **item 9**, followed by deaminative cleavage with nitrous acid as in **item 20** or by the action of N-deacetylase/N-sulfotransferase, C-5 epimerase, and various O-sulfotransferases to generate heparin-like sequences.
22. N-[$^3$H]Acetyl-labeled polysaccharide is obtained by partial N-deacetylation by hydrazinolysis of *E. coli* K5 capsular polysaccharide followed by reacetylation with [$^3$H]acetic anhydride *(35)*. Alternatively, it can be prepared by incubation of UDP-GlcA and UDP-GlcN[$^3$H]Ac (commercially available) together with microsomal systems *(36)*.
23. C-5-$^3$H-Labeled O-desulfated N-resulfated heparin/heparan substrate for measurement of GlcA to IdceA epimerization is prepared from heparin or N-sulfated *E. coli* K5 heparan by incubation with rat liver microsomes in the presence of $^3H_2O$ *(37)*.
24. Mixtures of glycosaminoglycans are treated with chondroitin B lyase followed by gel chromatography in order to obtain chondroitin sulfate without any IdceA-containing material (dermatan), or with chondroitin ABC lyase in order to obtain heparin/heparan sulfate without any chondroitin/dermatan sulfate material, or with heparanase to degrade heparan sulfate while leaving heparin mainly intact.

25. Some intact proteoglycans are obtained from commercial sources, but they are generally mixtures. Alternatively, they can be prepared from tissue sources by published techniques, and may be available from other investigators who have purified them.

### 2.4. Endogenous Acceptor Substrates

1. Nascent proteoglycans present in microsomal or Golgi preparations are the best substrates for synthesis of glycosaminoglycans.
2. The native endogenous acceptors are modified by treatment with various detergents to affect their capacity to serve as substrates by means of undefined changes in their juxtaposition to the enzymes of glycosaminoglycan synthesis *(38,39)*.
3. Specific modifications are made by use of testicular hyaluronidase to remove portions of glycosaminoglycan, resulting in changes in amounts of polymerization.
4. Bacterial lyases are used for limited periods to eliminate the capacity of the membrane-bound proteoglycans to act as acceptors by formation of Δ-uronic acid at the nonreducing ends of the partially stripped glycosaminoglycans. Resistance to this treatment provides a mechanism to reflect the degree that the nascent proteoglycans may be buried in vesicles acting as a barrier for access of the lyases *(38)*.

### 2.5. Enzyme Preparations

Microsomal preparations from the sources listed under **Subheading 2.1.** contain all the firmly associated appropriate glycosyl transferases, sulfotransferases, deacetylase, and epimerases for glycosaminoglycan synthesis plus firmly associated endogenous nascent proteoglycan primers on which the glycosaminoglycan chains can be initiated and/or extended and modified by sulfation and epimerization. Attempts to obtain similarly active preparations from many other tissues or cells have been unsuccessful, apparently due to limited levels of nascent proteoglycan primer, even though the glycosaminoglycan-synthesizing enzymes are present. Soluble enzymes that apparently lack the membrane-spanning region may also be present in supernatant fractions from tissue homogenates, serum, or media of cell cultures. Recombinant enzymes have proven to be valuable in identifying the specificities of some of the transferase reactions (*see* Chapter Shworak and Rosenberg).

1. Microsomes are prepared from tissues by first suspending 2–5 g of tissue in 4 volumes of 0.25 *M* sucrose containing protease inhibitors (1 m*M* phenylmethylsulfonylfluoride, 10 m*M* EDTA, 1 m*M* N-ethyl maleimide, 0.5 mg/mL leupeptin and 0.1 mg/mL pepstatin) in case there might be slight protease modifications of the enzymes (*see* **Note 10**), and homogenizing 3 to 5 times with a Polytron at the maximum speed for 1 min each. The homogenate is centrifuged at 12,000*g* (occasionally 10,000*g* or 20,000*g*) in a fixed-angle rotor for 20 min, and the supernatant fluid is then centrifuged at 105,000*g* for 1 h. The pellets are suspended in 0.25 M sucrose and recentrifuged at 105,000*g* for 1 h to wash the resulting microsomal fraction (*see* **Note 11**). Greatest experience has been with microsomal preparations from chick embryo epiphyseal cartilage (*see* **Note 12**) and mouse mastocytomas.
2. Similar procedures are used to obtain a microsomal fraction from lesser amounts of cultured cells except that sonication in short bursts (30 s) at maximal power with a microtip is utilized for cell disruption.
3. Fractionation of chick embryo epiphyseal cartilage microsomes to provide Golgi is performed by a sucrose density-gradient procedure that is a minor modification of that utilized by others for Golgi subfractionation with liver and cultured cells (*see* **Note 13**). A microsomal pellet

from fresh, unfrozen epiphyseal cartilage from 100 seventeen-day-old chick embryos is resuspended by using a Dounce homogenizer with a loose-fitting pestle in 0.5 mL of 0.25 *M* sucrose. A 0.2 mL portion of the fresh microsomal suspension is then fractionated on a six-step sucrose density gradient consisting of 4.0 mL of 55%, 1.4 mL of 40%, 2.2 mL of 35%, 2.2 mL of 30%, 2.0-mL of 25%, and 1.0 mL of 20% (wt/wt) sucrose containing 10 m*M* Tris-HCl, pH 8.0. After centrifugation for 40 h at 4°C in a Beckman SW-40 rotor, fractions are collected and used directly or frozen for further use in the assay of various enzyme activities (*see* **Note 14**). Protein is estimated by the method of Lowry et al. ***(40)***. Although Golgi fractions have not been prepared from other sources for investigation of proteoglycan synthesis, similar fractionations should be applicable.

4. For solubilization of membrane-bound enzymes, microsomal pellets are suspended in 10 mM HEPES buffer, pH 7.2, containing 10 m*M* $MgCl_2$, 2 m*M* $CaCl_2$, 1 m*M* DTT, 20% glycerol, and protease inhibitors. To this suspension an equal volume of the buffer solution containing 2% Triton X-100 is added. The suspension is gently mixed for 60 min and centrifuged at 105,000*g* for 90 min. The supernatant is saved and the pellets are resuspended in the original buffer. An equal volume of the buffer solution containing 2% Triton X-100 and 2.0 *M* NaCl is added. After incubation for 60 min with gentle mixing, the suspension is centrifuged at 105,000*g* for 90 min. The supernatant is combined with the first supernatant and used as the solubilized enzyme. By this two-step procedure, 90% or more of the glycosyl transferase and sulfotransferase activities are solubilized (*see* **Notes 15**, and **16**). Soluble glycosaminoglycan-synthesizing enzyme activities are found in cell culture medium, serum, and various body fluids, probably as a result of proteolytic cleavage from in vivo or intact cells presumably between the catalytic domain and transmembrane domain of the membrane-bound forms of these enzymes.

### 2.6. Other Reagents

1. Whatman #1 descending paper chromatography using:
   Solvent A: ethanol : 1 *M* ammonium acetate, pH 7.8 (5 : 2)
   Solvent B: 1-butanol : acetic acid : 1 *M* ammonium hydroxide (2 : 3 : 1)
   Solvent C: ethyl acetate : acetic acid : water (3 : 1 : 1)
2. Whatman #1 paper electrophoresis using:
   Solvent D: 0.08 *M* pyridine - 0.046 *M* acetic acid, pH 5.3

## 3. Methods

### 3.1. Use of Microsomal Preparations

Details in this section refer to synthesis of chondroitin/dermatan sulfate and to a lesser extent to synthesis of heparin/heparan sulfate. Similar microsomal reaction conditions with slight modifications have been used by U. Lindahl and his associates in their extensive examination of heparin/heparan sulfate biosynthesis ***(1,7)***.

1. Typically, reaction mixtures using endogenous acceptors are incubated with labeled or nonlabeled 0.05–0.2 m*M* UDP-GlcA, UDP-GalNAc, or UDP-GlcNAc with or without labeled or nonlabeled 0.01–0.2 m*M* PAPS (*see* **Note 17**) in 50 m*M* MES buffer, pH 6.5, and 15 m*M* $MnCl_2$ together with 1–10 µL of microsomal preparation, in a total volume of 15–25 µL (*see* **Note 18**). Incubations are at 37°C for as long as 24 h. Incorporation is usually linear for the first 3–5 h but then begins to diminish. True polymerization takes

place to form sulfated glycosaminoglycan (or nonsulfated glycosaminoglycan if PAPS is not present) by addition to nascent membrane-bound proteoglycan (*see* **Notes 19** and **20**).

2. Incubations for incorporation into exogenous oligosaccharide acceptors are similar except that better incorporation is obtained when Triton X-100 or other detergent is present (*see* **Notes 21–23**).
3. Incubations using exogenous glycosaminoglycan require detergent for any but minimal incorporation, and exogenous proteoglycan acceptors have an absolute requirement for detergent to provide any incorporation.

### 3.2. Use of Golgi Preparations

1. Reaction mixtures with Golgi fractions of chondroitin sulfate-producing tissues or cells are essentially the same as with microsomal preparations. However, total activities with these fractions is usually considerably greater than with the microsomal fractions, and there is less degradation of substrates.

### 3.3. Use of Solubilized and/or Purified Enzymes

1. Reaction mixtures with soluble and/or purified enzymes are the same as the reaction with microsomal preparations using exogenous acceptors.

### 3.4. Assay of Enzyme Activities

1. Each activity for synthesis of GalNAc/GlcNAc-GlcA-Gal-Gal-Xyl is determined by 2 to 5-h incubations at 37°C of particulate or soluble enzyme preparations in 0.05 *M* MES buffer, pH 6.5, 0.015 *M* $MnCl_2$ together with 0.2 m*M* of the appropriate $^{14}C$- or $^{3}H$-labeled sugar nucleotide and appropriate 0.2–2.0 m*M* acceptor (*see* **Note 24**) in a total volume of 15 - 35 µL. Thus Xyl transferase uses deglycosylated core protein (undetermined molarity of serine substituent) as acceptor; Xyl-methyl umbelliferyl for Gal I transferase; Gal-Xyl-methyl umbelliferyl or Gal-methyl for Gal II transferase; Gal-Gal-Xyl-methyl umbelliferyl or Gal-Gal-methyl for GlcA I transferase; and GlcA-Gal-Gal-Xyl-methyl umbelliferyl or GlcA-Gal-Gal-methyl for GalNAc I or GlcNAc I transferase. After incubation, each entire reaction mixture is spotted on Whatman No.1 paper and subjected to descending chromatography in solvent A or C (*see* **Note 25**) or paper electrophoresis in solvent D at 70 V/cm for 45 min *(41)*. When 4-methyl umbelliferyl-linked acceptors are used, the products move down the paper with solvent A or C, and their locations can be identified by fluorescence, followed by elution and assay by radioactivity. New radioactive spots corresponding to the oligosaccharide products are eluted with water and chromatographed on Sephadex G-50 for characterization by size and for separation from small amounts of labeled sugar nucleotide degradation products. Structure is confirmed by incubation with appropriate glycosidase.
2. Assay of microsomal or Golgi GlcA II, GalNAc II, and GlcNAc II polymerase activities are made as in **step 1** above with or without chondroitin or heparan pentasaccharide, hexasaccharide, or larger oligosaccharides. The total reaction mixtures are spotted on Whatman No. 1 paper and chromatographed overnight or longer in solvent B if oligosaccharide acceptors are present or in solvent A overnight if oligosaccharides are not present. Labeled products are eluted from the origins (*see* **Note 25**), and characterized by DEAE-cellulose, Sepharose, or Sephadex chromatography. Degradation with appropriate commercially available chondroitin, heparan, or heparin lyases is used to confirm structure (*see* **Note 26**).
3. Assay of soluble/purified enzymes is performed with oligosaccharide acceptors in the same fashion as above.

4. Assay of endogenous particulate or soluble sulfotransferases as well as separation and characterization of products is performed as above using the same incubation conditions with 3.0 m*M* PAP$^{35}$S as substrate. $^{35}$S-labeled products are subjected to nitrous acid degradation, and to heparan, heparin, or chondroitin lyases and sulfatases, and the disaccharide products then separated by paper chromatography (solvent B), electrophoresis, or HPLC in order to determine the degree and position of sulfation.
5. N-Deacetylase activity is assayed using [$^3$H]acetyl-labeled *E. coli* K5 capsular polysaccharide ***(35)*** or [$^3$H]acetyl-labeled heparan precursor as the exogenous substrate. Microsomal or solubilized enzyme protein is incubated at 37°C for 1–2 h with the [$^3$H]acetyl-labeled substrate in 0.05 *M* MES buffer, pH 6.5, 0.01 *M* $MnCl_2$ and 0.02% Triton X-100 in a total volume of 50 µL. When PAPS is added to the incubation mixtures, deacetylation is promoted along with N-sulfation ***(42)***. Reaction mixtures are spotted on Whatman No. 1 paper and subjected to descending chromatography in solvent A. [$^3$H]Acetate released moves down the paper and can be cut and counted in a scintillation counter.
6. The presence of chondroitin/dermatan sulfate uronyl epimerase is assayed by examination of the products of microsomal biosynthesis (**step 2**) for resistance to degradation by chondroitin AC lyase. However, this is not a quantitative assay. Alternatively the uronyl C-5 epimerase for chondroitin/dermatan sulfate is assayed based on the formation of $^3H_2O$ from [C-5-$^3$H]chondroitin ***(43)*** when this substrate is incubated with microsomal or solubilized enzyme and 50 m*M* MES buffer, pH 6.5, 15 m*M* $MnCl_2$, and 0.5% Nonidet P-40 in 50 µL total volume. After 2 h of incubation at 37°C, the reaction mixtures are extracted into a toluene-based scintillation fluid containing 10–25% isoamylalcohol. The use of such a biphasic liquid scintillation system allows the extraction of the radioactive water into the organic phase containing the scintillant, while the polysaccharide remains in the aqueous phase ***(37)***. The epimerase for heparin/heparan sulfate is measured by incorporation of $^3$H from $^3H_2O$ into the C-5 position of heparin/heparan sulfate or by the release of $^3H_2O$ from C-5-$^3$H-labeled N-sulfated heparan/heparan synthesizing systems.

## 4. Notes

1. Biosynthesis of keratan sulfate glycosaminoglycan has not received much attention. However, synthesis of the glycosaminoglycan probably occurs in a manner similar to that shown for other glycosaminoglycans, while synthesis of the linkage region apparently proceeds in a fashion similar to that of the N-linked or O-linked oligosaccharides found in glycoproteins. Biosynthesis of hyaluronic acid, a glycosaminoglycan that is not covalently linked to protein, is different from that of the glycosaminoglycan portions of proteoglycans. It is not formed within or attached to intracellular membrane structures ***(44)***, but is synthesized at the inner surface of the plasma membrane ***(45)*** by pathways quite distinct from those of proteoglycan glycosaminoglycan formation ***(46)***.
2. Incorporation of GalNAc onto linkage region GlcA with microsomal or soluble systems has been reported in all but one instance ***(19)*** to result in α-linked GalNAc rather than β-linked GalNAc ***(47)***. However, the addition of β-GalNAc to a naturally occurring protein that contains unsubstituted GlcA-Gal-Gal-Xyl structures has been reported to be catalyzed by the chondroitin-polymerizing GalNAc transferase ***(48)***. Thus addition of the first GlcNAc of heparin/heparan sulfate but perhaps not the first GalNAc of chondroitin, would appear to require an enzyme different from the hexosaminyl transferase enzyme for glycosaminoglycan formation.
3. Microsomal preparations from the chick cartilage make only chondroitin 6-sulfate, while those from skin fibroblast or bovine endothelial cell cultures also make chondroitin 4-sulfate, dermatan sulfate, and heparan sulfate.

4. Microsomal preparation from the rat chondrosarcoma and mast cells make no chondroitin 6-sulfate, dermatan sulfate, or heparan sulfate.
5. EHS tumors make heparan sulfate almost exclusively.
6. Smith-degraded proteoglycan ***(49)*** and Ser-Gly containing peptides ***(50)*** have also been utilized as substrates for Xyl transferase.
7. The β-glycosidases are commercially available and act sequentially to remove the non-reducing terminal sugar residues.
8. Commercial sources of chondroitin sulfate are invariably mixtures of chondroitin 4-sulfate and chondroitin 6-sulfate having different proportions of each. Chondroitin 4-sulfate that has no 6-sulfation or dermatan residues can be isolated from mouse mastocytoma cells or from Swarm rat chondrosarcoma. Chondroitin 6-sulfate free of any 4-sulfate is not available in any quantity, but can be prepared in small radioactively labeled amounts by biosynthesis with cartilage microsomal preparations.
9. Commercial heparin and heparan sulfate are mixtures, so that preparations of oligosaccharides will yield variable proportions of N-sulfated, N-acetylated, and O-sulfated products.
10. Microsomes prepared as above are stable for months or even years when stored at -20°C. Repeated freezing and thawing may modify or reduce but generally does not abolish activity. We have found no evidence of significant proteoglycan proteases, glycosidases, or sulfatases, since the products of synthesis with all microsomal preparations that we have used appear to be stable during incubations. The protease inhibitors are added only as a special precaution.
11. All the preparations have enzymatic activity for degradation of sugar nucleotides and particularly PAPS to varying degrees. Sodium fluoride has been used by others to inhibit these degrading enzymes, but we have found it to be of limited value. However, degradation can be reduced as much as 10-fold by use of 2,3-dimercaptopropan-1-ol without affecting glycosyl transferases ***(51)***. Nevertheless, it is not necessary to use these substances unless degradation is found to be particularly rapid in a specific microsomal preparation.
12. Approximately 400 fourteen- to seventeen-day old chick embryos are used for microsomal preparations from chick embryo epiphyses.
13. Density gradients have been widely utilized with preparations from liver and occasionally from other tissues in order to separate electron microscopically recognizable Golgi. Liver and other tissues as well as a variety of cultured cell lines have been used for preparation of Golgi fractions, leading to more specific and detailed characterizations of cis, medial, trans, and trans network subfractions with various enzymatic activities ***(52)***. The density of the compartments decreases from cis to trans, and the enzymes involved in glycoprotein ***(53)***, glycolipid ***(54)***, and glycosaminoglycan synthesis ***(55)*** have been shown to have a sequential spatial arrangement so that the early sugar additions occur in the denser cis regions of the Golgi proximal to the RER while the later reactions of oligosaccharide termination and glycosaminoglycan polymerization occur in more distal, less dense medial and trans regions of the Golgi.
14. Enzymatic activities and the products of synthesis are as stable in the Golgi preparations as they are in the microsomal preparations. Moreover, the degradation of sugar nucleotides and PAPS is much less.
15. The only microsomes that we have used for enzyme solubilization were from chick embryo epiphyseal cartilage. However, we would anticipate that effective solubilization from other tissues could be accomplished similarly.
16. The glycosyltransferases and sulfotransferases solubilized from microsomal preparations are much less stable than the naturally occurring soluble transferases. They can only be

stored for limited periods and have much greater losses of activities upon repeated freezing and thawing.

17. Apparent K*m* values for sugar nucleotides and PAPS in microsomal preparations are generally higher by at least 10-fold (0.01–0.2 m*M* range) than with the purified enzymes.
18. The small volumes are used in order to maximize the concentrations of expensive radioactive substrates, and to simplify assay.
19. Unlabeled PAPS can be added subsequent to formation of nonsulfated labeled glycosaminoglycan. In that case, a 100-fold unlabeled excess is added of whichever sugar nucleotide was labeled in the initial incubation. In this way, sulfation of the preformed glycosaminoglycan can be monitored, since only it will be labeled.
20. Sulfation of chondroitin or chondroitin oligosaccharide is usually in a slightly modified "all or nothing" pattern rather than random, so that in most cases sulfation yields long, completely sulfated areas with occasional short, nonsulfated areas and some glycosaminoglycan that remains completely nonsulfated.
21. The increased incorporation with detergent presumably is due to better substrate penetration to the microsomal vesicles sites of the biosynthetic enzymes.
22. The use of exogenous acceptors provides for the addition of only one or a few residues rather than true polymerization. The absence of polymerization on exogenous acceptors reinforces the concept that endogenous membrane-bound nascent proteoglycan substrates are juxtaposed firmly with the membrane-bound enzymes, thus permitting true polymerization.
23. Sulfation by addition of PAPS is not as extensive with exogenous acceptors, although it is still a modified "all or nothing" addition but with much greater areas that are not sulfated.
24. Exogenous substrates are needed for measuring the glycosyl transferases of the linkage region, since there is insufficient endogenous acceptor at these stages of synthesis.
25. Paper chromatography is the simplest and most rapid assay for multiple samples. After incubation of substrates and acceptors with purified enzymes, microsomes, or Golgi fractions, an aliquot or the entire reaction mixture is spotted on Whatman No. 1 paper for overnight descending chromatography in solvent A. With this system, nucleotide sugars, PAPS, and their degradation products move down the paper while radiolabeled proteoglycan and glycosaminoglycan products remain at the origin. When oligosaccharides are used as acceptors, descending chromatography in solvent B is used, since the oligosaccharides will move down the paper with solvent A. Incubation of the origins with proteases at 37°C or with 0.5 *M* NaOH at room temperature for a few hours will quantitatively provide cleavage and consequent solubilization of the glycosaminoglycan portion from endogenous proteoglycan acceptors. When glycosaminoglycans or oligosaccharides are used as exogenous acceptors, the products are eluted directly with water, thus providing a rapid means to separate the products of addition to exogenous acceptors from the products of addition to endogenous nascent proteoglycan. Eluants are then assayed for radioactivity. A single chromatography paper can be used for as many as 12 simultaneous assays, and a single large chromatography tank for as many as 12 sheets of chromatography paper, thus providing the potential for 144 simultaneous assays of addition to endogenous nascent proteoglycans and/or simultaneous assays of addition to exogenous glycosaminoglycans or oligosaccharides. Since all the solvent systems fix proteoglycan acceptors to the origin, the intact proteoglycan acceptors cannot be obtained by this technique. In order to obtain these, 4 *M* guanidine chloride and occasionally detergents are added to reaction mixtures that are then chromatographed directly on Sephadex G-50 to isolate labeled proteoglycans from the labeled substrates.

26. Use of these lyases and sulfatases as well as identification of products has been reviewed in detail *(6,32)*.

## References


1. Lindahl, U., Feingold, D. S., and Rodén, L. (1986) Biosynthesis of heparin. *TIBS* **11,** 221–225.
2. Esko, J. D. (1991) Genetic analysis of proteoglycan structure, function and metabolism. *Curr. Opin. Cell Biol.* **3,** 805–816.
3. Silbert, J. E. and Sugumaran, G. (1995) Intracellular membranes in the synthesis, transport and metabolism of proteoglycans. *Biochim. Biophys. Acta.* **1241,** 371–384.
4. Silbert, J. E. (1996) Organization of glycosaminoglycan sulfation in the biosynthesis of proteochondroitin sulfate and proteodermatan sulfate. *Glycoconj. J.* **13,** 907–912.
5. Silbert, J. E., Bernfield, M., and Kokenyesi, R. (1997) Proteoglycans: a special class of glycoproteins, in *Glycoproteins II* (Montreuil, J., Vliegenthart, J. F. G., and Schachter, H. eds.), Elsevier, Amsterdam, pp. 1–31.
6. Conrad, H. E., ed. (1998) *Heparin-Binding Proteins.* Academic, San Diego, CA.
7. Lindahl, U., Kusche-Gulberg, M., and Kjéllen, L. (1998) Regulated diversity of heparan sulfate. *J. Biol. Chem.* **273,** 24,979–24,982.
8. Habuchi, H., Habuchi, O., and Kimata, K. (1998) Biosynthesis of heparan sulfate and heparin: How are the multifunctional glycosaminoglycans built up? *Trends Glycosci. Glycotech.* **10,** 65–80.
9. Sugumaran, G. and Vertel, B. M. (2000) Biosynthesis of chondroitin sulfate and dermatan sulfate proteoglycans, in *Oligosaccharides in chemistry and biology: A comprehensive handbook* (Ernst, B., Hart, G., and Sinay, P. eds.), Wiley-VCH, Weinheim, Germany, pp. 375–394, in press.
10. Oegema, T. R., Kraft, E. L., Jourdian, G. W., and VanValen, T. R. (1984) Phosphorylation of chondroitin sulfate from the rat chondrosarcoma. *J. Biol. Chem.* **259,** 1720–1726.
11. Fransson, L.-Å., Silverberg, T., and Carlstedt, I. (1985) Structure of the heparan sulfate-protein linkage region: Demonstration of the sequence galactosyl-galactosyl-xylose 2-phosphate. *J. Biol. Chem.* **260,** 14,722–14,726.
12. Sugahara, K., Yamashina, I., De Waard, P., Van Halbeek, H., and Vliegenthart, J. F. G. (1988) Structural studies on sulfated glycopeptides from the carbohydrate-protein linkage region of chondroitin 4-sulfate proteoglycans of Swarm rat chondrosarcoma: demonstration of the structure Gal(4-O-sulfate)β1-3 Galβ1-4 Xylβ1-O-Ser. *J. Biol. Chem.* **263,** 10,168–10,174.
13. De Waard, P., Vliegenthart, J. F. G., Harada, T., and Sugahara, K. (1992) Structural studies on sulfated oligosaccharides derived from the carbohydrate-protein linkage region of chondroitin 6-sulfate proteoglycans of shark cartilage: II. Seven compounds containing 2 or 3 sulfate residues. *J. Biol. Chem.* **267,** 6036–6043.
14. Sugahara, K., Yamada, S., Yoshida, K., DeWaard, P. and Vliegenthart, J. F. G. (1992) A novel sulfated structure in the carbohydrate-protein linkage region isolated from porcine intestinal heparin. *J. Biol. Chem.* **267,** 1528–1533.
15. Silbert, J. E. and DeLuca, S. (1967) The synthesis of uridine diphosphate xylose by cell-free preparations from mouse mast cell tumors. *Biochim. Biophys. Acta* **141,** 193–196.
16. Abeijon, C., Mandon, E. C., and Hirschberg, C. B. (1997) Transporters of nucleotide sugars, nucleotide sulfate and ATP in the Golgi apparatus. *TIBS* **22,** 203–207.
17. Esko, J. D., Weinke, J. L., Taylor, W. H., Ekborg, G., Rodén, L., Anantharamaiah, G., and Gavish, A. (1987) Inhibition of chondroitin sulfate and heparan sulfate biosynthesis in Chinese hamster ovary cell mutants defective in galactosyltransferase I. *J. Biol. Chem.* **262,** 12,189–12,195.

18. Fritz, T. A., Gabb, M. M., Wei, G., and Esko, J. D. (1994) Two N-acetylglucosaminyl transferases catalyze the biosynthesis of heparan sulfate. *J. Biol. Chem.* **269,** 28,808–28,814.
19. Rohrmann, K., Niemann, R., and Buddecke, E. (1985) Two N-acetylgalactosaminyl-transferases are involved in the biosynthesis of chondroitin sulfate. *Eur. J. Biochem.* **148,** 463–469.
20. Sugumaran, unpublished results.
21. Lind, T., Lindahl, U., and Lidholt, K. (1993) Biosynthesis of heparin/heparan sulfate: identification of a 70 kDa protein catalyzing both the D-glucuronosyl and the N-acetylglucosaminyl transferase reactions. *J. Biol. Chem.* **268,** 20,705–20,708.
22. Malmström, A., Fransson, L.-Å., Höök, M., and Lindahl, U. (1975) Biosynthesis of dermatan sulfate: formation of L-iduronic acid residues. *J. Biol. Chem.* **250**, 3419–3425.
23. Kobayashi, M., Sugumaran, G., Liu, J., Shworak, N. W., Silbert, J. E., and Rosenberg, R. D. (1999) Molecular cloning and characterization of a human uronyl 2-sulfotransferase that sulfates iduronyl and glucuronyl residues in dermatan/chondroitin sulfate. *J. Biol. Chem.* **274**, 10,474–10,480.
24. Wei, Z., Sweidler, S. J., Ishihara, M., Orellana, A., and Hirschberg, C. B. (1993) A single protein catalyzes both N-deactylation and N-sulfation during biosynthesis of heparan sulfate. *Proc. Natl. Acad. Sci. (USA)* **90,** 3885–3888.
25. Sugumaran, G. and Silbert, J. E. (1990) Relationship of sulfation to ongoing chondroitin polymerization during biosynthesis of chondroitin 4-sulfate by microsomal preparations from cultured mouse mastocytoma cells. *J. Biol. Chem.* **265,** 18,284–18,288.
26. Vertel, B. M., Walters, L. M., Flay, N., Kearns, A. E., and Schwartz, N. B. (1993) Xylosylation is an endoplasmic reticulum to Golgi event. *J. Biol. Chem.* **268,** 11,105–11,112.
27. Silbert, J. E. and Reppucci, A. C., Jr. (1976) Biosynthesis of chondroitin sulfate: independent addition of glucuronic acid and N-acetylgalactosamine to oligosaccharides. *J. Biol. Chem.* **251**, 3942–3947.
28. Robbins, P. W. and Lipmann, F. (1957) Isolation and identification of active sulfate. *J. Biol. Chem.* **229,** 837–851.
29. Olson, C. A., Krueger, R., and Schwartz, N. B. (1985) Deglycosylation of chondroitin sulfate proteoglycans by hydrogen fluoride in pyridine. *Anal. Biochem.* **146,** 232–237.
30. Niemann, R. and Buddecke, E. (1982) Substrate specificity and regulation of activity of rat liver beta-D-glucuronidase. *Hoppe Seylers Z. Physiol. Chem.* **363,** 591–598.
31. Humphries, D. E., Silbert, C. K., and Silbert, J. E. (1988) Sulphation by cultured cells: cysteine, cysteinesulphinic acid, and sulphite as sources for proteoglycan sulphate. *Biochem. J.* **252,** 305–308.
32. Linhardt, R. J., Galliher, P. M., and Cooney, C. L. (1986) Polysaccharide lyases. *Appl. Biochem. Biotechnol.* **12,** 135–176.
33. Ludwigs, U., Elgavish, A., Esko, J. D., Meezan, E., and Rodén, L. (1987) Reaction of unsaturated uronic acid residues with mercuric salts. Cleavage of hyaluronic acid disaccharide 2-acetamido-2-deoxy-3-O-(β-D-gluco-4-ene-pyranosyl uronic acid)-D-glucose. *Biochem. J.* **245,** 795–804.
34. Malmstrom, A. and Aberg, L. (1982) Biosynthesis of dermatan sulphate: assay and properties of the uronosyl C-5 epimerase. *Biochem. J.* **201,** 489–493.
35. Pettersson, I. K., Kusche, M., Unger, E., Wlad, H., Nylund, L., Lindahl, U., and Kjellen, L. (1991) Biosynthesis of heparin: purification of a 110-kDa mouse mastocytoma protein required for both glucosaminyl N-deacetylation and N-sulfation. *J. Biol. Chem.* **266,** 8044–8049.
36. Silbert, J. E. (1963) Incorporation of $^{14}C$ and $^{3}H$ from nucleotide sugars into a polysaccharide in the presence of a cell-free preparation from mouse mast cell tumors. *J. Biol. Chem.* **238**, 3542–3546.

37. Campbell, P., Hannesson, H., Sandback, D., Roden, L., Lindahl U., and Li, J.-P. (1994) Biosynthesis of heparin/heparan sulfate: purification of the D-glucuronyl C-5 epimerase from bovine liver. *J. Biol. Chem.* **269,** 26,953–26,958.
38. Sugumaran, G. and Silbert, J. E. (1991) Formation of two species of nascent proteochondroitin in separate loci of a microsomal preparation from chick embryo epiphyseal cartilage. *Biochem. J.* **277,** 787–793.
39. Sugumaran, G. and Silbert, J. E. (1992) Effects of detergent on the sulphation of chondroitin by cell-free preparations from chick embryo epiphyseal cartilage. *Biochem. J.* **285,** 577–583.
40. Lowry, O. H., Rosebrough, N. J., Farr, A. L., and Randall, R. J. (1951) Protein measurement with the Folin phenol reagent. *J. Biol. Chem.* **193,** 265–275.
41. Sugumaran, G., Katsman, M., and Silbert, J.E. (1998) Subcellular co-localization and potential interaction of glucuronosyl-transferases with nascent proteochondroitin sulphate at Golgi sites of chondroitin synthesis. *Biochem. J.* **329,** 203–208.
42. Silbert, J.E. (1967) Biosynthesis of heparin: IV. N-Deacetylation of a precursor glycosaminoglycan. *J. Biol. Chem.* **242,** 5153–5157.
43. Malmström, A. (1984) Biosynthesis of dermatan sulfate: II. Substrate specificity of the C-5 uronosyl epimerase. *J. Biol. Chem.* **259,** 161–165.
44. Philipson, L. H. and Schwartz, N. B. (1984) Subcellular localization of hyaluronate synthetase in oligodendroglioma cells. *J. Biol. Chem.* **259,** 5017–5023.
45. Klewes, L., Turley, E. A., and Prehm, P. (1993) The hyaluronate synthase from a eukaryotic cell line. *Biochem. J.* **290,** 791–795.
46. Prehm, P., (1983) Synthesis of hyaluronate in differentiated teratocarcinoma cells: mechanism of chain growth. *Biochem. J.* **211,** 191–198.
47. Kitagawa, H., Tanaka, Y., Tsuchida, K., Goto, F., Ogawa, T., Lidholt, K., Lindahl, U., and Sugahara, K. (1995) N-Acetylgalactosamine transfer to the common carbohydrate-protein linkage region of sulfated glycosaminoglycans: identification of UDP-GalNAc: chondro-oligosaccharide α-N-acetylgalactosaminyltransferase in fetal bovine serum. *J. Biol. Chem.* **270,** 22,190–22,195.
48. Nadanaka, S., Kitagawa, H., Fumitaka, G., Tamura, J.-I., Neumann, K. W., Ogawa, T., and Sugahara, K. (1999) Involvement of the core protein in the first β-N-acetylgalactosamine transfer to the glycosaminoglycan-protein linkage-region tetrasaccharide and in the subsequent polymerization: the critical determining step for chondroitin sulphate biosynthesis. *Biochem. J.* **340,** 353–357.
49. Rodén, L., Baker, F. R., Helting, T., Schwartz, N. B., Stoolmiller, A. C., Yamagata, S., and Yamagata, T. (1972) Biosythesis of chondroitin sulfate, *Meth. Enzymol.* **28,** 662–676.
50. Helting, T. and Rodén, L. (1969) Biosynthesis of chondroitin sulfate. I. Galactosyl transfer in the formation of the carbohydrate-protein linkage region. *J. Biol. Chem.* **244,** 2790–2798.
51. Faltynek, C. R., Silbert, J. E., and Hof, L. (1981) Inhibition of the action of pyrophosphatase and phosphatase on sugar nucleotides. *J. Biol. Chem.* **256,** 7139–7141.
52. Pfeffer, S. R. and Rothman, J.E. (1987) Biosynthetic protein transport and sorting by the endoplasmic reticulum and Golgi. *Annu. Rev. Biochem.* **56,** 829–852.
53. Goldberg, D. E. and Kornfeld, S. (1983) Evidence for extensive subcellular organization of asparagine-linked oligosaccharide processing and lysosomal enzyme phosphorylation. *J. Biol. Chem.* **258,** 3159–3165.
54. Futerman A. H., Stieger B, Hubbard A. L., and Pagano R. E. (1990) Sphingomyelin synthesis in rat liver occurs predominantly at the cis and medial cisternae of the Golgi apparatus. *J. Biol. Chem.* **265,** 8650–8657.
55. Sugumaran, G. and Silbert, J. E. (1991) Subfractionation of chick embryo epiphyseal cartilage Golgi: localization of enzymes involved in the synthesis of the polysaccharide portion of proteochondroitin sulfate. *J. Biol. Chem.* **266,** 9565–9569.

# 12

# Disaccharide Composition of Hyaluronan and Chondroitin/Dermatan Sulfate

*Analysis with Fluorophore-Assisted Carbohydrate Electrophoresis*

**Anna H. K. Plaas, L. West, Ronald J. Midura, and Vincent C. Hascall**

## 1. Introduction

The glycosaminoglycan constituents of proteoglycans, heparan sulfate, chondroitin/dermatan sulfate, or keratan sulfate *(1)*, and the glycosaminoglycan hyaluronan *(2)* are important constituents of all tissue matrices. They provide tissue hydration for fluid flow and molecular transport, charge repulsions or attractions for intermolecular spacing *(3)* and polyanionic domains for growth factor or integrin signaling *(4,5)*, cell adhesion, and migration or proliferation *(6–8)*. As a result, studies of extracellular matrix metabolism in genetic, developmental, or tissue engineering experiments also include analyses of glycosaminoglycan components. The most widely used procedures for glycosaminoglycan analyses rely largely on the identification and quantitation of products generated by depolymerization with lyases *(9)* or hydrolases *(10)* by gel electrophoresis *(11)*, nuclear magnetic or mass spectrometry *(12–15)*, high-performance liquid chromatography *(16–21)* or capillary zone electrophoresis *(22–24)*. While sensitive and accurate, such procedures usually require costly or technically specialized equipment, and involve preparative steps that can lead to loss of products. We describe here an alternative procedure for the identification and quantitation of glycosaminoglycan lyase products using fluorophore-assisted carbohydrate electrophoresis or FACE *(25,26)*. It is especially applicable for studies when only small amounts of tissue or proteoglycans are available and/or a high sample throughout is needed. While we limit the illustration of this technically simple procedure to chondroitin/dermatan sulfate and hyaluronan in the chapter, it can also serve as an alternative sensitive and specific way to determine the identity and quantity of products of heparan sulfate-specific lyases *(9)* and keratan sulfate-specific hydrolases *(27)*.

The repeating disaccharide structures of chondroitin/dermatan sulfate and hyaluronan are illustrated in **Fig. 1**. In chondroitin and dermatan sulfate, the galNAc*β1,4*glcA repeats

From: *Methods in Molecular Biology, Vol. 171: Proteoglycan Protocols*
Edited by: R. V. Iozzo © Humana Press Inc., Totowa, NJ

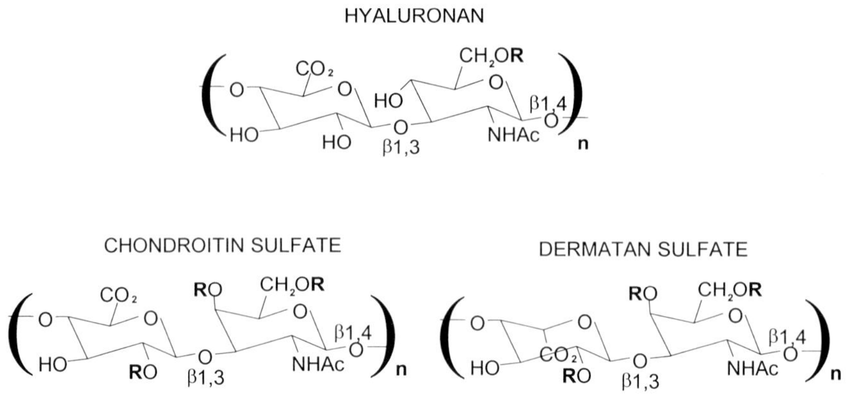

HYALURONAN

CHONDROITIN SULFATE

DERMATAN SULFATE

Fig. 1. Repeating disaccharide structures for hyaluronan, chondroitin sulfate, and dermatan sulfate. The sulfation at C4 and C6 of the hexosamine and at C2 of the hexuronic acid in chondroitin and dermatan sulfate are indicated by R. Dermatan sulfate is characterized by a variable proportion of L-iduronate residues in the chain, while chondroitin sulfate contains only D-glucuronate residues.

are extensively substituted with sulfate esters at C4 or C6 of the hexosamine residues, and occasionally also on C2 of an adjacent hexuronic acid residue. All three polymer sequences are cleaved at the *β1,4*-bonds by chondroitinase ABC ***(28)***, generating Δ-disaccharides from the chain interior, and mono- or disaccharides from the chain terminal (*see* **Fig. 2**). Each lyase product contains a free reducing end that can be stoichiometrically coupled to a fluorescent tag, such as 2-aminoacridone, via reductive amination ***(19,26,29)*** (*see* **Fig. 3**). The fluorotagged products can be resolved into discrete bands by electrophoresis on high percentage polyacrylamide gels ***(26,30–33)***. Each product can be detected down to the picomolar range by illuminating gels with ultraviolet light. Such images are electronically recorded and further processed using computerized image analysis software.

## 2. Materials

### 2.1. Glycosaminoglycan Preparation from Tissues and Proteoglycans

1. Proteinase K (fungal, > 20 U/mg, Gibco BRL (*see* **Notes 1–3**).
2. 100 m*M* ammonium acetate, pH 7.0, freshly prepared.
3. Absolute ethanol, precooled to –20°C.
4. Microcon centrifugal filter devices (YM-3) (Amicon).
5. Tabletop microcentrifuge.

### 2.2. Glycosaminoglycan Lyase Digestion

1. Chondroitinase ABC (*Proteus vulgaris*), (Seikagaku America), dissolved in water at 1.0 U/mL and stored in 10-µL aliquots at –80°C for up to 3 mo (*see* **Notes 4–6**).
2. 50 m*M* ammonium acetate, pH 7.0, freshly prepared.
3. Microcentrifuge tubes (1.5 mL capacity).

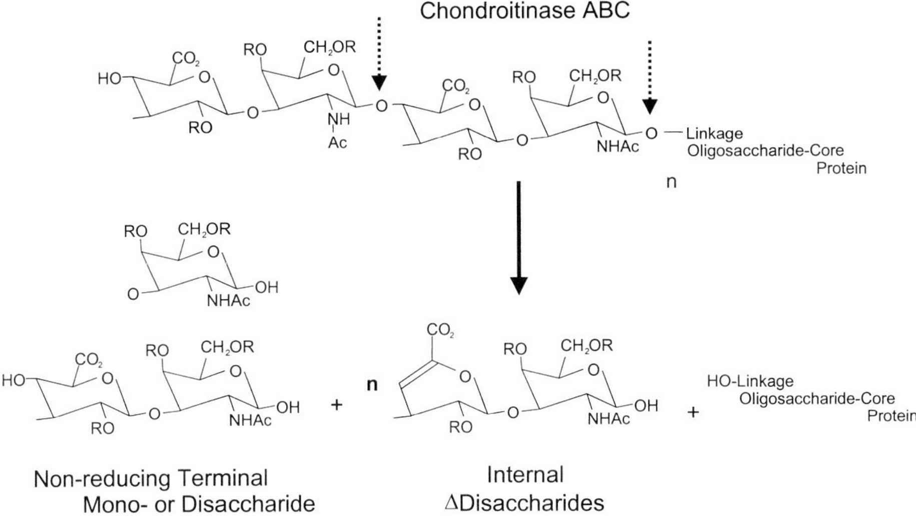


Fig. 2. Schematic representation of chondroitinase ABC cleavage in the glycosaminoglycan chain and the products generated from the interior and the nonreducing terminus.

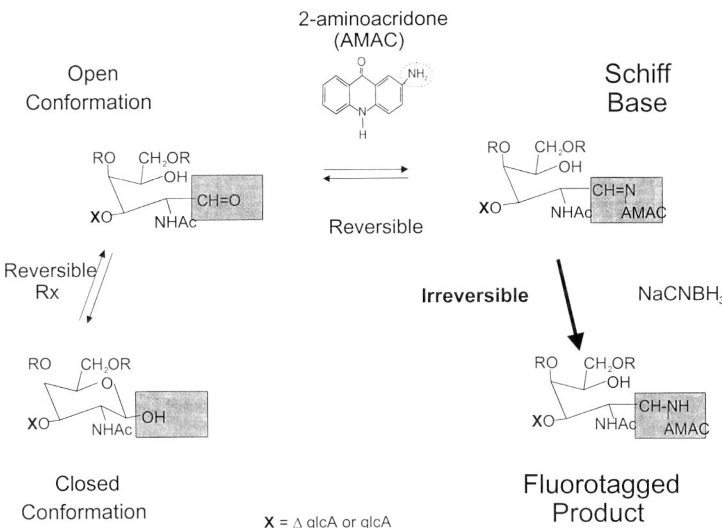


Fig. 3. Schematic representation of saccharide fluorotagging by reductive amination with 2-aminoacridone (AMAC) and sodium cyanoborohydride. In this reaction a Schiff's base is formed between the aldol group of the reducing saccharide and the amino group of a fluorotag, and this bond is then reduced to the stable secondary amine linkage to complete the formation of the fluorotagged products ***(26,30)***.

### 2.3. Fluorescent Derivitization of Lyase Digestion Products

1. Glucose (Sigma Chemical), stock solution of 10 mmol/mL, prepared in water and stored in 1-mL aliquots at –80°C and Δ-disaccharide standards (HPLC grade, Seikagaku America), stock solution of 1 mg/mL dissolved in water and stored in 10-µL aliquots at –20 °C (*see* **Note 7**).
2. Standard containing 1 nmol of glucose, and Δ-DiS (500–3000 pmol).
3. 2-aminoacridone (Molecular Probes).
4. Dimethylsulfoxide (99.7%) (Sigma-Aldrich Chemical).
5. Acetic acid (ACS grade) (Sigma-Aldrich Chemical).
6. Sodium cyanoborohydride (95 % w/v, in water) (Sigma-Aldrich Chemical) (*see* **Note 8**).
7. Glycerol (25% v/v, in water) (Sigma-Aldrich Chemical).
8. Oven or heating block at 37°C.

### 2.4. FACE Gel Electrophoresis of Fluorotagged Lyase Digestion Products

1. Mono™ Composition gel cassettes (Glyko) stored at 4°C until use (*see* **Notes 9** and **10**).
2. Mono™ gel running buffer (Glyko) freshly prepared in water and cooled to 4°C.
3. Electrophoresis apparatus (Glyko).
4. High voltage power supply (Bio-Rad PowerPac 1000 or Pharmacia Biotech EPS 1000).

### 2.5. Gel Imaging

1. Light-impermeable tape (*see* **Note 11**).
2. UV transilluminator (300–360 nm) (*see* **Note 12**).
3. Gel documentation system (Glyko Strategene Eagle Eye, or cooled CCD camera).

### 2.6. Product Identification

1. Image analysis sofware (NIH Image Analysis Version 1.7; Gel-Pro Analyzer 3.0) (*see* **Notes 13–15**).

## 3. Methods

### 3.1. Glycosaminoglycan Preparation from Tissues and Proteoglycans

1. Typically, 1 mg of tissue or 50 µg of purified proteoglycans are suspended in 300 µl of ammonium acetate buffer, pH 7.0, and digested at 60°C for 18–24 h with 50 µg of proteinase K.
2. The enzyme is inactivated by heating the samples at 100°C for 5 min, and undigested tissue debris is removed by centrifugation in a tabletop microcentrifuge (10,000*g*, 10 min at room temperature).
3. The clarified supernatants are cooled on ice, and glycosaminoglycan/peptides are precipitated by addition of 1.0 mL of ice-cold ethanol (to give a final concentration of 75% v/v) followed by storage at –20° C for 4 h (*see* **Notes 1** and **2**).
4. Precipitates are pelleted in a tabletop microcentrifuge at 10,000*g* for 15 min at 4°C.
5. Pellets are redissolved in water and portions are taken for estimation of total glycosaminoglycans by hexuronic acid content ***(34)*** or dimethymethylene blue dye binding ***(35)*** measurements (*see* **Notes 3, 4** and **7**).

### 3.2. Glycosaminoglycan Lyase Digestion

1. Glycosaminoglycan portions, 1–5 µg, are aliquoted into microcentrifuge tubes and dried in a vacuum concentrator.

2. They are resuspended in 100 μL of ammonium acetate buffer, pH 7.3, and digested at 37°C for 8–16 h with 5 mU of chondroitinase ABC (*see* **Notes 4–6**).
3. The digests are cooled on ice, and enzyme protein or undigested glycosaminoglycans are precipitated by addition of 900 μL of ice-cold ethanol at –20°C for 2 h.
4. Precipitates are removed in a microcentrifuge (10,000*g*, 20 min., at 4°C) and the Δ-disaccharides are recovered in the supernatant, which is dried by Speed Vac lyophilization.

### 3.3. Fluorescent Derivitization of Lyase Digestion Products

1. A 0.1 *M* solution of AMAC is prepared by dissolving 25 mg of the reagent in 1 mL of dimethylsulfoxide/acetic acid (85/15, v/v).
2. Then 5 μL of the freshly prepared reagent is added to the dried lyase products and the standards, dried in the bottom of the tube. Samples are vortex mixed carefully and spun birefly in the tabletop centrifuge to recover the mixture into the bottom of the tube. Samples are maintained at room temperature for 5–15 min, during which a 1 *M* solution of cyanoborohydride in water is prepared and a 5-μL aliquot immediately added to the AMAC-disaccharide mixture (*see* **Notes 7** and **8**).
3. Fluorotagging is carried out for 16 h at 37°C.
4. Samples are cooled to room temperature, and then mixed with 30 μL of 25% glycerol. These should be analyzed immediately by gel electrophoresis, or stored stably for up to 6 months at –80°C.

### 3.4. FACE Gel Electrophoresis of Fluorotagged Lyase Digestion Products

1. The electrophoresis tank is filled with precooled running buffer and set into a large container on ice to maintain tank and buffer temperatures at 4°C before and during the run.
2. The precast Mono composition gels are removed from the package, the loading wells extensively washed with the Mono gel running buffer, and the gel cassette placed into the electrophoresis tank.
3. A 4- to 6-μL portion (representing 10–15% of the total) of the standard mixture and the fluorotagged samples is loaded per well, and electrophoresis is done at 500 V, with a starting current of ~25 mA per gel and a final current of ~10 mA per gel (*see* **Note 9**).
4. Electrophoresis is terminated when the salt front reaches the bottom of the gel, which is usually achieved after ~80 min (*see* **Note 10**).

### 3.5. Gel Imaging and Product Quantitation (see Figs. 4–6)

1. After completion of the electrophoresis, one gel cassette at a time is removed from the tank. The outside of the glass plates as well as the loading wells are extensively washed with distilled water to remove running buffer and excess unreacted AMAC reagent (*see* **Note 11**).
2. The sample wells and the stacking gel are covered with light impermeable tape, and the gel cassette is placed on the transilluminator light box (*see* **Note 12**).
3. The gels are viewed and the images captured and stored in TIFF file formats.
4. Using standard image analysis software, an average pixel density per picomole of the Δ-disaccharide standards (based on the glucose standard) is determined. From that, Δ-disaccharide contents and composition in experimental samples are calculated (*see* **Notes 9, 13–15**).
5. A typical FACE gel of the chondroitinase ABC digestion products obtained from glycosaminoglycans prepared from fetal and adult bovine cartilages and adult human cornea

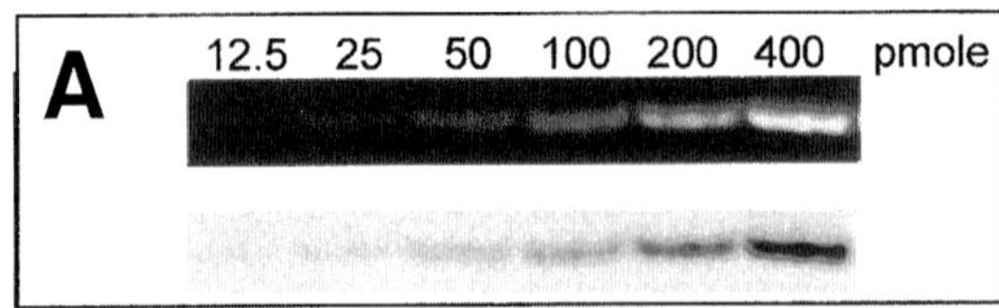


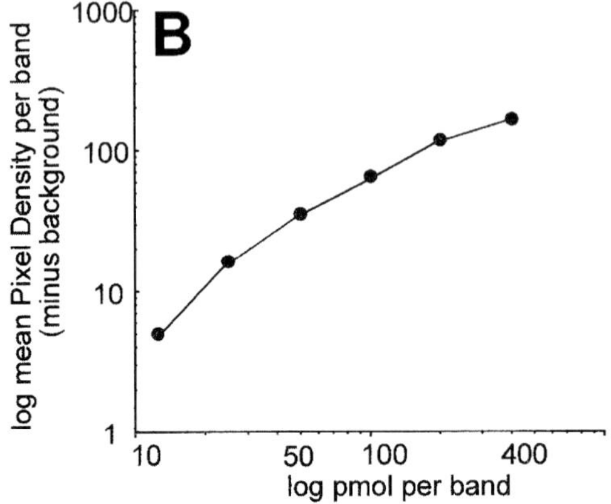


Fig. 4. Quantitation of a fluorotagged saccharide after FACE separation using pixel density measurements. **(A)** The fluorescent image and the inverted image used for determination of pixel density values of various concentrations of fluorotagged galactose. **(B)** Double log plot of pixel density per band vs. picomoles of saccharide per band, with near linearity between ~20–200 pmol of saccharide.

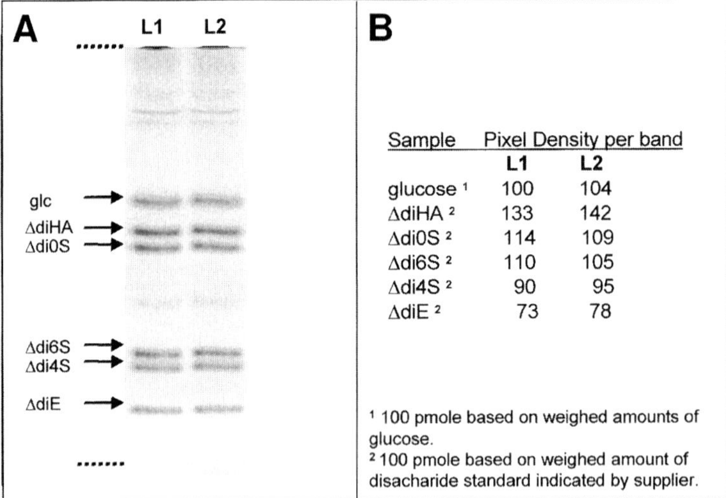


| Sample | Pixel Density per band | |
|---|---|---|
| | L1 | L2 |
| glucose [1] | 100 | 104 |
| ΔdiHA [2] | 133 | 142 |
| ΔdiOS [2] | 114 | 109 |
| Δdi6S [2] | 110 | 105 |
| Δdi4S [2] | 90 | 95 |
| ΔdiE [2] | 73 | 78 |

[1] 100 pmole based on weighed amounts of glucose.
[2] 100 pmole based on weighed amount of disacharide standard indicated by supplier.

Fig. 5. FACE analyses of Δ-disaccharide standards. Two separate preparations of 1 nmol each of Δ-di-HA, Δ-di-0S, Δ-di-6S, Δ-di-4S, and Δ-di-E (based on the quantity of disaccharide obtained from the supplier) and 1 nmol of glucose were fluorotagged and 10% portions of each reaction mixture (panel A, L1 and L2) were analyzed by FACE (*see* **Subheading 3.5.** in text for details).

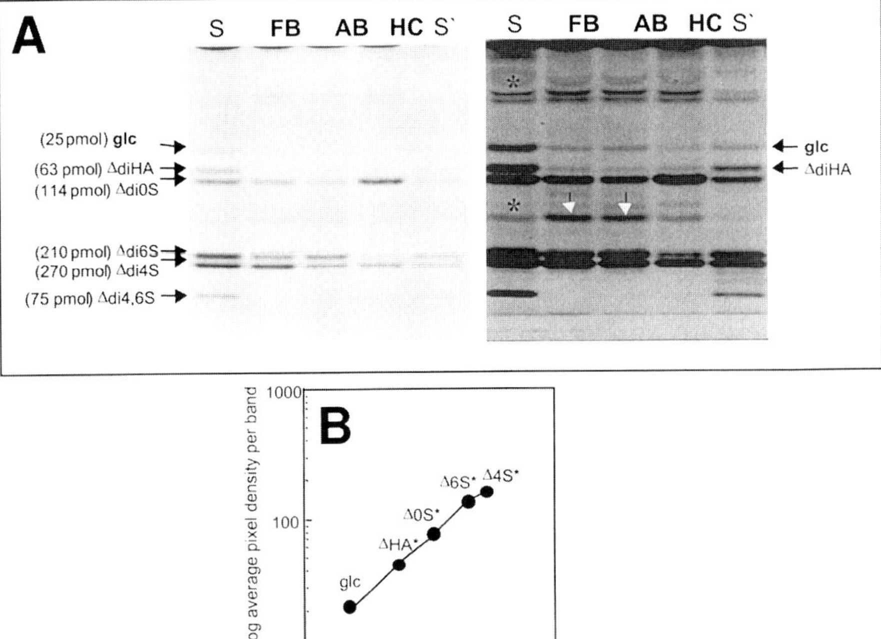


Fig. 6. FACE analyses of chondroitinase ABC-digested tissue glycosaminoglycans. 10 μg of glycosaminoglycans prepared from fetal bovine (FB) or adult bovine (AB) cartilages and from human cornea (HC) were digested with chondroitinase ABC, fluorotagged, and analyzed by FACE as outlined in the experimental procedure. The left-hand panel shows a short-exposure image that was used for quantitation of the major disaccharide products (Δ-di-0S, Δ-di-6S, and Δ-Di-4S). The right-hand panel shows a longer exposure that was used to visualize and quantitate minor products, such as Δ-di-HA, and the nonreducing terminal monosaccharide galNAc4S (indicated by white arrows). Nonspecific fluorescent bands from the reagents themselves are indicated by (*). A standard curve for product quantitation by pixel densities **(B)** was generated from the standard Δ-disaccharides and glucose shown in lane S (ranging from 25–270 pmol) and at a 1:4 dilution in lane S'.

is shown in **Fig. 6**. Two images representing different exposure times were used to visualize and quantitate the abundant Δ-disaccharides and minor products, respectively. The results are summarized in **Table 1**.

## 4. Notes

1. The separation of glycosaminoglycans from other protease products, such as collagen peptides, is usually required, since these can interfere with the lyase digestion and the recovery of products. Unsulfated chondroitin or short chains of hyaluronan and chondroitin/dermatan sulfate, however, are poorly recovered after ethanol precipitation. Alternatively, buffer salts and protein products can be removed by centrifugation (15 min at 9,500*g* at room temperature) of digests in MicroCon 3 devices. The glycosaminoglycan

**Table 1**
**Chonodroitin/Dermatan Sulfate Contents and Disaccharide Compositional Analyses of Cartilage and Cornea**

| | Glycosaminoglycan contents (mg/mg dry wt tissue) | | CS/CS disaccharide composition (% of total Δ-dis) | | |
|---|---|---|---|---|---|
| Tissue | CS/DS | HA[b] | Δ-di-0S: | Δ-di-4S: | Δ-di-6S |
| Fetal cartilage (FB[a]) | 130 | 1.8 | 19 | 54 | 27 |
| Adult cartilage (AB[b]) | 105 | 1.4 | 15 | 36 | 44 |
| Adult cornea (HC) | 5.75 | 0.11 | 63 | 31 | 6 |

[a] *See* **Fig. 6a.**
[b] For accurate hyaluronan quantitation, *see* **Subheading 3.5., item 4.**

peptides are retained on the filters, quantitatively recovered in water, and dried for further processing.

2. It should be noted that glucose is present in all tissues and biological fluids at ~0.8 mg/mL or less, which is solubilized with the protease digestion, but can be effectively removed from the glycosaminoglycans using the ethanol precipitation or MicroCon filtration procedure (*see* **Subheading 3.1.**). Conversely, fluorotagging and FACE analyses of a portion of the digests prior to the glycosaminoglycan purification step will provide a simple, yet accurate and sensitive, quantitation of the tissue or fluid glucose.
3. The recoveries of glycosaminoglycans as well as the completion of the lyase digestions can be determined by analyzing known amounts (1–5 μg) of commercially available standards, such as chondroitin, chondroitin sulfates (from Seikagaku America or Sigma-Aldrich Chemical) or hyaluronan (from Pharmacia or Sigma-Aldrich Chemical).
4. Chondroitinase digestions may not proceed to completion if the glycosaminoglycans are insufficiently dissolved before digestion. This can be improved by subjecting the glycosaminoglycan/water suspensions to several short cycles of agitation using a Vortex mixer and 10–15 min incubation at 37°C before the addition of the lyase. Attention should also be paid to maintaining the pH at 7–8 during the digestion, and it can be monitored by inclusion of 10 μL of 0.05% phenol red ***(26,33)*** in the digestion buffer.
5. Additional chondroitin lyases are commercially available. These include chondroitinase ACII (*Arthrobacter aurescens*), chondroitinase B (*Flavobacterium heparinum*), and proteinase-free ABC (*Proteus vulgaris*). It should be noted that each enzyme has distinct substrate specificity ***(9,10)*** and cleavage mode ***(36–39)*** that will result in a mixture of Δ-di- and oligosaccharides products from the chondroitin/dermatan sulfate chain interior. Conditions for optimal resolution of products generated by these enzymes by FACE will need to be established.
6. Chondroitinase ABC also digests hyaluronan, by cleavage at *β1,4* linkages between glcNAc and glcA, and Δ-di-HA product are therefore commonly present in lyase digests of total tissue glycosaminoglycans. Cleavage of hyaluronan, however, takes place at a significantly lower rate than that of dermatan and chondroitin sulfates. For the most accurate determination of tissue hyaluronan contents, samples should be digested sequentially with hyaluranidase SD (Seikagaku America) and chondroitinases ABC and ACII ***(26)***.
7. For quantitative fluorotagging of larger amounts (>5 μg) of Δ-disaccharide products increased volumes of both AMAC and cyanoborohydride solutions (up to 40 μL each) should be used. Reaction is carried out as described under **Subheading 3.3.** and glycerol added to a final concentration of 20% (v/v).

8. Completion of the reductive amination procedure also requires that the pH of the fluorotagging reaction mixture is maintained between 3.5 and 7.0. As a weak buffering in this pH range is provided by acetic acid and cyanoborohydride, only freshly prepared reagents should be used, to minimize decomposition of reducing agent or evaporation of acetic acid prior to fluorotagging.
9. Samples can be diluted before electrophoresis using a freshly prepared solution of dimethylsulfoxide (42%), acetic acid (7.5%), glycerol (20%) and 0.5 *M* cyanoborohydride.
10. Electrophoresis times should be monitored to adjust for variable rates of sample migration on different gel batches obtained from the supplier. For this, gels are briefly removed from the electrophoresis tank at 20- to 30-min intervals during the electrophoresis, and the migration of saccharides visualized with a hand-held UV light source.
11. Removal of excess AMAC reagent from the loading wells must be performed carefully, to avoid spillover of reagent onto the glass or into the gel. Its intense fluorescence otherwise masks the weaker fluorescent signals of individual saccharide bands.
12. If the transilluminator light source has a wavelength range of ~300 nm, the glass plates covering the gel may absorb incoming light and possibly emitted fluorescence. Fluorescence of lyase products can be intensified relative to the gel background by carefully removing one glass plate and placing the exposed gel surface directly on the transilluminator, with the remaining glass plate facing image-capture system.
13. Different image-capturing equipment and image analysis software can be used to quantitate lyase products. It is important, however, to calibrate each system for a near-linear range of product quantitation using pixel density measurements. A typical example of such a calibration is given in **Fig 4**. Various concentrations (12.5–400 pmol) of fluoratagged galactose were electrophoresed, the bands were visualized by illumination at 300 nm (panel A), and the image captured by a Strategene Eagle Eye II gel documentation system. The pixel densities of the bands were determined by NIH Image Analysis Software, version 1.57. The double-log plot of pixel density vs saccharide concentration shows linearity between ~20–200 pmole (panel B). A similar quantitation range is obtained when images are captured with a cooled CCD camera and pixel densities determined by Gel-Pro Image analyses software ***(26,33)***.
14. A sample containing a range of concentrations (20–300 pmol) of Δ-disaccharide standards is routinely included in one lane of each FACE gel, to generate a standard curve of pixel densities per picomole of saccharide for product identification and quantitation in experimental samples. It is recommended, however, that the nominal amount of the individual Δ-disaccharides as provided by the supplier, be validated first. This can be performed by FACE analyses, as illustrated in **Fig. 5**. Portions of Δ-disaccharide standards (1 nmol, based the amount [200 μg] of standard per vial from the supplier) and glucose (1 nmol, prepared from a stock solution of 1 mg/mL) are fluorotagged and electrophoresed (panel A). The pixel density values obtained for each Δ-disaccharide standard deviated by ~10–30 % from values for the glucose standard. In all subsequent runs (for example, *see* **Fig. 6**), the absolute amounts of each Δ-disaccharide standard can then be adjusted relative to the glucose standard.
15. Composition or identity of separated products should be verified by treatment of lyase products with mercuric acetate, sulfatases, and exoglycosidases ***(33)***. As these are generally carried out before the fluorotagging, the buffer conditions, salt removal, and product recoveries should be optimized for each treatment.

## Acknowledgment:

We would like to thank Dr. A. Calabro for his insightful comments during the preparation of the chapter and help in assembling **Figs. 1–3**.

## References

1. Iozzo, R. V., ed. (2000) *Proteoglycans: Structure, Biology and Molecular Interactions.* Marcel Dekker, New York.
2. Toole, B. P. (1997) Hyaluronan in morphogenesis. *J. Intern. Med.* **242,** 35–40
3. Iozzo, R. V. (1998) Matrix Proteoglycans: from molecular design to cellular function. *Annu. Rev. Biochem.* **67,** 609–652.
4. Lander, A., Stipp, C.,S., and Ivins, J. K. (1996) The glypican family of HSPGs: major cell surface proteoglycans in the developing nervous system. *Persp. Devel. Neurobiol.* **3,** 347–358.
5. Lin, X., Buff, E. M., Perrimon, N., and Michelson, A. M. (1998) HSPGs are essential for FGF receptor signaling during drosophila embryonic development. *Development* **126,** 3715–3723.
6. Cheung, W. F., Cruz, T. F., and Turley, E. A. (1999) Receptor for hyaluronan-mediated motility (RHAMM), a hyaladherin that regulates cell responses to growth factors. *Biochem. Soc. Trans.* **2,** 135–142.
7. Iida, J., Meijne, A. M., Knutson, J. R., Furcht, L. T., McCarthy, J. B., and Henderson, D. J. (1996) Cell surface chondroitin sulfate proteoglycans in tumor cell adhesion, motility and invasion. *Semin. Cancer. Biol.* **3,** 155–162.
8. Henderson, D. J. (1997) Versican expression is associated with chamber specification, septation and valvulogenesis in the developing mouse heart. *Circ. Res.* **83,** 523–532.
9. Linhardt, R. J., Galliher, P. M., and Cooney, C. L. (1986) Polysaccharide lyases. *Appl. Biochem.Biotechnol.* **12,** 135–176.
10. Ernst, S., Langer, R., Cooney, C. L., and Sasisekharan, R. (1995) Enzymatic degradation of glycosaminoglycans. *Crit. Rev. Biochem. Mol. Biol.* **30,** 387–444.
11. Lyon, M. and Gallagher, J. (1990) A general method for the detection and mapping of submicrogram quantities of glycosaminoglycan oligosaccharides on polyacrylamide gels by sequential staining with azure A and ammoniacal silver. *Anal. Biochem.* **185,** 63–70.
12. Yamanda, S., Yoshida, K., Sugiura, M., and Sugahara, K. (1992) One and 2D $^1$H NMR characterization of two series of sulfated disaccharides prepared from chondroitin sulfate and heparan sulfate/heparin by bacterial eliminase digestion. *J. Biochem. (Tokyo)* **112,** 440–447.
13. Linhardt, R. J., Wang, H. M., Loganathan, D., Lamb, D. J., and Mallis, L. M. (1992) Analyses of glycosaminoglycan derived oligosaccharides using fast atom bombardment mass-spectrometry. *Carbohydr. Res.* **225,** 137–145.
14. Tai, G. H., Huckerby, T. N., and Nieduszynski, I. A. (1996) Multiple non-reducing chain termini isolated from bovine corneal keratan sulfates. *J. Biol. Chem.* **271,** 23,535–23,546.
15. Pena, M., Williams, C., and Pfeiler, E. (1998) Structure of keratan sulfate from bone fish larvae deduced from NMR spectroscopy of keratanase derived oligosaccharides. *Carboh. Res.* **309,** 117–124.
16. Shinmei, M., Miyauchi, S., Machida, A., and Miyasaki, K. (1989) Quantitation of chondroitin 4- and chondroitin-6 sulfate in pathologic joint fluids. *Arth. & Rheum.* **35,** 1304–1308.
17. Midura, R., Salustri, A., Calabro, A., Yanagishita, M., and Hascall, V. C. (1994) High resolution separation of disaccharide and oligosaccharide alditols from chondroitin sul-

fate, dermatan sulfate and hyaluronan using CarboPac PA1 chromatography. *Glycobiology* **4,** 333–342.

18. Karamanos, N. K., Syroke, A., Vanky, P., Nurminem, M., and Hjerpe, A. (1994) Determination of 24 variously sulfated galactosaminoglycan and hyaluronan derived disaccharides by HPLC. *Anal. Biochem.* **221,** 189–199.
19. Plaas, A. H., Hascall, V. C., and Midura, R. (1996) Ion Exchange HPLC of chondroitin sulfate:quantitative derivitization of chondroitin lyase digestion products with 2-aminopyridine. *Glycobiology* **6,** 823–829.
20. Koshiishi, I, Takenouchi, M., and Imanari, T. (1999) Structural characteristics of oversulfated chondroitin/dermatan sulfates in the fibrous lesions of the liver with cirrhosis. *Arch. Biochem. Biophys.* **370,** 151–155.
21. Kinoshita, A. and Sugahara, K. (1999) Microanalyses of glycosaminoglycan derived oligosaccharides labeled with a fluorophore 2-aminobenzamide by HPLC: application to disaccharide compositional analyses and exosequencing of oligosaccharides. *Anal.Biochem.* **269,** 367–78.
22. al-Hakim, A. and Linhardt, R. J. (1991) Electrophoresis and detection of nanogram quantities of exogenous and endogenous glycosaminoglycans in biological fluids. *Appl. Theor. Electrophor.***1,** 305–312.
23. Kitagawa, H., Kinoshita, A., and Sugahara, K. (1997) Microanalyses of GAG derived disaccharides labeled with the fluorophore 2-aminoacridone by capillary zone electrophoresis and HPLC. *Anal. Biochem.* **232,** 114–121.
24. Lamari, F., Theocharis, A., Hjerpe, A., and Karamanos, N. K. (1999) Ultrasensitive CZE of sulfated disaccharides in CS/DS by laser induced fluorescence after derivitization with 2-aminoacridone. *J. Chrom. Biomed. Sci. Appl.* **730,** 129–133.
25. Jackson P. (1991) Polyacrylamide gel electrophoresis of reducing saccharides labeled with the fluorophore 2-aminoacridone: subpicomolar detection using an imaging system based on a cooled charge-coupled device. *Anal. Biochem.* **196,** 238–44.
26. Calabro, A., Hascall, V. C., and Midura, R.J. (2000) Adaptation of FACE methodology for microanalyses of total hyaluronan and chondroitin sulfate composition from cartilage. *Glycobiology* **10,** 283–293.
27. Nakazawa, K., Ito, M., Yamagata, T., and Suzuki, S. (1989) Substrate specificity of keratan sulfate degrading enzymes from microorganisms, in *Keratan Sulfate, Chemistry, Biology and Chemical Pathology* (Greiling, H. and Scott, J. E., eds.), Biochemical Society, London, U.K., pp. 99–110.
28. Suzuki, S., Saito, H., Yamagata, T., Anno, K., Seno, N., Kawai, Y., and Fuhashi, T. (1968) Formation of three types of disulfated dissacharides from chondroitin sulfates by chondroitinase digestion. *J. Biol. Chem.* **243,** 1543–1550.
29. Hase, S., Hara, S., and Matsushima, Y. (1979) Tagging of sugars with a fluorescent compound, 2-aminopyridine. *J. Biochem. (Tokyo)* **85,** 217–220.
30. Jackson, P. (1994) High resolution polyacrylamide gel electrophoresis of fluorophore-labeled reducing saccharides. *Meth. Enzymol.* **230,** 250–265.
31. Hu, J. (1994) Fluorophore assisted carbohydrate electrophoresis technology and applications. *J. Chromatogr.* **705,** 89–103.
32. Starr, C. M., Masada, R. I., Hague, C., Skop, E., and Klock, J. C. (1996) Fluorophore assisted carbohydrate electrophoresis in the separation, analysis, and sequencing of carbohydrates. *J. Chromatogr.* **720,** 295–321.
33. Calabro, A., Benavides, M., Tammi, M., Hascall, V. C., and Midura, R. J. (2000) Microanalyses of enzyme digests of hyaluronan and chondroitin/dermatan sulfate by fluorophore-assisted carbohydrate electrophoresis (FACE). *Glycobiology* **10**, 273–281.

34. Ludwigs, U., Elgavish, A., Esko, J. D., Meezan, E.. and Roden, L. (1987) Reaction of unsaturated uronic acid residues with mercuric salts. *Biochem. J.* **45,** 795–804.
35. Farndale, R. W., Buttle, D. J.. and Barrett, A. J. (1986) Improved quantitation and discrimination of sulfated glycosaminoglycans by use of dimethylmethylene blue. *Biochim. Biophys. Acta* **883,** 173–177.
36. Hamai, A., Hashimoto, N., Mochizuki, H., Kato, F., Makiguchi, Y., Horie, K., and Suzuki, S. (1997) Two distinct chondroitin sulfate ABC lyases. *J. Biol. Chem.* **272,** 9123–9130.
37. Huckerby, T. N., Lauder, R. M., and Nieduszynski, I. A. (1998) Structure determination for octasaccharides derived from the carbohydrate-protein linkage region of chondroitin sulfate chains in the proteoglycan aggrecan from bovine articular cartilage. *Eur. J. Biochem.* **258,** 669–676 .
38. West, L. A., Roughley, P., Nelson, F. R., and Plaas, A. H. (1999) Sulfation heterogeneity in the trisaccharide (GalNAcSbeta1,4GlcAbeta1,3GalNAcS) isolated from the non-reducing terminal of human aggrecan chondroitin sulfate. *Biochem. J.* **342,** 223–229
39. Aguiar, J. A. and Michelacci, Y. M. (1999) Preparation and purification of *Flavobacterium heparinum* chondroitinases AC and B by hydrophobic interaction chromatography. *Braz. J. Med. Biol. Res.* **32,** 45–50.

# 13

## Integral Glycan Sequencing of Heparan Sulfate and Heparin Saccharides

**Jeremy E. Turnbull**

### 1. Introduction

The functions of heparan sulfate (HS) are determined by specific saccharide motifs within HS chains. These sequences confer selective protein-binding properties and the ability to modulate protein bioactivities *(1,2)*. HS chains consist of an alternating disaccharide repeat of glucosamine (GlcN; N-acetylated or N-sulfated) and uronic acid (glucuronic [GlcA] or iduronic acid [IdoA]). The initial biosynthetic product containing N-acetylglucosamine (GlcNAc) and GlcA is modified by N-sulfation of the GlcN, ester (O)-sulfation (at positions 3 and 6 on the GlcN and at position 2 on the uronic acids) and by epimerization of GlcA to IdoA. The extent of these modifications is incomplete, and their degree and distribution varies in HS between different cell types. In HS chains, N- and O-sulfated sugars are predominantly clustered in sequences of up to 8 disaccharides separated by N-acetyl-rich regions with a low sulfate content *(3)*.

Sequence analysis of HS saccharides has presented a daunting analytical problem and until very recently sequence information had been obtained for only relatively short saccharides from HS and heparin. Gel chromatography and high-performance liquid chromatography (HPLC) methods have been employed to obtain information on disaccharide composition *(3,4)*. Other methods such as nuclear magnetic resonance (NMR) spectroscopy and mass spectroscopy *(5–9)* have provided direct sequence information, but are difficult for even moderately sized oligosaccharides and in the case of NMR requires large amounts of material (micro-moles). However, the scene has changed rapidly in the last few years with the availability of recombinant exolytic lysosomal enzymes. These exoglycosidases and exosulphatases remove specific sulfate groups or monosaccharide residues from the nonreducing end (NRE) of saccharides *(10)*. These can be used in combination with PAGE separations to derive direct information (based on band shifts) on the structures present at the nonreducing end of GAG saccharides (*11*; *see* **Fig. 1** for an example).

From: *Methods in Molecular Biology, Vol. 171: Proteoglycan Protocols*
Edited by: R. V. Iozzo © Humana Press Inc., Totowa, NJ

Integral glycan sequencing (IGS), a PAGE-based method using the exoenzymes, was recently developed as the first strategy for rapid and direct sequencing of heparan sulfate and heparin saccharides ***(11)***. Its introduction has been quickly followed by a variety of similar approaches using other separation methods including HPLC and MALDI mass spectrometry ***(12–14)***. In IGS, an oligosaccharide (previously obtained from the polysaccharide by partial chemical or enzymatic degradation and purification) is labeled at the reducing terminal with a fluorescent tag. This is subjected to partial nitrous acid degradation to give a ladder of evenly numbered oligosaccharides (di-, tetra-, hexa-, etc.), each bearing a fluorescent tag at its reducing-end terminus. Portions of this material are then treated with a variety of highly-specific exolytic lysosomal enzymes (exosulfatases and exoglycosidases), which act only at the nonreducing end of each saccharide if it is a suitable substrate. The various digests are then separated on a high-density polyacrylamide gel and the positions of the fragments are detected by excitation of the fluorescent tag with an ultraviolet (UV) transilluminator. Band shifts due to the different treatments permit the sequence to be read directly from the banding pattern (*see* **Fig. 1** for an example). This novel strategy allows direct readout sequencing of a saccharide in a single set of adjacent gel tracks in a manner analogous to DNA sequencing ***(11)***. IGS provides for the first time a rapid approach for sequencing HS saccharides, and it has proved invaluable in recent structure–function studies ***(15)***. It should be noted that this methodology is designed for sequencing purified saccharides, not whole HS preparations. Clearly, a critical factor in all sequencing methods is the availability of sufficiently pure oligosaccharide starting material. HS and heparin saccharides can be prepared following selective scission by enzymic (or chemical) reagents and isolation by methods such as affinity chromatography ***(4)***. Final purification usually requires the use of strong anion-exchange HPLC (***15***; *see* Chapter 14).

## 2. Materials

1. 2-aminobenzoic acid (2-AA; Fluka Chemicals).
2. Formamide.
3. Sodium cyanoborohydride (>98% purity).
4. Distilled water.
5. Oven or heating block at 37°C.
6. Desalting column (Sephadex G-25; e.g., HiTrap$^{TM}$ desalting columns, Pharmacia).
7. Centrifugal evaporator.
8. 200 m*M* HCl.
9. 20 m*M* sodium nitrite. 1.38 mg/mL in distilled water; prepare fresh.
10. 200m*M* sodium acetate, pH 6.0: 27.2 g/L sodium acetate trihydrate, pH to 6.0 using acetic acid.
11. Enzyme buffer: 0.2 *M* Na acetate, pH 4.5. Make 0.2 *M* sodium acetate (27.2g/L sodium acetate trihydrate) and 0.2 *M* acetic acid (11.6 mL/L) and mix in a ratio of 45 mL to 55 mL respectively.
12. Enzyme stock solutions (typically at concentrations of 500 mU/mL, where 1U = 1 µmol substrate hydrolyzed per minute). Available from Glyko (Novato, CA).
13. Vortex tube mixer.
14. Microcentrifuge.

15. Acrylamide stock solution (T50%/C5%). Caution: acrylamide is neurotoxic. Wear gloves (and a face mask when handling powdered forms). It is convenient to use pre-mixed acrylamide-*bis*, such as Sigma A-2917. Add 43 mL of distilled water to the 100 mL bottle containing the premixed chemicals and dissolve using a small stirrer bar (~2 h). Final volume should be ~80 mL. Store the stock solution at 4°C. Note that it is usually necessary to warm gently to redissolve the acrylamide after storage.
16. Resolving gel buffer stock solution: 2 *M* Tris-HCl, pH 8.8. 242.2 g/L Tris base, pH to 8.8 with HCl.
17. Stacking gel buffer stock solution: 1 *M* Tris-HCl, pH 6.8. 121.1 g/L Tris base, pH to 6.8 with HCl.
18. Electrophoresis buffer: 25 m*M* Tris-HC1, 192 m*M* glycine, pH 8.3. 3 g/L Tris base, 14.4 g/L glycine, pH to 8.3 if necessary with HCl.
19. 10% ammonium persulfate in water (made fresh or stored at –20°C in aliquots).
20. TEMED.
21. Vertical slab gel electrophoresis system (minigel or standard size).
22. DC power supply unit (to supply up to 500–1000 V and 200 mA).
23. UV tansilluminator (312-nm maximum emission wavelength).
24. Glass UV bandpass filter larger than gel size (type UG-11 or M-UG2).
25. CCD imaging camera fitted with a 450-nm (blue) band -pass filter.

## 3. Methods

### 3.1. Derivatization of Saccharides with the Fluorophore 2AA

HS and heparin saccharides can be labeled by reaction of their reducing aldehyde functional group with a primary amino group (reductive amination). For sulfated saccharides anthranilic acid (2-aminobenzoic acid; 2-AA; **ref. *11***) has been found to be effective for the IGS methodology. 2-AA conjugates typically display an excitation maxima in the range 300–320 nm, which is ideal for visualization with a commonly available 312 nm UV source (e.g., transilluminators used for visualizing ethidium bromide stained DNA). Emission maxima are typically in the range 410–420 nm (bright violet fluorescence). The approach described below allows rapid labeling and purification of tagged saccharide from free tagging reagent, gives quantitative recoveries, and the product is free of salts that might interfere with subsequent enzymic conditions. For saccharides in the size range hexa- to dodecasaccharides, approximately 2–3 nmol of purified starting material is the minimum required (~2–10 μg).

1. Dry down the purified saccharide (typically 2–20 nmol) in a microcentrifuge tube by centrifugal evaporation.
2. Dissolve directly in 10–25 μL of formamide containing freshly prepared 400 m*M* 2-AA (54.8 mg/mL) and 200 m*M* reductant (sodium cyanoborohydride; 12.6 mg/mL) and incubate at 37°C for 16–24 h in a heating block or oven. (Caution: the reductant is toxic and should be handled with care.) The volume used should be sufficient to provide a 500- to 1000-fold molar excess of 2-AA over saccharide (*see* **Note 1**).
3. Remove free 2-AA, reductant, and formamide from the labeled saccharides by gel filtration chromatography (Sephadex G-25 Superfine). Dilute the sample (maximum 250 μL of reaction mixture) to a total of 1 mL with distilled water (*see* **Note 2**).
4. Load onto two 5 mL HiTrap™ desalting columns (Pharmacia) connected in series. Alternatively, it is possible to use self-packed columns of other dimensions.

5. Elute with distilled water at a flow rate of 1 mL/min and collect fractions of 0.5 mL. Saccharides consisting of four or more monosaccharide units typically elute in the void volume (approximately fractions 7–12).
6. Pool and concentrate these fractions by centrifugal evaporation or freeze-drying.

### 3.2. Treatment of Saccharides with Nitrous Acid

Low-pH nitrous acid cleaves only at linkages between N-sulfated glucosamine and adjacent hexuronic acid residues (***16,17***; *see also* Chapter 34). Under controlled conditions nitrous acid cleavage can be used to create a ladder of bands that correspond to the positions of internal N-sulfated glucosamine residues in the original intact saccharide (***11***). To achieve this a series of different reaction stop points are pooled to produce a partial digest with a range of different fragment sizes.

1. Dry down 1–2 nmol of labeled saccharide by centrifugal evaporation.
2. Redissolve in 80 µl of distilled water and chill on ice.
3. Add 10 µl of 200 m*M* HCl and 10 µl of 20 m*M* sodium nitrite (both prechilled on ice) and incubate on ice.
4. At a series of individual time points (typically 0.5, 1, 2, and 3 h), remove an aliquot and stop the reaction by raising the pH to approximately 5.0 by the addition of 1/5 volume of 200 m*M* sodium acetate buffer, pH 6.0 (*see* **Note 3**).
5. Pool the set of aliquots and either use directly for enzyme digests or desalt as described under **Subheading 3.1.**

### 3.3. Treatment of Saccharides with Exoenzymes

The basic approach for treatment of HS/heparin samples with exoenzymes is described below. Details of the specificities of the exoenzymes are given in **Table 1**. Although these enzymes have differing optimal pH and buffer conditions, in general it is possible to use them under the single set of conditions given here, simplifying multiple enzyme treatments (*see* **Note 4**).

1. Dissolve the sample (typically 10–200 pmol of saccharide) in 10 µL of $H_2O$ in a microcentrifuge tube.
2. Add 5 µL of exoenzyme buffer, 1 µL of 0.5 mg/mL bovine serum albumin, 2 µL of appropriate exoenzyme (0.2–0.5 mU) and distilled water to bring the final volume to 20 µL.
3. Mix the contents well on a vortex mixer, and centrifuge briefly to ensure that the reactants are at the tip of the tube.
4. Incubate the samples at 37°C for 16 h in a heating block or oven.

### 3.4. PAGE Separation of Saccharides

Polyacrylamide gel electrophoresis (PAGE) is a high-resolution technique for the separation of HS and heparin saccharides of variable sulfate content and disposition. It provides a level of resolution for oligosaccharides larger than tetrasaccharides that is superior to gel filtration or anion-exchange HPLC (***18,19***). It is possible to obtain improved resolution using gradient gels. However, these are more difficult to prepare and use routinely and in most cases adequate resolution can be obtained with isocratic gels (*see* **Note 5**). Oligosaccharide mapping by PAGE is a rapid and reproducible method for the simultaneous comparison of multiple samples. It thus provides a simple but powerful approach for separating the saccharide products generated in the sequencing process.

**Table 1**
**Exoenzymes for Sequencing Heparan Sulfate and Heparin**

| Enzyme[a] | Substrate specificity[b] |
|---|---|
| Sulfatases | |
| Iduronate-2-sulfatase | IdoA(2S) |
| Glucosamine-6-sulfatase | GlcNAc(6S), $GlcNSO_3$(6S) |
| Sulphamidase (glucosamine N-sulfatase) | $GlcNSO_3$ |
| Glucuronate-2-sulfatase | GlcA(2S) |
| Glucosamine-3-sulphatase | $GlcNSO_3$(3S) |
| Glycosidases | |
| Iduronidase | IdoA |
| Glucuronidase | GlcA |
| α-N-Acetylglucosaminidase | GlcNAc |
| Bacterial exoenzymes | |
| Δ-4,5-Glycuronate-2-sulfatase | Δ-UA(2S) |
| Δ-4,5-Glycuronidase | Δ-UA |

[a]Enzyme availability: Glucuronidase is widely available commercially as purified enzyme. Recombinant iduronate-2-sulphatase, iduronidase, glucosamine-6-sulphatase, sulfamidase, and α-N-acetylglucosaminidase are available from Glyko (Novato, CA). Glucuronate-2-sulfatase and glucosamine-3-sulfatase have only been purified from cell and tissue sources to date. The bacterial exoenzymes are available from Grampian Enzymes, Nisthouse, Harray, Orkney, Scotland; e-mail grampenz@aol.com.

[b]The specificities are shown as the nonreducing terminal group recognised by the enzymes. Sulfatases remove only the sulfate group, whereas the glycosidases cleave the whole nonsulfated monosaccharide.

### *3.4.1. Preparing the PAGE Gel*

1. Assemble the gel unit (consisting of glass plates and spacers, etc.).
2. Prepare and degas the resolving gel acrylamide solution without ammonium persulfate or TEMED. To make a 30% acrylamide gel solution for a 16 cm ×12cm × 0.75 mm gel, 16 mL are required. Mix 9.6 mL of T50%/C5% acrylamide stock with 3 mL of 2 *M* Tris pH 8.8 and 3.4 mL of distilled water.
3. Add 10% ammonium persulfate (30 µl) and TEMED (10 µL) to the gel solution, mix well, and immediately pour into the gel unit.
4. Overlay the unpolymerized gel with resolving gel buffer (375 m*M* Tris-HCl, pH 8.8, diluted from the 2 *M* stock solution) or water-saturated butanol. Polymerization should occur within ~30–60 min. The gel can then be used immediately or stored at 4°C for 1–2 wk.

### *3.4.2. Electrophoresis*

1. Immediately before electrophoresis, rinse the resolving gel surface with stacking gel buffer (0.125 M Tris-HC1 buffer, pH 6.8, diluted from the 1 *M* stock solution).
2. Prepare and degas the stacking gel solution (for 5 ml, mix 0.5 mL of T50%/C5% acrylamide stock with 0.6 mL of 1 *M* Tris pH 6.8 and 3.9 mL of distilled water).

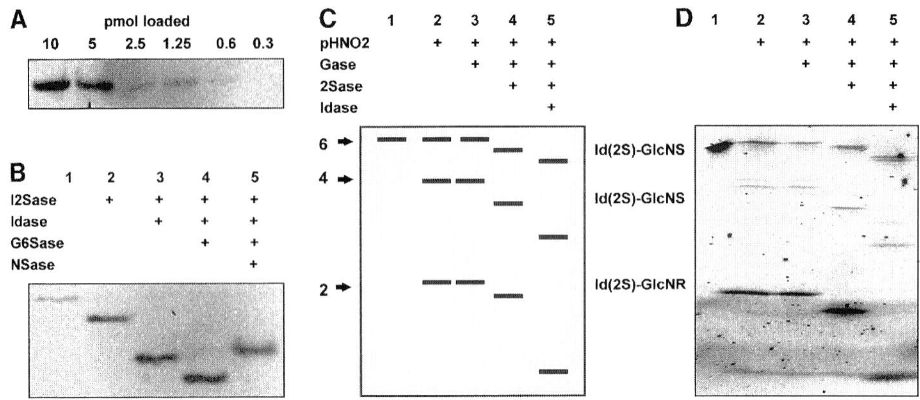


Fig. 1. Principles of integral glycan sequencing and an example. **(A)** Fluorescence detection of different amounts of a 2AA-tagged heparin tetrasaccharide run on a 33% minigel. **(B)** Exosequencing of a 2AA-tagged heparin tetrasaccharide with lysosomal enzymes and separation of the products on a 33% minigel (15 pmol per track). Band shifts following the exoenzyme treatments shown reveal the structure of the nonreducing end disaccharide unit (track 1, untreated). I2Sase, iduronate-2-sulfatase; Idase, iduronidase; G6Sase, glucosamine-6-sulfatase; Nsase, sulfamidase. **(C)** Schematic representation of IGS of a hexasaccharide (pHNO2, partial nitrous acid treatment). **(D)** Actual example of IGS performed on a purified heparin hexasaccharide, corresponding to the scheme in **(C)**, using the combinations of $pHNO_2$ and exoenzyme treatments indicated (track 1, untreated, 25 pmol; other tracks correspond to ~200 pmol/per track of starting sample for $pHNO_2$ digest). The hexasaccharide (purified from bovine lung heparin) has the putative structure IdoA(2S)-GlcNSO$_3$(6S)-IdoA(2S)-GlcNSO$_3$(6S)-IdoA(2S)-AMannR(6S). Electrophoresis was performed on a 16 cm 35% gel. Copyright © 1999 National Academy of Sciences, USA. From ref. ***(11)***.

3. Add 10% ammonium persulfate (10 µL) and TEMED (5 µL). Immediately pour on to the top of the resolving gel and insert the well-forming comb.
4. After polymerization (~15 min), remove the comb and rinse the wells thoroughly with electrophoresis buffer.
5. Place the gel unit into the electrophoresis tank and fill the buffer chambers with electrophoresis buffer.
6. Load the oligosaccharide samples (5–20 µL depending on well capacity, containing ~10% (v/v) glycerol or sucrose in 125 m*M* Tris-HCl, pH 6.8, carefully into the wells with a microsyringe. Marker samples containing bromophenol blue and phenol red should also be loaded into separate tracks.
7. Run the samples into the stacking gel at 150–200 V (typically, 20–30 mA) for 30–60 min, followed by electrophoresis at 300–400 V (typically 20–30 mA and decreasing during run) for approximately 5–8 h (for a 16 cm gel). Heat generated during the run should be dissipated using a heat exchanger with circulating tap water, or by running the gel in a cold room or in a refrigerator.
8. Electrophoresis should be terminated before the Phenol red marker dye is about 5 cm from the bottom of the gel. (At this point, disaccharides should be 3–4 cm from the bottom of the gel.)

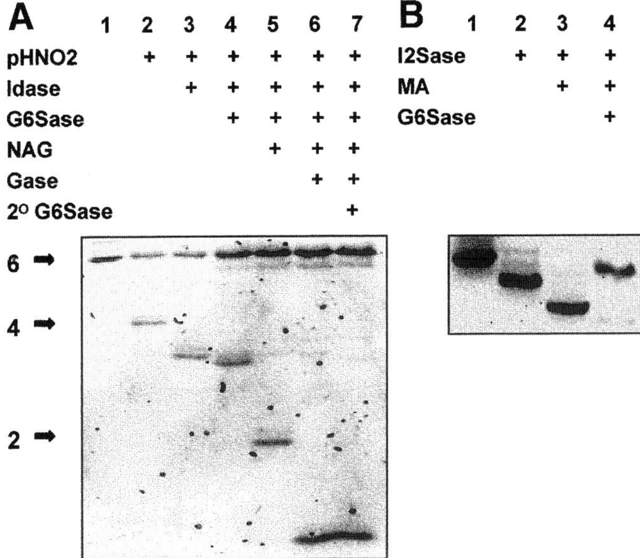


Fig. 2. IGS of a heparin hexasaccharide of known structure. A heparin hexasaccharide with the structure DHexA(2S)-GlcNSO$_3$(6S)-IdoA-GlcNAc (6S)-GlcA-GlcNSO$_3$(6S), was 2AA-tagged and subjected to sequencing on 16cm 33% gel. **(A)** IGS of hexasaccharide using the combinations of pHNO$_2$ and exoenzyme treatments indicated (track 1, untreated, 20 pmol; other tracks correspond to ~90 pmol/per track of starting sample for pHNO$_2$ digest). NAG, N-acetylglucosaminidase. **(B)** Determining the sequence of the nonreducing disaccharide unit of the hexasaccharide using the I2Sase, G6Sase, and mercuric acetate (MA) treatments shown (~20 pmol/track; track 1, untreated). Copyright © 1999 National Academy of Sciences, USA. From ref. *(11)*.

### 3.5. Imaging the Gels

Effective gel imaging requires a CCD camera that can detect faint fluorescent banding patterns by capturing multiple frames. Systems commonly used for detection of ethidium bromide stained DNA can usually be adapted with appropriate filters as described in **Note 6**.

1. Place a UV filter (UG-1, UG-11, or MUG-2) onto the transilluminator, and fit a 450-nm blue filter onto the camera lens.
2. Remove the gel carefully from the glass plates after completion of the run and place on the UV transilluminator surface wetted with electrophoresis buffer. Also wet the upper surface of the gel to prevent gel drying and curling.
3. Switch on transilluminator and capture image using CCD camera. Exposure times are typically 1–5 s depending on the amount of labeled saccharide (*see* **Note 7**).

### 3.6. Interpreting the Data

The sequence of saccharides subjected to IGS can be read directly from the banding pattern by interpreting the band shifts due to removal of specific sulfate or sugar

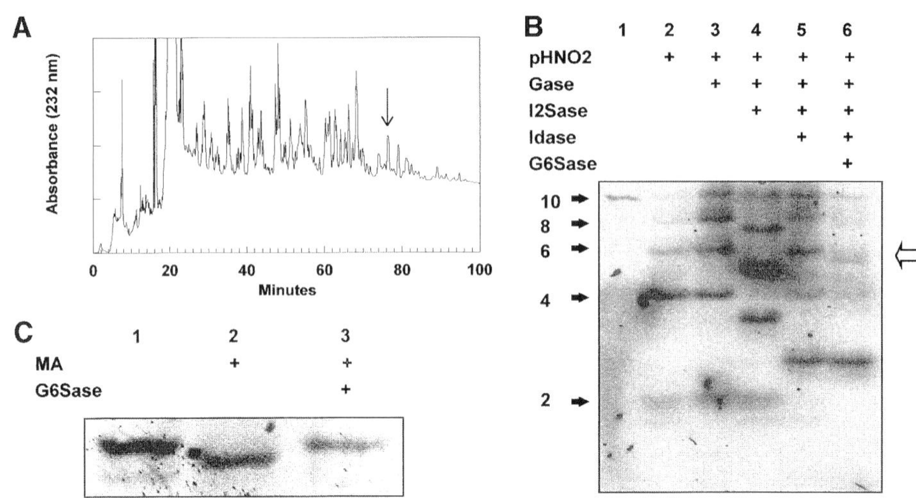


Fig. 3. Purification and IGS of a HS decasaccharide. **(A)** SAX-HPLC of a pool of HS decasaccharides derived by heparitinase treatment of porcine mucosal HS. For details of this technique, *see* Chapter 14. The arrowed peak was selected for sequencing. **(B)** IGS of the purified HS decasaccharide on a 16 cm 33% gel using the combinations of $pHNO_2$ and exoenzyme treatments indicated (track 1, untreated, 20 pmol; other tracks correspond to ~400 pmol/per track of starting sample for $pHNO_2$ digest). **(C)** Determining the sequence of the nonreducing disaccharide unit of the HS decasaccharide using the mercuric acetate (MA) and G6Sase treatments shown (approximately 40 pmol per track; track 1, untreated). Copyright© 1999 National Academy of Sciences, USA. From ref. *(11)*.

moieties. **Figure 1** shows an actual example and a schematic representation. First, bands generated by the partial nitrous acid treatment indicate the positions of N-sulfated glucosamine residues in the original saccharide (*see* **Fig. 1C**, track 2). Lack of a band at a particular position indicates the presence of an N-acetylated glucosamine residue (an example of this is shown in **Fig. 2**). Such saccharides can be sequenced with the additional use of the exoenzyme N-acetylglucosaminidase, which removes this residue and allows further sequencing of an otherwise "blocked" fragment. Following the nitrous acid treatment, the "ladder" of bands is then subjected to various exoenzyme digestions. The presence of specific sulfate or sugar residues can be deduced from the band shifts that occur (*see* **Fig. 1C**, tracks 3–5). **Figure 3** shows an example of a decasaccharide from HS which has been purified by SAX-HPLC and sequenced using IGS.

Usually the band shifts are downwards, due to the lower molecular mass and thus higher mobility of the product. However, it should be noted that occasionally upward shifts occur, probably due to subtle differences in charge/mass ratio (for examples, *see* **Figs. 1B**, **2B** and **3C**). Note also that minor "ghost" bands sometimes appear after the nitrous acid treatment. They are probably due to loss of an N-sulfate group, and normally these do not affect interpretation of the shifts in the major bands *(11)*.

If the saccharide being sequenced was derived by bacterial lyase treatment, it will have a Δ-4,5-unsaturated uronate residue at its nonreducing terminus. If this residue has a 2-O-sulfate attached, this can be detected by susceptibility to I2Sase (*see* **Fig. 2B**), but the sugar residue itself is resistant to both Idase and Gase. Its removal is required in order to confirm whether there is a 6-O-sulfate on the adjacent non-reducing end glucosamine (*see* **Figs. 2B** and **3C** for examples). However, bacterial enzymes which specifically remove the Δ-4,5-unsaturated uronate residues (and the 2-O-sulfate groups that may be present on them) are now available commercially (see **Table 1**). Alternatively, they can be removed chemically with mercuric acetate (***20***; *see* **Figs. 2B** and **3C**).

In addition to the basic sequencing experiment, it is wise to confirm agreement of the data with an independent analysis of the disaccharide composition of the saccharide (*see* Chapter 14). It can sometimes be difficult to sequence the reducing terminal monosaccharide, due to it being a poor substrate for the exoenzymes. In these cases it has proved more effective to analyze the terminal 2AA-labeled disaccharide unit in comparison to 2AA-labeled disaccharide standards ***(11)***.

## 4. Notes

1. Using large excesses of reagent as described, saccharides derived from HS and heparin by bacterial lyase scission generally couple with 2-AA with efficiencies in the range 60–70%. In contrast, saccharides derived from HS and heparin by low-pH nitrous acid scissioning (i.e., having an anhydromannose residue at their reducing ends) label more efficiently (~70–80% coupling efficiency).
2. Unwanted reactants and solvent can also be removed from labeled saccharides by methods such as dialysis, but the rapid gel filtration chromatography step described above using the HiTrap desalting columns is convenient and usually allows good recoveries of loaded sample, particularly for 2-AA-labeled saccharides (~80%).
3. It is useful to perform some trial incubations to test for optimal time points needed to generate a balance of all fragments in the partial nitrous acid digestion. With longer saccharides (octasaccharides and larger) it is observed that the largest products are generated quickly and thus a bias toward shorter incubations is required as saccharide length increases.
4. The enzyme conditions should provide for complete digestion of all susceptible residues. This is important to the sequencing process, since incomplete digestion would create a more complex banding pattern and would give a false indication of sequence heterogeneity. It is useful to run parallel controls with standard saccharides to enable monitoring of reaction conditions. When combinations of exoenzymes are required, these can be incubated simultaneously with the sample. If necessary, the activity of one enzyme can be destroyed before a secondary digestion with a different enzyme by heating the sample at 100°C for 2–5 min.
5. Adequate separations, particularly over limited size ranges of saccharides, can be obtained using single concentration gels, typically in the range 25–35% acrylamide. Improvements in resolution can be made by using longer gel sizes. Different voltage conditions (usually in the range 200–600 V) and running times are required for different gel formats, and should be established by trial and error with the particular samples being analyzed. Gels up to 24 cm in length can usually be run in 5–8 h using high voltages, whereas with longer gels it is more convenient to use lower voltage conditions and run overnight. We have also found that minigels can also be used effectively for separation of HS/heparin saccharides (*see* **Fig. 1**). Note that it is also possible to run Tris-acetate gels with a Tris-MES electrophoresis buffer (*see* **Fig. 1**; ***11***).

6. Because the emission wavelength is 410–420 nm, there is a need to filter out background visible wavelength light from the UV lamps. This can be done effectively with special glass filters that permit transmission of UV light but do not allow light of wavelengths >400 nm to pass. A blue bandpass filter on the camera also improves sensitivity. Suitable filters are available from HV Skan (Stratford Road, Solihull, UK; Tel: 0121 733 3003) or UVItec Ltd (St Johns Innovation Centre, Cowley Road, Cambridge, UK: www.uvitec.demon.co.uk).
7. Required exposure times are strongly dependent on sample loading and the level of detection required. Overly long exposures will result in excessive background signal. Note that negative images are usually better for band identification (*see* figures). Under the conditions described, the limit of sensitivity is ~1–2 pmol/band (*see* **Fig. 1**), with ~5–10 pmol being optimal. Recently it has been found that ~10-fold more sensitive detection is possible using an alternative fluorophore aminonaphthalenedisulfonic acid ***(21,22)***.

## References


1. Spillmann, D. and Lindahl, U. (1994) Gycosaminoglycan-protein interactions: a question of specificity. *Curr. Opin. Cell Biol.* **4,** 677–682.
2. Bernfield M., Gotte, M., Park, P. W., et al. Functions of Cell Surface Heparan Sulfate Proteoglycans. *Ann. Rev. Biochem.* (1999) **68,** 729–777.
3. Turnbull, J. E. and Gallagher, J. T. (1991) Distribution of Iduronate-2-sulfate residues in HS: evidence for an ordered polymeric structure. *Biochem. J.* **273,** 553–559.
4. Turnbull, J. E., Fernig, D., Ke, Y., Wilkinson, M. C., and Gallagher, J. T. (1992) Identification of the basic FGF binding sequence in fibroblast HS. *J. Biol. Chem.* **267,** 10,337–10,341.
5. Pervin, A., Gallo, C., Jandik, K., Han, X., and Linhardt, R., (1995) Preparation and structural characterisation of heparin-derived oligosaccharides. *Glycobiology* **5,** 83–95.
6. Yamada, S., Yamane, Y., Tsude, H., Yoshida, K., and Sugahara, K. (1998) A major common trisulfated hexasaccharide isolated from the low sulfated irregular region of porcine intestinal heparin. *J. Biol. Chem.*, **273,** 1863–1871.
7. Yamada, S., Yoshida, K., Sugiura, M., Sugahara, K., Khoo, K., Morris, H., and Dell, A. (1993) Structural studies on the bacterial lyase-resistant tetrasaccharides derived from the antithrombin binding site of porcine mucosal intestinal heparin. *J. Biol. Chem.* **268,** 4780–4787.
8. Mallis, L., Wang, H., Loganathan, D., and Linhardt, R. (1989) Sequence analysis of highly sulfated heparin-derived oligosaccharides using FAB-MS. *Anal. Chem.* **61,** 1453–1458.
9. Rhomberg A. J., Ernst, S., Sasisekharan, R., Biemann, K., et al. (1998) Mass spectrometric and capillary electrophoretic investigation of the enzymatic degradation of heparin-like glycosaminoglycans. *Proc. Natl. Acad. Sci. (USA)* **95,** 4176–4181.
10. Hopwood, J. (1989) Enzymes that degrade heparin and heparan sulfate. In: *Heparin* (Lane and Lindahl, eds.), Edward Arnold, London, UK, pp. 191–227.
11. Turnbull, J. E., Hopwood, J. J., and Gallagher, J. T. (1999) A strategy for rapid sequencing of heparan sulfate/heparin saccharides. *Proc. Natl. Acad. Sci. (USA)* **96,** 2698–2703.
12. Merry, C. L. R., Lyon, M., Deakin, J. A., Hopwood, J. J., and Gallagher, J. T. (1999) Highly sensitive sequencing of the sulfated domains of heparan sulfate. *J. Biol. Chem.* **274,** 18,455–18,462.
13. Vives, R. R., Pye, D. A., Samivirta, M., Hopwood, J. J., Lindahl, U., and Gallagher, J. T. (1999) Sequence analysis of heparan sulphate and heparin oligosaccharides. *Biochem. J.* **339,** 767–773.
14. Venkataraman, G., Shriver, Z., Raman, R., Sasisekharan, R., et al. (1999) Sequencing complex polysaccharides. *Science* **286,** 537–542.

15. Guimond, S. E. and Turnbull, J. E. (1999) Fibroblast growth factor receptor signalling is dictated by specific heparan sulfate saccharides. *Curr. Biol.* **9,** 1343–1346.
16. Shively, J. and Conrad, H. (1976) Formation of anhydrosugars in the chemical depolymerisation of heparin. *Biochemistry* **15,** 3932–3942.
17. Bienkowski, M. J. and Conrad, H. E. (1985) Structural characterisation of the oligosaccharides formed by depolymerisation of heparin with nitrous acid. *J. Biol. Chem.* **260,** 356–365.
18. Turnbull, J. E. and Gallagher, J. T. (1988) Oligosaccharide mapping of heparan sulfate by polyacrylamide-gradient-gel electrophoresis and electrotransfer to nylon membrane. *Biochem J* **251,** 597–608.
19. Rice, K., Rottink, M., and Linhardt, R. (1987) Fractionation of heparin-derived oligosaccharides by gradient PAGE. *Biochem. J.* **244,** 515–522.
20. Ludwigs, U., Elgavish, A., Esko, J., and Roden, L. (1987) Reaction of unsaturated uronic acid residues with mercuric salts. *Biochem. J.* **245,** 795–804.
21. Drummond, K. J., Yates, E. A., and Turnbull. J. E. (2001) Electrophoretic sequencing of heparin/heparin sulfate oligosaccharides using a highly sensitive fluorescent end label. Proteomics (in press).
22. Lee, K. B., Al-Hakim, A., Loganathan, D., and Linhardt, R. J. (1991) A new method for sequencing linear oligosaccharides on gels using charged fluorescent conjugates. *Carbohydr.Res.* (1991) **214,** 155–168

# 14

# Analytical and Preparative Strong Anion-Exchange HPLC of Heparan Sulfate and Heparin Saccharides

Jeremy E. Turnbull

## 1. Introduction

Studies on the structure–function relationships of the complex linear polysaccharides, known as glycosaminoglycans (GAGs), are becoming increasingly important as biological functions are established for them. However, structural analysis of GAGs presents a difficult technical problem, particularly in the case of the N-sulfated GAGs heparan sulfate (HS) and heparin, which display remarkable structural diversity *(1)*. A widely used and effective approach is to degrade the chains into smaller saccharide units that can then be separated and analyzed. In this regard, strong anion-exchange (SAX) HPLC techniques have proved particularly useful for both the analysis of disaccharide composition *(2,3)* and the separation of complex mixtures of larger saccharides *(3–6)*. However, in many methods the columns used have been silica-based and suffer from drawbacks related to poor stability of the support (e.g., inconsistency of run times, peak broadening, and short column life). There is clearly a need for improvements in column performance, especially for the purification of larger saccharides to homogeneity for sequencing and bioactivity testing. This chapter describes how a single type of polymer-based SAX column, ProPac PA1, can be used to provide high-resolution separations of both disaccharides and larger oligosaccharides derived from HS and heparin, with consistent elution times and excellent column performance characteristics (*see* **Notes 1** and **2**). Disaccharides from chondroitin sulfate and dermatan sulfate can also be separated (*see* **Note 3**). The improved resolution of saccharides compared to other SAX-HPLC methods, combined with the versatility and longevity of the columns, makes them a valuable tool for purification and structural analysis of HS/heparin and other GAG saccharides.

## 2. Materials

1. Gradient HPLC system (capable of mixing at least two solutions).
2. ProPac PA1 analytical columns (4 × 250 mm; Dionex).

From: *Methods in Molecular Biology, Vol. 171: Proteoglycan Protocols*
Edited by: R. V. Iozzo © Humana Press Inc., Totowa, NJ

3. Double-distilled water, pH 3.5: pH water to 3.5 using hydrochloric acid (high-purity HCl such as Aristar grade from BDH-Merck).
4. Sodium chloride solution (1 *M*): dissolve 58.4 g of HPLC-grade sodium chloride (e.g., HiPerSolv grade from BDH Merck) in 1L distilled water, and pH to 3.5 using HCl. Filter using a sintered glass filter or a 0.2 µm bottle-top filter.
5. Potassium dihydrogen phosphate solution (1 *M*, pH 4.6): dissolve 136.1 g of HPLC-grade $KH_2PO_4$ (e.g., HiPerSolv grade from BDH Merck) in 1L distilled water and filter using a sintered glass filter or a 0.2-µm bottle-top filter.
6. Unsaturated disaccharide standards (for both HS/heparin and chondroitin/dermatan sulfate) and heparitinases/chondroitinases are available from Seikagaku Kogyo (Tokyo).

## 3. Methods

### *3.1. Separation of Unsaturated HS/Heparin Disaccharides*

A simple and commonly used method to assess the structural composition of HS or heparin is to depolymerize the chains to disaccharides with a mixture of bacterial lyases. They can then be separated with reference to commercially available disaccharide standards of known structure (*see* **Fig. 1**). Note that the eliminative cleavage mechanism of the lyases results in unsaturated hexuronate residues in the resulting disaccharides (*see* **Note 4**).

1. Prepare unsaturated disaccharides from heparin/HS by bacterial lyase treatment with heparitinase I, heparitinase II, and heparinase; *see* **ref.** *7* for details.
2. After equilibration of the Pro-Pac PA1 column in mobile phase (double-distilled water adjusted to pH 3.5 with HCl) at 1 mL/min, inject the sample.
3. After 1 min injection time, elute the disaccharides with a linear gradient of sodium chloride (0–1 *M* over 45 min) in the same mobile phase.
4. Monitor the eluant in-line for UV absorbance at 232 nm (*see* **Note 5**).
5. Identify peaks by reference to standards separated under the same run conditions.

### *3.2. Separation of HS/heparin Disaccharides Derived by Nitrous Acid Degradation*

A further class of disaccharides, those derived by treatment of HS or heparin with nitrous acid, are more difficult to resolve by SAX-HPLC. However, in contrast to the lyase-derived saccharides, they have the advantage of containing intact hexuronate residues (*see* **Note 6**). The most widely reported method for their separation uses a silica-based SAX Partisil column with a $KH_2PO_4$ gradient separation system *(3)*. In our hands this method gives very broad and inadequately resolved peaks, which limits the sensitivity of detection and the accuracy of peak identification and quantitation. In marked contrast, these disaccharides can be resolved well on a ProPac PA1 column using appropriate shallow NaCl gradients. Resolution is improved slightly by use of two columns connected in series (*see* **Figs. 2A,B**), but adequate results can also be obtained using a single column. An alternative separation using a phosphate gradient is described under **Subheading 3.2.2.**

#### *3.2.1. Sodium Chloride Gradient*

1. Prepare reduced $^3$H-labeled disaccharides from heparin/HS by low-pH nitrous acid treatment, either from unlabeled samples using $^3$H-borohydride end-labeling *(3)* or from samples radiolabeled biosynthetically using $^3$H-glucosamine *(8)*.

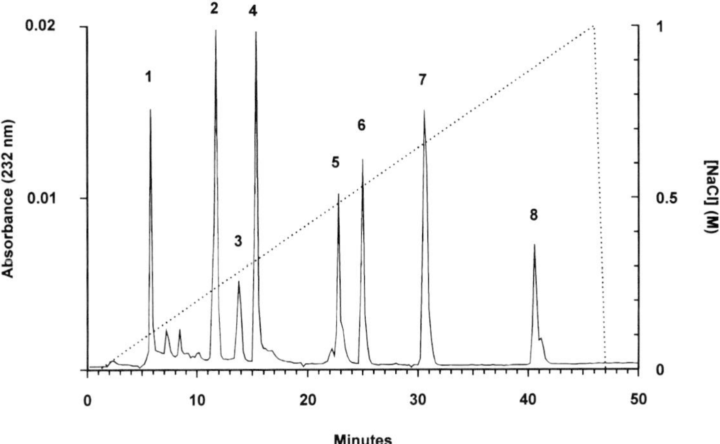


Fig. 1. Separation of lyase-derived HS/heparin disaccharides on a ProPac PA1 column. The profile shows the separation of the eight major unsaturated disaccharides released from these polysaccharides by treatment with a combination of the bacterial lyases heparitinase I, heparitinase II, and heparinase. The NaCl gradient is indicated by the dashed line. The structures of the standards and the amounts loaded were: 1, Δ-HexA-GlcNAc, 1 nmol; 2, Δ-HexA-$GlcNSO_3$ , 2 nmol; 3, Δ-HexA-GlcNAc(6S), 0.5 nmol; 4, Δ-HexA(2S)-GlcNAc, 1.5 nmol; 5, Δ-HexA-$GlcNSO_3$(6S), 1 nmol; 6, Δ-HexA(2S)-$GlcNSO_3$, 1.5 nmol; 7, Δ-HexA(2S)-GlcNAc(6S), 2 nmol; 8, Δ-HexA(2S)-$GlcNSO_3$(6S), 1.5 nmol. Abbreviations: GlcNAc, N-acetyl glucosamine; $GlcNSO_3$, N-sulfated glucosamine; 2S, 3S, and 6S are 2-O-, 3-O-, and 6-O-sulfate groups, respectively; Δ-HexA, unsaturated hexuronate residue formed at non-reducing end of disaccharides and oligosaccharides by eliminative lyase scission.

2. Separate samples on two ProPac PA1 columns in series, using the same mobile phase described under **Subheading 3.1.**, but with a two-step linear NaCl gradient (0–150 m*M* over 50 min followed by 150–500 m*M* over 70 min).
3. Monitor $^3H$ radioactivity either with an in-line radioactivity detector or by collecting fractions for scintillation counting (e.g., in Optiphase HiSafe III scintillant, Wallac). *See* **Note 7**.

### 3.2.2. Potassium Dihydrogen Phosphate Gradient

Improved separation of monosulfated disaccharides can be obtained using a gradient of phosphate (*see* **Figure 2c** and **Note 8**).

1. Using two ProPac PA1 columns in series, equilibrate in $H_2O$, pH 6.0.
2. Inject samples and elute with a linear gradient of 0–1 *M* $KH_2PO_4$ buffer, pH 4.6 over 90 min.
3. Monitor $^3H$ radioactivity as under **Subheading 3.2.1.**

## 3.3. Separation of HS/Heparin Oligosaccharides

In addition to separation of disaccharides, much larger HS and heparin oligosaccharides can also be resolved on ProPac PA1 using a linear NaCl gradient (*see* **Note 9**).

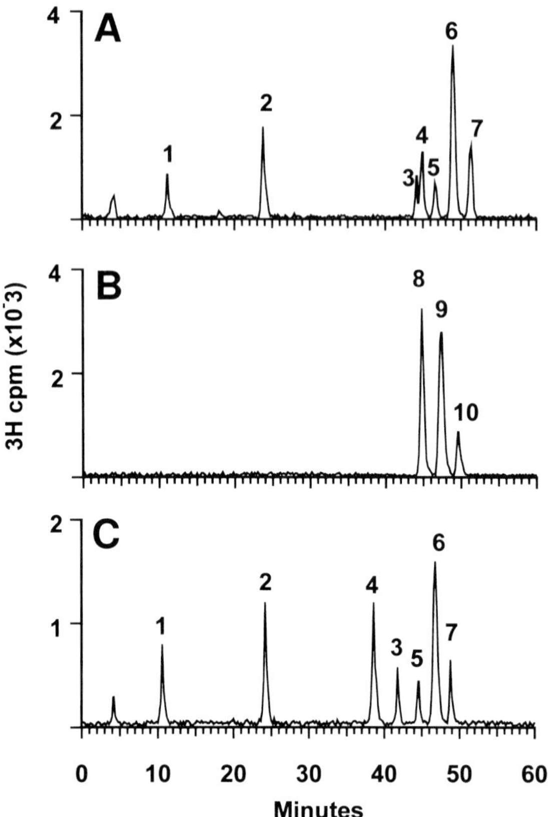


Fig. 2. Separation of $HNO_2$-derived HS/heparin disaccharides on a ProPac PA1 column. These profiles show the separation of the major disaccharides released from HS/heparin by low-pH nitrous acid treatment followed by $^3$H-borohydride end labeling *(3)*, or from $^3$H-biosynthetically labeled samples *(8)*. The structures of the $^3$H-labeled standards were: 1, IM/GM; 2, M6S (monosaccharide); 3, G2SM; 4, I2SM; 5, GM3S; 6, GM6S; 7, IM6S; 8, I2SM6S; 9, G2SM6S; and 10, GM3,6S (where G and I are glucuronic and iduronic acids, respectively; M is 2,5-anhydromannitol; and 2S, 3S, and 6S are 2-O-, 3-O-, and 6-O-sulfate groups, respectively). **(A, B)** show the first and last 60 min of the run time using a NaCl gradient as described under **Subheading 3.2.1.** **(C)** shows the separation of non- and monosulfated disaccharides using an alternative $KH_2PO_4$ gradient as described under **Subheading 3.2.2**.

**Figure 3** shows examples of separations performed on complex natural mixtures of hexa-, octa-, and decasaccharides derived from porcine mucosal HS by treatment with heparitinase I. It is noteworthy that the saccharides eluted in sets of resolved or partially resolved peaks, which presumably have similar overall levels of anionic charge but variant patterns of sulfation and disaccharide sequence. This method is particularly useful for purification of saccharides to the degree necessary for sequencing *(12)* or testing in biological assays *(13)*. Preparative scale separations are also possible (*see* **Note 10**).

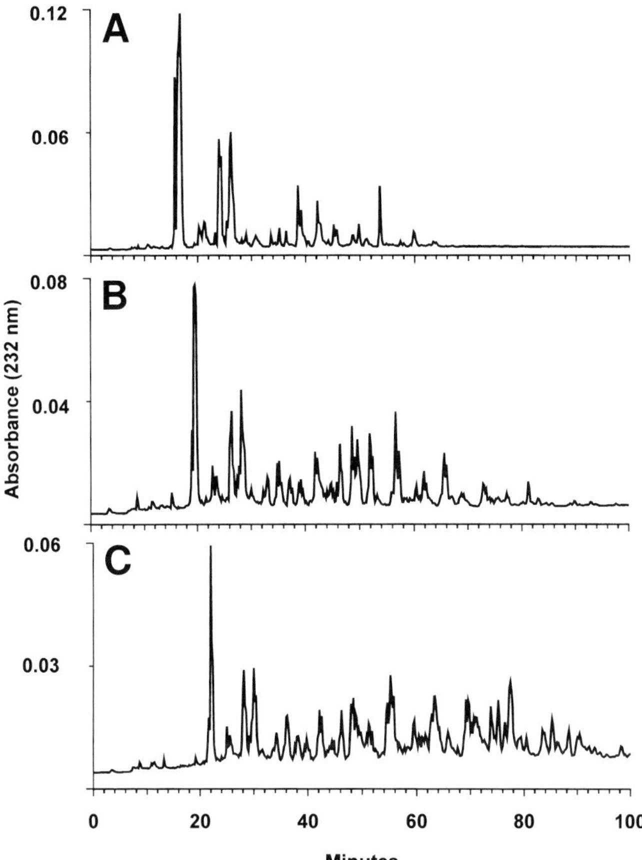


Fig. 3. Separation of HS oligosaccharides on a ProPac PA1 column. Porcine mucosal HS was depolymerised with heparitinase I and hexa-, octa-, and deca-saccharide pools prepared by gel filtration on Bio-Gel P6 *(9)* and resolved on two ProPac PA1 columns connected in series as described under Methods. **(A)**, hexasaccharides; **(B)**, octasaccharides; **(C)**, decasaccharides. In each case, 100 nmol of each saccharide mixture was loaded (approximately 150, 200, and 250 µg each, respectively).

1. Prepare HS and heparin oligosaccharides (tetrasaccharides and larger) as described previously—for example, by partial depolymerization with heparitinase I and gel filtration on Bio-Gel P6 *(7–9)*.
2. Inject samples onto two ProPac PA1 columns in series, using the same mobile phase described under **Subheading 3.1.**
3. Elute with a modified salt elution gradient (e.g., 0–1 *M* NaCl over 100 min).
4. Monitor the eluant in-line for UV absorbance at 232 nm.

## 4. Notes

1. During development of improved HPLC methods for separation of GAG saccharides we found that a polymer-based column, ProPac PA1, originally designed primarily for

SAX of proteins, provided excellent separations of disaccharides derived from the glucosaminoglycans HS and heparin. **Figure 1** shows the separation of the major unsaturated disaccharides (detected by absorbance at 232 nm) derived by digestion of these polysaccharides with a mixture of bacterial heparitinases. They elute as very sharp, symmetrical peaks at highly reproducible elution times, and good baselining between all the peaks was evident. These features of the separation provide more accurate identification of peaks and sensitive quantitation than possible with previous methods. Note that $^{3}H$ or $^{35}S$ radiolabeled disaccharides (derived from biosynthetically labeled polysaccharide) can also be separated using this method, using in-line monitoring of radioactivity (*see* **Fig. 2**) or counting of collected fractions.

2. Overall column performance is excellent. Elution times are highly consistent, and the columns exhibit very good stability and longevity, which outweighs their initial high cost compared to traditional silica columns. In one case a single column was used for more than 250 runs over a 2-yr period without significant loss of resolution. These properties are presumably dependent on the column resin, which is formed by 0.2m anion exchange microbeads being bonded to a 10-μm nonporous polystyrene/divinylbenzene polymeric resin.
3. Good separations of unsaturated disaccharides derived by chondroitinase ABC digestion of the galactosaminoglycans chondroitin sulfate and dermatan sulfate can also be obtained using the ProPac-PA1 column. Similar results have been reported for the separation of this class of disaccharides on another polymer-based column, the CarboPac PA1 ***(10)***.
4. The eliminative cleavage mechanism of the lyases results in unsaturated hexuronate residues (double bond between carbons 4 and 5 of the sugar ring) in the resulting disaccharides ***(14)***. It is thus not possible to distinguish directly whether they resulted originally from glucuronate or iduronate residues. In contrast, cleavage with nitrous acid leaves these residues intact (*see* **Note 6**).
5. The limit of detection of unsaturated disaccharides by UV absorbance is approximately 20–50 pmol.
6. Nitrous acid cleaves between N-sulfated glucosamine residues and hexuronate residues. The resulting saccharides have intact hexuronate residues but the reducing-end glucosamines are converted to an anhydromannose residues (which are reduced to anhydromannitol by reduction during preparation of the disaccharides).
7. When detecting radiolabeled samples, if accurate quantitation is required, be aware of the potential for salt-dependent variability in counting efficiencies, depending on the scintillant being used.
8. With in-line radioactivity monitoring, baselining between peaks is sufficient in most cases to allow accurate quantitation of the individual major species. The exception is the partial separation of I2SM from the rare disaccharide G2SM (*see* **Fig. 2a**). Note that slight changes in run conditions (e.g., pH) can reverse their order of elution. However, conditions using an alternative $KH_2PO_4$ buffer separation system give the best resolution of the monosulfated disaccharide standards, including excellent baseline separations of I2SM and G2SM when required (*see* **Fig. 2c**).
9. This approach is effective for separating saccharides up to approximately dp14-16 in size, although the patterns become inherently more complex and resolution less efficient with increasing size. Improved resolution can be obtained using shallower gradients than described under **Subheading 3.3.** Additional examples of applications can be found in **refs.** ***11–13***.
10. Preparative separations are possible with loadings up to approximately 0.5–1.0 mg per run on the 4 × 250 mm analytical columns, and a fivefold scale-up should be achievable on 9 × 250 mm preparative columns that are available from Dionex.

## References

1. Gallagher, J. T., Turnbull, J. E., and Lyon, M. (1992) Patterns of sulphation in heparan sulphate: polymorphism based on a common structural theme. *Int. J. Biochem.* **24,** 553–556.
2. Yoshida, K., Miyauchi, S., Kikuchi, H., Tawada, A., and Tokuyasu, K. (1989) Analysis of unsaturated disaccharides from glycosaminoglycuronan by HPLC. *Anal. Biochem.* **177,** 327–332.
3. Bienkowski, M. J. and Conrad, H. E. (1984) Structural characterisation of the oligosaccharides formed by depolymerisation of heparin with nitrous acid. *J. Biol. Chem.* **260,** 356–365.
4. Guo, Y. and Conrad, H. E. (1988) Analysis of oligosaccharides from heparin by reversed-phase ion pairing HPLC. *Anal. Biochem.* **168,** 54–62.
5. Linhardt, R. J., Gu, K., Loganathan, D., and Carter, S. R. (1989). Analysis of glycosaminoglycan-derived oligosaccharides using reverse-phase ion-pairing and ion-exchange chromatography with suppressed conductivity detection. *Anal. Biochem.* **181,** 288–296.
6. Rice, K. G., Kim, Y., Grant, A., Merchant, Z., and Linhardt, R. J. (1985) HPLC separation of heparin-derived oligosaccharides. *Anal. Biochem.* **150,** 325–331.
7. Turnbull, J. E. and Gallagher, J. T. (1991) Distribution of iduronate-2-sulphate residues in heparan sulphate: evidence for an ordered polymeric structure. *Biochem. J* . **273,** 553–559.
8. Turnbull, J. E., Fernig, D., Ke, Y., Wilkinson, M. C., and Gallagher, J. T. (1992) Identification of the basic FGF binding sequence in fibroblast HS. *J. Biol. Chem.* **267**, 10,337-10,341.
9. Walker, A., Turnbull, J. E., and Gallagher, J. T. (1994) Specific HS saccharides mediate the activity of basic FGF. *J. Biol. Chem.* **269,** 931–935.
10. Midura, R. J., Salustri, A., Calabro, A., Yanagashita, M., and Hascall, V. C. (1994) High-resolution separation of disaccharide and oligosaccharide alditols from chondroitin sulphate, dermatan sulphate and hyaluronan using CarboPac PA1 chromatography. *Glycobiology* **4,** 333–342.
11. Sanderson, R. D., Turnbull, J. E., Gallagher, J. T., and Lander, A.D. (1994) Fine structure of HS regulates cell syndecan-1 function and cell behaviour. *J. Biol. Chem.* **269,** 13,100–13,106.
12. Turnbull, J. E., Hopwood, J., and Gallagher, J. T. (1999) A strategy for rapid sequencing of heparan sulphate/heparin saccharides. *Proc. Nat. Acad. Sci. USA* **96,** 2698–2703
13. Guimond, S. E. and Turnbull, J. E. (1999) Fibroblast growth factor receptor signalling is dictated by specific heparan sulphate saccharides. *Current Biology* **9,** 1343–1346.
14. Linhardt R. J., Turnbull J. E., Wang H., Loganathan D., and Gallagher J. T. (1990) Examination of the substrate specificity of heparin and heparan sulphate lyases. *Biochemistry* **29,** 2611–2617.

# 15

## Proteoglycans Analyzed by Composite Gel Electrophoresis and Immunoblotting

George R. Dodge and Ralph Heimer

### 1. Introduction

Proteoglycans are the protein products of diverse genes posttranslationally modified with highly negatively charged side chains, commonly known as glycosaminoglycans. The latter consist of repeating disaccharides capable of forming polymers of varying size, which, depending on the specific disaccharide composition, are known as the chondroitin sulfates, the iduronate-containing dermatan sulfates, keratan sulfate, and heparan sulfate. Some of the more complex proteoglycans, such as aggrecan, are decorated by additional nonglycosaminoglycan oligosaccharides *(1)*. Aggrecan contains additionally a hyaluronan-binding domain that in the extracellular matrix facilitates formation of large noncovalently-bound complexes containing one hyaluronan molecule and numerous aggrecans *(1)*. Analysis of proteoglycans by standard SDS-PAGE is complicated by the presence of the negatively charged side chains, which prevent the linerarization of molecules usually achieved by treatment with a mercaptoethanol and SDS. Consequently, in SDS-PAGE most proteoglycans with multiple glycosaminoglycan side chains, if they enter the gel at all, migrate as broad bands and appear as smears due to size heterogeneity and large electroendosmotic effects particularly when the acrylamide composition of the PAGE gel is held to a minimum.

The studies of McDevitt and Muir, now more than three decades ago, made the surprising observation that the proteoglycans from an extract of cartilage resolved in a composite gel composed of 0.6% agarose and 1.2% acrylamide into two sharp bands *(2)*. The mechanism for the separation of the proteoglycans of cartilage into two bands appeared to be based not only on size but also on charge distribution of two subspecies of proteoglycans *(3)*. Gels such as these were originally designed to provide a larger pore size than 3.5% acrylamide alone, which is the limit concentration for gel formation. The technique of composite gel electrophoresis is best understood when the historical perspective is considered and its usefulness in previous applications discussed. These types of composite gels had been used for the separation of proteins *(4)*

From: *Methods in Molecular Biology, Vol. 171: Proteoglycan Protocols*
Edited by: R. V. Iozzo © Humana Press Inc., Totowa, NJ

**Table 1**
**Monoclonal and Polyclonal Antibodies Useful in Characterizing Proteoglycans**

| Antigen Name | Polyclonal/mAb | | Notes | Reference |
|---|---|---|---|---|
| Perlecan | 7B5 | mAb | Laminin-like region | ***19*** |
| Versican | 2B1 | mAb | | ***20*** |
| Keratan sulfate | AN9P1 | mAb | Bound to aggrecan | ***21*** |
| Aggrecan | BE123 | mAb | After c-ABC-treatment | ***22*** |
| Aggrecan, N-terminal | BC3 | mAb | After aggrecanase exposure | ***23*** |
| Aggrecan, C-terminal | BC4 | mAb | After aggrecanase exposure | ***23*** |
| Aggrecan | IC6 | mAb | HA-binding domain | ***24*** |
| Chondroitin sulfate | CS56 | mAb | Bound chondroitin 4 or 6 sulfate | ***25*** |
| Chondroitin-4 sulfate | 2B6 | mAb | After c-ABC digestion | ***26*** |
| Chondroitin-6 sulfate | 3B3 | mab | After c-ABC digestion | ***27*** |
| Unsulfated dissacharide | 1B5 | mAb | After c-ABC digestion | ***28*** |
| Biglycan | LF15 | Poly | *N*-terminus | ***29*** |
| Decorin | LF30 | Poly | *N*-terminus | ***29*** |

and RNA *(5)*. Improvement of this technique, originally done in tube gels, led inevitably to the use of slab gels ***(6)*** and also to improvement in sample preparation through use of SDS and reduction ***(6)*** or SDS only *(7)*. These modifications facilitated the analysis of tissue extracts and biologic fluids without prior fractionation. Originally the detection of proteoglycans was accomplished with the basic thiazine dye, toluidine blue. Increased sensitivity of the detection of proteoglycans was achieved by techniques such as biosynthetic labeling with $^{35}SO_4$ (*6*) transblotting to positively charged Nylon membranes followed by staining with Alcian blue *(7)*, and transblotting to unmodified Nylon and staining with the highly positively charged $^{125}$I-cytochrome c *(8)*. The detection achieved with Alcian blue staining of proteoglycans on positively charged Nylon-66 is nearly 100 times more sensitive than staining with toluidine blue *(7)*. With the availability of monoclonal and polyclonal antibodies specific for diverse proteoglycans (the core proteins and associated GAGs), standard Western blotting techniques could be utilized on proteoglycans transblotted from composite gels. **Table 1** contains a partial list of antibodies useful for the characterization of a wide range of known proteoglycans.

An interesting application of Western blotting involves sequential testing of a single blot with multiple antibodies. Antibodies bound to the proteoglycan antigen transblotted to nitrocellulose can be removed by exposure to sodium thiocyanate without materially affecting the binding of the immobilized material. The blot can then be probed again by another antibody. Densitometry scans of each blot can then be superimposed on one another, and by visual inspection this procedure can aid in resolving closely migrating species and also in determining the purity of a proteoglycan isolate or a standard ***(9)***.

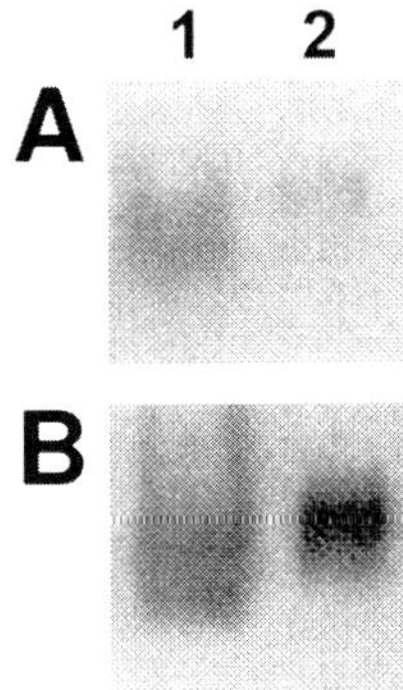


Fig. 1. Demonstration of different chondroitin/dermatan sulfate proteoglycans in myocardial extracts. Western blots **(A, B)** prepared from samples of primate myocardial extracts following composite agarose-acrylamide gel electrophoresis were incubated with a 1:200 dilution of the monoclonal CS-56 (ICN) (which reacts with intact chondroitin-4 and -6 sulfate chains) **(A)** or a 1:200 dilution of the monoclonal 2-B-6 (which reacts with terminal unsulfated disaccharides on chondroitinase ABC digestion) *(25)*. Lane 1 is the extract of the primate myocardium, lane 2 is a standard of 50 ng of bovine aggrecan from articular cartilage. **(B)** Western of duplicate samples that were exposed overnight at 37°C to 0.1 unit of chondroitinase ABC (ICN) in 5 mL of 0.2 *M* Tris buffer, pH 8.0, before being incubated with monoclonal 2-B-6. This demonstrates the presence of two clearly defined proteoglycan bands in the mycardial extract (lane 1), discernable only after chondroitinase digestion.

Results obtained with agarose/acrylamide gel electrophoresis have greatly enhanced our knowledge of proteoglycan biochemistry, synthesis, and degradation. First, avoiding time-consuming preparative steps has allowed the direct analysis of biologic specimens and tissue extracts. In general, such analyses have been useful in the simultaneous detection of proteoglycans with relatively large core proteins conjointly with detection of proteoglycans of smaller size. While large and small proteoglycans are easily separable, identification within each group can best be made by specific antibodies. The chief difficulty with separating proteoglycans with large core proteins is the presence of multiple glycosaminoglycan side chains and that the side chains often show size heterogeneity, which then affect overall mobility and electroendosmosis. An example of the number of glycosaminoglycan side chains affecting the mobility is illustrated when biglycan with its two glycosaminoglycan side chains has to be distinguished from decorin, with its single side chain. The core proteins of these proteoglycans have a rather similar molecular weight, although the presence of one additional (two) dermatan/chondroitin sulfate side chain biglycan results in biglycan having a molecular weight of approximately 100,000 daltons and decorin, with one such chain, only 70,000 daltons. One would assume, therefore, that this pair would be readily separable. **Figure 1** is an example of the power of using composite gels and Western blotting and demonstrates this by using two different antibodies, CS-56 (reactive with intact chondroitin 4- and 6-sulfate chains) and 2-B-6 (reactive with terminal dissacharides remaining on the core protein of proteoglycans after digestion with chondroitinase ABC *(26)*. One can see in **Fig. 1A** that, when using monoclonal CS-56,

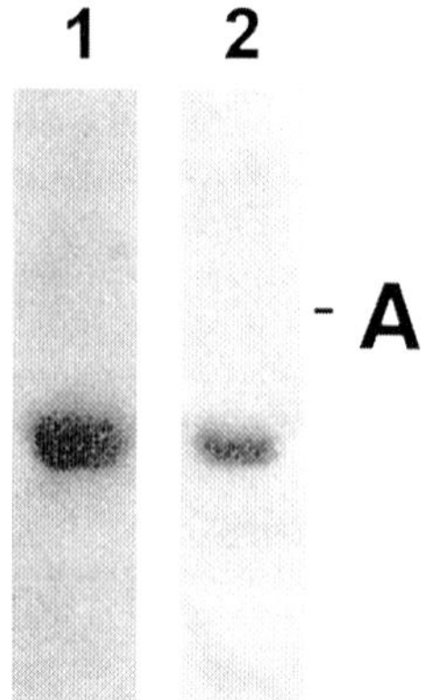


Fig. 2. Example of a composite agarose/acrylamide gel/Western blot that demonstrates the presence of bigylcan and decorin found in an extract of myocardium. Primate myocardium was extracted with 4 *M* guanidine-HCl, dialyzed against 6 *M* urea and aliquots electrophoresed directly into a composite gel as described in this chapter and elsewhere ***(10)***. The gel was blotted to nitrocellulose and lane 1 was incubated with LF-15 (anti-biglycan antiserum, 1:500) and lane 2 with LF-30 (anti-decorin antiserum, 1:500). The position of the migration of aggrecan isolated from bovine cartilage is noted (A).

one band was identified in the myocardial extract (*see* lane 1) that migrated ahead of the aggrecan standard (*see* lane 2). When an identically prepared blot (*see* **Fig. 1B**) was incubated first with chondroitinase ABC, then with a dilution of the monoclonal 2-B-6, two bands became evident in the myocardial extract, indicating the presence of two distinct species of chondroitin-4 sulfate-containing proteoglycans, presumably decorin and bigylcan. The precise identity of these proteoglycans would best be confirmed with specific decorin and biglycan antibodies (reactive with the core proteins). Shown in **Fig. 2**, using specific antibodies for the biglycan (*see* lane 1) and the decorin (*see* lane 2) core protein, one would be unable to distinguish them by just using monoclonal antibodies directed to the common dermatan/chondroitin sulfate side chain ***(10,11)***. It is remarkable that although biglycan and decorin have a similar mobility in composite gels, biglycan shows a greater polydispersity due to its more complex orientation and altered electroendosmotic effects.

Surprisingly, a search of the literature reveals that agarose/acrylamide gel electrophoresis has not been as widely used as this powerful method (*see* **Note 1**) would deserve. Historically, the major areas where agarose/acrylamide gel electrophoresis has been useful has been in the analysis of guanidine-HCl extracts of tissues, such as cartilage, and tendon. We summarize here in chronological order some of these findings.

Two types of aggregating proteoglycans were found by zonal rate centrifugation of guanidine-HCl extracts of bovine cartilage, one of *Mr* $3.5 \times 10^6$ and the other of *Mr* $1.3 \times 10^6$. The former did not contain a well-defined keratan sulfate-rich domain; the latter, however, contained the same percentage of hexosamine, and perversely the smaller species contained 20% more protein than the larger. Agarose/acrylamide gel electrophoresis showed one band for each proteoglycan and that the larger one had a lower mobility than the smaller ***(12)***. The electrophoresis, therefore, confirmed the result of the zonal rate electrophoresis, but it failed to explain the nearly threefold greater Mr of the keratan sulfate-poor species.

The guanidine-HCl extracts of adult bovine tendon subjected to CsCl density-gradient ultracentrifugation resolved into two populations of proteoglycan, one large and with a core protein of *Mr* 200,000 and one small with *Mr* 48,000. The large proteoglycan subjected to agarose-acrylamide gel electrophoresis showed three components, which could not be resolved further ***(13)***. Had Western blots for proteoglycans been perfected at that time, these components might have been more fully characterized.

Canine adult menisci contained two proteoglycans of high *Mr* that could be resolved into 2 fractions by agarose/acrylamide gel electrophoresis. These proteoglycans contain chondroitin sulfate and keratan sulfate, and they bound hyaluronan ***(14)***. This study suggested that there were tissue-specific high-*Mr* proteoglycans. By contrast to the proteoglycans of cartilage, which are keratan sulfate rich and keratan sulfate poor, menisci contain such molecules only in the keratan sulfate-rich form.

In a study of the proteoglycans of the human intervertebral disk, Jahnke and McDevitt ***(15)*** made extensive use of agarose/acrylamide gel electrophoresis. The hyaluronan aggregating proteoglycans resolved into two well-separated fractions, which retained their heterogeneity even after hydroxylamine fragmentation and removal of the hyaluronan-binding domain. The faster migrating species were resolvable into two components, one of which enriched in keratan sulfate. This work was extremely labor intensive, using associative and dissociative density-gradient ultracentrifugation as well as Sepharose 2B and Bio-Gel A-50 m column chromatography for preparation of samples to be subjected to the electrophoresis. Much of these labors might have been avoided had these authors used the modification of sample preparation (reduction and SDS treatment) introduced in 1985 by Heinegard et al. ***(6)***, which would have allowed instant characterization of the guanidine•HCl extract.

Agarose/acrylamide gel electrophoresis has also been used in characterizing the proteoglycans from clinical human specimens ***(16)***. In these studies Cs-Szabo et al. ***(16)*** studied the characteristic changes in aggrecan and the small proteoglycans in cartilage from patients with osteoarthritis and rheumatoid arthritis. Guanidine-HCl extracts of femoral condyle cartilage were subjected directly to agarose/acrylamide gel electrophoresis and Western blotting. Samples from patients with rheumatoid arthritis and some extent also from patients with osteoarthritis lacked certain aggrecan populations, indicating extensive proteoglycan degradation. The small proteoglycans biglycan and decorin appeared to be increased by the disease. Repair processes, particularly in osteoarthritis, were evident through production of fetal-type aggrecans. This represents an excellent example that takes full advantage of the powerful technique of agarose/acrylamide gel electrophoresis coupled with Western blotting.

The ability of articular cartilage to respond to exercise has been well documented, and this has led many investigators to analyze the proteoglycan content in defined anatomic sites in different species. One of the more interesting and successful studies, taking full advantage of agarose/acrylamide gel electrophoresis, was done with exercised and nonexercised race horses *(17)*. This study, from a technical point of view, demonstrated that a large number of samples can be analyzed simultaneously. The complex results, subject to various interpretations, however, showed that exercised horse had certain loci of articular cartilage that were capable of producing large polydispersed hyaluronan-binding proteoglycans. In these studies the data were gleaned from the unique ability of the composite gel to resolve large proteoglycans and subsequent immunoblotting for a specific domain of aggrecan (the hyaluronan-binding domain).

Versican, a proteoglycan of high *Mr* formed in the extracellular matrix, was surmised to be present through use of agarose/acrylamide gel electrophoresis and Western blots prepared from the gel. Without an antibody to canine versican, the authors *(18)* had to resort to monoclonal antibodies reacting with glycosaminoglycan epitopes. Chondroitin sulfate was the major glycosaminoglycan present in a proteoglycan of approximately 500,000 daltons, established by analysis of the core protein. As the tissue contained hyaluronan and also link protein, it was postulated that a domain of versican was able to react with hyaluronan and that the complex might be stabilized by link protein. The actual identification of canine versican, however, could only be established by partial amino acid sequencing.

Composite agarose/acrylamide gels have been instrumental in the characterization and study of various proteoglycans in a number of different cell types and species. The use of these specialized agarose/acrylamide gels has greatly enhanced our knowledge of proteoglycan biochemistry, synthesis, and degradation. These gels allow direct analysis of biological specimens and tissue extracts. Combined with the use of specific antibodies they potentially and simultaneously can detect distinct proteoglycan molecules of widely divergent molecular sizes.

## 2. Materials

1. Sample buffer: 6 M urea, 0.04 M Tris-acetate, pH 6.8, 0.1% Triton X-100, 1% sodium dodecylsulfate, 5% 2-mercaptoethanol, and 10% glycerol.
2. Agarose (*see* previous paragraph).
3. Electrode buffer: 0.04 *M* Tris-acetate, pH 6.8.
4. Triton X-100.
5. TEMED.
6. Large-format vertical protein gel electrophoresis chamber and accessories.
7. Power supply capable of supplying a constant 120 mA.
8. Glass plates with one side frosted (to fit the electrophoresis apparatus).
9. Spacers of 1.5 mm.
10. Transfer apparatus (suitable size to accommodate gel size).
11. Transfer membrane (positively charged Nylon, Immobilon-N [Millipore], or polyvinylidene fluoride [PVDF] Immobilon-P [Millipore] are recommended; supported nitrocellulose can also be used).
12. Filter paper (e.g., Whatman no. 3).

## 3. Method

### 3.1. Sample Preparation

1. Prepare samples in no more than 10 µL, containing 6 *M* urea, 0.04 *M* Tris-acetate, pH 6.8, 0.1% Triton X-100, 1% sodium dodecylsulfate, 5% 2-mercaptoethanol, and 10% glycerol. Boil samples for 5 min.

### 3.2. Gel Preparation

1. Before pouring the composite gel solution, a 0.5- to 1.0-cm plug of 10% acrylamide can be placed at the bottom of the plates after they are prepared. This ensures that the gel will not slip out during electrophoresis or handling.
2. Prepare the gel mixture in an Erlenmayer flask a final concentration of 0.6% agarose and 1.2% acrylamide. The agarose solution is prepared with 320 mg of agarose (*see* **Note 2**) and 30 mL of 0.04 *M* Tris-acetate, 1 m*M* sodium sulfate, and 1 m*M* EDTA, adjusted to pH 6.8.
3. Weigh the flask, add the agarose, and dissolve by heating the flask in a microwave oven.
4. Reweigh and restore to the original weight with distilled water. Keep the flask at 56°C to prevent gelling.
5. Make the acrylamide solution so that the total concentration of the acrylamide plus N,N'-methylene-bisacrylamide totals 2.4%, with an acrylamide to *bis* ratio of 20 to 1. These two reagents need to be dissolved in 0.04 *M* Tris-acetate, pH 6.8, containing 2% Triton X-100 and 3% TEMED.
6. Degas the solution and keep at 56°C.
7. Combine in a small beaker equal volumes of the two solutions, mix, and follow by the rapid addition of 1% ammonium persulfate.
8. Pour the solution into a preprepared chamber with frosted glass plates separated by 1.5-mm spacers (*see* **Notes 3** and **4**). The chamber should be preincubated at 45°C.
9. The comb is then inserted with care taken to minimize air bubbles and to keep the depth of the wells at ~0.5 cm.
10. One hour after the casting of the gel, electrode buffer is added to the upper electrode compartment. The assembly is stored in a cold room or refrigerator.
11. Withdrawing the comb without destroying the wells is the chief hazard of the preparation. This can be best accomplished by storing the gel in the cold for 24 h and keeping the depth of the wells to a 0.5-cm maximum. When ready to remove the comb, ease the comb out by adding a small amount of buffer along the top and pulling straight up.
12. The gel should be prerun for 1 h at 120 mA before loading the samples. When loading the gel, one well should be reserved for the bromphenol blue tracking dye. The electrophoresis should run in a cold room at 120 mA until the bromphenol blue marker has moved 3 cm. This may take about 3 h (for specifics on dismantling apparatus, *see* **Notes 5** and **6**). If not proceeding with immunodection of proteoglycans (transfer described below), after fixation the gel can be stained directly with toluidine blue. The gel can be left on one plate and placed in a tray containing fixative (50% methanol, 10% acetic acid) for 60 min at room temperature followed by a change to a solution of 0.025% toluidine blue in 3% acetic acid. This requires at least 2 h at room temperature although it can be protracted to overnight.
13. Destain with 3% acetic acid until the background is suitably clear. The gel can be stored by transferring to Gelbond film (FMC) as described by the manufacturer (note do not dry with heat higher than 55°C).

### 3.3. Western Blotting

1. Transblotting to nitrocellulose is carried out in 25 m*M* Tris and 192 mM glycine buffer adjusted to pH 8.3 and containing 5% methanol (reduced methanol improves efficiency of the transfer of large molecular molecules) (*see* **Note 7**). The plates need to be separated very carefully, usually using bent metal spatulas.
2. The gel adhering to one of the frosted glass plates is then overlaid with a nonsoaked filter paper and lifted off the plate by peeling backwards.
3. A small amount of transfer buffer should be placed on the top of the gel and the transfer membrane of choice (prewet) can then be laid on the top (*see* **Note 8**). It cannot be moved once it is placed on the gel, and the air bubbles should be eliminated by gently rolling a pipet over the membrane covering the gel.
4. A final piece of filter paper (prewet) should be laid on top after any visible bubbles are eliminated.
5. The transblot should be run with cold buffer in the cold at 75 mA on average for 5–6 h. It must be kept in mind, however, that a gel may contain both large and small proteoglycans such as decorin (*Mr* $1.0 \times 10^5$) and some species of aggrecan (*Mr* $3.5 \times 10^6$), therefore to effect a total transfer of each may be nearly impossible.
6. One partial solution to the dilemma is to use two sheets of nitrocellulose and examine the backup sheet for proteoglycans of lower *Mr*. Due to the problem of either incomplete transfer or loss of proteoglycans, one should attempt quantification of disparate species with caution.
7. There are no special problems with regard to immunological analysis of the blot and the remainder of the steps are identical to Western blotting of supports derived from SDS-PAGE.
8. Additionally, when positively charged Nylon is used, the membrane can be stained with Alcian blue (0.5% in 3% acetic acid) for 10–20 min followed by rinsing briefly with 5% acetic acid until the contrast of proteoglycans to background is satisfactory.

## 4. Notes

1. The technique has largely been underutilized. One reason for this might be that executing all the steps in the procedure requires practiced hands with some degree of technical patience, but in reality the method can be run routinely in almost any research laboratory.
2. The agarose to use to achieve the maximum resolving power must be linear (not branched), with none or low sulfate content and have a low EEO number. Most suppliers (e.g., FMC) carry a suitable agarose.
3. The cleanliness of the glass plates is imperative to a successful run. Plates should be cleaned with a concentrated detergent product (e.g., RBS-35 [Pierce]) and rinsed extensively with $H_2O$ and lastly ethanol.
4. To prevent the gel from slide-out, one or both of the glass plates must be frosted. Additionally, a 0.5- to 1.0-cm plug of ~10% acrylamide will ensure that the gel remains in place during the electrophoresis and handling.
5. Care should be taken to lift one plate off the gel; generally the nonfrosted plate will lift off more easily. In the event the gel is sticking, try to loosen it using a gentle stream of running buffer or water. Plates should be horizontal when trying to separate with the gel side down. Failure to separate plates without breaking the gel or when there is excessive sticking is typically due to imperfectly clean plates.
6. Particular care must be taken when handling the composite gel itself when taking the apparatus apart after the run. Gels composed of agarose and acrylamide are not at all the

consistency of standard Laemli SDS-polyacrylamide gels but are rather crumbly and easily broken or cracked. The best advice is to pick the gel from the glass plate with a dry piece of filter paper. Start by overlying the paper and gently peeling it back lifting the gel from the glass plate.

7. There are no special problems with regard to performing the immunological analysis of the blotted proteoglycans after the transfer is complete, although care must be taken when setting up the transfer. Placement of the membrane on the gel must be correct from the beginning since once placed on the gel it cannot be moved. Additionally, when preparing the sandwich of gel, membrane, and filter paper, care must be taken not to squeeze the package or use papers that are too thick, since it is possible to crush and destroy the gel.
8. The use of positively charged Nylon membranes for immunodetection can greatly enhance the sensitivity of detection (i.e., 10–50 ng of proteoglycan) over that of metachromatic dyes such as toluidine blue *(7)*. More recently, positively charged Immobilon-N (Millipore) has been used with nearly equal success.

## References


1. Heinegård, D. and Paulsson, M. (1984) Structure and metabolism of proteoglycans, in *Extracellular Matrix Biochemistry* (Piez, K. and Reddi, H., eds.), Elsevier, New York, NY, pp. 277–328.
2. McDevitt, C. A. and Muir, H. (1971) Gel electrophoresis of proteoglycans and glycosaminoglycans on large-pore composite polyacrylamide-agarose gels. *Anal. Biochem.* **44,** 612–622.
3. Uriel, J. (1966) Method of electrophoresis in acrylamide-agarose gels. *Bull. Soc. Chim. Biol.* **48,** 969–982.
4. Peacock, A. C. and Dingman, C. W. (1968) Molecular weight estimation and separation of ribonucleic acid by electrophoresis in agarose-acrylamide composite gels. *Biochemistry* **7,** 668–674.
5. Sweet, M. B., Thonar, E. J., and Immelman, A. R. (1978) Anatomically determined polydispersity of proteoglycans of immature articular cartilage. *Arch. Biochem. Biophys.* **189,** 28–36.
6. Heinegård, D., Sommarin, Y., Hedbom, E., Wieslander, J., and Larsson, B. (1985) Assay of proteoglycan populations using agarose-polyacrylamide gel electrophoresis. *Anal. Biochem.* **151,** 41–48.
7. Heimer, R. and Sampson, P. M. (1987) Detecting proteoglycans immoblized on positively charged nylon. *Anal. Biochem.* **162,** 330–336.
8. Heimer, R., Molinaro, L., Jr. and Sampson, P. M. (1987) Detection by $^{125}$I-cationized cytochrome c of proteoglycans and glycosaminoglycans immobilized on unmodified and on positively charged nylon 66. *Anal. Biochem.* **165,** 448–455.
9. Heimer, R. (1989) Proteoglycan profiles obtained by electrophoresis and triple immunoblotting. *Anal. Biochem.* **180,** 211–215.
10. Bashey, R. I., Sampson, P. M., Jimenez, S. A., and Heimer, R. (1993) Glycosaminoglycans and chondroitin/dermatan sulfate proteoglycans in the myocardium of a non-human primate. *Matrix* **13,** 363–371.
11. Roughley, P. J. and White, R. J. (1989) Dermatan sulphate proteoglycans of human articular cartilage. The properties of dermatan sulphate proteoglycans I and II. *Biochem. J.* **262,** 823–827.
12. Heinegård, D., Wieslander, J., Sheehan, J., Paulsson, M., and Sommarin, Y. (1985) Separation and characterization of two populations of aggregating proteoglycans from cartilage. *Biochem. J.* **225,** 95–106.

13. Vogel, K. G. and Heinegård, D. (1985) Characterization of proteoglycans from adult 1bovine tendon. *J. Biol. Chem.* **260,** 9298–9306.
14. Adams, M. E., McDevitt, C. A., Ho, A. and Muir, H. (1986) Isolation and characterization of high-buoyant-density proteoglycans from semilunar menisci. *J. Bone Joint Surg.* **68,** 55–64.
15. Jahnke, M. R. and McDevitt, C. A. (1988) Proteoglycans of the human intervertebral disc. Electrophoretic heterogeneity of the aggregating proteoglycans of the nucleus pulposus. *Biochem. J.* **251,** 347–356.
16. Cs-Szabo, G., Roughley, P. J., Plaas, A. H., and Glant, T. T. (1995) Large and small proteoglycans of osteoarthritic and rheumatoid articular cartilage. *Arthritis Rheum.* **38**, 660–668.
17. Palmer, J. L., Bertone, A.L., Malemud, C, J., Carter, B. G., Papay, R. S., and Mansour, J. (1995) Site-specific proteoglycan characteristics of third carpal articular cartilage in exercised and nonexercised horses. *Am. J. Vet. Res.* **56,** 1570–1576.
18. Yamauchi, S., Cheng, H., Neame, P., Caterson, B., and Yamauchi, M. (1997) Identification, partial characterization, and distribution of versican and link protein in bovine dental pulp. *J. Dental Res.* **76,** 1730–1736.
19. Murdoch, A. D., Lliu, B. Schwarting, R., Tuan, R. S., and Iozzo, R. V. (1994) Widespread expression of perlecan proteoglycan in basement membranes and extracellular matrices of human tissues as detected by a novel monoclonal antibody against domain III and by in situ hybridization. *J. Histocem. Cytochem.* **42,** 239–249.
20. Isogai, Z., Shinomura, T., Yamakawa, N., Takeuchi, J., Tsuji, T., Heinegård, D., and Kimata, K. (1996) 2B1 antigen characteristically expressed on extracellular matrices of human malignant tumors is a large chondroitin sulfate proteoglycan, PG-M/versican. *Cancer Res.* **56,** 3902–3908.
21. Poole, A. R., Webber, C., Reiner, A., and Roughley, P. J. (1989) Studies of a monocloncal antibody to skeletal keratan sulphate. Importance of antibody valency. *Biochem. J.* **260,** 849–856.
22. Caterson, B., Christner, J. E., and Baker, J. R. (1983) Identification of a monoclonal antibody that specifically recognizes corneal and skeletal keratan sulfate. Monoclonal antibodies to cartilage proteoglycan. *J. Biol. Chem.* **258,** 8848–8854.
23. Hughes, C. E., Caterson, B., Fosang, A. J., Roughley, P. J., and Mort, J. S. (1995). Monoclonal antibodies that specifically recognize neoepitope sequences generated by "aggrecanase" and matrix metalloproteinase cleavage of aggrecan: application to catabolism in situ and in vitro. *Biochem. J.* **305,** 799–804.
24. Fosang, A. J. and Hardingham, T. E. (1991) 1-C-6 epitope in cartilage proteoglycan G2 domain is masked by keratan sulphate. *Biochem. J.* **273,** 369–373.
25. Avnar, Z. and Geiger, B. (1984) Immunocytochemical localization of native chondroitin-sulfate in tissues and cultured cells using specific monoclonal antibody. *Cell* **38,** 811–822.
26. Caterson, B., Christner, J. E., Baker, J. R., and Couchman, J. R. (1985) Production and characterization of monoclonal antibodies directed against connective tissue proteoglycans. *Fed. Proc.* **44,** 386–393.
27. Couchman, J. R., Caterson, B., Christner, J. E., and Baker, J. R. (1984) Mapping by monoclonal antibody detection of glycosaminoglycans in connective tissues. *Nature* **307,** 650–652.
28. Christner, J. E., Caterson, B., and Baker, J. R. (1980) Immunological determinants of proteoglycans. Antibodies against the unsaturated oligosaccharide products of chondroitinase ABC-digested cartilage proteoglycans. *J. Biol. Chem.* **255,** 7102–7105.
29. Fisher, L. W., Hawkins, G. R., Tuross, N., and Termine, J. D. (1987) Purification and partial characterization of small proteoglycans I and II, bone sialoproteins I and II, and osteonectin from the mineral compartment of developing human bone. *J. Biol. Chem.* **262,** 9702–9708.

# 16

# Quantitation of Proteoglycans in Biological Fluids Using Alcian Blue

**Madeleine Karlsson and Sven Björnsson**

## 1. Introduction

Alcian blue is a tetravalent cation with a hydrophobic core that contains a copper atom, hence its blue color *(1)*. The four charges allow Alcian Blue to bind to glycosaminoglycans (GAGs) at high ionic strength, in contrast to other cationic dyes, which are all monovalent. The molecular structure, that is, the plane tetragonal hydrophobic core with positive charges attached at its corners, may facilitate formation of aggregates of several molecules side by side rather than micelle formation. The ionic strength and presence of other detergents will affect the size of these aggregates in solution. Such positively charged linear aggregates will bind strongly to some negatively charged polymers such as GAGs, but not to others such as nucleic acids reflecting the conformation of both the polymer and the dye. Alcian blue is not a well defined substance. There are large differences in solubility and binding characteristics among different manufacturers but also between batches from the same manufacturer. In addition, the stain has a limited shelf life and is slowly degraded both in solution and as dry substance.

Glycosaminoglycans are negatively charged polymers built by repeating units of disaccharides containing one uronic acid or galactose and one N-acetylglycosamine. The variation in charge may be very large, since each disaccharide is more or less sulfated. The ionic bonding between cationic dyes and the negatively charged GAGs is generally thought to be proportional to the number of negative charges present on the GAG chain—that is, both sulfate and carboxyl groups. In contrast, studies of the interaction at different pH and ionic strength *(2)* revealed that the carboxyl groups are not involved in the binding of Alcian blue to GAGs. Moreover, the presence of charged carboxyl groups interferes with binding of Alcian blue to the GAGs. One species of GAG, keratan sulfate, contains not uronic acid but galactose, and thus lacks the

From: *Methods in Molecular Biology, Vol. 171: Proteoglycan Protocols*
Edited by: R. V. Iozzo © Humana Press Inc., Totowa, NJ

carboxyl group. Accordingly, the binding of Alcian blue to keratan sulfate is not pH dependent, and the amount of dye bound per milligram keratan sulfate is equal to that of other types of GAGs *(2)*.

The principle is based on the specific interaction between sulfated polymers and the tetravalent cationic dye Alcian blue at a pH low enough to neutralize all carboxylic and phosphoric acid groups and at an ionic strength great enough to eliminate ionic interactions other than those between Alcian blue and sulfated GAGs *(2)*. Hyaluronan, a nonsulfated GAG, does not react in this assay. There is no interference from proteins or nucleic acids in this method (*see* **Figs 1** and **9**, later, and **refs.** *2* and *4*), in contrast to the dimethylmethyl blue (DMMB)-method *(5,6)* or other dye-binding methods. At 0.4 M guanidine-HCl, binding of Alcian blue to GAG chains in proteoglycans is the same as binding to isolated GAG chains. Thus, the method is equally applicable to proteoglycans samples as to GAG samples.

A method for simultaneous quantitation and preparation of intact proteoglycans in biological fluids (blood plasma, synovial fluid) or 4 *M* guanidine extracts of tissues has been published *(2)*. The tube/absorbance assay has a measuring range of 1–20 μg of GAG. Proteoglycans prepared by Alcian blue precipitation are suitable for electrophoresis *(3)* or biochemical analysis. The assay principle can be applied at any scale, from a 50 μL sample in a quantitative assay to a 600-mL sample in large-scale preparation.

Quantitation may also be performed using the dot blot format. In the dot blot/reflectance assay, the Alcian blue–GAG complexes are collected on a polyvinylidene diflouride (PVDF)-membrane by filtration in a dot-blot apparatus, and the stain is quantitated as reflectance by scanning densitometry. The dot-blot/reflectance assay has a measuring range of 10–800 ng of GAG, which is enough to measure the low contents of proteoglycans in plasma, urine, or wound fluid.

## 2. Materials

1. Alcian blue, certified for GAG/PG quantitation, Wieslab, Lund, Sweden.
2. Chondroitin-6-sulphate (C-4384) from shark cartilage, keratan sulphate (K3001) from bovine cornea, heparan sulphate (H7640) from bovine kidney, from Sigma (USA).
3. Capped polypropylene vials (1- or 2-mL), Sarstedt, Sweden.
4. Centrifuge with a fixed-angle rotor for 2-mL vials, and a minimal centrifugal force 12,000*g*.
5. ELISA photometer equipped with a 605- or 620-nm filter.
6. Syringe (5–10 mL) with a large-bore needle (1.5 mm).
7. MilliBlotD system (#MBBDDD 960) dot-blot apparatus, Biogenex, San Ramon, CA, USA.
8. PVDF membrane (Immobilon-P #IVPH00010), Millipore, USA.
9. A flat-bed scanner, Scanmaker II (Mikrotek Lab, USA) and a PowerMacintosh with Adobe Photoshop™ (Adobe Systems, USA) was used for imageing.

### 2.1. Stock Solutions

1. 10% (v/v) $H_2SO_4$: Add 50 mL of $H_2SO_4$ slowly with stirring to 450 mL of $H_2O$ to a final volume of 500 mL. The solution will be stable for for 1 yr at room temperature.
2. 10% (w/v) Triton X-100 : Dissolve 50g of Triton X-100 in 500 mL of $H_2O$, stir until in solution. The solution will be stable for for 1 yr at room temperature.

3. Alcian blue stock solution (Alcian blue in 0.1% $H_2SO_4$/0.4 *M* guanidine-HCl ): Add 0.1 mL of 10% $H_2SO_4$ to 0.5 mL of 8 *M* guanidine-HCl. Add $H_2O$ to 10 mL. Dissolve 0.1-0.2g (depending on batch) of Alcian blue 8 GS (Wieslab, Sweden). Mix for 2 h to overnight. Store in the dark at 4°C. The solution will be stable for for 1 mo.
   The guanidine-insoluble portion of the stain is removed by centrifugation (12,000*g*, 15 min) before each preparation of AB reagent. Dilute 100 µL of the supernatant to 10 mL and read absorbance at 600 nm. The concentration of Alcian blue in the stock solution should be adjusted, by addition of 0.4 *M* guanidine-HCl/0.1% $H_2SO_4$, corresponding to an absorbance of 1.4 of the diluted stock.
4. 8.0 *M* guanidine-HCl (Gu-HCl): Dissolve 764 g of guanidine-HCl p.a. (*Mw* 95.53) in $H_2O$ and add to 1000 mL. The solution will be stable for for 1 yr at room temperature. If technical-grade guanidine-HCl is used, it must be stirred overnight with active charcoal and then filtered. The concentration of a guanidine-HCl may be checked by measuring the density of the solution: 8.0 *M* guanidine-HCl has a density of 1.192 g/L.
5. 3.0 *M* $MgCl_2$: Dissolve 305 g of $MgCl_2 \times 6H_2O$ (*Mw* 203.30) in $H_2O$ and add $H_2O$ to 500 mL.The solution will be stable for for 1 yr at room temperature.
6. CsC calibrator stock solution (chondroitin-6-sulfate 10mg/mL): Dissolve 20 mg of chondroitin-6-sulfate (Sigma C4384) in 2 mL of $H_2O$. The solution will be stable for several years at –20°C.
7. TG × 10 (0.25 *M* Tris/1.92 *M* glycine pH 8.3): Dissolve 12.1 g Tris base and 57.6g of glycine in a final volume of 400 mL of $H_2O$.The pH should be approximately 8.3, but need not be adjusted. The solution will be stable for for 1 yr at room temperature.
8. 10% (w/v) SDS: Dissolve 25 g of SDS in a final volume of 250 mL of $H_2O$. The solution will be stable for for 1 yr at room temperature.
9. 1.0M Tris-acetate, pH 7.4: Dissolve 121 g of Tris base in $H_2O$. Adjust pH to 7.4 with acetic acid. Add $H_2O$ to 1000 mL. The solution will be stable for for 1 yr at room temperature.

### 2.2. Reagents

1. SAT-reagent (0.3% $H_2SO_4$/0.75% Triton X-100): Take 0.3 mL of 10% $H_2SO_4$ and 0.75 mL of 10% Triton X-100. Add $H_2O$ to 10 mL. The solution will be stable for 1 wk at room temperature.
2. AB reagent (Alcian blue in 0.1% $H_2SO_4$/0.25% Triton): Take 5.0 mL of Alcian blue stock solution and add 1.0 mL of 10% $H_2SO_4$ and 2.5 mL of 10% Triton. Add $H_2O$ to a final volume of 100 mL. The solution will be stable for 1 wk at 4°C.
3. DMSO washing solution (40% (v/v) DMSO/0.05M $MgCl_2$): Add 40 mL of DMSO to 40 mL of $H_20$ and add 1.7 mL of 3.0 M $MgCl_2$. Add $H_20$ to 100 mL. Store at room temperature. The solution will be stable for 1 mo.
4. Ethanol washing solution (50% (v/v) ethanol/0.05M $MgCl_2$): Take 52 mL of ethanol (96%, v/v) and add 1.7 mL of 3.0 *M* $MgCl_2$. Add $H_2O$ to 100 mL. Store at room temperature. The solution will be stable for 1 mo.
5. Gu-prop-$H_2O$: 4 *M* guanidine-HCl/33% 1-propanol/0.25% Triton/0.1% Ficoll.) Take 50 mL of 8.0 *M* guanidine-HCl, 33 mL of 1-propanol p.a., and 2.5 mL of 10% Triton. Add 0.1 g of Ficoll. Mix and add $H_2O$ to 100 mL. Do not use 2-propanol (isopropanol)! Store at room temperature. The solution will be stable for 1 yr.
6. Prop-gu: 71% 1-propanol/2.3 *M* guanidine-HCl): Mix 71 mL of 1-propanol p.a. and 29 mL of 8.0 *M* guanidine-HCl. Do not use 2-propanol (isopropanol)! Store at room temperature. The solution will be stable for 1 yr.

**Table 1**
**Preparation of Calibrators, Tube Absorbance Assay**

| Calibrator (mg/mL) | Dilution | H2O (μL) | μg GAG in 50μL |
|---|---|---|---|
| 0.4 | 80μl CsC stock | 1920 | 20 μg |
| 0.2 | 1000μl 0.4mg/mL | 1000μl | 10 μg |
| 0.1 | 1000μl 0.2mg/mL | 1000μl | 5 μg |
| 0.05 | 1000μl 0.1mg/mL | 1000μl | 2.5 μg |
| 0.025 | 1000μl 0.05mg/mL | 1000μl | 1.25 μg |
| 0.0125 | 1000μl 0.025mg/mL | 1000μl | 0.0625 μg |

7. Prop-Tris: 75% 1-propanol/0.1 *M* Tris-Ac. Mix 75 mL of 1-propanol and 10 mL of 1.0 *M* Tris/Ac, pH 7.4. Add $H_20$ to 100 mL. Store at room termperature. The solution will be stable for 1 yr.

### 2.3. Sample Preparation

GAGs are stable at 4°C, in the absence of cells. The protein cores of proteoglycans may be degraded by proteolytic enzymes. The sample must not contain any particles that sediment during centrifugation or that are insoluble in 0.4 *M* guanidine-HCl at pH 1.5. Cell debris and insoluble material should be removed by centrifugation (12,000*g* for 15 min). If the sample does contain material that precipitates in 0.4 *M* guanidine-HCl at pH 1.5, it must be removed by a preceding step (*see* **Subheading 3.4.**).

Samples such as plasma or synovial fluid with a high protein and a low proteoglycan content may be concentrated and purified with an extra preceding Alcian blue precipitation. (*see* **Subheading 3.5.**)

## 3. Methods

### 3.1. Quantitation, Tube Absorbance Assay (Table 1)

The GAG/PG are allowed to precipitate in 0.4 *M* guanidine-HCl at pH 1.5 and collected by centrifugation. Excess stain in the pellet is removed by washing in DMSO. The GAG/PG–Alcian blue complexes are dissolved and dissociated in a 4 *M* guanidine-HCl/propanol mixture. The amount of GAG/PG is directly proportional to the Alcian blue concentration measured as absorbance at 605 or 620 nm. The tube/absorbance assay has a measuring range of 1–20 μg of GAG. Sample volumes given are only suggestive. The method can be applied at any scale as long as the final concentration of guanidine-HCl is 0.4 mol/l at pH 1.5 - 2.0.

1. Put 50 μL of either blank (water), sample, calibrator, or control in 2-mL polypropylene centrifuge tubes.
2. Add 50 μL of 8 *M* Gu-HCl. Mix and leave for 15 min. Samples in 4 *M* guanidine-HCl are analyzed without further addition of guanidine-HCl. Take 100 μL of the sample and go directly to **step 3**.
3. Add 50 μL of SAT reagent. Mix and leave for 15 min.
4. Add 750 μL of cold AB reagent. Mix and leave for 15 min to overnight at 4°C.
5. Centrifuge for 15min at 12,000*g*. Remove supernatant and discard. Care should be taken that pellets are left intact when supernatants are removed by suction. The use of a manu-

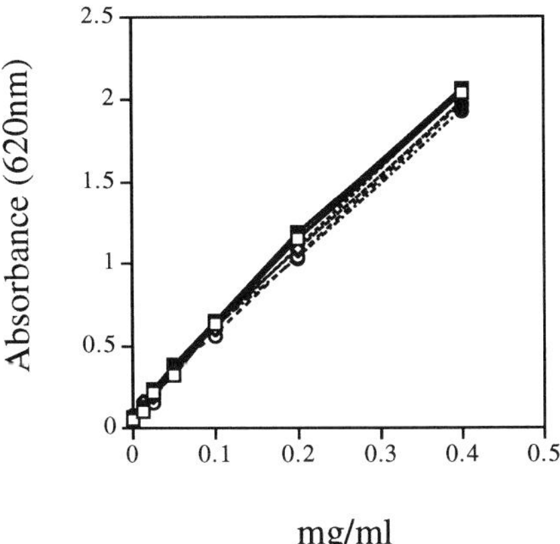


Fig 1. Tube assay of different GAG in blood plasma. Commercial preparations of glycosaminoglycans were dissolved in water or in blood plasma (0–0.4mg/mL) and was quantitated with Alcian Blue using the tube assay. -❑- Chondroitin-6-sulphate; -◇- Keratan sulphate; -○- Heparan sulphate; Open symbols: GAG dissolved in water. Closed symbols: GAG dissolved in blood plasma.

ally operated syringe for suction and immediate removal of the supernatants after centrifugation is therefore recommended. The tube should be held at an angle and in such a way that the pellet and needle are visible during the entire operation. The needle should be gradually lowered with the meniscus during suction and never be allowed to touch the pellet.

6. Add 500 µL of DMSO solution to the pellet. Mix thoroughly until the pellet is suspended in the washing solution. Mix for 15 min on a shaker.
7. Centrifuge as in **step 5**. Remove supernatant and discard.
8. Add 500 µL Gu-prop-$H_2O$ solution to the pellet. Mix for 15 min on a shaker. Check that the pellet is completely dissolved.
9. Dispense 2 × 240 µL of the supernatant into an ELISA-microplate. Read absorbance at 605 nm.
10. Plot the absorbances against amount of GAG in each calibrator (0–20 µg). The plot should be a perfectly straight line with an absorbance of approximately 2.0 at 20 µg and blank values below 0.05 (**Fig. 1**). High blank values may be caused by insoluble, degraded stain. The dry Alcian blue stain has a limited shelf life, and some batches are not suitable for biochemical work. The stock solution also has a limited shelf life, and it is important that the AB stock solution be centrifuged in order to remove the insoluble portion of Alcian blue. Fit a linear equation and calculate the amount of GAG in each sample. Alternatively, if only a single calibrator concentration is used, a factor is calculated by dividing the calibrator (µg) by the corresponding absorbance. The absorbance of the unknown samples is then multiplied by the factor to obtain micrograms of GAG in each sample.

    All commercial sulfated GAG samples (CsA, CsC, Ds, Ks , Hs) should give a similar colour yield. HA, Hyaluronan, DNA, and RNA do not react with Alcian blue at all. There is no interference from proteins in blood plasma (**Fig. 1**).

### 3.2. Preparation for Electrophoresis Without Removal of Alcian Blue

The GAG/PGs are allowed to precipitate in 0.4 *M* guanidine-HCl at pH 1.5 and collected by centrifugation. Excess stain in the pellet is removed by washing in DMSO.

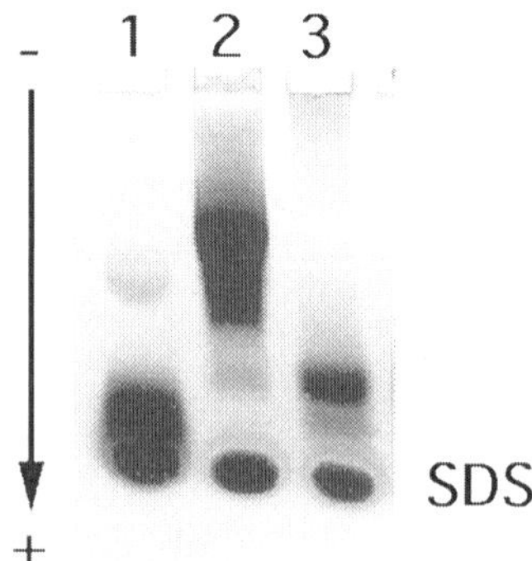


Fig. 2. Alcian blue precipitation without stain removal. Proteoglycans in 4 *M* Gu-HCL extracts of cartilage and skin were precipitated with by Alcian blue as described under **Subheading 3.2.** The samples were analyzed by electrophoresis in 1% agarose and stained with AB reagent containing 0.4 *M* GuHCl. Lane 1; chondroitin-6-sulfate. Lane 2; 4 *M* Gu-HCl extract of bovine nasal cartilage; Lane 3; 4 *M* Gu-HCl extract of human skin. The position of AB–SDS complexes is indicated in the figure.

The GAG/PG/Alcian blue complexes are dissolved directly in an SDS buffer and analyzed by electrophoresis.

The SDS used to dissolve the GAG/PG–Alcian blue complexes will move in front of any GAG/PG, **Fig. 2**. The relative mobility of the SDS–Alcian blue complexes will depend on the ratio of SDS bound per Alcian blue molecule.

This method is useful for revealing losses during the propanol precipitation step (*see* **step 9** and **10** under **Subheading 3.3.**). Any losses can be detected by comparing the electrophoresis patterns of the same samples obtained as described under **Subheading 3.2.** and **3.3.**

1. Take 100 µL of either blank (water), sample, calibrator or control to 2 mL polypropylene centrifuge tubes.
2. Add 100 µL of 8M GuHCl. Mix and leave for 15 min. Samples in 4 *M* guanidine-HCl are analyzed without further addition of guanidine-HCl. Take 200 µL of the sample and go directly to **step 3**.
3. Add 100 µL of SAT reagent. Mix and leave for 15 min.
4. Add 1500 µL of cold AB reagent. Mix and leave for 15min - overnight at 4°C.
5. Centrifuge for 15min at 12000*g*. Remove supernatant and discard.
6. Add 750 µL of DMSO solution to the pellet. Mix thoroughly; the pellets should be suspended in the washing solution. Mix for 15 min.
7. Centrifuge as in **step 5**. Remove supernatant and discard.
8. Add 50 µL of TG × 3/2.5% SDS to pellets (sample, calibrator and control).
9. Quantitation may be performed as follows. Put 5 µL of each dissolved pellet on a microtiter plate. Add 100 µL TG × 3/0.5% SDS to each well and mix. Use reverse pipetting technique to minimize formation of bubbles. Read absorbance at 600 nm.
10. The amount of sample to be applied to each lane is calculated as follows. Divide the calibrator (µg) by the corresponding absorbance. The absorbance of the unknown sample is then multiplied by the factor to obtain micrograms of GAG in each sample. About 3–10 µg are suitable for electrophoresis. Add TG × 3/0.5% SDS to a final volume of 50 µL.

11. Add 25 µL of liquified 1% agarose with 15% glycerol. Heat at 105°C for 5 min and apply 75 µL of the hot sample on each lane on a 1% agarose gel in 0.1 *M* Tris-Ac. Electrophorese for 1 h at 100 V, until the BFB front has moved 60 mm. Stain in AB-reagent with 0.4 *M* GuHCl (*see* **Fig. 2.**).

### 3.3. Preparation with Removal of Alcian Blue

The GAG/PGs are allowed to precipitate in 0.4 *M* guanidine-HCl at pH 1.5 and collected by centrifugation. Excess stain in the pellet is removed by washing in DMSO. The GAG/PG–Alcian Blue complexes are dissociated and dissolved in a 4 *M* guanidine-HCl/propanol mixture. The GAG/PGs are precipitated by increasing the propanol concentration, with addition of a 71% propanol/guanidine-HCl mixture, and are then recovered by centrifugation. The amount of GAG/PG is directly proportional to the Alcian blue concentration in the supernatant, measured as absorbance at 605 nm.

1. Put 100 µL of either blank (water), sample, calibrator, or control in 2-mL polypropylene centrifuge tubes.
2. Add 100 µL of 8 *M* Gu-HCl. Mix and leave for 15 min. Samples in 4 *M* guanidine-HCl are analyzed without further addition of guanidine-HCl. Take 200 µL of the sample and go directly to **step 3**.
3. Add 100 µL of SAT reagent. Mix and leave for 15 min.
4. Add 1500 µL of cold AB reagent. Mix and leave for 15min to overnight at 4°C.
5. Centrifuge for 15 min at 12000*g*. Remove supernatant and discard.
6. Add 750 µL of DMSO solution to the pellet. Mix thoroughly; the pellets should be suspended in the washing solution. Mix for 15 min.
7. Centrifuge as in **step 5**. Remove supernatant and discard.
8. Add 250 µL of Gu-prop-$H_2O$ to the pellet. Mix for 15 min or until the pellet is dissolved.
9. Add 1750 µL of Prop-gu. Mix and leave for 60 min . Loss of GAG/PG may occur during propanol precipitation/centrifugation. Most proteoglycans will precipitate quantitatively at a rather low propanol concentration but isolated GAG chains may not precipitate quantitatively. Thus, keratan sulfate require higher propanol concentrations than other GAGs, **Fig. 3**. Any losses can be detected by comparing the electrophoresis patterns of the same samples without propanol precipitation (*see* **Subheading 3.2.**) and with propanol precipitation (*see* **Subheading 3.3.**).
10. Centrifuge for 15 min at 12000*g*. Remove the supernatant using a syringe. Read absorbance at 605 nm of the supernatant. The expected absorbance of the supernatant is 1.0 at 0.4 mg/mL (40 µg).
11. Add 750 µL of Prop-Tris to the pellet. Mix.
12. Centrifuge for 5 min at 12,000*g*. Remove the supernatant and discard. The pellet will always contain some insoluble degraded Alcian blue. The dye in the pellet is not water soluble, and can be sedimented by centrifugation. The concentration of propanol in the Prop-gu solution, used for precipitation of GAG/PG, will affect the amount of Alcian blue remaining in the pellet, **Fig. 4**. The amount of Alcian blue in the pellet also depends on the GAG and is especially high with keratan sulfate, **Fig. 4**.
13. Dissolve the pellet in 50 µL TG × 3/0.1% SDS. The pellet may also be dissolved in water or a buffer without SDS.
14. The amount of proteoglycan needed for electrophoresis may be calculated as above (*see* **Subheading 3.2., step10**) and analyzed by electrophoresis (*see* **Fig. 5**).

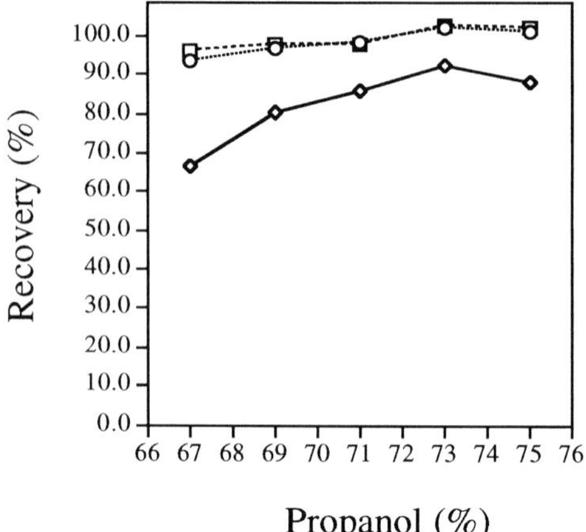


Fig. 3. Removal of Alcian blue stain: precipitation of GAG at different concentrations of propanol. GAGs were precipitated by Alcian blue and the precipitates were dissolved in 250 µL of a 4 *M* GuHCl/33% propanol mixture as described under **Subheading 3.3.** GAGs were precipitated by adding 1750 µL of a propanol/GuHCl mixture. The following propanol/GuHCl mixtures were used: 67% propanol 2.66 *M* GuHCl; 69% propanol/2.48 *M* GuHCl; 71% propanol/2.32 *M* GuHCl; 73% propanol/2.16 *M* GuHCl; 75% propanol/2.0 *M* GuHCl. The precipitated GAGs, recovered by centrifugation, were again quantitated using the method of **Subsection 3.1.** and the recovery calculated. -❑- Chondroitin-6-sulfate; -◇- Keratan sulfate; -❍- Heparan sulfate.

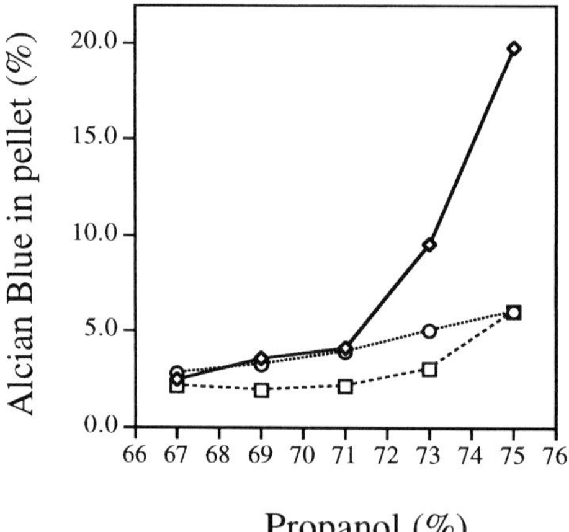


Fig. 4. Removal of Alcian blue stain: coprecipitation of Alcian blue with GAG at different concentrations of propanol. Same experiment as in Fig. 3. The amount of Alcian blue that co-precipitated with GAGs after centrifugation was quantitated as absorbance at 620 nm of the solubilized pellet. -❑- Chondroitin-6-sulfate; -◇- Keratan sulfate; -❍- Heparan sulfate.

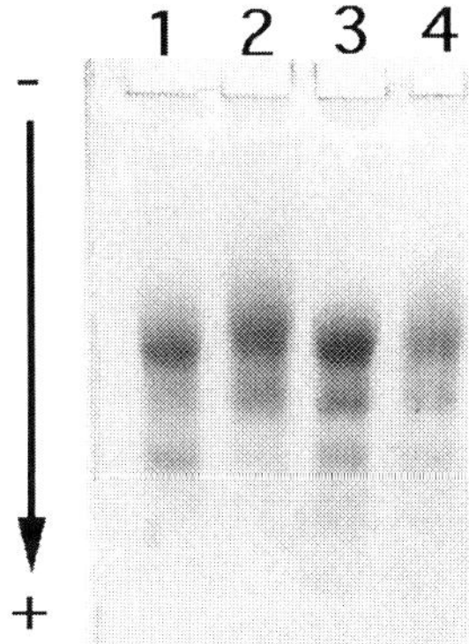


Fig. 5. Alcian blue precipitation with stain removal. Proteoglycans in four different human synovial fluids were precipitated by Alcian blue as described under **Subheading 3.3.** The samples were analyzed by electrophoresis in 1% agarose and stained with AB-reagent containing 0.4 *M* Gu-HCl.

### 3.4. Purification of Difficult Samples

Some samples may contain material that precipitates at a low pH or that binds unspecifically to Alcian blue. Such material may be removed by an extra Alcian blue precipitation step as described below.

1. Put 100 µL of either blank (water), sample, calibrator, or control into 2-mL polypropylene centrifuge tubes.
2. Add 100 µL of 8 *M* Gu-HCl. Mix and leave for 15 min.
3. Add 100 µL of SAT reagent. Mix and leave for 15 min.
4. Add 1500 µL of cold AB reagent. Mix and leave for 15 min to overnight at 4°C.
5. Centrifuge for 15 min at 12000*g*. Remove supernatant and discard.
6. Dissolve the pellet in 100 µL of $H_2O$ + 100 µL of 8 *M* GuHCl + 100 µL of SAT reagent. Mix for 15 min.
7. Centrifuge for 15 min at 12000*g*. Transfer the supernatant to a new vial. Discard the pellet. Alternatively, the pellet may be washed in DMSO-washing solution and dissolved in TG × 3/2.5% SDS and analyzed by electrophoresis to verify that no GAG/PG are present in the pellet, **Fig. 6**.
8. Continue at **step 4** under **Subheading 3.2.** or **3.3.**

### 3.5. Large-Scale Preparation

Large-scale preparation consists of two successive precipitations with Alcian blue and is suitable for dilute samples such as plasma and urine. In the first precipitation step the GAG/PGs are concentrated. When the first precipitate is dissolved and centrifuged, any acid-insoluble material is left behind in the pellet. The supernatant will contain the proteoglycans, which are further purified by a second Alcian blue precipitation.

The GAG/PG–Alcian blue complexes recovered after the second precipitation are dissolved and dissociated in a 4 *M* guanidine-HCl/propanol mixture. The GAG/PG is recovered free from stain by selective propanol precipitation.

**Table 2**
**Sample of Results with Large Scale Preparation Method**

| Fraction | Absorbance | Dispensed volume (mL) | Total volume (mL) | Total absorbance | Approx. amount of GAG/PG μg |
|---|---|---|---|---|---|
| Step 5—supernatant | 1.86 | 0.1 | 3.75 | 69.7 | 348 |
| Step 8—supernatant | 0.51 | 0.2 | 10 | 25.5 | 127 |
| Step 11—supernatant | 1.97 | 0.24 | 6.8 | 55.8 | 279 |
| Final quantitation of 100 μL | 1.75 × 5 | 0.24 | 0.5 | 45.5 | 227 |

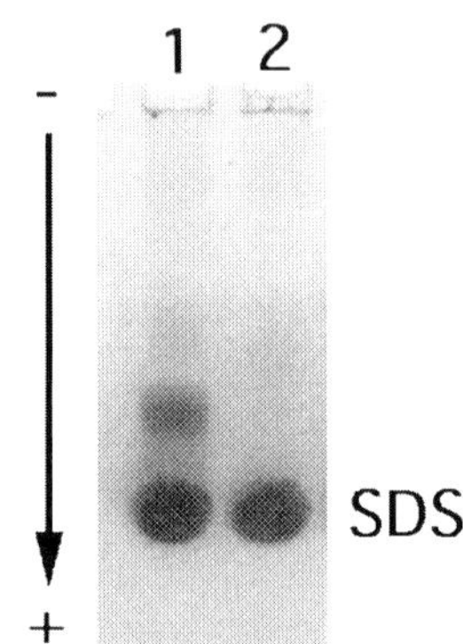


Fig 6. Purification of difficult samples. Proteoglycans in human blood plasma were precipitated with Alcian blue and the pellets extracted with a mixture of 4 M Gu-HCl/SAT-reagent and centrifuged as described under **Subheading 3.4.** The extracted proteoglycans in the supernatant were reprecipitated as described under **Subheading 3.2.** The pellet was washed in DMSO washing solution and dissolved in TG × 3-2.5% SDS. The samples were analyzed by electrophoresis in 1% agarose and stained with AB-reagent containing 0.4 *M* GuHCl. Lane 1; supernatant from **Subheading 3.4., step 7**; Lane 2; pellet from **Subheading 3.4., step 7**. The position of AB - SDS complexes is indicated in the figure.

The amount of GAG is directly proportional to the Alcian blue concentration in the supernatant, measured as absorbance at 605 or 620 nm.

Recovery of GAG/PG throughout the procedure is approximately 70%. The main losses are in the propanol precipitation/centrifugation step (*see* **Table 2**). The size distribution and purity of the purified preteoglycans could be analyzed by agarose electrophoresis and parallel lanes stained with Coomassie Brilliant blue and Alcian blue, **Fig. 7.** The proportions given below can be applied at any scale.

1. Put a 25 mL sample in a 50 mL propylene tube. Add 2.5 mL 8 *M* Gu-HCl and 1.25 mL 10% Triton. Mix and leave for 15 min.
2. Add 2.5 mL of the Alcian blue stock solution. Mix carefully and leave for 15 min.
3. Add 0.67% (v/v) $H_2SO_4$ up to 50 mL. Leave overnight at 4°C.
4. Centrifuge at 3300*g*, 60 min. Remove the supernatant and discard. Let the tube drain upside down.

Fig. 7. Large scale preparation. Proteoglycans were prepared from human blood platelets with two successive precipitations with Alcian Blue as described under **Subheading 3.5**. The purified proteoglycans were analyzed by electrophoresis in 1% agarose.Lane A: stained with AB - reagent with 0,4 M GuHCl. Lane B: stained with Coomassie Brilliant Blue in 5% (v/v) acetic acid - 25% (v/v) ethanol.

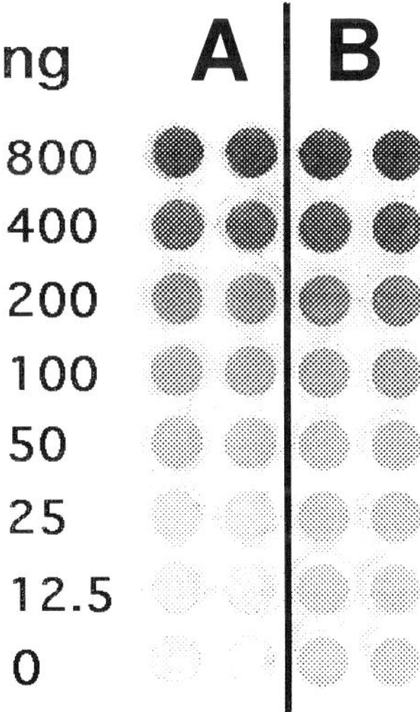


Fig. 8. Dot-blot assay. Heparan sulfate calibrator (0–80 mg/L), dissolved in water (**A**) or in blood plasma (**B**), was precipitated with Alcian blue and the GAG precipitates collected on a PVDF membrane by filtration in a dot-blot apparatus. The membrane was scanned on a single layer of clear plastic.

5. Add 1.25 mL of 8 *M* Gu-HCl, and 1.25 mL of $H_2O$, and 1.25 mL of SAT reagent. Mix for 15 min. Centrifuge at 3300*g* for 15min. Transfer the supernatant to another tube. Determine absorbance of a suitable volume at 605 nm.
6. Add 22.5 mL of AB reagent. Leave for 2 h. Centrifuge at 3300*g* for 1h. Remove the supernatant and discard. Let the tube drain upside down.
7. Add 10 mL of DMSO washing solution and mix until the pellet is suspended. Mix for 15 min.
8. Centrifuge at 3300*g* for 20 min. Remove the supernatant and determine absorbance of a suitable volume at 605 nm. Let the tube drain upside down.
9. Dissolve the pellet in 1000 µL of Gu-prop-$H_2O$ solution (prepared without Ficoll). Mix for 30 min or until the pellet is completely dissolved. Divide into four 2 mL propylene vials (approximately 250 µL/vial).
10. Add 1750 µL of Prop-gu solution. Leave for 1h to precipitate the GAG/PG.
11. Centrifuge at 12,000*g* for 15 min. Remove the supernatants carefully, using a syringe, to new tubes. Determine absorbance of a suitable volume at 605 nm. Loss of GAG/PG may occur during the propanol precipitation/centrifugation in **steps 10–11**. Such losses are revealed by comparison of the total absorbances calculated in **step 11** and **13** (*see* **Table 2**).
12. Add 750 µL of Prop-Tris solution to the pellet. Mix. Centrifuge at 12,000*g* for 15 min. Remove the supernatant and discard. Let the tube drain upside down. Enzymatic degradation of GAG/PG is inhibited if propanol is present in the pellet. Propanol may be evaporated by drying at 37°C. Too intensive drying should be avoided, since the GAG/PG may be difficult to redissolve.
13. Dissolve the pellet in a suitable buffer (*see* **Note 1**). Take a suitable volume for quantitation (*see* **Subheading 3.1.**). Calculate total absorbance and the amount of GAG/PG.

**Calculation.** The recovery of GAG/PG in each step may be monitored by taking the absorbance of a suitable portion and calculating the total absorbance. Care should be taken not to exceed the actual absorbance range of the instrument since most ELISA photometers are only reliable below an absorbance of 2.0. Calculate the total absorbance as follows: total volume of supernatant /volume in ELISA well × $A_{605}$. The approximate amount of GAG/PG is obtained by multiplying by 5.

The following results in **Table 2** were obtained when 250 µg of shark CsC were dissolved in 25 mL of human blood plasma. The absorbance was measured in each of the steps indicated above. The purified GAG/PG was dissolved in 500 µL 4 *M* guanidine-HCl and 100 µL were taken for final quantitation (*see* **Subheading 3.1.**).

Comments: The figure obtained for **step 5**—supernatant overestimates the amount of GAG/PG since it also contains some co-precipitating dye. This dye is removed by washing in DMSO. The figure obtained for **step 8**—supernatant overestimates the amount of dye washed away since the molar absorbance coefficient of Alcian blue is higher in DMSO than in guanidine-HCL solution. The figure obtained for **step 11**—supernatant is always higher than that obtained by final quantitation of the redissolved purified GAG/PG sample. This is presumably because of GAG/PG losses during the propanol precipitation/centrifugation in **step 11** or by incomplete solubilization in **step 13**.

### 3.6. Dot-Blot Reflectance Assay (see Table 3)

The Alcian blue–GAG complexes are collected on a PVDF membrane, by filtration in a dot-blot apparatus, and the stain quantitated as reflectance by scanning and densitometry. The assay requires 10 µL of sample and has a measuring range of 10–800 ng

**Table 3**
**Dot-Blot Reflectance Assay**

| Calibrator | Dilution | | H2O | GAG in 10μl |
|---|---|---|---|---|
| 80 | 16 μL | CsC stock | 1984 | 800 |
| 40 | 1000 μL | 80 mg/L | 1000 | 400 |
| 20 | 1000 μL | 40 mg/L | 1000 | 200 |
| 10 | 1000 μL | 20 mg/L | 1000 | 100 |
| 5 | 1000 μL | 10 mg/L | 1000 | 50 |
| 2.5 | 1000 μL | 5 mg/L | 1000 | 25 |
| 1.25 | 1000 μL | 2.5 mg/L | 1000 | 12.5 |

of GAG. No interference from plasma proteins is evident when chondroitin sulfate, keratan sulfate or heparan sulfate is dissolved in a blood plasma with low endogenous GAG content (*see* **Fig. 8**). The dot-blot assay is sensitive enough to measure the amount of GAG released upon coagulation; that is serum values (*see* **Fig. 10**) are much higher than plasma values (*see* **Fig. 9**).

1. Put 10 μL of either blank (water), sample, calibrator, or control in duplicate in a 96-well polystyrene microplate. All six calibrator levels, plus blank, should be used, as the calibration curve is nonlinear.
2. Add 20 μL of a 1/1 mixture of 8 *M* GuHCl and SAT reagent.
3. Mix on a microplate shaker at 200 rpm for 15 min.
4. Add 200 μL of cold AB reagent. Mix as above for 60 min.
5. Assemble a 96-well MilliBlotD apparatus according to the instructions of the manufacturer, using a PVDF membrane wetted and blocked for 1 h in 1% (v/v) Triton X-100.
6. Add 200 μL of 0.4 *M* Gu-HCl/0.1% (v/v) sulfuric acid/0.25% (v/v) Triton X-100 to each well immediately upon assembly (*see* **Note 2**). Evacuate approximately 100 μL by suction. Close the outlet tubing with a rubber stopper.
7. Transfer 200 μL of the samples from the microplate using an eight-channel pipet and the reverse pipetting technique. Precipitates of GAG/PG–Alcian blue may form during incubation in **step 4**. Any precipitates present at this point can be dispersed by repetitive pipetting in the microplate well before transfer.
8. Evacuate the wells by suction.
9. Add 200 μL of 50% (v/v) ethanol in 0.05 *M* $MgCl_2$ and evacuate the wells by suction. Repeat once
10. Remove the membrane from the apparatus and wash briefly in distilled water before air drying. High blank values are caused by degraded Alcian blue stain. Degradation may occur both in dry form and in liquid preparations (stock or AB reagent).
11. *Scanning*. Scanning is performed in the reflectance color mode with the gamma curve set at 1.0 and the dynamic range adjusted by positioning the 256 gray scales over the actual range displayed in the histogram window. The membrane is placed on top of one transparent plastic sheet in order to increase the sensitivity. The green and blue channels are discarded. The red channel is saved in gray-scale mode, **Fig. 8**.
12. *Densitometry*. The Scan Analysis software for Macintosh from Biosoft® (Cambridge, UK) is used for densitometry (*see* **Note 3**). The manual mode is used without background subtraction, and the densitometry is performed on a rectangle encompassing two dots of a

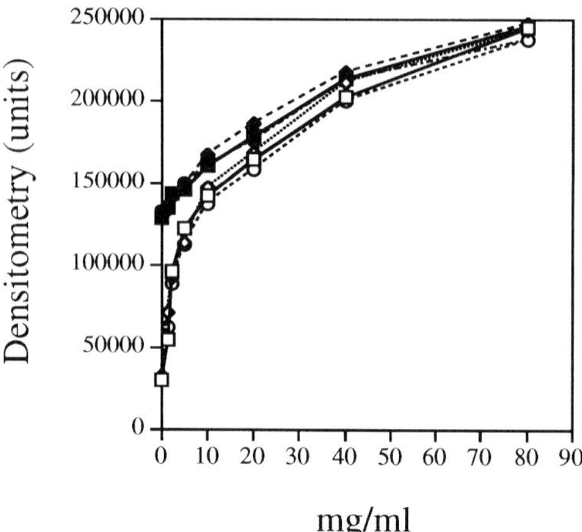


Fig. 9. Dot-blot assay of different GAGs in serum. Commercial preparations of glycosaminoglycans were dissolved in water or in serum (0-80 mg/L) and quantitated with Alcian blue using the dot-blot assay-❑- Chondroitin-6-sulfate; ◆ Keratan sulfate; -O- Heparan sulfate; open symbols, GAG dissolved in water. Closed symbols, GAG dissolved in serum.

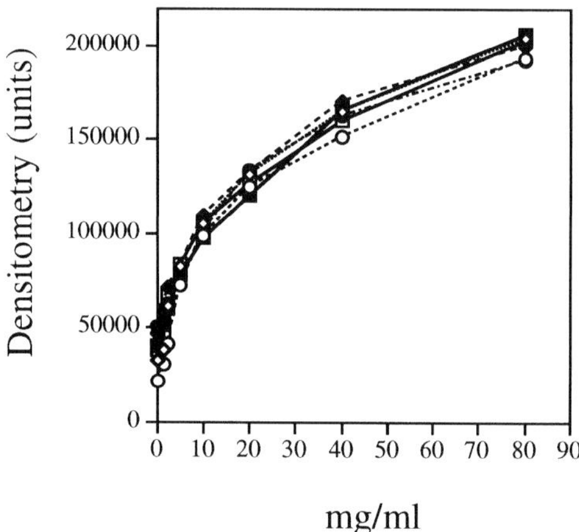


Fig. 10. Dot-blot assay of different GAGs in blood plasma. Commercial preparations of glycosaminoglycans were dissolved in water or in blood plasma (0-80 mg/L) and quantitated with Alcian blue using the dot blot assay. -❑- Chondroitin-6-sulphate; ◆ Keratan sulphate; -O- Heparan sulphate; open symbols, GAG dissolved in water; closed symbols, GAG dissolved in blood plasma.

duplicate sample. The non linear relationship between reflectance and concentration is evident (*see* **Figs. 9** and **10**).

13. *Curve fitting.* The arbitrary units obtained in Scan Analysis are pasted into the Ultra Fit software from Biosoft® and multiplied by 0.01 in order to get manageably sized factors (*see* **Note 3**). The densitometry units (x axis) are plotted against the calibrator values (y axis). The Cubic equation (a third-degree polynomial) is used for curve fitting. In order to prevent minima appearing within the fitted curve, parameter c should be restricted to values $\geq 0$. The fitted equation is used to calculate the concentration of unknown samples.

## 4. Notes

1. The pellet will always contain some degraded Alcian blue, which is not water soluble and can be sedimented by centrifugation.
2. Uneven spots caused by air bubbles present underneath the membrane are prevented by always having excess fluid beneath and above the membrane when assembling the apparatus and applying the samples. Therefore, buffer should be added to each well immediately upon assembly.
3. Densitometry and curve fitting may be performed with any suitable software. A third-degree polynomial or any other suitable equation may be used for curve fitting.

## References

1. Scott. J. E. (1970) Histochemistry of Alcian blue I. Metachromasia of Alcian Blue, Astrablau and other cationics and other phthalocyanin dyes. *Histochemie* **21,** 277–285
2. Björnsson, S. (1993) Simultaneous preparation and quantitation of proteoglycans by precipitation with Alcian Blue. *Anal. Biochem.* **210,** 282–291.
3. Björnsson, S. (1993) Size dependent separation of proteoglycans by electrophoresis in gels of pure agarose. *Anal. Biochem.* **210,** 292–298.
4. Björnsson, S. (1998) Quantitation of proteoglycans as glycosaminoglycans in biological fluids using an Alcian Blue dot blot analysis. *Anal. Biochem.* **256,** 229–237.
5. Farndale, R. W., Sayers, C. A., and Barrett A. J. (1982) A direct spectrophotometric microassay for sulfated glycosaminoglycans in cartilage cultures. *Connect. Tissue Res.* **9,** 247–248.
6. Farndale, R. W., Buttle, D. J., and Barrett A. J. (1986) Improved quantitaion and discrimination of sulphated glycosaminoglycans by use of dimethylmetylene blue. *Biochim. Biophys. Acta.* **883,** 173–177.

# 17

## Cellulose Acetate Electrophoresis of Glycosaminoglycans

**Yanusz Wegrowski and Francois-Xavier Maquart**

### 1. Introduction

Electrophoresis on cellulose acetate membrane (zone electrophoresis) is a common method for qualitative and semiquantitative analysis of glycosaminoglycan (GAG) mixtures. The advantage of this method is its simplicity, rapidity, the possibility of processing several samples at the same time, and the low cost of analysis. Apart from cellulose acetate strips and electrophoresis apparatus, usually applied in diagnostic laboratories for serum protein electrophoresis, no special equipment is needed (*see* **Note 1**). Several original papers and book chapters describe this technique in different running conditions; the most common is a monodimensional electrophoresis in bivalent cation buffer or pyridine/formiate buffer *(1–3)*. However, no simple system can separate all known GAGs in one run. The separation of different GAGs in one dimension can be done by a several steps method described by Hopwood and Harrison *(4)*, requiring careful temperature control and selective ethanol precipitation after each run. Alternatively, the GAGs can be separated by a two-dimensional method *(5)*, but only one sample at once may be applied on the cellulose acetate sheet. The scope of this chapter is to describe a simple method for rapid separation and visualization of the most common GAGs from tissue or cell culture. Comparison of electrophoretic profiles before and after selective enzymatic treatment (*see* Chapters 32-36) or $HNO_2$ depolymerization of heparin/heparan sulfate *(6)* allows characterization in a single run of the major fractions of GAGs in a given sample.

Glycosaminoglycans should be displaced from the protein core of proteoglycans before analysis. This can be done by exhaustive digestion of proteins with nonspecific proteases such as pronase or papain *(7)*. Apart from corneal keratan sulfate, which is bound to protein via an N-glycosyl linkage between N-acetylglucosamine and the amide group of asparagine, O-glycosyl covalent bonding can be alternatively disrupted by a beta-elimination reaction, which liberates all linked GAGs from the protein core *(8)*. The GAG samples have to be desalted and sufficiently concentrated (*see* **Note 2**). Tissue

From: *Methods in Molecular Biology, Vol. 171: Proteoglycan Protocols*
Edited by: R. V. Iozzo © Humana Press Inc., Totowa, NJ

or body fluids usually contain sufficient amounts of GAGs to be detected by cationic dye staining without radioactive labeling (*see* Chapter 16). The limit of Alcian blue staining in the case of electrophoresis is about 100 ng of individual GAG. The techniques that use Safranin O ***(9)*** or ruthenium red ***(10)*** staining have a detection limit even 100-fold lover, but they are complicated to use. When working with cell culture, one can easily radiolabel the GAGs with $^{35}$S-sulfate and/or $^{3}$H-glucosamine. Tritium labeling also permits the study of nonsulfated GAGs, such as hyaluronan or chondroitin. The methods for $^{35}$S-sulfate labeling as well as beta-elimination are described elsewhere (*8*).

We present here two simple variants of cellulose acetate electrophoresis. The first one, published originally by Wessler ***(11)***, permits the separation of hyaluronan, heparan sulfate, galactosaminoglycans (dermatan and chondroitin sulfates), and heparin in 0.1 *M* HCl. The second one was used to separate galactosaminoglycans in 0.1 *M* zinc acetate, pH 5.1. If the electrophoresis is performed first in zinc acetate and then in HCl, the separation of five glycosaminoglycans can be done. The simple method of selective detection of $^{35}$S-sulfate *vs* $^{3}$H-glucosamine is also described.

## 2. Materials

### 2.1. Electrophoresis of Glycosaminoglycans

1. Cellulose acetate strips and electrophoresis apparatus (e.g., Titan III cellulose acetate plates and Titan Zip Zone Chamber from Helena Research Laboratories, Beaumont, TX, USA, or Sebiagel® and Electrophoresis Apparatus from Sebia, Issy-les-Moulineaux, France).
2. Glycosaminoglycan standards (Sigma, St. Louis, MO, USA). Dissolve each glycosaminoglycan in deionized water at concentration 1 mg/mL. This is stable indefinitely if frozen.
3. 0.1 *M* zinc acetate buffer, acidified to pH 5.1 with acetic acid.
4. 0.1 *M* HCl.
5. Absolute ethanol.
6. Phenol red (Sigma). Saturated aqueous solution.
7. A microsyringue for sample application (e.g., Hamilton).

### 2.2. Staining with Alcian Blue

1. Stock solution of Alcian blue 8GX (Sigma): Prepare 0.4% (m/v) solution in absolute ethanol. Filter through cotton or Kleenex or Büchner funnel. Stable indefinitely in the dark.
2. Staining buffer: 0.05 *M* Natrium acetate containing 0.1 *M* $MgCl_2$, pH 5.8. Sodium azide 0.05 % (m/v), may be added to prevent bacterial development. Store at 4°C.
3. Staining solution: Mix 1-to-1 stock Alcian solution with staining buffer. May be reused several fold until the formation of precipitate.
4. Destaining solution: Mix 1-to-1 staining buffer with ethanol.

### 2.3. Autoradiography of Glycosaminoglycans

1. Glycerol 2% (v/v) in absolute ethanol.
2. PPO solution: Dissolve PPO (2,5-diphenyloxasol, Merck, Darmstadt, Germany), 2% (m/v) in glycerol-containing ethanol.
3. Autoradiography cassette with intensifying screens and autoradiography film.
4. Saran Wrap foil.
5. Developing facilities.

## 3. Method

### 3.1. Electrophoresis

1. Soak cellulose acetate membrane for at least 10 min in electrophoresis buffer with gentle agitation (*see* **Note 3**).
2. Fill the electrode chambers with the same buffer.
3. Blot the membrane in filter paper to remove the excess of liquid. Do not dry. Immediately install in electrophoresis apparatus (*see* **Note 4**).
4. Starting 8 mm from one border and about 1 cm from cathode (–) end, mark with a phenol red a series of 5-mm traits indicating the places for sample loading. Leave a space of 5 mm between the traits. Let the liquid dry.
5. Apply samples or standards by portions not exceeding 3 µL. Let the liquid dry every time. Optimal loading quantity for staining is the equivalent of 1 µg of standard GAG and, for autoradiography, 40,000 cpm as $^{35}$S-sulfate.
6. Run the electrophoresis at room temperature for 2 h at constant voltage. The current at the beginning should not exceed 1 mA/cm of width for zinc acetate or 1.5 mA/cm for HCl (*see* **Note 5**).
7. If the electrophoresis is performed in 0.1 *M* HCl, fix the membrane (*see* **step 9**). If it is done in zinc acetate buffer, remove the membrane and soak it for about 1 min in 0.1 *M* HCl. At the same time, fill the buffer tanks of apparatus with 0.1 *M* HCl.
8. Blot and reinstall the membrane. Continue the electrophoresis for 1 h at the same current.
9. After electrophoresis, the GAGs are fixed by soaking the membrane for about 1 min in absolute ethanol.

### 3.2. Staining

1. Soak the membrane in Alcian blue staining solution for 30 min with gentle agitation.
2. Wash the membranes in three baths of destaining solution (*see* **Note 6**).
3. Wash the membranes in water if scanning or photography have to be performed. The color is stable for several days. **Figure 1a** shows an example of electrophoresis of tissue GAGs performed in zinc acetate/HCl solutions, followed by Alcian blue staining.
4. To dry the membranes, place them in anhydrous methanol for 1 min with gentle agitation, then transfer for exactly 1 min to freshly prepared 18% (v/v) acetic acid in methanol. The membranes are removed, deposited on glass plates, and dried at 80°C for 10 min. After cooling at room temperature, the membranes are unstuck from the glass plate with a spatula or scalpel. They may be stored indefinitely.

### 3.3. Autoradiography

1. Wash the membrane containing the radiolabeled GAGs in ethanolic glycerol. If the labeling was performed with only one isotope, wash the membrane directly in PPO solution (*see* **step 2**). Dry between Whatman filter papers. Install in cassette. Cover with Saran foil and autoradiography film in a dark room. Keep at –80°C for 24 h for 40,000 cpm as $^{35}$S-sulfate deposited. Less isotope needs longer exposition time. Develop the film. **Figure 1b** shows an example of electrophoresis in 0.1 *M* HCl and direct autoradiography.
2. For tritium autoradiography, wash the membrane in PPO solution. Dry and reexpose as above. (**Fig. 1c**).
3. Optionally, after autoradiography, the bands can be excised, dissolved in xylene, and counted by liquid scintillation.

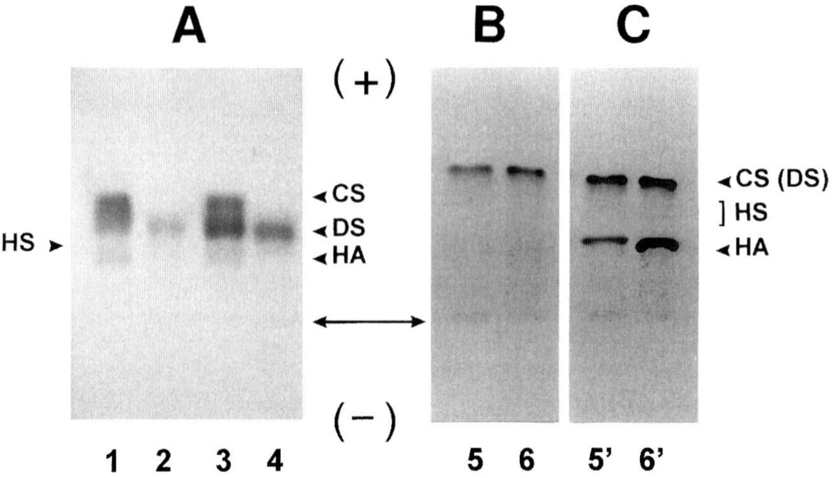


Fig. 1. Examples of electrophoresis in zinc acetate and HCl of tissue GAGs (**A**) and electrophoresis in HCl of $^{35}$S-sulfate and $^{3}$H-glucosamine labeled GAGs extracted from MRC5 fibroblast's culture medium (**B**,**C**). Panel (**A**) shows the separation of 1 µg (as uronic acid) of GAGs from fibrous tissue obtained from patient with type IV Ehlers-Danlos syndrome ***(1,2)*** and control fibrous tissue ***(3,4)*** before ***(1,3)*** and after ***(2,4)*** chondroitinase AC treatment ***(12)***. The GAGs were stained with Alcian blue. Note the higher proportion of dermatan sulfate in control tissue. The panels at right (**B**,**C**) show the electrophoresis of the secreted, radiolabeled GAGs from control cells (5,5') and the cells stimulated with 5 ng/mL each of TGFβ and EGF (6,6'). Ten thousand cpm of $^{35}$S-sulfate were deposited on each lane. The electrophoresis was autoradiographied before (**B**) and after (**C**) impregnation with 2% (m/v) PPO. The migration position of chondroitin sulfate (CS), dermatan sulfate (DS), heparan sulfate (HS), and hyaluronan (HA) standards are indicated on the margins. The bi-directional arrow indicates the application points. Anode and cathode extremities are indicated by (+) and (–), respectively.

## 4. Notes

1. Any horizontal electrophoresis apparatus may be easily adapted to support cellulose acetate strips. The strips are lying in the gel platform and are connected with the electrode chambers (buffer tanks) by Whatman no. 1 filter paper. Do not submerge the membrane.
2. The salts provoke lateral diffusion of GAGs. It can be easily eliminated by dialysis or gel-filtration. Precipitated GAGs can be desalted by successive washing with 90% ethanol and absolute ethanol, drying, and dissolving in water. For practical reasons, maximal volume for electrophoresis should not excess 10 µL.
3. Use zinc acetate or HCl as electrophoresis buffer. Keep the liquid at 4°C before electrophoresis.
4. The membranes do not dry in the electrophoresis tank when the extremities sink in the buffer.
5. The electrophoresis in zinc acetate buffer is performed at 80 V for a membrane of 10 cm length and that performed in 0.1 M HCl at 40 V for the same membrane dimension.
6. If necessary, the membrane can be destained overnight at room temperature in a covered tank, to avoid ethanol evaporation.

## References


1. Morris, J. E., Canoy, D. W., and Rynd, L. S. (1981) Electrophoresis with two buffers in one dimension in the analysis of glycosaminoglycans on cellulose acetate strips. *J. Chromatogr.* **224,** 407–413.
2. Beeley, J. C. (1985) Characterization of charge, in *Glycoprotein and Proteoglycan Techniques* (Burdon, R. H. and van Knippenberg, P. H. eds.), Elsevier, Amsterdam, The Netherlands, pp. 88–94.
3. Kodama, C., Kodama, T., and Yosizawa, Z. (1988) Methods for analysis of urinary glycosaminoglycans. *J. Chromatogr.* **429,** 293–313.
4. Hopwood, J. J. and Harrison, J. R. (1982) High-resolution electrophoresis of urinary glycosaminoglycans: an improved screening test for the mucopolysaccharidoses. *Anal. Biochem.* **119,** 120–127.
5. Hata, R. and Nagai, Y. (1973) A micro-colorimetric determination of acidic glycosaminoglycans by two dimensional electrophoresis on a cellulose acetate strip. *Anal. Biochem.* **52,** 652–656.
6. Cifonelli, J. A. (1976) Nitrous acid depolymerisation of glycosaminoglycans. *Meth. Carbohydr. Chem.* **7,** 139–142.
7. Breen, M., Weinstein, H. G., Blacik, L. J., Borcherding, M. S., and Sittig, R. A. (1976) Microanalysis and characterization of glycosaminoglycans from human tissue via zone electrophoresis. *Meth. Carbohydr. Chem.* **7,** 101–115.
8. Rider, C. C. (1998) Analysis of glycosaminoglycans and proteoglycans, in *Glycoanalysis Protocols* (Hounsell, E. F., ed.), Humana, Totowa, NJ, pp. 131–143.
9. Lammi, M. and Tammi, M. (1988) Densitometric assay of nanogram quantities of proteoglycans precipitated on nitrocellulose membrane with Safranin O. *Anal. Biochem.* **168,** 352–357.
10. Gaffen, J. D., Price, F. M., Bayliss, M. T., and Mason, R. M. (1994) A ruthenium-103 red dot blot assay specific for nanogram quantities of sulfated glycosaminoglycans. *Anal. Biochem.* **218,** 124–130.
11. Wessler, E. (1971) Electrophoresis of acidic glycosaminoglycans in hydrochloric acid: a micro-method for sulfate determination. *Anal. Biochem.* **41,** 67–69.
12. Wegrowski, Y., Bellon, G., Quereux, C., and Maquart, F. X. (1999) Biochemical alterations of uterine leiomyoma extracellular matrix in type IV Ehlers-Danlos syndrome. *Am. J. Obstet. Gynecol.* **180,** 1032–1034.

# 18

# Disaccharide Composition in Glycosaminoglycans/Proteoglycans Analyzed by Capillary Zone Electrophoresis

**Nikos K. Karamanos and Anders Hjerpe**

## 1. Introduction

During the past decade, the use of fully automated equipment for capillary electrophoresis (CE) has made this a routine method for the study of soluble analytes. The separation of such compounds in capillaries (20–200 μm id) and in strong electric fields (around 50 kV/m) seems to be exceptionally efficient for separating both large and small molecules. The most common type of CE is capillary zone electrophoresis (CZE), which is done with the separation buffer free in an otherwise empty capillary. These separations, in principle, share some features not only with gel electrophoresis but also with high-performance liquid chromatography (HPLC), and they provide a unique combination of both analytical techniques. CZE is therefore an alternative to HPLC and it is sometimes advantageous because there are no problems related to laminar flow and other wall effects.

Other advantages of CZE, compared to HPLC and gel electrophoresis, are that it is more friendly to the user and the environment. Only minute amounts of solvents are needed, and the use of aqueous separation buffers in open capillaries eliminates the need for toxic organic solvents and acrylamide.

The migration of analytes toward the detector depends on two main factors: electrophoretic mobility (EM), due to the net charge of the analyte; and electroosomotic flow (EOF) of the free solution, caused mainly by dissociation of silanol groups in the capillary glass wall and migration of resultant $H_3O^+$ toward the anode. Both these effects can be modulated to change the separation. The net charge of the analyte is readily modified by ion pairing or ion suppression. The EOF of the solution depends on the pH and can be completely blocked by the inclusion of detergents in the separation buffer or by coating the inner capillary surface with hydrophobic agents or surfactants.

From: *Methods in Molecular Biology, Vol. 171: Proteoglycan Protocols*
Edited by: R. V. Iozzo © Humana Press Inc., Totowa, NJ

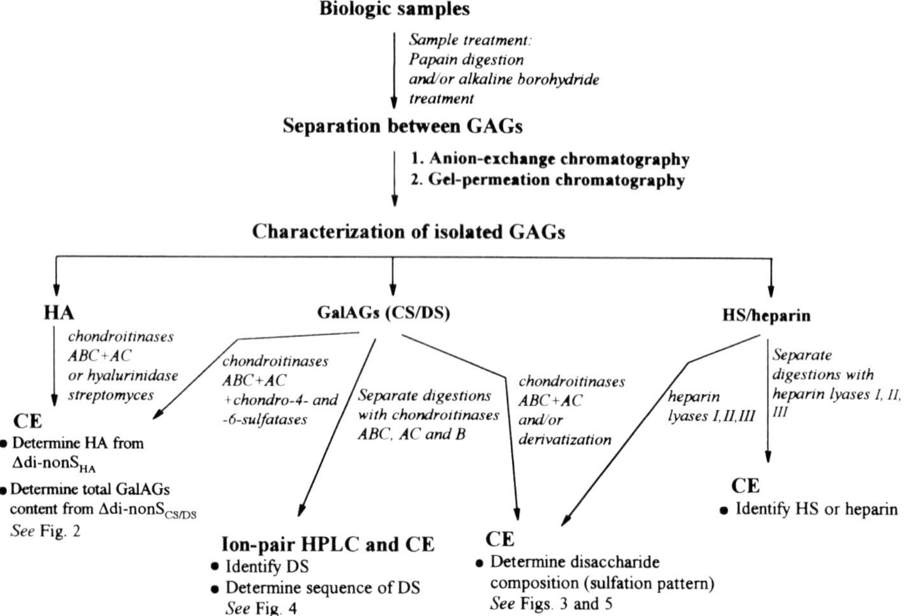


Fig. 1. A possible strategy to identify and analyze GAGs.

Modern detectors yield an electropherogram that is similar to a chromatogram. The ideal flow characteristics and resultant high resolution provide a considerably higher sensitivity than HPLC in terms of analyte amounts, while the short light paths through the detection window of the capillary often necessitate concentrations similar to those of HPLC. Stacking of injected material, however, sometimes creates deviations from this general rule to give exceptional sensitivity *(1)*. Another way to improve the sensitivity is to derivatize the analytes with fluorescent tags, allowing the use of laser-induced fluorescence (LIF) detection *(2,3)*.

The proteoglycans (PGs) with their glycosaminoglycan (GAG) side chains are involved in a wide range of biological processes. Although the GAGs are synthesized to yield only a few main types, the possible variations in structure are enormous *(4)*. Our knowledge of the structural background of specific GAG interactions and their biological importance increases with improved methods for elucidation of the fine structure. In such studies, CE can be powerful as an alternative tool or complement to other analytical techniques. The various methods developed in this context have recently been reviewed *(5,6)*.

In this chapter we present protocols for the analysis of Δ-disaccharides obtained from digestion of GAGs with specific lyases. These separations can be used to obtain information on the total amounts of hyaluronan (HA) and galactosaminoglycans (GalAGs)—i.e., chondroitin sulfate (CS) and dermatan sulfate (DS) *(7)*—and the disaccharide composition of these GAGs, the latter including the type of uronic acid in DS, sulfation patterns in CS/DS *(1)* or heparan sulfate (HS)/heparin, and the extent of N-acetylation in heparin/HS *(8,9)*. A strategy for analyzing all these GAGs is shown in **Fig. 1**.

To our knowledge, there are no CZE-based methods for analyzing keratan sulfate disaccharide composition, although the availability of specific keratanases would make such separations possible.

## 2. Materials

### 2.1. Total Content of HA and GalAGs

1. Digestion buffer: 25 m*M* Tris-HCl, pH 7.5. The buffer is passed through a 0.2-μm membrane filter and kept at –20°C pending use.
2. Standard HA/CS solution: Prepare stock solutions with HA and CS (preferably use CSA of mammal origin), each containing 1.0 mg/mL. Determine the exact GAG content in each solution colorimetrically (for example, by the carbazole reaction). Make a standard solution with 1 mL of each stock solution and 8 mL of the digestion buffer. Divide in portions and store at –20°C until use.
3. HA/GalAGs degradation buffer: Dissolve chondroitinases ABC and AC as well as chondrosulfatases-4 and -6 in the digestion buffer, so as to give 1 unit/mL. An equi-unit mixture of all lyases is then prepared by mixing 10 μL of each enzyme to produce 40 μL of the buffer containing 0.01 unit of each enzyme (*see* **Note 1**).
4. Capillary: Uncoated fused-silica (75-μm id, effective length 50 cm) (*see* **Note 2**).
5. Operating buffer: 15 m*M* sodium orthophosphate buffer, pH 3.0. The buffer is passed through a 0.2-μm membrane filter, divided into portions of 1 mL, and kept at –20°C pending use (*see* **Note 3**).
6. 0.1 *M* NaOH, prepared in 2 × distilled water and passed through a 0.2-μm membrane filter.
7. Centricon 3 membrane (cutoff 3000 daltons) microfuge tubes.

### 2.2. Structural Characterization of GalAGs

#### 2.2.1. Analysis of Δ-Disaccharides Using Ultraviolet Detection

1. Digestion buffer I: 50 m*M* Tris-HCl, pH 7.5. The buffer is passed through a 0.2-μm membrane filter and kept at –20°C pending use.
2. Standard Δ-disaccharides: Prepare a stock solution (1.0 mg/mL) of nonsulfated (Δdi-nonS$_{CS/DS}$), monosulfated (Δdi-mono2S, Δdi-mono4S, and Δdi-mono6S), disulfated [Δdi-di(4,6)S, Δdi-di(2,4)S, Δdi-di(2,6)S, previously referred to as E, B, and D types] and trisulfated [Δdi-tri(2,4,6)S, also referred to as Δdi-triS] chondro-/dermato-derived Δ-disaccharides by dissolving the various disaccharides in digestion buffer I. Make serial dilutions (1/10,000, 1/5,000, and 1/1,000) in digestion buffer I so as to prepare standard Δ-disaccharide solutions of 0.1, 0.2, and 1.0 μg/mL.
3. GalAGs degradation buffer: Dissolve chondroitinases ABC and AC in digestion buffer I to give 1 unit/mL. An equi-unit mixture of both lyases is then prepared by mixing 10 μL of each enzyme solution with 20 μL of digestion buffer I to produce 40 μL of the GalAG degradation buffer containing 0.01 unit of each chondroitinase (*see* **Note 1**).
4. Operating buffer: 15 m*M* sodium orthophosphate buffer, pH 3.0. The buffer is passed through a 0.2-μm membrane filter, divided in portions of 1 mL, and kept at –20°C, pending use (*see* **Notes 3 and 4**).

#### 2.2.2. Analysis of Δ-Disaccharides Using Laser-Induced Fluorescence

1. Derivatizing agent: 0.1 *M* 2-aminoacridone (AMAC) is dissolved in glacial acetic acid/DMSO (3/17 v/v).
2. Reductive agent: 1 *M* $NaCNBH_3$ dissolved in 2 × distilled water.

### 2.2.3. Structural Characterization of DS

1. DS degradation buffer I: Chondroitinase AC is prepared by dissolving the enzyme in digestion buffer I to give 1 unit/mL. Then 10 µL of the enzyme solution is mixed with 30 µL of digestion buffer I to produce 40 µL of DS degradation buffer I containing 0.01 unit of the enzyme.
2. Digestion buffer II: 50 m*M* Tris-HCl, pH 8.0. The buffer is passed through a 0.2-µm membrane filter and kept at –20°C, pending use.
3. DS degradation buffer II: Chondroitinase B is prepared by dissolving the enzyme in digestion buffer II to give 1 unit/mL. Then 10 µL of the enzyme solution is mixed with 30 µL of digestion buffer II to produce 40 µL of DS degradation buffer II, containing 0.01 unit of the enzyme (*see* **Note 1**).

## 2.3. Disaccharide Composition of heparin/HS

1. Digestion buffer III: 20 m*M* sodium acetate/acetic acid buffer, pH 7.0. The buffer is passed through a 0.2-µm membrane filter and kept at –20°C pending use.
2. Standard Δ-disaccharides: Prepare a stock solution (1.0 mg/mL) of nonsulfated (acetylated $\Delta$di-non$S_{HS}$, also referred to as a$\Delta$di-non$S_{HS}$ or nonacetylated $\Delta$di-non$S_{HS}$), monosulfated (a$\Delta$di-mono6$S_{HS}$, $\Delta$di-mono6$S_{HS}$, a$\Delta$di-mono2$S_{HS}$, $\Delta$di-mono2$S_{HS}$, and $\Delta$di-mono$NS_{HS}$), disulfated [a$\Delta$di-di(2,6)$S_{HS}$, $\Delta$di-di(2,6)$S_{HS}$, $\Delta$di-di(2,N)$S_{HS}$, and $\Delta$di-di(6,N)$S_{HS}$], as well as trisulfated [$\Delta$di-tri(2,6,N)$S_{HS}$, also referred to as $\Delta$di-tri$S_{HS}$] heparin-/HS-derived Δ-disaccharides, by dissolving the various disaccharides in digestion buffer III. Make serial dilutions (1/10,000, 1/5,000 and 1/1,000) in digestion buffer III to prepare standard Δ-disaccharide solutions of 0.1, 0.2, and 1.0 µg/mL.
3. Heparin/HS degradation buffer: Dissolve heparin lyases I (EC 4.2.2.7), II (no EC number), and III (EC 4.2.2.8) in digestion buffer III to give 1 unit/mL. An equi-unit mixture of all three lyases is then prepared by mixing 10 µL of each enzyme solution with 10 µL of digestion buffer III to produce 40 µL of the heparin/HS degradation buffer containing 0.01 unit of each heparin lyase (*see* **Notes 1 and 5**).
4. Operating buffer: 15 m*M* sodium orthophosphate buffer, pH 3.5. The buffer is passed through a 0.2-µm membrane filter, divided in portions of 1 mL, and kept at –20°C pending use (*see* **Notes 3 and 6**).

# 3. Methods

## 3.1. Total Content of HA and GalAGs

Biological samples or isolated GAGs can be easily analyzed for their content of HA and total GalAGs, that is, CS and DS, by complete depolymerization of these GAGs with a combination of specific lyases and chondrosulfatases. Combining chondroitinases ABC and AC, both HA and GalAGs are converted to Δ-disaccharides. Those derived from HA are nonsulfated and those from GalAGs contain various numbers of sulfates. Most sulfate groups in mammals are esterified at C-4 and C-6 of the GalNAc and are eliminated by using digestion with both chondrosulfatases-4 and -6. Therefore, all HA is recovered in the $\Delta$di-non$S_{HA}$ peak and both CS and DS in the $\Delta$di-non$S_{CS/DS}$ peak (*see* **Fig. 1**). These two peaks are completely resolved and the contents of HA and total GalAGs are estimated using the following protocol. (*See* **Note 7.**)

1. Dissolve the samples containing approximately 0.1–10 µg of HA and/or GalAGs in 10 µL of digestion buffer. Place 10 µL of the HA/CS standard solution in separate vials.

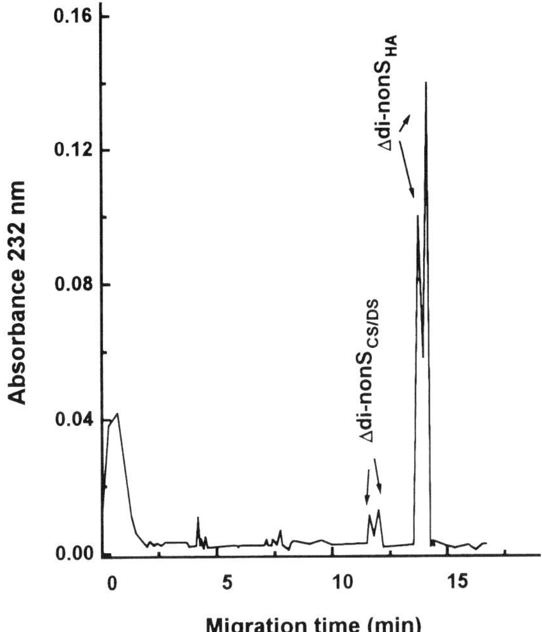


Fig. 2. Typical CZE profile obtained by combined digestion of HA and CS and/or DS with chondroitinases ABC and AC and chondrosulfatases-4 and -6. (Reprinted with permission from *ref. 7.*)

2. Add 40 μL of HA/GalAGs degradation buffer to the samples and standards. Digest for 90 min at 37°C (*see* **Notes 8 and 9**).
3. To remove nondegraded HS and protein/proteoglycans, centrifuge the digestion mixture on a Centricon 3 membrane at 11,000*g* for 5 min (*see* **Note 10**).
4. Place a 5 - 10-μL portion of the filtrate in the electrophoresis vials and keep at 4°C pending use.
5. Start the CE instrument (*see* **Note 11**) and set the detector wavelength at 232 nm.
6. Prepare the software of CE instrument as follows:
   a. Rinse the capillary with 0.1 *M* NaOH for 1 min.
   b. Condition the capillary with operating buffer for 4 min (*see* **Note 12**).
   c. Inject the sample at the cathode (*reversed polarity*), using the pressure mode (500 mbar × s) (*see* **Note 13**).
   d. Run the samples for 15 min at 40 kV/m and 25°C (*see* **Notes 2 and 14**).
   e. Before each run, rinse and condition the capillary as in steps **6a, b**.
7. After every four injections of the samples, inject the standard solution to check the performance of the electrophoresis.
8. Calculate the amount of HA and total GalAGs, using the peak areas estimated and the standard curve (*see* **Fig. 2**) (*see* **Notes 8, 15 and 16**).

## 3.2. Disaccharide Composition of GalAGs

### 3.2.1. Analysis of Δ-Disaccharides Using Ultraviolet Detection

Following digestion of samples containing CS and/or DS with both chondroitinases ABC and AC, all constituent nonsulfated and variously sulfated disaccharides of these

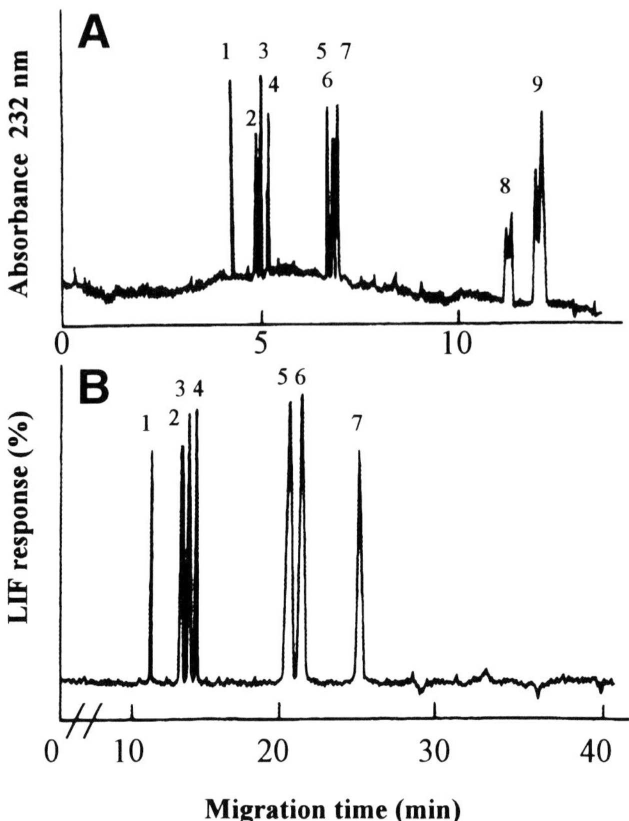


Fig. 3. Electropherograms showing the resolution of GalAG-derived Δ-disaccharides following degradation of CS/DS with chondroitinases ABC and AC. (**A**) Direct UV detection at 232 nm and (**B**) derivatization of the Δ-disaccharides obtained with fluorescent tag 2-aminoacridone and detection using LIF ($\lambda_{exc}$ 488 nm). In both cases, analysis is performed using 15 mM phosphate buffer, pH 3.0. Peaks: 1 = Δdi-tri(2,4,6)S; 2 = Δdi-di(2,6)S; 3 = Δdi-di(2,4)S; 4 = Δdi-di(4,6)S; 5 = Δdi-mono2S; 6 = Δdi-mono4S; 7 = Δdi-mono6S; 8 = Δdi-non$S_{CS/DS}$; 9 = Δdi-non$S_{HA}$. Reprinted from **refs.** ***1*** and *2*, copyright 1995, 1999 Elsevier Science, with permission.

GAGs are recovered as Δ-disaccharides. All CS/DS-derived Δ-disaccharides are completely resolved and determined using direct UV detection with the following protocol.

1. Dissolve the samples containing approximately 0.1–10 µg of CS and/or DS in 10 µL of digestion buffer I.
2. Add 40 µL of GalAGs degradation buffer to the samples and digest for 90 min at 37°C (*see* **Note 7**).
3. Centrifuge the digestion mixture in a microfuge tube at 11,000*g* for 5 min (*see* **Note 8**).
4. Place aliquots of the supernatant in the electrophoresis vials and keep at 4°C pending use.
5. Start the CE instrument, adjust the detector wavelength at 232 nm, prepare to use the method and analyze the samples, as described under **Subheading 3.1, steps 6–8**. (*see* **Fig. 3A**).

### *3.2.2. Analysis of Δ-Disaccharides Using Laser-Induced Fluorescence (LIF)*

To obtain higher sensitivities, LIF detection can be used to determine CS- and/or DS-derived Δ-disaccharides. Therefore, following the same digestion scheme as in **Subheading 3.2.1.**, the variously sulfated CS- and/or DS-derived Δ-disaccharides can be easily converted to fluorescent derivatives by reductive amination with AMAC. CS/DS-derived sulfated Δ-disaccharides are completely separated by reversed-polarity CZE and determined with Ar-laser source LIF detection, using the following protocol.

1. Evaporate standard Δ-disaccharides and digestion mixtures in a microfuge tube at low temperature.
2. Add 5 μL of each of the derivatizing and reductive agents to the dry residues.
3. Incubate the mixture at 45°C for 2 h (*see* **Note 7**).
4. Evaporate the mixtures to dryness, reconstitute in 50% DMSO, transfer to the electrophoresis vials and keep at 4°C pending use.
5. Start the CE instrument (*see* **Note 11**) and adjust the LIF detector at an excitation wavelength of 488 nm.
6. Prepare the software of the CE instrument, as follows:

   a. Rinse the capillary with 0.1 *M* NaOH for 1 min and 2 × distilled water for 0.5 min.
   b. Condition the capillary with operating buffer for 4 min.
   c. Introduce the samples at the cathode (*reversed polarity*), using the pressure mode (500 mbar × s) (*see* **Note 13**).

7. Run the samples for 30 min at 40 kV/m and 25°C (*see* **Notes 2 and 14**). Estimate the amount, as described under **Subheading 3.1., steps 7 and 8.** (*see* **Fig. 3B**; *see* **Note 16**).

### *3.2.3. Structural Characterization of DS*

DS is a copolymer GAG constructed by both GlcA- and IdoA-containing repeating disaccharide units. From a biological point of view, it is often important to know whether DS is present in a GAG preparation. Differential digestion with chondroitinase ABC and chondroitinase AC or B may provide useful information on this aspect. When the total amount of Δ-disaccharides recovered following digestion by chondroitinase AC is less than that obtained by ABC, this is evidence of the presence of DS. Furthermore, comparing the amount of Δ-disaccharides recovered, using separate digestions with chondroitinase AC and B, we may well determine whether DS is rich in IdoA or GlcA and, at the same time, the sulfation pattern of IdoA- and GlcA-containing disaccharides (*see* **Fig. 4**). The protocols are presented below.

1. Dissolve separately the samples containing DS in 10 μL of digestion buffers I and II (*see* **Note 18**).
2. To each one of them, add 40 μL of DS degradation buffers I and II, respectively, and incubate for 90 min at 37°C.
3. Centrifuge the digestion mixture in a microfuge tube at 11,000*g* for 5 min.
4. Place aliquots of the supernatant in the electrophoresis vials and keep at 4°C pending use.
5. Start the CE instrument (*see* **Note 11**), adjust the detector wavelength to 232 nm, prepare the method, and analyze the samples as described under **Subheading 3.1, steps 6–8**.

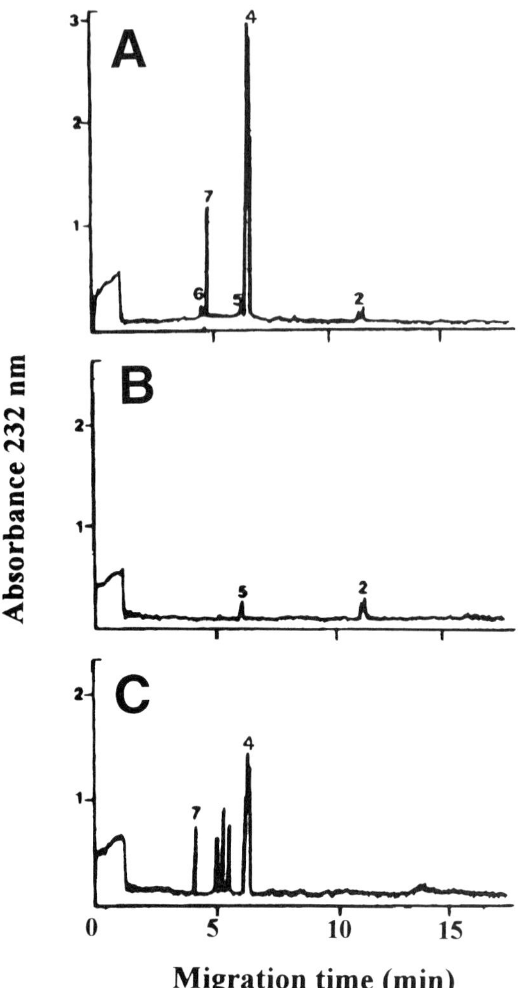


Fig. 4. Analysis of DS isolated from porcine skin following separate digestion with chondroitinase ABC (**A**), AC (**B**) and B (**C**). The electropherograms obtained show that DS is rich in IdoA-containing disaccharides (very low susceptibility to chondroitinase AC in contrast to that obtained with chondroitinase B) and that most of the GlcA-containing disaccharides of the DS chain occur in short sequences (extra peaks representing Δ-oligosaccharides are obtained when DS is treated with chondroitinase B). For identity of peaks (*see* Figure 3). Reprinted from **ref.** ***1***, copyright 1995, Elsevier Science, with permission.

### 3.3. Disaccharide Composition of heparin/HS

Heparin and HS can be almost quantitatively (>90%) degraded to Δ-disaccharides by using all three heparin lyases (I, II, and III) in combination (*See* **Note 5**). All variously sulfated Δ-disaccharides and those containing N-acetylated and unsubstituted GlcN at the amino group are completely resolved, using reversed-polarity CZE, according to the following protocol.

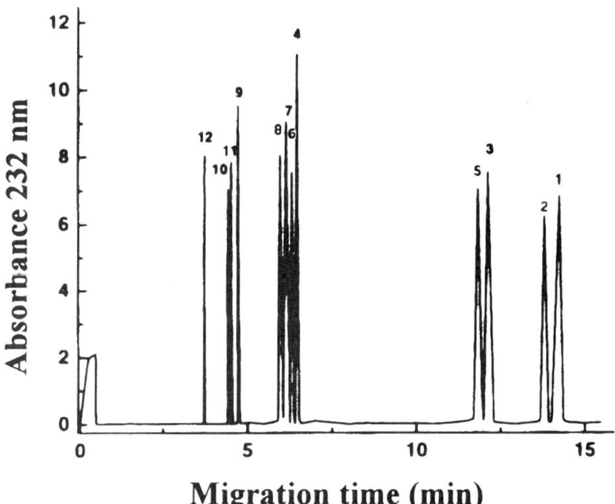


Fig. 5. CZE profile showing the resolution of all 12 heparin-/HS-derived Δ-disaccharides. Analysis is performed with 15 mM phosphate buffer, pH 3.50. Peaks: 1 = aΔdi-nonS$_{HS}$; 2 = Δdi-nonS$_{HS}$; 3 = aΔdi-mono6S$_{HS}$; 4 = Δdi-mono6S$_{HS}$; 5 = aΔdi-mono2S$_{HS}$; 6 = Δdi-mono2S$_{HS}$; 7 = Δdi-monoNS$_{HS}$; 8 = aΔdi-di(2,6)S$_{HS}$; 9 = Δdi-di(2,6)S$_{HS}$; 10 = Δdi-di(2,N)S$_{HS}$; 11 = Δdi-di(6,N)S$_{HS}$; 12 = Δdi-tri(2,6,N)S$_{HS}$. Reprinted from **ref.** *8*, with permission of Wiley - VCH, STM Copyright of Licenses.

1. Dissolve the samples containing approximately 0.1–10 µg of heparin and/or HS in 10 µL of digestion buffer III.
2. Add 40 µL of heparin/HS degradation buffer to the samples and add 1 µmol calcium acetate from a 2 *M* stock solution. Digest at 37°C overnight (*see* **Note 8**).
3. Centrifuge the digestion mixture in a microfuge tube at 11,000*g* for 5 min (*see* **Note 9**).
4. Place aliquots of the supernatant in the electrophoresis vials and keep at 4°C pending use.
5. Start the CE instrument (*see* **Note 11**), adjust the detector wavelength to 232 nm, prepare the method, and analyze the samples as described under **Subheading 3.1., steps 6–8** (*see* **Fig. 5**).

## 4. Notes

1. Digestion buffers containing the various lyases and chondrosulfatases should be divided into small portions of 50–200 µL and kept in sealed plastic tubes at –20°C. Enzyme-containing solutions should not be frozen and used again after thawing.
2. In each protocol, a certain capillary diameter and length are given. When capillaries with different characteristics are available, the resolution may be tested using external standards. Voltage can also be modified (±5–10 kV) to affect the rate of migration and resolution. However, if no complete resolution is obtained, use the proposed capillary and the recommended conditions. Large amounts of injected material may cause peak broadening. Optimal peak shape can be obtained with £100 ng disaccharide injected.
3. Buffers and samples used in capillary electrophoresis should be handled carefully so as to protect the capillary from particles and air bubbles. Therefore, all solutions should be passed through a 0.2-µm membrane filter and degassed in an ultrasonic bath for 5–10 min. Samples dissolved in low volumes are preferably centrifuged at 11,000*g* for 5 min.
4. A pH of 3.0 in the operating buffer is important for the resolution of the variously sulfated Δ-disaccharides and should be carefully adjusted. The separation is partly based on ion

suppression ($pK_a$ of the carboxyl groups is just above 3 and slightly different in the various disaccharides). Slightly modified pH values (±0.1 pH unit) therefore affect the net charge and the separation of Δdi-mono4S and Δdi-mono6S and decreased pH may cause close to infinite retardation of nonsulfated Δ-disaccharides.

5. Due to the different specificities of the three heparin lyases (type of uronic acid and position of GlcN sulfate), it is possible to obtain separate information on HS- and heparin like sequences in the analyte ***(9)***. The amounts of enzyme recommended represents an excess to warrant maximal yield of disaccharides, but can be expensive when analyzing large series of samples. Less than 10% of these amounts is sufficient, but one should then be sure that the amounts of GAG is 1 mg or less.
6. The pH of 3.50 in the operating buffer and reversed polarity are critical for the resolution of the heparin- and HS-derived Δ-disaccharides carrying N-acetylated, N-sulfated, or unsubstituted GlcN. The pH should be checked carefully, since slightly modified pH values during electrophoresis (±0.1 pH) affect the separation (*see* **Note 4**). Nonsulfated, nonacetylated Δ-disaccharide migrates very slowly and can sometimes be difficult to recover at all. In this case, it can be recommended to run one separation in reversed polarity as above, followed by a second separation with normal polarity, moving this Δ-disaccharide by EOF.
7. Biological samples or tissue extracts taken for analysis of HA and total GalAGs content can be concentrated by precipitation with 4 volumes of 90% (v/v) ethanol, also containing 2.5% (w/v) sodium acetate. Small amounts of dextran can be used as carrier. Digestion buffer containing the lyases and chondrosulfatases is then added directly to the precipitate. Following heating to stop the enzymic effects, the digests should be centrifuged at 11,000*g* for 5 min.
8. Incubation times and the units of the enzymes indicated should not be exceeded, since some enzyme preparations may contain contaminating lyase activities. Enzymic digestions should be terminated by boiling the incubation mixtures in a water bath for 1 min.
9. Evaporation during incubation necessitates the use of sealed, capped tubes. Drops on the tube walls are recovered at the same time as particles are removed by centrifugation at 11,000*g* for 5 min.
10. Removal of chondroitinase-resistant macromolecules is obtained by ultrafiltration in Centricon 3 membrane tubes. However, overly long centrifugation times may cause the membrane to dry and break and therefore should be avoided. In the absence of experience with the handling of these membrane tubes, use them only once.
11. When the capillary is mounted into the capillary cassette, care should be taken to keep the detection window clean. Preferably, use gloves when handling the capillary!
12. Due to some electrolysis during the electrophoretic separation, the operating buffer characteristics (ionic strength and pH) may be affected. Therefore, replace the operating buffers frequently (at least every 5 runs) and run frequent standards (once for each new buffer vial) to ensure uniform performance.
13. When using pressure injection, these protocols give suitable parameters for pressure (mbar) × time (s). When working with low concentrations, it is always possible to increase the injection times without significantly affecting the resolution. However, use the same injection conditions for samples and standards.
14. Electrical current and temperature should be constant during electrophoresis, since changes may give extra peaks or baseline drift. The current should therefore be monitored throughout the run. Temperature also influences the viscosity of the solution and may therefore affect resolution and quantification. Performance can be considerably improved, by creating a "concentration zone" during injection. If the analyte is dissolved in water with as little electrolyte as possible, the conductivity in the zone of injected material will be low, hence

increasing the electric field through the injected solvent. This causes the analyte to move rapidly through this zone and it becomes concentrated when entering the running buffer (similar to what happens in TLC when using a concentration zone). To obtain such conditions one can use volatile buffers for digestion (for example, ammonium formate instead of those conventionally recommended), followed by evaporation of the digested material and redissolving it in pure water. The EOF will push the injected low-electrolyte solvent backward out of the capillary, not interfering at the detection window.

15. The amounts of HA and GalAGs are preferably given in terms of uronic acid per milliliter, since the presence of crystal water and the poorly defined weights of the counterions make it difficult to determine the weight of a GAG preparation correctly. Therefore, GAG standards with a well-defined amount of uronic acid (carbazole reaction) can be used. This also monitors the efficiency of the enzymic digestion. Alternatively, the quantity of HA and total GalAGs in samples can be determined, based on standard curves and using commercially available Δ-disaccharides, as under **Subheading 3.2**.
16. Disaccharides sulfated in uronic acid migrate earlier than the main nonsulfated peak (mainly in the region of monosulfated disaccharides). However, in most CS preparations from mammalian tissues this peak will be minute and the calculation more readily made when relating to a mammalian CS standard. When analyzing CS/DS preparations with significant amounts of uronic acid sulfation, also include a Δdi-di(2,4)S standard (see **Subheading 2.2.1**) when digesting the standard mixture.
16. LIF detection of AMAC-conjugated Δ-disaccharides increases the sensitivity at least 100 times more than that obtained by UV detection at 232 nm of underivatized Δ-disaccharides. However, the AMAC derivatives of the nonsulfated Δ-disaccharides do not appear on the electropherogram, due to the effect of the fluorochrome. The advantage of this derivatization is its sensitivity, which permits the study of the sulfation pattern with less sample present. The AMAC labeling also allows the non reducing end of the GAG chain to be studied. This fragment does not carry the UV-absorbing $\Delta^{4,5}$-structure. Recent studies indicate that this part of the chain may be of particular biological importance ***(10,11)***.
18. When DS is digested with only one of the chondroitinases AC or B, these digests will also contain Δ-oligosaccharide fragments. They can be identified in the electropherogram, thereby providing information also regarding longer DS sequences.

## References


1. Karamanos, N. K., Axelsson, S., Vanky, P., Tzanakakis, G. N., and Hjerpe, A. (1995) Determination of hyaluronan- and galactosaminoglycan-derived disaccharides by high-performance capillary electrophoresis at the attomole level. Applications to analyses of tissue and cell culture proteoglycans. *J. Chromatogr. A* **696(2)**, 295–305.
2. Lamari, F., Theocharis, A., Hjerpe, A., and Karamanos, N. K. (1999) Ultrasensitive capillary electrophoresis of sulfated disaccharides in chondroitin/dermatan sulfates by laser-induced fluorescence after derivatization with 2-aminoacridone. *J. Chromatogr. B.* **730,** 129–133.
3. Lamari, F. and Karamanos, N. K. (1999) High-performance capillary electrophoresis as a powerful analytical tool of glycoconjugates. *J. Liq. Chromatogr. Rel. Technol.* **22,** 1295–1317.
4. Karamanos, N. K. (1999) Proteoglycans: biological roles and strategies for isolation and determination of their glycan constituents, in *Proteome and Protein Analysis* (Kamp, M., Kyriakides, D., and Choli-Papadopoulou, T. eds.), Springer-Verlag, Heidelberg, Germany, pp. 341–363.

5. Karamanos, N. K. and Hjerpe, A. (1998) A survey of methodological challenges for glycosaminoglycan/proteoglycan analysis and structural characterization by capillary electrophoresis. *Electrophoresis* **19,** 2561–2571.
6. Karamanos, N. K. and Hjerpe, A. (1999) Strategies for analysis and structure characterization of glycan/proteoglycans by capillary electrophoresis. Their diagnostic and biopharmaceutical importance. *Biomed. Chromatogr.* **13,** 507–512.
7. Karamanos, N. K., and Hjerpe, A. (1997) High-performance capillary electrophoretic analysis of hyaluronan in effusions from human malignant mesothelioma. *J. Chromotogr. B.* **696,** 277–281.
8. Karamanos, N. K., Vanky, P., Tzanakakis, G. N., and Hjerpe, A. (1996) High-performance capillary electrophoresis method to characterize heparin and heparan sulfate disaccharides. *Electrophoresis* **17,** 391–395.
9. Karamanos, N. K., Vanky, P., Tzanakakis, G. N., Tsegenidis, T., and Hjerpe, A. (1997) High-performance liquid chromatography for determining disaccharide composition in heparin and heparan sulfate. *J. Chromatogr. A* **765,** 169–179.
10. Plaas, A. H., West, L. A., Wong-Palms, S., and Nelson, F. R. (1998) Glycosaminoglycan sulfation in human osteoarthritis. Disease-related alterations at the non-reducing termini of chondroitin and dermatan sulfate. *J. Biol. Chem.* **273,** 12,642–12,649.
11. Plaas, A. H., Wong-Palms, S., Roughley, P. J., Midura, R. J., and Hascal, V. C. (1997) Chemical and immunological assay of the nonreducing terminal residues of chondroitin sulfate from human aggrecan. *J. Biol. Chem.* **272** (33)**,** 20,603–20,610.

# 19

## Intact and Oligomeric Glycosaminoglycans Analyzed with Capillary Electrophoresis

**Nikos K. Karamanos and Anders Hjerpe**

### 1. Introduction

Various types of capillary electrophoresis (CE) can be used to study native and oligomeric glycan structures. For a recent review, *see* **ref. 1.** These separations include not only capillary zone electrophoresis (CZE), discussed in Chapter 18, but also capillary gel electrophoresis (CGE). This can be performed with the capillary filled with a cross-linked chemical gel as in common polyacrylamide electrophoresis. Alternatively, a neutral polymer (pullulan, PEG, dextran, etc.) can be added to the operating buffer, creating a matrix that retards the analyte in a similar way, depending on its molecular size. Although it is not described in this chapter, micellar electrokinetic capillary chromatography (MECC) may also be valuable for the study of glycosaminoglycan (GAG) fragments *(2)*. In MECC, the separation buffer contains surfactant in a concentration sufficient for the formation of micelles that migrate more slowly than the electroosmotic (EOF) or in the opposite direction. This creates a situation similar to that in reversed-phase chromatography, with the micelles as the stationary phase. The separation of analytes depends on differences in the partition between micelles and buffer.

Several biologically important interactions of proteoglycans (PGs) require GAG structures longer than a disaccharide, the size of this sequence in several cases being around 5 monosaccharides or more. A limitation in establishing such functional oligomeric sequences and their fine structures has been a lack of techniques that sufficiently separates various larger GAG fragments. The high-resolution efficiency of the CE-based techniques permits such separations *(3)*, and this technique can, after further methodological development, become an important tool in such a context.

In the present chapter, we outline protocols for the separation of different GAGs, depending on charge density (hyaluronan [HA], chondroitin sulfate [CS], dermatan sulfate [DS], heparan sulfate [HS], and heparin) *(4)* and visualization of molecular

From: *Methods in Molecular Biology, Vol. 171: Proteoglycan Protocols*
Edited by: R. V. Iozzo © Humana Press Inc., Totowa, NJ

size and polydispersity (HA) *(5,6)*. CE can also be used to demonstrate and identify specific oligosaccharide sequences *(3)*. This is of value for the structural characterization of GAGs, and a procedure for such separations is provided in the Chapter 18, **Subheading 3.2.**

## 2. Material

### 2.1. CZE of Different GAGs Related to Charge Density

1. Standard GAGs: Prepare a stock solution (1.0 mg/mL) of HA, CSA (CS sulfated mainly at C-4), or CSC (CS sulfated mainly at C-6), DS, HS, heparin and/or Fragmin® (a fragment of heparin with a molecular size of 5000 daltons) by dissolving the substances in the operating buffer. Make serial dilutions (1/1,000, 1/300, and 1/100) in the operating buffer so as to prepare standard GAG solutions of 1.0, 3.3, and 10.0 µg/mL.
2. Capillary: Uncoated fused-silica (75 µm id, 50 cm effective length).
3. Operating buffer: 20 m*M* orthophosphate buffer, pH 3.0. The buffer is passed through a 0.2-µm membrane filter, divided in portions of 1 mL, and kept at –20°C.
4. 0.1 *M* NaOH, prepared in 2 × distilled water and passed through a 0.2-µm membrane filter.

### 2.2. CGE of HA Using Pullulan Matrix

1. Operating buffer: 50 m*M* phosphate buffer, pH 4.0, is passed through a 0.2-µm membrane filter, whereafter 0.10% (w/v) pullulan with a molecular weight of 1,600,000 is added (*see* **Note 1**).
2. Capillary: the separation is performed in a 75-µm-id uncoated fused-silica capillary with an effective length of 50 cm.

### 2.3. CGE of HA Oligomers Using PEG Matrix

1. Operating buffer: 0.1 *M* Tris-HCI/0.25 *M* borate, pH 8.5, containing 10% (w/v) polyethylene glycol (PEG) with average molecular weight of 70,000 (PEG70000).
2. The separation is performed in a 100-µm-id fused-silica capillary coated with (50% phenyl)methyl-polysiloxane, effective length 20 cm.

## 3. Methods

### 3.1. CZE of Different GAGs Related to Charge Density

GAG preparations can be analyzed for the homogeneity of their charge density by simple CZE using detection at low wavelength. Following the protocol described below, HA, CS/DS, HS, and HA are separated at low pH using reversed polarity. CSA, CSB, and DS migrate almost identically since they have similar charge densities. Heparin and Fragmin migrate first due to their high content of sulfates, whereas HA migrates last since it is a nonsulfated GAG (*see* **Fig. 1**).

1. Dissolve the samples containing approximately 0.1–10 µg of GAGs in 100 µL of operating buffer.
2. Start the CE instrument and adjust the detector wavelength at 190 nm (*see* **Note 2**).
3. Prepare the software of the CE instrument as follows.
   a. Rinse the capillary with 0.1 *M* NaOH for 1 min.
   b. Condition the capillary with operating buffer for 4 min.
   c. Inject the sample at the cathode *(reversed polarity)*, using the pressure mode (500–700 mbar × s).

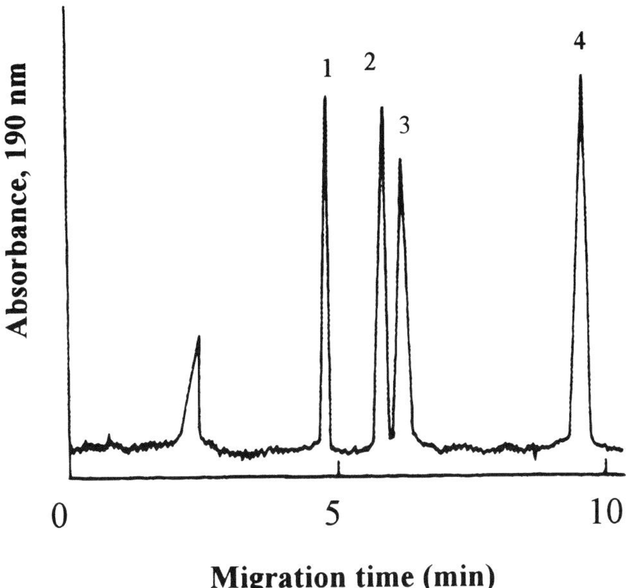


Fig. 1. CZE profile showing the resolution of variously sulfated intact GAGs and HA. CE was performed on an uncoated fused-silica capillary at reversed polarity using 20 m*M* phosphate, pH 3.0, as operating buffer and detection of separated GAGs at 190 nm. Peaks: 1 = fragmin, an oversulfated fragment of heparin, bearing on the average 2.5–3 sulfate groups per disaccharide residue; 2 = CSA, CSC, or DS bearing on average 0.8–1 sulfate group per disaccharide unit; 3 = HS containing on average 0.6–0.7 sulfates per unit; 4 = HA containing no sulfate groups. Reprinted from **ref.** ***4***.

   d. Run the samples for 12 min at 34 kV/m.
   e. Before each run, rinse and condition the capillary, as in **steps a** and **b**.

4. Change the operating buffer and run a standard mixture after every four injections of the samples to ensure the performance of the separation (*see* **Notes 3** and **4**).

### 3.2. CGE of HA Using Pullulan Matrix

A simple open capillary separation allows determination of the amount and estimation of the molecular mass and polydispersity of high-molecular-size HA preparations *(5)*. The addition of the polysaccharide pullulan to the elution buffer creates a gel matrix in which larger molecules become retarded (*see* **Fig. 2**). With a separation buffer at a pH of 4.0, most of the glucuronic carboxyl groups are dissociated while the EOF is low, and the electrophoretic mobility (EM) will make the HA move toward the anode. The molecular size and working concentration of the matrix-forming pullulan are important, and the protocol below describes one setup that permits analysis in the 50-to 2000-kDa range. The detection is achieved with low ultraviolet and the specificity can be improved by repeating the analysis after hyaluronidase digestion.

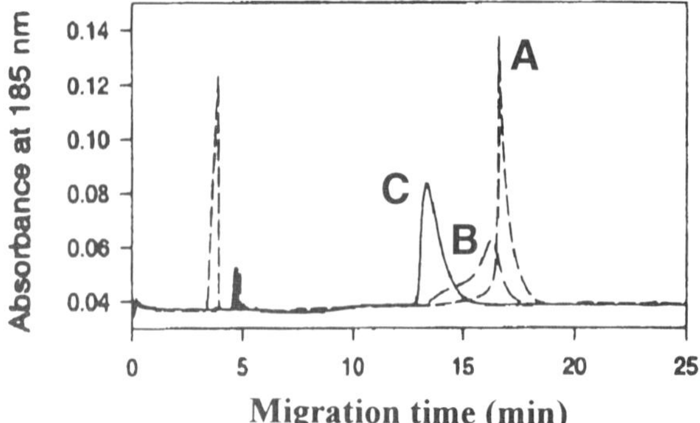


Fig. 2. Separation of variously sized HA in the presence of 0.10% pullulan (molecular mass 1600 kDa) in the operating buffer as a matrix-forming agent. **(a)** HA-H (molecular mass 1500–2100 kDa); **(b)** HA-M (800–1200 kDa) and **(c)** HA-L (molecular mass 40–60 kDa). Reprinted from *Journal of Chromatography,* volume 768, Hayase, S., et al., High perfomrance capillary electrophoresis of hyaluronic acid: determination of its amount and molecular mass, 1997, pp. 290–305, with permission from Elseveir Science.

1. Dissolve the sample in water at an approximate concentration of 0.25 mg/mL and add 0.025 volume of 0.1% w/v of naphthalene trisulfonic acid trisodium salt (NTS) as internal standard (*see* **Note 5**). The mixture is kept at 4°C pending use.
2. Start the CE instrument. Adjust the detector to monitor absorbance at 190 nm.
3. Prepare the software of the CE instrument as follows:
   a. Rinse the capillary with 0.1 *M* NaOH for 5 min.
   b. Rinse the capillary with 2 × distilled water for 3 min.
   c. Rinse the capillary with the operating buffer for 5 min.
   d. Inject the sample by pressure (300 mbar × s).
   e. Run the separation for 25 min at 34 kV/m (20 kV for a total capillary length of 58 cm).
4. The total amount of HA is calculated from the peak area, comparing with standard HA samples also containing the NTS internal standard. All macromolecular HA should disappear after hyaluronidase digestion (*see* **Note 6**). The molecular size and polydispersity are estimated from the shape of the HA peak, although accurate calculation of the *Mw* and *Mn* values is difficult (*see* **Note 7**).

### 3.3. CGE of HA Oligomers Using PEG Matrix

When studying shorter fragments of HA, the CGE procedure reported by Kakehi et al. *(6)* and presented here gives an exceptional separation, permitting the resolution of fragments from 2 to 50 disaccharides (*see* **Fig. 3**). Running the separation under the same conditions in a longer capillary (50-cm effective length) and for a longer period allows the identification of individual fragments close to 100 disaccharides. Such size is still far from that of many intact HA preparations, which normally consist of several thousand disaccharides per molecule. Information on the molecular size distribution of longer fragments (> 100 disaccharides, i.e., molecular mass > 40 kDa) can be obtained, using the protocol described under **Subheading 3.2.** The capillary coating abolishes the EOF and migration is therefore caused only by the EM (*see* **Note 8**). The

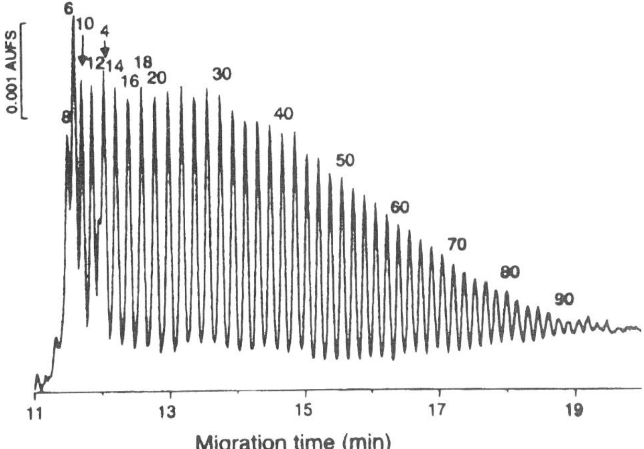


Fig. 3. Separation of HA oligomers in the presence of PEG70000 as a neutral polymer. The fused capillary is coated with 50% phenyl(methyl)polysiloxane (effective length 20 cm, 100 μm id). Operating buffer is 0.1 *M* Tris-0.25 M borate, pH 8.5, containing 10% PEG70000. The numbers indicate the degree of polymerization (number of monosaccharide units). Reprinted from **ref. *6***, © American Chemical Society.

operating buffer also contains high-molecular-weight polyethylene glycol which forms a gel through which the HA fragments migrate.

1. Dissolve the samples (1.0 mg/mL) in an aqueous 15 m*M* citric buffer, pH 5.3, or in the operating buffer.
2. Start the CE instrument. Adjust the detector to monitor absorbance at 200 nm (*see* **Note 9**).
3. Prepare the software of the CE instrument as follows.
   a. Rinse the capillary with operating buffer for 5 min.
   b. Introduce the sample by the electrokinetic method, using 5 kV for 10 s.
   c. Run the separation for 25 min at 7.4 kV/m (10 kV for a total capillary length of 27 cm).
4. Compare elution patterns with external HA standards. The HA oligosaccharide fragments (GlcAβ1–3GlcNAcβ1)*n* should elute in the following order: *n* = 4, 3, 5, 6, 2, 7, 8, 9, 10, etc. (*see* **Fig. 3**; **Note 10**).

## 4. Notes

1. To obtain a homogeneous operating solution, the buffer should be kept at room temperature for at least 12 h after dissolution of the polysaccharide. The buffer is degassed under reduced pressure immediately prior to use.
2. This method employs low-UV detection, and it must be borne in mind that a variety of substances will cause absorption. To avoid some of this uncertainty, the spectrum (190–600 nm) may be monitored. Substances with absorbance maximum at wavelengths higher than 190 nm do not represent GAGs (aromatic amino acids of PGs absorb at 280 nm).
3. The electropherograms monitor the homogeneity of the charge density in the GAG preparations. Homogeneous sulfation of the polysaccharide results in sharp peaks. Broad and split peaks indicate that the GAG in question is contaminated or degraded.

4. For alternative CE analyses of charge density and purity of GAGs (*see* **ref. 7**).
5. This method was originally described for the analysis of pure HA preparations. Presence of interacting compounds, such as hyalectans and lectins, may interfere with the electrophoretic performance of HA. The addition of 0.1% (w/v) SDS to the operating buffer may prevent such interactions *(2,8)*.
6. All polyanions detected at 190 nm are not HA, and this lack of specificity still requires purification of the analyte HA and/or a comparison with hyaluronidase-digested aliquots.
7. For alternative CE analyses of high-molecular-weight HA (*see* **refs.** ***1,8*** and ***9***).
8. Most fragments migrate according to their molecular size, the largest being more retarded. The smallest fragments (2–4 disaccharides), however, differ in this respect, slower migration being obtained with the smallest ones. This effect is difficult to explain, but could be related to the complex system with alkaline borate that may interact with both the reducing end monosaccharide and the PEG matrix.
9. This determination also employs low UV detection, and the concerns for specificity presented in **Notes 2, 4,** and **6** should also be carefully thought about here.
10. For alternative CE analyses of HA fragments, (*see* **refs.** ***8*** and ***9***).

## References


1. Karamanos, N. K. and Hjerpe, A. (1998) A survey of methodological challenges for glycosaminoglycan/proteoglycan analysis and structural characterization by capillary electrophoresis. *Electrophoresis* **19,** 2561–2571.
2. Payan, E., Presle, N., Lapicque, F., Jouzeau, J. Y., Bordji, K., Oerther, S., Miralles, G., and Netter, P. (1998) Separation and quantitation by ion-association capillary zone electrophoresis of unsaturated disaccharide units of chondroitin sulfates and oligosaccharides derived from hyaluronan. *Anal. Chem.* **70,** 4780–4786.
3. Karamanos, N. K., Axelsson, S., Vanky, P., Tzanakakis, G. N., and Hjerpe, A. (1995) Determination of hyaluronan and galactosaminoglycan-derived disaccharides by high-performance capillary electrophoresis at the attomole level. Applications to analyses of tissue and cell culture proteoglycans. *J. Chromatogr. A* **696,** 295–305.
4. Karamanos, N. K. and Hjerpe, A. (1999) Strategies for analysis and structure characterization of glycan/proteoglycans by capillary electrophoresis. Their diagnostic and biopharmaceutical importance. *Biomed. Chromatogr.* **13,** 507–512.
5. Hayase, S., Oda, Y., Honda, S., and Kakehi, K. (1997) High-performance capillary electrophoresis of hyaluronic acid: determination of its amount and molecular mass. *J. Chromatogr. A.* **768,** 295–305.
6. Kakehi, K., Kinoshita, M., Hayase, S., and Oda, Y. (1999) Capillary electrophoresis of N-acetylneuraminic acid polymers and hyaluronic acid: correlation between migration order reversal and biological functions. *Anal. Chem.* **71,** 1592–1596.
7. Toida, T. and Linhardt, R. J. (1996) Detection of glycosaminoglycans as a copper (II) complex in capillary electrophoresis. *Electrophoresis* **17,** 341–346.
8. Grimshaw, J., Kane, A., Trocha-Grimshaw, J., Douglas, A., Chakravarthy, U., and Archer, D. (1994) Quantitative analysis of hyaluronan in vitreous humor using capillary electrophoresis. *Electrophoresis* **18,** 2408–2414.
9. Hong, M., Sudor, J., Stefansson, M., and Novotny, M. V. (1998) High-resolution studies of hyaluronic acid mixtures through capillary electrophoresis. *Anal. Chem.* **70,** 568–583.

# II

## Expression, Detection, and Degradation

# 20

# Recombinant Expression of Proteoglycans in Mammalian Cells

## *Utility and Advantages of the Vaccinia Virus/T7 Bacteriophage Hybrid Expression System*

**David J. McQuillan, Neung-Seon Seo, Anne M. Hocking, and Camille I. McQuillan**

## 1. Introduction

### *1.1. Only Mammalian Cells Will Make a Proteoglycan*

Proteoglycans are molecules that comprise a *core protein to which at least one glycosaminoglycan (GAG) chain is attached*, and a consideration of recombinant proteoglycan expression must operate under this simple definition. There are numerous systems currently available to express proteoglycans (and other complex glycoconjugates), but this chapter will focus on a system adopted by us to express proteoglycans and other extracellular matrix proteins in mammalian cells ***(1–5)***.

With few reported exceptions, proteoglycans are exclusively products of eukaryotic organisms, generally of mammalian origin. Although there is ample evidence that core protein homologs exist throughout evolution in many members of the phylogenetic tree, the capacity to synthesize glycosaminoglycans (chondroitin/dermatan, keratan, and heparin/heparan sulfates) appears to be exclusive to higher organisms. Hence, when embarking on a strategy to express heterologous proteoglycan genes, the host cell must possess the appropriate machinery (glycosyltransferases, sulfotransferases, epimerases, deacetylases, etc.) to synthesize and modify a GAG chain onto a recognizable acceptor site in the core protein substrate. In general, this requirement is satisfied by most mammalian cells. Although the capacity for GAG synthesis by different cells and tissues will vary, phenotypically stable cell lines *will* have the posttranslational machinery required for synthesis of some form of GAG.

From: *Methods in Molecular Biology, Vol. 171: Proteoglycan Protocols*
Edited by: R. V. Iozzo © Humana Press Inc., Totowa, NJ

Structural and functional studies of proteoglycan domains can be facilitated by the isolation and purification of native proteoglycan. Most current procedures for isolation of proteoglycans from tissue require the use of denaturing solvents. An alternative method for the generation of native proteoglycan is the use of a recombinant expression system. Due to the extensive posttranslational modifications of proteoglycans, an expression system capable of complex modifications, especially addition of glycosaminoglycan chains, is essential. Prokaryotic expression systems are capable of generating high yields of protein, but these products lack the posttranslational modifications that may regulate folding, solubility, and biological activity. For example, in *Escherichia coli* expression systems, biglycan and decorin are synthesized as core proteins devoid of glycosaminoglycan chains and *N*-linked oligosaccharides. The recombinant core protein is often insoluble, requiring the use of denaturing solvents for efficient extraction from inclusion bodies ***(6–9)***. The baculoviral expression system is also capable of high-level production of processed protein. However, conflicting data makes it difficult to assess whether *Spodoptera frugiperda* (Sf21) cells have the appropriate posttranslational machinery for effective processing of a proteoglycan. Perlecan domains expressed using the baculoviral system have been reported to be substituted with chondroitin sulfate ***(10)***, but, decorin produced in Sf21 cells was secreted devoid of glycosaminoglycan chains ***(11)***.

Recombinant proteoglycans expressed in stably transfected mammalian cells are likely to be folded and glycosylated correctly, but effective purification of the recombinant proteoglycan often involves the use of denaturing solvents. Therefore, it is also important to consider the method of purification of the proteoglycan subsequent to its expression so that it will be representative of the molecule found in vivo. This is most easily achieved by generation of a fusion protein in which the molecule of interest is expressed in complex with a tag (peptide or protein) that can be isolated by a high-affinity, nondenaturing purification procedure. Again, a variety of examples are available, including lectin domains, peptide–antibody pairs, protein–protein ligand complexes, and metal chelating peptides. In our laboratory we have utilized several fusion protein systems but, like numerous other investigators, have found many advantages in the divalent cation-binding properties of the hexa-histidine tag ***(12)***.

### 1.2. Vaccinia Virus as an Expression Vector in Mammalian Cells—Expression by All Cells Achieved with Natures Own Transfection Agent

Introduction of naked DNA into mammalian cells is generally inefficient, despite the increasing number of reagents available from vendors purporting to yield "high-efficiency" expression. In general, the use of calcium phosphate, DEAE-dextran, cationic lipid formulations, or electroporation yield less that 20% transfection efficiency (depending on the cell line) and can be both cumbersome and expensive. It has also been estimated that of the DNA that finds its way into a cell, only 10% will transit to the nucleus and be available for transcription. Therefore, a system whereby efficiency of uptake of DNA is optimized and circumvention of inefficiency of nuclear processing is achieved is likely to be optimal for overexpression of heterologous genes. The use of vaccinia virus as an expression vector satisfies these requirements.

Vaccinia virus, a member of the poxvirus family, has a number of unique properties that has made it very popular for a number of years as a vector for the expression of foreign genes ***(13)***. Vaccinia replicates entirely within the cytoplasm of eukaryotic cells and encodes a complete transcription system, including RNA polymerase, capping and methylating enzymes, and poly A polymerase. As a eukaryotic expression vector, vaccinia virus has a number of useful characteristics, such as a large capacity to incorporate foreign DNA (> 20 kb), retention of infectivity, a wide host range, high level of protein synthesis, and appropriate transport, processing, and posttranslational modification of proteins.

Detailed methods for the propagation, manipulation, and construction of recombinant vaccinia viruses will not be provided here. The reader is referred to several excellent reviews where this information is readily available ***(13,14)***. However, the protocols used in our laboratory for the design, construction, and selection of recombinant viruses used to express proteoglycans will be provided in the following sections. The primary disadvantage of the use of vaccinia virus is that it has virulence for humans and animals. Laboratory workers who work with viral cultures (or other infective materials) should always observe appropriate biosafety guidelines and adhere to published infection control procedures. In the United States, National Institutes of Health guidelines recommend that workers likely to come in direct contact with vaccinia virus be vaccinated with smallpox vaccine every 3–10 yr. Vaccination with smallpox vaccine is generally a simple and safe procedure, but investigators are advised to become well informed of all potential hazards before embarking on the use of this expression system.

### 1.3. Vaccinia Virus/T7 Phage Expression System—A Powerful System for Overexpression of Heterologous Genes

The vaccinia/T7 phage eukaryotic expression system was developed by Moss and co-workers ***(15,16)*** to exploit the high transcriptase activity, stringent promoter specificity, and excellent processivity of the RNA polymerase from the T7 bacteriophage. The main steps involved in the expression of proteoglycans using this system are outlined in **Fig. 1**. Eukaryotic cells infected with the recombinant virus, vTF7-3 (available from the ATTC, accession number VR-2153) express high cytoplasmic levels of T7 bacteriophage RNA polymerase ***(17)***. These cells are co-infected with a recombinant virus containing a cDNA under the control of the T7 promoter. This expression construct also includes an encephalomyocarditis virus (EMC) untranslated region (UTR) that facilitates cap-independent ribosome binding and hence increases translation efficiency up to 10-fold ***(15)***. Transcription of the PG cDNA are driven by phageT7 RNA polymerase in the cytoplasm, and the resultant high levels of PG mRNA are then available to the host cell machinery. Targeting to the secretory pathway by a signal sequence, and subsequent processing including addition of N- and O-linked oligosaccharides, GAGs, sulfation, phosphorylation, folding, and secretion, should all occur appropriately.

### 1.4. Expression Vectors—The Cam Series Works in Vitro and In Vivo

We have modified the basic vaccinia/T7 cloning and expression plasmid, pTM1 ***(15)*** to facilitate targeted secretion and nondenaturing purification of recombinant proteoglycans.

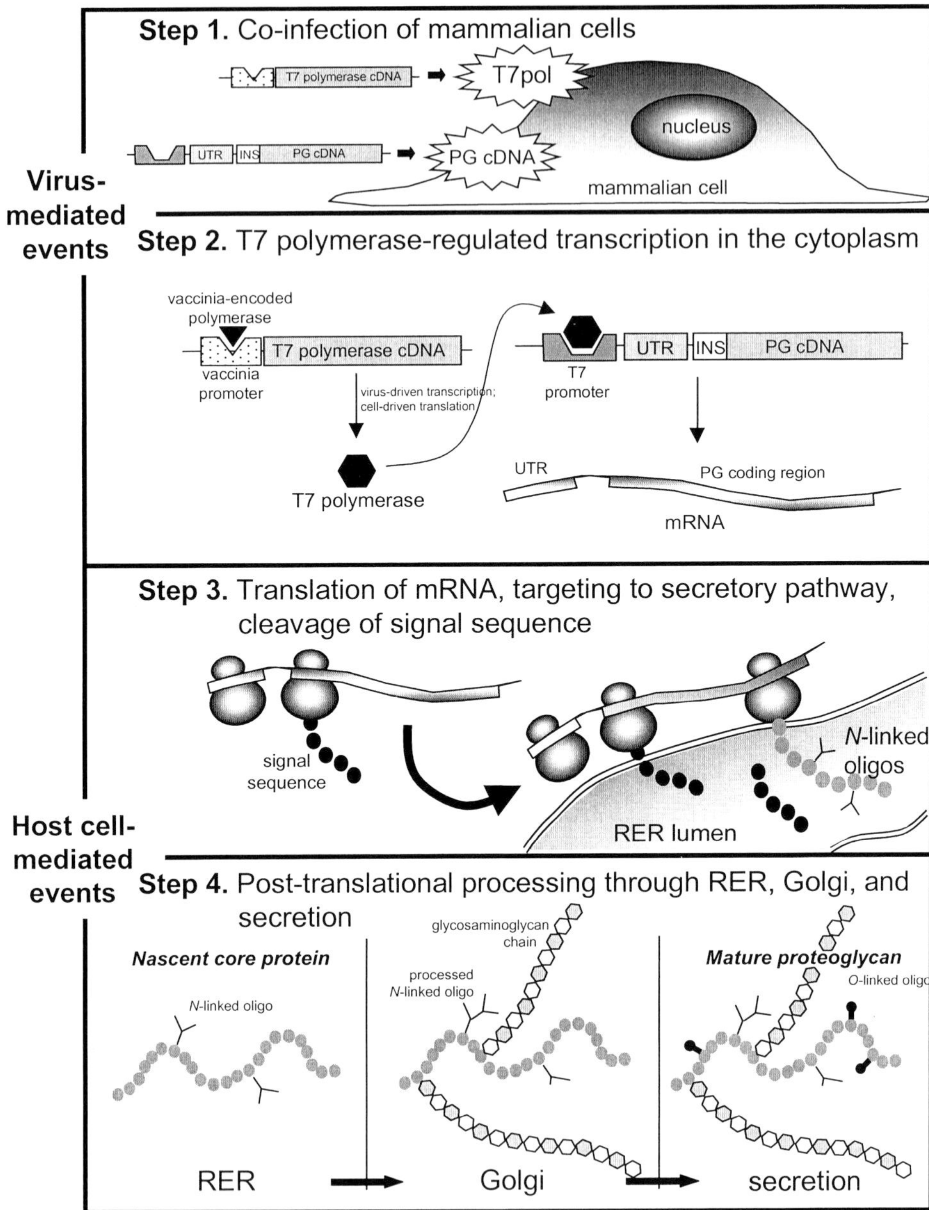


Fig. 1. Schematic representation of vaccinia virus/T7 phage expression system. The major steps of this powerful eukaryotic recombinant protein expression system are shown. Vaccinia virus mediates the initial steps 1 and 2. *Step 1*, co-infection of mammalian cells by a recombinant virus containing the gene encoding phage T7 RNA ploymerase under control of the early/late vaccinia virus promotor; and a recombinant virus containing the gene encoding the target protein under control of the phage T7 promotor. *Step 2*, cytoplasmic expression of T7 RNA

The important domains of these vectors are shown in **Fig. 2**. The most versatile of the series is pT-cam1 (*see* **Fig. 2a, b**), and comprises (1) a bacteriophage T7 promoter; (2) an ATG start codon at a precise distance downstream of an encephalomyocarditis virus ribosome-binding site (EMC UTR); (3) a canine insulin signal sequence, for targeting to the secretory pathway; (4) a hexa-histidine (6 × His) sequence, for nondenaturing purification by metal-chelating affinity chromatography; (5) a factor Xa recognition cleavage sequence, for removal of the histidine tag after purification; (6) a versatile multicloning site, for in-frame insertion of the cDNA of interest (preferably devoid of endogenous signal sequence); and (7) stop codons in three reading frames upstream of a polyadenylation signal.

The other vectors shown are essentially modifications of pT-cam1, for situations where all the elements of the parent vector are not required. If one has an alternative method of purification (e.g., antibody affinity column or interaction with hyaluronan), pT-cam2 does not have the hexa-histidine tag nor the factor Xa cleavage sequence. If potential bioactive domains are predicted to be located in the vicinity of the N-terminus, pT-cam3 has the hexa-histidine tag located at the C-terminus of the secreted protein. Our experience indicates that the insulin signal sequence can enhance expression of some proteins compared to the endogenous sequence, but pT-cam4 is designed for insertion of the cDNA inclusive of endogenous signal sequence with the hexa-histidine tag at the C-terminus. In many instances, removal of the small hexa-histidine sequence following purification is likely to be no advantage and potentially cumbersome, and in these circumstances pT-cam5, which is devoid of the factor Xa cleavage sequence, may be the vector of choice. All vectors and complete sequences are available upon request from the authors.

## 2. Materials

1. Cell lines, shown in **Table 1**.
2. Wild-type vaccinia virus, WR (ATCC # VR-119), titer should be about $5 \times 10^9$ pfu/mL.
3. DMEM alone (DMEM SF), containing 2.5% FCS (DMEM 2.5%), containing 10% FCS (DMEM 10%).
4. Lipofectin™ transfection kit (GIBCO-BRL #18292-011) or other efficient transfection reagents.
5. Plasmid containing cDNA (pT-cam series) at 1 µg/µL.
6. Sterile polystyrene tubes.
7. Bromodeoxyuridine (BrdU) 200-fold concentrated stock solution, 5 mg/mL in water, filter sterilized.
8. 143B (TK–) cells in log phase, passaged at least once in the presence of BrdU.
9. All culture media for the plaque assays should be supplemented with 25 µg/mL BrdU.
10. Neutral red solution, 3.33 mg/mL, tissue culture grade (Gibco-BRL #15330-079).
11. Low-melting-point agarose (Gibco-BRL #15517-014), 2% solution in PBS, autoclave and mix well. Can be stored at 45°C for 1–2 wk.

---

polymerase and transcription of mRNA encoding the target protein. The host mammalian cell machinery is utilized for steps 3 and 4. *Step 3*, ribosome directed translation of the target protein mRNA that is enhanced by inclusion of a ribosome "landing pad" (UTR); targeting to the rough endoplasmic reticulum by specific signal sequence; signal sequence cleavage; and addition of *N*-linked oligosaccharides to the nascent chain. *Step 4*, posttranslational modifications as appropriate for the target protein, including processing of *N*-linked oligosaccharides, addition of *O*-linked oligosaccharides, synthesis of glycosaminoglycan chains, and finally secretion of a proteoglycan.

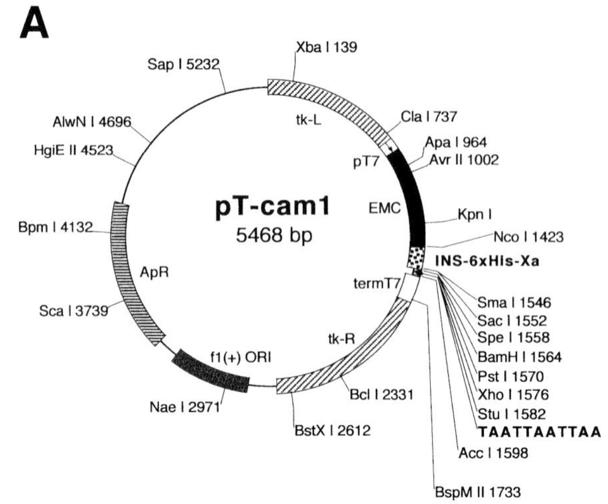


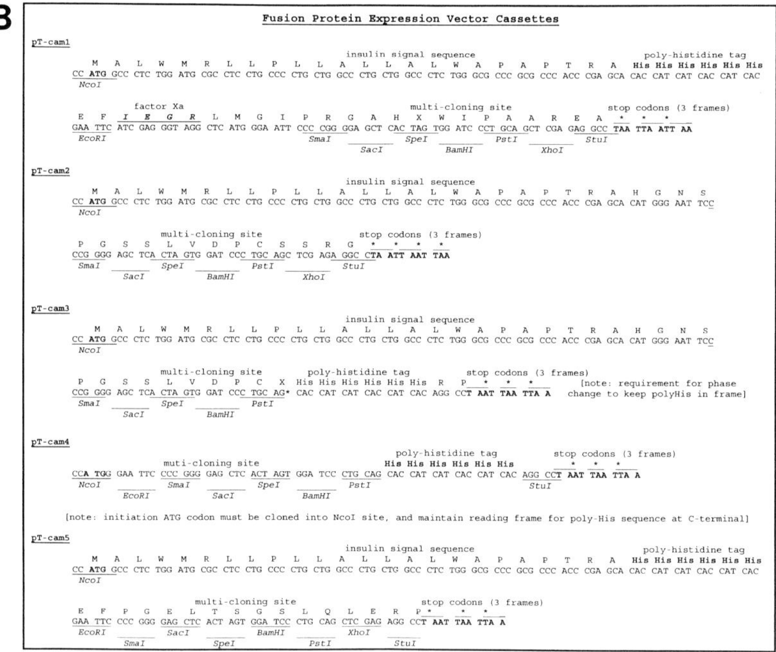


Fig. 2. pT-cam series of expression vectors. (**A**) Main structural features of the pT-cam series of vectors. tk-L and tk-R, thymidine kinase flanking sequences to facilitate insertion of the expression cassette via homologous recombination into wild-type vaccinia virus genome; EMC, encephalomyocarditis untranslated region (UTR) that allows for cap-independent ribosome binding; INS, insulin signal sequence; 6xHis, hexa-histidine sequence; Xa, factor Xa cleavage site. (**B**) cDNA and protein sequences of pT-cam series expression cassettes, showing unique restriction sites available in the multicloning site (MCS), and positioning of other structural features in relation to the MCS.

**Table 1**
**Cell Lines Used for Homologous Recombination, Plaque Assays, Amplification, Titration, and Protein Expression**

| Cell line (origin, morphology) | Culture medium[a] | ATTC catalog number |
|---|---|---|
| CV-1 (monkey kidney cell, fibroblastic) | DMEM + 2.5% FCS[b]<br>5.0% FCS<br>10.0% FCS | CCL-70 |
| 143B (TK–) (human osteosarcoma, fibroblastic) | MEM + 5% FCS<br>10% FCS | CRL-8303 |
| HeLa (human adeno-carcinoma, epithelial) | DMEM + 10% FCS | CCL-2 |
| UMR-106 (rat osteosarcoma, fibroblastic) | DMEM + 10% FCS | CRL-1661 |
| HT-1080[c] (human fibrosarcoma, epithelial) | DMEM + 10% FCS | CCL-121 |

[a]Inclusion of antibiotics is optional, except in the case of the initial transfection for the homologous recombination. If using Lipofectin, then antibiotics should be omitted, as it can inhibit transfection efficiency.

[b]Fetal calf serum, or serum substitutes.

[c]Any other cell line that is likely to process proteoglycans appropriately can be substituted here, with the exception of CHO cells.

12. Proteinase K stock solution (20 mg/mL, Gibco-BRL #25530-031).
13. TE-saturated phenol.
14. Reagents or kit for standard PCR reaction.
15. Thermal cycler.
16. Crystal violet, 0.1% in 20% ethanol.
17. vTF7-3, recombinant vaccinia virus encoding phage T7 RNA polymerase.
18. Methionine and cysteine-free DMEM.
19. Trans-$^{35}$S-label, a mixture of $^{35}$S-methionine and $^{35}$S-cysteine (ICN Biochemicals, #51006).
20. Sephadex G-50 (available from Pharmacia-Amersham Biotech).
21. Sepharose 6B conjugated to iminodiacetic acid functional group (slurry or prepacked 1-mL columns available from Pharmacia-Amersham Biotech).
22. Column buffers: For optimal recovery, column solvents (with the exception of the "charge buffer") should contain a detergent (either 0.1% or greater Triton X-100, or 0.2% or greater CHAPS).
    a. Charge buffer: 100 m*M* $NiCl_2.6H_2O$.
    b. Sample and loading buffer: 5 m*M* imidazole, 0.5 *M* NaCl, 20 m*M* Tris-HCl, pH 8.0.

c. Wash buffer: 20 m*M* imidazole, 0.5 *M* NaCl, 20 m*M* Tris-HCl, pH 8.0.
d. Elution #1: 60 m*M* imidazole, 0.5 *M* NaCl, 20 m*M* Tris-HCl, pH 8.0.
e. Elution #2: 250 m*M* imidazole, 0.5 *M* NaCl, 20 m*M* Tris-HCl, pH 8.0.
f. Strip buffer: 100 m*M* EDTA, 0.5 *M* NaCl, 20 m*M* Tris-HCl, pH 8.0.

23. Sonicator, either probe or cup (e.g., Fisher Sonic Dismembranator 550).
24. Phase-contrast light microscope.
25. Biohazard containment.

## 3. Methods

### 3.1. In Vitro Transcription and Translation (see Note 1)

We routinely use the TNT® coupled reticulocyte lysate system available from Promega (#L4610), and essentially follow the manufacturer's instructions. In a typical cell-free transcription/translation reaction, we would use the following protocol.

1. Mix in an eppendorf tube for a final reaction volume of 50 μL: 25 μL rabbit reticulocyte lysate, 2 μL TNT reaction buffer, 1 μL T7 RNA polymerase, 1 μL methionine-free amino acid mixture, 4 μL Trans$^{35}$S-label, a mixture of $^{35}$S-methionine and $^{35}$S-cysteine (ICN Biochemicals), 1 μL RNAsin® ribonuclease inhibitor, 1 μL plasmid, containing cDNA (1 μg/μL CsCl-purified, or equivalent quality), and 15 μL nuclease-free water
2. Incubate mixture at 30°C for 90 min.
3. Analyze reaction products by (a) direct application to SDS-PAGE; (b) immunoprecipitation followed by SDS-PAGE; or (c) purification on small metal chelating column followed by SDS-PAGE (reaction mix must be exchanged into appropriate binding buffer, e.g., PD-10 column from Amersham-Pharmacia Biotech).
4. Bands are visualized by standard autoradiography/fluorography. The predominant signal should be the target protein of interest migrating at the predicted size for the unsubstituted core protein.
5. If DNA sequencing does not show any errors, and cell-free translation generates a protein product of the predicted size, then one can confidently move to the next step of generating a recombinant virus from the plasmid.

### 3.2. Generating a Recombinant Virus

In this section, several protocols are outlined. The homologous recombination allows transfer of the cDNA under control of the T7 promoter to be inserted into the thymidine kinase (TK) locus of wild-type vaccinia virus. Disruption of the TK gene allows for selection of positive recombinants by culture in the presence of bromodeoxyuridine, a lethal analog of deoxyuridine that is incorporated into replicating DNA by an active TK gene. Plaques are selected and undergo three rounds of purification to ensure clonal selection and remove contaminanting dormant wild-type virus. These procedures require standard cell culture ware and tissue culture facilities. As indicated above, these protocols assume a basic knowledge of virus manipulation. Extensive protocols relating to vaccinia virus are available elsewhere ***(13)***. Cells required are shown in **Table 1**, indicating the ATTC catalog number and the recommended culture medium. Specialized equipment includes a sonicator (either probe or cup, e.g., Fisher Sonic Dismembranator 550), phase-contrast light microscope, and biohazard containment.

### 3.2.1. Homologous Recombination

1. *Day 1:* Initiate culture of CV-1 cells. Seed a T-25 flask (one flask per construct) with CV-1 cells at $1 \times 10^6$ cells/flask, in a total volume of 10 mL of DMEM + 10% FCS. Use the next day when ~80% confluent (should be ~$3 \times 10^6$ cells/flask).
2. *Day 2:* Infection and transfection. Thaw WR virus stock (titer of stock should be ~$5 \times 10^9$ pfu/mL). Sonicate for 30 s, chill on ice for 2 min, and vortex vigorously for 30 s. Repeat once.
3. Trypsinize virus by mixing 20 µL of virus with an equal volume of trypsin (cell culture grade). Incubate at 37°C for 30 min, vortexing briefly every 5 min. This is tube A, a 1/1 dilution (~$2.5 \times 10^9$ pfu/mL).
4. Dilute A 1/100 in 2.5% DMEM (10 µL A + 990 µL 2.5% DMEM). This is tube B, a 1/200 dilution (~$2.5 \times 10^7$ pfu/mL).
5. Aliquot a volume of B containing $1.5 \times 10^5$ pfu (~6 µL) and add to 1 mL of 2.5% DMEM. This is tube C, final concentration of $1.5 \times 10^5$ pfu/mL.
6. Aspirate medium from the T-25 flask of CV-1 cells, and replace with 1 mL of tube C (represents ~0.05 pfu per cell).
7. Incubate for 2 h at 37°C in 5% $CO_2$, with gentle rocking every 15 min.
8. Set up transfection reagents 30 min before end of **step 7**. Note that all plasmids should be at 1 µg/µL, and lipofectin is inhibited by serum proteins and penicillin.
9. Set up two sterile polystyrene tubes:
   a. Tube A: 10 µL DNA + 200 µL DMEM SF
   b. Tube B: 30 µLlipofectin (or lipofectamine) + 200 µL DMEM SF
10. Gently mix A and B. Incubate for 30 min at room temperature. Add 1.6 mL of DMEM SF to the AB mixture.
11. Aspirate the virus inoculum from the T-25 flask. Add the DNA/lipofectin transfection mixture (volume 2 mL) to the cells, ensuring good coverage of the cell layer.
12. Incubate for 5–6 h in a cell culture incubator.
13. Overlay transfection with 3 mL of 10% DMEM (*do not* aspirate transfection cocktail). Continue culture for 2 d.
14. *Day 4:* Harvest homologous recombination. Cytopathic effect (CPE) should be clearly visible by light microscopy. Scrape cell layer with a rubber policeman into a 15 mL conical tube. Pellet cells by centrifugation at 300*g* (3000 rpm), 10 min in a bench-top centrifuge. Aspirate and discard supernatant. Resuspend pellet in 0.5 mL of 2.5% DMEM.
15. Freeze and thaw lysate three times in a Dry Ice/ethanol bath.
16. Store at –80°C, or proceed with plaque assay (*see* **Subheading 3.2.2.**).

### 3.2.2. Plaque Assay (First Round)

1. *Day 1:* Initiate culture of 143B cells. Seed two 6-well plates (35-mm-diameter wells) per recombination with 143B cells at a density of $5 \times 10^5$ cells/well in 10% MEM. Use the next day when greater than 90% confluent (~$1 \times 10^6$ cells per well). Note that a well-defined monolayer is critical to the recognition of clearly defined plaques.
2. *Day 2:* Infection of 143B cells. Thaw the homologous recombination lysate (**Subheading 5.1.2.**) on ice. Sonicate for 30 s, chill on ice for 2 min, and vortex vigorously for 30 s. Repeat once.
3. Incubate 100 µL of lysate with an equal volume of trypsin and incubate at 37°C for 30 min, vortex briefly every 5 min.
4. The trypsinized lysate is diluted into 2.5% MEM, and tested at a range of dilutions from 1/1000 through 1/10 000. An example of a serial dilution protocol is as follows:

| | Virus solution | | Diluent | | Stock | Dilution |
|---|---|---|---|---|---|---|
| 200 µL | trypsinized lysate | + | 1.8 mL 2.5% MEM | = | A | (1/10) |
| 200 µL | A | + | 1.8 mL 2.5% MEM | = | B | (1/100) |
| 400 µL | B | + | 3.6 mL 2.5% MEM | = | **C** | (1/1000) |
| 500 µL | C | + | 1.0 mL 2.5% MEM | = | **D** | (1/2000) |
| 400 µL | C | + | 1.6 mL 2.5% MEM | = | **E** | (1/5000) |
| 200 µL | C | + | 1.8 mL 2.5% MEM | = | **F** | (1/10 000) |

5. Aspirate media from 6-well plates, and infect with 0.5 mL of diluted lysate per well. Infect with stock dilutions C, D, E, and F in triplicate.
6. Incubate for 2 h in the culture incubator, with gentle rocking every 15 min to ensure the monolayer remains wet.
7. Prepare enough LMP agarose overlay to allow 2.5 mL per well. For example, for four plates (i.e., one construct = four dilutions in triplicate = four 6-well plates), 32.0 mL 2% LMP agarose (~45°C), 32.0 mL 5% MEM, and 0.36 mL BrdU (5 mg/mL).
8. Add 2.5 mL of agarose overlay (~45°C) solution per well. *Do not* aspirate virus inoculum. Place plates at 4°C for 15 min to allow agarose to solidify, and then incubate plates in culture incubator for 2 d.
9. *Day 4:* To visualize plaques easily, the 143B monolayer is stained overnight with neutral red (a relatively nontoxic reagent). Prepare enough neutral red/agarose overlay to allow 2.0 mL per well. For example, for four plates, 32.0 mL 2% LMP agarose (~45°C), 32.0 mL 5% MEM, and 2.0 mL neutral red.
10. Add 2.0 mL of neutral red/agarose overlay (~45°C) solution per well. Place plates at 4°C for 15 min to allow agarose to solidify, and then incubate plates in the culture incubator overnight.
11. The first-round plaque assay will be complete on day 5; therefore, to reduce "down-time", 143B cell cultures should be set up at this time for the second round (*see* **Subheading 3.2.3.**).
12. *Day 5:* Inspect plates by holding up to a light, and circle clearly defined, well-separated plaques (these appear as a weakly stained "hole" in a red background). Wells from dilution C should contain an excessive number of plaques, and wells from dilution F are likely to contain very few to no plaques. If this is not the case, then the homologous recombination step has not worked and should be repeated.
13. Pick plaques using sterile glass Pasteur pipets to pull the agarose plug (containing virus) with slight suction. The plug is then expelled into a sterile Eppendorf tube containing 500 µL of pre-warmed 2.5% MEM. Rinse the pasteur with the medium to recover all the agarose. It is recommended that 24 plaques be picked.
14. Rapidly freeze and thaw the plaques three times in a Dry Ice/ethanol bath.
15. Store at –80°C, or proceed directly with the second-round plaque assay.

### *3.2.3. Second- and Third-Round Plaque Assays (*see ***Note 2**)

1. *Day 1:* Initiate culture of 143B cells. Seed one 6-well plate (35-mm-diameter wells) per plaque, with 143B cells at a density of $5 \times 10^5$ cells/well in MEM 10%. Use the next day when greater than 90% confluent (~$1 \times 10^6$ cells per well).
2. *Day 2:* Infection of 143B cells. Thaw lysate from first-round plaque assay. Sonicate for 30 s, chill on ice for 2 min, and vortex vigorously for 30 s. Repeat once.
3. Incubate 250 µL of lysate with an equal volume of trypsin and incubate at 37°C for 30 min, vortex briefly every 5 min.

4. The trypsinized lysate is diluted into 2.5% MEM, and tested at a range of dilutions of 1/5, 1/50, and 1/500. An example of a serial dilution protocol is as follows :

| | Virus solution | | Diluent | | Stock | Dilution |
|---|---|---|---|---|---|---|
| 500 µL | lysate-trypsin | + | 2.0 mL 2.5% MEM | = | **A** | (1/5) |
| 200 µL | A | + | 1.8 mL 2.5% MEM | = | **B** | (1/50) |
| 200 µL | B | + | 1.8 mL 2.5% MEM | = | **C** | (1/500) |

5. Aspirate media from 6 well plates, and infect with 0.5 mL of diluted lysate per well. Infect with stock dilutions A, B, and C in duplicate.
6. Incubate for 2 h in the culture incubator, with gentle rocking every 15 min to ensure that the monolayer remains wet.
7. Prepare enough LMP agarose overlay to allow 2.5 mL per well. For example, for eight plates (i.e., one plate per first round plaque), 64.0 mL 2% LMP agarose (~45°C), 64.0 mL 5% MEM, and 0.72 mLBrdU (5 mg/mL),
8. Add 2.5 mL agarose overlay (≤ 45°C) solution per well. Place plates at 4°C for 15 min to allow agarose to solidify, and then incubate plates in culture incubator for 2 d.
9. *Day 4:* The 143B monolayer is stained overnight with neutral red. Prepare enough neutral red/agarose overlay to allow 2.0 mL per well. For example, for eight plates, 64.0 mL 2% LMP agarose (~45°C), 64.0 mL 5% MEM, and 4.0 mL neutral Red.
10. Add 2.0 mL neutral red/agarose overlay (~45°C) solution per well. Place plates at 4°C for 15 min to allow agarose to solidify, and then incubate plates in the culture incubator overnight.
11. The second-round plaque assay will be complete on d 5, therefore, to reduce "downtime", 143B cell cultures should be set up at this time for the third round.
12. *Day 5:* Inspect plates by holding up to a light, and circle clearly defined, well-separated plaques. Plaques are likely to be present in all wells if the first-round plaque is a genuine TK- recombinant.
13. Pick at least 4 plaques per plate, and place into 500 µL of 2.5% MEM, as for the first round.
14. Rapidly freeze and thaw the plaques three times in a Dry Ice/ethanol bath.
15. Store at –80°C, or proceed directly with the third-round plaque assay.
16. The third round is identical to the second round. Although you have picked four plaques at the second round, take one plaque from each plate to the third round. After the third round you should have eight recombinants that have been derived from eight distinct plaques in the first-round plaque assay.

### 3.3. A Convenient Nomenclature: Keeping Track of All Those Plaques

There are a large number of samples that must be archived in the event that a putative positive drops out during selection, or a catastrophe (such as bacterial contamination) wipes out a third-round positive. Storage of original and subsequent rounds of selection will allow one to rescreen at any stage. We have found the following numbering system to be useful. By example, selection for recombinant virus expressing the proteoglycan "pg":

1. Recombinantion step: pg recombination, 03-04-2000 (i.e., date of recombination).
2. Select first-round plaques: pg 1, pg 2, pg 3, …, pg 24.

3. Select second-round plaques derived from pg 1, pg 2, pg 3, pg 4: pg 1.1, pg 1.2, pg 1.3, pg 1.4, pg 2.1, pg 2.2, …, pg 4.1, pg 4.2, pg 4.3, pg 4.4.
4. Select third-round plaques derived from pg 1.1, pg 2.1, pg 3.1, pg 4.1: pg 1.1.1, pg 1.1.2, pg 1.1.3, pg 1.1.4, …, pg 4.1.1, pg 4.1.2, pg 4.1.3, pg 4.1.4 (*see* **Note 3**).

### 3.4. Screening Recombinants by PCR (see Note 4)

1. *Day 1:* Seed 6-well plates with 143B cells at a density of $5 \times 10^5$ cells/well in 10%MEM. You will need one well per plaque. Use the next day when greater than 90% confluent (~$1 \times 10^6$ cells per well).
2. *Day 2:* Thaw third round plaque lysate on ice. Sonicate for 30 s, chill on ice for 2 min, and vortex vigorously for 30 s. Repeat once.
3. Add 200 µL of lysate is added to 300 µL of 2.5% MEM to make the virus inoculum. Note that there is no requirement for trypsinization at this step.
4. Add virus inoculum directly to the cell layer, followed by incubation for 2 h with gentle rocking every 15 min.
5. Overlay cultures with 2 mL of 2.5% MEM and incubate a further 2 d.
6. *Day 4:* Harvest cells and media by scraping with a rubber policeman and transfer to a centrifuge tube.
7. Pellet cells by centrifugation at 300*g* (3000 rpm) for 10 min. Resuspend pellet in 200 µL of 1 *M* Tris-HCl, pH 9.0.
8. Set up proteinase K digestion for a final volume of 333.6 µL: 200 µL cell pellet, 16.7 µL 1 *M* Tris. HCl pH 7.8, 16.7 µL 10% SDS, 33.4 µL 60% sucrose, and 66.8 µL 10 mg/mL proteinase K. Digest ~5 h at 37°C.
9. Extract two times with TE-saturated phenol.
10. Extract one time with phenol/chloroform
11. Ethanol-precipitate DNA overnight at –20°C
12. Resuspend pellet in 30 µL of TE buffer
13. Perform PCR reaction, including a "no DNA" and "plasmid DNA" controls. A typical reaction mix with a final volume of 100 µL is: 1 µL of extract, or 100 ng of control DNA, DNA, dNTP's (10 m*M*) 8 µL, forward primer (0.1 µg/µL) 1 µL, reverse primer (0.1 µg/µL) 1 µL, 10 X Pfu buffer 10 µL, sterile water 78 µL, and Pfu polymerase 1 µL. Mix, spin briefly, and set in PCR machine. Do PCR as required for specific primers.

### 3.5. Amplification of Recombinant Virus (see Note 5)

#### 3.5.1. Plaque Amplification into 6-Well Plates

1. *Day 1*: Seed 6-well plates with 143B cells at a density of $5 \times 10^5$ cells/well in 10% MEM. You will need one well per plaque. Use the next day, when greater than 90% confluent.
2. *Day 2:* Thaw lysate from third-round plaque assay on ice. Sonicate for 30 s, chill on ice for 2 min, and vortex vigorously for 30 s. Repeat once.
3. Add 250 µL of lysate to 250 µL 2.5% MEM to make virus inoculum. Make sure BrdU is present at a final concentration of 25 µg/mL.
4. Add 0.5 mL virus inoculum directly to cell layer, followed by incubation for 2 h with gentle rocking every 15 min.
5. Overlay cultures with 1.5 mL of 2.5% MEM and 25 µg/mL BrdU.
6. *Day 4*: Cytopathic effect (CPE) should be clearly visible when held under the light. Harvest cells and media by scraping into a 15-mL conical tube. Pellet cells by centrifugation

at 300*g* (3000 rpm) for 10 min. Aspirate and discard the supernatant. Resuspend cell pellet in 500 µL 2.5% MEM.

7. Rapidly freeze and thaw plaques three times in a Dry Ice/ethanol bath. Store at –80°C.

### 3.5.2. Plaque Amplification into T-25 Flasks

1. *Day 1:* Seed T-25 flasks with 143B cells at $1 \times 10^6$ cells per flask. Use the next day, when confluent. A single flask is needed for each individual clone.
2. *Day 2*: Thaw lysate from 6-well plate amplification (*see* **Subheading 5.5.1.**). Sonicate for 30 s, chill on ice for 2 min, and vortex vigorously for 30 s. Repeat once.
3. Add 300 µL of lysate to 700 µL of 2.5% MEM. Make sure BrdU is present at a final concentration of 25 µg/mL.
4. Aspirate media from cells and infect with 1.0 mL of diluted lysate per flask.
5. Incubate for 2 h at 37°C with gentle rocking every 15 min.
6. Overlay each flask with 3.0 mL of 2.5% MEM and 25 µg/mL BrdU.
7. *Day 4:* CPE is clearly visible. Scrape cell layer into 15-mL conical tube. Pellet cells at 300*g* (3000 rpm) for 10 min. Aspirate supernatant and resuspend pellet in 500 µl of 2.5% MEM.
8. Rapidly freeze and thaw plaques three times in a Dry Ice/ethanol bath. Store at –80°C if required.

### 3.5.3. Plaque Amplification into T-175 Flasks

1. *Day 1*: Seed T-175 flasks with HeLa cells. Note that selection agent (BrdU) is no longer required. A single flask is needed for each individual putative recombinant.
2. *Day 2*: Thaw lysate from T-25 amplification (*see* **Subheading 5.5.2.**). Sonicate for 30 s, chill on ice for 2 min, and vortex vigorously for 30 s. Repeat once.
3. Add 250 µL of lysate to 1.75 mL of 2.5% DMEM.
4. Aspirate media from cells and infect with 2.0 mL of diluted lysate per flask.
5. Incubate for 2 h with gentle rocking every 15 min.
6. Overlay each flask with 25 mL of 2.5% DMEM.
7. *Day 4*: Cytopathic effects should be clearly visible. Scrape cell layer and media into a 50-mL conical tube. Pellet cells by centrifugation at 300*g* (3000 rpm) for 10 min. Aspirate supernatant and resuspend pellet in 1.5 mL of 2.5% DMEM.
8. Rapidly freeze and thaw plaques three times in a Dry Ice/ethanol bath. Store at –80°C if required.

## 3.6. Titration of Virus Stock (see Note 6)

1. *Day 1:* Seed 6-well plates with either UMR-106 or CV1 cells at $1 \times 10^6$ cells/well. (One plate per recombinant virus). Use the next day, when confluent.
2. *Day 2:* Thaw virus stock (from **Subheading 5.5.3.**). Sonicate for 30 s, chill on ice for 2 min, and vortex vigorously for 30 s. Repeat once.
3. Incubate 10 µL of lysate with an equal volume of trypsin and incubate at 37°C for 30 min; vortex briefly every 5 min.
4. The trypsinized lysate is diluted into 2.5% MEM and tested at a range of dilutions as follows:

| Virus solution | | | Diluent | | Stock | Dilution |
|---|---|---|---|---|---|---|
| 20 µL | trypsinized lysate | + | 1.8 mL 2.5% MEM | = | A | ($10^{-3}$) |
| 200 µL | A | + | 1.8 mL 2.5% MEM | = | B | ($10^{-4}$) |

| | | | | | | |
|---|---|---|---|---|---|---|
| 200 μL | B | + | 3.6 mL 2.5% MEM | = | C | ($10^{-5}$) |
| 200 μL | C | + | 1.0 mL 2.5% MEM | = | D | ($10^{-6}$) |
| 200 μL | D | + | 1.6 mL 2.5% MEM | = | E | ($10^{-7}$) |
| 200 μL | E | + | 1.8 mL 2.5% MEM | = | F | ($10^{-8}$) |
| 200 μL | F | + | 1.8 mL 2.5% MEM | = | G | ($10^{-9}$) |

5. Aspirate media from cells and infect with 0.5 mL of diluted lysate per well (plate dilutions **E**, **F**, **G** in duplicate).
6. Incubate for 2 h at 37°C, rocking every 15 min.
7. Overlay with 1.5 mL of 2.5% DMEM per well. Incubate for 2 d.
7. *Day 4:* Aspirate medium and add 0.5 mL of crystal violet for each well. Incubate at room temperature for 5 min. Aspirate and allow wells to air dry.
8. Count plaques. There should be 30–90 plaques per well to give an accurate titration. The titer of the virus is calculated and expressed as "plaque-forming units" (pfu) per milliliter of virus stock.

### 3.7 Amplification of a virus stock (see Note 7)

1. *Day 1*: Seed T-175 flasks with HeLa cells at $5 \times 10^7$ cells per flask. The number of flasks is determined by the quantity of virus desired.
2. *Day 2*: Infect each flask, following a standard infection protocol (see following) at 3 pfu/cell. That is, for $10 \times 10^7$ cells per flask, infect with $30 \times 10^7$ pfu.
   Standard infection protocol:
   a. Combine equal volumes of virus and trypsin (volume determined by titer of virus stock). Incubate 37°C for 30 min, vortex every 5 min.
   b. Add 5.0 mL of 2.5% DMEM to the virus mixture.
   c. Remove media from flask and add the 5.0 mL of virus-media mix.
   d. Incubate 2 h at 37°C, with gentle rocking every 15 min.
   e. Overlay with 25 mL 2.5% DMEM, and incubate 2–3 d.
3. Scrape cells into 50-mL conical tube and centrifuge at 300*g* (3000 rpm) for 10 min. Discard the supernatant and resuspend the cell pellet in 2.0 mL of 2.5% DMEM.
4. Rapidly freeze and thaw plaques three times in a Dry Ice/ethanol bath. Store at –80°C if required.
5. Do a virus titration (*see* **Subheading 3.5.**), and aliquot as appropriate.

### 3.8. Screening for Positive Recombinants by Protein Expression (see Note 8)

1. *Day 1:* set up cells. Seed 6-well plate with cells at $1 \times 10^6$ cells per well. Use 2 wells per recombinant screen (i.e., perform in duplicate).
2. *Day 2*: Co-infection should be done with 10–20 pfu/cell for each virus; infection at doses outside this range will result in poor to no detectable expression. Control cultures should include (a) uninfected and (b) infected with vTF7-3 alone (at 20–40 pfu/cell to maintain constant virus load). Also, for UMR-106 or HT-1080 cells, cell density can be assumed to be $3 \times 10^6$ cells per plate at the time of infection. For other cell lines, this needs to be determined.
3. Thaw recombinant viruses on ice, and add an aliquot to an equal volume of trypsin. Incubate at 37°C for 30 min, vortexing every 5 min.
4. Mix an appropriate amount of each virus (i.e., titered recombinant and vTF7-3) into 0.5 mL of 2.5% DMEM for each well to be infected.

5. Remove media from cells and wash cell layer twice with PBS.
6. Overlay cells with virus–DMEM cocktail and incubate at 37°C for 2 h, with gentle rocking every 15 min.
7. Add 2 mL of 2.5% DMEM, and incubate a further 4–6 h. This "recovery" period generally results in a higher rate of protein production, possibly because it allows the sells to recover somewhat from the conditions of minimal media, as well as allowing time for viral-related events to occur in the host cell. The optimal rate of protein secretion is generally seen at 12–16 h postinfection.
8. Biosynthetically label proteins by incubating in the presence of 20–100 μCi/well of $^{35}$S-Trans label in a final volume of 400 μl of methionine- and cysteine-free DMEM, serum free (*see* **Note 9**).
9. Incubate cells overnight, up to 30 h postinfection.
10. Harvest conditioned media and process by metal chelating chromatography (*see* **Subheading 3.9.**). Due to the nature of the fusion protein cassette, the recombinant protein should be secreted. However, particularly in the case of mutants, it is possible that the target protein will not escape quality control systems in the cell. This can be tested by harvesting the cell layer with a detergent-containing solvent (e.g., 20 m*M* Tris-HCl, pH 8.0, 20 m*M* imidazole, 1% Triton X-100), followed by purification as described above.

## 3.9 Metal Chelating Affinity Chromatography ($Ni^{2+}$ Column)

This section brifely outlines one strategy for purifying a hexa-histidine-tagged protein by metal-chelating affinity chromatography. Many similar methods are described elsewhere, in addition to a number of "kits" and different resins available to achieve the same goal (*see* **Note 10**).

### 3.9.1. Column Preparation

1. Transfer the desired amount of slurry (for screening, a 0.5-mL resin volume is ample) to a polypropylene column or, alternatively, use a prepacked column (Amersham Pharmacia Biotech).
2. Wash resin with dd$H_2O$ (5 vol).
3. Charge resin 1× charge buffer (5 vol).
4. Wash resin 1× loading buffer (3 vol).

### 3.9.2 Sample Preparation

1. Harvest the supernatant from 6-well plates.
2. De-salt the sample on Sephadex G-50 (or PD-10 columns, Pharmacia) column to remove unincorporated label and exchange sample into load buffer.
   Disposable G-50 column preparation:
   a. Prepare G-50 slurry in water.
   b. Generate a column in a disposable 10-mL pipet, with the end plugged with glass wool.
   c. Wash resin with 3 column volumes of load buffer.
   d. Behavior of macromolecular elution is essentially linear up to 20 mL of resin.
      i. For 8-mL column, load 2-mL sample and wash in with 0.5 mL, and elute macromolecular fraction with 2.5 mL.
      ii. For 20-mL column, load 5-mL sample and wash in with 1.25 mL, and elute macromolecular fraction with 6.25 mL.

### 3.9.3. Column Chromatography

1. Apply the exchanged sample to the $Ni^{2+}$-charged column.
2. Wash the column with 5 vol of loading buffer.
3. Wash the column with 5 vol of wash buffer.
4. Elute weakly bound material with 3 vol of elution #1.
5. Elute tightly bound proteins with 3 vol of elution #2.

### 3.9.4. Concentration of Sample and SDS-PAGE

1. Concentrate about one-quarter of each eluted fractions with 10,000-MW cutoff membrane (Centricon, Millipore).
2. Wash concentrate one time with excess TE buffer (10 m*M* Tris-HCl, 10 m*M* EDTA), and reduce volume to ≤ 20 mL.
3. Add SDS-PAGE sample buffer and process as usual for SDS-PAGE analysis, followed by fixation and fluorography.
4. Core protein and/or proteoglycan should be apparent in positive lanes (*see* **Fig. 3**).

## 3.10. Mass Production

The following protocols briefly describe methods for production of larger amounts of protein (milligrams and higher), and essentially comprise simple scale-up of the above protocols. The system lends itself to high-level batch production of recombinant proteins, and purification under native conditions.

1. *Day 1*: Set up cells. Seed roller bottles (or cell factory) with cells at $5 \times 10^7$ cells per bottle (or per layer of the cell factory). Use 200 mL of 10% DMEM per roller (100 mL per layer).
2. Allow 1–2 d to reach 80–90% confluence.
3. *Day 2*: Co-infection. Thaw recombinant viruses and add an equal volume of trypsin. Incubate at 37°C for 30 min, vortexing every 5 min.
4. Mix an appropriate amount of each virus into 10 mL of 2.5% DMEM for each roller to be infected.
5. Remove media from cells and wash cell layer twice with PBS.
6. Overlay cells with virus–DMEM cocktail and incubate at 37°C for 2 h in roller apparatus.
7. Add 50 mL of 2.5% DMEM for each roller, and incubate a further 2–4 h.
8. Optional: Biosynthetically label one bottle to use as a trace during purification, by incubating in the presence of 1 mCi per roller of $^{35}$S-Trans label in a final volume of 30 mL of methionine- and cysteine-free DMEM. For nonlabeled bottles, use 30 mL of DMEM SF per roller.
9. Incubate cells overnight, up to 30 h postinfection.
10. Harvest conditioned media and process by metal-chelating chromatography
11. Harvest the supernatant from the roller bottles and add solid imidazole up to a final concentration of 5 m*M*.
11. Add a detergent to reduce nonspecific binding (concentration should be the same as for the column buffer).
12. Adjust pH to 8.0 using concentrated NaOH.
13. Purify hexa-histidine tagged proteoglycan by affinity purification on $Ni^{2+}$-charged metal-chelating column (*see* **Subheading 3.9.**), using imidazole gradient elution as shown in **Fig. 3**.

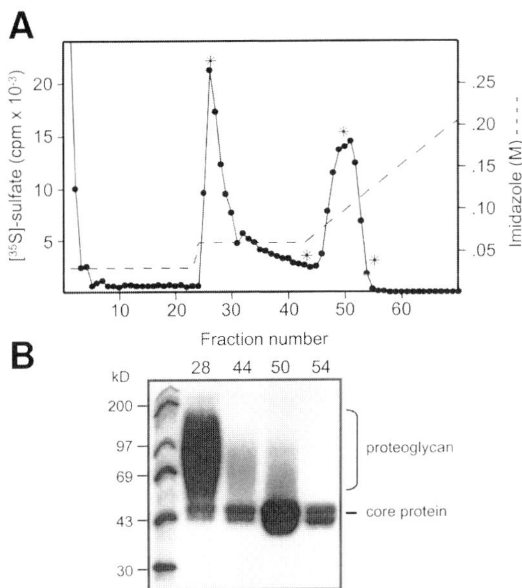


Fig. 3. Purification of recombinant proteoglycan by metal-chelating affinity chromatography. (**a**) $^{35}S$-labeled proteoglycan (decorin) in conditioned media was applied to a $Ni^{2+}$-charged iminodiacetic acid column (2 mL) in 20 m*M* imidazole buffer. Bound material was eluted by a stepwise increase of imidazole to 60 m*M* (proteoglycan elutes), and a linear gradient of imidazole from 60 to 200 m*M* (core protein elutes); see text for discussion. (**b**) Analysis of selected fractions (indicated by the asterisk in *panel a*) by SDS-PAGE and visualized by fluorography. Note that the separation of proteoglycan and core protein glycoforms is likely to be specific to decorin (and biglycan) due to the proximity of the glycosaminoglycan chain to the hexa-histidine tag resulting in a lower affinity interaction.

### 3.11. Characterization of Recombinant Proteoglycan

Following purification of recombinant proteoglycan and core protein by the procedures described above, further steps are dictated by the ultimate disposition of the protein. Simple carbohydrate analysis should yield information about glycosaminoglycan type, number, and size; more detailed analysis will show substitution with *N*- and *O*-linked oligosaccharides; and finally, study of the structure (e.g., CD spectroscopy, NMR analysis, crystallization) and biological function (e.g., interaction with other matrix components, role in ECM assembly, in-vivo activity) will take advantage of the high-level expression and purification of native proteoglycan that is likely to closely resemble the in-vivo product. Many of these methods are detailed elsewhere in this book.

## 4. Notes

1. Constructs developed for expression in the vaccinia/T7 phage hybrid system have a T7 promoter driving expression of the target protein. Following construction of an expres-

sion vector in any of the pT-cam series of vectors, we recommend doing in-vitro transcription directly from the plasmid vector with T7 RNA polymerase, and in-vitro translation with a reticulocyte lysate system. This is in addition to direct sequencing through cloning sites or, optimally, through the entire coding region. A significant amount of wasted time can be avoided by following these two simple steps.

2. Since there is likely to be background wild-type virus as well as spontaneous deletion of the TK gene activity, it is recommended to take 8 plaques through to the second round. Some of these may not survive a second round of selection, and others may not express protein. Ideally, 50–75% of the first round plaques should be the desired recombinant; however, this efficiency can be somewhat less (ultimately, you need only one positive!).
3. If pg 1.1.3 is a positive, the name indicates it is derived from plaque number pg 1 from the pg recombination, and that it has gone through three rounds of selection. If pg 1.1.3 is lost through a laboratory accident, then it is likely that pg 1.1.1, pg 1.1.2, and pg 1.1.4 are identical clones and the positive can be recovered.
4. It is possible to screen plaques at this stage by PCR analysis. Our experience suggests that it is difficult to design appropriate positive and negative controls for this procedure. Since the ultimate test of a recombinant virus will always be expression of the protein of interest, we recommend amplifying the eight recombinants, determining the titration, and screening for protein expression. However, PCR screening can be useful, particularly when generating and screening a large number of constructs (truncations, deletions, mutations, etc.). Therefore this protocol describes the isolation of viral DNA for subsequent PCR by standard methods using appropriately designed oligonucleotides.
5. These procedures are followed to generate enough virus for an accurate titration and to screen for protein expression. We recommend amplifying 4–8 recombinants from the third-round plaque assay.
6. An accurate titration is required before any attempts to screen for protein expression. Titers can vary by 1-2 orders of magnitude, and optimal protein expression is achieved only within a discrete range of infection (5–30 pfu/cell). Infection with too little or too much virus will not yield detectable levels of protein expression.
7. Following identification of a positive recombinant, large stocks of virus are made and can be aliquoted and stored at –70°C for extended periods. For generation of milligram amounts of protein, and to avoid significant batch-to-batch variation, it is recommended to make as much virus stock as is practical.
8. After generating up to eight putative positive recombinant viruses that have undergone at least three rounds of selection in the presence of BrdU, a simple proteoglycan/protein expression screen is done to identify genuine positives. There are a number of ways to screen for the protein of interest, including Western blotting with specific antibodies or even sensitive bioassays if available. However, in our laboratory we almost exclusively utilize biosynthetic labeling of the core protein and purification with the hexa-histidine tag, followed by predicted migration on SDS-PAGE.
9. Most cells will tolerate serum-free conditions for the period of labeling, and this has the advantage of reducing total protein in the harvested conditioned medium. However, for cell lines that are particularly sensitive to serum-free conditions, one can add minimal serum or some other form of defined medium.
10. The behavior of a particular protein is not easily predicted, and better performance can often be achieved by an empirical approach, testing different resins, different divalent cautions, and changes in solvent conditions. However, for the purposes of an initial screen, this method is appropriate.

## References


1. Chan, D., Lamande, S. R., McQuillan, D. J., and Bateman, J. F. 1997. In vitro expression analysis of collagen biosynthesis and assembly. *J. Biochem. Biophys. Meth.* **36,** 11–29.
2. Chan, D., Weng, Y. M., Hocking, A. M., Golub, S., McQuillan, D. J. and Bateman, J. F. 1996. Site-directed mutagenesis of human type X collagen. Expression of alpha1X) NC1, NC2, and helical mutations in vitro and in transfected cells. *J. Biol. Chem.* **271**, 13,566–13,572.
3. Hocking, A. M., Strugnell, R., Ramamurthy, A. P., and McQuillan, D. J. 1996. Eukaryotic expression of recombinant biglycan. Post-translational processing and the importance of secondary structure for biological activity. *J. Biol. Chem.* **271**, 19,571–19,577.
4. Krishnan, P., Hocking, A. M., Scholtz, J. M., Pace, C. N., Holik, K. K., and McQuillan, D. J. 1999. Distinct secondary structures of the leucine-rich repeat proteoglycans decorin and biglycan. Glycosylation-dependent conformational stability. *J. Biol. Chem.* **274**,10,945–10,950.
5. Ramamurthy, P., Hocking, A. M., and McQuillan, D. J. 1996. Recombinant decorin glycoforms. Purification and structure. *J. Biol. Chem.* **271**, 19,578–19,584.
6. Hering, T. M., Kollar, J., Huynh, T. D., and Varelas, J. B. 1996. Purification and characterization of decorin core protein expressed in Escherichia coli as a maltose-binding protein fusion. *Anal. Biochem.* **240**, 98–108.
7. Hildebrand, A., Romaris, M., Rasmussen, L. M., Heinegard, D., Twardzik, D. R., Border, W. A., and Ruoslahti, E. 1994. Interaction of the small interstitial proteoglycans biglycan, decorin and fibromodulin with transforming growth factor beta. *Biochem. J.* **302**, 527–534.
8. Schonherr, E., Hausser, H., Beavan, L., and Kresse, H. 1995. Decorin-type I collagen interaction. Presence of separate core protein-binding domains. *J. Biol. Chem.* **270**, 8877–8883.
9. Schonherr, E., Witsch-Prehm, P., Harrach, B., Robenek, H., Rauterberg, J., and Kresse, H. 1995. Interaction of biglycan with type I collagen. *J. Biol. Chem.* **270**, 2776–2783.
10. Groffen, A. J., Buskens, C. A., Tryggvason, K., Veerkamp, J. H., Monnens, L. A., and van den Heuvel, L. P. 1996. Expression and characterization of human perlecan domains I and II synthesized by baculovirus-infected insect cells. *Eur. J. Biochem.* **241**, 827–834.
11. Gu, J., Nakayama, Y., Nagai, K., and Wada, Y. 1997. The production and purification of functional decorin in a baculovirus system. *Biochem. Biophys. Res. Commun.* **232**, 91–95.
12. Hochuli, E., Dobeli, H., and Schacher, A. 1987. New metal chelate adsorbent selective for proteins and peptides containing neighbouring histidine residues. *J. Chromatogr.* **411**, 177–184.
13. Mackett, M., Smith, G. L., and Moss, B. 1985. The construction and characterization of vaccinia virus recombinants expressing foreign genes; in *DNA Cloning*, volume II (Glover, D. M., ed.), IRL, Oxford, UK, pp.191–212.
14. Ausubel, F. M., Brent, R., Kingston, R. E., Moore, D. M., Seidman, J. G., Smith, J. A., and Struhl, K., eds. 1987. *Current Protocols in Molecular Biology*. Wiley, New York, NY.
15. Elroy-Stein, O., Fuerst, T. R., and Moss, B. 1989. Cap-independent translation of mRNA conferred by encephalomyocarditis virus 5' sequence improves the performance of the vaccinia virus/bacteriophage T7 hybrid expression system. *Proc. Natl. Acad. Sci. (USA)* **86**, 6126–6130.
16. Fuerst, T. R., Niles, E. G., Studier, F. W., and Moss, B. 1986. Eukaryotic transient-expression system based on recombinant vaccinia virus that synthesizes bacteriophage T7 RNA polymerase. *Proc. Natl. Acad. Sci. (USA)* **83**, 8122–8126.
17. Fuerst, T. R., Earl, P. L., and Moss, B. 1987. Use of a hybrid vaccinia virus-T7 RNA polymerase system for expression of target genes. *Mol. Cell. Biol.* **7**, 2538–2544.

# 21

# Purification of Recombinant Human Decorin and Its Subdomains

**Anna L. McBain and David M. Mann**

## 1. Introduction

Conventional approaches to purifying most connective tissue proteoglycans (PGs) generally involve treatments that consequently alter the binding properties of the purified PG (e.g., exposure to strong detergents or harsh denaturants such as guanidinium). Herein we describe a protocol that we use to isolate human decorin in which the PG remains functional in its binding to various ligands ***(1,2)***. This method utilizes a eukaryotic expression system that produces high quantities of recombinant human decorin as a soluble component of the transfected cell culture medium together with a combination of conventional ion-exchange and hydrophobic-interaction chromatography. Here we provide detailed instructions for (1) the production of conditioned medium containing secreted recombinant human decorin, (2) the initial ion-exchange chromatography step using DEAE-Sepharose, and (3) the final hydrophobic-interaction chromatography step.

In addition, we have also developed a prokaryotic system for the expression of subdomains of human decorin. This system utilizes a derivative of the pATH plasmid series ***(3)*** and was selected because our preliminary studies indicated that it yields high levels of recombinant decorin core protein with minimal degradation. Decorin subdomains are expressed in *Escherichia coli* as fusion hybrids with the amino-terminal 323 residues of anthranilate synthase (*trpE*), which tends to stabilize the fusion protein and cause it to form protease-resistant inclusion bodies within the cytosol. Inclusion body formation also facilitates isolation of the decorin-subdomain fusion proteins.

We used a cassette mutagenesis approach to engineer a derivative of the pATH3 expression vector to introduce several desirable features. For example, our modified pATH3 vector encodes a Factor Xa cleavage site between the *trpE* encoded sequence and the inserted decorin sequence, and a polyhistidine tag at the carboxy-terminus of the decorin sequence. This tag provides an affinity handle to facilitate a rapid and highly specific method of purifying the expressed protein from bacterial cell extracts

From: *Methods in Molecular Biology, Vol. 171: Proteoglycan Protocols*
Edited by: R. V. Iozzo © Humana Press Inc., Totowa, NJ

if needed. We have used *E. coli* WM6 as the host cell for expression of the decorin fusion proteins because these cells are deficient in protease activity *(4)*. They are a derivative of the *E. coli* strain CAG 456, which carry a mutation in the RNA polymerase sigma subunit specific for heat-shock promoters. Thus, in addition to a reduced proteolysis, there is a potential to accumulate the expressed protein for longer periods of time in these cells. In this chapter, we describe the general outline for subcloning a decorin core protein domain into our modified pATH3 vector and provide a protocol for expressing this subdomain as an anthranilate synthase hybrid protein. We also have provided our protocol for the isolation and solubilization of the resulting inclusion bodies.

## 2. Materials

### 2.1. Cell Culture and Conditioned Medium Production

1. Chinese hamster ovary cells stably transfected to express the proteoglycan form of human decorin *(5)*.
2. Supplemented alpha(–)MEM: Alpha(–)MEM (without nucleosides) supplemented with final concentrations of 25 m*M* glucose, 0.8 m*M* sodium sulfate, 0.02% Tween 80 (Sigma, St. Louis, MO), 100 units/mL penicillin G sodium, 100 units/mL streptomycin sulfate, and 0.292 mg/mL L-glutamine.
3. Serum-containing supplemented alpha(–)MEM: As described above, with 10% defined and supplemented bovine calf serum (Hyclone Laboratories, Logan, UT) added.
4. Trypsin EDTA 1× solution in Hanks' balanced salts (0.05% Trypsin, 0.53 m*M* EDTA, without calcium or magnesium).
5. 850 $cm^2$ polystyrene roller bottles (Corning, Corning, NY).
6. Cell wash buffer: Dulbecco's phosphate–buffered saline containing 1 m*M* calcium chloride and 0.5 m*M* magnesium chloride.
7. Protease inhibitor stocks: 1 mg/mL aprotinin, 1 mg/mL leupeptin, 1 mg/mL pepstatin A.
8. Sodium azide stock solution (20% w/v).

### 2.2. DEAE-Sepharose Chromatography

1. Buffer A: 300 m*M* NaCl in phosphate buffer, pH 7.4 (6.5m*M* $Na_2HPO_4.7H_2O$, 1.1 m*M* $KH_2PO_4$, 2.7 m*M* KCl) with 0.02% sodium azide.
2. 5 *M* NaCl in water.
3. Buffer B: 6 *M* urea containing 6 m*M* CHAPS and 300 m*M* NaCl in phosphate buffer, pH 7.4 (as defined above), with 0.02% sodium azide. Prepare this urea buffer fresh from a 9 *M* urea stock that has been maintained over a mixed-bed ion-exchange resin, TMD-8 (Sigma, St. Louis, MO) to remove cyanate ions.
4. Buffer C: 1 *M* NaCl in phosphate buffer (as defined above), pH 7.4, with 0.02% sodium azide.
5. Two 2.5 cm (diameter) by 20 cm (length) chromatography columns (e.g., Kontes Flex), each packed with 80-mL DEAE-Sepharose Fast Flow (Pharmacia).
6. A 2.5 cm by 20 cm guard column packed with 30-mL Sepharose-4B (Pharmacia).

### 2.3. Hydrophobic Interaction Chromatography (HIC)

1. Saturated ammonium sulfate (4.1 *M* at 25°C) in 50 m*M* phosphate buffer, pH 6.3.
2. A 50 mm by 4.6 mm Hydrocell C4-1000 HIC column, 10 µm by 1000 Å (prepacked by BioChrom Labs, Terre Haute, IN).

3. 2.5 *M* ammonium sulfate in 50 m*M* phosphate buffer, pH 6.3.
4. 50% Ethylene glycol in 50 m*M* phosphate buffer, pH 8.3.

### 2.4. Prokaryotic Expression of Decorin Subdomains as Anthranilate Synthase Hybrid Proteins

1. pATH3 prokaryotic expression plasmid (American Type Culture Collection [ATCC], Manassas, VA, product #37697), modified within the multiple cloning site by cassette mutagenesis to contain the following: Factor Xa protease cleavage site (Ile-Glu-Gly-Arg), a directional cloning site (BamHI-XhoI-XbaI) for PCR-generated decorin domains, an acid-labile cleavage site (Asp-Pro), and a polyhistidine affinity tag (six consecutive histidine residues) followed by two consecutive lysine residues that can be used for chemical cross-linking to a cyanogen bromide-activated agarose matrix.
2. PCR product encoding the desired subdomain of human decorin, terminating with an appropriate restriction site for subcloning into the expression vector above such that the decorin codons remain *in frame* with those of the anthranilate synthase sequence.
3. *E. coli* WM6 (ATCC product #47020, genotype *F- supC(Ts) lac(Am) mal trp(Am) rpsL htpR165 pho(Am) lambda-*).
4. Luria-Bertani medium (LB medium).
5. M9 minimal medium with casamino acids.
6. Ampicillin: 50 mg/mL stock in water.
7. Indoleacrylic acid stock of 10 mg/mL in ethanol (this should be prepared immediately before use).
8. Phenylmethylsulfonylfluoride (PMSF): 200 m*M* stock in 95% ethanol.
9. Lysis buffer: 50 m*M* Tris-HCl (pH 8.0), 1 m*M* EDTA, 0.1 *M* NaCl.
10. Lysis buffer with 0.5% Triton X-100 and 10 m*M* EDTA.
11. Lysis buffer with 8 *M* Urea and 2 m*M* PMSF.
12. Deoxycholic acid (sodium salt, Sigma).
13. Deoxyribonuclease I (DNaseI) from bovine pancreas (Sigma), dissolved at 1 mg/mL in Tris-buffered saline.

## 3. Methods

### 3.1. Cell Culture and Conditioned Medium Production

1. Chinese hamster ovary cells stably transfected with the proteoglycan form of human decorin are grown in serum-containing alpha(–)MEM medium at 37°C in 5% $CO_2$. We typically grow these cells in 75 $cm^2$ tissue culture flasks until confluent and then expand them into 850 $cm^2$ roller bottles. After detaching cells from one confluent 75 $cm^2$ flask using trypsin/EDTA, the cells are gently pelleted by centrifugation at 320*g*, 25°C. The trypsin/EDTA supernatant is discarded and the resuspended cell pellet is added to 150 mL of serum-containing medium in an 850 $cm^2$ roller bottle. Cells are maintained in serum-containing supplemented alpha(–)MEM until they have reached confluency. Roller bottle cultures should be grown at 37°C in 5% $CO_2$ with a rotational speed of 0.5 rpm.
2. Remove the serum-containing conditioned medium (*see* **Note 1**) and wash the adherent cell layer three times each with approximately 150 mL of cell wash buffer at 37°C.
3. Add 50 mL of supplemented alpha(–)MEM (serum-free) to the roller bottle and rotate the cells in this medium at 0.5 rpm and 37°C for 8–24 h to allow them to condition this medium with secreted decorin (*see* **Note 2**).
4. Collect the serum-free conditioned medium and refeed cells with 150 mL of serum-containing medium. Culture cells in this medium for at least 48 h to allow them to recover from the serum deprivation before producing serum-free conditioned medium again.

5. Add protease inhibitors (aprotinin, leupeptin, pepstatin A) to the conditioned medium to a final concentration of 1 μg/mL for each (*see* **Note 3**).
6. Sodium azide should be added to the conditioned medium to a final concentration of 0.02% to prevent microbial contamination during subsequent handling steps.
7. Remove cell debris by centrifugation for 10 min at 430*g* and store clarified conditioned medium at –80°C (preferred) or –20°C.
8. Cultures maintained in roller bottles should be harvested and reseeded into fresh roller bottles at approximately $1 \times 10^7$ cells per roller bottle every 2–3 wk or whenever significant sloughing can be observed macroscopically (i.e., greater than 20% of the surface area is cell free). Production of conditioned medium from roller bottle cultures that are overgrown will result in increased histone and DNA contamination of the decorin due to significant cell lysis.

### 3.2. DEAE-Sepharose Chromatography

1. Thaw conditioned medium at 37°C.
2. Bring 20–30 L of pooled medium to a final concentration of 300 m*M* NaCl by adding 31 mL of 5 *M* NaCl stock per liter of conditioned medium (*see* **Notes 4** and **5**).
3. Apply the conditioned medium over two 80-mL DEAE-Sepharose fast flow columns connected and preceded by a 30-mL Sepharose-4B precolumn guard to remove nonspecific Sepharose-binding molecules (columns should be previously equilibrated with buffer A). This should be performed at 4°C with a flow rate not exceeding 3 mL/min. Typically we pass 15 L over these columns in about 72 h using gravity flow without pumping (*see* **Note 6**).
4. Detach the preguard column and individually wash the 80 mL DEAE-Sepharose columns each with 2–3 L of buffer A until the absorbance at 280 nm of this wash is below 0.010. This step can be performed at room temperature with a gravity-driven flow rate of approx 750 mL/h (*see* **Note 7**).
5. Wash each DEAE-Sepharose column seperately with 250 mL of buffer B. The presence of 6 m*M* CHAPS in this buffer removes low-molecular-weight decorin-binding contaminants *(**6**)*.
6. Remove the urea/CHAPS thoroughly by washing each column with approx 0.5–1.0 L of buffer A.
7. Place the two DEAE-Sepharose columns back in series and remove the bound decorin with a linear NaCl gradient generated between 100% buffer A and 100% buffer C (300 m*M* to 1 *M* NaCl, in phosphate buffer, pH 7.4). We recommend performing this step at 4°C and pumping a total elution volume of 500 mL at a constant flow rate of 20–30 mL/h while collecting 50 × 10 mL fractions.
8. Measure the absorbance of each fraction at 280 and 260 nm and plot these values vs the fraction number. Typically, the elution profile will contain two partially overlapping peaks, the first composed predominately of nucleic acid contaminants and the second, smaller, peak containing recombinant human decorin. The bound decorin will dissociate from the DEAE columns in approx 600–800 m*M* NaCl.
9. Analyze the purity of decorin in the fractions by SDS-PAGE (*see* **Note 8**). Pool the decorin-containing fractions that lack significant DNA contamination (determined by absorbance at 260 nm). Our typical yields of recombinant human decorin are approx 2.5–3.0 mg/L from serum-free conditioned medium (*see* **Note 9**). Greater yields (5–10 mg/L) are obtained from serum-containing medium that has been conditioned for 48 h or longer.

### 3.3. Hydrophobic Interaction Chromatography

1. Dilute the pooled decorin-containing fractions with an equal volume of saturated ammonium sulfate in 50 m*M* phosphate buffer, pH 6.3.

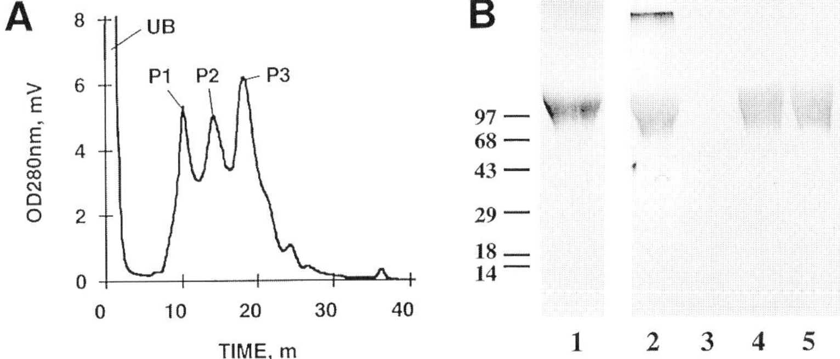


Fig. 1. Purification of human recombinant decorin by hydrophobic interaction chromatography. **(A)** Representative chromatogram resulting from C4-hydrophobic interaction column loaded with pooled DEAE-Sepharose purified decorin. Note the three predominant peaks detected in the eluted material, P1-P3. **(B)** SDS-PAGE analysis of three peaks fractionated by HIC: *lane 1*, Coomassie blue toluidine blue staining of the starting material loaded on the HIC column; *lanes 2–5,* immunoblot analysis of the starting material *(lane 2)* and the eluted peaks, P1-P3 *(lanes 3–5, respectively)*, using a polyclonal antibody directed against the amimo terminal end of human decorin *(7)*.

2. Apply this material to a C4 hydrophobic interaction column (previously equilibrated with 2.5 *M* ammonium sulfate in 50 m*M* phosphate buffer, pH 6.3) at room temperature with a slow flow rate (e.g., 1.0 mL/min for a 50 by 4.6 mm column) and wash the column with the equilibration buffer until a baseline signal is achieved (*see* **Note 10**).
3. Remove bound decorin with a reverse linear gradient of ammonium sulfate from 2.5 to 0 *M* in 50 m*M* phosphate buffer, pH 6.3, while simultaneously applying an increasing linear gradient of ethylene glycol from 0 to 50% in 50 m*M* phosphate buffer, pH 8.3. When monitored by absorbance at 280 nm, we typically observe three major peaks eluting from the column (**Fig. 1A;** *see* **Note 11**). Coomassie blue/toluidine blue-staining SDS-polyacrylamide gels and immunoblotting using a polyclonal antibody directed against the amino-terminal end of human decorin *(7)* reveal that the second and third peaks in the chromatogram contain highly purified decorin, with no other molecules detected (**Fig. 1B**).
4. Collect decorin containing peaks and remove ethylene glycol and ammonium sulfate by dialysis using a 30-kDa MWCO dialysis membrane (Spectrapor). We typically dialyze against phosphate-buffered saline before storing the final decorin product.

### 3.4. Prokaryotic Expression of Decorin Sequences as Anthranilate Synthase Fusion Proteins

1. *Cloning of human decorin subdomains into modified pATH3 vector*: PCR amplify the desired decorin core protein sequence(s) using synthetic oligonucleotide primers that are designed to incorporate appropriate restriction sites for subcloning the product into the modified pATH3 vector in the correct reading frame. Restrict PCR product with appropriate restriction enzymes (e.g., BamHI and XbaI) and subclone into the modified pATH3 vector. **Figure 2** illustrates the boundaries of the decorin subdomains that we have subcloned into our modified pATH3 vector and expressed as fusion proteins, as well as the sequence of the salient features of this expression vector and resultant fusion polypeptides.

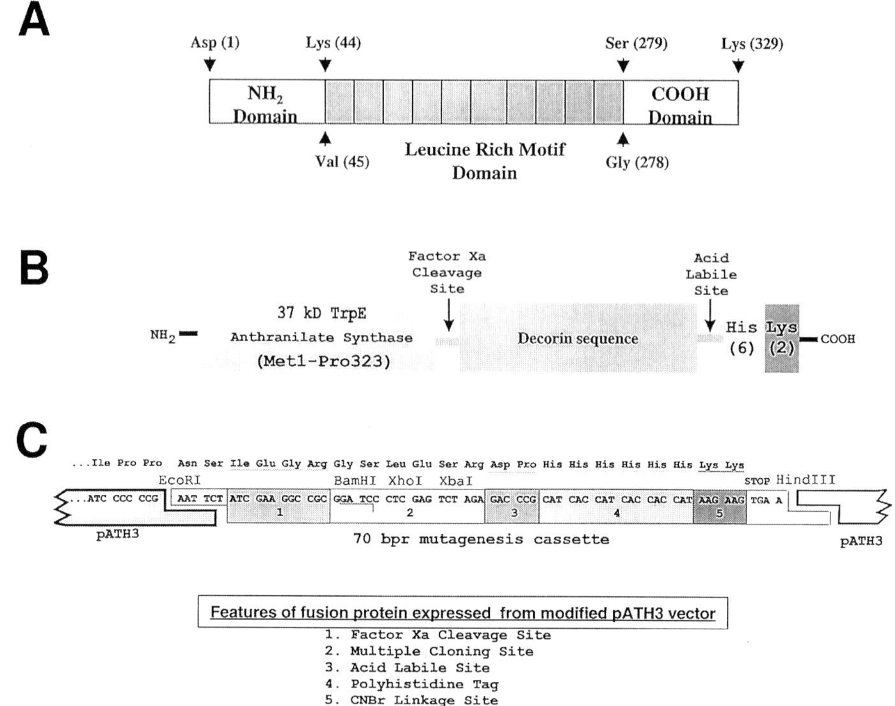


Fig. 2. Structure of decorin core protein and prokaryotic expression vector used to produce anthranilate synthase-decorin hybrid proteins. **(A)** Delineation of human decorin subdomains that have been expressed as hybrid proteins. **(B)** Model of hybrid fusion protein illustrating the salient features introduced by expression from our modified pATH3 vector. **(C)** Sequence of the fusion site in modified pATH3 vector illustrating position of specific engineered features.

2. *Expression of fusion proteins in E. coli WM6*: Use *E. coli* WM6 cells to express the recombinant fusion protein. Prepare a freshly seeded plate with WM6 transformants and pick a single colony for overnight expansion in LB medium containing 100 µg/mL ampicillin (37°C while shaking at 225 rpm). Dilute the overnight culture with 4 volumes of M9 minimal medium containing casamino acids supplemented with 100 µg/mL ampicillin (i.e., cell density diluted to 20% that of the overnight culture) (*see* **Note 12**). Incubate the diluted culture for 3 h at 37°C with shaking and then induce fusion protein expression by adding indoylacrylic acid (IAA) to a final concentration of 10 µg/mL. Incubate cultures in IAA containing medium for 5 h at 37°C with shaking at 225 rpm to accumulate high levels of expressed fusion protein (*see* **Note 13**). Add PMSF to the cultures to a final concentration of 0.2 m*M*. At this stage the total bacterial culture can be analyzed for the expression of the fusion protein. Remove 10 µL of culture, dilute with 2× SDS-PAGE sample buffer (reducing), and analyze total protein content by SDS-PAGE (*see* **Note 14**). **Figure 3** shows the profile of total proteins present in the WM6 cultures (cell + medium) and their immunoreactivity with rabbit polyclonal antisera raised against either human decorin *(7)* or a carboxy-terminal peptide of human decorin *(8)*.

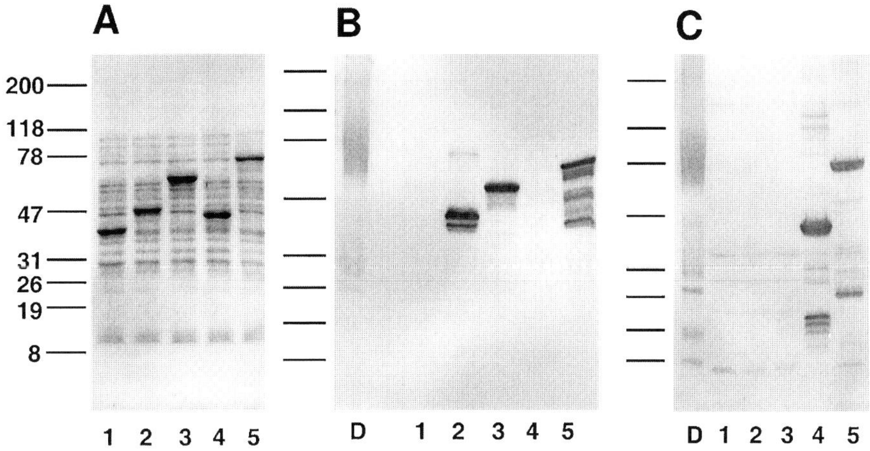


Fig. 3. Expression of anthranilate synthase-decorin hybrid proteins in E. coli WM6. Fractionation of 5–20% SDS-polyacrylamide gel of total proteins present in combined media and cell pellet: **(A)** Ponceau S staining of a nitrocellulose membrane before immunodetection; **(B,C)** immunoblot analysis using rabbit polyclonal antisera raised angainst either human decorin proteoglycan or a carboxy-terminal peptide of the core protein of human decorin *(8)*, respectively. Total proteins derived from WM6 cultures transformed with *(lane 1)* modified pATH3 vector only (no decorin sequence inserted); *(lane 2)* modified pATH3 vector with decorin $NH_2$ domain insert (Asp1-Lys44); *(lane 3)* modified pATH3 vector with decorin LRM domain insert (Val45-Gly278); *(lane 4)* modified pATH3 vector with decorin COOH domain insert (Ser279-Lys329); *(lane 5)* modified pATH3 vector with whole decorin sequence insert (Asp1-Lys329). As a positive control, *lane D* contains the proteoglycan from of human decorin.

3. *Isolation and solubilization of fusion protein inclusion bodies*: Pellet WM6 cells by centrifugation at 5000*g* for 15 min at 4°C and determine the weight of the cell pellet. Resuspend the cell pellet in lysis buffer using 3 mL of lysis buffer per gram of wet cell pellet weight. While shaking cell suspension at 25°C, add 4 mg of deoxycholic acid per gram of original cell pellet weight. Incubate lysate at 37°C until it becomes viscous (10–30 min) and then add 20 µL of DNase I solution per gram cell pellet weight. Incubate at 25°C until no longer viscous (30–60 min). Centrifuge at 12,000*g* for 15 min at 4°C. Decant supernatant and resuspend pellet in 9 vol of lysis buffer containing 0.5% Triton X-100 and 10 m*M* EDTA, pH 8.0. Incubate for 5 min at 25°C before centrifuging for 15 min at 12,000*g*, 4°C. Decant supernatant and solubilize the residual pellet in lysis buffer containing 8 *M* urea and 2 m*M* PMSF. Analyze all supernatants and urea insoluble pellet by SDS-PAGE to determine the distribution of the fusion protein.

## 4. Notes

1. Serum-containing medium can be processed as per **steps 5–7** under **Subheading 3.1.**, but, the decorin produced from this material can contain low levels of bovine proteoglycans derived from the serum. By collecting and processing conditioned medium made under serum-free conditions, the levels of contaminating proteoglycans from the bovine serum are undetectable.

2. We have found that 16 h is an optimal period to allow cultures to condition their serum-free medium with decorin. Longer incubations tend to increase DNA contamination in the decorin preparations, due to increased cell lysis, while shorter incubations yield lower concentrations of decorin.
3. It is important to add protease inhibitors before the centrifugation of conditioned medium in order to reduce degradation of decorin core protein caused by proteases released from any lysed cells.
4. Increasing the ionic strength of the conditioned medium prior to performing ion-exchange chromatography reduces binding of weak polyanionic molecules to the DEAE-Sepharose matrix and thus increases the binding capacity of the column and the purity of the isolated decorin.
5. For handling of large volumes of conditioned medium we typically use the 5-gal plastic bottles from common water coolers (each will hold 18–19 L of conditioned medium).
6. Occasionally it may be necessary to siphon off the top 1–2 mm of the matrix from the guard column, as this may become clogged with a floculant precipitate.
7. To reduce loss of decorin in the wash buffer of the DEAE-Sepharose column, the total wash time should not exceed 4 h. The fast flow rates necessary can be achieved under a gravity-driven flow by positioning the connected wash bottle approximately 7 ft (2 m) above the column.
8. We typically analyze the purity of eukaryotic expressed decorin during the purification scheme by electrophoresing samples under reducing conditions on 5–20% SDS-polyacrylamide gradient gels. These gels are subsequently stained with Coomassie blue to detect any protein contaminants and then with 0.1% toluidine blue (w/v) in 0.1 *M* acetic acid, to detect proteoglycans.
9. The DEAE column flow-through conditioned medium can be reapplied to the ion-exchange columns for an approx 10% additional yield of decorin.
10. We use the Hydrocell C4-1000, 50 mm × 4.6 mm (BioChrom Labs) on a TOSOH HPLC system (TSK 6011) equipped with a TSK 6041 UV detector and a GM8010 gradient monitor; however, a Pharmacia FPLC system (or other equivalent) can be used.
11. Preliminary preparative runs using the larger, 150 mm x 21 mm, Hydrocell C4-1000 column indicate that approx 600 mg of recombinant human decorin can be bound and recovered from this column.
12. Although the pATH3 expression vector uses a fairly tight promoter, if low yields of the fusion protein are observed expression can sometimes be improved by depleting tryptophan levels in the cultures for 2–3 h before inducing expression with IAA. For example, the overnight LB cultures are separated from the tryptophan-containing spent medium by centrifugation and resuspended at 20%, their overnight density in tryptophan-free M9 medium containing casamino acids.
13. The optimal expression period must be determined emperically for each particular fusion molecule. For example, depending on the stability and accumulation of individual proteins, the induction period can range from 2 to 12 h.
14. To determine whether the induction step has been successful, it is advised that a duplicate culture not induced by IAA be prepared and analyzed by SDS-PAGE.

## References

1. Moscatello, D. K., Santra, M., Mann, D. M., McQuillan, D. J., Wong, A. J., and Iozzo, R. V. (1998) Decorin suppresses tumor cell growth by activating the EGF-receptor. *J. Clin. Invest.* **101**, 406–412.

2. Yamaguchi, Y., Mann, D. M., and Ruoslahti, E. (1990) Transforming growth factor is negatively regulated by a proteoglycan. *Nature* **346**, 281–284.
3. Koerner, T. J., Hill, J. E., Myers, A. M., and Tzagoloff, A. (1991) High-expression vectors with multiples cloning sites for construction of *trpE* fusion genes: pATH vectors. *Meth. Enzymol.* **94**, 477–490.
4. Mandecki, W., Powell, B. S., Mollison, K. W., Carter, G. W., and Fox, J. L. (1986) High-level expression of a gene encoding the human complement factor C5a in *Escherichia coli. Gene* **43**, 131–138.
5. Yamaguchi, Y. and Ruoslahti, E. (1988) Expression of human proteoglycan in Chinese hamster ovary cells inhibits cell proliferation. *Nature* **336**, 244–246.
6. Choi, H. U., Johnson, T. L., Pal, S., Tang, L-H, Rosenberg, L., and Neame, P. J. (1989) Characterization of the dermatan sulfate proteoglycans, DS-PGI and DS-PGII, from bovine articular cartilage and skin isolated by octyl-sepharose chromatography. *J. Biol. Chem.* **264**, 2876–2884.
7. Mann, D. M., Yamaguchi, Y., Bourdon, M. A., and Ruoslahti, E. (1990) Analysis of glycosaminoglycan substitution in decorin by site-directed mutagenesis. *J. Biol. Chem.* **265**, 5317–5323.
8. Roughly, P. J., White, R. J., Magny, M-C., Liu, J., Pearce, R. H., and Mort, J. S. (1993) Non-proteoglycan forms of biglycan increase with age in human articular cartilage. *Biochem. J.* **295**, 421–426.

# 22

## Prokaryotic Expression of Proteoglycans

**Alan D. Murdoch and Renato V. Iozzo**

### 1. Introduction

For molecules with such extensive posttranslational modifications, it may not be immediately obvious why one would wish to express the protein cores of proteoglycans in prokaryotic systems, where none of this modification can occur. However, there are several cases where this is not a problem, and some where there is a positive advantage to expressing a core protein with none of the normal eukaryotic modifications.

Bacterially expressed proteoglycan core proteins have been used successfully to raise polyclonal antisera ***(1,2)*** and as both immunization and screening agents in monoclonal antibody production, notably in the production of domain specific monoclonals ***(3,4)***. Bacterial fusion proteins have also been used to probe core protein domain–carbohydrate interactions ***(5)***, as substrate for modification enzymes such as core protein–UDP-xylose xylosyltransferase ***(6,7)***, and to gain at least some structural and functional information about important extracellular matrix molecules ***(8,9)***.

Bacterial fusion systems can be divided roughly into two main groups depending on the size of the vector-encoded fusion partner. Those with small tags such as polyhistidine (6 × His), which allow purification on immobilized metal ions (e.g., Qiagen, Clontech), FLAG (Sigma), or c-myc peptide tags (e.g., Invitrogen), which allow detection and purification with specific antibodies, and the small (4-kDa) calmodulin-binding peptide, which allows purification on calmodulin resin (Stratagene), have the advantage that the tag can be left on the protein of interest with minimal or no interference with the structure or function. Larger tags such as glutathione S-transferase, protein A, maltose-binding protein, or cellulose-binding domain (Amersham Pharmacia Biotech, Novagen, New England Biolabs) may help with fusion proteins that have solubility problems, but the large fusion partner can present its own problems. There are many methods available for enzymatic or autocatalytic cleavage of fusion partners, but in many cases, such as monoclonal antibody production or binding studies, it is not necessary to remove the fusion partner, as

From: *Methods in Molecular Biology, Vol. 171: Proteoglycan Protocols*
Edited by: R. V. Iozzo © Humana Press Inc., Totowa, NJ

experimental design can control for the presence of the partner. The choice of fusion partner may be largely empirical, since not all proteins will express in the first system of choice, and several may need to be tested before satisfactory expression levels are achieved.

The pMAL system from New England Biolabs fuses the protein of interest to maltose-binding protein (MBP), the product of the *Escherichia coli malE* gene ***(10,11)***. MBP is normally secreted into the bacterial periplasmic space; the pMAL-p2 vector can be used to take advantage of this and produce secreted periplasmic fusion proteins. The pMAL-c2 vector has the signal peptide sequence of the *malE* gene removed and can be used for high-level cytoplasmic expression of fusion proteins. In both cases, the fusions can be purified by virtue of the affinity of MBP for maltose using a column of immobilized amylose (a maltose polymer). In this chapter we present details of pilot-scale experiments for cytoplasmic and periplasmic expression of MBP fusion proteins, with an example of cytoplasmic expression, and a protocol for affinity purification. Further details about the system can be found in the pMAL system handbook (available on the NEB website at **http://www.neb.com**). In addition, since many fusion proteins will form insoluble inclusion bodies in many of the common *E.coli* strains, we include a convenient method for solubilization of protein from these inclusion bodies in a form suitable for affinity purification.

## 2. Materials

1. pMAL-c2 and pMAL-p2 vectors, components of the pMAL system. General molecular biology supplies: restriction endonucleases, T4 DNA ligase, etc., facilities for running horizontal agarose gels, water baths, DNA sequencing facilities. PCR reagents, including proofreading polymerases such as Pfu or Tli polymerases if necessary. SDS-PAGE gel equipment, columns, pumps, and fraction collector for affinity chromatography.
2. Luria-Bertani (LB) growth medium: 10 g of tryptone, 5 g of yeast extract, 5 g of NaCl per liter. Adjust pH to 7 and autoclave.
3. Glucose. 1 *M* glucose stock: Dissolve 18 g of glucose and make up to 100 mL. Filter sterilize and add 10 mL to cooled sterile LB for rich growth medium (*see* **Note 1**).
4. Ampicillin 1000 × stock: Dissolve 0.5 g of ampicillin in 5 mL of $H_2O$. Filter sterilize into aliquots and store at –20°C.
5. X-gal: Dissolve at 40 mg/mL in dimethylformamide. IPTG 1 *M* stock: Dissolve 2.38 g of IPTG in 10 mL of $H_2O$, filter sterilize into aliquots. Store both X-gal and IPTG at –20°C.
6. Column buffer: 20 m*M* Tris-HCl, pH 7.4, 200 m*M* NaCl, 1 m*M* EDTA. Dissolve 2.42 g of Tris base, and 11.7 g of NaCl per liter and add 2 mL of 0.5 *M* EDTA stock. Adjust pH to 7.4. In addition, the column buffer used in the example shown in **Fig. 1** also contained 1 m*M* sodium azide (1 mL of 1 *M* stock per liter) and 10 m*M* β-mercaptoethanol (0.7 mL/L).
7. Cold osmotic shock wash buffer: 30 m*M* Tris-HCl, pH 8.0, 20% sucrose. Dissolve 3.63 g of Tris base and 200 g of sucrose per liter and adjust pH to 8.0.
8. For the cold osmotic shock: 5 m*M* $MgSO_4$. Dissolve 60 mg of anhydrous $MgSO_4$ in 100 mL of $H_2O$.
9. Anti-MBP serum, component of the pMAL system.
10. Amylose resin, component of the pMAL system.
11. Inclusion body lysis buffer: 50 m*M* Tris-HCl, pH 8.0, 100 m*M* NaCl, 1 m*M* EDTA. Dissolve 3.03 g of Tris base, and 2.92 g of NaCl and add 1 mL of 0.5 M EDTA for 500 mL of buffer and adjust the pH to 8.

12. Inclusion body wash buffer: as per lysis buffer above, but add 10 mL of 0.5 *M* EDTA and 2.5 mL of Triton X-100 per 500 mL.
13. Inclusion-body solubilization buffer: Lysis buffer containing 8 *M* urea and 0.1 m*M* PMSF. Dissolve 48 g of urea by adding 50 mL of $H_2O$ and then making the solution up to 100 mL. Deionize by adding 1 g of mixed-bed resin TMD-8 (Sigma) and stir for 1 h. Filter the beads out of the solution through a sintered glass filter. Add to the solution 0.61 g of Tris base, 0.58 g of NaCl, 0.2 mL of 0.5 *M* EDTA and adjust pH to 8. Add PMSF just before use to 0.1 m*M* (*see* **Note 2**).
14. Alkaline pH buffer: 50 m*M* $KH_2PO_4$, pH 10.7, 50 m*M* NaCl, 1 m*M* EDTA. Dissolve 3.4 g of $KH_2PO_4$ and 1.46 g of NaCl and add 1 mL of 0.5 *M* EDTA per 500 mL. Adjust the pH to 10.7.

## 3. Methods

### 3.1. Subcloning and Screening

1. Subclone the cDNA encoding the gene or open reading frame of interest into both the pMAL-c2 and pMAL-p2 vectors for small-scale expression. The translational reading frame of the insert must be the same as the *malE* gene from the vector. If no vector-derived amino acids are wanted in a factor Xa-cleaved protein, then the open reading frame should be cloned into the *XmnI* site in the vector. Bear in mind that the first encoded amino acid in this case should not be proline or arginine, since this will inhibit factor Xa cleavage. Otherwise, the remaining restriction sites can be used, preferably for directional cloning. If using the PCR to generate insert, proofreading polymerases will produce suitable blunt-ended fragments for *XmnI* cloning, and downstream primers should include a translational stop codon. A number of cloning strategies are also presented in the pMAL system manual.
2. Ligate 25–50 ng of suitably digested pMAL-c2 and -p2 with 1–4 times the molar amount of insert in a standard ligation. For directionally cloned inserts, include a vector-only control reaction.
3. Make competent TB1 or other *E. coli* strain (or purchase ready competent cells). We have found the $CaCl_2$ method works well for TB1.
4. Mix the ligation reaction with 50 µL of competent cells, incubate on ice for 30 min, and heat shock at 42°C for the length of time suitable for the strain used (2 min for TB1).
5. Recover the cells by adding 100–200 µL of LB and incubating at 37°C for 30 min. Spread all of the transformation on 2 or 3 LB plates containing 100 µg/mL of ampicillin and incubate overnight at 37°C.
6. If the insert was directionally cloned into the vector, and the vector control plate has few or no colonies, pick colonies into 3 mL of LB with 100 mg/mL of ampicillin and shake overnight at 37°C for plasmid minipreps (*see* **Note 3**).
7. If directional cloning was not possible and blue/white screening is necessary, either replica plate the colonies onto a fresh plate containing 100 µg/mL amp, 80 µg/mL of X-gal, and 0.1 m*M* IPTG, or pick colonies with sterile toothpicks and stab or patch them onto an LB amp plate and an LB amp/X-gal/IPTG plate and incubate at 37°C overnight. White colonies on the X-gal plate show which colonies should be picked from the LB amp plate for overnight liquid culture in 3 mL of LB with 100 µg/mL amp (*see* **Note 3**). Do not pick colonies directly from plates containing IPTG.
8. Prepare miniprep DNA from the overnight cultures. Cut the DNA with suitable restriction enzyme(s) and run on agarose gels to check for insert. Plasmids that appear to contain insert should be sequenced to check for appropriate insert sequence and that the reading frame across the MBP/fusion partner junction has been retained. Glycerol stocks

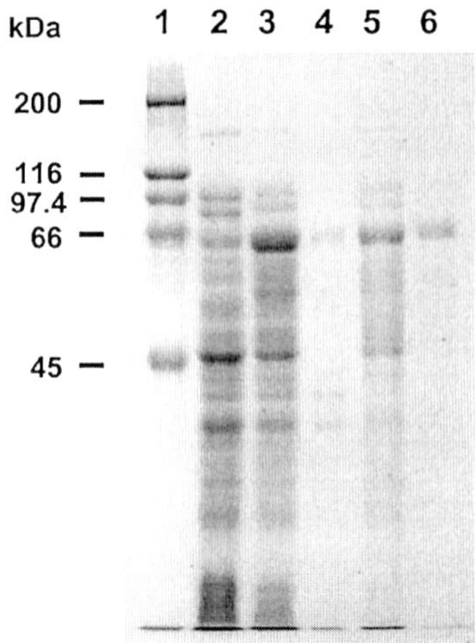


Fig. 1. Purification scheme for a maltose-binding protein/perlecan domain II fusion protein. A cDNA encoding the entire low-density lipoprotein-like region (domain II) of the basement membrane heparan sulfate proteoglycan perlecan was subcloned into the pMAL-c2 vector (3) and a pilot-scale expression was performed as described under **Subheading 3.1.** Samples were run on a 10% SDS-PAGE gel and stained with Coomassie brilliant blue. *Lane 1*, molecular weight markers. *Lane 2*, uninduced crude total lysate. *Lane 3*, induced crude total lysate. *Lane 4*, crude soluble extract. *Lane 5*, crude insoluble material. *Lane 6*, amylose resin sample: material from the crude soluble extract that bound to the amylose resin in the small batch binding procedure detailed under **Subheading 3.1.6.**

should be prepared for archival purposes and cultures restreaked on LB plates containing 100 µg/mL ampicillin for small-scale expression testing.

9. Grow a small (5-mL) culture from a single bacterial colony in LB + glucose + ampicillin, shaking with good aeration at 37°C, to an $A_{600}$ of ~0.5. Take a 1-mL sample and pellet the cells in a microfuge (top speed, 1 min). Aspirate the supernatant, resuspend the pellet in 50 µL of SDS-PAGE sample buffer and store at –20°C until needed (uninduced sample).
10. Add IPTG to the remaining 4 mL to 0.3 m*M*, and continue shaking at 37°C for 2 h. Remove a 0.5-mL sample, microfuge as before and resuspend in 100-µL SDS-PAGE sample buffer (induced sample).
11. Boil the uninduced and induced samples for 5 min and run 20-µL samples on an SDS-PAGE gel of a suitable percentage for the expected fusion protein size and check for an induced band of the correct size following staining with Coomassie blue (*see* **Fig.1**, lanes 2 and 3).

### 3.2. Pilot-Scale Expression—Cytoplasmic Expression

1. Grow an overnight culture of cells (5 mL) in LB + glucose + amp from a single colony of a stock identified as expressing the fusion protein (above).
2. Add 0.8 mL of overnight culture to 80 mL (1/100) of fresh growth medium and grow as before to an $A_{600}$ of ~0.5. Take an uninduced sample as before, add IPTG to a final concentration of 0.3 m*M*, and shake for a further 2 h (*see* **Note 4**). Take an induced sample as before and centrifuge the remaining culture at 4000 g for 10 min. Resuspend the pellet in 10 mL of column buffer (*see* **Note 5**).

3. Freeze the suspension at –20°C (–70/80°C can also be used). Thaw in cold water.
4. Sonicate the cells to disrupt them and release protein. Prevent the suspension from overheating by using an ice-water bath and sonicating in short pulses. Protein release can be monitored with the Bradford assay. Sonicate until no more protein is being released.
5. Centrifuge the sonicated suspension at 9000 g for 20 min. Remove the supernatant (crude soluble extract) and store on ice. Resuspend the pellet in 10 mL of column buffer (crude insoluble material).
6. Add approx 200 µL of settled amylose resin as supplied and microfuge briefly. Remove supernatant and wash the resin twice by resuspension in 1.5 mL of column buffer and microfuging. Resuspend the resin in 200 µL of column buffer and add 50 µL of resuspended slurry to 50 µL of the crude soluble extract. Incubate on ice for 15 min, microfuge for 1 min, wash pellet with 1 mL of column buffer, microfuge again, and resuspend the resin pellet in 50 µL of SDS-PAGE sample buffer.
7. Take 5 µL of the crude soluble extract and the crude insoluble material and add 5 µL 2× SDS-PAGE sample buffer to each. Boil these and the uninduced, induced, and amylose resin samples for 5 min. Microfuge the tubes to pellet the amylose resin and run 20 µL of the supernatant along with 20 µL of the uninduced and induced total lysate samples and all of the crude soluble and insoluble samples on an SDS-PAGE gel (*see* **Fig.1**).

### 3.3. Pilot-Scale Expression—Periplasmic Expression

1. Proceed as above to grow and induce a culture of cells. After centrifuging for 10 min at 4000 g (*see* **Subheading 3.1.**, **step 2**) resuspend the pellet in 10 mL of 30 m*M* Tris-HCl, pH 8.0; 20% sucrose.
2. Add EDTA to a final concentration of 1 m*M* (20 µL of 0.5 *M* EDTA to 10 mL) and incubate at room temperature with shaking for 5–10 min.
3. Centrifuge at 8000 g at 4°C for 10 min. Decant all the supernatant and resuspend the pellet in 10 mL of ice-cold 5 m*M* $MgSO_4$. Shake for 10 min in an ice-water bath.
4. Centrifuge again at 8000 g for 10 min at 4°C. The supernatant is the cold osmotic shock extract. Samples of this extract (10 µL plus 10 µL 2× SDS-PAGE sample buffer) can be run on SDS-PAGE gels. Depending on the level of expression, immunoblotting with the anti-MBP serum supplied with the kit may be necessary.

### 3.4. Affinity Chromatography

1. The maltose-binding protein will bind to the amylose resin in a variety of buffer systems at around pH 7 and in varying ionic strengths. Intracellular proteins extracted in column buffer are ready to apply to the resin, following dilution if necessary (*see* below). Proteins in cold osmotic shock extracts should be adjusted to 20 m*M* Tris-HCl, pH 7.4 by the addition of 1 *M* Tris-HCl pH 7.4. Proteins from cytoplasmic extracts that contain intramolecular disulfide bridges (such as the perlecan domain II fusion in **Fig. 1**) may need to be purified over the column in a reducing environment (e.g., 10 m*M* β-mercaptoethanol) to prevent loss from incorrect disulfide bonding and aggregation. A refolding protocol may then be necessary for the downstream application (*see* **Note 6**).
2. Pour an amylose resin column suitable for the amount of protein to be purified. The pMAL system handbook suggests a 2.5 × 10 cm column; we have found K9/15 and XK16/20 columns from Pharmacia and 1 × 10 cm econocolumns from Bio-Rad to be satisfactory. Column size and bed volume should be adjusted to the expected yield given a binding capacity of ~3 mg/mL packed resin. We have generally used 10-mL columns (~30 mg binding capacity is more than enough for many purposes, and larger bed volumes only need longer washing steps). Wash the column with 8 vol of column buffer.

3. Dilute the crude extract to a total protein content of 2.5 mg/mL or less with column buffer. Cold osmotic shock extracts may not need diluting. Load the extract onto the column at [10 × (diameter of column in cm)$^2$] mL/h (~0.5 mL/min for an XK16 column).
4. Wash the column with 12 volumes of column buffer (or with an excess overnight).
5. Elute the bound fusion protein in column buffer containing 10 m*M* maltose, collecting fractions of one-fifth column size (usually 2–3 mL). Fusion protein will usually start to elute within the first column volume and should be quite a tight peak; collecting beyond 3 column volumes is usually not necessary.
6. Determine the protein content of fractions with the Bradford assay and pool the fractions containing protein.
7. Check the protein concentration of the pooled material and run samples on SDS-PAGE gels to verify the integrity of the protein.
8. Protein can be dialyzed, or concentrated and buffer-exchanged in Centricon concentrators (Amicon) if it is necessary for your application.

### 3.5. Inclusion Body Solubilization

1. This method is taken from Sambook et al. ***(12)***, based on a method of Marston ***(13)*** and uses a combination of 8 *M* urea and alkaline pH for solubilisation. Unless stated, all procedures should be carried out on ice or in a cold room.
2. Pellet up to 1 L of an induced culture of cells by centrifugation at 4000 g for 10 min at 4°C.
3. Decant the supernatant and weigh the pellet. Resuspend the pellet in 3 mL of lysis buffer per gram wet weight.
4. Add 8 µL of 50 m*M* PMSF per gram of cells (*see* **Note 2**) and then 80 µL of 10 mg/mL lysozyme per gram. Incubate for 20 min with occasional mixing.
5. Stir the lysate with a glass rod and, while stirring, add 4 mg of deoxycholate per gram of original cell weight. Move the lysate to a 37°C water bath and continue to stir. When the lysate becomes viscous, add 20 µL of 1-mg/mL DNAse I per gram of cells, mix well, and incubate at room temperature.
6. Check the lysate periodically for viscosity. After approx 30 min, the mixture should be no longer viscous.
7. Centrifuge the lysate at 12,000 g for 15 min at 4°C.
8. Resuspend the pellet in 9 vol of inclusion body wash buffer and incubate at room temperature for 5 min (first wash).
9. Centrifuge at 12,000 g for 15 min at 4°C.
10. Remove the supernatant and keep. The pellet can be rewashed several times. At each wash, save the supernatant and analyze small (10-µL) samples along with a small sample of the final pellet resuspended in 100 µL of $H_2O$ by SDS-PAGE. If large amounts of the fusion protein are washing out of the pellet, the washes can be kept and pooled with the solubilised and dialyzed material from the pellet (below). The small amount of Triton X-100 in a diluted sample applied to an amylose column may interfere with MBP binding, but will have to be determined on an empirical basis.
11. Resuspend the pellet in 100 µL of inclusion body solubilization buffer. Store at room temperature for 30 min.
12. Add the mixture to 9 vol of alkaline pH buffer. Incubate at room temperature for 30 min. Check the pH by removing very small samples to pH strips at regular intervals. If the pH begins to drop, add KOH to maintain it at 10.7.
13. Adjust the pH to 8.0 with HCl and incubate for a further 30 min at room temperature. It is important during these alterations in the pH that the temperature is not allowed to rise above room temperature.

14. Centrifuge at 12000 g for 15 min at room temperature. Decant the supernatant and store. Resuspend the pellet in 100 μL of SDS-PAGE sample buffer. Run 10 μL samples of supernatant and resuspended pellet on an SDS-PAGE gel to check the amount of solubilized material.
15. The supernatant should be dialysed into pMAL column buffer before affinity chromatography (above). We have found a single-step dialysis in the presence of β-mercaptoethanol to be satisfactory, but depending on the fusion partner and the downstream application a stepwise dialysis from 8 *M* urea or dropwise dilution into an excess of column buffer may be tried.

## 4. Notes

1. The glucose in the growth medium represses the expression of *E. coli* amylase. Expression of amylase can lead to problems with the affinity purification due to degradation of the amylose resin.
2. Some of the protocols listed here use PMSF as a serine protease inhibitor. In many cases this compound can now be replaced with an equimolar concentration of the much safer and water-soluble AEBSF (Calbiochem).
3. In our hands the pMAL-c2 and -p2 vectors are not as stable as other plasmids such as pBluescript. In order to generate plasmids containing insert with no obvious deletions or rearrangements of the vector, it is normally necessary to pick at least 10 colonies at a time for overnight liquid culture. More than one set of 10 may need to be screened by restriction digestion.
4. The conditions for IPTG induction may need to be optimized for the particular fusion. Lower or higher IPTG concentrations, induction for shorter or longer times, and induction at higher cell densities can all be tested. Problems with fusion protein solubility can also be improved by altering the growth and induction conditions—for example, expressing the protein at a lower temperature—but often it is easier to optimize for maximum expression and then solubilize the protein inclusions, especially if the application does not require native folded protein.
5. Adding additional protease inhibitors at this step may help with protein stability if this is a problem.
6. Many refolding protocols are available. A good selection to start from can be found in the pMAL handbook (at **http://www.neb.com**).

## References


1. Dudhia, J., Davidson, C. M., Wells, T. M., Vynios, D. H., Hardingham, T. E., and Bayliss, M. T. (1996) Age-related changes in the content of the C-terminal region of aggrecan in human articular cartilage. *Biochem. J.* **313**, 933–940.
2. Zimmerman, D. R., Dours-Zimmerman, M. T., Schubert, M., and Bruckner-Tuderman, L. (1994) Versican is expressed in the proliferating zone in the epidermis and in association with the elastic network of the dermis. *J. Cell Biol.* **124**, 817–825.
3. Murdoch, A. D., Liu, B., Schwarting, R., Tuan, R. S., and Iozzo, R. V. (1994) Widespread expression of perlecan proteoglycan in basement membranes and extracellular matrices of human tissues as detected by a novel monoclonal antibody against domain III and by in situ hybridization. *J. Histochem. Cytochem.* **42**, 239–249.
4. Whitelock, J. M., Murdoch, A. D., Iozzo, R. V., and Underwood, P. A. (1996) The degradation of human endothelial cell-derived perlecan and release of bound basic fibroblast growth factor by stromelysin, collagenase, plasmin, and heparanases. *J. Biol. Chem.* **271**, 10,079–10,086.

5. Ujita, M., Shinomura, T., Ito, K., Kitagawa, Y., and Kimata, K. (1994) Expression and binding activity of the carboxyl-terminal portion of the core protein of PG-M, a large chondroitin sulfate proteoglycan. *J. Biol. Chem.* **269**, 27,603–27,609.
6. Brinkmann, T., Weilke, C., and Kleesiek, K. (1997) Recognition of acceptor proteins by UDP-D-xylose proteoglycan core protein beta-D xylosyltransferase. *J. Biol. Chem.* **272**, 11,171–11,175.
7. Weilke, C., Brinkmann, T., and Kleesiek, K. (1997) Determination of xylosyltransferase activity in serum with recombinant human bikunin as acceptor. *Clin. Chem.* **43**, 45–51.
8. Williamson, R. A., Natalia, D., Gee, C. K., Murphy, G., Carr, M. D., and Freedman, R. B. (1996) Chemically and conformationally authentic active domain of human tissue inhibitor of metalloproteinases-2 refolded from bacterial inclusion bodies. *Eur. J. Biochem.* **241**, 476–483.
9. Ruppert, R., Hoffman, E., and Sebald, W. (1996) Human bone morphogenetic protein 2 contains a heparin-binding site which modifies its biological activity. *Eur. J. Biochem.* **237**, 295–302.
10. Guan, C., Li, P., Riggs, P. D., and Inouye, H. (1987) Vectors that facilitate the expression and purification of foreign peptides in *Escherichia coli* by fusion to maltose-binding protein. *Gene* **67**, 21–30.
11. Maina, C. V., Riggs, P. D., Grandea, A. G. III, Slatko, B. E., Moran, L. S., Tagliamonte, J. A., McReynolds, L. A., and Guan, C. (1988) A vector to express and purify foreign proteins in *Escherichia coli* by fusion to, and separation from, maltose binding protein. *Gene* **74**, 365–373.
12. Sambrook, J., Fritsch, E. F., and Maniatis, T., ed (1989) *Molecular Cloning: A Laboratory Manual.* Cold Spring Harbor Laboratory Press, Cold Spring Harbor, MA.
13. Marston, F. A. O., Lowe, P. A., Doel, M. T., Schoemaker, J. M., White, S., and Angal, S. (1984) Purification of calf prochymosin (prorennin) synthesised in *Escherichia coli. Bio/Technology* **2**, 800–804.

# 23

# Proteoglycan Gene Targeting in Somatic Cells

Giancarlo Ghiselli and Renato V. Iozzo

## 1. Introduction

### *1.1. Overview of Gene Targeting in Mammalian Cells*

Gene targeting allows the generation of specifically designed mutations in cells by the mean of homologous recombination between exogenous DNA and endogenous genomic sequences *(1–3)*. This is possible because a fragment of genomic DNA introduced into a cell can locate and recombine with the chromosomal homologous sequences *(4)*. Although the mechanisms of exogenous DNA transfer to the nucleus and of homologous recombination are still poorly understood, important advances have been made in identifying the requirements for the targeting vector and the desirable features of the targeted locus that maximize gene targeting *(5)*. Presently, various gene targeting techniques can be successfully applied to a variety of organisms. In mammals, gene targeting strategies have been developed mainly for the genetic manipulation of mouse embryonic stem (ES) cells. Mating of the transgenes leads to the generation of animals in which the functional significance of a specific gene silencing can be investigated in different conditions. Cell lines can also be established from mouse organ biopsies. There are, however, important limitations to the attainability of cell lines of interest from transgenic animals. First, ES gene knockout techniques are currently limited to the mouse. Because there are major interspecies differences with regard to gene pattern of expression and implication in disease, extrapolation from the results obtained with mice cells to the pathophysiological mechanisms in humans is not warranted. This is particularly the case when genes involved in malignant transformation and development are investigated. Second, the availability of established cell lines bearing specific mutations makes somatic gene targeting highly desirable. In this case the implication of the expression of a particular gene on the cell behavior can be assessed directly against a well-defined genetic background. Third, the establishment of an organ-specific cell line involves complex genetic rearrangements leading to immortalization. In other words, cell lines, even if

From: *Methods in Molecular Biology, Vol. 171: Proteoglycan Protocols*
Edited by: R. V. Iozzo © Humana Press Inc., Totowa, NJ

derived from the same animal cannot be considered isogenic in the sense that the pattern of expression of several genes — as opposed to the single targeted gene—is affected. In addition, the genetic mutation that has been introduced through homologous recombination in ES cells may be lethal even at a stage preceding embryonic development, thus precluding the possibility of establishing a cell line. Finally, the cost of originating and maintaining a transgenic animal colony is high.

These potential drawbacks are particularly relevant when considering the gene knockout of proteoglycan core proteins. There is in fact ample evidence that proteoglycans play a crucial role, both in development as well as in modulating the phenotypic expression of somatic cells. This is a case where the ES cell knockout strategy is of dubious value, as the generated transgenic animals will likely display a phenotype that is the result of a constellation of effects ensuing at different stages of life. A mechanistic interpretation of the function of a gene at a cellular level is thus difficult. This is more so in the case of diseases in which an array of genes rather than a single gene play a role, such as in cancer.

The subject of homologous recombination and gene targeting has been reviewed extensively in the recent past *(5,6)*. The reader is directed to that literature for a critical examination of the various strategies devised for achieving successful gene targeting and knockout. In this chapter, we will discuss those aspects of the targeting vector design that are pertinent to somatic gene knockout and the methodology for vector transfer into established cell lines other than ES cells. The design of an effective screening strategy for the identification of homologous recombinants by genotyping and functional assays is also discussed. As proteoglycans play a role in cell differentiation and growth *(7)* their gene knockout can significantly affect cell behavior. Consequently, care should be taken in devising a screening strategy that is not negatively biased toward the selection of clones with a rate of growth different from that of the parent cell line, as this would lead to poor recovery of the clones in which homologous recombination has occurred.

### 1.2. Targeting Strategies for Gene Knockout

A major limitation of gene targeting is that DNA introduced into the mammalian genome integrates by random, nonhomologous recombination at a rate 100 to 100,000 times more frequently than by homologous recombination. In order to identify and recover the small fraction of cells in which a homologous recombination occurs, several strategies have been developed. Through these strategies, successful homologous recombination have been reported to occur in at about 1% of stable transfected ES clones. Two of these strategies, the so-called positive–negative selection (PNS) *(8)* and the promotorless strategy *(9,10)* have been found to be particularly advantageous. The PNS approach exploits the use of two selectable markers in the targeting vector. The first marker gene harboring a drug resistance cassette is inserted within the region of homology of the vector and is driven by an exogenous promoter. This marker serves to disrupt the gene translation and at the same time to select the cells that have incorporated the homologous DNA region of the targeting vector within the chromosomal DNA. Examples of selectable markers are *neo,* which confer resistance to G418, an analog of neomycin that is toxic to mammalian cells, and *hydro* and *puro,* which confer resistance to the hygromycin and the puromycin antibiotic, respectively. The second marker's gene, also driven by an exogenous promoter, is placed outside the homology region of the targeting vector and confers sensitivity to a toxic agent (e.g., Herpes simplex virus thymidine kinase [HSV-*tk*] which impart sensitivity to purine analog such as *ganciclovir*, FIAU, and *acyclovir* through phosphorylation of these agents). The rationale is that, unless a double reciprocal crossover (i.e., a replacement) is taking place and the HSV-*tk* gene is excised out of the inserted vector, cells become sensitive to the purine toxic agent as the negative-selection gene is incorporated into the chromosomal DNA. The main advantage of the PNS vector strategy is that there is no requirement for the targeted gene to be expressed, as both markers gene are driven by their own promoters. This strategy has been adopted extensively in ES cells. The occurrence of homologous recombination is somatic cell lines is, however, two to three orders of magnitude lower than in ES cells (*5*) and targeting strategies have been developed that allow the identification of homologous recombinant events with higher efficiency than that allowed by the PNS strategy. The attainment of high screening efficiency is crucial for somatic gene knockout inasmuch as a double round of gene targeting is required. The so-called promotorless selectable marker strategy relies on a single selectable marker that becomes activated if the targeting vector is incorporated in a chromosomal region that allows the gene to be driven by an endogenous promoter. The expected transcriptional product is represented by a fusion protein comprising the N-terminal region of the targeted protein and the selectable marker. Since activation by cellular promoters occurs in about 1% of random integration events, the use of promotorless vectors provides a 100-fold enrichment for homologous recombination by reducing the background of nonhomologous recombination. By careful manipulation of the concentration of the selection agent it has been possible to increase the rate of legitimate recombination to values approaching or even exceeding 10% ***(11)***. A recent survey of the literature ***(12)*** of 23 experiments of gene targeting with a promotorless vector has revealed that gene targeting frequency in human cell lines is highly variable, ranging from 1/3 for the p21 gene targeted in HCT116 colon carcinoma cells to 1/940 for the p53 gene in the same cell line, suggesting important locus-to-locus variability. A significant advantage of the promotorless strategy with reference to somatic gene knockout is that since the end point of the experiment is represented by the suppression of a cell phenotypic trait, assays relying on the expression of the targeted gene product can be considered for screening purposes. This can greatly facilitate the screening of the stable transfected clones, bypassing the problem of identifying the modality of gene mutation through genomic characterization of the targeted locus.

### 1.3. Design of a Promotorless Targeting Vectors for Somatic Knockout of Proteoglycan Genes

Although the description of the method is general, this approach can be applied to the somatic knockout of any given proteoglycan gene. A targeting vector is designed to recombine with and mutate a specific chromosomal locus. The construction of a promotorless targeting vector involves the selection of a suitable genomic fragment harboring an exon into which a positive selection cassette can be inserted in frame. The positive selection marker serves two functions: (1) to isolate the rare transfected cells among which there are those that have properly integrated DNA, and (2) to silence the gene by mutation. The selected genomic fragment must be free of any functional promoter/enhancer elements and of repeating sequences that might increase the rate of detectable spurious recombinations. The desirable features of the homologous sequence to be incorporated in the targeting vector are:

1. The presence of regions coding for the N-terminus of the protein such that the expression of a proteins with functional activity is rigorously precluded.
2. The absence of repetitive sequences in order to reduce the chances of scattered illegitimate recombination in the chromosomal DNA ***(4)***.
3. The coding region harboring the marker should be located 5 kb or more from the transcriptional starting site. This is to prevent the gene promoter influencing the expression of selectable marker ***(5,11)***.
4. The selectable marker should be preferably inserted within an exon initiating and terminating in different splicing phase. This to avoid that splicing of the mutated exon might give rise to a functional gene product ***(13)***.
5. The size of the homologous sequence should be between 3 and 6 kb. Efficiency of homologous recombination is dependent on the length of the homologous region, but there is little evidence that beyond 6 kb the frequency of homologous recombination occurs at significantly higher rate ***(14–16)***. On the other hand, limitation in the size of the homologous region facilitates the construction of the vector
6. The rate of recombination is affected by the length of homologous end regions of the targeting vector ***(17)***. A nonmutated region of 500 bp at either end of the homologous region should provide sufficient linearized DNA area such that hybridization of exogeneous and chromosomal DNA at the site of crossover might occur efficiently.
7. The targeting vector needs to be linearized such that a single crossover event (insertion targeting) or rather a double crossover event (replacement targeting) is favored ***(18)***. Gene knockout by insertion occurs at a rate 10 times higher than by replacement ***(13)***. In order to favor insertion targeting, the targeting vector is linearized at a restriction site located within the region of homology. On the other hand, a replacement vector requires linearization outside the region of homology, within the vector backbone where unique cutting sites are already mapped. In this sense, the design of a replacement vector is facilitated compared to that of an insertion vector.

Detailed mapping of the region of homology is required, especially when the use of an insertion vector is contemplated. In-depth knowledge of the targeting site and its restriction fragment mapping is also required in order to generate DNA probes that can provide informative results on the successful integration of the targeting vector by Southern hybridization screening. Insertion and replacement targeting vectors give raise to different integration products ***(6)***. The genotype screening strategies should take into account that although some of the crossover events are favored over others, there is the possibility that a rather wide range of integration products is obtained, especially when using an insertion vector. This is because the pattern of integration is largely dependent on the DNA recombination and repair machinery of the host cell

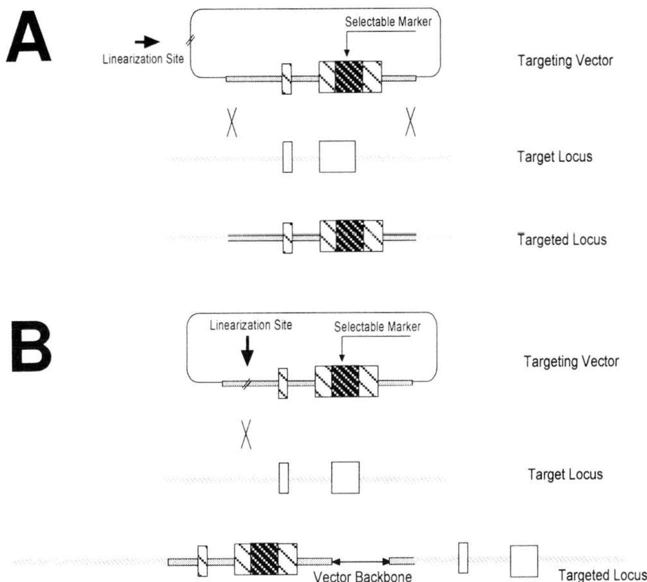


**Fig. 1.** Integration pattern of the targeting vector into a genomic locus. A promotorless targeting vector is constructed by inserting a mutagen-selectable marker (illustrated by the heavy shaded box) into the coding region of the gene of interest (shown as lightly shaded boxes) such that a fusion protein terminated by the stop codon of the selectable marker is generated. The expression of the selectable marker (usually a gene whose product confer-resistance toward a cytotoxic drug) is driven by an endogenous cellular promoter coinciding with that of the targeted gene when a legitimate recombination takes place. Targeting vectors linearized at a site external to the region of homology integrate into the chromosomal DNA preferentially by replacing the endogenous DNA through a double crossover event **(A)**. Gene disruption by insertion is favored when cells are transfected with a targeting vector that has been linearized within the region of homology **(B)**. In this case the vector backbone becomes part of the mutated chromosomal locus. Because a single crossover event is necessary for the integration of foreign DNA, mutagenesis by insertion occurs at higher frequency than by replacement.

line *(4)*. Furthermore, there is clear evidence for a locus-to-locus susceptibility for integration of foreign DNA that affects both the rate of homologous integration as well as the pattern of integration *(12)*. Generally, the final recovery product of a replacement vector is equivalent to the replacement of the homologous chromosomal sequence with all the components of the vector except for the regions flanking the homologous sequence, as all the heterologous part of the vector is excised and lost *(17)* (*see* **Fig. 1A**). The favored final product of an insertion vector on the other hand, is an increase in length of chromosomal DNA corresponding to the full length of the targeting vector including the vector backbone (*see* **Fig. 1B**).

### 1.4. Somatic Gene Knockout: Overall Strategy

After a suitable cell line and the targeting strategy have been selected, the somatic gene targeting involves the following general steps: (1) electroporation of the targeting vector into the host cell line, (2) selection of stable transfected cell colonies by drug selection, (3) rescue and growth of the cell colonies, (4) preparation of frozen stock and of duplicate plates for identification of the targeted clones by Southern blot analysis, PCR, or RT-PCR, (5) new round of targeting in the heterozygous clones utilizing a targeting vector harboring a new selectable marker, and (6) cloning, cell rescue, and gene analysis. The knockout is then confirmed by genotyping, Northern blot hybridization, and immunoassay techniques. For the construction of the promotorless targeting vector in p-Bluescript (Stratagene) or another suitable cloning vector, the neomycin cassette harboring the Neomycin phosphotransferase gene (*neo*) and its polyA tail is generated by PCR from a pSV2-Neo template using primers that allow for the in-frame ligation of the cassette into the selected coding region of the targeting vector. Following cloning and linearization of the construct, the targeting vector is introduced into the selected human cell line by electroporation using 50–100 μg DNA per $10^7$ cells. Human colon carcinoma HCT116 cells has been frequently used for this purpose ***(12)***. This cell line offers some key advantages over other normal or malignant cell lines. First, HCT116 carries a DNA mismatch repair-deficient gene, thereby enhancing the stability of the inserted foreign DNA ***(19)***. Second, these cells are sensitive to the cytotoxic effect of G418 over a relative wide range of concentrations (between 0.4 and up to 10 mg/mL), which is useful when attempting to improve the rate of recovery of successful transfectants by increasing the drug concentration ***(11)***. Third, unlike most transformed cell lines, HCT116 has a normal euploid, which implies that somatic knockout is completed in two rounds of gene targeting. Within 2 wk after electroporation and selection in G418, the drug-resistant colonies are ring-cloned and expanded. The targeting of the second allele is pursued by a second round of transfection with a targeting vector carrying a different drug resistance gene, usually *hygro* or *puro*. After genotyping of the rescued clones by Southern hybridization, final confirmation of the functional destruction of the gene of interest is performed by immunoassays or RNA-based techniques ***(20)***. The targeting strategy and the analysis of the integration products by Southern blot hybridization of the human perlecan gene in HCT116 colon carcinoma cells is illustrated in **Fig. 2**. For the targeting of this gene, a knockout replacenment strategy was considered.

## 2. Materials

1. Low-electroendoosmosis agarose for electrophoresis (from Fisher).
2. TE buffer: 10 m*M* Tris-HCl, pH 7.4, 1 m*M* $Na_2EDTA$.
3. Absolute ethanol (95%), reagent grade.
4. DNA restriction enzymes and related reaction buffers can be purchased from various vendors.
5. Cell electroporation apparatus (example: Hoefer's Progenetor II Electroporation unit).
6. DMEM/FCS: Dulbecco's Minimal Essential Medium with 10% fetal calf serum.
7. DPBS without Ca/Mg: Dulbecco's phosphate buffer saline without calcium and magnesium.
8. Trypsin solution: 0.05% trypsin, 0.53 m*M* $Na_4EDTA$ in Hank's phosphate buffer without Ca/Mg.

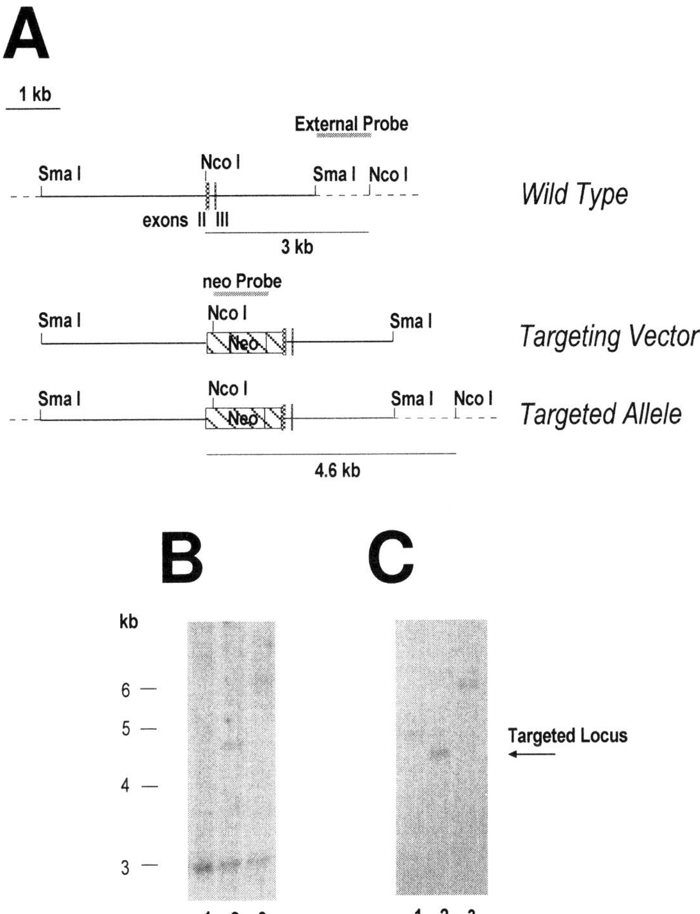


**Fig. 2. (A)** Strategy for targeting of the human perlecan gene in HCT 116 colon carcinoma cells by replacement. The restriction site map of the wild-type allelic locus is illustrated at the top. The solid lines represent the targeted DNA, whereas the dotted line correspond to its flanking regions. An in-scale diagram of the targeting vector constructed by in-frame insertion of the *neo* selectable marker at the *Nco*I restriction site of exon II of perlecan is also illustrated. Note the presence of the *Nco*I site within the *neo* insert that facilitates the RFLP analysis of the recombination products. The genomic DNA of G418-resistant HCT116 cell clones was digested with *Nco*I, the products separated by agarose gel electrophoresis, and transferred to a nitrocellulose membrane for hybridization with two $^{32}$P-DNA probes identified by the shadowed bars in the figure. Panel **(B)** illustrates the autoradiogram obtained by using the "external" probe flanking the 3' region of the targeted locus whereas panel **(C)** illustrates the results with the same blot using a "neo" probe spanning the *neo* DNA. Lane 1, unsuccessful targeting; lane 2, successful targeting identified by the presence of a *Nco*I restriction fragment of the expected size (4.6 kb) that hybridizes with both probes; lane 3, evidence for a third integration product giving a larger-than-expected restriction fragment likely generated by the rearrangement of the targeting vector and/or of the targeted locus.

9. Plasmid DNA purification kit from Qiagen or Promega.
10. *DH5a*-competent *Escherichia coli* bacteria for transfection (from GIBCO-BRL).
11. Sterile cloning rings (from Scienceware).
12. Geneticin G418 (Neomycin sulfate analog) (Life-Technologies).
13. Hygromycin B (Sigma H 0654).
14. TEA buffer (50×): 2 *M* Tris base, 1 *M* glacial acetic acid, 0.05 *M* EDTA (pH 8.0) to 1 L with water.
15. Ethydium bromide: 1% solution in water.
16. Tris-HCl (pH 7.6) saturated phenol: Mix equal volumes of the phenol and the Tris-buffered solution and let the phases separate at room temperature. Store phenol overlayed with the Tris buffer in a dark bottle.
17. SSC buffer (20×): 3 *M* NaCl, 0.3 *M* Na-citrate (pH 7.2).
18. SSPE buffer (20×): 3 *M* NaCl, 0.25 *M* $NaH_2PO_4$, 0.025 *M* EDTA (pH 7.4).
19. Denhart's hybridization solution (50×): 10 mg/mL Ficoll 400, 10 mg/mL polyvynil-pyrrolidone, 10 mg/mL BSA fraction V.
20. Thermostatic water baths set at 56 and 65°C. Heating blocks set at 37°C.
21. Thermostatic shakers at 55°C and 65°C.
22. Heath blocks set at 55°C and 100°C.

## 3. Methods

### 3.1. Vector Cloning and Linearization

Vector DNA is cloned by standard techniques in quantities to achieve a yield of 200–500 μg. Following purification by ion-exchange chromatography, the vector is linearized by digestion with a suitable restriction enzyme. The linearized vector is then freed of contaminating proteins and recovered by phenol–chloroform extraction followed by ethanol precipitation. The resulting DNA is solubilized in water and, following linearization is used directly for the transfection into mammalian cells. The linearized DNA can be safely stored at –70°C for prolonged periods of time.

1. Transform *rec*A1 mutated *E.Coli* bacteria (example DH5α) with the targeting vector and plate on agar using a suitable antibiotic for selection. Pick a formed colony and transfer to 250 mL of LB medium. Grow overnight at 35–37°C or until $OD_{600}$ reaches ~2.0.
2. Collect the bacteria by centrifugation and extract DNA by the alkaline lysis method followed by DNA purification by ion-exchange chromatography. Several companies (including Qiagen and Promega) sell suitable kits for this purpose.
3. Recover the DNA pellet by ethanol or isopropanol precipitation. Carefully evaporate the excess solvent under a sterile hood but do not allow the pellet to desiccate, as this causes DNA to become insoluble.
4. Dissolve the DNA in 500 μL of sterile water. Measure the DNA amount by spectrophotometry at 260/280 nm. Assess the quality of the purified DNA by running 1 μg on 0.7% agarose in TEA buffer. Ethydium bromide-stained DNA should appear as a single band corresponding to supercoiled DNA, with minimal or no nick DNA product evident.
5. Aliquot 100 μg of DNA into a sterile tube. Bring the final volume to 500 μL with water and 10× incubation buffer, and incubate for a minimum of 3 h with 50 U of a suitable restriction enzyme at the manufacturer's suggested temperature (*see* **Note 1**).
6. Stop the incubation by heat-inactivating the enzyme. Bring the sample to room temperature and add 250 μL of saturated phenol solution. Mix by tube inversion several times. Avoid

vortexing as this may cause DNA shearing. Add 250 µL of chloroform and mix carefully. Centrifuge at 5000*g* for 5 min at room temperature and transfer the supernatant to a new tube.

7. Add 500 µL of chloroform and mix gently. Centrifuge and transfer the supernatant to a new tube.
8. Add one-tenth of the volume of 3 *M* Na-acetate (pH 7.6) and 1 mL of ice-cold ethanol. Mix by tube inversion and place the tube at –20°C for no less then 1 h.
9. Recover the precipitated DNA by centrifugation at 10,000*g* for 20 min in a refrigerated centrifuge. Discard the supernatant and carefully remove the residual solvent.
10. Dry the DNA pellet under a sterile hood and solubilize in sterile water by incubating overnight at 4°C.
11. Measure the DNA concentration by spectrophorometry and assess the quality of the linearized DNA by agarose electrophoresis. Store at –20°C for up to a week.

### 3.2. DNA Transfection by Electroporation

The initial step of the gene targeting process is the introduction of DNA into the recipient cells. This can be achieved through several means, including DNA electroporation, lipofectamine-mediated, and calcium phosphate-based procedures. In our hands, DNA electroporation achieves the highest degree of transfection efficiency, as evidenced by the number of drug-resistant colonies recovered.

1. Prepare the linearized targeting vector by the procedure described above.
2. One day before the electroporation, passage 1:2 the actively growing cell (at this point ~80% conflent).
3. Feed the cells with fresh medium 4 h before harvesting and electroporation.
4. Wash the plates twice with PBS and detach the cells by treatment with trypsin solution for 10 min at 37°C.
5. Stop the action of the trypsin solution by adding 1 volume of fresh medium and dissociate the cell clumps by pipetting.
6. Centrifuge the cell at 1100*g(*1000 rpm) for 5 min in a clinical centrifuge and discard the supernatant. Resuspend the cell in 10 mL of electroporation buffer and determine the number of cells.
7. Withdraw and recover by centrifugation $10^7$ cells. Discard the supernatant and resuspend the cell in 1 mL of electroporation buffer.
8. Mix 50 µg of the linearized targeting vector with the cell suspension in an electroporation cuvette. Incubate for 5 min at room temperature (*see* **Note 2**).
9. If HCT116 cells are used, electroporate at 230 V, 1080 µF, 1 sec. Incubate for 5 min at room temperature.
10. Plate the entire content of the cuvette into 10-cm tissue culture plate with DMEM.
11. Apply G418 selection 24 h after the electroporation.
12. Refeed the cell when the medium starts to turn yellow, usually daily for the first 5 d.
13. Ten days after the electroporation, most of the colonies have reached a size suitable for subcloning (*see* **Note 3**).

### 3.3 Recovery and Expansion of Stable Transfected Colonies

Colonies that have reached suitable degree of growth are processed for DNA extraction and subsequent analysis of the restriction digestion fragments by Southern blot hybridization analysis. A main consideration in the design of the screening procedure is the possible differential growth of the clones. This implies that the colony harvesting be performed at subsequent times (*see* **Note 4**).

1. Wash the plate containing the colonies with Ca/Mg-free PBS.
2. By holding the plastic dish against a light source, visually inspect the bottom. Colonies appears as translucent areas of 2–4 mm in diameter. Circle them with a marker for easy identification.
3. Remove the medium and wash the plates once with DPBS without Ca/Mg. Carefully remove the washing buffer.
4. Streak the bottom of a sterile cloning ring into silicone grease and place it around the colony of interest. The purpose is to seal out the colony.
5. Add 20 µL of trypsin solution to each cloning ring and place in a humidified incubator for 15 min.
6. Add 50 µL of media to each ring and transfer the cells to a 48-well plate into which 250 µL of media per well had been added in advance.
7. Carefully aspirate all the remaining cells. Remove the cloning ring and replenish the dishes with fresh medium containing the selection agent. This allows slowly growing colonies to reach suitable size. Perform a second round of harvesting a week later.
8. Grow the cells that had been transferred to 48-well plates in the presence of selection medium.
9. When the cells are approaching confluence, wash with Ca/Mg-free PBS and dissociate with 50 µL of trypsin. Add 350 µL of fresh medium and pipet vigorously to dissociate the cells. Transfer 200 µL to a 6-well plate and add 2 mL of medium. Once grown, these cells may be stocked frozen. The remaining of the cells are transferred to 12-well plates and expanded. Upon reaching confluence, the clones are processed for DNA extraction.

### 3.4. Genomic DNA Isolation and Southern Blot Analysis

Cell growth is examined under a light microscope and clones are processed when they have reached 75–100% confluence. Following cell lysis, DNA is precipitated with ethanol and digested with the selected restriction enzyme. The generated restriction fragments are isolated by agarose gel electrophoresis and hybridized with a suitable $^{32}$P-labeled probe.

1. Grow the cells to confluence. Cells at this stage are also recognizable by the yellow medium appearing 24 h after medium replacement.
2. Wash the cells with PBS lacking Ca/Mg and add 300 µL of cell lysis buffer.
3. Incubate at 37°C in an incubator for 10 min and then transfer to an Eppendorf tube. Cell lysate can be stored at –70°C without appreciable degradation of DNA for several years (*see* **Note 5**).
4. When sufficient numbers of samples have been collected, add 10 µL of protease (10 mg/mL) and place the sample in 65°C heat-block overnight.
5. Perform a phenol–chloroform extraction with 0.5 mL of phenol followed by a final extraction with chloroform.
6. Precipitate DNA by adding 2 vol of absolute ethanol. Turn the tube upside down several times until a filamentous DNA precipitate appears. Collect the DNA by swirling a glass rod inside the tube and transfer the DNA to a new Eppendorf tube containing 100 µL of TE (*see* **Note 6**).
7. Let DNA dissolve at 4°C overnight and then store at –70°C.

8. Withdraw a 20-μL aliquot of DNA solution and digest with 20 U of restriction enzyme in a final volume of 60 μL at the proper temperature overnight.
9. Heat the sample at 65°C for 15 min and then add 10 μL of sample buffer.
10. If the expected size of the informative restriction fragment is between 2 and 7 kb, apply the sample to a 0.75% agarose gel containing ethydium bromide. For resolution of fragments of different size, adjust the agarose content accordingly within the 0.5% to 1.5% limits. Load the sample in 0.8-cm-wide wells and run at 1.5 V/cm of running gel overnight (*see* **Note 7**).
11. Stop the electrophoretic run and take a picture of the gel under a UV light source, taking care to place a fluorescent ruler to the side of the gel to identify the relative migration of the DNA standard ladder.
12. Process the agarose gel for Southern blot and $^{32}$P-probe hybridization with cross-linking to the nitrocellulose membrane by UV. Prehybridization is performed at 65°C in 2× SSPE, 1% SDS, and 10% Denhart's solution for 4–16 h. Hybridization is carried out by exchanging the prehybridization cocktail with a fresh solution containing 50–100 × 106 cpm $^{32}$P-labeled DNA probe and incubating for 16 h at 65°C. Excess probe is washed out by incubation in 2% SSC, 1% SDS solution at 65°C twice for 15 min each, followed by autoradiography.

## 4. Notes

1. If the process of linearization gives rise to two DNA fragments, the targeting vector can be recovered following separation by agarose gel electrophoresis. For this purpose use low-temperature melting agarose and recover the DNA by agarase treatment or by ion exchange using glass beads available as part of a kit from Bio101 or other suppliers.
2. Carefully collect all the cells that remain attached to the electrodes, after the electric discharge. This is done by pipetting the medium against the electrodes and collecting the cells in a dish. Cell clumps are dissolved by vigorous pipetting and distributed in four 10-cm plastic dishes containing 13 mL of medium. It is recommended that the dishes with the medium be preincubated for at least 1 h before the addition of the cells, in order to allow the deposition of the collagen and other extracellular matrix macromolecules at the bottom of the dish to facilitate cell adherence.
3. Drug-resistant colonies should be collected at a sufficient stage of growth to maximize the chance of attachment and growth in the cloning dishes. A minimum size of 2–4 mm is recommended. With time, colony density become too high, with the risk of contamination during ring cloning. Furthermore, some overgrown colonies may undergo apoptosis.
4. Cell colonies should be expanded gradually by initial grow in 48-well plates. Grow rate can vary considerably, and care should be taken to trypsinize and replate the cells if signs of growth resting become evident. Usually, cells reaching confluence in 48-well plates are passed to 12-well plates and later to 6-well plates in 1–2 wk.
5. Cells extracts are stored at –70°C for up to 1 yr. Protease is added only at the time of incubation and DNA precipitation.
6. If the DNA amount is low, the addition of ethanol will not result in the formation of visible filamentous material. In this case DNA can be recovered by mild centrifugation at room temperature.
7. Digested DNA is separated onto a 20-cm long agarose gel in TBE buffer at 1.5 V/cm. Forced cooling is not required. Completion of the electrophoretic run requires 16–20 h.

## References


1. Zimmer A. (1992) Manipulating the genome by homologus recombination in embryonic stem cells. *Annu. Rev. Biochem.* **15**, 115–137.
2. Waldman, A. S. (1992) Targeted homologous recombination in mammalian cells. *Crit. Rev. Oncol. Hematol.* **12**, 49–64.
3. Capecchi, M. R. (1989) Altering the genome by homologous recombination. *Science* **244**, 1288–1292.
4. Bollag, R., Waldman, A., and Liskay, R. (1989) Homologous recombination in mammalian cells. *Annu. Rev. Genet.* **23**, 199–225.
5. Sedivy, J. M. and Joyner, A. L. (1992) *Gene Targeting*, W. H. Freeman, New York, NY.
6. Hasty, P. and Bradley, A. (1993) Gene targeting vectors for mammalian cells, in *Gene Targeting: A Practical Approach*, Joyner, A.L., eds. IRL Press, pp. 1–32.
7. Iozzo, R. V. (1998) Matrix proteoglycans: from molecular design to cellular function. *Annu. Rev. Biochem.* **67,** 609–652.
8. Mansour, S. L., Thomas, K. R., and Capecchi, M. R. (1988) Disruption of the proto-oncogene *int-2* in mouse embryo-derived stem cells: a general strategy for targeting mutations to non-selectable genes. *Nature* **336**, 348–352.
9. Schwartzberg, P., Robertson, E., and Goff, S. (1990) Targeted disruption of the endogenous c-*abl* locus by homologous recombination with DNA encoding a selectable fusion protein. *Proc. Natl. Acad. Sci. (USA)* **87,** 3210–3214.
10. Sedivy, J. M. and Sharp, P.A. (1989) Positive genetic selection for gene disruption in mammalian cells by homologous recombination. *Proc. Natl. Acad. Sci. (USA)* **86**, 227–231.
11. Hanson, K. D. and Sedivy, J. M. (1995) Analysis of biological selections for high-efficiency gene targeting. *Mol. Cell Biol.* **15**, 45–51.
12. Sedivy, J. M., Vogelstein, B., Liber, H., Hendrickson, E., and Rosmarin, A. (1999) Gene targeting in human cells without isogenic DNA. *Science* **283**, 9.
13. Hasty, P., Rivera-Pérez, J., Chang, C., and Bradley, A. (1991) Target frequency and integration pattern for insertion and replacement vectors in embryonic stem cells. *Mol. Cell Biol.* **11,** 4509–4517
14. Hasty, P., Rivera-Pérez, J., and Bradley, A. (1991) The length of homology required for gene targeting in embryonic stem cells. *Mol. Cell Biol.* **11**, 5586–5591.
15. Thomas, A. P. and Capecchi, M. R. (1987) Site-directed mutagenesis by gene targeting in mouse embryo-derived stem cells. *Cell* **51**, 503–512.
16. te Riele, H., Maandag, E. R., and Berns, A. (1992) Highly efficient gene targeting in embryonic stem cells through homologous recombination with isogenic DNA constructs. *Proc. Natl. Acad. Sci. (USA)* **89**, 5128–5132.
17. Hasty, P., Rivera-Pérez, J., and Bradley, A. (1992) The role and fate of DNA ends for homologous recombination in embryonic stem cells. *Mol. Cell Biol.* **12**, 2464–2474.
18. Jeannotte, L., Ruiz, J., and Robertson, E. J. (1991) Low level of Hox1.3 gene expression does not preclude the use of promotorless vectors to generate gene discruption. *Biol. Mol. Cell* **11**, 5578–5585.
19. Parsons, R., Li, G.-M., Longley, M. J., Fang, W. H., Papadopoulos, N., Jen, J., et al. (1993) Hypermutability and mismatch repair deficiency in RER+ tumor cells. *Cell* **75**, 1227–1236.
20. Kontgen, F. and Steward, C. (1993) Simple screening procedure to detect gene targeting events in embryonic stem cells in: *Guide to Techniques in Mouse Development*, Wassarman, P. and DePanphilis, M., eds. Academic Press, 878–889.

# 24

## Constitutive and Inducible Antisense Expression

**Melissa Handler and Renato V. Iozzo**

### 1. Introduction

The use of antisense technology has been applied to a number of diverse organisms, including *Drosophila melanogaster*, plants, and mammals. The purpose is to take advantage of the base-pair specificity of both RNA:RNA and RNA:DNA interactions, ultimately to block protein production. This technique requires the production of exogenous antisense mRNA, generated either constitutively or inducibly, from a functional promoter. Classically, a constitutive promoter, such as the cytomegalovirus (CMV) promoter, is utilized to express high levels of antisense mRNA. This mRNA then binds irreversibly to endogenous sense mRNA, thus preventing its efficient translation, which results in attenuation of the protein product. Synthetic antisense oligonucleotides are also commonly used for constitutive antisense targeting. These DNA oligomers bind to endogenous RNA to generate RNA:DNA hybrids that are targeted by RNase H and degraded. This method will not be described at length in this chapter.

One problem often associated with constitutive antisense systems is low levels of expression, such that the system does not completely block translation of the targeted transcript. Another disadvantage is that the investigator has no temporal control over the antisense levels when using a constitutive expression system, because the promoter is always active. Therefore, the Tet-On and Tet-Off system is commonly utilized to circumvent these problems. This system allows researchers to exploit the tetracycline-regulated expressions systems originally described by Gossen and Bujard (Tet-Off) and Gossen et al. (Tet-On) and to regulate both spatially and temporally the levels of inductive antisense expression ***(1–6)***.

In order to obtain a Tet-Off or Tet-On cell line, the generation of a double stable cell line is required ***(3–6)***. First, the cell line is transfected with a regulatory plasmid. This vector expresses a transactivator element (tTA) for the Tet-On system, or a reverse transactivator element (rtTA), for the Tet-Off system, which is produced from the fusion of a tet repressor with the VP16 activation domain of the Herpes simplex virus,

From: *Methods in Molecular Biology, Vol. 171: Proteoglycan Protocols*
Edited by: R. V. Iozzo © Humana Press Inc., Totowa, NJ

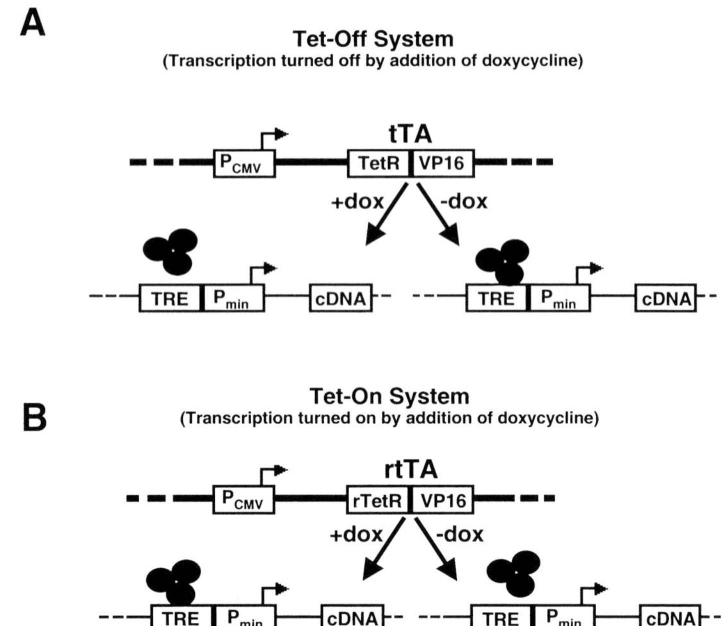


Fig. 1. Schematic representation of gene regulation in the Tet-Off and Tet-On Systems. **(A)** Under the control of the CMV promoter, the tet-tesponsive transcriptional activator (tTA), which is a fusion protein of the tetracycline repressor (TetR) and VP16 activation domain of the Herpes simplex virus, will bind to a tetracycline responsive element (TRE) in the absence of tetracycline or its derivative doxycycline (dox). The TRE contains seven copies of the 42-bp tet operator and is located upstream of the minimal CMV promoter ($P_{min}$) which drives expression of a specific cDNA. **(B)** The reverse tet-responsive transcriptional activator (rtTA) differs from the tTA by 4 amino acid changes, which result in a reversal of its response to tetracyline and its derivates. Therefore, transcription is turned on in the presence of doxycyline through the binding of rtTA to the TRE. Reprinted with permission from Gossen, M., Freundlieb, S., Bender, G., Muller, G., Hillen, W., and Bujard, H. (1995) Transcriptional activation by tetracyclines in mammalian cells *Science* **268**, 1766–1769. Copyright ©1995 American Association for the Advancement of Science.

under the control of the CMV promoter (**Fig. 1**). When tetracycline, or its more commonly used derivative doxycycline, is added to the culture medium, the transactivator is induced to bind the tetracycline response element (TRE) located on the response plasmid (**Fig. 1**). This plasmid contains a multiple cloning site (MCS) located immediately downstream of the Tet-responsive minimal promoter. The promoter consists of the TRE, which contains seven copies of the 42-bp tet operator sequence (tetO) residing upstream of the minimal CMV promoter ($P_{minCMV}$), which lacks the enhancer region of the full-length CMV promoter. Therefore, the TRE, in conjunction with the minimal CMV promoter, drives expression of a specific cDNA only upon binding of the tetR or rTetR to the tetO sequence (**Fig. 2**). Both Tet-On and Tet-Off utilize the

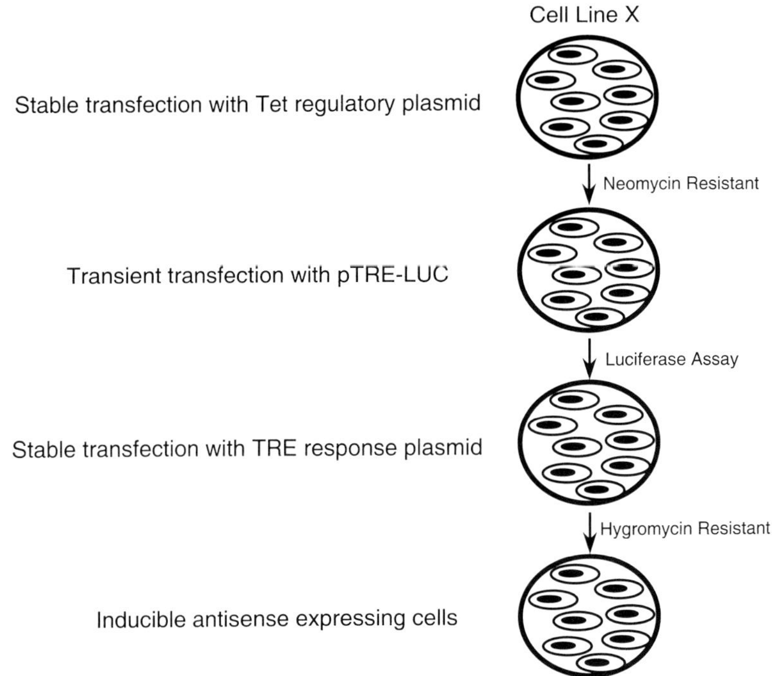


Fig. 2. Schematic representation of generation of double stable cell line. Cells are grown to confluence and transfected with a pTet-On or pTet-Off regulatory plasmid. Neomycin-resistant cells are cloned and expanded. Transient transfection with a luciferase construct (pTet-Luc) is performed to determine low-background and high-inducibility of the clones. The most optimal clone is then transfected with the response plasmid expressing specific antisense under the control of the TRE and the minimal CMV promoter. Hygromycin-resistant clones are then isolated and expanded.

pTRE response plasmid. However, for the Tet-On system, the tTA activates transcription in the presence of doxycycline. Conversely, in the Tet-Off system, the rtTA, which has four amino acids mutated in the tet repressor, blocks transcription in the presence of doxycycline.

One advantage of the tetracycline inducible system is that it allows for the tight regulation of antisense expression with the presence of the TRE on the response plasmid. By incubating the clones with a minimal dose of doxycycline, it is possible to control when antisense expression is induced. Expression levels can then be assayed by Western blotting with an antibody specific to your protein, RT-PCR using gene-specific primers, Northern blot analysis, or a functional assay specific for your system. Once a stably transfected cell line has been generated, it is then possible to do more in-vitro and in-vivo analyses, including injection of inducible cells into mice followed by oral administration of doxycycline to induce expression *(7–8)*. A second advantage of this system is that doxycycline is not toxic to the cells or to animals, and it is only slightly cytotoxic at relatively high levels (milligram amounts). Recent data also show evidence for the controlled expression of two genes in a mutually exclusive manner by utilizing this technology *(9)*. Overall, this system has proven to be extremely beneficial in specific areas of research, and it ultimately has the potential to be utilized for therapeutic purposes.

## 2. Materials

### 2.1. General

1. DMEM (Life Technologies, Inc.).
2. Doxycycline (Sigma).
3. Fetal bovine serum (HyClone Laboratories, Inc.).
4. Glutamine (Mediatech, Inc.).
5. G418 (Geneticin, Sigma).
6. Hygromycin (Sigma).
7. Luciferase Assay Kit (Clontech).
8. Penicillin/streptomycin (Life Technologies, Inc.).
9. Serum-free defined media (Life Technologies, Inc.).
10. 50× TAE electrophoresis buffer, pH 8.5: 242 g Tris base, 57.1 mL glacial acetic acid, 37.2 g $Na_2EDTA{\cdot}2H_2O$, made up to 1 L with distilled $H_2O$.
11. TE buffer, pH 7.5: 10 m*M* Tris-HCl, pH 7.5, 1 m*M* EDTA, pH 8.0.
12. Trypsin (Life Technologies, Inc.).
13. Tet-On/Tet-Off expression system (Clontech).
14. Tissue culture dishes (12 well, 6 well, 100 mm).
15. Loading buffer: 50% glycerol, 0.2 *M* EDTA, pH 8.3, 0.05% (w/v) bromophenol blue.

## 3. Methods

### 3.1. General

As mentioned earlier, the overall concept of this system is to generate a double stable cell line that will inducibly express specific antisense mRNA (*see* **Fig. 2** for a general scheme). First, cells are stably transfected with a response plasmid expressing either the tTA or rtTA. Second, neomycin-resistant clones from the first transfection are then transiently transfected with a reporter plasmid, which is under the control of the TRE and the minimal CMV promoter, to measure the inducibility and background levels of the selected clones. In this case, we describe the utilization of the luciferase gene as a reporter and subsequent measurement of its activity with the luciferase assay. Third, after a suitable clone with low background and high inducibility has been selected, it is again stably transfected with the response plasmid, which contains a specific cDNA driven by the TRE and the minimal CMV promoter. Transcription is then regulated by the presence of tetracycline and its derivatives. These protocols will be described in more detail below.

### 3.2. Construction of Antisense Response Plasmid Vector

The most critical aspect of this procedure is determining the best region of the gene to target with the antisense mRNA. Often, several regions of the gene need to be tested before finding the most optimal. In general, investigators typically utilize the most conserved region in an effort to inhibit nonspecific hybridization. Cloning of cDNA can be accomplished by PCR amplification from a cDNA library, RT-PCR with gene-specific primers, or screening of a cDNA library.

1. In a sterile Eppendorf tube, digest 1–2 µg of cDNA and response plasmid with compatible enzymes. If cDNA is cloned from a library, the sequence must be scanned for restriction sites that are located in the MCS of the response plasmid. There are several computer programs available, such as DNA Strider and PC Gene, which can be very helpful. If the cDNA is generated either by PCR from a cDNA library or RT-PCR from total RNA, it is possible to introduce novel restriction sites into the cloned sequences with the specifically designed primers.
2. Run digests on agarose gel. Agarose gel can be made by weighing out the necessary amount of agarose (typically a 1% gel will suffice) and add to TAE buffer. For these purposes, a small agarose gel can be utilized (~150 mL). Microwave until the agarose is dissolved, add ethidium bromide (0.5 µg/mL), pour into gel tray and cool at room temperature until polymerized.
3. Add 2 µL of 10× loading buffer (50% glycerol, 0.2 *M* EDTA, pH 8.3, 0.05% (w/v) bromophenol blue) to each sample and run on agarose gel along with the appropriate DNA ladder marker for 1.5–2 h at ~90 V.
4. When the gel is finished running, visualize the DNA by long-wave UV light and isolate the correct size cDNA fragment and linearized response plasmid with a clean razor blade. Purify the DNA and resuspend the DNA in an appropriate volume of TE.
5. Ligate the cDNA insert (0.5–1 µg) into response plasmid (0.1 µg) with T4 DNA ligase for 6 h at 13°C. Transform 1 µL of ligation into *Escherichia coli* bacteria according to standard protocol and spread onto ampicillin-containing LB agar plates. Allow to grow at 37°C for at least 12 h.
6. Select 10–15 ampicillin colonies from the transformed plates and grow in 3 mLs of LB with ampicillin (50 mg/mL). Purify the plasmid by standard miniprep protocol. Correct orientation can be determined by appropriate restriction endonuclease digestion.

### 3.3. Generation of Stable Neomycin-Resistant Clones

The first stable transfection is with the regulator plasmid, containing either the tTA or the rtTA driven by the CMV promoter, as well as the neomycin resistance gene. Following transfection, clones are selected based on their ability to survive in G418.

1. One vial of cells is thawed and plated onto a 100-mm tissue culture dish containing suitable culture medium. Cells should be grown and/or passaged until they reach a density of approx $10^8$ (*see* **Note 1**).
2. Cells are then trypsinized and then mixed with 40 µg of the pTet-On regulatory plasmid DNA (Clontech) in a 0.4-mL cuvette. Cells are then transfected by electroporation in a Bio-Rad Gene Pulsar according to the manufacturer's suggested conditions (*see* **Note 2**).
3. Following electroporation, cells are allowed to recover for ~10 min on ice and are then plated evenly on ten 100-mm tissue culture dishes. Use 10 mL of medium per dish (*see* **Note 3**).
4. Allow cells to attach and grow in a 37°C incubator (5% $CO_2$).
5. After 48 h, G418 (400 µg/mL) can be added to the culture medium. Media should be changed every 3–4 d (*see* **Note 4** and **Note 5**).
6. After G418-resistant clones have reached a suitable size, 60–70 should be isolated by ring cloning. They can be plated directly into 12-well dishes. When confluent, cells can be expanded to 35-mm dishes (*see* **Note 6**).

### 3.4. Luciferase Assay

Following the first stable transfection, neomycin-resistant clones are transiently transfected with the plasmid pTRE-LUC. This plasmid is similar to the response plasmid except that the luciferase gene is expressed in place of a specific cDNA under the control of the TRE and the minimal CMV promoter. By measuring the level of luciferase activity with a functional assay, it is possible to determine which individual clones display the highest level of inducibility as well as the lowest background. This allows the investigator to screen the clones quickly before performing a second stable transfection.

1. Transiently transfect each neomycin resistant clone with 10 µg of pTRE-LUC plasmid according to calcium phosphate protocol. We recommend using the CalPhos Maximizer (Clontech) for maximal transfection efficiency. Although we have suggested introducing the luciferase vector via calcium phosphate transfection, electroporation can also be utilized (*see* **Note 5**).
2. Cells are grown in 35-mm dishes ± doxycycline (2 µg/mL) until 90% confluent.
3. Growth medium is removed from the cells and then rinsed twice with 1× PBS. Then 500 µL of lysis buffer (25 m*M* Tris-phosphate, pH 7.8, 2 m*M* DTT, 2 m*M* 1,2-diaminocyclohexane-N,N,N,N-tetracetic acid, 10% glycerol, 1% Triton X-100) are added to each dish and the cells are scraped off and placed in a sterile Eppendorf tube.
4. Tubes are then spun briefly (~5 s) in a Beckman microcentrifuge at 12,000*g* to pellet large debris.
5. Next, 20 µL of room-temperature cell extract is mixed with 100 µL of room-temperature Luciferase Assay Reagent (Promega).
6. The sample is placed in a luminometer to measure the release of light as the luciferin is being oxidized (*see* **Note 7**).

### 3.5. Generation of Antisense Inducible Clones

Once a clone has been selected to have low background levels and high inducibility based on the luciferase assay results, it undergoes a second stable transfection. It is transfected with two plasmids: a response plasmid containing specific cDNA under the control of the TRE and minimal CMV promoter, as well as a plasmid expressing the hygromycin resistance gene. Because the cells are already neomycin resistant, it is necessary to use a different antibiotic resistance marker to select clones.

1. The most optimal stably transfected clone, as determined by luciferase assay, is cotransfected with 40 µg of the pTRE construct and 2 µg of the hygromycin plasmid by electroporation as described previously. Because the clone is already G418 resistant, hygromycin is needed as a second antibiotic resistance marker.
2. After 48 h, 200 µg/mL hygromycin was added to the culture medium (*see* **Note 4**).
3. Then 30–40 hygromycin-resistant clones are isolated by ring cloning and expanded as described previously.

### 3.6. Southern Blot Analysis

Although the clones have been selected through antibiotic resistance, it is also important to show that the regulatory and response plasmids have in fact integrated into the genome. This can be achieved by PCR with primers that amplify a specific region of the regulatory and response plasmids or by Southern analysis utilizing these plasmids as probes. We have found that Southern analysis is a much more reliable although labor-intensive method. It is preferable to perform Southern analysis after the first stable transfection as well before proceeding with the second stable transfection.

1. Extract genomic DNA from a 35-mm dish of each clone to be screened. DNA should be phenol:chloroformed and ethanol precipitated prior to restriction digestion.
2. Separate endonuclease digestions should be performed with an enzyme(s) that will liberate a distinct region within either the response plasmid and/or the regulatory plasmid. Digest 10 µg of genomic DNA overnight at 37°C, followed by inactivation with 0.5 µL of 0.5 *M* EDTA, phenol:chloroform extraction and ethanol precipitation.
3. Genomic digestions are run on a 0.8% agarose gel in TAE buffer at 80 V for 4–5 h. Gel percentage and duration of electrophoresis will depend on the size of the fragments that are expected.
4. Plasmid DNA should be digested with the same enzymes as genomic DNA and inserts purified as described previously.
5. After gel is finished running, it is denatured in 1.5 *M* NaCl and 0.5 *M* NaOH for 30 min at room temperature and then neutralized in 1 *M* ammonium acetate and 0.5 *M* NaOH for 30 min at room temperature. The DNA is transferred onto a nylon membrane in 10× SSC overnight and then UV cross-linked in a Stratalinker.
6. The plasmid probes should be labeled with $^{32}$P dCTP or, if preferred, a nonradioactive isotope such as digoxigenin, by random priming with Klenow according to the manufacturer's protocol.
7. Genomic DNA filters are prehybridized at 65°C for 3 h and then hybridized with the corresponding DNA probes at 65°C overnight. Membranes are then washed and exposed to autoradiography film. Development of the film should yield the expected results.

### 3.7. Northern Blot Analysis

Northern analysis is needed to determine that the tTA or rtTA and the antisense cDNA are expressed. This will confirm expression of both the regulatory and response transcripts. Proper controls include extraction of total RNA in both the presence and absence of doxycycline. Once again, Northern analysis should ideally also be performed after the first stable transfection.

1. Total RNA is extracted from 35-mm confluent dishes of the clones to be screened by commonly used guanidium thiocyanate protocols (e.g., TRI-Reagent, Sigma) (*see* **Note 8**).
2. Then 20 µg of total RNA are run on a 1% agarose/formaldehyde denaturing gel in MOPS buffer at 70 V for 4–5 h.
3. Total RNA is transferred onto a nylon filter and cross-linked with a UV Stratalinker. Filters are prehybridized at 42°C for 3 h and hybridized at 42°C overnight with the probes labeled the same as described for Southern analysis.
4. Membranes are then washed and exposed to autoradiography film. Development of the film should yield the expected results.

### 3.8. Western Immunoblotting Analyses

Western analysis is necessary to determine that there is a decrease in specific protein expression upon addition or removal of doxycycline. Which system is being used (Tet-On or Tet-Off) will dictate whether expression should be blocked with or without doxycycline. It is also possible to perform functional assays for the protein as well.

1. Antisense transfected clones are expanded into 35-mm dishes and then incubated for 48 h in serum-free defined media with or without 2 μg/mL doxycycline.
2. Serial dilutions of the conditioned media from each clone are vacuum blotted onto a nylon filter (Schleicher and Scheull).
3. The filter is then blocked in 3% milk for 3 h and then incubated overnight at room temperature with a primary antibody to Protein X.
4. After several washes, the blot is incubated with an appropriate horseradish peroxidase-conjugated secondary antibody.
5. Enhanced chemiluminescence is performed according to the manufacturer's protocol (Amersham).
6. Autoradiographs are then scanned by laser densitometry and quantitated to determine the relative protein levels.

## 4. Notes

1. When working with tissue cultures, sterility is of concern. Although antibiotics are added to the culture media, be sure to wipe down the hood with 70% ethanol before and after use, and wear gloves throughout the procedure.
2. Any DNA to be transfected into cells should be of highest quality. That is, after plasmid preparation, DNA should be phenol:chloroformed and ethanol precipitated under a sterile tissue culture hood. Also, as a final check, it is a good idea to run an aliquot on an agarose gel prior to transfection.
3. Be sure to determine the correct capacitance and voltage for the cell line and electroporator.
4. Due to variations in potency between different lots, it may be necessary to titrate the proper concentrations of G418 and hygromycin for each cell line.
5. It is important to mention that certain cell lines can alter the effectiveness of the tetracycline controlled gene expression ***(11)***. The type of transfection method as well as the amount of plasmid that is transfected can also alter considerably the background levels and inducibility of each clone. Therefore, it may be necessary to test several experimental parameters for each cell line and screen many clones in order to find one that is highly inducible but also gives very low levels of background expression. Although labor intensive, the results will be much more trustworthy.
6. When ring cloning, we have found that plastic rings work better than metal rings and require less vacuum grease to attach to the plate. Also, while picking clones, try to work rapidly, because the plates can dry out quickly and the cells will die.
7. We have noticed that often there are extremely high levels of background luciferase expression in many of the selected clones. Therefore, it may be necessary to select more than the recommended number of clones until one of low background can be isolated.
8. Be certain when working with RNA that gloves are always worn and are changed frequently. RNA is very susceptible to degradation from RNases that are found on most surfaces, including skin.

## References


1. Gossen, M. and Bujard, H. (1992) Tight control of gene expression in mammalian cells by tetracycline-responsive promoters. *Proc. Natl. Acad. Sci. (USA)* **89**, 5547–5551.
2. Gossen, M., Bonin, A. L., and Bujard, H. (1993) Control of gene activity in higher eukaryotic cells by prokaryotic regulatory elements. *TIBS* **18**, 471–475.
3. Gossen, M., Freundlieb, S., Bender, G., Muller, G., Hillen, W., and Bujard, H. (1995) Transcriptional activation by tetracyclines in mammalian cells. *Science* **268**, 1766–1769.

4. Baron, U., Freundlieb, S., Gossen, M., and Bujard, H. (1995) Co-regulation of two gene activities by tetracycline via a bidirectional promoter. *Nucleic Acids Res.* **23**, 3605–3606.
5. Baron, U., Gossen, M., and Bujard, H. (1997) Tetracycline-controlled transcription in eukaryotes: novel transactivators with graded transactivation potential. *Nucleic Acids Res.* **25**, 2723–2729.
6. Freundlieb, S., Baron, U., Bonin A.L., Gossen, M., and Bujard, H. (1997) Use of tetracycline-controlled gene expression systems to study mammalian cell cycle. *Meth. Enzymol.* **283**, 159–173.
7. Furth, P. A., Onge, L. S., Boger, H., Gruss, P., Gossen, M., Kistner, A., Bujard, H., and Henninghausen, L. (1994) Temporal control of gene expression in transgenic mice by a tetracycline-responsive promoter. *Proc. Natl. Acad. Sci. (USA)* **91**, 9302–9306.
8. Kistner, A., Gossen, M., Zimmermann, F., Jerecic, J., Ullmer, C., Lubbert, H., and Bujard, H. (1996) Doxycycline-mediated quantitative and tissue-specific control of gene expression in transgenic mice. *Proc. Natl. Acad. Sci. (USA)* **93**, 10,933–10,945.
9. Baron, U., Schnappinger, D., Helbl, V., Gossen, M., Hillen, W., and Bujard, H. (1999) Generation of conditional mutants in higher eukaryotes by switching between the expression of two genes. *Proc. Natl. Acad. Sci. (USA)* **96**, 1013–1018.
10. Bonin, A. L., Gossen, M., and Bujard, H. (1994) Photinus pyralis luciferase: vectors that contain a modified luc coding sequence allowing convenient transfer into other systems. *Gene* 141, 75–77.
11. Gossen, M. and Bujard, H. (1995) Efficacy of tetracycline-controlled gene expression is influenced by cell type: commentary. *BioTechniques* **19**, 213–216.

25

# Cell-Mediated Transfer of Proteoglycan Genes

Jens W. Fischer, Michael G. Kinsella, David Hasenstab, Alexander W. Clowes, and Thomas N. Wight

## 1. Introduction

Replication-defective retroviral systems for transduction of genes into target eukaryotic cells have grown recently in popularity. In general, the use of a retroviral packaging system involves the preparation and transfection by standard molecular biological techniques of the retroviral vector into packaging cell lines for the production of replication-defective retrovirus *(1,2)*. Virus-laden medium from packaging cells is then used to infect target cells. Retroviral transduction of cDNA sequences has distinct advantages and disadvantages when compared to other gene transfer technologies. Foremost among the advantages of viral gene delivery approaches are the high cellular infection rates. This efficiency allows the preparation and rapid selection of large pools of transduced cells without the added step of cell cloning. This is critical in the preparation of sufficient numbers of low-passage primary cells for both characterization of cellular phenotype and expression of the transgene in vitro, as well as for experimental use in vivo. Moreover, because pools of transduced cells can be used rather than clonal cell lines, clonal variability unrelated to transgene expression is not a major issue in interpretation of experimental results. A major advantage in the use of the retroviral delivery system is that the transduced gene is rapidly incorporated into the host genome of dividing cells, thus ensuring stable expression of the gene product, although typically a single copy of the target gene is inserted. However, because the insertion of the viral sequences requires mitosis of the target cell, retroviral vectors cannot be used effectively to transfer genes to nondividing cells. Thus, retroviral transduction is usually performed on cells in culture, and gene transfer in animal models can be performed by implantation of cells that express the transgene, a process referred to as cell-mediated gene transfer. One advantage of the cell-mediated gene transfer approach is that long-term expression of the transgene in vivo can be achieved *(3)*. Moreover, local expression of the gene can be induced in an adult animal without concern that adaptive processes that compromise function or viability might occur

From: *Methods in Molecular Biology, Vol. 171: Proteoglycan Protocols*
Edited by: R. V. Iozzo © Humana Press Inc., Totowa, NJ

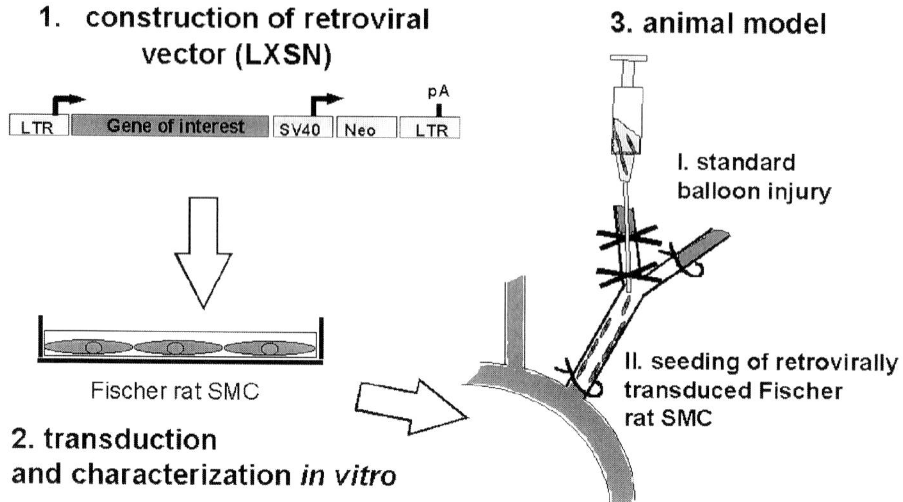


Fig. 1. Flow diagram of protocols involved in cell-mediated proteoglycan gene transfer. In step 1 the dark panel indicates the gene of interst.

during development, such as is possible in transgenic and knockout models. However, the preimplantation culture of the target cells in vitro may induce a phenotype that differs from that of endogenous cells. The approach of cell-mediated gene transfer has been used both to model disease processes and to evaluate therapeutic agents in vivo. For example, the role of tissue factor in the thrombotic occlusion of restenotic vessels has been examined by local overexpression of tissue factor by cell-mediated gene transfer *(4)*. Retrovirally transduced cells that overexpress the human granulocyte colony stimulating factor (G-CSF) gene have been used to examine the effects of the systemic delivery of that factor on circulating neutrophil levels *(5)*. Several studies have used cell-mediated gene transfer to express locally potential therapeutic agents in vivo. For example, tissue inhibitor of metalloproteinases has been assessed as an approach to limiting vessel aneurysms in an animal model *(6)*, and the local expression of decorin induced by cell-mediated gene transfer has been assessed as an agent to limit the growth of the neointima *(7)*. We include protocols that we have used for cell-mediated gene transfer to the injured rat carotid artery as examples of this approach (**Fig. 1**).

## 2. Materials

### *2.1. Retroviral Transfection of Target Cells*

1. PE501 cells.
2. PA317 cells.
3. NIH 3T3 TK– cells.
4. LXSN vector (~20 µg/transfection).
5. 2.5 *M* $CaCl_2$, pH 7.2.

6. 2× HEPES buffered saline (2× HBS): 50 m*M* HEPES, 280 m*M* NaCl, 1.5 m*M* $Na_2HPO_4$, pH 7.2.
7. Polybrene (hexadimethrine bromide) stock solution, 4 mg/mL, store at 4°C. Polybrene is added to the media of cultures to be infected in order to increase infection rates.
8. G-418 (Geneticin) stock solution, 50 mg/mL in PBS. G-418 is a neomycin analog used for cell selection after transduction of vector.
9. Standard tissue culture supplies.
10. Culture medium: Dulbecco's Minimal Essential Medium (DMEM) with 10% fetal bovine serum.

### 2.2. Internal Seeding of Retrovirally Transfected Rat Smooth Muscle Cells into the Injured Rat Carotid Artery

The following is an example of how cell mediated gene transfer, using retrovirally transduced smooth muscle cells *(3,8)*, can be used to overexpress proteoglycans in the vessel wall and to investigate their effect on neointima formation *(7)*. For this purpose smooth muscle cell from Fischer 344 rats were retrovirally transduced using the protocol described above. Since Fischer 344 rats are an inbred rat strain, Fischer 344 rat smooth muscle cells can be transferred into any Fischer 344 rat without inducing an immune response (*see* **Note 1**).

1. Fischer 344 rats, male, 250–300 g.
2. Anesthetic: 50 mg/kg ketamine, 5 mg/kg xylazine, 1mg/kg acepromazine.
3. Heparin, 10U/kg.
4. 2F Fogarty balloon embolectomy catheter (Baxter) filled with water, attached to 1-$cm^2$ syringe filled with lactate ringer's solution.
5. 1-$cm^3$ syringe attached to silicone elastomer medical grade tubing (0.012 in. id × 0.025 in. od).
6. Lactated Ringer's solution.
7. Gauze sponges.
8. Cotton-tipped applicators.
9. 4.0 silk sutures.
10. Scalpel.
11. Ethanol, betadine.
12. Instruments: fine point (watchmaker's) forceps; blunt-curved forceps, fine curved forceps, retractor, 3 mosquito clamps, arterial (vanna) scissors, regular scissors.

### 2.3. Peri-adventitial Seeding on Decellularized Carotids

In some circumstances, it may be advantageous not to have endogenous cells present in the media and neointima and to repopulate a vessel entirely with exogenously seeded cells. Rapid freeze/thaw cycles can be used to decellularize the artery before seeding the cells in the peri-adventitial region. Transduced smooth muscle cells can be cultured in 5'-Bromo-2'-deoxyuridine (BrdU)(Roche, 24 h in 0.06 µg/mL), which will be incorporated in the DNA and serve as a genomic tag for identifying seeded cells at later times. Smooth muscle cells seeded in the peri-adventitial region will migrate through the media and form a neointima by 2 wk. This protocol is well suited for measuring the effects of gene expression on migration in vivo *(4,10)*.

1. As under **Subheading 2.2.1.**
2. Liquid nitrogen.

3. 5'-Bromo-2'-deoxyuridine (BrdU).
4. Teflon sheet.
5. Anti-BrdU antibodies.
6. Secondary antibody and standard supplies for immunocytochemistry (ICC).

## 3. Methods

### 3.1. Retroviral Transfection of Target Cells

1. Insert the gene of interest into the LXSN vector using standard procedures (*see* **Note 2**).
2. Transfection of primary ecotropic (PE501) cells with the retroviral vector (LXSN and L *gene of interest* SN) and transduction of secondary amphotropic (PA317) packaging cells with the replication-defective retrovirus (*see* **Note 3**):

   *Day 1*: Plate PE501 cells at $5 \times 10^5$ cells/60mm dish, using 4 mL of medium.

   *Day 2*: Prepare tube A with 20 µg of vector DNA, and 50 µL of 2.5 *M* $CaCl_2$. Add $H_2O$ to make 500 µL. Prepare tube B with 500 µL of 2× HBS solution. Add contents of tube A dropwise to tube B, mixing by constant moderate aeration of tube B with a Pasteur pipet. Incubate at ambient temperature for 30 min (a precipitate will form). Add 1 mL of the mixture to the medium in each plate and incubate cells overnight at 37°C in the incubator.

   *Day 3*: Change medium of transfected PE501 cells. Plate secondary packaging cell line (PA317) at $5 \times 10^5$ cells/60-mm dish; include an extra dish with PA317 cells to monitor the G-418 selection (*see* **Note 4**). *Note*: After transfection, cells, media and associated culture materials must be considered *Biohazardous!*

   *Day 4*: Collect 16-h-conditioned medium from PE501 cells. Centrifuge collected culture supernatant for 10 min at 4800*g* (*see* **Note 5**). Prepare medium for infection of PA317 cells by mixing fresh medium 1/1 with the conditioned medium of PE501 cells and add Polybrene from 1000× stock at final concentration of 4 µg/mL. Replace the culture medium of the target PA317 cells with the freshly prepared viral infection medium and incubate overnight.

   *Day 5*: Replate infected and control PA317 cells at various dilutions (1/10, 1/100, 1/1000) into 100-mm culture dishes with medium containing 800 µg of G-418/mL for selection of virally transduced cells (*see* **Note 4**).
3. Selection and cloning of secondary packaging cell line (PA317): Replace selection medium every 2–3 d until all cells in the control dish die (approximately 7–10 d). Pick clones with the use of cloning rings following standard procedures and pass individual clones into 60-mm plates. Continue growth in selection media for 3 d more. Freeze clones back before you determine the virus titer of PA317 clones (**step 4**).
4. Titer virus produced by PA317 clones. In this step, the relative number of virus produced by the different clones is determined, in order to select the PA317 clone that produces the highest titer for transduction of target cell lines.

   *Day 1*: Plate cells from the PA317 clone to be titered at $5 \times 10^5$ cells/60-mm dish.

   *Day 2*: Replace medium of the PA317 cells with 4 mL of fresh medium and plate one dish of NIH 3T3 TK– cells at $5 \times 10^5$ cells/60-mm dish for each clone to be titered. Plate one additional dish to use as a selection control.

   *Day 3*: Collect 16-h-conditioned medium of PA317 cells, centrifuge for 10 min at 4800*g*, prepare medium for infection of NIH 3T3 TK– cells by mixing fresh medium with 10 µL of the virus-containing medium of PA317 cells and add Polybrene at a final concentration of 4 µg/mL. Replace the medium of the NIH 3T3 TK– cells with the freshly prepared infection medium. Add fresh medium only to the selection control dish. Incubate overnight at 37°C.

*Day 4*: Trypsinize 3T3 $TK^-$ cells and replate three different dilutions (1/20, 1/100, 1/500) of the cells into 6-well plates in medium containing G-418 (600 µg/mL). Grow the cells for 7–10 d until the cells in the selection control dish have died and the colonies in the titering plates are big enough to be counted. To count clones, remove media, rinse with PBS, fix for 2 min with 100% methanol, rinse with distilled water, and stain cells with 0.1% methylene blue in 50% ethanol for 5 min. After staining, dishes are rinsed with water until excess stain is removed and clones are visible, then dishes are air-dried, and the colonies counted by visual inspection. The number of clones and the dilution factors allow the calculation of the number of virus particles produced by the different clones. For example, 10 clones at the 1/500 dilution equals 5000 infectious units (cfu) in 10 µL of medium or $5 \times 10^5$ cfu/mL.

After virus-titering, selected clones are expanded and frozen back under liquid nitrogen.

5. Infection of the target cells (*see* **Note 6**).

   *Day 1*: Plate the PA317 clone with the highest titer at $5 \times 10^5$ cells/60-mm dish (*see* **Note 7**).

   *Day 2*: Change medium of PA317 cells (4 mL), plate target cells at $5 \times 10^5$ cells/60-mm dish. Plate a control dish of target cells to monitor the G-418 selection.

   *Day 3*: Collect 16-h-conditioned medium from PA317 cells (*see* **Note 7**) and centrifuge for 10 min at 4800*g*. Prepare infection medium for the target cells by mixing fresh medium 1/1 with the conditioned medium of PA317 cells and adding Polybrene at 4 µg/mL. Remove medium from the target cells and apply the freshly prepared infection medium.

   *Day 4*: Replate cultures of transduced target cells into 100-mm dishes and add G-418 at the previously determined concentration for selection.

   *Days 5–14*: Continue selection in G-418, with medium changes every 3 d, until all of the cells in nontransduced control cultures die. Target cells are now stably transduced and can be passed and frozen back following standard tissue culture procedures.

6. Characterization of the transduced cells: The characterization procedures typically include Northern and Western blot analysis to demonstrate mRNA and protein expression of the transgene. The DNA sequences that encode the gene of interest and the neomycin phosphotransferase gene are expressed in a single contiguous mRNA. Therefore, the mRNA for the transgene is ~3.1 kb larger than that of the endogenous gene, allowing the retrovirally expressed gene product to be easily distinguished from the endogenous mRNA. Further characterization after the Western blotting depends largely on the gene of interest. If the target gene is a proteoglycan, it is important to determine the extent to which glycosaminoglycan chains are added and elongated on the core protein. The appropriate procedures are described elsewhere.

### 3.2. Internal Seeding Of Retrovirally Transfected Rat Smooth Muscle Cells into the Injured Rat Carotid Artery

1. Preparation of cells for the seeding procedure (*see* **Note 9**): After characterization of the transduced smooth muscle cells in vitro, grow cells nearly to confluence and trypsinize. Suspend the cells in DMEM/10% FCS to inactivate the trypsin, centrifuge cells at 960*g* (1000 rpm) and wash with DMEM without serum, repeat the washing step once, count the cells, and resuspend at $2.5 \times 10^6$ cells/mL in serum-free DMEM. The estimated amount of cell suspension per carotid is 40 µL; to allow for waste, prepare 150 µL per seeded carotid (*see* **Note 10**).
2. Anesthetize and weigh animals (*see* **Note 11**).

3. Shave throat.
4. Restrain rat lying on its back (make sure that the tongue does not obstruct the airway).
5. Disinfect the shaved throat area.
6. Make a midline incision from breast bone to chin bone.
7. Blunt dissect until the carotid artery is visible.
8. Set retractor to open neck cavity.
9. Dissect the common carotid artery from chin bone (distal to the bifurcation) to approx 1 cm proximal of the bifurcation.
10. Put a double, temporary loop around the common carotid artery as far proximally as possible.
11. Install a single loop around the external carotid just distal of the bifurcation.
12. Ligate the external carotid as distal as possible.
13. Put a double, temporary loop around the internal carotid.
14. Attach hemostats to all ties for better control.
15. Shut off blood flow by tightening the double loop around the proximal part of the common carotid artery.
16. Make an arteriotomy between the distal tie and the single temporary loop aroung the external carotid artery (**step 11**). Use tips of the Vanna scissors at a 45° angle, close and insert the tips of the scissors into the arteriotomy, and carefully open the scissors to spread the arteriotomy, then close the scissors again before pulling it out.
17. Hold the arteriotomy open with the cut tip of a 30-g needle and insert tip of the catheter filled with lactated Ringer's and flush to remove blood.
18. Insert balloon catheter and perform balloon injury to the carotid artery (Clowes et al. [1983], *Lab Invest* **49,** 208), afterwards tighten loop to prevent blood flow.
19. Flush again with lactated Ringer's.
20. Resuspend cells thoroughly (no clumps) and draw about 150 μL into the $1 cm^3$ syringe, attach the catheter, remove air, and insert the catheter into the ateriotomy, proximal of the middle tie at the bifurcation. Tie the catheter in with the middle loop (make sure that it seals) and instill the cells into the carotid artery while pulling the catheter back toward the arteriotomy. Be careful not to pull the catheter tip out. Invert the rat for 2 min, then return the rat to its original position for 13 min (cover the wound with lactated Ringer's-soaked gauze).
21. To close, remove the catheter, loosen the proximal tie of the common carotid artery and flush the artery with blood to remove the cells that did not adhere to the denuded vessel wall.
22. Ligate the middle loop around the external carotid artery around the bifurcation.
23. Restore blood flow by loosening the proximal tie around the common carotid artery.
24. Remove the loop on the internal carotid artery, and check blood flow through the internal and common carotid artery.
25. Cut the ends of the ties; leave the proximal loop loosely around the common carotid artery as a marker.
26. Close the incision.
27. At the end of the experimental period, seeded carotid arteries can be harvested as fresh tissue for further examination by methods of molecular biology or protein and proteoglycan biochemistry, or the animal can be perfusion fixed at physiological pressures for morphological examination.

### 3.3. Peri-adventitial Seeding on Decellularized Carotids

1. Cells are prepared as described previously except that 24 h before trypsinization, 0.06 μg/mL of BrdU are added to the culture medium. This dose should be determined for each cell

type to ensure that it does not decrease proliferation and gives complete labeling as detected by immunocytochemistry.
2. The carotid is injured as described previously but not recannulated for seeding (omit **step 20**) (*see* **Note 12**).
3. A small Teflon sheet (1.5 × 3.0 cm) is placed underneath the carotid to protect surrounding tissue. Loose connective tissue surrounding the carotid artery should be removed.
4. The carotid artery is decellularized by gently touching liquid nitrogen-cooled forceps to the carotid artery until frozen (several seconds). The vessel is allowed to thaw and the cycle repeated a total of three times.
5. The Teflon sheet is removed, blood flow restored, and the decellularized carotid artery immersed in the suspension of BrdU-labeled cells (1.25–2.5 × $10^6$ cells/mL in 400 μL). The cells are allowed to incubate in the peri-adventitial region of the carotid for 15 min and then the neck incision is closed.
6. Seeded cells are identified in histological sections using anti-BrdU antibodies (Roche) and standard immunocytochemical stains (*see* **Note 13**).

## 4. Notes

1. Non-species-matched genes can be used in the procedure as well. However, the development of antibodies against the gene of interest is to be expected and might abrogate the biological function *(11)*.
2. In the protocols described above, we reference the LXSN vector, developed in the laboratory of A. D. Miller, Fred Hutchinson Cancer Research Center, Seattle, WA, USA, although procedures for transduction of related replication-defective retroviruses are similar. This vector uses the viral long terminal repeat (L) promoter to drive constitutive expression of the gene of interest, which is inserted into the multicloning site of the vector (X). Expression of the neomycin phosphotransferase gene (N) is driven by an internal SV40 promoter (S) and is responsible for the survival of the cells in the presence of the neomycin analog, G-418. After insertion of the cDNA for the gene of interest, the orientation of the insert in the LXSN vector should be determined by PCR or restriction digestion. Related vectors with different promoter elements are available, including retroviral vectors that are designed to express the gene of interest under control of a regulatable promoter.
3. PE501 and PA317 virus packaging cells *(12)* are grown in Dulbecco's Minimal Essential Medium supplemented with 10% fetal bovine serum. We suggest the sequential use of both the mouse 3T3 cell-derived ecotropic packaging line (PE 501), which is designed for packaging of virus that will efficiently infect mouse cells, such as the amphotropic packaging line, PA317. Amphotropic virus packaging lines allow the production of virus that will infect cells from a wide variety of different species. Well-designed packaging cell lines are capable of synthesizing the retroviral proteins necessary for the assembly of infectious virus, but do not release replication-competent virus.
4. Control dishes that are not infected with the virus should always be included to monitor the selection of cells expressing neomycin phosphotransferase by G-418. In advance of attempting viral transductions, each cell type to be used as a target must be titered against different concentrations of G-418 to determine the concentration that will kill the cells during the course of 10–14 d. Note that the sensitivity to G-418 varies substantially among cells of different tissue origin or species and that, moreover, different lots of G-418 contain different concentrations of active inhibitor. Note also that cells must divide in order to incorporate the retroviral sequences into their genome. Passage of cells to relatively low density after transduction also aids in selection, as rapidly growing cells are more

readily killed by the inhibition of protein synthesis by G-418. During passage, be careful not to over-trypsinize the PA317 cells.

5. Centrifugation of the conditioned media is a precaution to avoid transfer of the virus-producing cells (PE501 or PA317) to the next cell type. Alternatively (or additionally), a 0.45-μm filter can be used to process virus-laden medium. Media of retrovirally transduced PA317 and PE501 cell cultures contains virus particles and, although the virus is modified to be replication-defective, all culture products must be considered *Biohazardous*.
6. Use the target primary cell lines at as a low passage number as is feasible. Passage 3–4 has been appropriate for endothelial cells and aortic smooth muscle cells. Typically, transfection efficiency of >80% can be achieved, and the resultant pools of transduced cells can be used without cloning. The high transduction efficiency is an advantage of this method because it allows the establishment of transfected cell lines before cells transform or become senescent. For experimental controls, it is always desirable to infect the target cell line with vector alone, and to passage this control cell line in parallel with the target cells carrying the gene of interest.
7. The procedures can be modified for the infection of larger quantities of target cells.
8. Use only freshly prepared the virus and the infection media. It is possible to repeat the infection of target cells several times in sequence if transduction efficiency for single infections is low. Note, however, that target cells must continue to divide.
9. Cells stay viable for a couple of hours at room temperature. Ensure that cells are resuspended before they are instilled into the balloon-injured rat carotid artery.
10. Note that only one carotid per rat can be seeded with cells, due to vagus nerve damage, which results in breathing difficulties for the animal.
11. The application of heparin directly before surgery can be helpful to reduce the risk of thrombosis associated with the ballooning and cell seeding procedure.
12. Injury is not required for peri-adventitial cells to migrate through the media, but balloon injury increases the rate of migration.
13. Cell divisions will decrease the intensity of BrdU staining.

## References


1. Miller, A. D. and Rosman, G. J. (1989) Improved retroviral vectors for gene transfer and expression. *Biotechniques* **7**, 980–990.
2. Linial, M. L. and Miller, A. D. (1990) Retroviral RNA packaging: sequence requirements and implications. In *Current Topic in Microbiology and Immunology. Vol. 157 Retrovirues: Strategies of Replication.* (Swanstrom, R. and Vogt, P. K., eds.), Springer-Verlag, Berlin, pp.125–152.
3. Lynch, C. M., Clowes, M. M., Osborne, W. R. A., Clowes, A. W., and Miller, D. A. (1992) Long-term expression of human adenosine deaminase in vascular smooth muscle cells of rats: a model for gene therapy. *Proc. Natl. Acad. Sci. (USA)* **89**, 1138–1142.
4. Hasenstab, D., Lea, H., Hart, C. E., Lok, S., and Clowes, A. W. (2000) Tissue factor overexpression in rat arterial neointima models thrombosis and progression of advential atherosclerosis. *Circulation*, **101**, 2651–2657.
5. Bonham, L., Palmer, T., and Miller, A. D. (1996) Prolonged expression of therapeutic levels of human granulocyte colony-stimulating factor in rats following gene transfer to skeletal muscle. *Hum. Gene Ther.* **7**, 1423–1429.
6. Allaire, E., Forough, R., Clowes, M., Starcher, B., and Clowes, A. W. (1998) Local overexpression of TIMP-1 prevents aortic aneurysm degeneration and rupture in a rat model. *J. Clin. Invest.* **102**, 1413–1420.

7. Fischer, J. W., Kinsella, M. G., Clowes, M. M., Lara, S., Clowes, A. W., and Wight, T. N. (2000) Local expression of bovine decorin by cell-mediated gene transfer reduces neointima formation after balloon injury in rats. *Circ. Res.*, **86**, 676–683.
8. Clowes, M. M., Lynch, C. M., Miller, A. D., Miller, D. G., Osborne, W. R. A., and Clowes, A. W. (1994) Long-term biological response of injured rat carotid artery seeded with smooth muscle cells expressing retrovirally introduced human genes. *J. Clin. Invest.* **93**, 644–651.
11. Miller, A. D., and Buttimore, C. (1986) Redesign of retrovirus packaging cell lines to avoid recombination leading to helper virus production. *Mol. Cell. Biol.* **6**, 2895–2902.
11. Forough, R., Hasenstab, D., Koyama, N., Lea, H., Clowes, M., and Clowes, A. W. (1996) Generating antibodies against secreted proteins using vascular smooth muscle cells transduced with replication-defective retrovirus. *Biotechniques* **20**, 694–701.
12. Mason, D. P., Kenagy, R. D., Hasenstab, D., Bowen-Pope, D. F., Seifert, R. A., Coats, S., Hawkins, S. M., and Clowes, A. W. (1999) Matrix metalloproteinase-9 overexpression enhances vascular smooth muscle cell migration and alters remodeling in the injured rat carotid artery. *Circ. Res.* **85**, 1179–1185.

# 26

# Morphological Evaluation of Proteoglycans in Cells and Tissues

**Stephanie L. Lara, Stephen P. Evanko, and Thomas N. Wight**

## 1. Introduction

A variety of techniques are available to assess the structure of proteoglycans, their interactions, and their locations and distribution in cells and tissues. In this chapter, we will review the procedures that we have used over the years to examine the morphology of proteoglycans and cite examples of how these techniques have been used to address questions related to the role of these molecules in health and disease. At the outset, we want to stress that all of the procedures we describe were developed by other investigators, and we are merely sharing our experiences and presenting a compilation of our favorite techniques. It should also be stressed that what follows is a small sampling of a number of different techniques available to investigators who wish to study proteoglycans using morphological approaches.

## 2. Materials

### *2.1. Molecular Spreads*

In an effort to resolve the structural features of purified proteoglycans and the complex structures that occur as these molecules interact with themselves and other molecules, we describe a spreading technique that was first used to visualize DNA *(1)* and later adapted to visualizing purified proteoglycans from tissues and cells *(2–4)*. This method utilizes a basic protein film spread technique, which is one of several possible approaches to preparing biochemically isolated and purified matrix molecules for ultrastructural visualization. The approach depends on the adsorption of negatively charged matrix molecules to a positively charged globular protein, cytochrome c, within a hyperphase solution, and the subsequent layering of the complex onto the surface of a hypophase. The globular protein denatures, collapses around, and coats the negatively charged molecules and forms a strong, laterally cohesive monolayer in which is embedded the protein-coated, negatively charged molecule. Contrast

From: *Methods in Molecular Biology, Vol. 171: Proteoglycan Protocols*
Edited by: R. V. Iozzo © Humana Press Inc., Totowa, NJ

enhancement is achieved by staining with the heavy metals uranyl acetate and phosphotungstate and by rotary shadowing with evaporated metal. We have used this technique to evaluate the structure of individual proteoglycan monomers as well as proteoglycan aggregates from human intestine ***(5)***, cultured arterial smooth muscle cells ***(6)***, bovine aorta ***(7)***, and nervous tissue ***(8–10)***.

1. Glassware: Cleaned and EtOH rinsed (*see* **Note 1**).
   a. Acid-washed, thoroughly DI-$H_2O$-rinsed glass microscope slides, stored in 90% EtOH.
   b. 9-cm glass Petri dish, acid washed and thoroughly rinsed, siliconized with Prosil-28, 1/100.
2. Samples: Buffer possibilities include 1 *M* $NH_4AC$, pH 5; 0.01, 0.05 *M*, and 0.3 *M* $NH_4Ac$, pH 5; 0.5 *M* guanidine/HCl, 0.005 *M* benzamidine, 0.1 *M* 6-aminehexanoic acid, 0.01 *M* EDTA, 0.0005 *M* phenylmethylsulfonylfloride (*see* **Note 2**).
3. Grids: Freshly prepared (2- to 3-d), very clean (acetic acid, water, EtOH rinsed), 300-mesh, Cu grids, coated with parlodion and light carbon (*see* **Note 3**).
4. Spreading solutions
   a. Hyperphase solution (prepare just before spreading): Sample diluted in appropriate buffer, pH 5, 4°C; 1 m*M* Tris-HCl, 1 m*M* EDTA, pH 8.5; cytochrome C stock: equal parts a and b (a. 4 *M* Tris-HCl, 0.1 *M* EDTA, pH 8.5 and b. 5 mg Cytochrome C (purest grade)/mL $H_2O$) (Store at 4°C up to a year.)
   b. Hypophase solution may be one of the following: Deionized $H_2O$, or 0.3 *M* $NH_4Ac$, pH 5.0.
5. Staining solutions
   a. 0.001% phosphotungstic acid in 90% EtOH, prepare fresh daily (stock: 0.1% PTA in $H_2O$).
   b. 10 m*M* uranyl acetate (UA) in 90% EtOH, prepare fresh daily (stock: $5 \times 10^{-2}$ *M* UA in 50 m*M* HCl) (Store in dark at 4°C ).
6. Rotary shadowing metal: 10 mm, 8 mil gage, platinum/paladium wire.

### 2.2. Histochemistry of Proteoglycans—Light Microscopy

Since a majority of our work involves distinguishing regions of the extracellular matrix (ECM) that are enriched in proteoglycans from those regions of the ECM that contain collagen and elastic fibers, we use a modification of the Movat pentachrome stain ***(11)*** developed by Schmidt and Wirtala ***(12)*** at the University of Washington, (USA). This protocol involves the use of a modified Verhoeff elastic tissue stain, which is more tolerant of minor variations in technique and does not require differentiation to a subjective end point ***(13)***. This procedure is particularly useful for evaluating ECM in lung and vascular tissue (*see* **Note 4**) ***(14,15)***.

#### 2.2.1. Modified Movat's Pentachrome

1. Alcian blue, 1%: Alcian Blue 8GX (1 g) in distilled water (DW) (100 mL), stable for 6 mo.
2. Alkaline alcohol: 95% EtOH (180 mL), conc. ammonium hydroxide (20 mL), stable for 6 mo.
3. Musto elastin stain
   a. Alcoholic hematoxylin, 2% stock: Hematoxylin (2 g) in 95% EtOH (100 mL), stable for 3 mo.

   b. Ferric Chloride, 1.48% stock: Ferric chloride hexahydrate (2.48 m) in DW (99 mL), conc. HCl (5 mL), stable for 6 mo.
   c. Iodine stock: Potassium iodide (4 g) in DW (20 mL), iodine (2 g), DW (80 mL); stable for 6 mo.
   d. Working solution: Mix in order immediately before use 30 mL 2% hematoxylin, 20 mL 1.48% ferric chloride, and 10 mL iodine solution. Discard after use.
4. Crocein scarlet–Acid fuchsin stain
   a. Crocein scarlet stock: 1 g Brilliant Crocein MOO in 99.5 mL DW, 0.5 mL conc. acetic acid, stable for 6 mo.
   b. Acid fuchsin stock: 0.1 g acid fuchsin in 99.5 mL DW, 0.5 mL conc. acetic acid, stable for 6 mo.
   c. Working solution: 4 parts crocein scarlet, 1 part acid fuchsin, stable for 3 mo.
5. Acetic acid, 1%: 1 mL glacial acetic acid, 99 mL DW, stable for 3 mo.
6. Phosphotungstic acid, 5%: 5 g Phosphotungstic acid, 100 mL DW, stable for 6 mo.
7. Saffron, alcoholic: 6 g Spanish saffron stamens, 100 mL absolute EtOH in an airtight jar. Incubate at 60°C for 48 h to extract saffron, leave stamens in stain solution, and periodically top up alcohol over the life of the stain (1 yr).
8. Bouin's fixative: 750 mL Saturated picric acid/$H_2O$, 37–40% formaldehyde (Formalin) (250 mL), 50 mL glacial acetic acid.
9. 5 µm Paraffin sections of tissue fixed in 10% neutral buffered Formalin and processed routinely (*see* **Note 5**).

### 2.3. Histochemistry of Hyaluronan—Light Microscopy

This technique utilizes a highly specific peptide that interacts with hyaluronan in tissue sections ***(16–18)***. The peptide is derived from the proteoglycan aggrecan and can be prepared from tryptic digests of bovine cartilage by hyaluronan affinity columns and then biotinylated ***(19)***. We have used this procedure to detect hyaluronan in vascular lesions ***(20,21)***.

1. 5 µm Paraffin sections mounted on Superfrost + slides, baked 1 h at 50°C (*see* **Note 6**).
2. Xylene.
3. Graded EtOH series for rehydration: 100, 95, 70, 50, and 35% .
4. 0.7% $H_2O_2$ in absolute methanol.
5. Phosphate buffered saline (PBS), pH 7.3: 8 g NaCl, 0.2 g KCl, 1.44 g $Na_2HPO_4$, 0.24 g $KH_2PO_4$ , 800 mL distilled $H_2O$. Adjust pH to 7.3 with HCl and bring volume to 1 L with distilled $H_2O$.
6. PBS/0.1% BSA: Add 1 mg of BSA (globulin-free)/mL of PBS.
7. Blocking solution: 1% BSA, globulin-free in TBS.
8. Biotinylated hyaluronan binding protein, 2–4 µg/mL PBSA (HABP).
9. Streptavidin/HRP 1:400 in PBS.
10. Tris-HCl buffer (TB), 0.05 *M*, pH 7.6 with 1 *N* HCL.
11. Diaminobenzidine (DAB).
12. Nickel chloride.
13. 30% $H_2O_2$.
14. Counterstain: 2% methyl green in NaAc buffer, pH 4.2.
15. *Streptomyces* hyaluronidase (1 U/µL stock).
16. Hyaluronan (for use as probe-absorbing control), add 100 µg/mL of diluted HABP and allow to absorb overnight at 4°C.

### 2.4. Intracellular Localization of Hyaluronan in Cultured Cells

We recently found that intracellular hyaluronan can be detected inside a number of cell types *(22)*. The cells must first be pretreated with hyaluronidase to remove pericellular and cell surface hyaluronan and then permeabilized. The source of the intracellular hyaluronan, whether taken up from the pericellular matrix or synthesized within an endosomal membrane compartment, is not yet clear.

1. Cells grown on glass coverslips in 35-mm tissue culture dishes (*see* **Note 7**).
2. 10% neutral buffered formalin.
3. PBS and PBS/1% BSA as described above for tissue sections.
4. Biotinylated HABP (100 µg/mL stock) as above.
5. *Streptomyces* hyaluronidase, 1 µg/µL stock solution.
6. Streptavidin–Texas red conjugate.
7. Gel-Mount fluorescence mounting medium or suitable equivalent.

### 2.5. Ultrastructural Localization of Hyaluronan—Scanning Electron Microscopy

A hyaluronan-dependent pericellular matrix can be visualized around cultured cells at the light level using the particle-exclusion method. However, the hyaluronan network is highly soluble and fragile, and is usually lost during preparation for scanning electron microscopy (SEM) when standard techniques are employed. Therefore, we provide a simple method for preserving the hyaluronan coat for viewing by SEM to visualize the hyaluronan filaments, which extend from the cell surface into the pericellular matrix *(23)*. Proteoglycan granules present in the cell coat can also be visualized. (Also *see* **Subheading 2.6.**)

1. Cells grown on glass coverslips (as described above for intracellular localization of hyaluronan).
2. Karnovsky's fixative (half-strength): 2% paraformaldehyde and 2.5% EM-grade glutaraldehyde in 0.1 *M* sodium cacodylate, pH 7.3, containing 3 m*M* $CaCl_2$, 5% sucrose, and 0.2% ruthenium red (RR).
3. 0.1 *M* Sodium cacodylate, pH 7.3, containing 3 m*M* $CaCl_2$, 5% sucrose, and 0.1% RR.
4. 1% $OsO_4$ in cacodylate buffer containing 0.05% RR.
5. *Streptomyces* hyaluronidase (1 U/µL stock).

### 2.6. Histochemistry of Proteoglycans—Transmission Electron Microscopy

A critical element in preserving proteoglycans in tissues and cells is to prevent their solubility by including within the fixative cationic dyes that will neutralize the negative charges and make them less water soluble. One such dye that has proven to be very effective is ruthenium red *(24,25)*. Neutralizing the negative charges on the glycosaminoglycans of the proteoglycans results in collapse of the glycosaminoglycan chains onto the core protein and the formation of a granule that is of a size that reflects the size of the proteoglycan monomer *(4,26)*. An essential component of the ultrastructural visualization of polyanionic glycosaminoglycans with cationic dyes is the oxidation of bound ruthenium red and reduction of $OsO_4$ to an insoluble product to provide electron density. Thus it is possible to apply the techniques of morphologic stereology to estimate quantitatively the content of proteoglycans that are spatially preserved in tissue sections ***(26–29)***.

1. 0.2% ruthenium red (RR) in half-strength Karnovsky's fixative (2% paraformaldehyde and 2.5% EM-grade glutaraldehyde in 0.1 *M* NaCacodylate, 3 m*M* $CaCl_2$, pH 7.3) ***(4,26)***.
2. 0.1% ruthenium red in 0.1 *M* NaCacodylate, 3 m*M* $CaCl_2$, 3% sucrose, pH 7.3.
3. 0.05% ruthenium red in 1% $OsO_4$ in 0.1 *M* Cacodylate, 3 m*M* $CaCl_2$, pH 7.3 (*see* **Note 8**).
4. 0.1 *M* NaCacodylate, 3 m*M* $CaCl_2$, pH 7.3.
5. Graded EtOH series: 35, 50, 70, 90, 95, and 100%.
6. 3% uranyl acetate in 70% EtOH, stored in dark, filtered before use.
7. Propylene oxide.
8. Epoxy resin: Mix thoroughly in this order: 29.0 g Eponate; 16.g DDSA; 14.3 g NMA. Add and mix thoroughly, BDMA 1.18 g. Store frozen in air-tight 20-mL aliquots (*see* **Note 9**).
9. 7% (saturated) uranyl acetate in dd-$H_2O$.
10. Reynolds lead citrate: 2.66 g Lead nitrate, 3.52 g Sodium citrate, 60 mL fresh dd-$H_2O$, shake 5 min until dissolved, then invert every 5 min over the next 0.5 h. Add 1 *N* NaOH (freshly made, 1 g/25 mL dd-H2O) (16 mL), bring volume to 100 mL with dd-$H_2O$.

### 2.7. Immunocytochemistry of Proteoglycans—Light Microscopy

During the last 5 yr there has been an explosion of available antibodies to virtually all of the core proteins of the known proteoglycans as well as to several glycosaminoglycan epitopes. This has provided the opportunity to "map" the specific location of particular proteoglycans to defined regions in tissues and associated with cells. Defining the precise location of a proteoglycan in tissue sections aids in identifying different functions of specific proteoglycans ***(21,30–33)***. In addition to the use of paraffin processed tissue and avidin–biotin detection systems for the immunolocalization of proteoglycans, we have found it useful to process tissue with the goal of visualizing tissue antigens at both the light and electron microscopic level on the same section or on adjacent sections. This can be done by processing tissue for electron microscopy and removing the plastic resin following sectioning for light microscopy ***(15,34)***. Below we describe two different procedures we use to localize proteoglycans in tissue sections.

#### 2.7.1. Paraffin

1. 5 µm paraffin sections mounted on Superfrost + slides, baked 1 h, 50°C (*see* **Note 10** and **Note 11**).
2. Xylene.
3. Graded EtOH series for rehydration: 100, 95, 70, 50, and 35%.
4. 0.7% $H_2O_2$ in absolute methanol.
5. Tris-HCl buffer (TB), 0.05 *M*, pH 7.6 with 1 *N* HCL.
6. Tris-HCl buffered saline (TBS), 0.9% NaCl, pH 7.6.
7. Phosphate–buffered saline (PBS), pH 7.3: 8 g NaCl, 0.2 g KCl, 1.44 g $Na_2HPO_4$, 0.24 g $KH_2PO_4$, 800 mL distilled $H_2O$. Adjust pH to 7.3 with HCl and bring volume to 1 L with distilled $H_2O$.
8. PBS/0.1% bovine serum albumin: Add 1 mg BSA (globulin-free)/mL PBS (*see* **Note 12**).
9. Blocking solution: 10% normal serum (corresponding to secondary antibody species), 1% bovine serum albumin, globulin-free (BSA) in TBS.

10. Diluent: 0.1% BSA in TBS.
11. Primary antibody: Diluent, dilution determined by titration on positive control tissue.
12. Biotinylated secondary antibody, 1/200 diluent.
13. Vectastain Elite ABC reagent, vector, solutions A and B.
14. Diaminobenzidine.
15. Nickel chloride.
16. 30% $H_2O_2$.
17. Counterstain: 2% methyl green in sodium acetate buffer, pH 4.2.
18. Enriched Tris buffer, pH 8.0: 0.6 g Tris base, 0.8 g sodium acetate, 0.3 g NaCl, 100 mL distilled $H_2O$, adjust pH with 10 *N* HCl, add BSA (1 mg/mL).
19. Enzymes: Chondroitin ABC lyase (0.2 unit/mL enriched Tris buffer) trypsin (1 mg/mL 0.2 *M* Tris, 4 m*M* $CaCl_2$, pH 7.7).

### *2.7.2. Plastic Removal*

Tissue Fixation and Embedding in Plastic

1. 0.2 *M* $PO_4$ buffer: To 405 mL of 0.2 *M* $Na_2HPO_4$ (28.39 g/L), add approx 95 mL of 0.2 *M* $NaH_2PO_4{\bullet}H_2O$ (27.6 g/L) to give a pH of 7.4. Add distilled $H_2O$ to give final volume of 500 mL.
2. Fixative: 3% paraformaldehyde, 0.25% EM-grade glutaraldehyde in 0.05 *M* $PO_4$ buffer: In a fume hood stir 3 g of paraformaldehyde into 20 mL of 65°C distilled $H_2O$, add 2–3 drops of 1*N* NaOH and stir until clear, then cool. Mix 25 mL of 0.2 *M* phosphate buffer, 50 mL distilled $H_2O$, 0.5 mL of 50% EM-grade glutaraldehyde, and cooled paraformaldehyde. Adjust pH to 7.4 and bring final volume to 100 mL with d-$H_2O$.
3. Phosphate–buffered saline (PBS): 0.05 *M* phosphate buffer, 0.9% NaCl, pH 7.3.
4. 0.1 *M* glycine in PBS.
5. Graded ethanol series: 35, 50, 70, 90, 95, and 100% (absolute).
6. 3% uranyl acetate in 70% EtOH, stored in the dark and filtered with #1 Whatman filter paper before use.
7. Epoxy resin: Mix equal parts of solution A (50 mL of DDSA and 37.5 mL of EPON 812, well mixed) and solution B (50 mL of EPON 812 and 36.7 mL of NMA, well mixed). Add 12 drops of DMP-30 for each 10 mL of resin and stir thoroughly (*see* **Note 13**).

Sectioning, Plastic Removal, and Immunostaining

1. Superfrost plus microscope slides.
2. Saturated Na ethoxide: Put 15 g of NaOH in a 100 mL plastic bottle, fill with absolute EtOH and allow to dissolve and age 2–3 d on a shaker. The resulting solution will be a light tea color.
3. Absolute EtOH.
4. Graded ethanol series for hydration.
5. 0.3% $H_2O_2$ in absolute methanol.
6. Immunoreagents as in Procedure 1—Paraffin.

## *2.8. Immunocytochemistry—Transmission Electron Microscopy*

The successful immunodetection and localization of specific molecules within tissue prepared for ultrastructural examination is subject to a number of factors, including epitope maintenance during fixation and tissue/reagent penetration limitations. The physical and chemical nature of individual epitopes within the microenvironment of the tissue or culture condition of interest determine the impact of these factors and ultimately the choice of detection protocols. No single processing procedure can be recommended, but preliminary light microscopy establishes some of the limitations to be encountered and provides direction. If a molecule cannot be detected using light microscopy, ultrastructural localization is unlikely to be achieved. There are a number of different ways to prepare tissue and cells for the ultrastructural examination of proteoglycan location. Below, we describe two different methods that we have used with considerable success to examine proteoglycan localization at the ultrastructural level. In Procedure 1 the acrylic monomer resin LR White is used because its hydrophilic nature allows the penetration of immunochemicals without plastic removal or etching, in contrast to epoxy resins, avoiding possible alteration of antigenic sites with removal or etching agents ***(15,35,36)***. In Procedure 2 the removal of plastic from sections of epoxy resin-processed tissue has the effect of "unmasking" previously inaccessible epitopes and allowing a precise correlation of light and EM immunostaining ***(37,38)***.

### *2.8.1. Procedure 1: LR White (LRW) Resin*

#### Tissue Fixation and Embedding (*see* **Note 14**)

1. Phosphate buffered saline (PBS), pH 7.3: 8 g NaCl, 0.2 g KCl, 1.44 g $Na_2HPO_4$, 0.24 g $KH_2PO_4$, 800 mL distilled $H_2O$. Adjust pH to 7.3 with HCl and bring volume to 1 L with distilled $H_2O$.
2. Fixative: 3% paraformaldehyde, 0.25% EM-grade glutaraldehyde in PBS, pH 7.3, freshly prepared.
3. 0.3 *M* glycine in PBS.
4. 30%, 50%, and 70% methanol and 80% EtOH (*see* **Note 15**).
5. 2% uranyl acetate (UA) in 70% methanol, store in the dark ***(36)*** (*see* **Note 16**).
6. 2/1 mixture LR White/80% EtOH.
7. LR White acrylic resin (medium or hard).
8. Gelatin capsules.
9. Reverse carbon-coated formvar substrated Ni grids, 200 mesh.

#### Immunolabeling and Poststaining

1. 80 to 100-nm LRW-embedded sections mounted on formvar-coated Ni grids (*see* **Note 17**).
2. Tris-HCl buffered saline (TBS): 50 m*M* Trizma base, 0.15 *M* NaCl, pH 7.6 with 1 *N* HCl.
3. Blocking buffer:10% normal serum, 1% BSA in TBS.
4. Diluent/wash buffer: 0.1% Tween 20, 0.1% BSA, 0.1% $NaN_3$, TBS, pH 8.2.
5. Primary antibody diluted in diluent/wash buffer and microfuged for 1 min just before use (*see* **Note 18**).
6. 10-nm colloidal gold-conjugated secondary antibody diluted in diluent/wash buffer and microfuged 1 min just before use.
7. 3% glutaraldehyde in PBS.
8. 7% uranyl acetate in dd-$H_2O$.
9. Reynolds lead citrate: 2.66 g Lead nitrate, 3.52 g Sodium citrate, 60 mL $H_2O$; shake 5 min until dissolved, then invert every 5 min over the next 0.5 h, add 16 mL 1 *N* NaOH (freshly made, 1 g/25 mL dd-H2O), bring volume to 100 mL with dd-$H_2O$.
10. Aqueous 2% $OsO_4$.

### 2.8.2. Procedure 2: Plastic Removal

#### Tissue Fixation and Embedding in Plastic

1. 0.2 *M* phosphate buffer: To 405 mL of 0.2 *M* $Na_2HPO_4$ (28.39 g/L), add approx 95 mL of 0.2 *M* $NaH_2PO_4 \cdot H_2O$ (27.6 g/L) to give a pH of 7.4. Add distilled $H_2O$ to give a final volume of 500 mL.
2. Fixative: 3% paraformaldehyde, 0.25% EM-grade glutaraldehyde in 0.05 *M* $PO_4$ buffer: In a fume hood, stir 3 g of paraformaldehyde into 20 mL of 65°C distilled $H_2O$. Add 2–3 drops of 1 *N* NaOH and stir until clear, then cool. Mix 25 mL of 0.2 *M* $PO_4$ buffer, 50 mL of distilled $H_2O$, 0.5 mL of 50% EM-grade glutaraldehyde, and cooled paraformaldehyde. Adjust pH to 7.4 and bring final volume to 100 mL with d-$H_2O$.
3. 0.1 *M* glycine in PBS.
4. Graded ethanol series, 35%, 50%, 70%, 90%, 95%, and 100% (absolute).
5. 3% uranyl acetate in 70% EtOH, stored in the dark and filtered with #1 Whatman filter paper before use.
6. Epoxy resin: Mix equal parts of solution A (50 mL of DDSA and 37.5 mL of EPON 812, well mixed) and solution B (50 mL of EPON 812 and 36.7 mL of NMA, well mixed). Add 12 drops of DMP-30 for each 10 mL of resin and stir thoroughly (*see* **Note 19**).

#### Sectioning, Plastic Removal, Immunostaining, and Reembedding Thin Sections

1. Reverse carbon-coated formvar substrated Ni grids, 200 mesh.
2. Saturated Na ethoxide: Put 15 g of NaOH in a 100-mL plastic bottle, fill with absolute EtOH and allow to dissolve, and age 2–3 d on a shaker. The resulting solution will be a light tea color.
3. Absolute ethanol.
4. Graded ethanol series for hydration and dehydration (including fresh absolute ethanol).
5. Parafilm.
6. Covered chamber for humidified incubations.
7. Tris-HCl-buffered saline (TBS): 0.05 *M* Trizma base, 0.15 *M* NaCl; pH 7.6 with 1 *N* HCl (*see* **Note 19**).
8. Blocking buffer: 10% normal serum, 1% BSA in TBS.
9. Diluent/wash buffer: 0.1% Tween 20, 0.1% BSA, 0.1% $NaN_3$, TBS, pH 8.2.
10. Primary antibody diluted in filtered diluent.
11. 10-nm colloidal gold conjugated secondary antibody diluted in diluent/wash buffer and microfuged 1 min just before use.
12. Phosphate-buffered saline (PBS): 0.05 *M* phosphate buffer, 0.9% NaCl, pH 7.3.
13. 2% glutaraldehyde in PBS.
14. 1% $OsO_4$ in PBS.
15. 2% aqueous uranyl acetate.
16. Epoxy resin: Mix equal parts of solution A (50 mL of DDSA and 37.5 mL of EPON 812, well mixed) and solution B (50 mL of EPON 812 and 36.7 mL of NMA, well mixed). Add 12 drops of DMP-30 for each 10 mL of resin and stir thoroughly.

## 3. Methods

### 3.1. Molecular Spreads

1. Fill siliconized Petri dish with hypophase solution to give a convex meniscus and clean by skimming it with optical lens tissue.
2. Thoroughly rinse with water and drain a glass slide, handling with forceps. Immerse the slide at a 30–45° in the hypophase and support on the edge of the dish.

3. Prepare the final spreading solution by gently mixing: 50 µL of diluted sample; 50 µL of 10 m*M* Tris-HCl, 1 m*M* EDTA, pH 8.5; 2 µL cytochrome C stock (2.5 µg/µL) (add just before spreading preparation).
4. With a micropipet, draw a small amount of hypophase onto the slide to wet it.
5. Slowly (over 10–30 s) deliver 75 µL of the hyperphase spreading solution to the slide at 3–6 mm above the hypophase meniscus. Maintain a continuous sheet of liquid between the delivery pipet and the hypophase.
6. Allow the hyperphase solution to spread and denature for 15–60 s (*see* **Note 20**).
7. Pick up a flat, supported grid with a pair of clean self-locking forceps. Lay it flat, carbon side down, onto a smooth filter paper. Lift the forceps up to a 20–30° to bend the edge of the grid and set the forceps/grid aside. Prepare 2–3 such grids in advance of sampling the monolayer.
8. Gently lower a grid, substrate down, onto the monolayer.
9. Stain the sample for approximately 5 s by immersing the entire grid in 0.001% PTA without draining after pickup. Rinse in 90% EtOH for 10 s.
10. Stain for about 30 s in 5 µ*M* uranyl acetate, rinse in 90% EtOH for 10 s, and air dry.
11. Mount grids on a rotating stage in vacuum evaporator (*see* **Note 21**).
12. Load with 10-mm platinum/palladium wire coiled to fit down in tip if using a tungsten basket.
14. Position aperture to assure proper alignment of path from metal source to sample.
15. Insert shield between metal source and sample and pump down evaporator to below $5 \times 10^{-5}$ torr. Use $LN_2$ trap if evaporator is equipped with one (*see* **Note 22**).
16. After achieving vacuum, heat tungsten basket to a red glow to drive off contaminants, begin specimen rotation, and retract shield.
17. Heat metal slowly and just enough to achieve metal evaporation. After completely exhausting metal, continue specimen rotation for 30 s before venting and removing grids (*see* **Note 23**).

### 3.2. Histochemistry of Proteoglycans - Light Microscopy

1. Deparaffinize and hydrate tissue sections.
2. Preheat alkaline alcohol in 60°C oven.
3. Mordant tissue sections in Bouin's in microwave 45 min (staining tub volume), cool 10 min (*see* **Note 24**).
4. Rinse sections in running water for 10 min.
5. Immerse sections in 1% Alcian blue for 20 min.
6. Rinse sections in running water for 2–5 min.
7. Place sections in 60°C alkaline alcohol for 10 min.
8. Rinse sections in running water for 2–5 min.
9. Immerse sections in Musto working solution for 30 min.
10. Wash briefly in running water, rinse well in distilled water.
11. Place sections in crocein scarlet/acid fuchsin working solution for 1–2 min (*see* **Note 25**).
12. Rinse in distilled water.
13. Immerse slides in 5% phoshotungstic acid for 5 min (*see* **Note 26**).
14. Place directly in 1% acetic acid for 5 min (*see* **Note 27**).
15. Rinse slides well in distilled water.
16. Dehydrate tissue by rinsing briefly in 95% EtOH, twice in absolute EtOH, 1 min each.
17. Immerse slides in alcoholic saffron for 60 min, keep container covered.
18. Rinse slides in two changes absolute EtOH, 1 min each (*see* **Note 28**).
19. Clear stained slides in xylene, and mount with xylene soluble mounting media.

### 3.3. Histochemistry of Hyaluronan—Light Microscopy

1. Deparaffinize tissue sections in three changes of xylene, 7 min each on slow shaker.
2. Rinse tissue in 100% EtOH.
3. Quench endogenous tissue peroxidase by incubating tissue in 0.7% $H_2O_2$ in absolute methanol for 20 min.
4. Rinse tissue in 100% EtOH, and rehydrate through graded EtOH, 1 min each.
5. Rinse for 10 min in PBS.
6. Treat control slides with hyaluronidase, 20 U/mL NaAc buffer, 1 h, 37°C.
7. Rinse 3 times over 10 min with PBS.
8. Block nonspecific protein interactions by incubating tissue in 10% serum, 1% BSA in TBS for 1 h.
9. Apply biotinylated HABP (4 μg/mL PBSA) to experimental and HAdase control sections and apply HA-absorbed HABP control to sections. Incubate overnight at 4°C in a humidified chamber.
10. Rinse tissue sections with PBS, 3 × 5 min on a slow shaker.
11. Apply Streptavidin/HRP, 1/400 in PBS, 1 h, in a humid atmosphere.
12. Rinse tissue sections with PBS, 3 × 5 min on a slow shaker.
13. TB, 10 min, 37°C.
14. Prepare developing solution: Dissolve 64 mg of diaminobenzadine (DAB) in 200 mL of TB, 37°C. Add 0.08 g of $NiCl_2$ if black reaction product is desired. Just before use, add 164 μL of 30% $H_2O_2$ and mix.
15. Incubate sections in developing solution, 10 min, 37°C. Stop reaction by washing in TB, then rinse with $H_2O$.
16. Counterstain with 2% methyl green for 5 min.
17. Dehydrate through 95% and 100% EtOH (2×), clear in xylene, and cover slip with xylene-soluble mounting media.

### 3.4. Intracellular Localization of Hyaluronan in Cultured Cells

1. Pretreat cells with *Streptomyces* hyaluronidase (2 U/mL in the culture medium) to remove pericellular and cell surface hyaluronan for 1 h, 37°C (*see* **Note 29**).
2. Fix cells by addition of 0.5 mL of 10% neutral formalin directly to the medium (2 mL) to give a final concentration of 2% formalin (this is roughly equivalent to 0.75% formaldehyde) for 10 min, at 22°C (*see* **Note 30**).
3. Rinse three times, 2 min each, with PBS. The rinses are conveniently done with the cover slips still in the 35-mm dishes.
4. Permeabilize with 0.5% Triton X-100 in PBS for 10 min at 22°C. (*see* **Note 31**).
5. Rinse three times, 2 min each, with PBS.
6. Using fine forceps to handle cover slips, gently blot the back of the coverslip and lay face up (cells on top) on parafilm on a hard, flat surface. Drying the back helps prevent wicking to the underside of the cover slip. (*see* **Notes 32, 33**)
7. Incubate with 100–200 μL of biotinylated HABP (at 4 μg/mL in PBS/1% BSA) for 1 h at 22°C, covering with the lid from the 35-mm dish to minimize evaporation.
8. Transfer cover slip back to the 35-mm dish and rinse three times, 5 min each.
9. Blot the back of the coverslip and transfer back to parafilm, cell side up.
10. Incubate with 200 μL of streptavidin-Texas red (1/500) in PBS/1% BSA, for 1 h at 22°C, covering with the lid from the 35-mm dish.

11. Transfer the cover slips back to the culture dishes, and rinse three times, 5 min each, with PBS.
12. Mount cover slip face down on a glass slide using Gel-mount.

### 3.5. Ultrastructural Localization of Hyaluronan—Scanning Electron Microscopy

1. Rinse cells in serum free medium. Control cells can be pretreated before fixation with *Streptomyces* hyaluronidase to remove hyaluronan. Digest the cells using 2 U/mL in culture medium for 1 h at 37°C.
2. Fix with Karnovsky's fixative in 0.1 *M* sodium cacodylate containing 2 m*M* $CaCl_2$, 5% sucrose, and 0.2% RR for 1 h at 22°C.
3. Rinse cells with 0.1 *M* sodium cacodylate, pH 7.3, containing 3 m*M* $CaCl_2$, 5% sucrose, and 0.1% RR.
4. Postfix in 1% $OsO_4$ in cacodylate buffer containing 0.05% RR for 1 h at 22°C.
5. Rinse cover slips gently by dipping several times in phosphate-buffered saline and then $H_2O$.
6. Let the coverslips air dry in a dust-free environment (*see* **Note 34**).
7. Cut the cover slip to an appropriate size with a diamond pencil.
8. Sputter coat with gold/palladium (*see* **Note 35**).

### 3.6. Histochemistry of Proteoglycans—Transmission Electron Microscopy

1. Fix 1-mm$^3$ tissue pieces in 0.2% RR in half-strength Karnovsky's for 3 h (*see* **Notes 36–39**).
2. Wash tissue three times, 10 min each, with 0.1% RR in 0.1 *M* Na cacodylate buffer.
3. Postfix with 0.05% RR in 1.0% $OsO_4$ in 0.1 *M* cacodylate buffer for 3h, at room temperature.
4. Wash for 10 min with 0.1 *M* NaCacodylate buffer on a shaker.
5. Dehydrate with 35%, 50% and 70% EtOH, 10 min each, on a shaker.
6. *En bloc* stain with 3% uranyl acetate/70% EtOH, 1 h in dark.
7. Continue dehydration with 80%, 90%, and 95% EtOH, 10 min each, on the shaker, absolute EtOH three times, 10 min.
8. Pass tissue through the transition fluid, propylene oxide two times for 10 min.
9. Infiltrate tissue with 1/2, 1/1, and 2/1 mixtures of complete epoxy resin and propylene oxide, 1 h each, on shaker.
10. Infiltrate with complete resin, uncovered on the shaker.
11. Infiltrate with fresh complete resin, uncovered, overnight, under vacuum.
12. Drain tissue pieces of excess epoxy and place in fresh complete resin in beem capsules or silicon rubber molds with labels.
13. Allow to equilibrate for 1 h under vacuum and polymerize in a 60–65°C oven for 48 h.
14. Cut 80- to 100-nm sections and mount on cleaned Cu grids.
15. Poststain with 7% uranyl acetate for 8–10 min. Wash with filtered (0.4 µm) dd-$H_2O$. Stain with Reynolds lead citrate, for 8–10 min, and wash as above (*see* **Note 40**).
16. Examine in transmission electron microscope.

### 3.7. Immunocytochemistry of Proteoglycans—Light Microscopy

#### 3.7.1. Paraffin

1. Deparaffinize tissue sections in three changes of xylene, 7 min each, on slow shaker.
2. Rinse tissue, 2 × 1 min, in 100% EtOH.
3. Quench endogenous tissue peroxidase by incubating tissue in 0.7% $H_2O_2$ in absolute methanol for 20 min.(*see* **Note 41**).
4. Rinse tissue 2 × 1 min in 100% EtOH, and rehydrate through graded EtOH, 1 min each.

5. Rinse, 2 × 1 min, in $H_2O$, then 10 min in TBS.
6. Epitope unmasking treatment as necessary with enzyme digestion (*see* **Note 43**): Chondroitin ABC lyase 1 h, 37°C; Trypsin 15 min, 37°C (*see* **Notes 42** and **44**).
7. Rinse in TBS, 2 × 5 min.
8. Block non-specific protein interactions by incubating tissue in 10% serum, 1% BSA in TBS for 1 h.
9. Blot off blocking serum on filter paper and apply primary antibody diluted in TBS/0.1% BSA and incubate in a humidified atmosphere overnight at 4°C.
10. Segregate sections with the same primary antibody into the same washing tub. Rinse tissue sections with TBS, 3 × 5 min, on slow a shaker.
11. Apply biotinylated secondary antibody, diluted 1/200 in TBS/0.1% BSA, to tissue sections and incubate 1 h, in a humidified atmosphere.
12. During secondary antibody incubation, prepare Vectastain avidin-biotin-peroxidase-complex (ABC) reagent: Add 15 µL of reagent A and 15 µL of reagent B to each 1 mL of PBS/0.1% BSA and allow to stand for 30 min before use (*see* **Note 45**).
13. Rinse tissue sections with TBS, 3 × 5 min on slow a shaker.
14. Apply ABC reagent to sections and incubate for 30 min in a humidified atmosphere.
15. Rinse sections with PBS, 3 × 5 min, TB 10 min, 37°C.
16. Prepare developing solution: Dissolve 64 mg of diaminobenzadine (DAB) in 200 mL of TB at 37°C. Add 0.08 g of $NiCl_2$ if black reaction product is desired. Just before use, add 164 µL of 30% $H_2O_2$ and mix.
17. Incubate sections in developing solution, 10 min, 37°C. Stop reaction by washing in TB, then rinse with $H_2O$ (*see* **Note 41**).
18. Counterstain with 2% methyl green for 5 min.
19. Dehydrate through 95% and 100% EtOH (2×), clear in xylene, and cover slip with xylene-soluble mounting media.

### *3.7.2. Plastic Removal*

#### Tissue Fixation and Embedding in Plastic (*see* **Note 46**)

1. Dissect tissue to 1-$mm^3$ cubes or smaller and fix for 2 h at room temperature.
2. Wash tissue 2 × 5 min with PBS.
3. Quench free aldehyde groups with 0.1 *M* glycine/PBS 2 × 10 min.
4. Wash once with PBS, 5 min.
5. Dehydrate with 35%, 50%, 70% EtOH, 10 min each on shaker.
6. *En bloc* stain with 3% uranyl acetate/70% EtOH 1 h.
7. Continue dehydration through 90%, 95%, and 100% ethanol, 10 min each, on shaker.
8. Prepare epoxy resin and use in 1:1 mixture EPON/ethanol. Infiltrate tissue for 1 h on the shaker with EPON/ethanol in a capped vial.
9. Prepare 2/1 mixture of EPON/ethanol and infiltrate tissue for 1 h on the shaker with the cap removed.
10. Prepare fresh epoxy resin and infiltrate the tissue on the shaker, for 1 h, in an uncapped vial.
11. Change the epoxy and infiltrate overnight under vacuum.
12. Prepare fresh epoxy resin and embed the tissue in Beem capsules or coffin molds.
13. Polymerize blocks in vacuum oven at 55°C for 48 h.

#### Sectioning, Plastic Removal and Immunostaining

1. Cut 1- to 2-mm sections and float on a small pool of filtered water on slides.
2. Dry overnight on a 45°C hot plate.

3. Circle sections on slide underside with diamond etching pencil.
4. Place slides in plastic staining rack and immerse in saturated Na ethoxide solution for 15–20 min. To ensure adequate removal of plastic, at about 15 min, remove one slide, rinse in 100% EtOH, and examine under the dissecting scope, being careful not to allow the tissue to dry. Continue at 1-min intervals to ensure removal.
5. Rinse in 4 changes of 100% EtOH, 5 min each.
6. Place slides in 0.3% $H_2O_2$/methanol for 10 min to block endogenous peroxide.
7. Rinse once in 100% EtOH for 1 min, and through graded ethanol series to $H_2O$ for 1–2 min each.
8. Soak in TBS for 10 min.
9. Proceed with immunostaining as in **step 2** of Procedure 1 - Paraffin.

## 3.8. Immunocytochemistry—Transmission Electron Microscopy

### 3.8.1. Procedure 1: LR White Resin

#### Tissue Fixation and Embedding

1. Cut tissue into small pieces (1 mm$^3$) and rinse briefly tissue with PBS at room temperature.
2. Fix tissue pieces with 3%/0.25% fixative for 1h at room temperature.
3. Rinse fixed tissue with 0.3 *M* glycine/PBS 3 × 10 min, on a shaker.
4. Rinse with PBS for 10 min or overnight at 4°C.
5. Dehydrate with 30%, 50%, and 70% methanol, 10 min each, on shaker.
6. *En bloc* stain with 2% UA/70% methanol, 1 h, dark, room temperature.
7. Dehydrate with 80% EtOH, 10 min, on shaker.
8. Infiltrate with 2/1 LR White/80% EtOH, 1 h, on shaker.
9. Continue infiltration with 100% LR White, two changes, 1 h each, on shaker, 4°C.
10. Infiltrate overnight at 4°C on shaker in fresh LR White.
11. Transfer tissue to gelatin capsules, add fresh LR White, cap, and polymerize in vacuum, 50–52°C, 24 h.
12. Cut 100 nm (gold) sections and mount on the non-carbon side of formvar-coated nickel grids.

#### Immunolabeling and post-staining

1. Float grids (section surface down) on droplets of dd-$H_2O$, 5 min (*see* **Note 47**).
2. Block tissue/secondary antibody cross-reactivity and nonspecific protein interactions by incubating sections on droplets of blocking buffer in humidified chamber, 1 h, room temperature.
3. Drain excess blocking buffer with filter paper at the grid/forceps junction.
4. Place grids directly on droplet of diluted primary antibody in a humidified chamber and incubate overnight at 4°C (*see* **Note 48**).
5. Drain antibody with filter paper and soak grids on 15 successive droplets of diluent/wash buffer over 15 min.
6. Incubate sections on droplet of gold-conjugated secondary antibody diluted 1:20–1:50, 1 h, room temperature.
7. Drain secondary with filter paper and soak grids on 15 successive droplets of diluent/wash buffer or TBS.
8. Fix on droplet of 3% glutaraldehyde/PBS, 10 min, room temperature.
9. Soak grids on five successive droplets of dd$H_2O$ over 5 min.
10. The grids may be air dried at this point until poststain.
11. Working in a hood, place grids face up on hard filter paper in a 9-cm glass Petri dish. On a piece of parafilm in the Petri dish, place a 100- to 200-μL droplet of aqueous 2% $OsO_4$.

Cover and allow the sections to postfix in osmium vapor for 10 min to 1 h, depending on desired contrast. Drain with filter paper and air dry. Avoid contact with eyes, skin, or respiratory membranes.

12. Poststain sections by floating grids face down on droplets of aqueous 7% UA (microfuge 1 min before use) for 10 min. Wash by placing on 10 successive drops of filtered dd$H_2O$. Float grids on drops of lead citrate for 2–10 min and wash as before. Drain with filter paper and air dry.
13. Examine sections in TEM.

### 3.8.2. Procedure 2: Plastic Removal

#### Tissue Fixation and Embedding in Plastic

1. Dissect tissue to 1-mm cubes or smaller and fix for 2 h at room temperature.
2. Wash tissue 2 × 5 min with PBS.
3. Quench free aldehyde groups with 0.1 *M* glycine/PBS, 2 × 10 min.
4. Wash once with PBS, 5 min.
5. Dehydrate with 35%, 50%, and 70% EtOH, 10 min each, on shaker.
6. *En bloc* stain with 3% uranyl acetate/70% EtOH, 1 h.
7. Continue dehydration through 90%, 95%, and 100% ethanol, 10 min each, on shaker.
8. Prepare epoxy resin and use in 1/1 mixture EPON/ethanol. Infiltrate tissue for 1 h on shaker with EPON/ethanol in capped vial.
9. Prepare 2/1 mixture of EPON/ethanol and infiltrate tissue for 1 h on shaker with cap removed.
10. Prepare fresh epoxy resin, infiltrate tissue on shaker 1 h in uncapped vial.
11. Change epoxy and infiltrate overnight under vacuum.
12. Prepare fresh epoxy resin and embed tissue in Beem capsules or coffin molds.
13. Polymerize blocks in vacuum oven at 55°C for 48 h.

#### Sectioning, Plastic Removal, Immunostaining, and Re-Embedding Thin Sections

1. Prepare grids for collecting sectioned material: Acetic acid wash nickel grids in a water bath sonicator, rinse with dd-$H_2O$ three times in the sonicator, and finally rinse with absolute EtOh and air dry. Prepare formvar film by stripping 0.25% formvar off a precleaned glass microscope slide onto a cleaned dd-$H_2O$ surface. Place the grids, matte side down, onto to the film and pick up formvar with grids attached with a sheet of parafilm. Allow film to air dry protected from dust. Carefully remove individual grids from the parafilm with forceps, and place, formvar side down, onto a clean microscope slide. Evaporate a medium coat of carbon onto this "reverse" side of the grids (*see* **Notes 49** and **50**).
2. Collect 80- to 100-nm sections on the formvar side of the prepared grids and air dry (*see* **Note 51**).
3. Prepare a 1:3 dilution of saturated Na ethoxide with absolute ethanol. When not in use, keep this solution covered to retard evaporation and surface NaOH crystal formation.
4. Immerse individual grids, using anticapillary forceps, in the removal solution for 2–5 min (*see* **Note 52**). Wash with three consecutive 1 min immersions in absolute EtOH and hydrate through a graded alcohol series, 1 min each, to filtered dd-$H_2O$.
5. The "reverse" carbon side is blotted with filter paper and the grid is floated on an individual droplet of filtered 50 m*M* TBS on parafilm, where it remains as subsequent grids are prepared for immunostaining (*see* **Note 53**).

6. Using a humidified chamber for all incubations, transfer grids from TBS to individual filtered blocking buffer droplets for 10 min, and wash by floating grids on three consecutive droplets of diluent/washing buffer, 1 min each.
7. Transfer grids to droplets of primary antibody diluted in filtered diluent (*see* **Note 54**).
8. Incubate overnight at 4°C in a humidified chamber.
9. Wash the grids by floating on six successive drops of filtered wash buffer over 15 min.
10. Incubate on colloidal gold-conjugated secondary antibody diluted 1/50 in filtered diluent 1 h in the humidified chamber at room temperature.
11. Wash the grids by floating on eight successive drops of filtered wash buffer over 15 min, followed by three successive drops of filtered PBS.
12. Fix the antigen/antibody complex with 2% glutaraldehyde in PBS for 10 min and rinse on 4 drops of PBS.
13. Postfix the tissue with 1% $OsO_4$ in PBS for 10 min and wash on 3 droplets of filtered dd-$H_2O$.
14. Proceed to reembedding the thin sections by immersing each forceps-held grid through the graded alcohol series, 1 min each, and in absolute EtOH 3 × 1 min.
15. Immerse in 2% EPON in absolute ethanol 2 min and carefully blot between two pieces of No. 50 Whatman filter paper.
16. Position grids upright in razor blade slits in the back of a silicon rubber embedding mold and polymerize overnight to 48 h at 60°C.
17. Post-stain with 2% aqueous UA and lead citrate.
18. Examine in the transmission electron microscope.

## 4. Notes

1. Surfactants such as detergents, oil vapors, and grease from fingerprints interfere with formation and continuity of protein monolayers.
2. Something to consider in choosing sample buffer conditions is that, in the final cytochrome C/sample solution, at pH 9 and above, basic residues on cytochrome C become neutralized. This may result in less protein coating of the negatively charged molecules. At higher salt concentration (above 0.2 *M*), electrostatic interactions between polyanions and polycations become significantly reduced, resulting in less cytochrome C binding to the negatively charged molecule.
3. Preparation of collodion support films is probably the most important source of day-to-day variation in quality of the final molecular spread image. The goal is to prepare a hydrophobic surface to come in contact with the hydrophobic protein monolayer. The light carbon coat is useful in achieving an appropriate collection surface. A light carbon coat has been shown to yield proteoglycan monomers with extended disaccharide side chains, while an uncoated collodion surface yields mononers with side chains condensed onto the protein core. A quick indicator of the efficiency of protein pickup is a very flat meniscus of the collected droplet.
4. Results of the stain combination include the discrimination of various tissue components: coarse and delicate elastin fibers are black, collagens appear bright yellow, sulfated macromolecules (proteoglycans, glycosaminoglycans, and secreted mucins) are blue/blue-green, nuclei are dull red/black, erythrocytes are bright red, while the myofibrils (contractile apparatus of smooth muscle cells and striated muscle) are red/orange.
5. While tissue fixed in methacarn (acid/alcohol) can be used, it results in less satisfactory staining.
6. Prepared biotinylated probe is stored unaliquoted at approx 100 µg/mL dissolved in a 50/50 (v/v) mixture of glycerol and 0.15 *M* NaCl at –20°C. This stock preparation does

not lose detectable amounts of binding activity when stored for more than a year in this manner.

7. We achieve the best results when cover slips are incubated face up on the parafilm and the solutions are applied directly to the cells. We do not recommend putting parafilm squares on top of the incubation solutions. Evaporation is negligible if covered with the lid from the culture dish.
8. It is of interest that a radioactive isotope of ruthenium red, $^{103}Ru$, has been used to estimate glycosaminoglycans quantitatively after separation by electrophoresis ***(29)***.
9. We performed radiolabeling experiments to assess the loss of sulfated proteoglycans from fixed cultures of arterial smooth muscle cells and found that approx 40% of the total radiolabeled proteoglycans were lost during routine processing for electron microscopy ***(30)***. Inclusion of ruthenium red in the fixatives reduced the losses to less than 1%!
10. There is no one ideal tissue fixative, the goals being to maintain adequate tissue integrity for interpretation while preserving antigen/antibody recognition. Buffered 4% formaldehyde or methyl carnoys (10 mL acetic acid, 30 mL chloroform, 60 mL methanol) for 2 h, are common first choices. During tissue processing into paraffin, avoid temperatures exceeding 56–58°C.
11. Never allow sections to dry out.
12. Bovine serum albumin should be globulin free (immunohistochemical grade) when used to block nonspecific protein interactions.
13. Solutions A and B may be premixed and stored frozen as stocks. Complete resin should be made fresh immediately before each use by mixing room temperature A and B, adding DMP-30, and mixing thoroughly again. Bubbles can be pumped out in a vacuum chamber
14. This fixation and embedding scheme has proven useful as a standard first choice because the degree of molecular cross-linking by the low glutaraldehyde content results in the retention of recognizable epitopes and reasonable ultrastructural morphology. The low viscosity, tolerance of small amounts of water, and low temperature of polymerization also contribute to its being chosen when no ultrastructural localization of a particular antigen has been previously tried in our lab.
15. Use of methanol in the dehydration scheme is an attempt to reduce the amount of extraction during the dehydration process.
16. Use 50- to 100-µL reagent droplets on parafilm sheets for incubations.
17. Millipore (0.22-µm) filter all diluents, buffer washes, and water.
18. Handle grids with nonmagnetic #3 Dumont forceps on the grid rim only. LR White sections tend to swell during staining because of the resin's hydrophillic nature and become extremely fragile, so that touching them with forceps or filter paper can easily cause folds or damage.
19. All buffers and $H_2O$ for section washes are 0.22-µm millipore filtered.
20. One microgram of protein denatures to cover 9 $cm^2$, so the approx 3.7 µg delivered here should spread easily across the entire surface in front of the slide. One source of excessive graininess and low sample contrast is overcompression of the cytochrome C film.
21. The substrate surface should be as flat as possible, so that it receives a uniform metal coating.
22. Vacuum cleanliness and ultimate bell-jar pressure contribute to the final metal coating quality. If the vacuum is not at least $5 \times 10^{-5}$ torr during evaporation, the mean free path of evaporated metal will be less than the distance from the filament to the sample. Metal vapor colliding with gases will cool off, slow down, and may aggregate into coarse particles, resulting in an excessively grainy shadow.
23. Slowly heating metal just to the evaporation point results in the finest granularity (because of local vapor pressure characteristics).

24. The Bouin's-fix mordant step improves the color quality of primarily the crocein scarlet-stained components.
25. As staining time is increased, there is a decrease in the differentiation between crocein scarlet (erythrocytes, fibrin) and acid fuchsin (myofibrils, smooth muscle) stained tissue components.
26. The phosphotungstic acid step that removes excess Musto elastin stain is progressive and will, if extended, begin to decolorize the fine elastic fibers.
27. Acetic acid removes excess PTA
28. Limit the duration of the final ethanol rinses, as saffron (collagen) will destain.
29. If extracellular hyaluronan is to be visualized, omit the hyaluronidase pretreatment.
30. In some cells, overfixation (i.e., 4% paraformaldehyde for several hours) can result in unintentional permeabilization and/or higher background staining of mitochondria with the Streptavidin–Texas red.
31. Controls for intracellular staining and specificity include omitting the permeabilization step; digestion with hyaluronidase a second time, following permeabilization; and preincubation of the bHABP with excess hyaluronan.
32. If it is necessary to conserve reagents, cover slips can be cut in half using a diamond pencil to score the coverslip. 100 μL is the minimum volume that will cover a full 22-mm cover slip without excessive evaporation.
33. Cover slips are preferable to chamber slides if they will be viewed with a 100× oil objective.
34. Although cellular morphology is somewhat compromised, air drying preserves the fragile hyaluronan network. The relatively rigorous critical point drying techniques tend to wash away the hyaluronan filaments, as well as other extracellular matrix components, leaving a very smooth cell surface.
35. The coating may cover some of the hyaluronan filaments, especially farther away from the cell. Thus, coating times may have to be determined empirically.
36. The presence of phosphate ions in buffer vehicles with ruthenium red results in the precipitation of $OsO_4$-RR before the critical $OsO_4$/RR/polyanion reaction takes place.
37. Thaw and bring to room temperature.
38. Ruthenium red penetration is dependent on the histological structure of a particular tissue defining the diffusion pathway into the interior and the progressive concentration change of dye at different depths along this pathway. Extended fixation time allows the binding of ruthenium red to ionizable carboxylic acid groups and subsequent oxidation to ruthenium brown at increasing tissue depths as well as the oxidation of the polysaccharide substrate generating new carboxyl groups for RR binding.
39. Inclusion of RR in the primary fixative and intervening washing buffer is not necessary for the critical reaction but is included to limit the possible leaching away of bindable substances.
40. The greatest contrast of RR staining is apparent in sections with no poststain, but fine detail of other tissue structures is revealed when uranyl acetate and lead citrate are used.
41. Hydrogen peroxide potency must be considered; bubbles on the inside of the bottle indicate adequate activity in 30% solutions.
42. Plan on 50–150 μL of solution/tissue, depending on the size of the section.
43. Including 0.04% NiCl in the developing reagent changes the color of the reaction product to blue/black.
44. The specific enzymes and digestion time required must be worked out for each primary antibody. Excessive or inadequate digestion will result in an altered epitope or an inadequately unmasked antigen respectively.

45. Sodium azide is an inhibitor of peroxidase activity and should not be included in buffers used to make the peroxidase substrate or the ABC reagent.
46. Tissue may be harvested after perfusion fixation with 3% paraformaldehyde if desired and then continue fixation in 3%/0.25%.
47. Never allow the section surface to dry between droplet changes.
48. Include appropriate buffer/no-primary, irrelevant antibody, and positive and negative tissue controls.
49. Extreme care should be used in handling coated grids, as an intact support film limits difficulties later with contamination and section deformation. The use of clean, well-aligned anticapillary forceps to pick up the grids on the rimmed edge will minimize mechanical damage to the sections and support film.
50. The carbon coat will not only strengthen the coat, but will render this "reverse" side somewhat hydrophobic, facilitating the subsequent restriction of reagents to the section side of the grid.
51. It is sometimes helpful for later interpretation to collect sections consecutively.
52. Times can vary with block hardness and section thickness, and can be predetermined with test grids examined in the electron microscope for a loss of defined section edge.
53. Ideally, the carbon side of the grid remains dry during the entire immunostaining process. If the grid sinks at some point, it is best to proceed through each droplet immersed.
54. Antibody dilution is best determined by bracketing the dilution that gave the best light level localization.

## References


1. Broker, T. R. and Chow, L. T. (1976) Electron Microscopy of Nucleic Acids, Cold Spring Harbor Laboratory, Cold Spring Harbor, NY.
2. Rosenberg, L., Hellman, W., and Kleinschmidt, A. K., (1970) Macromolecular models of proteinpolysaccharides from bovine nasal cartilage based on electron microscopic studies. *J. Biol. Chem.* **245**, 4123–4130.
3. Heinegard, D., Lohmander, S., and Thyberg, J. (1978) Cartilage proteoglycan aggregates. Electron microscopic studies of native and fragmented molecules. *Biochem. J.* **175**, 913–919.
4. Hascall, G. K. (1980) Cartilage proteoglycans: comparison of sectioned and spread whole molecules. *J. Ultrastruct. Res.* **70**, 369.
5. Iozzo, R. V., Marroquin, R., and Wight, T. N. (1982) Analysis of proteoglycans by high-performance liquid chromatography. A rapid micromethod for the separation of proteoglycans from tissue and cell culture. *Anal. Biochem.* **126**, 190–199.
6. Wight, T. N. and Hascall, V. C. (1983) Proteoglycans in primate arteries. III. Characterization of the proteoglycans synthesized by arterial smooth muscle cells in culture. *J. Cell Biol.* **96**, 176–176.
7. Kapoor, R., Phelps, C. F., and Wight, T. N. (1986) Physical properties of chondroitin sulfate-dermatan sulfate proteoglycans from bovine aorta. *Biochem. J.* **240**, 575–583.
8. Carlson, S. S. and Wight, T. N. (1987) Nerve terminal anchorage protein one (TAP-1) is a chondroitin sulfate proteoglycan: biochemical and electron microscopic characterization. *J. Cell Biol.* **105**, 3075–3086.
9. Iwata, M., Wight, T. N., and Carlson, S. (1993) A brain extracellular matrix proteoglycan forms aggregates with hyaluronan. *J. Biol. Chem.* **268**, 15,061–15,069.
10. Carlson, S. S., Iwata, M., and Wight, T. N. (1996) A chondroitin sulfate/keratan sulfate proteoglycan, PG-1000 forms complexes which are concentrated in the reticular laminae of electric organ basement membranes. *Matrix Biol.* **15**, 281–292.

11. Movat, H. Z. (1955) Demonstration of all connective tissue elements in a single section. *Arch. Pathol.* **60,** 289–295.
12. Schmidt, R. and Wirtla, J. (1996) Modification of movat pentachrome stain with improved reliability of elastin staining *J. Histotechnol.*19, 325–327.
13. Musto, L. (1981) Improved iron-hematoxylin stain for elastic fibers. *Stain Technol*, **56**, 185–187.
14. Sheenhan, D. C. and Hrapchak, B. B. (1980) Theory and practice of histotechnology, 2nd ed. C. V. Mosby, St Louis, MO.
15. Wight, T. N., Lara, S., Riessen, R., Le Baron, R., and Isner, J. (1997) Selective deposits of versican in the extracellular matrix of restenotic lesions from human peripheral arteries. *Am. J. Pathol.* **151**, 963–973.
16. Tengblad, A. (1979) Affinity chromatography on immobilized hyaluronate and its application to the isolation of hyaluronate binding proteins from cartilage. *Biochem. Biophys. Acta* **5781**, 281–289.
17. Ripellino, J. A., Klinger, M. M., Margolis, R. U., and Margolis, R. K. (1985) The hyaluronic acid binding region as a specific probe for the localization of hyaluronic acid in tissue sections. *J. Histochem. Cytochem.* **33**, 1060–1066.
18. Knudson, C. B. and Toole, B. P. (1985) Fluorescent morphological probe for hyaluronate. *J. Cell Biol.* **100**, 1753–1758.
19. Underhill, C., Nguyen, H., Shizari, M., and Culty, M., (1993) CD44 positive macrophages take up hyaluronan during lung development. *Dev. Biol.* **155**, 324–336.
20. Riessen, R., Wight, T. N., Pastore, C., Henley, C., and Isner, J. M. (1996) Distributions of hyaluronan during extracellular remodeling in human restenotic arteries and balloon-injured rat carotid arteries. *Circulation* **93**, 1141–1147.
21. Evanko, S., Raines, E. W., Ross, R., Gold, L., and Wight, T. N. (1998) Proteoglycan distribution in lesions of atherosclerosis depends on lesion severity, structural characteristics and the proximity of PDGF and TGF-β. *Am. J. Pathol.* **152,** 533–546.
22. Evanko, S. P. and Wight, T. N. (1999) Intracellular localization of hyaluronan in proliferating cells. *J. Histochem. Cytochem.* **47**, 1331–1341.
23. Evanko, S. P., Angello, J. C., and Wight, T. N. (1999) Formation of hyaluronan- and versican-rich pericellular matrix is required for proliferation and migration of vascular smooth muscle cells. *Arterioscler. Thromb. Vasc. Biol.* **19**, 1004–1013.
24. Luft, J. H. (1971) Ruthenium red and violet. I. Chemistry, purification, methods of use, electron microscopy and mechanism of action. *Anat. Rec.* **171**, 347–368.
25. Wight, T. N. and Ross, R. (1975) Proteoglycans in primate arteries. II. Synthesis and secretion of glycosaminoglycans by smooth muscle cells *in vitro*. *J. Cell Biol.* **67**, 686.
26. Iozzo, R. V., Bolender, R. P., and Wight, T. N. (1982) Proteoglycan changes in the intercellular matrix of human colon carcinoma: an integrated biochemical and stereological analysis. *Lab. Invest.* **47**, 124–138.
27. Snow, A. D., Bolender, R. P., Wight, T. N., and Clowes, A. W. (1989) Heparin modulates the composition of the extracellular matrix domain surrounding arterial smooth muscle cells. *Am. J. Pathol.* **137**, 313–330.
28. Bolender, R. P., Hyde, D. M., and Dehoff, R. T. (1993) Lung morphometry: a new generation of tools and experiments for organ, tissue, cell and molecular biology. *Am. J. Physiol.* **265**, 521–548.
29. Carlson, S. S. (1982) $^{103}$Ruthenium red, a reagent for detecting glycosaminoglycans at the nanogram level. *Anal. Biochem.* **122**, 364–367.

30. Chen, K. and Wight,T. (1984) The ultrastructural identification of proteoglycans in arterial smooth muscle cell cultures using cationic dyes. *J. Histochem. Cytochem.* **32**, 347–357.
31. Lark, M.W., Yeo, T-K., Mar, H., Lara, S., Hellström, I., Hellström, K. E., and Wight, T. N. (1988) Arterial chondroitin sulfate proteoglycan: localization with a monoclonal antibody. *J. Histochem. Cytochem.* **36**, 1211–1221.
32. Gutierrez, P., O'Brien, K. D., Ferguson, M., Nikkari, S., Alpers, C. E., and Wight, T. N. (1997) Differences in the distribution of versican, decorin, and biglycan in atherosclerotic human coronary arteries. *Cardiovasc. Pathol.* **6**, 271–278.
33. O'Brien, K. D., Alpers, C. E., Chiu, W., Hudkins, K., Wight, T. N. ,and Chait, A. (1998) A comparison of apolipoprotein and proteoglycan deposits in human coronary atherosclerotic plaques: colocalization of apolipoprotein E and biglycan. *Circulation* **98**, 519–527.
34. Mar, H., Tsukada, T., Gown, A. M., Wight, T. N., and Baskin, D. G. (1987) Correlative light and electron microscopic immunocytochemistry on the same section with colloidal gold. *J. Histochem. Cytochem.* **35**, 419–425.
35. LR White product instructions.
36. Erikson, P. Anderson, D., and Fisher, S. (1987) Use of UA en bloc to improve tissue preservation and labeling for immunoelectron microscopy. *J. Elect. Micro. Tech.* **5,** 303–314.
37. Mar, H. and Wight, T.N. (1989) Correlative light and electron microscope immunocytochemistry on reembedded resin sections with colloidal gold, in: *Colloidal Gold: Principles, Methods and Applications, Vol. 2*, (Hyatt M. A., ed.), Academic, Orlando, FL.
38. Mar, H. and Wight, T.N. (1988) Colloidal gold immunostaining on deplasticized ultrathin section. *J. Histochem. Cytochem.* **36**:1386–1395

27

# Expression and Characterization of Engineered Proteoglycans

Wei Luo, Chunxia Guo, Jing Zheng, and Marvin L. Tanzer

## 1. Introduction

Target proteoglycan gene(s) can be rearranged to fulfill special designs for expression, purification, or characterization purposes. These are achieved through several procedures, including polymerase chain reaction (PCR) amplification of a target gene, cloning of PCR products into cloning vectors, subcloning of a correct gene fragment into an expression vector, purification of target gene DNA via a large preparation, transfection into host cells for expression, purification of candidate proteins, and identification of the proteins. Characterization can be done of core proteins or of the glycosaminoglycan chains attached to the core proteins. A summary of the expression of an engineered protien is shown in **Fig. 1**.

### *1.1. Construct Assembly of Target Proteoglycans*

Gene structure arrangements are accomplished through PCR with specifically designed oligonucleotides and target template(s) *(1)*. Appropriate PCR products are engineered into a suitable eukaryotic expression vector using appropriate polylinkers. The vector containing target gene(s) can express (transiently) proteoglycans through transfection into candidate eukaryotic cell lines (*see* **Subheading 1.2.**).

### *1.2. Expression of Target Proteoglycans*

Large amounts of DNA with molecular-level purity are needed for transfection and are prepared by a maxiprep kit (Qiagen). The maxiprep-prepared DNA is then used to transfect eukaryotic cell lines to express/obtain the corresponding proteins. Cell lines are chosen based on two major criteria: (1) the cell lines themselves do not express the same target endogenous genes; (2) the cell lines have the capability (e.g., machinery to add glycosaminoglycan chains) to express proteoglycans.

From: *Methods in Molecular Biology, Vol. 171: Proteoglycan Protocols*
Edited by: R. V. Iozzo © Humana Press Inc., Totowa, NJ

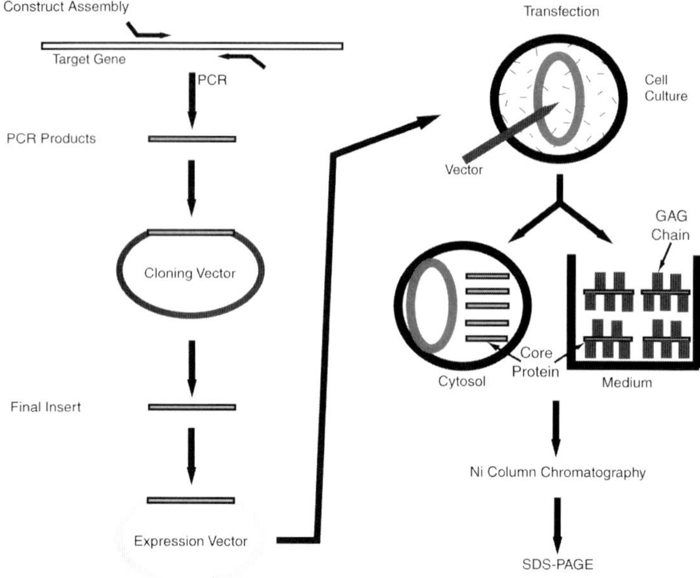


Fig 1. Summary diagram of expression of an engineered proteoglycan.

### 1.3. Purification of Target Proteoglycans

Expressed products are purified using a nickel column with stringent washing procedures to eliminate background proteins. A histidine-6 (His6) tag is engineered in the construct assembly (*see* **Subheading 1.1.**) to allow such a purification procedure.

### 1.4. Identification of Target Proteoglycans

The purified proteins can be run on sodium dodecyl sulfate-polyacrylamide gel electrophoresis (SDS-PAGE) directly to separate the products from background proteins or processed further for other experiments. Purified products can be characterized by various procedures, e.g., quantitation by radioactivity, classification of polysacchrides by enzymatic digests, and identification of core proteins by immunocapture.

### 1.5. Characterization of Expressed Proteoglycans

Proteoglycans have unique features that can be individually characterized. Chondroitin sulfate, keratan sulfate, dematan sulfate, and heparan sulfate can be digested with appropriate enzymes to confirm their presence. The example protocol is for chondroitin sulfate digestion (*see* **Subheading 3.5.**).

### 1.6. Detection of Proteoglycan Interaction with Heat-Shock Proteins Part I

Expressed proteoglycans may interact with certain types of heat shock proteins (Hsp) *(2)*. For example, aggrecan's G3 domain interacts with Hsp25. The example protocol is for immunocapture of Hsp25 using anti-Hsp25 antibody (*see* **Subheading 3.6.**).

### 1.7. Detection of Proteoglycan Interaction with Heat-Shock Proteins Part II

Interaction between heat-shock proteins and proteoglycans can also be detected by *in situ* cross-linking. This method is suitable only when the interacting candidate Hsp and proteoglycan are adjacent to each other and they have suitable amino group(s) in proximity that can be cross-linked. The procedure here uses 3,3'-dithiobis sulfosuccinimidylpropionate (DTSSP) as a cross-linking reagent that requires lysines or an N-terminal amino group for reaction. Other reagents may be used for other chemical groups.

## 2. Materials (*see* Note 1)

### 2.1. Construct Assembly of Target Proteoglycans

1. Target proteoglycan gene.
2. Specifically designed oligonucleotides with certain features, e.g., restriction sites, His6 tag, etc., as desired.
3. PCR-Script SK(+) cloning vector kit (Stratagene).
4. *Escherichia coli* (*E. coli* ) XL-blue MRF' Kan supercompetent cells.
5. X-gal (5-bromo-4-chloro-3-indolyl β-D-galactopyranoside), IPTG (isopropyl β-D-thio-galactopyranoside), and N,N-dimethyl formamide (Sigma).
6. Restriction enzymes and matching buffers corresponding to the engineered restriction sites.
7. DNA purification system Minipreps (Wizard Plus from Promega).
8. LB broth base (Life Technology).
9. Eukaryotic expression vector (e.g., pcDNA3 from Invitrogen).
10. T4 DNA ligase (Invitrogen).
11. *E. coli* TOP10F' competent cells (Invitrogen).
12. Ampicillin-containing agar plates (only if ampicillin is the selecting reagent).
13. QiaFilter DNA purification systems (for Maxipreps from Qiagen).
14. Sequencing oligonucleotides (pairing to expression vector or cloning vector or templates).
15. Agarose (Life Technology).
16. NanoSep 0.2-μm DNA filters (Fisher Scientific/Pall Gelman).
17. 125-mL glass flask.
18. 15-mL Falcon tube.
19. DNA sequence verification software (optional).

### 2.2. Expression of Target Proteoglycans

1. Eukaryotic cell line(s) (e.g., Chinese Hamster Ovary K1 cells, COS cells).
2. Lipofectamine + PLUS reagent (Life Technology).
3. Regular minimal essential culture medium (MEM), regular OPTI-MEM (Life Technology), and methionine-free and cysteine-free OPTI-MEM (Life Technology, special order).
4. Fetal bovine serum (certified grade).
5. Antibiotic–antimycotic solution (containing100 U/mL penicillin, 100 μg/mL streptomycin, 250 ng/mL amphotericin, Life Technology).
6. Radioactive isotopes $^{35}$S-Pro-Mix (methionine and cysteine, Amersham, 7.15 mCi, for 4–5 plates of cells) for methionine + cysteine labeling (labeling core proteins) or $^{35}S$-$Na_2SO_4$ (sodium sulfate, Dupont NEN, 5 mCi, for 4–5 plates of cells) for sulfate labeling (labeling sulfated glycosaminoglycan chains).
7. Decolorizing NORIT-A carbon (ACROS Organics).
8. Sterilized 9-in. glass pipets (Fisher Scientific).

9. 10-mL individually wrapped plastic pipets (sterilized).
10. Cell culture dishes (100 mm).
11. Eppendorf tubes (regular and boiling tubes).
12. Table-top centrifuge (up to 14,000*g* speed).
13. Nonidet P-40 detergent.
14. Phenylmethylsulfonyl fluoride (PMSF, *see* **Subheading 3.2.3.**), leupeptin, antipain, benzamidine, aprotinin, chymostatin, pepstatin, dimethyl sulfoxide (DMSO) (Sigma).
15. Hanks' Balanced Salt Medium (Life Technology).

### 2.3. Purification of Target Proteoglycans

1. Ni NTA agarose (Ni resin, Qiagen).
2. Equilibration buffer: 0.1 *M* NaCl, 44 m*M* NaHCO3, 1 m*M* imidazole, 0.5% Triton X-100 in $H_2O$.
3. Met-free and Cys-free OPTI-MEM medium.
4. 10% Triton X-100 solution.
5. pH indicator paper.
6. Stringent wash buffer A:10 m*M* HEPES, 2 *M* NaCl, 10% glycerol, 0.1 m*M* PMSF, 2 m*M* imidazole, 0.5% Triton X-100, 6 *M* urea, 250 m*M* dithiothreitol in $H_2O$, pH 8.0.
7. Stringent wash buffer B: 10 m*M* HEPES, 750 m*M* NaCl, 10% glycerol, 0.1 m*M* PMSF, 10 m*M* imidazole, 0.5% Triton X-100, 1 mg/mL bovine serum albumin in $H_2O$, pH 8.0.
8. Stringent wash buffer C: 10 m*M* HEPES, 750 m*M* NaCl, 10% glycerol, 0.1 m*M* PMSF, 10 m*M* imidazole in $H_2O$, pH 8.0.
9. Elution buffer: 10 m*M* HEPES, 750 m*M* NaCl,10% glycerol, 100 m*M* EDTA, 0.1 m*M* PMSF, 250 m*M* imidazole in $H_2O$, pH 8.0.
10. Roto Torque heavy–duty rotator (Cole Parmer Instruments).

### 2.4. Identification of Target Proteoglycans

1. 5–15% linear gradient SDS-containing acrylamide gel (can be self-prepared or purchased from manufacturer).
2. Dithiothreitol (DTT), final concentration 38 mg/mL in loading buffer.
3. Gyrotory shaker (New Brunswick Scientific).
4. Gel fixing solution: 25% Methanol, 10% acetic acid in $H_2O$.
5. Square Petri dish (100 × 100 × 15 mm, Nalge Nunc #4021).
6. Entensify Universal Autoradiography Enhancer Part A and Part B (Dupont).
7. Radiogram cassette and X-ray film (Kodak, Bio-Max films are best).

### 2.5. Characterization of Expressed Proteoglycans

1. Chondroitinase ABC (Seikagagu) (dissolved in buffer conditions recommended by the manufacturer).
2. 20× Chondroitinase ABC digest buffer: 2 *M* Tris-HCl, 600 m*M* sodium acetate, pH 8.0 (final conditions = 100 m*M* Tris-HCl, 30 m*M* sodium acetate, pH 8.0).
3. 10% Triton X-100 solution.
4. 100 m*M* N-ethylmaleimide (NEM) solution.

### 2.6. Detection of Proteoglycan Interaction with Heat-Shock Proteins Part I

1. Hsp25 antibody (Stressgen).
2. Protein A beads (Pierce).
3. Washing buffer D: 50 m*M* Tris-HCl, 1 *M* NaCl, 1% Triton X-100, pH 7.5.

### 2.7. Detection of Proteoglycan Interaction with Heat-Shock Proteins Part II

1. DTSSP (Pierce).
2. Washing buffer E: 130 m*M* NaCl, 20 m*M* Bicine, pH 8.0, ice cold.
3. Digitonin lysis buffer: 50 m*M* Bicine, pH 8.0, 150 m*M* NaCl, 5 m*M* KCl, 5 m*M* $MgCl_2$, 0.2% digitonin, 1.5 m*M* DTSSP, ice cold.
4. 10 m*M* glycine solution, ice cold.

## 3. Methods

### 3.1. Construct Assembly of Target Proteoglycans

1. The target sequence was amplified by PCR using an upstream oligonucleotide (oligo 1) and a downstream oligonucleotide (oligo 2). Oligo 1 should contain a specific restriction site, which corresponds to one of the sites in the polylinker region in the eukaryotic expression vector. Oligo 2 should contain a specific restriction site, a His6 sequence (engineered for purification purposes), and a stop codon. The restriction site in oligo 2 can be the same as in oligo 1 or can be different, but it must correspond to a site in the polylinker region in the eukaryotic expression vector in correct orientation. Usually it is easier to have two different restriction sites so that you don't have to verify the orientation of the sequence after its insertion (*see* **step 22**).
2. The PCR reaction is usually a Hot-Start program (*see* **Note 2**). A portion of the PCR products (usually 5 μL from a 50-μL PCR reaction) is denatured and electrophoresed in an ethidium bromide agarose gel to verify the size of the products.
3. Correct PCR products (the right size) are digested with SrfI and ligated to predigested pCR-Script SK(+) vector (Stratagene), which is then used to transform *E. coli* XL1-blue MRF' Kan supercompetent cells (following the manufacturer's manual).
4. Prepare LB broth following the manufacturer's instructions.
5. Dissolve X-gal in N,N-dimethyl formamide (20 mg/mL) and IPTG in water (50 mg/mL).
6. Place transformed products on an ampicillin-containing agar plate (containing X-gal and IPTG) and incubate at 37°C for 16 h. Blue and white colonies should appear on the plate.
7. Select white *E. coli* colonies, inoculate one colony into a 15-mL Falcon tube containing 3 mL of LB broth supplemented with ampicillin (50 ng/mL final concentration using 50 mg/mL 1000× stock), shake at 250 rpm at 37°C for 16 h (overnight).
8. Purify miniprep DNA using Wizard Plus Minipreps (Promega) following the manufacturer's instructions.
9. Digest purified DNA (5 μL from 50 μL) with appropriate restriction enzymes at 37°C for 1–2 h.
10. Electrophorese digested DNA in a 0.8–1% agarose gel to detect and select suitable DNA fragments, affirming the correct size of DNA fragments.
11. Digest the DNA fragments (16 μL) again with the same restriction enzymes as in **step 8**.
12. Digest the pcDNA3 vector (Invitrogen) with the same restriction enzymes.
13. Electrophorese (10) and (11) in a 0.8–1% agarose gel.
14. Cut out the agarose bands containing the correct DNA fragments.
15. Purify the DNA using NanoSep filters following the manufacturer's instructions.
16. Ligate purified DNA fragments and pcDNA3 vector using Invitrogen T4 DNA ligase following the manufacturer's instructions.
17. Transform *E. coli* TOP10F' supercompetent cells using (16) following the manufacturer's instructions.

18. Repeat **steps 7–10** to verify DNA inserts. Store bacteria-containing liquid broth in an 80/20 ratio glycerol at –80°C.
19. Strip bacteria onto an ampicillin-containing agar plate from glycerol stock and incubate overnight at 37°C for 16 h (overnight).
20. Inoculate one colony into a 125-mL glass flask containing 50 mL of LB broth supplemented with ampicillin (50 ng/mL final concentration) and shake at 250 rpm on a bacterial incubation shaker at 37°C for 16 h (overnight).
21. Prepare a maxiprep using QiaFilter DNA purification systems (Qiagen) following the manufacturer's instructions.
22. Quantify the DNA concentration and sequence the DNA to verify the correct sequence. The sequence must be 100% identical to the original design at the amino acid level.
23. The verified DNA is ready for the next step (transfection).

## 3.2. Expression of Target Proteoglycans

### 3.2.1. Cell Culture

1. (All procedures are performed in sterile-hood cell culture conditions.) Prepare MEM according to the manufacturer's instructions (supplemented with 5–10% fetal bovine serum and 100 U/mL penicillin, 100 µg/mL streptomycin, and 250 ng/mL amphotericin, filtered).
2. Maintain eukaryotic cells (e.g., CHO) at 37°C in humidified air with 5% $CO_2$.
3. Seed 1.8 million cells per 100-mm dish in MEM medium 2 d before transfection. Cells should be 70–80% confluent when transfected.

### 3.2.2. Transfection Procedure

1. All procedures are performed in sterile-hood cell culture conditions. Mix 9 µg DNA (in less than 20 µL volume), 460 µL OPTI-MEM, and 30 µL PLUS reagent and invert 5× to mix well (refer to Life Technology protocol).
2. Incubate at room temperature (RT) for 15 min.
3. Mix 30 µL Lipofectamine and 470 µL OPTI-MEM and invert 5× to mix well.
4. Combine (2) + (3), invert 10×, incubate at RT for 15 min (*see* **Note 3**).
5. Aspirate MEM from the cell culture (regular culture medium).
6. Wash cells with 10 mL of regular OPTI-MEM, incubate at 37°C for 20 min and aspirate.
7. Wash cells with 5 mL of regular OPTI-MEM and aspirate.
8. Add 4 mL of regular OPTI-MEM.
9. Add (4) onto cells in a plate-tilted position and swirl medium gently until well mixed.
10. Incubate at 37°C for 5 h. Do not agitate cells when handling dishes.
11. Place twenty 100-mm dishes containing 2 tablespoons of decolorizing NORIT-A carbon each into another incubator (same conditions as **Subheading 3.2.1., step 2**) designated for radioactive isotope usage. Carbon is placed to absorb free radioactive materials released by metabolizing cells in the incubator.
12. Aspirate medium from (10).
13. Add 5 mL of Met- and Cys-free OPTI-MEM.
14. Add 100 µL of $^{35}$S-Pro-Mix labeling mix to each dish (about 1.40–1.80 mCi/dish).
15. Incubate at 37°C for 16 h in the incubator supplemented with decolorizing NORIT-A carbon (**step 11**). The cells and media are ready for collection.

(The above protocol is for CHO cells.)

### *3.2.3. Sample Collection*

1. Dissolve 0.5 g of Nonidet P-40 in 100 mL of $dH_2O$ to prepare 0.5% Nonidet lysis buffer, which can be stored at 4°C.
2. Protease inhibitor I: 2 mg/mL leupeptin, 0.4 mg/mL antipain, 2 mg/mL benzamidine, and 2 mg/mL aprotinin dissolved in $dH_2O$.
3. Protease inhibitor II: 1 mg/mL chymostatin and 1 mg/mL pepstatin dissolved in DMSO.
4. 250 m*M* PMSF: 250 m*M* PMSF dissolved in 100% ethanol.
5. Prepare 10 mL of Nonidet P-40 lysis buffer (from **step 1**) with 10 µL of protease inhibitor I, 10 µL of protease inhibitor II, and 20 µL of 250 m*M* PMSF.
6. Add 20 µL of 250 m*M* PMSF, 5 µL of protease inhibitor I, and 5 µL of protease inhibitor II to a 15-mL tube.
7. Transfer medium (from **Subheading 3.2.2., step 15**) to the tube.
8. Spin tube for 60s at 925*g* (1000 rpm) in cell culture centrifuge to eliminate dead cells.
9. Transfer medium (*see* **step 8**) to a new 15-mL tube, store at –20°C. Do not disturb the pellet at the bottom of the tube. The medium samples are ready for purification.
10. Simultaneously, wash cells with 5 mL of Hanks' medium/dish and discard medium as radioactive Waste.
11. Add 1 mL of Nonidet P-40 lysis buffer (*see* **step 5**) to each dish. Swirl buffer to wet the entire surface.
12. Completely scrape the cells from the dish and transfer to a 1.5-mL Eppendorf tube; tilt the dish to decant the cell lysate liquid.
13. Incubate on ice for 10 min and then spin at 14,000*g* (table-top centrifuge) for 5 min.
14. Transfer supernatant to a new tube, and store the supernatant and debris separately at –20°C. The supernatant is ready for purification.

## ***3.3. Purification of Target Proteoglycans***

All procedures are to be done with extreme care because of the high energy of radioactive isotope in use.

### *3.3.1. Ni Resin Equilibration*

1. Transfer 100 µL of dispersed resin beads (well-mixed beads with buffer) to a 1.5-mL Eppendorf tube.
2. Spin at 14,000*g* for 15 s, then remove liquid.
3. Add 500 µL of equilibration buffer to beads.
4. Place the tube on a Rotator and rotate at RT for 30 min (low speed at 6.5).
5. Spin at 14,000*g* for 15 s and remove liquid. The resin is ready for sample loading.

### *3.3.2. Ni Resin Loading*

1. Medium loading: 900 µL of medium (from **Subheading 3.2.3., step 9**) and 47.5 µL of 10% Triton X-100 for final 0.5% Triton-X concentration; adjust pH to 8.0 (usually it should be pH 7.5–8.0 by itself after collection, using pH paper to monitor).
2. Lysate loading: 700 µL of Met-free medium, 200 µL of lysate (from **Subheading 3.2.3., step 14**), and 47.5 µL of 10% Triton X-100; adjust pH to 8.0.
3. Rotate for 1 h at RT.
4. Spin at 14,000*g* for 15 s, then transfer supernatant to waste.

### *3.3.3. Stringent Washes*

1. Wash 1: Add 1000 µL of stringent wash buffer A (*see* **Note 4**) to each pellet and rotate at RT for 30 min (notice that color changes from clear to brown at RT).

2. Spin at 14,000*g* for 15 s, then remove supernatant to waste.
3. Repeat (1) and (2) twice.
4. Wash 2: Add 1000 µL stringent wash buffer B to each pellet and rotate at RT for 30 min.
5. Spin at 14,000*g* for 15 s, then remove supernatant to waste.
6. Wash 3: Add 250 µL of stringent wash buffer C to each pellet and rotate at RT for 15 min.
7. Spin at 14,000*g* for 15 s, then remove supernatant to waste.
8. Carefully and completely remove all liquid from **step 7**.
9. Elution: Add 50 µL of elution buffer to each pellet and rotate at RT for 1 h.
10. Spin at 14,000*g* for 15 s.
11. Collect 50 µL of supernatant into a new tube and take 2 µL for radioactive β-monitoring. The eluate is ready for electrophoresis by SDS-PAGE or other process.
12. Store samples at –20°C.

## 3.4. Identification of Target Proteoglycans

### 3.4.1. SDS-PAGE

1. Transfer 20 µL of medium eluate (or 10 µL lysis eluate) to a 1.5-mL boiling tube.
2. Add 5 µL 5× sample loading buffer (non-DTT for nondenatured condition or DTT-containing for disulfide cleavage condition).
3. Boil tube at 100°C in a water bath for 5 min.
4. Cool tube using running water (not ice water).
5. Spin at 14,000*g* for 10 s.
6. Load sample into an individual well of the SDS gel.
7. Run the gel at constant 38 mA (set voltage to highest, 500 V) for about 90 min. Monitor the gel to avoid sample overrun.

### 3.4.2. Processing of SDS-PAGE and Radiography

8. Transfer gel to 300 mL of fixing solution in a glass container.
9. Place the container onto a shaker and shake at 62.5 rpm for 30 min.
10. Drain solution and replace with a fresh 300 mL of fixing solution and shake at 62.5 rpm for 30 min.
11. Transfer gel to a square Petri dish.
12. Add 15 mL of Entensify Enhancer Part A to dish.
13. Shake on shaker at 125 rpm for 45 min.
14. Add 15 mL of Entensify Enhancer Part B
15. Shake on shaker at 125 rpm for 45 min.
16. Dry gel in a Gel Dryer at 60°C for 120 min, until completely dry.
17. Place the dried gel on a radiogram cassette and expose an X-ray film to the gel in a dark room.
18. Place the cassette in a –80°C freezer for hours to days (depending on the intensity of radioactivity) (*see* **Note 4**).
19. Develop the film in an X-OMAT (Kodak) machine. Alternatively, bands in the gel can be quantified on an Instant Imager (Packard). The methionine-labeled (Pro-Mix) proteins are shown in **Fig 2**.

## 3.5. Characterization of Expressed Proteoglycans

1. Transfer 30 µL of eluate (from **Subheading 3.3.3., step 11**) to a 1.5-mL boiling Eppendorf tube (For a 60-µL digest reaction).

**Fig. 2. Methionine labeling experiment.** (**A**) Spent medium: autoradiography of empty vector (control, lane 1) and aggrecan G3 construct (lane 2) from wild-type CHO cells (*1*). Note that the diffuse band in lane 2 shows a typical proteoglycan pattern (bracket). (**B**) Cell lysates: autoradiography of empty vector (lane 1) and aggrecan G1 construct (lane 2) from wild-type CHO cells (*1*). Note the intracellular G1 core protein (arrow, lane 2).

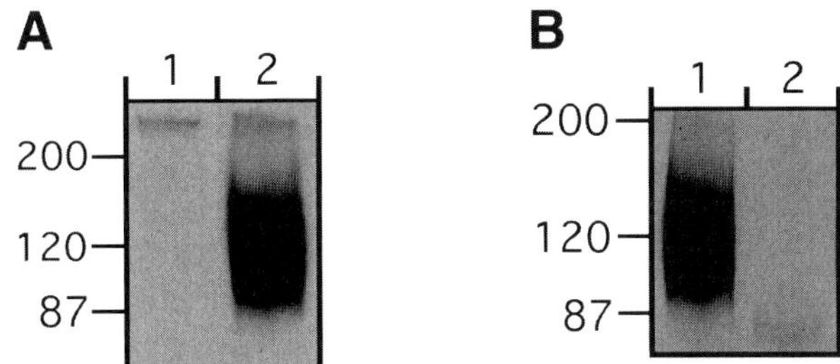


**Fig. 3. Sulfate labeling experiment and GAG chain characterization.** (**A**) Autoradiography of sulfate-labeled products of empty vector (lane 1) and aggrecan G3 construct (lane 2) from wild-type CHO cells (*1*). Note that only chondroitin/heparan sulfate GAGs are labeled (not the core protein). (**B**) Autoradiography of sulfate-labeled products of aggrecan G3 construct in Chondroitinase buffer (lane 1) and G3 digested by Chondroitinase ABC (lane 2). Note that the enzyme digestion almost completely eliminates the diffuse band.

2. Add 3 µL of 20× Chondroitinase ABC digest buffer, 5 µL of 10% Triton X-100, 5 µL of 0.1 *M* NEM, and 2 µl of $H_2O$.
3. Add 15 µL of Chondroitinase ABC solution (0.3 Units of Chondroitinase ABC).
4. Mix well, incubate at 37°C for 120 min.
5. Run samples in a SDS-PAGE gel, comparing to nondigested eluate (same as **Subheading 3.4.**). The Chondroitinase ABC-digested samples (sulfate labeled) are shown in **Fig 3**.

### 3.6. Detection of Proteoglycan Interaction with Heat-Shock Proteins Part I

1. (**Steps 1–7** at 4°C) Combine spent medium (or cell lysate) with protein A beads and Hsp25 antibody, rotating for 5 h.
2. Spin at 14,000*g* for 15 s, then transfer supernatant to waste.
3. Add 1000 µL of washing buffer D to each tube and rotate for 60 min.
4. Spin at 14,000*g* for 15 s, then transfer supernatant to waste.
5. Repeat **steps 3 and 4** twice.

6. Spin at 14,000*g* for 15 s.
7. Carefully and completely remove all liquid from **step 6**.
8. (From here on, at RT) Add appropriate amount (e.g., 30 μL) of 1× loading buffer with or without DTT.
9. Boil tube at 100°C in a water bath for 5 min.
10. Cool tube using running water.
11. Spin at 14,000*g* for 10 s.
12. Load supernatant onto SDS-PAGE.
13. Run the gel and process the gel as under **Subheading 3.4.**

### 3.7. Detection of Proteoglycan Interaction with Heat-Shock Proteins Part II

1. Transfect and label cells as under **Subheading 2.** (before sample collection).
2. (**Steps 2–5** in cold room) Wash labeled cells (live on dish) twice with 5 mL of washing buffer E.
3. Lyse cells with 1 mL of digitonin lysis buffer (containing DTSSP), then scrape cells off dish.
4. Transfer lysate to a 1.5-mL Eppendorf tube and incubate on ice for 30 min.
5. Add 200 μL of 10 m*M* glycine to sample (to deactivate excess cross-linker), then incubate on ice for 10 min.
6. Ni chromatography as under **Subheading 3.3.**
7. Electrophorese samples with dithiothreitol loading buffer (to release cross-linked proteins) as under **Subheading 3.4.**

## 4. Notes

1. All ingredients are sterilized, molecular biology grade.
2. A PCR hot-start program typically contains one denature cycle (94°C for 4 min, 72°C for 2 min and 55°C for 1 min) and 30 cycles of amplification (94°C for 45 s, 55°C for 45 s and 72°C for 3 min). Annealing temperature and elongation temperature need to be adjusted for different oligo lengths and templates. See a PCR publication for details.
3. Low speed spin for 15 s can be used to collect all liquid in the tube after inversion.
4. Stringent washes are necessary to eliminate background proteins and for clear visibility of the candidate proteins. Stringent washing buffer (wash 1) can be increased with higher molar concentrations of urea (utmost 8 *M*). If more stringency is needed, radioimmune precipitation (RIPA) components can be added to the buffer and subsequent washes. A repeat wash 3 is recommended if RIPA is added to wash 2 and wash 3, since deoxy cholate will allow different charges on the proteins, converting in one regular band into two bands on a gel (thereby making it hard to judge whether the correct proteins are expressed).
5. Film exposure time has to be adjusted (from 1 h to several days) so that the appropriate exposure is obtained. The criterion is that you can see the desired bands clear enough. Or you can use an imager to obtain the appropriate exposure through computer imaging.

## References

1. Luo, W., Kuwada, T. S., Chandrasekaran, L., Zheng, J., and Tanzer, M. L. (1996) Divergent secretory behavior of the opposite ends of aggrecan. *J. Biol. Chem.* **271**, 16,447–16,450.
2. Zheng, J., Luo, W., and Tanzer, M. L. (1998) Aggrecan synthesis and secretion: a paradigm for molecular and cellular coordination of multiglobular protein folding and intracellular trafficking. *J. Biol. Chem.* **273**, 12,999–13,006.

# 28

# Intracellular Localization of Engineered Proteoglycans

**Tung-Ling L. Chen, Peiyin Wang, Seung S. Gwon, Nina W. Flay, and Barbara M. Vertel**

## 1. Introduction

Proteoglycans undergo numerous synthetic and processing events as they progress through the exocytic pathway. In cells that express genetically engineered constructs encoding proteoglycans, immunolocalization is a useful approach in identifying specific intracellular compartments involved in their processing and trafficking.

## 2. Materials

### 2.1. Expression of Target Proteoglycan (Cell Culture and Transfection)

1. Proteoglycan constructs packaged into vectors suitable for expression in target cells (*1,2*). In our experiments, pcDNA3 and its derivatives (Invitrogen) were used. FLAG and 6xHis epitope tags were engineered at the C-terminus.
2. Mammalian cell lines: Chinese hamster ovary (CHO) cells, COS-1 cells, and HeLa cells were used.
3. Culture medium: Ham's F12 medium (for CHO, Life Technologies) or Dulbecco's Modified Eagle's Medim High Glucose Pyruvate (HG-DMEM, for COS-1 and HeLa, Irvine Scientific) with 10% fetal bovine serum (Atlanta Biologicals) and 1% antibiotic-antimycotic solution (Life Technologies).
4. Trypsin-EDTA solution (Sigma, cell culture grade).
5. Tissue culture dishes.
6. Sterilized cover slips: No. 1 cover slips were prepared by soaking in 95% ethanol for at least 10 min, followed by rinsing with sterile distilled water.
7. SuperFect transfection reagent (Qiagen).
8. Opti-MEM (Life Technologies).
9. Phosphate buffered saline (PBS).

### 2.2. Immunofluorescence Localization of Expressed Proteoglycan

1. Fixatives: 100% methanol, stored at –20°C or 2–4 % paraformaldehyde in PBS.
2. Nonidet P40 (NP-40), 0.1% in PBS.

From: *Methods in Molecular Biology, Vol. 171: Proteoglycan Protocols*
Edited by: R. V. Iozzo © Humana Press Inc., Totowa, NJ

3. Normal goat serum (NGS).
4. Primary antibody: M2 anti-FLAG antibody (Eastman Kodak), Penta-His or Tetra-His antibody (Qiagen).
5. Secondary antibody: FITC-conjugated goat IgG anti-mouse IgG (Jackson Labs).
6. PBS.
7. Mounting medium *(3)*: 15% Vinol 205 polyvinyl alcohol (w/v, *see* **Note 1**), 33% glycerol (v/v), 0.1% azide, pH 8.5. Dissolve 20 g of Vinol 205 polyvinyl alcohol in 80 mL of 0.1 *M* Tris, 0.1 *M* NaCl, pH 8.5, for 16 h with stirring. Add 40 mL of glycerol and 1.2 mL of 10% Na azide with continued stirring for another 16 h. Pellet undissolved particles at 20,000*g* for 20 min. Aliquot the viscous supernatant and store at –20°C. Use a defrosted aliquot stored at 4°C as your working solution.

### 2.3. Ultrastructural Localization of Expressed Proteoglycan

1. Paraformaldehyde-lysine-periodate fixative *(4)*:
   a. Solution A: 0.1 *M* lysine–0.05 *M* phosphate buffer, final pH = 7.4. Dissolve 1.827 g of lysine HCl in 50 mL of distilled water. Adjust to pH 7.4 with 0.1 *M* $Na_2HPO_4$. Bring the final volume to 100 mL with 0.1 *M* phosphate buffer, pH 7.4. Osmolarity should be approx 300 mo. Store at 4°C.
   b. Solution B: 20% paraformaldehyde. Mix 10 g of paraformaldehyde in 50 mL of distilled water, heating in a 60°C water bath with stirring. Slowly add 1–3 drops of 1 *N* NaOH until the solution clears. Store at –20°C. Spin or filter to remove debris before use.

   To make 10 mL of fixative, mix 7.5 mL of solution A with 1 mL of solution B and 1.5 mL of distilled water. Add 21.4 mg of $NaIO_4$. Final composition is 0.01 *M* $NaIO_4$, 0.075 *M* lysine, 0.0375 *M* phosphate buffer, 2% paraformaldehyde. Upon mixing solutions A and B, the pH will decrease from 7.4 to approx 6.2. The fixative is used at this lower pH and needs to be made fresh directly before use.
2. Wash and permeabilization solution: 0.05% saponin in PBS.
3. Blocking solution: 10% NGS/0.06% glycine/0.05% saponin/PBS.
4. Primary antibody: M2 anti-FLAG, diluted 1/100 in PBS with 5% NGS and 0.05% saponin.
5. Rabbit IgG anti-mouse IgG, diluted 1/25 in PBS with 5% NGS and 0.05% saponin.
6. Peroxidase-conjugated goat IgG anti-rabbit IgG Fab fragments (Jackson Labs), diluted 1/50 in PBS with 5% NGS and 0.05% saponin.
7. Diaminobenzidine (DAB): Tare a 15-mL polystyrene tube, weigh out DAB, and dissolve in DAB buffer to generate a 0.2% solution—e.g., weigh out 10 mg of DAB and dissolve in 5 mL of DAB buffer. (*see* **Note 2**.)
8. DAB buffer: 0.05 *M* Tris-HCl, pH 7.4.
9. Hydrogen peroxide (30% solution, Sigma).
10. Sucrose-containing cacodylate buffer: Mix 0.2 *M* Na cacodylate, pH 7.4, and 60% sucrose to make 0.1 *M* Na cacodylate containing 6% sucrose.
11. Glutaraldehyde (grade I, 25% aqueous solution, Sigma). (*see* **Note 2**.)
12. Osmium tetroxide (2% aqueous solution, Electron Microscopy Sciences) (*see* **Note 2**).
13. Potassium ferrocyanide [$K_4Fe(CN)_6 \cdot 3H_2O$, Sigma]. (*see* **Note 2**.)
14. Ethanol.
15. Hydroxypropyl methacrylate (HPMA, Electron Microscopy Sciences).
16. Epon with catalyst:
    a. tEpon 812 (Tousimis Research Corp.).
    b. Nadic methyl anhydride (NMA, Tousimis Research Corp.).

c. Dodecenylsuccinic anhydride (DDSA, Tousimis Research Corp.).
d. Tri-dimethylamino methyl phenol (DMP-30, Tousimis Research Corp.).

To make 50 mL of mixture, pour 21 mL of NMA into a 100 mL plastic beaker with a volume marker, add 3 mL of DDSA using a syringe, then add tEpon 812 to the 50-mL mark. Stir with a wood stick for 15 min, taking caution to avoid bubbles. Add 0.75 mL of DMP-30 with a syringe, and stir well for another 10 min. Unused portions may be stored in syringes at –20°C for later use. (*see* **Note 2**.)

17. Lead citrate: The lead citrate solution is prepared and used according to Reynolds *(5)*. All glassware must be washed with 3% HCl and thoroughly rinsed with double-distilled (dd) water in advance. Fill a 50-mL volumetric flask with 30 mL of dd water. Add 1.33 g of $Pb(NO_3)_2$ and 1.76 g of trisodium citrate. Shake continuously for 1 min. The contents will appear milky. Let stand for an additional 30 min with intermittent shaking. This is necessary for the complete conversion of reagents into lead citrate. Adjust to pH 12 by the slow addition (with inversion of the flask) of 1.0 *N* NaOH. The proper pH is reached just at the point when the solution clears (after the addition of about 8 mL of 1.0 *N* NaOH). Adjust to the final volume of 50 mL using dd water. Filter with Whatman's #2 filter paper. Store the solution at room temperature in a tightly capped amber glass bottle. The solution should not be jarred; allowing it to stand undisturbed enables small particles to settle to the bottom and improves the quality of the stain.

## 3. Methods

### 3.1. Expression of Target Proteoglycan (Cell Culture and Transfection)

1. Seed $1.1 \times 10^6$ CHO cells onto a 100-mm culture dish containing sterilized cover slips 36–40 h before transfection (*see* **Notes 3** and **4**). For ultrastructural studies, seed $1 \times 10^5$ cells onto a 35-mm culture dish.
2. Incubate in a 37°C incubator with 5% $CO_2$. For CHO cells, the dishes should be 80% confluent on the day of transfection.
3. In a polystyrene tube, dilute 5 µg of DNA into 150 µL of Opti-MEM (contains no serum or antibiotics). Mix solution.
4. Add 10 µL of SuperFect transfection reagent to the DNA solution (*see* **Note 5**). Mix solution.
5. Incubate the samples for 5–10 min at room temperature (20–25°C) to allow complex formation.
6. While complex formation takes place, set up 35-mm dishes with 1 mL of Opti-MEM and transfer one cover slip with attached cells to each dish.
7. Add 1 mL of complete medium containing serum and antibiotics to the reaction tube containing the transfection complexes. Mix by pipetting up and down twice, and immediately transfer the total volume to the cover slips in the 35-mm dishes.
8. Incubate the cells with the complexes for 2 h at 37°C in an atmosphere of 5% $CO_2$.
9. Remove medium containing the remaining complexes from the cells. Wash cells once with PBS.
10. Add new complete medium. Incubate cells for additional hours as each experiment requires (usually 24–48 h).

### 3.2. Immunofluorescence Localization of Expressed Proteoglycan

1. Prior to fixation, wash cells with PBS 2 times.
2. Fix with cold methanol for at least 20 min, or with room-temperature paraformaldehyde for 15 min.

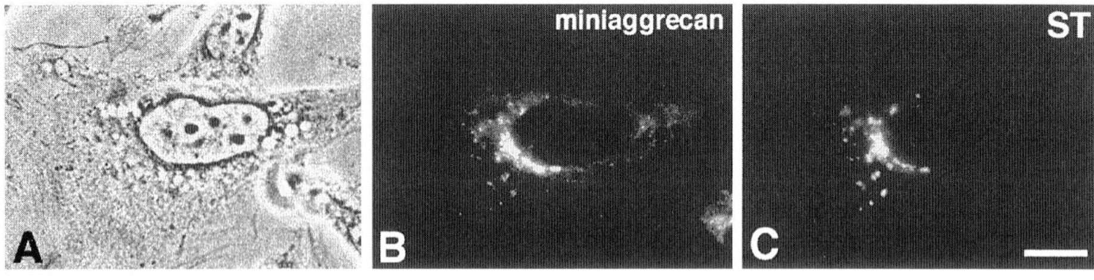


Fig. 1. Immunofluorescence localization of expressed "miniaggrecan" (containing the chicken aggrecan N-terminus with the signal sequence and G1 domain, a segment of the chondroitin sulfate attachment region, the G3 domain, and a 6xhis epitope tag) and the Golgi enzyme, ST. Transfected CHO cells exhibit miniaggrecan localization within the perinuclear Golgi complex **(B)**. The region is further identified as the Golgi complex by the co-transfection and localization of the ST Golgi enzyme **(C)**. The corresponding phase micrograph shown in **(A)** contains an expressing cell that exhibits positive immunofluorescence localized to the Golgi subcellular compartment, and nonexpressing cells that serve as useful negative controls. Miniaggrecan was detected using the mouse monoclonal Tetra-His Antibody and FITC-goat IgG anti-mouse IgG. ST was localized using polyclonal anti-ST antibodies and Texas Red-goat IgG anti-rabbit IgG. The calibration bars of **Figs. 1** and **2** represent 10 μm.

3. Permeabilize the paraformaldehyde-fixed cells with 0.1% NP-40 for 15 min. (Skip this step if the cells were fixed with methanol.)
4. Rinse with PBS.
5. Block the fixed samples with 15% NGS at 37°C for 15 min.
6. After removing NGS, add primary antibody to the sample (anti-FLAG was used at 0.04 mg/mL, Tetra-and Penta-His were used at 0.01 mg/mL) and incubate for 2 h at 37°C.
7. Wash with PBS for 1 h with 5–6 changes.
8. Incubate with FITC-conjugated goat IgG anti-mouse IgG (0.2 mg/mL) for 1 h at 37°C.
9. Wash with PBS for 1 h with 5–6 changes.
10. If double-fluorescence localization is desired, repeat **steps 4–7** using selected rabbit polyclonal antibodies and Texas Red-conjugated goat IgG anti-rabbit IgG (*see* **Note 6**).
11. Mount samples on microscope slides with mounting medium.
12. Observe samples with a microscope equipped with phase-contrast (or differential interference-contrast) and incident-light fluorescence optics (*see* **Notes 7** and **8**). Document with conventional photographic camera or capture images using digital camera (*see* **Figs. 1** and **2**).

### 3.3. Ultrastructural Localization of Expressed Proteoglycan

1. Wash cells with PBS 2 times.
2. Fix with paraformaldehyde-lysine-periodate fixative for 45 min at room temperature (*see* **Notes 9** and **10**).
3. Wash with PBS 2 times.
4. Incubate 15 min with 0.05% saponin/PBS to permeabilize cells.
5. After most of the wash solution is removed, use Kimwipes to dry the edge of culture dish. This step generates surface tension around the central area of the dish so that as little as 50 μL of blocking reagents (and later, antibody solutions) is required for covering the area during incubation. Incubate in blocking reagent for 20 min.

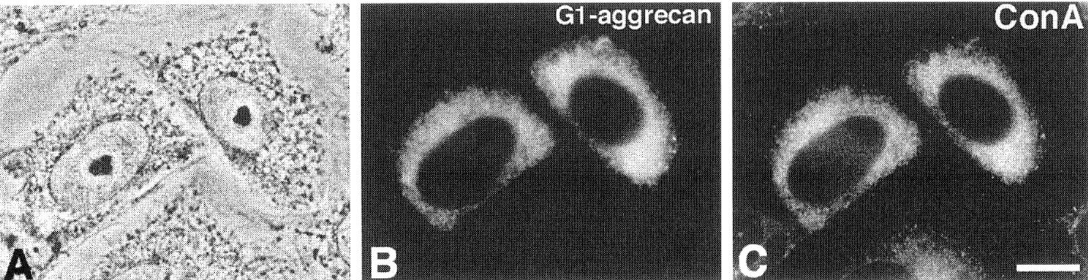


Fig. 2. Colocalization of G1-aggrecan (containing the chicken aggrecan N-terminus with the signal sequence and G1 domain, a segment of the chondroitin sulfate attachment region, and a 6xhis epitope tag) and concanavalin A. CHO cells transfected with G1-aggrecan exhibit protein localized principally in regions throughout the cytoplasm (**B**). Identification of these cytoplasmic regions as the ER is established by the colocalization of FITC-concanavalin A, a lectin that recognizes mannose-rich oligosaccharides in the ER (**C**). Two expressing cells and segments of several nonexpressing cells are revealed in the corresponding phase micrograph (**A**) and by the concanavalin A localization (**C**). The nonexpressing cells serve as negative controls for G1-aggrecan localization, which was accomplished using the mouse monoclonal Penta-His Antibody and Texas Red-goat IgG anti-mouse IgG. Reproduced with permission from Wang, P. W., Chen, T.-L., Luo, W., Zheng, J., Qian, R., Tanzer, M. L., Colley, K., and Vertel, B. M. (1999) Immunolocalization of 6xHis-tagged proteins in CHO cells with QIA express Anit-His Antibodies. *Qiagen News* **1**, 3–6. *(6)*. Used with permission.

6. Wash with 0.05% saponin/PBS 2 times before adding primary antibody.
7. Run Kimwipes around edge of culture dish to prepare the surface for antibody addition and incubate in primary antibody for 1.5 h at 37°C.
8. Wash 6–7 times over 45 min with 0.05% saponin/PBS.
9. Run Kimwipes around edge of culture dish to prepare the surface for antibody addition and incubate in rabbit IgG anti-mouse IgG for 1.5 h at 37°C.
10. Wash 6–7 times over 45 min with 0.05% saponin/PBS.
11. Run Kimwipes around edge of culture dish to prepare the surface for antibody addition and incubate in peroxidase-conjugated goat anti-rabbit FAB for 1.5 h at 37°C.
12. Wash 6–7 times over 45 min with 0.05% saponin/PBS.
13. Wash 3 times over 5 min in DAB buffer.
14. During wash, prepare DAB solution.
15. Add 1.3 mL of DAB solution per dish and incubate 10 min on a shaker.
16. Add 1.5 µL of hydrogen peroxide for each dish, cover the dish, and incubate for an additional 5–10 min. Monitor color change using an inverted microscope and when reaction is sufficient, stop by removing DAB solution (*see* **Note 11**).
17. Wash 3 times over 10 min with DAB buffer.
18. Wash for 10 min with 1/1 mixture of DAB buffer and sucrose-containing cacodylate buffer.
19. Wash with sucrose-containing cacodylate buffer 2 times, 10 min each.
20. Fix with 2% glutaraldehyde in 0.1 *M* cacodylate buffer containing 6% sucrose for 30 min at room temperature.
21. Wash with sucrose-containing cacodylate buffer 2 times, 10 min each.
22. Fix with 1% $OsO_4$/1.5% $KFe_4(CN)_6$ in 0.1 *M* cacodylate buffer containing 6% sucrose for 45 min in the dark and cold.

23. Wash with sucrose-containing cacodylate buffer 4 times, 15 min each.
24. Fix with 1% tannic acid in 0.1 *M* cacodylate buffer containing 6% sucrose for 15 min.
25. Wash with sucrose-containing cacodylate buffer 2 times, 10 min each.
26. Wash with water 3 times over 10 min.
27. Dehydrate through 50% ethanol, 10 min; 75% ethanol, 10 min; 90% ethanol, 10 min; 90% HPMA, 3 times over 15 min; 95% HPMA, 15 min; 97% HPMA, 15 min (*see* **Note 12**).
28. Exchange through a series of HPMA/tEpon solutions: 2/1 HPMA/tEpon, 15 min; 1/1 HPMA/tEpon, 30 min; 1/2 HPMA/tEpon, 30 min. Add HPMA/tEpon mixture to dish, cover, and mix continuously.
29. Exchange into Epon with catalyst over 30 min with 3 changes.
30. Drain off excess Epon with catalyst to leave just enough to cover the cell layer.
31. Infiltrate overnight at 37°C in oven with dessicant. Be sure dishes in the oven are flat, and cover them with a single layer of foil into which small holes have been poked to allow residual dehydrating agents to evaporate off.
32. Transfer dishes to 60°C oven for polymerization, keeping the dishes flat to maintain a uniform depth of Epon polymer on the cells. Two days are required for complete polymerization.
33. Upon completion of polymerization, break the culture dish to recover the embedded cells in a thin layer of polymerized Epon (*see* **Note 13**). Cut out a small piece (<1 mm each side) to be mounted (using epoxy or Krazy Glue) on a block suitable for sectioning. A light microscope can be used to help in selecting optimal areas for mounting.
34. Ultrathin sections are collected onto electron microscope grids. Cells may be counterstained briefly with lead citrate before viewing under the electron microscope. For lead citrate counterstaining, prepare a Petri dish chamber with parafilm on the bottom. Add moist NaOH pellets to the chamber in order to sequester $CO_2$ and prevent the formation of lead carbonate precipitate. For counterstaining, drop the lead citrate solution from a syringe with a filter onto the parafilm. Float each grid, sample side down, onto separate drops. Cover the chamber with aluminum foil to protect against light and $CO_2$ contamination. Remove excess stain from the grids by dipping them through a series of double-distilled water rinses and allow the grids to dry before viewing under the electron microscope.

## 4. Notes

1. Vinol 205 polyvinyl alcohol (also called Airvol 205) is available as a sample on request from Air Products and Chemicals, Inc. (Allentown, PA). Defrosted working solutions are stable for 6 mo at 4°C.
2. Toxic compounds used in this procedures include DAB, cacodylate, osmium tetroxide, glutaraldehyde, and the dehydration solutions and nonpolymerized embedding materials. Handle the solutions in the hood with gloved hands. Dispose as chemical waste. DAB can be detoxified by treatment with chlorox.
3. The initial cell density required to seed the cover slips varies among cell types used for transfection. The major factors to be considered are cell growth rate, the time required for cells to attach to the cover slips, and the survival rate after transfection. For example, CHO cells can grow directly on the glass surface of cover slips, but require 36–40 h to attach and spread well, while COS and HeLa cells attach and spread better on gelatinized carbon-coated cover slips, and do so by 24 h. The COS and HeLa cells are less sensitive to SuperFect transfection reagent, and so 70% confluence is the optimal density at the time of transfection. Cell density adjustments may need to be made to allow for differences in the toxicity of SuperFect lots.
4. Gelatinized carbon-coated coverslips can be used for cells that do not attach well to uncoated glass surfaces. A carbon coat is applied to glass cover slips using carbon rods

(Ted Pella) in a vacuum evaporator (Denton, DV502) run under standard conditions. Carbon-coated cover slips are stored in 95% ethanol and treated with 1% gelatin before use.

5. The optimal volume (μL) of SuperFect reagent and the ratio of SuperFect volume to the quantity of DNA (μg) may vary depending on the specific cell type and DNA construct.
6. Compartment-specific antibodies and fluorophore-conjugated lectins can be used to help identify intracellular compartments involved in trafficking of proteoglycans. For example, concanavalin A, a lectin that reacts with mannose-rich oligosaccharides added co-translationally to glycoproteins in the endoplasmic reticulum (ER), can be used as a marker for the ER. Co-transfection with a construct that encodes the Golgi enzyme sialyltransferase (ST) has been used in our experiments to identify the Golgi complex *(6)*. In this case, expressed ST was localized with polyclonal antibodies against ST. Alternatively, antibodies specific for Golgi complex proteins, such as TGN38, may be used to identify the trans-Golgi network.
7. When observing immunostained cells under the fluorescence microscope, it is important to distinguish real signals from nonspecific background. An overall high level of cellular fluorescence usually suggests a background problem. In transfection experiments, the nontransfected (therefore, nonexpressing) cells serve as a convenient internal negative control for immunostaining (*see* **Figs. 1** and **2**).
8. The background problem in immunostaining reactions can usually be reduced by diluting the primary and/or secondary antibody concentrations and by modifying incubation times.
9. If the final concentration of paraformaldehyde used in the fixative is increased to improve ultrastructure, the effect on antigenicity must also be determined.
10. The protocol for preembedment immunoperoxidase localization is a modification *(7)* of the method described by Brown and Farquhar *(8)*.
11. DAB color reaction should be stopped when the reaction becomes saturated and starts to extend into peripheral structures.
12. The protocol for embedding monolayer cell cultures is according to Brinkley et al. *(9)*.
13. For ultrastructural localization studies, the monolayer of cells is best cultured in dishes with thin walls because these plastic dishes can be pried off easily to release the polymerized Epon-containing embedded cells.

## Acknowledgements

Research support was provided by National Institutes of Health grants DK28433, AR45909, and grants from the Arthritis Foundation. We are grateful to Drs. Marvin Tanzer and Wei Luo (Univ. Connecticut Health Center) for providing aggrecan constructs and Dr. Karen Colley (Univ. Illinois College of Medicine, Chicago) for providing antibodies and constructs for sialyltransferase.

## References

1. Zheng, J., Luo, W. and Tanzer, M. L. (1998). Aggrecan synthesis and secretion. A paradigm for molecular and cellular coordination of multiglobular protein folding and intracellular trafficking. *J. Biol. Chem.* **273**, 12,999–13,006.
2. Luo, W., Kuwada, T. S., Chandrasekaran, L., Zheng, J., and Tanzer, M. L. (1996). Divergent secretory behavior of the opposite ends of aggrecan. *J. Biol. Chem.* **271**, 16447–16450.
3. Dahdal, R. Y. and Colley, K. L. (1993) Specific sequences in the signal anchor of the galactoside-2,6-sialyltransferase are not essential for Golgi localization. *J. Biol. Chem.* **268**, 26,310–26,319.

4. McLean, I. W. and Nakane, P. K. (1974) Periodate-lysine-paraformaldehyde fixative: a new fixative for immunoelectron microscopy. *J. Histochem. Cytochem.* **22**, 1077–1083.
5. Reynolds, E. S. (1963) The use of lead citrate at high pH as an electron-opaque stain in electron microscopy. *J. Cell Biol.* **17**, 208–212.
6. Wang, P. W., Chen, T.-L., Luo, W., Zheng, J., Qian, R., Tanzer, M. L., Colley, K., and Vertel, B. M. (1999) Immunolocalization of 6xHis-tagged proteins in CHO cells with QI *A* express Anti-His Antibodies. *Qiagen News* **1**, 3–6.
7. Vertel, B. M., Velasco, A., LaFrance, S., Walters, L., and Kaczman-Daniel, K. (1989) Precursors of chondroitin sulfate proteoglycan are segregated within a subcompartment of the chondrocyte endoplasmic reticulum. *J. Cell Biol.* **109**, 1827–1836.
8. Brown, W. J. and Farquhar, M. G. (1984) The mannose-6-phosphate receptor for lysosomal enzymes is concentrated in *cis* Golgi cisterna. *Cell* **36**, 295–307.
9. Brinkley, B. R., Murphy, P., and Richardson, C. L. (1967) Procedure for embedding *in situ* selected cells cultured in vitro. *J. Cell Biol.* **35**, 279–283.

# 29

# Selection of Glycosaminoglycan-Deficient Mutants

**Xiaomei Bai, Brett Crawford, and Jeffrey D. Esko**

## 1. Introduction

Mutant cell lines provide an excellent model for studying the structure, assembly and function of proteoglycans under the controlled conditions of tissue culture. Numerous proteoglycan-deficient strains have been isolated, mostly in Chinese hamster ovary cells, and in many cases the defects have been characterized both genetically and biochemically (*see* **Table 1**). Biochemical analysis of the mutants has confirmed that various enzyme activities detected in cell-free extracts using synthetic substrates actually play a role in proteoglycan assembly in vivo. The cell lines have allowed investigators to study how altering the composition of proteoglycans affects fundamental properties of cells, such as adhesion and signaling. Moreover, animal cell mutants provide the background for predicting the phenotype of organismal mutants defective in proteoglycan assembly.

Until recently, high-capacity selection methods for isolating glycosaminoglycan-deficient cell mutants have not been available. Instead, most mutants have been identified by indirect screening methods using a modified form of Lederberg-style replica plating, as originally devised for the isolation of microbial mutants *(1)*. As applied to animal cells, replica plating involves the transfer of colonies growing on plastic culture dishes onto overlying disks of polyester cloth. The colonies on the replicas are then used to screen for mutants by metabolic labeling schemes or a biochemical assay. Mutants identified as lacking a particular activity are recovered from the original set of colonies present on the plate. Although the technique has moderate capacity compared to selection schemes (~$10^5$ colonies can be screened at one time), the frequency of mutations is sufficiently high after mutagenesis that a large collection of strains have been identified (*see* **Table 1**).

One limitation of replica plating is that it is not amenable to all types of cells, and mutations in some steps appear to be relatively rare. To circumvent this problem, direct selection techniques have been developed. One procedure employed repeated rounds

From: *Methods in Molecular Biology, Vol. 171: Proteoglycan Protocols*
Edited by: R. V. Iozzo © Humana Press Inc., Totowa, NJ

**Table 1**
**Cell Mutants with Defined Defects in Glycosaminoglycan Biosynthesis**

| ComplementationGroup | Biochemical Defect | Phenotype |
|---|---|---|
| pgsA (CHO) ***(28)*** | Xylosyltransferase | Glycosaminoglycan-deficient |
| pgsB (CHO) ***(29)*** | Galactosyltransferase I | Glycosaminoglycan-deficient |
| pgsG (CHO) ***(20)*** | Glucuronosyltransferase I | Glycosaminoglycan-deficient |
| pgsD (CHO) ***(30)*** | N-acetylglucosaminyl/ glucuronosyltransferase (EXT-1) | Heparan sulfate-deficient |
| Gro2C (mouse L-cells) ***(3,31)*** | N-acetylglucosaminyl/ glucuronosyltransferase (EXT-1) | Heparan sulfate-deficient |
| *ldl*D (CHO) ***(32,33)*** | UDP-glucose/galactose (GlcNAc/GalNAc) 4-epimerase | Chondroitin sulfate-deficient when starved for GalNAc; GAG-deficient when starved for galactose |
| pgsC (CHO) ***(34)*** | Sulfate transporter | Normal glycosaminoglycans; deficient labeling with $^{35}SO_4$ |
| pgsE (CHO) ***(35)*** | N-deacetylase/ N-sulfotransferase 1 (NDST-1) | Undersulfated heparan sulfate |
| CM-15 (COS cells) ***(36)*** | N-deacetylase/N-sulfotransferase (undefined locus) | Undersulfated heparan sulfate |
| pgsF (CHO) ***(26)*** | 2-O-sulfotransferase | Deficient 2-O-sulfation of heparan sulfate |

of cell lysis mediated by guinea pig complement and an antibody against a carbohydrate domain of a surface proteoglycan ***(2)***. More recently, Tufaro and co-workers have exploited the dependence of Herpes simplex virus on cell surface proteoglycans to identify strains resistant to the viral cytopathic effect ***(3,4)***. One class of mutants isolated in this way lacks cell surface proteoglycans required for viral attachment and invasion. Since Herpes viruses in general depend on cell surface proteoglycans acting as co-receptors ***(5–9)***, these methods could be applicable to a variety of cell types.

Compared to replica plating, direct selection methods have high capacity since they can be applied to large populations of cells (~$10^9$). To expand this approach, it would be desirable to have a variety of agents targeted to specific glycosaminoglycan sequences that make up binding sites for ligands. In Asn-linked glycosylation, numerous mutants have been isolated by taking advantage of the specificity and cytotoxicity of plant lectins ***(10)***. Plant lectins bind to oligosaccharides, often detecting subtle differences in anomeric linkage or in the stereochemistry of one or more sugars *(11)*. Binding results in cell death or in the release of the cells from their substratum, depending on the nature of the lectin. Direct selections based on lectin resistance have yielded recessive, loss-of-function mutations in various glycosyltransferases *(10,12)*, as well as dominant, gain-of-function mutants expressing novel activities *(13)*. These cell lines have helped to unravel the branching pathway of Asn-linked glycosylation, the assembly and transport of lipid and nucleotide precursors, and the effects of altered glycosylation on glycoprotein function *(10,14)*.

Unfortunately, plant lectins that bind to glycosaminoglycans have not yet been identified. Monoclonal antibodies could potentially fulfill this role, but few antibodies with defined sequence specificity have been described or they do not fix complement *(15–17)*. To circumvent this problem, chimeric toxins consisting of a GAG-binding protein coupled to a cytotoxin have been made. The prototype GAG-dependent cytotoxin consists of basic fibroblast growth factor (FGF-2) fused to a ribosome-inactivating protein, saporin (SAP) *(18)*. Application of this chimeric toxin to cultured cells results in cell death, mediated through high-affinity FGF signaling receptors and/or low-affinity heparan sulfate co-receptors *(19,20)*. CHO cells express very few high-affinity FGF receptors, and therefore the cytotoxic activity depends critically on the expression of cell surface heparan sulfate chains. Although the use of chimeric toxins targeted to glycosaminoglycans is a relatively new concept, novel groups of mutants have already been identified *(20)*. Thus, the technique should have broad impact since a variety of cytotoxins can be made with different specificities dependent on the GAG-binding moiety of the chimera *(21)*.

The following technical description takes advantage of the cytotoxin FGF-Saporin (*see* **Note 1**). The latter can be produced by chemical cross-linking of saporin to FGF-2 *(19,22–24)* or by expression of a recombinant chimera in *Escherichia coli (18,25)*. The generation and use of other chimeras should follow the same principles. Combining this technique with replica plating provides both the capacity and biochemical specificity needed to identify desirable mutant cell lines.

## 2. Materials

### 2.1. Selection

1. Culture medium appropriate for specific cell line under study. For CHO cells, use Ham's F12 medium supplemented with 10% (v/v) fetal bovine serum, 100 U/mL of penicillin, and 100 µg/mL of streptomycin.
2. FGF-SAP. The original preparation was obtained from Selective Genetics, Inc. (San Diego, CA), but the company no longer produces the material for commercial distribution. Recombinant material can be generated using published procedures for expression and purification of the chimera in *E. coli (18,25)* (*see* **Note 2**).
3. 10% trichloroacetic acid (TCA). *Caution: TCA is caustic and will cause severe burns.* Coomassie brilliant blue G (0.05% ) in methanol/water/acetic acid, 45/45/10 (v/v). The destaining solution consists of methanol/water/acetic acid, 45/45/10 (v/v).

### 2.2. Replica Plating

1. Polyester cloth disks. Polyester cloth of different pore sizes can be obtained from Tetko, Inc. (Elmsford, NY). The optimal pore size should be determined empirically, but for

many cells 5- to 17-μm-pore-diameter cloth works well (PeCap 7-5, PeCap 7-17). The disks should be prepared as described *(1)*.
2. Sterile 4-mm-diameter glass beads (Pyrex, Fisher) prepared as described *(1)*.
3. Sterile Whatman #1 filter paper cut to fit the culture dish prepared as described *(1)*.

## 3. Methods

### 3.1. Determine the Dose–Response Curve for the Cytotoxin

1. Plate wild-type cells in a 96-well dish at a density of ~$1 \times 10^3$ cells/well in 0.2 mL of growth medium supplemented with serum and antibiotics. Add 0.05, 0.1, 0.2, 0.4, 0.6, 0.8. 1.0, 2.0, 4.0, and 8.0 μg/mL FGF-SAP to individual wells.
2. After 4–5 d, decant the spent medium and rinse the wells with buffered saline to remove dead cells. Fix the remaining attached cells with 10% TCA for 15 min at room temperature and then rinse the plate 3 times with water. Stain the cells with Coomassie blue and remove excess dye with destaining solution.
3. Examine the staining intensity in each well and select the concentration of cytotoxin that completely inhibited cell growth.

### 3.2. Selection of Resistant Mutants

1. Plate cells in growth medium containing the optimal concentration of cytotoxin. A pilot experiment is necessary to determine the incidence of mutants in the population. Seed multiple 100-mm-diameter plates at $10^2$, $10^3$, $10^4$, and $10^5$ cells. If cells are plated at too high a density, the incidence of resistant strains may be too great to pick individual clones easily. Plating the cells at too low of a density wastes the cytotoxin and other materials (*see* **Note 3**).
2. With direct selection, it may not be necessary to treat cells with a mutagen to find the desired mutants, since the capacity of the technique is high (~$10^9$ cells). However, if resistant mutants are not observed, then a mutagen should be employed. The details of mutagen-treatment have been described elsewhere *(1)*. Typically one can expect a $10^2$- to $10^4$-fold increase in the incidence of toxin-resistant mutants after chemical mutagenesis, but the increased likelihood of finding mutants should be weighed against the enhanced probability of inducing DNA damage and multiple mutations (*see* **Note 4**).
3. Since saporin is a ribosome-inactivating agent (RNA N-glycosidase), an immediate cessation of growth does not occur, and detachment of dead cells from the plate takes a few days. After 4 d, remove dead cells and the spent medium, and add fresh medium containing cytotoxin. After ~10 d in culture, visible colonies arise that can be picked with a glass cloning cylinder or a Pasteur pipet *(1)*. To assure their purity and stability, these clones should be repurified by serial dilution in microtiter plates in the presence of selection medium.
4. Examine the glycosaminoglycan composition of the resistant colonies by radiolabeling the cells with $^{35}SO_4$ or [6-$^3$H]glucosamine following established procedures (Chapters 1,9,30).

### 3.3. Selection and Replica Plating

In some cases it may be desirable to couple selection with replica plating in order to subsort the mutants into different biochemical groups. The details of replica plating have been described previously *(1)* and are presented here in abbreviated form (**Fig. 1**; **Subheadings 3.1.** and **3.2.**).

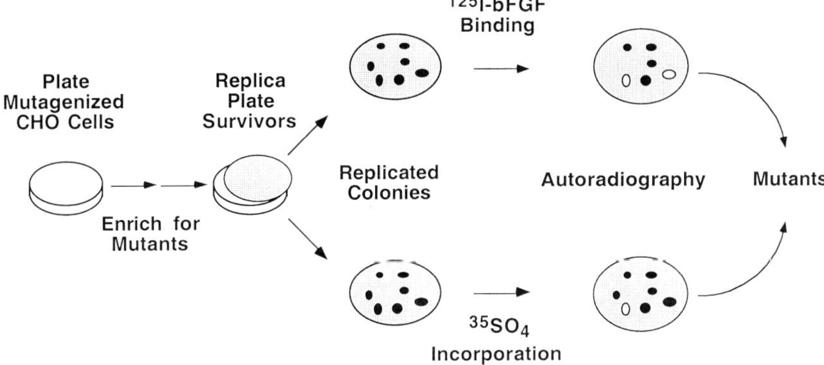


Fig. 1. Selection and replica plating of animal cells for glycosaminoglycan-deficient mutants.

1. From the pilot experiments described above, choose the appropriate number of cells to seed in 100-mm-diameter tissue culture plates so that selection with the cytotoxic agent will result in ~100 resistant colonies/dish. Under these conditions only 5–10 plates are needed to screen 500–1000 resistant colonies.
2. After 4 d, remove the dead cells by replacing the growth medium. Float two layers of polyester cloth in each dish and add glass beads as a ballast to hold the disk against the bottom of the plate ***(1)***.
3. Change the medium on day 8. After 12 d, remove the media, decant the glass beads, and remove the top disk with a pair of tweezers. Gently remove the bottom disk and place it into fresh (bacterial) dish with 5 mL of medium.
4. Gently rinse the master dish with growth medium and overlay it with Whatman #1 filter paper and fresh glass beads. This step prevents satellite colonies from growing while screening of the replica disks is underway. The master dish should be placed at 33°C in a $CO_2$ incubator.
5. Transfer the disk to growth medium containing 10 µCi of $H_2{}^{35}SO_4$ in order to radiolabel newly made proteoglycans. Precipitate radioactive proteoglycans on the disks with 10% TCA, and wash the disk three times by dipping them in 2% TCA followed by one dip in water. Stain the disk with Coomassie blue to visualize the colonies. Dry the disk and image the colonies on film or a phosphorimager. As an alternative to measuring proteoglycan biosynthesis with $^{35}SO_4$, the colonies can be screened by blotting methods using a suitably labeled GAG-binding protein (e.g., $^{125}$I-FGF) ***(20)***.
6. Compare the autoradiographic image to the Coomassie blue staining, and circle the colonies that are present on the disk but lack an autoradiographic halo.
7. To recover the colony, remove the glass beads and Whatman paper from the master dish. Rinse the plate once with saline solution, orient the stained disk in the plate, and circle the desired colony on the bottom of the plate. Pick the colony with glass cloning cylinders or a Pasteur pipet ***(1)***.

## 4. Notes

1. Saporin is a type I ribosome-inactivating inhibitor that enzymatically depurinates ribosomal RNA. By itself, the toxin is inactive against cells due to limited uptake, but its fusion

to a ligand for a cell surface receptor greatly enhances its potency. Toxicity presumably requires uptake of the cytotoxin by endocytosis and delivery to the cytoplasm, but the mechanism underlying escape from the endocytic pathway is unknown. Because of the complex mechanism underlying its action, one might expect to derive resistant mutants with altered endocytic properties or ribosomal structure. For unknown reasons, all of the CHO cell mutants derived to date have defects in proteoglycan assembly, suggesting that these alternate modes of resistance may be relatively rare events.

2. Optimizing the concentration of the cytotoxin is essential for each cell line. FGF-SAP binds to high-affinity, tyrosine kinase receptors as well as low-affinity heparan sulfate co-receptors, and evidence suggests that the killing effect may be a combination of both receptor subtypes ***(19)***. Thus, resistance could be due to loss of either class of receptors, depending on their relative number and the concentration of ligand that is used.
3. There are two key characteristics of an ideal cell line for selection. First, the cells must give rise to colonies from single cells at high efficiency. Second, it is favorable if the cell line is hypodiploid or aneuploid, since having multiple copies of the targeted gene will drastically reduce the probability of isolating the desired mutants. In CHO cells, many loci appear to be functionally and/or physically hemizygous, thus increasing the likelihood of finding mutants. However, the selection strategy may work even in diploid lines as well since its capacity is very high ($\sim 10^9$ cells).
4. The combination of replica plating with direct selection provides a powerful method for identifying mutants in proteoglycan biosynthesis. One advantage of the combined procedure is that it provides the capacity to find rare mutants and the selectivity to find strains with specific biochemical characteristics. For example, the method described here can be used with $^{35}SO_4$ or other radioactive metabolic precursors as a global screen for GAG deficiency. Alternatively, a radioactive or fluorescently labeled ligand or antibody can be used to pick out defects affecting a specific binding sequence ***(26,27)***. Colonies can also be screened by direct enzymatic assay to find variants in a specific biochemical step ***(1)***. Thus, the combination of selection and screening provides the order of magnitude and specificity required for identifying even rare mutants in proteoglycan biosynthesis.

## References


1. Esko, J. D. (1989) Replica plating of animal cells. *Meth. Cell Biol.* **32**, 387–422.
2. Inestrosa, N. C., Matthew, W. D., Reiness, C. G., Hall, Z. W., and Reichardt, L. F. (1985) Atypical distribution of asymmetric acetylcholinesterase in mutant PC12 pheochromocytoma cells lacking a cell surface heparan sulfate proteoglycan. *J. Neurochem.* **45**, 86–94.
3. Banfield, B. W., Leduc, Y., Esford, L., Schubert, K., and Tufaro, F. (1995) Sequential isolation of proteoglycan synthesis mutants by using herpes simplex virus as a selective agent: Evidence for a proteoglycan-independent virus entry pathway. *J. Virol.* **69**, 3290–3298.
4. Gruenheid, S., Gatzke, L., Meadows, H., and Tufaro, F. (1993) Herpes simplex virus infection and propagation in a mouse L cell mutant lacking heparan sulfate proteoglycans. *J.Virol.* **67**, 93–100.
5. WuDunn, D. and Spear, P.G. (1989) Initial interaction of herpes simplex virus with cells is binding to heparan sulfate. *J. Virol.* **63**, 52–58.
6. Shieh, M.-T., WuDunn, D., Montgomery, R. I., Esko, J. D., and Spear, P. G. (1992) Cell surface receptors for herpes simplex virus are heparan sulfate proteoglycans. *J. Cell Biol.* **116**, 1273–1281.
7. Spear, P. G. (1993) Entry of alphaherpesviruses into cells. *Virology* **4**, 167–180.

8. Shieh, M. T. and Spear, P. G. (1994) Herpesvirus-induced cell fusion that is dependent on cell surface heparan sulfate or soluble heparin. *J. Virol.* **68**, 1224–1228.
9. Shukla, D., Liu, J., Blaiklock, P., Shworak, N. W., Bai, X. M., Esko, J. D., et al. (1999) A novel role for 3-*O*-sulfated heparan sulfate in herpes simplex virus 1 entry. *Cell* **99**, 13–22.
10. Stanley, P. and Ioffe, E. (1995) Glycosyltransferase mutants: Key to new insights in glycobiology. *FASEB J.* **9**, 1436–1444.
11. Cummings, R. D. (1999) Plant Lectins, in *Essentials of Glycobiology* (Varki, A., Cummings, R. D., Esko, J. D., Freeze, H. H., Hart, G. W., and Marth, J. D., eds.) Cold Spring Harbor Laboratory Press, Cold Spring Harbor, NY, pp. 455–467.
12. Stanley, P. (1993) Use of mammalian cell mutants to study the functions of N- and O-linked glycosylation, in *Cell Surface and Extracellular Glycoconjugates: Structure and Function* (Roberts, D. D. and Mecham, R. P., eds.) Academic Press, San Diego, pp. 181–222.
13. Stanley, P., Raju, T. S. and Bhaumik, M. (1996) CHO cells provide access to novel *N*-glycans and developmentally regulated glycosyltransferases. *Glycobiology* **6**, 695–699.
14. Esko, J. D. (1999) Genetic Disorders of Glycosylation in Cultured Cells, in *Essentials of Glycobiology* (Varki, A., Cummings, R. D., Esko, J. D., Freeze, H. H., Hart, G. W., and Marth, J. D., eds.) Cold Spring Harbor Laboratory Press, Cold Spring Harbor, NY, pp. 469–477.
15. Van Den Born, J., Gunnarsson, K., Bakker, M. A. H., Kjellén, L., Kusche-Gullberg, M., Maccarana, M., et al. (1996) Presence of N-unsubstituted glucosamine units in native heparan sulfate revealed by a monoclonal antibody. *J. Biol. Chem.* **270**, 31,303–31,309.
16. Van Den Born, J., Jann, K., Assmann, K. J. M., Lindahl, U., and Berden, J. H. M. (1996) *N*-acetylated domains in heparan sulfates revealed by a monoclonal antibody against the *Escherichia coli* K5 capsular polysaccharide—Distribution of the cognate epitope in normal human kidney and transplant kidney with chronic vascular rejection. *J. Biol. Chem.* **271**, 22,802–22,809.
17. Van Kuppevelt, T. H., Dennissen, M. A. B. A., Van Venrooij, W. J., Hoet, R. M. A., and Veerkamp, J. H. (1998) Generation and application of type-specific anti-heparan sulfate antibodies using phage display technology—Further evidence for heparan sulfate heterogeneity in the kidney. *J. Biol. Chem.* **273**, 12,960–12,966.
18. Lappi, D. A., Ying, W., Barthelemy, I., Martineau, D., Prieto, I., Benatti, L., Soria, M., and Baird, A. (1994) Expression and activities of a recombinant basic fibroblast growth factor-saporin fusion protein. *J. Biol. Chem.* **269**, 12,552–12,558.
19. Lappi, D. A., Maher, P. A., Martineau, D., and Baird, A. (1991) The basic fibroblast growth factor-saporin mitotoxin acts through the basic fibroblast growth factor receptor. *J. Cell Physiol.* **147**, 17–26.
20. Bai, X. M., Wei, G., Sinha, A., and Esko, J. D. (1999) Chinese hamster ovary cell mutants defective in glycosaminoglycan assembly and glucuronosyltransferase I. *J. Biol. Chem.* **274**, 13,017–13,024.
21. Jackson, R. L., Busch, S. J., and Cardin, A. D. (1991) Glycosaminoglycans: Molecular properties, protein interactions, and role in physiological processes. *Physiol. Rev.* **71**, 481–539.
22. Lappi, D. A., Martineau, D., and Baird, A. (1989) Biological and chemical characterization of basic FGF-saporin mitotoxin. *Biochem. Biophys. Res. Commun.* **160**, 917–923.
23. Lappi, D. A., Matsunami, R., Martineau, D., and Baird, A. (1993) Reducing the heterogeneity of chemically conjugated targeted toxins: homogeneous basic FGF-saporin. *Anal. Biochem.* **212**, 446–451.
24. Buechler, Y. J., Sosnowski, B. A., Victor, K. D., Parandoosh, Z., Bussell, S. J., Shen, C., et al. (1995) Synthesis and characterization of a homogeneous chemical conjugate between basic fibroblast growth factor and saporin. *Eur. J. Biochem.* **234**, 706–713.

25. McDonald, J. R., Ong, M., Shen, C., Parandoosh, Z., Sosnowski, B., Bussell, S., and Houston, L. L. (1996) Large-scale purification and characterization of recombinant fibroblast growth factor-saporin mitotoxin. *Protein. Expr. Purif.* **8**, 97–108.
26. Bai, X. M. and Esko, J. D. (1996) An animal cell mutant defective in heparan sulfate hexuronic acid 2-O-sulfation. *J. Biol.C hem.* **271**, 17,711–17,717.
27. de Agostini, A. L., Lau, H. K., Leone, C., Youssoufian, H., and Rosenberg, R. D. (1990) Cell mutants defective in synthesizing a heparan sulfate proteoglycan with regions of defined monosaccharide sequence. *Proc. Natl. Acad. Sci. (USA)* **87**, 9784–9788.
28. Esko, J. D., Stewart, T. E., and Taylor, W. H. (1985) Animal cell mutants defective in glycosaminoglycan biosynthesis. *Proc. Natl. Acad. Sci. (USA)* **82**, 3197–3201.
29. Esko, J. D., Weinke, J. L., Taylor, W. H., Ekborg, G., Rodén, L., Anantharamaiah, G., and Gawish, A. (1987) Inhibition of chondroitin and heparan sulfate biosynthesis in Chinese hamster ovary cell mutants defective in galactosyltransferase I. *J. Biol. Chem.* **262**, 12,189–12,195.
30. Lidholt, K., Weinke, J. L., Kiser, C. S., Lugemwa, F. N., Bame, K. J., Cheifetz, S., et al. (1992) A single mutation affects both N-acetylglucosaminyltransferase and glucuronosyltransferase activities in a Chinese hamster ovary cell mutant defective in heparan sulfate biosynthesis. *Proc. Natl. Acad. Sci. (USA)* **89**, 2267–2271.
31. McCormick, C., Leduc, Y., Martindale, D., Mattison, K., Esford, L.E., Dyer, A. P., and Tufaro, F. (1998) The putative tumour suppressor EXT1 alters the expression of cell-surface heparan sulfate. *Nat. Genet.* **19**, 158–161.
32. Kingsley, D. M., Kozarsky, K. F., Hobbie, L., and Krieger, M. (1986) Reversible defects in O-linked glycosylation and LDL receptor expression in a UDP-Gal/UDP-GalNAc 4-epimerase deficient mutant. *Cell* **44**, 749–759.
33. Esko, J. D., Rostand, K. S., and Weinke, J. L. (1988) Tumor formation dependent on proteoglycan biosynthesis. *Science* **241**, 1092–1096.
34. Esko, J. D., Elgavish, A., Prasthofer, T., Taylor, W. H., and Weinke, J. L. (1986) Sulfate transport-deficient mutants of Chinese hamster ovary cells. Sulfation of glycosaminoglycans dependent on cysteine. *J. Biol. Chem.* **261**, 15,725–15,733.
35. Bame, K. J. and Esko, J. D. (1989) Undersulfated heparan sulfate in a Chinese hamster ovary cell mutant defective in heparan sulfate N-sulfotransferase. *J. Biol. Chem.* **264**, 8059–8065.
36. Ishihara, M., Kiefer, M. C., Barr, P. J., Guo, Y., and Swiedler, S. J. (1992) Selection of COS cell mutants defective in the biosynthesis of heparan sulfate proteoglycan. *Anal. Biochem.* **206**, 400–407.

# 30

# Xyloside Priming of Glycosaminoglycan Biosynthesis and Inhibition of Proteoglycan Assembly

**Timothy A. Fritz and Jeffrey D. Esko**

## 1. Introduction

A powerful approach for studying the relationship of proteoglycan (PG) structure to function employs inhibitors to block glycosaminoglycan (GAG) biosynthesis. Although true enzyme-based, active site-directed inhibitors of the glycosyltransferases and sulfotransferases have not yet been described, decoys consisting of β-D-xylose linked to hydrophobic aglycones have been available for some time ***(1)***. As shown over 25 years ago ***(2)***, xylosides block PG assembly by serving as alternate substrates, thereby diverting GAG assembly from xylosylated proteoglycan core proteins onto the exogenous xyloside primer. This method of derailing PG biosynthesis has been used to explore PG function in cells, tissues, and animals. The priming of oligosaccharides on xylosides has also been used to define the nature of mutations in cell lines deficient in PG biosynthesis ***(3–5)***, to co-localize glycosyltransferases in Golgi subcompartments ***(6–8)***, and as a model for glycoside primers that affect other kinds of glycoconjugates ***(9–12)***.

Beta-D-xylosides consist of xylose in beta linkage to an aglycone (*see* **Fig. 1**). The aglycones typically consist of a hydrophobic compound which aids in carrying the polar sugar moiety across cell membranes into the Golgi apparatus, where GAG biosynthesis takes place. While all β-D-xylosides prime chondroitin/dermatan sulfate, studies have shown that their ability to prime heparan sulfate depends on the structure of the aglycone ***(13,14)***. Xylosides that prime heparan sulfate require an aglycone containing fused rings that assume a more or less planar configuration. Priming of heparan sulfate also depends on concentration, and usually requires a higher dose than is needed for priming chondroitin sulfate. This structural specificity is thought to arise by recognition of the aglycone by the α-GlcNAc transferase that initiates the formation of the repeating disaccharide units of heparan sulfate ***(15,16)***.

From: *Methods in Molecular Biology, Vol. 171: Proteoglycan Protocols*
Edited by: R. V. Iozzo © Humana Press Inc., Totowa, NJ

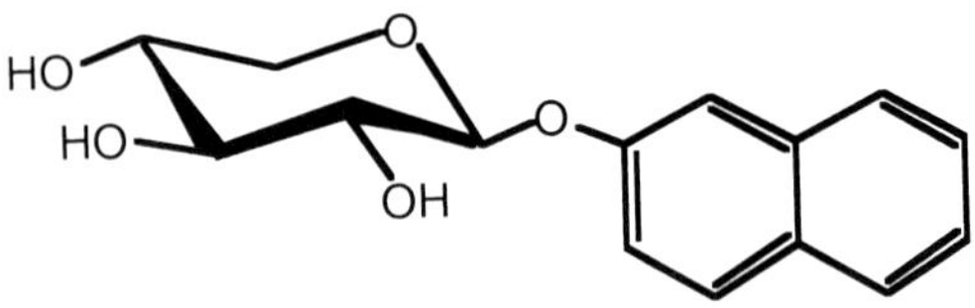

Fig. 1. Structure of 2-naphthol-β-D-xyloside.

Inhibition of GAG assembly on PGs starts at xyloside concentrations as low as 10–30 μ*M*, concentrations not significantly higher than those needed for maximum stimulation of GAG priming ***(13,17–19)***. Inhibition of heparan sulfate PG formation typically requires higher concentrations of xyloside than needed for chondroitin/dermatan sulfate PG inhibition, consistent with the higher dose requirements for priming heparan sulfate described above. As might be expected, xylosides that prime heparan sulfate inhibit heparan sulfate PG synthesis at lower doses ***(13)***.

The following procedures describe the use of β-D-xylosides to inhibit PG assembly in both cultured cells and tissues. The protocol consists of three steps: (i) addition of the compounds to the growth/incubation medium, (ii) PG/GAG extraction and isolation, and (iii) analysis of the extracted PGs/GAGs.

## 2. Materials

### 2.1. GAG Priming and Inhibition of PG Assembly

1. 0.2 *M p*-nitrophenyl, 4-methylumbelliferyl, or 2-napthol-β-D-xyloside in dimethylsulfoxide (DMSO).
2. 0.2 *M* control glycoside (e.g., *p*-nitrophenyl-α-D-xyloside or *p*-nitrophenyl-β-D-arabinoside) in DMSO.
3. Culture medium appropriate for the cells or tissue being studied.
4. Radioactive precursors ($H_2{}^{35}SO_4$, [6-$^3$H]glucosamine or [1-$^3$H]galactose).

### 2.2. PG/GAG Extraction

1. Guanidine extraction buffer. 4.0 *M* guanidine-HCl, 0.2% (w/v) Zwittergent 3-12, 50 m*M* sodium acetate (pH 6.0), 10 m*M* EDTA, 10 m*M* N-ethylmaleimide (NEM), 1 m*M* phenylmethylsulfonyl fluoride (PMSF), 1mg/mL pepstatin A, 0.5 mg/mL leupeptin. Add protease inhibitors from 200× stocks (2 *M* NE*M* in ethanol, 0.2 *M* PMSF in ethanol, 0.2 mg/mL pepstatin A in ethanol, 0.1 mg/mL leupeptin in water). Store stocks of protease inhibitors at –20°C.
2. Triton X-100 extraction buffer: 20 m*M* Tris-HCl, pH 7.4, 0.5% (w/v) Triton X-100, 0.15 *M* NaCl, 10 m*M* EDTA, 10 m*M* NEM, 1 m*M* PMSF, 1 μg/mL pepstatin A, 0.5 μg/mL leupeptin.
3. Dialysis membranes, 6- to 8-kDa cutoff.

### 2.3. PG/GAG Isolation

1. Dialysis buffer: Same as the guanidine extraction buffer but with 4.0 *M* urea substituted for the guanidine-HCl.

2. DEAE-Sephacel (Amersham Pharmacia Biotech): Equilibrate the resin with 0.2 *M* NaCl, 50 m*M* sodium acetate, pH 6.0.
3. DEAE wash buffer: Add solid NaCl to dialysis buffer for a final NaCl concentration of 0.2 *M*.
4. DEAE elution buffer: Add solid NaCl to dialysis buffer for a final NaCl concentration of 1.0 *M*.
5. 20 mg/mL chondroitin sulfate A or heparin in 20 m*M* Tris-HCl, pH 7.0.
6. 20 m*M* Tris-HCl, pH 7.4, 0.2 *M* NaCl.

### 2.4. Analysis of PGs/GAGs

1. TSK 4000 SW HPLC gel filtration column (30 cm x 7.5 mm inner diameter).
2. Gel filtration buffer: 0.1 *M* $KH_2PO_4$, pH 6.0, 0.5 *M* NaCl, 0.2% (w/v) Zwittergent 3-12.

## 3. Methods

### 3.1. GAG Priming and Inhibition of PG Assembly

1. Prepare serial dilutions of the stock xyloside in cell culture medium to achieve final concentrations of 0, 0.01, 0.03, 0.1, 0.3, and 1 m*M*. Prepare similar, separate dilutions of the control glycoside. Add DMSO as necessary to maintain a constant vehicle concentration for all dilutions (*see* **Note 1**).
2. Add radioactive precursors to the diluted xylosides (10 µCi/mL [$^{35}$S]$H_2SO_4$, 20 µCi/mL [6-$^3$H]glucosamine, or 50 µCi/mL [1-$^3$H]galactose should be sufficient, but this depends on the sulfate and glucose content of the growth medium. Warm the supplemented medium under culture conditions before adding it to cells or tissue. If sufficient levels (~50 µg of uronic acid) of PGs/GAGs will be produced, the material can be quantitated chemically, and the radioactive precursors may be omitted (*see* **Note 2**).
3. Replace the culture medium in the culture dish or flask with the medium containing xyloside and radioactive precursors, and incubate the samples under appropriate conditions for the cells or tissue under study. Priming occurs rapidly and continues throughout the incubation, since the xyloside is not significantly depleted during the incubation. The number of cells and/or labeling time may need to be varied depending on the cell or tissue type and the sensitivity of the assays employed. If the content of PGs/GAGs is to be measured chemically, the incubation may need to be extended since preexisting molecules must turn over in order to see significant inhibition of PG glycosylation.

### 3.2. PG/GAG Extraction

1. Extraction of PGs/GAGs from cells and culture medium *(20)*. Chill the cells and medium to 4°C. For most cells, adding Triton X-100 extraction buffer (2 mL/$10^7$ cells) will suffice to solubilize membrane and cell-associated PGs and GAGs. The extent of solubilization can be assessed by treating the monolayer (or residual cell pellet) with 0.5 mL/$10^7$ cells of 0.1 *M* NaOH at room temperature for 10 min. Neutralize the solution with 10 *M* acetic acid, clarify the sample by centrifugation, and measure the amount of residual material in the supernatant. If the Triton extraction fails to solubilize all of the material from the plate, a guanidine extraction procedure should be used. Add 1–2 mL/$10^7$ cells of guanidine extraction buffer. Alternatively, add solid guanidine-HCl to achieve a final concentration of 4.0 *M*, Zwittergent 3–12 to 0.2% w/v, EDTA to 10 m*M*, and sodium acetate to 50 m*M*. Add protease inhibitors from 200× stock solutions to achieve final concentrations of 10 m*M* NEM, 1 m*M* PMSF, 1 µg/mL pepstatin A and 0.5 µg/mL leupeptin. Incubate at 4°C for 1 h. If tissue samples are under study, mince or homogenize the tissue in 5–10 vol of chilled guanidine extraction buffer per gram (wet weight) of tissue. Stir overnight at 4°C.

2. Clarify the extracts by centrifugation at 4°C for 20 min at 12,000 *g* or 10 min at maximum speed in a microcentrifuge.

### 3.3. PG/GAG Isolation

1. Samples extracted with Triton X-100 solution may be analyzed by anion-exchange chromatography without further processing. If guanidine-HCl buffer was used, the sample must be dialyzed to lower the ionic strength. Dialyze samples at 4°C against dialysis buffer using membranes with a 6- to 8-kDa cutoff. Alternatively, gel filtration chromatography (e.g., PD-10 columns, Amersham Pharmacia Biotech) or ultrafiltration may be used to exchange buffers.
2. Add 1–2 mg of chondroitin sulfate A or heparin as a carrier to the PG/GAG extract if the samples are radioactively labeled (*see* **Note 3**). Note: If PGs/GAGs are to be quantitated chemically by uronic acid assay, omit this step.
3. Prepare a 0.5- to 1.0-mL DEAE-Sephacel column in a disposable pipet tip plugged with glass wool. Remove the bottom few millimeters of the tip to enlarge the opening and improve flow rates. Equilibrate the resin with 5 column volumes of DEAE wash buffer.
4. Apply the PG/GAG extract to the column and wash with 10–20 column volumes of DEAE wash buffer. Elute the PGs/GAGs with 5 column volumes of DEAE elution buffer and collect the eluate as a single fraction.
5. Desalt samples using Sephadex G25 columns or PD-10 columns. These columns can be run in 10% ethanol, which allows the samples to be concentrated.
6. Lyophilize the sample and resuspend it in 0.2 mL of 20 m*M* Tris-HCl, pH 7.4. Store at 4°C.

### 3.4. Analysis of Isolated PGs/GAGs

1. Apply samples (≤0.2 mL) to a TSK G4000 SW gel filtration column equilibrated in gel filtration buffer. Adjust the flow to 0.5 mL/min, collect fractions (1 min), and assay aliquots for radioactivity. PGs will elute before primed GAGs because of their large size, but baseline resolution may not be achieved. Other FPLC or HPLC columns can be used as well to optimize separation. These columns can be internally calibrated by comparing the elution of intact PGs and free chains liberated from the core proteins by beta-elimination ***(13)***. If enough PGs/GAGs are present, their distribution can be measured by the carbazole assay for uronic acid ***(21)***.
2. The inhibition of GAG assembly on PG core proteins can be monitored by measuring the decrease in high-molecular weight material corresponding to the intact proteoglycans (*see* **Fig. 2**). To separate the effects on heparan sulfate or chondroitin/dermatan sulfate PGs, the samples must be treated with low-pH nitrous acid to depolymerize the heparan sulfate chains (Chapter 34) or by treatment with heparinase(s) or chondroitinase(s) (Chapter 35 and 36).

## 4. Notes

1. Stock solutions of xylosides in neat DMSO should not be added directly to cells in culture, since DMSO at high concentration can lyse cells. DMSO is also known to cause differentiation of certain cells, which may be associated with a change in PG expression. As an alternative approach, a xyloside stock may be prepared in ethanol or the compounds can be dissolved directly in culture medium up to ~5 m*M* (with warming if necessary). DMSO stocks of xylosides are stable at –20°C and should be protected from light, since the aglycones absorb in the ultraviolet. Cell monolayers should be visually inspected for signs of toxicity.
2. Recent findings have shown that xyloside priming is not as specific as once thought. A variety of unexpected, small oligosaccharides not necessarily related to GAGs are also

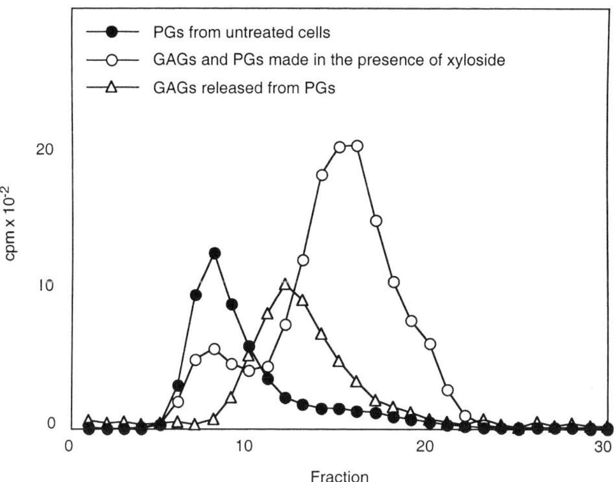


Fig. 2. Inhibition of PG and priming of free GAG chains on a xyloside. (*Source:* Lugemwa, F. N. and Esko, J. D. (1991) Estradiol-β-D-xyloside, an efficient primer of heparan sulfate biosynthesis. *J. Biol. Chem.* **266**, 6674–6677.)

primed and the array of oligosaccharides depends on the cell type ***(22–27)***. These unusual oligosaccharides can constitute the majority of the primed material, even at relatively low concentrations of primer ***(26)***. Thus, it is important to correlate any change of cellular phenotype with the degree of GAG priming and the actual extent of PG inhibition.

3. Alpha-D-xylosides show little or no ability to stimulate GAG synthesis ***(17,28,29)*** but serve as substrates for galactosyltransferase I in vivo ***(22)***. High concentrations of both α- and β-xylosides can inhibit glycolipid biosynthesis, and this inhibition may be related to the unusual oligosaccharides produced ***(22)***. At high xyloside concentrations the large mass of material generated on the exogenous primer may deplete an internal pool of a nucleotide sugar, although this has yet to been determined. Thus, it is important to work at the minimal xyloside dose required for the desired level of PG inhibition.

   A variety of xylosides with different aglycones are available commercially (Sigma or Aldrich) or can easily be synthesized by coupling protected and activated xylose to different aglycones ***(14)***. Different aglycones can affect the extent of GAG priming and the composition of the chains, as described above. These differences also may depend on cell type. Thus, it may be important to test more than one xyloside to optimize priming and PG inhibition. By comparing the effects of a xyloside that primes only chondroitin sulfate to one that primes both heparan sulfate and chondroitin sulfate, the role of each type of chain can be potentially assessed ***(14,30)***.

   GAGs primed by xylosides may have the activities associated with PGs. For example, the heparan sulfate chains primed on 2-naphthol-β-D-xyloside will bind to basic fibroblast growth factor (FGF-2) and enhance growth factor association with high-affinity tyrosine kinase receptors ***(30)***. Although the chains differ somewhat in structure from those assembled on natural core proteins ***(14)***, they possess the binding sequence for the growth factor and apparently the receptor ***(31)***. Therefore, even though PG assembly may be inhibited, a particular proteoglycan-dependent activity may not be altered if the primed GAGs are able to substitute for the PGs.

4. If PGs/GAGs are to be quantitated chemically by uronic acid assay, omit this step.

## References


1. Esko, J. D. and Montgomery, R. I. (1996) Synthetic glycosides as primers of oligosaccharide biosynthesis and inhibitors of glycoprotein and proteoglycan assembly, in *Current protocols in molecular biology: Preparation and analysis of glycoconjugates* (Ausubel, R., Brent, R., Kingston, B., Moore, D., Seidman, J., Smith, J., et al., eds.), Greene Publishing and Wiley-Interscience, New York, pp. 17.11.1–17.11.6
2. Okayama, M., Kimata, K., and Suzuki, S. (1973) The influence of *p*-nitrophenyl β-D-xyloside on the synthesis of proteochondroitin sulfate by slices of embryonic chick cartilage. *J. Biochem.(Tokyo)* **74**, 1069–1073.
3. Esko, J. D., Weinke, J. L., Taylor, W. H., Ekborg, G., Rodén, L., Anantharamaiah, G., and Gawish, A. (1987) Inhibition of chondroitin and heparan sulfate biosynthesis in Chinese hamster ovary cell mutants defective in galactosyltransferase I. *J. Biol. Chem.* **262**, 12,189–12,195.
4. Esko, J. D., Stewart, T. E., and Taylor, W. H. (1985) Animal cell mutants defective in glycosaminoglycan biosynthesis. *Proc. Natl. Acad. Sci. USA* **82**, 3197–3201.
5. Bai, X. M., Wei, G., Sinha, A., and Esko, J. D. (1999) Chinese hamster ovary cell mutants defective in glycosaminoglycan assembly and glucuronosyltransferase I. *J.Biol.Chem.* **274**, 13,017–13,024.
6. Etchison, J. R., Srikrishna, G., and Freeze, H. H. (1995) A novel method to co-localize glycosaminoglycan-core oligosaccharide glycosyltransferases in rat liver Golgi. Co-localization of galactosyltransferase I with a sialyltransferase. *J. Biol. Chem.* **270**, 756–764.
7. Etchison, J. R. and Freeze, H. H. (1996) A new approach to mapping co-localization of multiple glycosyl transferases in functional Golgi preparations. *Glycobiology* **6**, 177–189.
8. Freeze, H. H. and Etchison, J. R. (1996) A new side of xylosides and their close relatives: Co-localization mapping glycosyltransferases in the functional Golgi. *Trends Glycosci.Glyotech.* **8**, 65–77.
9. Kuan, S. F., Byrd, J. C., Basbaum, C., and Kim, Y. S. (1989) Inhibition of mucin glycosylation by aryl-N-acetyl-α-galactosaminides in human colon cancer cells. *J. Biol. Chem.* **264**, 19,271–19,277.
10. Sarkar, A. K., Fritz, T. A., Taylor, W. H., and Esko, J. D. (1995) Disaccharide uptake and priming in animal cells: Inhibition of sialyl Lewis X by acetylated Galb1-4GlcNAcβ-*O*-naphthalenemethanol. *Proc. Natl. Acad. Sci. USA* **92**, 3323–3327.
11. Neville, D. C. A., Field, R. A., and Ferguson, M. A. J. (1995) Hydrophobic glycosides of *N*-acetylglucosamine can act as primers for polylactosamine synthesis and can affect glycolipid synthesis *in vivo*. *Biochem.J.* **307**, 791–797.
12. Sarkar, A. K., Rostand, K. S., Jain, R. K., Matta, K. L., and Esko, J. D. (1997) Fucosylation of disaccharide precursors of sialyl Lewis$^X$ inhibit selectin-mediated cell adhesion. *J. Biol. Chem.* **272**, 25,608–25,616.
13. Lugemwa, F. N. and Esko, J. D. (1991) Estradiol β-D-xyloside, an efficient primer for heparan sulfate biosynthesis. *J. Biol. Chem.* **266**, 6674–6677.
14. Fritz, T. A., Lugemwa, F. N., Sarkar, A. K., and Esko, J. D. (1994) Biosynthesis of heparan sulfate on β-D-xylosides depends on aglycone structure. *J. Biol. Chem.* **269** , 300–307.
15. Fritz, T. A., Gabb, M. M., Wei, G., and Esko, J. D. (1994) Two N-acetylglucosaminyltransferases catalyze the biosynthesis of heparan sulfate. *J. Biol. Chem.* **269**, 28,809–28,814.
16. Fritz, T. A., Agrawal, P. K., Esko, J. D., and Krishna, N. R. (1997) Partial purification and substrate specificity of heparan sulfate α-N-acetylglucosaminyltransferase . 1. Synthesis, NMR spectroscopic characterization and *in vitro* assays of two aryl tetrasaccharides. *Glycobiology* **7**, 587–595.

17. Robinson, H. C., Brett, M. J., Tralaggan, P. J., Lowther, D. A., and Okayama, M. (1975) The effect of D-xylose, β-D-xylosides, and β-D-galactosides on chondrotin sulphate biosynthesis in embryonic chicken cartilage. *Biochem. J.* **148**, 25–34.
18. Kolset, S. O., Ehlorsson, J., Kjellén, L., and Lindahl, U. (1986) Effect of benzyl β-D-xyloside on the biosynthesis of chondroitin sulphate proteoglycan in cultured human monocytes. *Biochem. J.* **238**, 209–216.
19. Sobue, M., Habuchi, H., Ito, K., Yonekura, H., Oguri, K., Sakurai, K., Kamohara, S., Ueno, Y., Noyori, R., and Suzuki, S. (1987) β-D-xylosides and their analogues as artificial initiators of glycosaminoglycan chain synthesis: Aglycone-related variation in their effectiveness *in vitro* and *in ovo*. *Biochem. J.* **241**, 591–601.
20. Esko, J.D. (1993) Special considerations for proteoglycans and glycosaminoglycans and their purification, in *Current protocols in molecular biology* (Ausubel, F., Brent, R., Kingston, B., Moore, D., Seidman, J., Smith, J., et al., eds.) Greene Publishing and Wiley-Interscience, New York, pp. 17.2.1–17.2.9
21. Esko, J. D. and Manzi, A. (1996) Measurement of uronic acids, in *Current protocols in molecular biology* (Ausubel, F., Brent, R., Kingston, B., Moore, D., Seidman, J., Smith, J., et al., eds.) Greene Publishing and Wiley-Interscience, New York, pp.17.9.8–17.9.11
22. Freeze, H. H., Sampath, D., and Varki, A. (1993) α- and β- xylosides alter glycolipid synthesis in human melanoma and Chinese hamster ovary cells. *J. Biol. Chem.* **268**, 1618–1627.
23. Izumi, J., Takagaki, K., Nakamura, T., Shibata, S., Kojima, K., Kato, I., and Endo, M. (1994) A novel oligosaccharide, xylosylβ1-4xylosylβ1-(4-methylumbelliferone), synthesized by cultured human skin fibroblasts in the presence of 4-methylumbelliferyl-β-D-xyloside. *J. Biochem. (Tokyo)* **116**, 524–529.
24. Nakamura, T., Izumi, J., Takagaki, K., Shibata, S., Kojima, K., Kato, I., and Endo, M. (1994) A novel oligosaccharide, GlcAβ1-4Xylβ1-(4-methylumbelliferone), synthesized by human cultured skin fibroblasts. *Biochem. J.* **304**, 731–736.
25. Manzi, A., Salimath, P. V., Spiro, R. C., Keifer, P. A., and Freeze, H. H. (1995) Identification of a novel glycosaminoglycan core-like molecule I. 500 MHz $^1$H NMR analysis using a nano-NMR probe indicates the presence of a terminal α-GalNAc residue capping 4-methylumbelliferyl-β-D-xylosides. *J. Biol. Chem.* **270**, 9154–9163.
26. Salimath, P. V., Spiro, R. C., and Freeze, H. H. (1995) Identification of a novel glycosaminoglycan core-like molecule II. α-GalNAc-capped xylosides can be made by many cell types. *J. Biol. Chem.* **270**, 9164–9168.
27. Miura, Y. and Freeze, H. H. (1998) α-*N*-Acetylgalactosamine-capping of chondroitin sulfate core region oligosaccharides primed on xylosides. *Glycobiology* **8**, 813–819.
28. Schwartz, N. B., Galligani, L., Ho, P.-L., and Dorfman, A. (1974) Stimulation of synthesis of free chondroitin sulfate chains by β-D-xylosides in cultured cells. *Proc. Natl. Acad. Sci. USA* **71**, 4047–4051.
29. Sudhakaran, P. R., Sinn, W., and Von Figura, K. (1981) Initiation of altered heparan sulphate on β-D-xyloside in rat hepatocytes. *Hoppe-Seyler's Z. Physiol. Chem.* **362**, 39–46.
30. Miao, H.-Q., Fritz, T. A., Esko, J. D., Zimmermann, J., Yayon, A., and Vlodavsky, I. (1995) Heparan sulfate primed on β-D-xylosides restores binding of basic fibroblast growth factor. *J. Cell. Biochem.* **57**, 173–184.
31. Plotnikov, A. N., Schlessinger, J., Hubbard, S. R., and Mohammadi, M. (1999) Structural basis for FGF receptor dimerization and activation. *Cell* **98**, 641–650.

# 31

## Inhibition of Heparan Sulfate Synthesis by Chlorate

**H. Edward Conrad**

### 1. Introduction

With the exception of hyaluronic acid, all glycosaminoglycans are O-sulfated. Heparins and heparan sulfates are, in addition, N-sulfated on many of their glucosamine residues. Other naturally occurring structures are also sulfated. For example, some proteins contain sulfotyrosine residues *(1)*, and some complex lipids are O-sulfated. In addition, certain N-linked oligosaccharides are sulfated *(2)*. In all of these cases, the enzymes that add sulfate residues to these structures utilize 3'-phosphoadenosine-5'-phosphosulfate (PAPS) as the sulfate donor. If PAPS cannot be synthesized, these biological sulfation reactions will not proceed. Chlorate is a competitive inhibitor of PAPS synthesis and, as such, can block all sulfation reactions.

PAPS is synthesized by a two-step process utilizing ATP sulfurylase (reaction 1) and adenosine-5'-phosphosulfate kinase (reaction 2), activities that are in some cases catalyzed by a single bifunctional enzyme that may localize to the nucleus in mammalian cells *(3,4)*.

$$ATP + SO_4^{2-} \rightarrow APS + PP_i \qquad (1)$$

$$APS + ATP \rightarrow PAPS + ADP \qquad (2)$$

Chlorate competes with the $SO_4^{2-}$ in reaction 1. Consequently, as the concentration of $ClO_4^-$ in the cells is increased, the amount of PAPS formed is reduced. Under these circumstances, the sulfotransferase reactions, which require PAPS as a substrate, will proceed progressively more slowly as the concentration of PAPS drops. Furthermore, the sulfotransferases with the highest $K_m$ for PAPS will be selectively inhibited compared to the sulfotransferases with lower $K_m$. Such selectivity has been reported in the biosynthesis of heparan sulfate *(5)* wherein, with increasing $ClO_4^-$ concentrations, 6-O-sulfation of glucosamine residues is inhibited strongly, 2-O-sulfation of uronic acids is inhibited moderately, and N-sulfation of glucosamine residues is inhibited

From: *Methods in Molecular Biology, Vol. 171: Proteoglycan Protocols*
Edited by: R. V. Iozzo © Humana Press Inc., Totowa, NJ

modestly. In the heparan sulfate case, there is a requirement that N-sulfation proceed before the various O-sulfation reactions, so that the types of sulfation that occur may not be controlled totally by the $K_m$. Of course, at high $ClO_4^-$ concentrations, all sulfation reactions are inhibited.

Sulfation of glycosaminoglycans occurs in the trans Golgi; i.e., the sulfation is the final stage in the processing of glycosaminoglycans in the secretory pathway. Thus, when cells are labeled with $^{35}SO_4^{2-}$, $^{35}SO_4$-labeled glycosaminoglycans appear outside of the cell within minutes. Consequently, when cells are treated with $ClO_4^-$, changes in the structures of secreted glycosaminoglycans can be observed in a few minutes, and the full effect of the $ClO_4^-$ treatment is usually seen in about 30 min. Since the binding of many growth factors, enzymes, etc., requires sulfated heparan sulfate, $ClO_4^-$ treatment will result in the loss of protein binding to the cell surface. Thus, $ClO_4^-$ is a useful reagent for determining whether various proteins require sulfated glycosaminoglycans for cell binding and activity.

## 2. Materials

1. $NaClO_4$.
2. Cells in desired culture medium.

## 3. Methods

### 3.1. Defining the Concentration of Chlorate Required to Cause Total Inhibition of Sulfation (see Note 1)

Chlorate has been added to cultured cells at concentrations up to 50 m*M* for 24 h or more with no apparent adverse effects on the cells ***(5–10)***. The concentration of chlorate required to cause total inhibition of sulfation differs in different cell types. Because of variations in cell types and culture media compositions, the following general approach may used to determine the effects of varying $ClO_4^-$ concentrations on glycosaminoglycan sulfation.

1. Incubate cultured cells with increasing concentrations of $NaClO_4$ (e.g., 5–50 m*M*) for 1 h.
2. Then incubate the cells for an additional 1 h with a sufficient quantity of $^{35}SO_4^{2-}$ to produce ~10,000 cpm of cell surface $^{35}SO_4$–glycosaminoglycan. The amount of $^{35}SO_4^{2-}$ required will vary with the cell type and the type of culture medium used.
4. Remove the culture medium and wash the cells with fresh medium lacking $ClO_4^-$ and $^{35}SO_4^{2-}$.
5. Treat the cells with trypsin in phosphate-buffered saline to release the cells from the culture dish. Centrifuge out the cells and determine the number of $^{35}SO_4$–labeled glycosaminoglycan in the supernatant.
6. Repeat these steps for each concentration of $ClO_4^-$.
7. Choose the concentration of $ClO_4^-$ required to give the desired degree of inhibition of sulfate incorporation.

### 3.2. Determination of the Effect of Chlorate on the Activity of Heparan Sulfate-Binding Proteins

Inhibition of sulfate incorporation into heparan sulfate will prevent the binding of proteins to cell surface heparan sulfate proteoglycans (HSPG) and will block the normal changes in the activities of the cells that result from the protein binding. To study these changes, the approach used depends on the heparan sulfate-binding protein that is being studied, the cells that are used, and the activities of the heparan sulfate-binding protein on the cells. Thus, there is no standard protocol. However, several literature examples illustrate the general approach to be used.

1. It is known that cell surface heparan sulfate proteoglycan must interact with basic fibroblast growth factor in order for the growth factor to bind to its receptor and exhibit its mitogenic effects ***(11–13)***. In the presence of chlorate, basic fibroblast growth factor cannot bind or exert these effects on cells.
2. Chlorate treatment of rat hepatoma cells prevents the stimulation of β-very-low-density lipoprotein binding to rat hepatoma cells transfected with hepatic lipase cDNA ***(14)***.
3. In cells that produce lipoprotein lipase, newly synthesized lipoprotein lipase normally adheres to the surface of cells that produce it, but in chlorate-treated cells, there is a direct secretion of the enzyme into the culture medium ***(15,16)***.

## 4. Notes

1. Heparan sulfate structures that are synthesized by cells in the presence of chlorate are quite different from any structures that are normally synthesized. When the chlorate concentration is high enough, cells produce a completely unsulfated heparan proteoglycan. The polysaccharide chains that accumulate in the proteoglycan are [GlcA→GlcNAc]*n* chains in which a significant proportion of the GlcNAc residues are de-N-acetylated and ready to receive a sulfate group. Such products have been identified in chlorate-treated cells in culture. These products, which have anionic carboxyl groups on the GlcA residues, cationic amino groups on many of the GlcN residues, and very few IdoA residues, apparently appear on the cell surface but lack the ability to bind proteins. Thus, although chlorate "disarms" the cell so that it cannot respond to the growth factors or other heparan sulfate-binding proteins, it also results in the synthesis of an unnatural heparan sulfate structure that appears on the cell surface. Since heparan sulfate binds to many structures on and near the surface of cells (collagen, fibronectin, laminin, thrombospondin, etc.), it is a major contributor to the organization of the extracellular matrix. Consequently, in the presence of $ClO_4^-$, the structure of the cell surface and the extracellular matrix may be altered significantly. Furthermore, all secondary cellular responses that depend on the primary responses to altered heparan sulfate structure will be altered, even though one observes only those changes that are measured.

## References

1. Mintz, K. P., Fisher, L. W., Grzesik, W. J., Hascall, V. C., and Midura, R. J. (1994) Chlorate-induced inhibition of tyrosine sulfation on bone sialoprotein synthesized by a rat osteoblast-like cell line (UMR 106-01 BSP). *J. Biol. Chem.* **269,** 4845–4852.
2. Baenziger, J. U., Kumar, S., Bradbeck, R. M., Smith, P. L., and Beranek, M. C. (1992) Circulatory half-life but not interaction with the lutropin/chorionic gonadotropin receptor is modulated by sulfation of bovine lutropin oligosaccharides. *Proc. Natl. Acad. Sci. (USA)* **89,** 334–338.
3. Lyle, S., Stanczak, J., Ng, K., and Schwartz, N. B. (1994) Rat chondrosarcoma ATP sulfurylase and adenosine 5'-phosphosulfate kinase reside on a single bifunctional protein, *Biochemistry* **33,** 5920–5925.

4. Besset, S., Vincourt, J.-B., Amalric, F., and Girard, J. P. (2000) Nuclear localization of PAPS synthetase 1: a sulfate activation pathway in the nucleus of eukaryotic cells. *FASEB J.* **14,** 345–354.
5. Safaiyan, F., Lindahl, U., and Salmivirta, M. (1998) Selective reduction of 6-O-sulfation in heparan sulfate from transformed mammary epithelial cells. *Eur. J. Biochem.* **252,** 576–582.
6. Baeuerle, P. A. and Huttner, W. B. (1986) Chlorate—a potent inhibitor of protein sulfation in intact cells. *Biochem. Biophys. Res. Commun.* **141,** 870–877.
7. Greve, H., Cully, Z., Brumborg, P., and Kresse, H. (1988) Influence of chlorate on proteoglycan biosynthesis by cultured human fibroblasts. *J. Biol. Chem.* **263,** 128,865–128,892.
8. Humphries, D. E. and Silbert, J. E. (1988) Chlorate: a reversible inhibitor of proteoglycan sulfation. *Biochem. Biophys. Res. Commun.* **14,** 365–371.
9. Keller, K. M., Brauer, P. B., and Keller, J. M. (1989) Modulation of cell surface heparan sulfate structure by growth of cells in the presence of chlorate. *Biochemistry* **28,** 8100–8107.
10. Hoogewerf, A. J., Cisar, L. A., Evans, D. C., and Bensadoun, A. (1991) Effect of chlorate on the sulfation of lipoprotein lipase and heparan sulfate proteoglycans: sulfation of heparan sulfate proteoglycans affect lipoprotein lipase degradation. *J. Biol. Chem.* **266,** 16,564–16,571.
11. Yayon, A., Klagsbrun, M., Esko, J. D., Leder, P., and Ornitz, D. M. (1991) Cell surface, heparin-like molecules are required for binding of basic fibroblast growth factor to its high affinity receptor. *Cell* **64,** 841–848.
12. Ornitz, D. M., Yayon, A., Flanagan, J. G., Svahn, C. M., Levi, E., and Leder, P. (1992) Heparin is required for cell-free binding of basic fibroblast growth factor to a soluble receptor and for mitogenesis in whole cells. *Mol. Cell. Biol.* **12,** 240–247.
13. Ishihara, M., Tyrrell, D. J., Stauber, G. B., Brown, S., Cousens, L. S., and Stack, R. J. (1993) Preparation of affinity-fractionated, heparin-derived oligosaccharides and their effects on selected biological activities mediated by basic fibroblast growth factor. *J. Biol. Chem.* **268,** 4675–4683.
14. Ji, Z.-S., Lauer, S. J., Fazio, S., Bensadoun, A., Taylor, J. M., and Mahley, R. W. (1994) Enhanced binding and uptake of remnant lipoproteins by hepatic lipase-secreting hepatoma cells in culture. *J. Biol. Chem.* **269,** 13,429–13,436.
15. Berryman, D. E., and Bensadoun, A. (1995) Heparan sulfate proteoglycans are primarily responsible for the maintenance of enzyme activity, binding, and degradation of lipoprotein lipase in Chinese hamster ovary cells. *J. Biol. Chem.* **270,** 24,525–24,531.
16. Liu, G., Bengtsson-Olivecrona, G., and Olivecrona, T. (1993) Assembly of lipoprotein lipase in perfused guinea pig hearts. *Biochem. J.* **292,** 277–282.

# 32

# Detection of Proteoglycan Core Proteins with Glycosaminoglycan Lyases and Antibodies

**John R. Couchman and Pairath Tapanadechopone**

## 1. Introduction

Proteoglycans are quite abundant components of many extracellular matrices, while most cell surfaces also bear these macromolecules. Frequently the profiles are complex. For example, several members of the syndecan and glypican families of cell surface heparan sulfate proteoglycans may be present on a single cell type *(1,2)*. Some extracellular matrices, e.g., from brain, may also contain a variety of proteoglycans including several members of the hyalectans or aggregating proteoglycans such as neurocan, brevican, and versican *(3)*. Frequently it is useful to monitor the nature and variety of proteoglycans in a pool from tissues or cell cultures in a simple manner, before moving on to further purification steps, use of core protein-specific antibodies, or pursuit of a potentially novel core protein.

While proteoglycans require some specialized techniques for analysis, advantage can be taken of their glycanation to identify core proteins even when their precise characteristics remain unresolved. Specific enzymes are readily available, first from bacterial sources but more recently of recombinant origin, which selectively degrade glycosaminoglycans. Chondroitinase ABC will degrade virtually all chondroitin and dermatan sulfates, while leaving heparan and keratan sulfate chains intact. Conversely, heparitinase enzymes will degrade nearly all forms of heparan sulfate, but are unable to degrade chondroitin, dermatan, or keratan sulfate (*see* **Fig. 1**). Further, the consequences of glycosaminoglycan removal can be monitored by sodium dodecyl sulfate-polyacrylamide gel electrophoresis (SDS-PAGE). The heterogeneous nature of proteoglycans ensues largely from the variable number, length and charge of glycosaminoglycans in a pool of a single core protein (e.g., aggrecan from cartilage, or perlecan from a basement membrane preparation). Proteoglycans are frequently seen as broad smears, or sometimes, when large, may not even penetrate a 3% resolving gel (*see* **Fig. 2A**). Once glycosaminoglycan lyases have removed most of the chains, the core proteins become much more readily resolved by SDS-PAGE as discrete polypeptides (*see* **Fig. 2**).

From: *Methods in Molecular Biology, Vol. 171: Proteoglycan Protocols*
Edited by: R. V. Iozzo © Humana Press Inc., Totowa, NJ

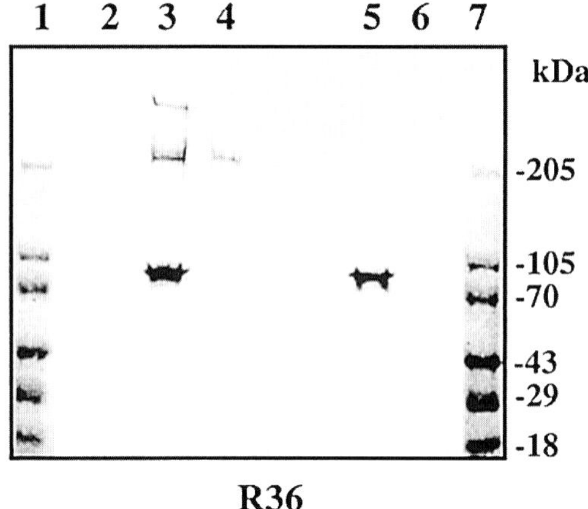


Fig. 1. Detection of chondroitin/dermatan sulfate proteoglycans core proteins from EHS tumor with R36, a polyclonal antibody recognizing chondroitin/dermatan sulfate stubs remaining after chondroitinase ABC treatment. Lanes 1 and 7 are standards whose molecular weight in kilodaltons is indicated. Lane 2, untreated sample; lane 3, sample treated with chondroitinase ABC and heparinase II and III; lane 4, sample treated with chondroitinase ABC only; lane 5 contains heparinase II and III only, while lane 6 contains chondroitinase ABC alone. The common polypeptide seen in lanes 3 and 5 is present in the heparinase II preparation. The data show that a CS/DS proteoglycan with a core protein of *Mr* ~ 200,000 (in lanes 3 and 4) is accompanied by a second, large core protein that is revealed only after additional heparinase treatment. This, therefore, represents a hybrid form of perlecan bearing both HS and CS/DS chains.

Chondroitinases and heparitinases are eliminases, so that the remaining core proteins have serine (usually) residues bearing not only the stem oligosaccharide (xylose-galactose-galactose-uronic acid) but also a disaccharide or larger oligosaccharide with a terminal unsaturated uronic acid residue. This, it turns out, is quite antigenic, and monoclonal ***(4,5)*** as well as polyclonal antibodies have been raised ***(6)*** which recognize the carbohydrate "stubs" remaining after chondroitinase or heparitinase treatments. They are also very specific. An antibody recognizing a heparan sulfate "stub" with a terminal unsaturated uronic acid residue will not recognize the equivalent "stub" generated by a chondroitinase enzyme, and vice versa. Therefore, the combined use of enzymes and antibodies can be used, for example in Western blotting, to estimate the sizes of core proteins, and the type of glycosaminoglycan present. This can be particularly useful where a particular core protein, e.g., perlecan, can be substituted with heparan and/or chondroitin sulfate chains (*see* **Fig. 1**). It can provide evidence of hybrid proteoglycans that bear more than one glycosaminoglycan type. Further, since the antibodies do not recognize core protein epitopes, a mixed population of heparan and/or chondroitin and dermatan sulfate proteoglycan can be quickly analyzed for their number, size, and glycanation profiles. Such evidence can be supported by more traditional metabolic labeling methods, combined with chemical or enzymatic degradation techniques followed by gel filtration analysis.

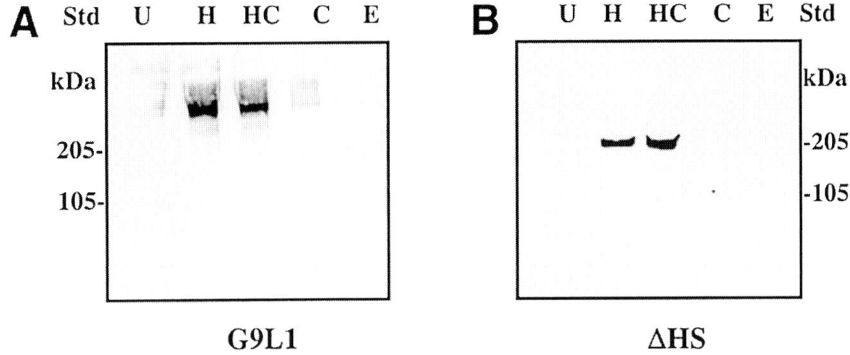


Fig. 2. Detection of intact murine perlecan **(A)** and recombinant domain IV-V of mouse perlecan transfected into COS-7 cells **(B)** with a rat monoclonal antibody specific to perlecan core protein **(A)** and monoclonal Δ-heparan sulfate antibody recognizing HS stub after heparinase III (heparitinase) treatments **(B)**. **(A)** shows that the intact perlecan core protein is not clearly visible until the HS chains have been removed, while (B) shows HS substitution on the recombinant perlecan. In each blot, the proteoglycans are untreated (U), heparitinase-pretreated (H), heparitinase- and chondroitinase-pretreated (HC), or chondroitinase-pretreated (C). Lanes E contain heparitinase and chondroitinase enzymes only.

## 2. Materials

1. Samples for analysis dissolved in suitable buffers (*see* **Note 1**).
2. Heparitinase buffer: 0.1 *M* sodium acetate, 0.1 m*M* calcium acetate, pH 7.0.
3. Chondroitinase buffer: 50 m*M* Tris-HCl, 30 m*M* sodium acetate, 20 m*M* ethylemediamintetraacetic acid (EDTA), 10 m*M* NEM, 0.2 m*M* phenylmethyl sulfonyl flouride (PMSF), and 0.02 sodium azide, pH 8.0 (*see* **Note 1**).
4. Protease inhibitor for all heparitinase treatments (*see* **Note 2**): 10× trypsin inhibitor type III-0 (ovomucoid 100 µg/mL, Sigma).
5. Glycosaminoglycan lyases (Seikagaku or Sigma):
   a. Heparan sulfate: heparinase III (EC 4.2.2.8). This enzyme is also known as heparitinase and heparitinase I.
   b. Chondroitin sulfate and dermatan sulfate: chondroitin ABC lyase (EC 4.2.2.4).
   c. Chondroitin sulfate: chondroitin AC II lyase (EC 4.2.2.5).
   d. Dermatan sulfate: chondroitin B lyase (no EC number).
6. Glycosaminoglycan carriers (optional): chondroitin sulfate type A (chondroitin 4-sulfate) or type C (chondroitin 6-sulfate), heparan sulfate (Sigma).
7. 2× SDS-PAGE sample buffer (Sigma), with or without reducing agent (e.g., 40 m*M* dithioerythreitol).
8. Prestained protein molecular-weight standards (Sigma or Bio-Rad).
9. SDS-PAGE gels. If a wide size range of core proteins is suspected, 3–15% gradient gels can be useful.
10. Electroblotting and transfer buffers and apparatus.
11. Transfer membrane: 0.45-µm nitrocellulose (Bio-Rad Trans-Blot® transfer medium or Schleicher & Schuell #BA85), PVDF (Millipore Immobilon P), or positively charged nylon (Bio-Rad Zetabind) membranes.

12. Blocking buffer: 5% nonfat dried milk in 0.1% Tween 20 is optional in phospate-buffered saline, PBS (TPBS).
13. Diluting buffer: 1% nonfat dried milk, 0.1% bovine serum albumin, and 0.1% (v/v) Tween 20 in PBS (for monoclonal antibodies) or triphosphate-buffered saline, TBS (for polyclonal antibodies).
14. Primary antibodies recognizing carbohydrate "stubs" (Seikagaku):
    a. Monoclonal Δ-heparan sulfate (for heparan sulfate GAG).
    b. Monoclonal anti proteoglycan Δ-di-0S, -4S, -6S (for chondroitin/ dermatan sulfate GAGs; *see* **Note 3**).
    These antibodies are also available as biotin conjugates.
    c. Equivalent antibodies recognizing protein of interest.
15. Secondary antibodies: horseradish peroxidase- or alkaline phosphatase-anti-Ig conjugate. Alternately, streptavidin-horseradish peroxidase conjugate should be used where the primary antibodies are biotin conjugates.
16. Chromogenic and chemiluminescence visualization system, e.g., ECL™ Western blotting detection reagents (Amersham Pharmacia Biotech) for peroxidase conjugates or alkaline phosphatase-conjugate substrate kit (Bio-Rad).

## 3. Methods

1. Divide the proteoglycan sample to be analyzed into equal aliquots. The number depends on the enzyme treatments to be performed. For example, if a proteoglycan pool is suspected to contain chondroitin and heparan sulfate proteoglycans, four aliquots should be used. One sample is left untreated, while others receive chondroitinase ABC, heparitinase, or both enzyme treatments. Ideally, each sample should contain 1–10 μg of proteoglycan. The choice of buffer depends on the enzymes to be used (*see* **Note 1**). Further controls contain buffer with enzyme only (no proteoglycan).
2. Treat samples with appropriate enzymes at 37°C. The amount and duration of enzyme treatment depend on the proteoglycan concentration. For 1–10 μg of proteoglycan, 2–3 h of incubation with 0.5–1 mU chondroitinase ABC or 1–2 mU heparitinase in the presence of protease inhibitor (1× ovomucoid, *see* **Note 2**) is suggested. Where concentrations of proteoglycan are higher, adding a second aliquot of enzyme after 2 h, for a further incubation can be beneficial. Where proteoglycan concentrations are very low (below 100–200 ng per sample), adding approximately 0.5 μg of appropriate free glycosaminoglycan can be added as carrier to aid efficiency and recovery (but *see* **Note 4**).
3. If enzyme activity needs to be verified, set up samples of free glycosaminoglycan (approx 0.5 mg/mL) in buffer, to which the enzymes are added, and incubate simultaneously. Enzyme activity is monitored spectrophotometrically at 232 nm.
4. Enzyme treatments are terminated by adding SDS-PAGE sample buffer (with or without reducing agent, **Subheading 2.**) and heating to 100°C, if required. Samples can be frozen at –20°C or immediately resolved by SDS-PAGE.
5. Samples are applied to SDS-PAGE gels for conventional electrophoresis and transfer to nitrocellulose or other medium (*see* **Note 5**). If a range of core protein masses is suspected, or not known, acrylamide gradients are preferable (e.g., 3–15%).
6. Membranes are blocked conventionally, for example in 5% dried milk powder in phosphate-buffered saline for at least 1 h. They are then probed with monoclonal or polyclonal antibodies recognizing carbohydrate "stubs" created by glycosaminoglycan lyases. These are available as purified IgG, and sometimes in biotinylated form, and should be used at 10–25 μg/mL. Incubation can be at 4°C overnight, or shorter periods at room temperature or 37°C, but for at least 1 h. Constant gentle agitation is advised.

7. Thorough washing is followed by secondary antibody (e.g., affinity purified goat anti-mouse IgG conjugated to horseradish peroxidase) in the same buffer for 1 h at room temperature. Antibody concentrations should accord with manufacturer's instructions. Extensive washes are then followed by visualization as preferred, such as chemiluminescence.

## 4. Notes

1. A suitable buffer for chondroitinase ABC or AC II is listed under **Subheading 2.**, as is one suitable for heparitinase (also known as heparinase III). However, where samples are to be treated with both enzymes, we have found the heparitinase buffer to be suitable. It should be noted that chondroitinase B is inhibited by phosphate. Heparinase activity is increased by the presence of calcium ions, but it is reported that the activity of heparinase III is not much decreased by its absence.
2. Polysaccharide lyases are primarily of microbial origin. Protease contamination can be present in the enzyme preparation, especially all heparinase enzymes. This can cause misleading results. Thus, protease inhibitor should be added in case of heparitinase treatments. Chondroitinase ABC is available in protease-free form.
3. Separate, specific antibodies are available that, while all recognizing the terminal unsaturated uronic acid residue, as described, have specificity for the presence and position of sulfate on the adjacent galactosamine residue. The three antibodies can be used combined. At the current time they are only available separately. Most commonly, the prevalence of sulfation is 4S > 6S > 0S.
4. We have found that the use of chondroitinase ABC and heparinase III together leads to a less efficient identification of chondroitin sulfate proteoglycan core proteins than the use of the former enzyme alone. The reasons are not clear, but it may be that products of heparan sulfate lyases are slightly inhibitory to chondroitinase enzymes. It is known that heparin will inhibit chondroitinases, and should therefore not be used as a carrier.
5. Intact proteoglycans transfer poorly to nitrocellulose or similar membrane. The more glycosaminoglycan present on a core protein, the more difficult it becomes. This is a result of high mass as well as charge. Therefore, while decorin with one chain can be quite efficiently transferred, aggrecan with >100 chains may not. Transfer to cationic membranes can enhance proteoglycan capture.

## References

1. Bernfield, M., Götte, M., Park, P. W., Reizes, O., Fitzgerald, M. L., Lincecum, J., and Zako, M. (1999) Functions of cell surface heparan sulfate proteoglycan. *Annu. Rev. Biochem.* **68**, 729–777.
2. David, G., Bai, X. M., Van der Schueren, B., Cassiman, J-J, and Van der Berghe, H. (1992) Developmental changes in heparan sulfate expression: in situ detection with mAbs. *J. Cell Biol.* **119**, 961–975.
3. Iozzo, R. V. (1998) Matrix proteoglycans: from molecular design to cellular function. *Annu. Rev. Biochem.* **67**, 609–652.
4. Couchman, J. R., Caterson, B., Christner, J. E., and Baker, J. R. (1984) Mapping by monoclonal antibody detection of glycosaminoglycans in connective tissues. *Nature* **307**, 650–652.
5. Tapanadechopone, P., Hassell, J. R., Rigatti, B., and Couchman, J. R. (1999) Localization of glycosaminoglycan substitution sites on domain V of mouse perlecan. *Biochem. Biophys. Res. Commun.* **265**, 680–690.
6. Couchman, J. R., Kapoor, R., Sthanam, M., and Wu, R. R. (1996) Perlecan and basement membrane-chondroitin sulfate proteoglycan (bamacan) are two basement membrane chondroitin/dermatan sulfate proteoglycans in the Engelbreth-Holm-Swarm tumor matrix. *J. Biol. Chem.* **271**, 9595–9602.

# 33

# Proteoglycan Core Proteins and Catabolic Fragments Present in Tissues and Fluids

**John D. Sandy**

## 1. Introduction

The full- or partial-length c-DNA and deduced core protein sequence is now available for at least 38 distinct proteoglycans [for review, *see* (*1*)]. Many of these can be placed into several large family groupings, such as the 10 members of the small leucine-rich repeat proteoglycans (decorin, biglycan, fibromodulin, lumican, keratocan, PRELP, epiphycan, mimecan, oculoglycan and osteoadherin), the glypicans (GPC-1, cerebroglycan, OCI-5, K-glypican, GPC-5, and GPC-6), the hyaluronan-binding proteoglycans (aggrecan, brevican, neurocan, and versican), the syndecans (SYN-1, fibroglycan, neuroglycan/N-syndecan, and amphiglycan/ryudocan), and the glycosaminoglycan-substituted collagens (types IX, XII, and XVIII). In addition, there is a group of apparently unrelated species, which includes agrin, perlecan, leprecan, bamacan, betaglycan, serglycin, phosphacan, NG2, CD44/epican, and testican.

Since most extracellular matrix proteins, and presumably also proteoglycans, undergo some degree of proteolytic modification during biosynthesis or catabolism, each of the core species described above probably exists in vivo in both the intact form and in one or more fragmented forms. Proteolysis may indeed be necessary for the conversion of the intact proteoglycan to a product, or a series of products, that can serve one or more functions in the cell or tissue where it is located. Proteolysis will almost certainly be involved in the removal of proteoglycans from cells or tissues, whether this be part of the normal turnover process or in pathological states where degradative pathways may be altered or accelerated. Despite the wealth of knowledge on core structures, tissue distribution, and function of proteoglycans, there is still very limited information on the role of proteolysis in the biology of these molecules. On the other hand, there are now a number of examples of what appear to be physiologically important proteolytic processing events for proteoglycans, and for aggrecan (*2*) and

From: *Methods in Molecular Biology, Vol. 171: Proteoglycan Protocols*
Edited by: R. V. Iozzo © Humana Press Inc., Totowa, NJ

brevican *(3)* the precise cleavage sites and the family of proteinases responsible (ADAMTS) appears to have been established.

A 17-kDa N-terminal fragment of decorin accumulates in human skin with aging *(4,5)* and a 20-kDa N-terminal biglycan fragment is generated by bFGF treatment of bovine aortic endothelial cells *(6)*, but the details of cleavage of these proteoglycans are unknown in both cases. The ectodomain of syndecan-1 is shed from cell surfaces into wound fluids by proteolysis in vivo *(7,8)*, and similar cell -associated proteolysis appears to generate fragments of betaglycan *(9)*, NG2 *(10)*, testican *(11)*, and perlecan *(12)*, but the structural details have not been reported either. The most detailed analyses of proteoglycan core protein degradation in vivo have been done with the family of hyaluronan-binding proteoglycans, aggrecan, brevican, neurocan, and versican. At least two much-studied HA-binding proteins, hyaluronectin *(13)* and glial hyaluronate-binding protein (GHAP) *(14)* appear to represent the N-terminal globular domain of versican, although the versican isoform involved, the precise cleavage sites and the protease(s) responsible for the cleavage(s) are unclear. Neurocan has been isolated from rat brain in two fragments *(15)* which are generated by cleavage by an unknown proteinase near the middle of the core protein (apparently at the methionine 638–leucine 639 bond), and immunostaining with monoclonal antibodies that recognize only the N-terminal fragment (1F6) or the C-terminal fragment (1D1) suggests that the two fragments have very different functions *(16)*.

Studies on brevican/BEHAB *(17)* and particularly aggrecan *(18–20)* have provided the majority of the molecular details on the pathways and enzymes involved in vivo in the proteolysis of proteoglycans. Brevican is present in the brain as the full-length protein (about 145 kDa) and a proteolytic fragment (about 80 kDa), which represents the C-terminal portion. The sequence at the cleavage responsible for this fragment (glutamate 400– serine 401) shows a striking similarity to the five cleavage sites that have been identified in aggrecan degradation in cartilage. The enzyme(s) responsible (aggrecanases) have now been identified as members of the ADAMTS family of metalloproteinases. The detailed studies on cartilage aggrecan degradation leading to the discovery of aggrecanase activity *(18)* and their cloning as ADAMTS-4 *(2)* and ADAMTS-11 (alias ADAMTS-5) *(20)* have provided some proven methodological approaches to the study of proteoglycan fragmentation, and these methods would appear to have general applicability to other proteoglycans. Indeed, the protocols to be described here have largely been developed *(18,21–23)* for analysis of small quantities of aggrecan and fragments (5- to 250-mg glycosaminoglycan [GAG]) such as those present in small tissue biopsies, biological fluids, and cell or tissue explant culture medium.

## 2. Materials

### 2.1. Proteoglycan and Core Protein Preparation from Fluids, Culture Medium, and Collagen-Rich Tissues

1. *Streptomyces* hyaluronidase (*Str. hyalurolyticus*), chondroitinase ABC (protease-free, *Proteus vulgaris*), Endobetagalactosidase (*Escherichia freundii*), and keratanase II(*Bacillus sp.*) (Seikagaku).
2. DE 52 cellulose, preswollen microgranular (Whatman).

3. Polyprep chromatography columns (Bio-Rad).
4. Dialysis of small volumes in Slide-a-Lyzer minidialysis units, 10,000 MWCO (Pierce) and large volumes (greater than 200 µL) in Specta/Por dialysis tubing, 12,000–14,000 MWCO (Fisher).

### 2.2. Proteoglycan and Core Protein Preparation from Brain and Spinal Chord

1. 10-mL glass homogenizer with Teflon plunger (Thomas Scientific).

### 2.3. Preparation of Tissue Proteoglycans with No Glycosaminoglycan Substitution

1. Eppendorf microfuge (model 5413 or 5415C).

### 2.4. N-Terminal Sequencing of Proteoglycan Core Proteins

1. Protein assayed by bicinchoninic acid assay kit (Pierce).
2. Pharmacia FPLC Superose 12 (HR-10/30) and a fast desalting column (HR-10/10).
3. PVDF membranes (Hybond PVDF from Amersham or Immobilon from Millipore).

### 2.5. Western Blot Analysis and Core Identification

1. Sample preparation in 0.6-mL Snap-cap microcentrifuge tubes (Continental Lab Products).
2. 2× Tris-glycine sodium dodecyl sulfate (SDS) sample buffer as a premixed buffer (Novex).
3. Gel-loading pipet tips (200 µL) (Labsource, Chicago, IL).
4. Mini-Protean II gel assembly (Bio-Rad) and E19001-XCELL II Mini Cell (Novex).
5. Trans-Blot transfer medium (roll of pure nitrocellulose membrane) (0.45 µ, Bio-Rad).
6. Dry nonfat milk, blotting-grade blocker (Bio-Rad), and polyvinyl chloride laboratory wrap (Fisher).
7. Primary antibodies to most proteoglycans are described in the literature, and aggrecan antibodies, including those which detect neoepitopes on specific cleavage products, have recently been reviewed ***(24)***. Secondary antibodies for rabbit primary antibodies are HRP-conjugated goat anti-rabbit IgG (Chemicon), and for mouse monoclonals are generally HRP-conjugated goat anti-mouse IgG (Sigma), but HRP-conjugated goat anti-mouse IgM (Sigma) for antibody 3-B-3.
8. Chemiluminescent substrates, ECL (Amersham) or SuperSignal West Pico (Pierce), and high-performance chemiluminescence film, Hyperfilm ECL (Amersham).

### 2.6. Image Capture and Quantitation

1. ScanJet 3c/T, with DeskScan II software for PC or Mac (Hewlett Packard.) NIH Image (obtainable online from wayne@helix.nih.gov(for Mac) or from Scioncorp.com (for Windows). Adobe Photoshop for presentation.

## 3. Method

### 3.1. Proteoglycan and Core Protein Preparation from Fluids and Culture Medium (Note 1)

1. Typically, biological fluids (synovial fluid) or medium from cell or tissue culture, which contains at least 5 µg of proteoglycan core protein, will be required. A desirable starting amount for aggrecan analyses is 250 µg of GAG (as chondroitin sulfate [CS] and keratan sulfate [KS]) or about 25µg of core protein.

2. If the fluid is viscous due to a high content of hyaluronan (such as is found in synovial fluid or fibroblast-conditioned medium), it should be predigested as follows: Add 1/5 vol of 0.5 *M* ammonium acetate, 20 m*M* ethylenediaminetetraacetic acid (EDTA), 0.5 m*M* 4-(2-aminoethyl)benzenesulfonyl flouride (AEBSF), pH 6.0, and 2 TRU of streptomyces hyaluronidase and incubate at 60°C for 3 h.
3. The digest is dried to remove the ammonium acetate, and may be deglycosylated directly for Western analysis (**Subheading 3.6.**) as in **steps 5 and 6** to follow. If the sample contains a high concentration of nonproteoglycan proteins (such as for serum-containing medium), the dried sample is taken up in about 2 mL of 7 *M* urea, 50 m*M* Tris-acetate, pH 8.0, and applied to a 0.5-mL-bed-volume column of DE 52 cellulose in Poly-Prep chromatography columns. The DE 52 cellulose is prepared by washing 3 times in distilled water and stored as a 50% slurry in water at 4°C. The 0.5-mL-bed-volume column is equilibrated in at least 2 mL of 7 *M* urea, 50 m*M* Tris-acetate, pH 8.0, before the sample is loaded. After sample loading and collection of unbound material (flow-through), the column is eluted as follows: (a) 2 mL of 7 *M* urea, 50 m*M* Tris-acetate, pH 8.0, which is added to the flow-through (b) 8 mL of 0.1 *M* NaCl, 7 *M* urea, 50 m*M* Tris-acetate, pH 8.0, (c) 4 mL of of 0.2 *M* NaCl, 7 *M* urea, 50 m*M* Tris-acetate, pH 8.0, (d) 1.5 mL of 0.8 *M* NaCl, 7 *M* urea, 50 m*M* Tris-acetate, pH 8.0, and (e) 1.5 mL of 1.5 *M* NaCl, 7 *M* urea, 50 m*M* Tris-acetate, pH 8.0.
4. All fractions (b, c, d, e) are dialyzed exhaustively against distilled water (at least 24 h at 4°C in 4 L of water with multiple changes) and the retentates are dried.
5. If the samples contain chondroitin sulfate/dermatan sulfate (CS/DS), the dried retentates are dissolved in 50 m*M* sodium acetate, 50 m*M* Tris, 10 m*M* EDTA,pH 7.6, and 25 mU (per 100 µg GAG) chondroitinase ABC (protease-free) is added, followed by incubation at 37°C for 1–2 h.
6. If the samples contain KS alone or in addition to CS/DS, the above digestion condition is adjusted to pH 6.0 with acetic acid and 0.5 mU (per 100 µg GAG) of endo-betagalactosidase and 0.5 mU (per 100 µg GAG) of Keratanase II are added and incubated at 37°C for 2–4 h.

### 3.2. Proteoglycan and Core Protein Preparation from Collagen-Rich Tissues

1. Typically, a tissue sample containing at least 5 µg of proteoglycan core protein will be required.
2. The tissue (cartilage, tendon, ligament, meniscus, aorta, etc.) is rinsed in cold PBS, chopped finely, and extracted by rocking at 4°C (15 mL of extractant per gram wet weight) for 48 h in 4 *M* guanidine-HCl, 10 m*M* MES, 50 m*M* sodium acetate, 5 m*M* EDTA, 0.1 m*M* AEBSF, 5 m*M* iodoacetic acid, 0.3 *M* aminohexanoic acid, 15 m*M* benzamidine,1µg/mL pepstatin, pH 6.8.
3. A clear extract is obtained after centrifugation to pellet the tissue and the extract is dialyzed exhaustively against distilled water (at least 24 h at 4°C in 4 L of water with multiple changes), and the retentate is either dried and deglycosylated for Western analysis (**Subheading 3.6.**) as in **steps 5** and **6** under **Subheading 3.1.**, or, if purification is required, it is adjusted to 7 *M* urea, 50 m*M* Tris-acetate, pH 8.0, and processed for analysis as for proteoglycans in **steps 3–6** under **Subheading 3.1.**

### 3.3. Proteoglycan and Core Protein Preparation from Brain and Spinal Chord

1. The tissue (typically 50–300 mg wet weight of brain or spinal chord) is finely sliced and added to ice-cold 0.3 *M* sucrose, 4 m*M* HEPES, 0.15 *M* NaCl, 5 m*M* EDTA, 0.1 m*M*

AEBSF, 5 m*M* iodoacetic acid, 0.3 *M* aminohexanoic acid, 15 m*M* benzamidine,1 µg/mL pepstatin, pH 6.8 (about 9 mL of extractant per gram wet weight of tissue).

2. After a brief (3 × 1 min), cold homogenization, the sample is clarified by centrifugation at maximum speed in an Eppendorf microfuge for 30 min at 4°C and the supernatant is applied to a 0.5-mL-bed-volume column of DE 52 cellulose in Poly-Prep chromatography columns which is equilibrated in 50 m*M* Tris-HCl, 0.15 *M* NaCl, 0.1% Triton X-100, pH 8.0. After sample loading and collection of unbound material (flow-through) the column is eluted as follows: (a) 4 mL of 50 m*M* Tris-HCl, 0.15 *M* NaCl, 0.1% Triton X-100, pH 8.0, which is added to the flow-through,(b) 4 mL of 50 m*M* Tris-HCl, 6 *M* urea, 0.25 *M* NaCl, 0.1% Triton X-100, pH 8.0, and (c) 50 m*M* Tris-HCl, 1.5 *M* NaCl, 0.5% CHAPS, pH 8.0.
3. The flow-through pool, the 0.25 *M* NaCl wash and the 1.5 *M* NaCl wash are each prepared for Western analysis (**Subheading 3.6.**) as in **steps 4–6** under **Subheading 3.1.**

### 3.4. Preparation of Tissue Proteoglycans with No Glycosaminoglycan Substitution (Notes 2–4)

1. Some proteoglycans, and/or fragments, such as the "free" G1 domains of the different hyaluronan-binding proteoglycans, are present in tissues without GAG substitution. Solubilization of these proteins for SDS-PAGE and Western analysis may be achieved by direct extraction in detergent-containing buffer.
2. Tissue (cartilage, tendon, ligament, meniscus, spinal chord, aorta, sclera, cornea, brain, etc.) is washed in cold PBS, sliced finely, and suspended (about 6 mL of extractant per gram wet weight tissue) by mild agitation in 1% Triton X-100, 50 m*M* Tris-HCl, 150 m*M* NaCl, 5 m*M* CaCl2, 0.05%(v/v) BRIJ-35, 0.025% NaAzide, 5 m*M* EDTA, 0.1 m*M* AEBSF, 5 m*M* iodoacetic acid, 15 m*M* benzamidine, 1 µg/mL pepstatin, pH 6.8. After 30 min at room temperature the tissue is removed by centrifugation (microfuge at maximum speed) at 4°C and a portion of the extract taken directly for SDS-PAGE and Western blot. Since deglycosylation is not used, only proteoglycans without GAG substitution will be detected.

### 3.5. N-Terminal Sequencing of Proteoglycan Core Proteins (Note 5)

1. Deglycosylated core proteins from **step 6** under **Subheading 3.1.** can be prepared for N-terminal sequencing as follows: A portion of the deglycosylated product (containing at least 15 µg of proteoglycan core protein) is fractionated on Superose 12, eluted in 0.5 *M* guanidine·HCl at 0.5 mL/min/fraction. The eluant is monitored at 214 nm and the high molecular-weight-pool (fractions 4–8) and low-molecular-weight pool (fractions 9–13) are concentrated to 0.5 mL and desalted on a fast-desalting column run in water at 1 mL/min and monitored at 214 nm. The desalted protein is dried and taken directly for N-terminal analysis.
2. If multiple N-terminal sequences are obtained, the proteins can be further separated on SDS-PAGE, electroblotted to PVDF membrane, stained with Coomassie blue, and individual stained bands cut out with a scalpel for direct N-terminal analysis.

### 3.6. Western Blot Analysis and Core Identification (Figs 1–3; Note 6)

1. Dry samples (0.1–5 µg of protein) are dissolved in 20–40 µL of gel sample buffer (prepared fresh with 12 mg of dithiothreitol, 200 µL of 6 *M* urea, and 200 µL of 2× Tris-glycine SDS sample buffer, pH 6.8) heated in a heating block at 100°C for 5–10 min, followed by centrifugation to spin down the liquid.
2. Samples (up to 40 µL) are applied with gel-loading pipet tips to Novex precast gels (4–12% or 4–20% Tris-glycine gels, 1.0 mm × 10 wells) and run in electrode buffer (50 m*M* Tris base, 384 m*M* glycine, 0.2% SDS, pH 8.8) at 200 V for about 45 min at 4°C.

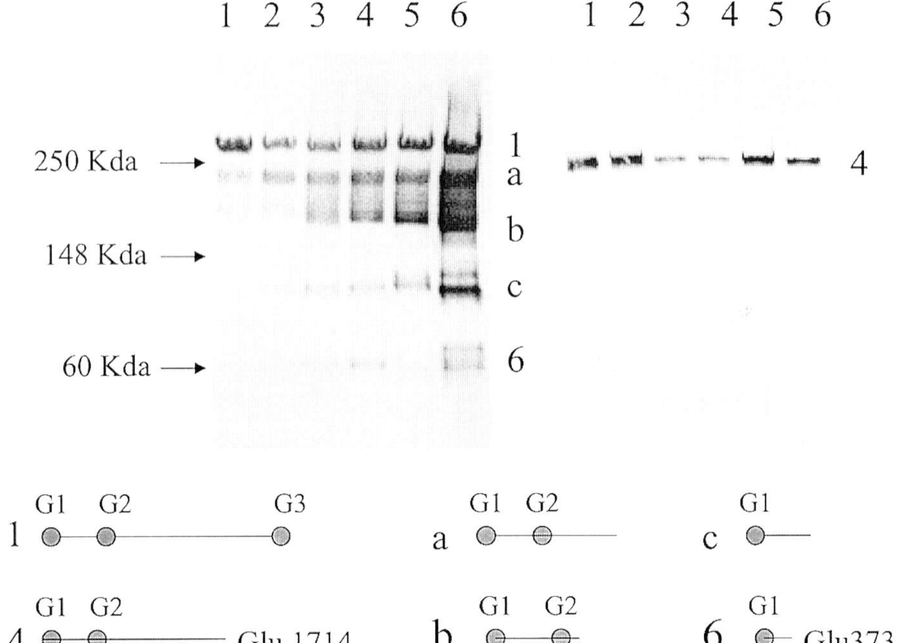


Fig. 1. Western analysis of aggrecan species isolated by guanidine extraction from normal human articular cartilages of different ages. Lanes 1–6 in both panels show analysis of human aggrecan from fetal, 2 mo, 1-y, 5-y, 15-y and 68-yr samples.The left panel is probed with a general aggrecan antiserum (anti-G1-2) raised against bovine aggrecan G1 domain.The right panel shows the same samples probed with the neo-epitope antiserum (anti-KEEE) to the ADAMTS-generated C-terminal neoepitope at Glu1714. The structure of the six peptides (1, a–c, 4, and 6) is shown diagrammatically, although the C-terminals of some major human aggrecan fragments (peptides a, b, and c) have yet to be identified. The figure clearly illustrates the well-established age-dependent C-terminal truncation of aggrecan in articular cartilage. In addition peptide 4, which migrates between peptide 1 and peptide a, is of very low relative abundance and is not detected readily with the anti-G1 antiserum. This illustrates the extreme sensitivity of the anti-KEEE (anti-peptide) antiserum relative to the anti-G1 antiserum for aggrecan. It should be noted that, in general, aggrecan peptides migrate with molecular sizes that are about twice the size predicted from the peptide size.

3. Proteins are transferred onto nitrocellulose membrane in transfer buffer (25 m*M* Tris base, 192 m*M* glycine, 20% (v/v) methanol, pH 8.4) at 100 V for 1 h at 4°C.
4. The membrane is blocked by incubation for at least 10 min at room temperature in 200 m*M* Tris-HCl, 1.37 *M* NaCl, 0.1% (v/v) Tween-20, 1% (w/v) dry nonfat milk, pH 7.6, and then incubated for between 1 h and 16 h at 4°C in 200 m*M* Tris-HCl, 1.37 *M* NaCl , 0.1% (v/v) Tween-20, 5% (w/v) dry nonfat milk, pH 7.6, containing the primary antibody at a dilution that is generally about 1/3000 .
5. The membrane is washed (3 × 2 min) in 200 m*M* Tris-HCl, 1.37 *M* NaCl, 0.1% (v/v) Tween-20, pH 7.6, and then incubated for about 1 h at room temperature in 200 m*M*

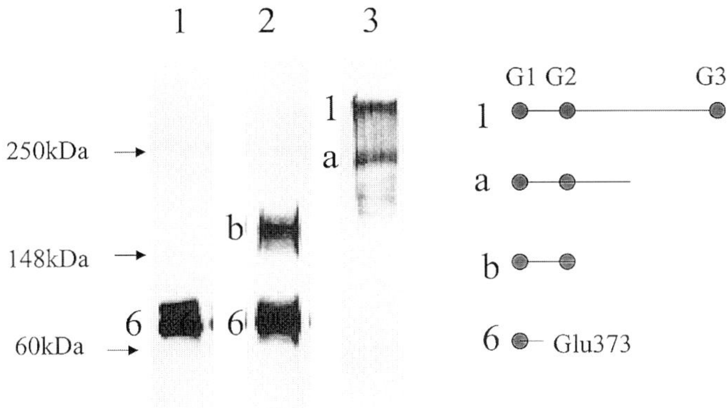


Fig. 2. Western analysis of aggrecan species present in immature pig proximal femoral epiphyseal cartilage and separated by DE 52 chromatography. A guanidine extract of pig cartilage was processed by DE 52 chromatography as detailed in **steps 3–6** under **Subheading 3.1.** The samples shown in lanes 1, 2, and 3 were recovered from the 0.1 *M* NaCl, 0.2 *M* NaCl, and 0.8 *M* NaCl eluants, respectively, and the blot was probed with the general aggrecan antiserum, anti-G1-2. The structure of the four peptides (1, a, b, and 6) is shown diagrammatically, although the precise C-terminal sequence for peptides a and b have not been identified. Peptides b and 6 elute from DE 52 with low-salt buffer because they are not substituted with CS.

Tris-HCl, 1.37 *M* NaCl, 0.1% (v/v) Tween-20, 5% (w/v) dry nonfat milk (Bio-Rad), pH 7.6, containing the secondary antibody at about 1/3000.

6. The membrane is washed (3 × 10 min) in 200 m*M* Tris-HCl, 1.37 *M* NaCl, 0.1% (v/v) Tween-20, pH 7.6, and developed with Amersham or Pierce chemiluminescence reagents as follows: mix 5 mL each of distilled water, reagent 1, and reagent 2 from the ECL kit. Submerge the membrane in mix for about 1 min, protein side up. Remove the membrane with tweezers and allow excess liquid to drip off. Wrap the membrane in clear laboratory wrap and expose on Hyperfilm ECL.
7. Multiple exposures (for example, 5 s,30 s,1 min, 5 min) should be developed to optimize signal intensity and separation for major and minor species.

### 3.7. Image Capture and Quantitation (Note 7)

1. The film images are captured on an HP ScanJet 3c/T with DeskScan II software (set to Black and White Photo) and opened for quantitation in NIH Image and/or presentation in Adobe Photoshop.
2. For quantitation, individual bands are selected and integrated pixel density values obtained with preset value of pixel aspect ratio (generally about 750). Standardization with known loadings of core protein or fragment (where available) can be used to establish the linear detection range for the assay *(23)*.

## 4. Notes

1. Treatment with *Streptomyces* hyaluronidase should totally eliminate viscosity due to hyaluronan before application of samples to DE 52, and the concentration of GAG in the applied samples should not exceed 100 μg/mL. The DE52 flow-through will contain pro-

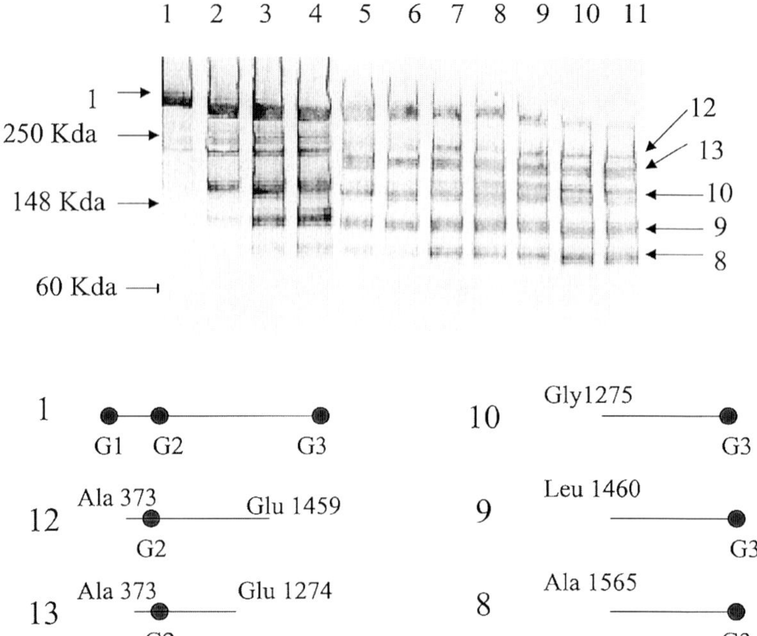


Fig. 3. Western analysis of aggrecan species generated in rat chondrosarcoma cell cultures treated with IL-1b. Cultures of rat chondrosarcoma cells (each containing about 30 µg of GAG) were treated with IL-1b and at intervals (0, 24, 42, 48, 66, 72, 93, 98, 110, 114, and 138 h) the total culture was terminated by addition of buffer and digestion with chondroitinase ABC (**step 5** under **Subheading 3.1.**) After removal of cells by centrifugation, the supernatants (1–11, respectively) were taken directly for Western analysis with monoclonal Ab 2-B-6 (reactive with chondroitin-4-sulfate stubs left on the core protein after Chase ABC digestion). The figure illustrates the time-dependent change in composition of immunoreactive fragments. The structure of six of the peptides is shown diagrammatically. In each case the N-terminal and C-terminal of the individual species were established either by N-terminal sequencing of bands or immunoreactivity with polyclonal antisera to G1 and G3 and neo-epitope antisera to the terminals at Ala 374 (anti-ARGSV), Glu1459 (anti-KEEE), and Glu1274 (anti-SELE).

tein that is not GAG-substituted whereas the washes with 0.1, 0.2, 0.8, and 1.5 *M* NaCl will contain proteoglycans of different compositions. The bed volume of the DE 52 can be increased for isolation of larger amounts (capacity is about 500 µg of GAG-substituted proteoglycan per milliliter of DE 52) but the concentration loaded should not exceed 100 µg/mL. Also, the volume of wash solutions should be increased proportionately for larger amounts.

2. An alternative approach to isolation of proteoglycans from all tissues described above is extraction in guanidine-HCl as under **Subheading 3.2.**, followed by ethanol precipitation as follows: To a portion of the guanidine extract add 3 vol of ice-cold ethanol (sodium acetate saturated) and let stand at –20°C for 16 h. Centrifuge at 4°C in the Eppendorf microfuge at maximum speed, remove, and discard the ethanol. Dry the pellet and sus-

pend it in chondroitinase buffer at 37°C and proceed as from **step 5** under **Subheading 3.1.** This protocol isolates both protein and proteoglycan and is most successful with tissues, such as cartilage, that do not contain abundant guanidine-extractable nonproteoglycan proteins.

3. High-yield purification of proteoglycans and fragments substituted only with KS, such as fibromodulin and lumican, will require the addition of 0.5% CHAPS as detergent. Fibromodulin can be purified on MonoQ anion exchanger in buffers containing 6 *M* urea and 0.5% CHAPS *(25)* and lumican from corneas on Q-Sepharose in buffers containing 8 *M* urea and 0.5% CHAPS *(26)*.
4. Reference *(27)* gives more detail on the isolation of specific proteoglycans and fragments from nervous tissues.
5. Clear N-terminal sequencing requires 10–100 pmol of protein, which is about 1–10 μg for most proteoglycan core species. This is similar in sensitivity to Coomassie staining of proteins and therefore, if the species of interest can be identified by Coomassie staining, it should be possible to obtain an N-terminal sequence.
6. Optimal separation and detection of proteoglycan core proteins by Western analysis requires strict control of the completeness of the deglycosylation steps (**steps 5** and **6** under **Subheading 3.1.**). This can be monitored by measuring the loss of reactivity in the dimethylmethylene blue assay *(28)*. For all CS/DS- and KS-substituted proteoglycans, greater than 85% loss of the DMMB reactivity should be achieved on deglycosylation before Western analysis is attempted.
7. Since the chemiluminescence signal obtained with each species and antibody is highly dependent on epitope presentation on the nitrocellulose and the reactivity of the antibody in use, each species requires independent standardization. Because of the unavailability of standard preparations of many proteoglycan core proteins and fragments, quantitation by Western analysis is not possible in most cases. Other methods, such as radioimmunoassays and N-terminal quantitation by chemical means, are needed for stricter quantitation of these products.

## References


1. Iozzo R. V. (1998) Matrix proteoglycans: from molecular design to cellular function. *Annu. Rev. Biochem.* **67**, 609–652.
2. Tortorella, M. D., Burn, T. C., Pratta, M. A., Abbaszade, I., Hollis, J. M., Liu R., et al. (1999). Purification and cloning of aggrecanase-1: a member of the ADAMTS family of proteins. *Science* **284**, 1664–1666.
3. Yamaguchi, Y. (1996) Brevican: a major proteoglycan in adult brain *Perspect. Dev. Neurobiol.* **3**, 307–317.
4. Carrino, D. A., Sorrell, J. M., and Caplan, A. I. (2000) Age-related changes in the proteoglycans of human skin. *Arch. Biochem. Biophys.* **373**, 91–92.
5. Sorrell, J. M., Carrino, D. A., Baber, M. A., and Caplan, A. I. (1999) Versican in human fetal skin development. *Anat Embryol (Berl.)* **199**, 45–56.
6. Kinsella, M. G., Tsoi, C. K., Jarvelainen, H. T., and Wight T. N. (1997) Selective expression and processing of biglycan during migration of bovine aortic endothelial cells. The role of endogenous basic fibroblast growth factor. *J. Biol. Chem.* **272**, 318–325.
7. Dhodapkar, M. V., Kelly, T., Theus, A., Athota, A. B., Barlogie, B., and Sanderson, R. D. (1997) Elevated levels of shed syndecan-1 correlate with tumour mass and decreased matrix metalloproteinase-9 activity in the serum of patients with multiple myeloma. *Br. J. Haematol.* **99**, 368–371.

8. Kato, M., Wang, H., Kainulainen, V., Fitzgerald, M. L., Ledbetter, S., Ornitz, D. M., and Bernfield, M. (1998) Physiological degradation converts the soluble syndecan-1 ectodomain from an inhibitor to a potent activator of FGF-2. *Nat. Med.* **4**, 691–697.
9. Philip, A., Hannah, R., and O'Connor-McCourt, M. (1999) Ectodomain cleavage and shedding of the type III transforming growth factor-beta receptor in lung membranes effect of temperature, ligand binding and membrane solubilization. *Eur. J. Biochem.* **261**, 618–628.
10. Nishiyama, A., Lin, X. H., and Stallcup, W. B. (1995) Generation of truncated forms of the NG2 proteoglycan by cell surface proteolysis. *Mol. Biol. Cell.* **12**, 1819–1832.
11. Alliel, P. M., Perin, J. P., Jolles, P., and Bonnet, F. J. (1993) Testican, a multidomain testicular proteoglycan resembling modulators of cell social behaviour. *Eur. J. Biochem.* **214**, 347–350.
12. Eriksen, G. V., Carlstedt, I., Morgelin, M., Uldbjerg, N., and Malmstrom, A (1999) Isolation and characterization of proteoglycans from human follicular fluid. *Biochem. J.* **340**, 613–620.
13. Delpech, B., Girard, N., Olivier, A., Maingonnat, C., van Driessche, G., van Beeumen, J., Bertrand, P., Duval, C., Delpech, A., and Bourguignon, J. (1997) The origin of hyaluronectin in human tumors. *Int. J. Cancer* **72**, 942–948.
14. Perides, G., Asher, R. A., Lark, M. W., Lane, W. S., Robinson, R. A., and Bignami, A. (1995) Glial hyaluronate-binding protein: a product of metalloproteinase digestion of versican? Biochem. J. **312**, 377–384.
15. Rauch, U., Karthikeyan, L., Maurel, P., Margolis, R. U., and Margolis, R. K. (1992) Cloning and primary structure of neurocan, a developmentally regulated, aggregating chondroitin sulfate proteoglycan of brain. *J. Biol. Chem.* **267**, 19,536–19,546.
16. Matsui, F., Nishizuka, M., Yasuda, Y., Aono, S., Watanabe, E. and Oohira, A. (1998) Occurrence of a N-terminal proteolytic fragment of neurocan, not a C-terminal half, in a perineuronal net in the adult rat cerebrum. *Brain Res.* **790**, 45–51.
17. Yamada, H., Watanabe, K., Shimonaka, M., Yamasaki, M., and Yamaguchi, Y. (1995) cDNA cloning and the identification of an aggrecanase-like cleavage site in rat brevican. *Biochem. Biophys. Res. Commun.*, **216**, 957–963.
18. Sandy, J. D., Neame, P. J., Boynton, R. E., and Flannery, C. R. (1991) Catabolism of aggrecan in cartilage explants. Identification of a major cleavage site within the interglobular domain. *J. Biol. Chem.* **266**, 8683–8685.
19. Arner, E. C., Pratta, M. A., Trzaskos, J. M., Decicco, C. P., and Tortorella, M. D. (1999) Generation and characterization of aggrecanase. A soluble, cartilage-derived aggrecan-degrading activity. *J. Biol. Chem.* **274**, 6594–6601.
20. Abbaszade, I., Liu, R. Q., Yang, F., Rosenfeld, S. A., Ross, O. H., Link, J. R., et al. (1999) Cloning and characterization of ADAMTS11, an aggrecanase from the ADAMTS family. *J. Biol. Chem.* **274**, 23,443–23,450.
21. Sandy, J. D., Plaas, A. H., and Koob, T. J. (1995) Pathways of aggrecan processing in joint tissues. Implications for disease mechanism and monitoring. *Acta Orthop. Scand. Suppl.* **266**, 26–32.
22. Sandy, J. D., Flannery, C. R., Neame, P. J., and Lohmander, L. S. (1992) The structure of aggrecan fragments in human synovial fluid. Evidence for the involvement in osteoarthritis of a novel proteinase which cleaves the Glu 373-Ala 374 bond of the interglobular domain. *J. Clin. Invest.* **89**, 1512–1516.
23. Sandy, J. D., Gamett, D., Thompson, V., and Verscharen, C. (1998) Chondrocyte-mediated catabolism of aggrecan: aggrecanase-dependent cleavage induced by interleukin-1 or retinoic acid can be inhibited by glucosamine. *Biochem. J.* **335**, 59–66.

24. Sandy, J. D. and Lark, M. W. (1998) Proteolytic degradation of normal and osteoarthritic cartilage matrix, in *Osteoarthritis* (Brandt, K. D., Doherty, M., Lohmander, S. L., eds.), Oxford University Press, pp. 84–93.
25. Plaas, A. H., Neame, P. J., Nivens, C. M., and Reiss, L. (1990) Identification of the keratan sulfate attachment sites on bovine fibromodulin. *J. Biol. Chem.* **265**, 20,634–20,640.
26. Midura, R. J., Hascall, V. C., MacCallum, D. K., Meyer, R. F., Thonar, E. J., Hassell, J. R., Smith, C. F., and Klintworth, G. K. (1990) Proteoglycan biosynthesis by human corneas from patients with types 1 and 2 macular corneal dystrophy. *J Biol Chem.* **265**, 15,947–15,955.
27. Margolis, R. U. and Margolis, R. K. (1994) Aggrecan-versican-neurocan family proteoglycans. *Meth. Enzymol.* **245**, 105–126.
28. Farndale, R. W., Buttle, D. J., and Barrett, A. J. (1986) Improved quantitation and discrimination of sulphated glycosaminoglycans by use of dimethylmethylene blue. *Biochem. Biophys. Acta* **883**, 173–177.

# 34

## Degradation of Heparan Sulfate by Nitrous Acid

**H. Edward Conrad**

### 1. Introduction

All glycosaminoglycans contain an amino sugar in every second position in their linear chains. In most cases these amino sugars are N-acetylated, e.g., N-acetyl-D-galactosamine (GalNAc) in chondroitin sulfates and dermatan sulfates, or N-acetyl-D-glucosamine (GlcNAc) in heparins, heparan sulfates, and hyaluronic acids. Heparins and heparan sulfates contain only glucosamine residues, and, in a high percentage of these, the N-acetyl substituents are replaced by N-sulfate groups. The N-sulfated glucosamines ($GlcNSO_3$), found only in heparin-like glycosaminoglycans, are sites that are unique in their susceptibility to facile cleavage by nitrous acid at room temperature and at low pH (~1.5). Thus, when heparins or heparan sulfates are treated with nitrous acid, they are specifically cleaved into fragments with ranges of molecular weights that depend on the distributions of the $GlcNSO_3$ residues in the chain. Since N-acetylated amino sugars are not affected by the nitrous acid treatment, only the heparin-like structures are cleaved when they are present in mixtures containing other glycosaminoglycans.

Interestingly, glycosaminoglycans containing amino sugars that have unsubstituted amino groups can be cleaved specifically with nitrous acid at a higher pH (~4), again at room temperature. Under these conditions, the $GlcNSO_3$ residues do not react. Since a small number of N-unsubstituted glucosamine residues (GlcN) occur in heparan sulfates *(1)*, cleavage occurs at these positions.

N-acetylated amino sugars do not react with nitrous acid under any conditions. However, by hydrazinolysis, it is possible to remove N-acetyl groups from N-acetylated amino sugars under conditions that do not remove N-sulfate groups or otherwise alter the glycosaminoglycan structures *(2)*. Thus, following hydrazinolysis, nitrous acid (at pH 4) cleaves heparins and heparan sulfates specifically at the glucosamine residues that were originally N-acetylated, yielding fragments with ranges of molecular weights that depend on the distributions of the GlcNAc residues in the chains. In a

From: *Methods in Molecular Biology, Vol. 171: Proteoglycan Protocols*
Edited by: R. V. Iozzo © Humana Press Inc., Totowa, NJ

mixture of glycosaminoglycans, the hydrazinolysis/nitrous acid procedure yields the variously sized fragments from the heparin-like glycosaminoglycans, but converts all other glycosaminoglycans completely to disaccharides (since all other glycosaminoglycans contain N-acetylated amino sugars at every second residue).

The cleavage of the glycosidic bonds of the amino sugars with nitrous acid is an elimination reaction not a hydrolysis reaction. In all cases, the reaction yields fragments in which the amino sugar at the site of cleavage is converted to a reducing terminal anhydrosugar—in the case of D-glucosamine, 2,5-anhydro-D-mannose is formed; in the case of D-galactosamine, 2,5-anhydro-D-talose is formed. In order to stabilize the products, it is helpful to reduce the aldehyde groups of these anhydrosugars to alditols using $NaBH_4$. The nitrous acid, hydrazinolysis, and $NaBH_4$ reactions have been discussed in detail elsewhere ***(3)***.

## 2. Materials

### 2.1. Chemicals

1. Heparin.
2. Heparan sulfate.
3. 0.1 *M* NaOH.
4. 0.2 *M* $Na_2CO_3$.
5. 1 *M* $Na_2CO_3$.
6. Reacti-Vials (Pierce Chemical Co., Rockford, IL).
7. 0.5 *M* $H_2SO_4$.
8. 0.5 *M* $Ba(NO_2)_2$ (A. D. Mackay, Darien, CT).
9. 5.5 *M* $NaNO_2$.
10. Water bath.
11. Hydrazine, anhydrous.
12. Hydrazine $SO_4$.
13. 3 *M* $H_2SO_4$.

### 2.2. Reference Standards

Heparin and heparan sulfate can be obtained from Sigma Chemical Co. Stock solutions of these glycosaminoglycans contain 20 mg/mL water and are stored frozen.

### 2.3. pH 1.5 Nitrous Acid Reagent

Solutions of 0.5 *M* $H_2SO_4$ and 0.5 *M* $Ba(NO_2)_2$ (114 mg/mL) are prepared and cooled separately to 0°C in an ice bath. A mixture containing 1 mL of each solution (0.5 mmol of each reagent) is prepared at 0°C, and the mixture is centrifuged in a clinical centrifuge to pellet the $BaSO_4$ precipitate. The supernatant is drawn off with a Pasteur pipet, and stored on ice. This reagent should be prepared when needed and used immediately (*see* **Note 1**).

### 2.4. pH 4.0 Nitrous Acid Reagent

pH 4 Nitrous acid is generated by adding 5 mL of 5.5 *M* $NaNO_2$ to 2 mL of 0.5 *M* $H_2SO_4$. This reagent should be prepared when needed and used immediately. When glycosaminoglycans are hydrolyzed in 0.5 *M* $H_2SO_4$, the pH 4 nitrous acid is generated *in situ* by adding 5 µL of the 5.5 *M* $NaNO_2$ solution to 2 µL of the hydrolysate (*see* **Notes 2** and **3**).

### 2.5. Hydrazine Reagent

Hydrazine sulfate (100 mg) is dissolved in 3 mL of distilled water. Anhydrous hydrazine (7 mL) is added to this solution to give a solution containing 1% hydrazine sulfate in 30% aqueous hydrazine. Hydrazine is a toxic and corrosive reagent and should be handled accordingly.

### 2.6. Sodium Borohydride Reagent

1. 0.5 *M* $NaBH_4$ in 0.1 *M* NaOH.

## 3. Methods

### 3.1. Cleavage of N-Sulfated Glycosaminoglycans with Nitrous Acid at pH 1.5

1. Cool a 5-mL aliquot of a solution of heparin or heparan sulfate (20 mg/mL of water) to 0°C and add 20 mL of the cold pH 1.5 nitrous acid reagent.
2. Let the mixture warm to room temperature; deamination is complete within 10 min following nitrous acid addition.
3. Adjust the pH of the deaminated product to 8.5 with 1 *M* $Na_2CO_3$.
4. Reduce the sample with $NaBH_4$ as described under **Subheading 3.3.**
5. The reduced sample may be separated into its individual components by gel filtration to separate di-, tetra-, hexasaccharides, etc., and then by ion-exchange chromatography, high-pressure liquid chromatography, or capillary electrophoresis to separate the oligosaccharides according to charge ***(2,4–14)***. Since the nitrous acid cleavage products are complex mixtures of oligosaccharides, multiple separation steps are required to obtain individual components in a pure form.

### 3.2. Hydrazinolysis of N-Acetylated Glycosaminoglycans and Cleavage with pH 4 Nitrous Acid

1. Place 15 μL of a solution of heparin or heparan sulfate (20 mg/mL) in a 100-μL Reacti-Vial and dry the sample in a stream of air.
2. Redissolve the dried sample in 20 μL of hydrazine reagent.
3. Cap the vial and place it in a 100°C water bath for 4 h.
4. Cool the sample, dry it in a stream of air; and lyophilize the partially dried sample to remove as much of the hydrazine as possible.
5. Due to residual hydrazine $SO_4$ and hydrazine, the pH of this solution is actually alkaline (pH 8–10). Add 5–10 μL of 3 *M* $H_2SO_4$ to the sample to bring the pH to 4.0, as measured with pH paper.
6. Dry the pH-adjusted sample in a stream of air.
7. Add 20 μL of the pH 4 nitrous acid reagent to the dried sample.
8. After 15 min, the cleavage reaction is complete.
9. Adjust the pH of the reaction mixture to 8.5
10. Reduce the cleavage products with $NaBH_4$ as described below and separate the individual components by gel filtration and ion-exchange chromatography (above).

### 3.3. Reduction of Cleavage Products with $NaBH_4$

1. Treat the pH 8.5 solutions of the nitrous acid-cleaved products with 10 mL of the sodium borohydride reagent and incubate at room temperature for 15 min. Since $H_2$ gas is evolved in this reaction, all $NaBH_4$ reductions should be carried out in a fume hood.

2. Destroy excess $NaBH_4$ by addition of ~5 mL of 3 *M* $H_2SO_4$ to give a slightly acidic solution. Evaporate the sample to dryness. Redissolve the sample in water and again evaporate the solution to dryness to remove as much $H_2$ as possible.
3. Redissolve the sample in water for separation of the individual components.

## 4. Notes

1. Once the nitrous acid is prepared, there is a series of complex reactions of the resulting oxides of nitrogen. Within a short time, the active species of "nitrous acid" undergoes changes that result in the loss of the capacity of the reagent to cleave the glycosidic bonds of the amino sugars [for a discussion, *see* **ref. *15***)]. Consequently, the nitrous acid reagents must be used within a few minutes after their preparation. This is of less concern for the pH 4 reagent than for the pH 1.5 reagent, because the pH 4 reagent is more highly concentrated in nitrite.
2. The pH of these reactions is important for maintaining the selectivity of the cleavage. Although there is good selectivity for N-unsubstituted GlcN's and N-sulfated GlcN's at pH 4 and 1.5, respectively, the glycosidic bonds of both types of GlcN residues are cleaved at a pH between these values. Thus, it is desirable to let the reactions at the respective pH proceed only for 10–15 min. Also, when samples are derived from buffered solutions, it is necessary to check the pH with pH paper before addition of the nitrous acid reagent. In fact, it is desirable to dialyze the glycosaminoglycan solution to remove *all* salts before beginning the cleavage step. An important reason for the predialysis is that $NaBH_4$ is catalytically destroyed by oxyanions, such as $PO_4^3$ ***(16)***.
3. Although the nitrous acid reactions cleave the bonds of β-linked amino sugars virtually stoichiometrically, the reaction of nitrous acid with α-linked amino sugars takes two pathways. In both cases treatment with nitrous acid leads to the loss of the amino group and the formation of a carbonium ion at carbon 2. In the most prominent further conversion, the α-glycosidic bond is cleaved with the formation of the reducing terminal anhydrosugar as described above. However, a significant proportion (~10%) of the carbonium ion undergoes a reaction in which the ring is contracted to a furanose ring without glycosidic bond cleavage. This "ring contraction reaction" is described elsewhere ***(3)***.

## References


1. van den Born, J., Gunnarsson, K., Bakker, M. A. H., Kjellén, L., Kusche-Gullberg, M., Maccarana, M., Berden, J. H. M., and Lindahl, U. (1995) Presence of N-unsubstituted glucosamine units in native heparan sulfate revealed by a monoclonal antibody. *J. Biol. Chem.* **270**, 31,303–31,309.
2. Guo, Y. and Conrad, H. E. (1989) The disaccharide composition of heparins and heparan sulfates. *Anal. Biochem.* **176**, 96–104.
3. Conrad, H. E. (1998) *Heparin-Binding Proteins*, Academic Press, San Diego, CA.
4. Ampofo, S. A., Wang, H. M., and Linhardt, R. J. (1991) Disaccharide compositional analysis of heparin and heparan sulfate using capillary zone electrophoresis. *Anal. Biochem.* **199**, 249–255.
5. Bienkowski, M. J. (1984) Structure and metabolism of heparin and heparan sulfate, Ph.D. dissertation, University of Illinois, Urbana, IL.
6. Delaney, S. R., Leger, M., and Conrad, H. E. (1980) Quantitation of the sulfated disaccharides of heparin by high performance liquid chromatography. *Anal. Biochem.* **106**, 253–261.

7. Desai, U. R., Wang, H.-M., Ampofo, S. A., and Linhardt, L. J. (1993) Oligosaccharide composition of heparin and low-molecular-weight heparins by capillary electrophoresis. *Anal. Biochem.* **213**, 120–127.
8. Desai, U. R., Wang, H.-M., Kelly, T. R., and Linhardt, R. J. (1993) Structure elucidation of a novel acidic tetrasaccharide and hexasaccharide derived from a chemically modified heparin. *Carbohydr. Res.* **241**, 249–259.
9. Guo, Y. and Conrad, H. E. (1988) Analysis of oligosaccharides from heparin by reversed phase ion-pairing high pressure liquid chromatography. *Anal. Biochem.* **168**, 54–62.
10. Kitagawa, H., Kinoshita, A., and Sugahara, K. (1995) Microanalysis of glycosaminoglycan-derived disaccharides labeled with the fluorophore 2-aminoacridone by capillary electrophoresis and high-performance liquid chromatography. *Anal. Biochem.* **232**, 114–121.
11. Liu, J., Shworak, N. W., Fritze, L. M., Edelberg, J. M., and Rosenberg, R. D. (1996) Purification of heparan sulfate D-glucosaminyl 3-O-sulfotransferase. *J. Biol. Chem.* **271**, 27,072–27,082.
12. Merchant, Z. M., Kim, Y. S., Rice, K. G., and Linhardt, R. J. (1985) Structure of heparin-derived tetrasaccharides. *Biochem. J.* **229**, 369–377.
13. Murata, K., Murata, A., and Yosida, K. (1995) High-performance liquid chromatographic identification of eight constitutional disaccharides from heparan sulfate isomers digested with heparitinases. *J. Chromatogr.* B **670**, 3–10.
14. Pervin, A., Al-Hakim, A., and Linhardt, R. J. (1994) Separation of glycosaminoglycan-derived oligosaccharides by capillary electrophoresis using reverse polarity. *Anal. Biochem.* **221**, 182–188.
15. Shively, J. E., and Conrad, H. E. (1976) Formation of anhydrosugars in the chemical depolymerization of heparin. *Biochemistry* **15**, 3932–3942.
16. Conrad, H. E., James, M. E., and Varboncoeur, E. (1973) Qualitative and quantitative analysis of reducing carbohydrates by radiochromatography on ion exchange papers. *Anal. Biochem.* **51**, 486–500.

35

# Degradation of Heparan Sulfate with Heparin Lyases

**Laurie A. LeBrun and Robert J. Linhardt**

## 1. Introduction

Glycosaminoglycan (GAG), heparan sulfate (HS), and heparin are a polydisperse mixture of linear polysaccharides composed of glucosamine residues 1→ 4 linked to uronic acid residues. The major repeating unit in heparin is → 4)-α-D-*N*-sulfoglucosamine-6-sulfate (1→ 4)-α-L-iduronic acid-2-sulfate (1→, corresponds to 75–90% of its sequence *(1)* (*see* **Fig. 1A**), whereas heparan sulfate consists of 50–75% → 4)-α-D-*N*-acetylglucosamine (1→ 4)-β-glucuronic acid (1→ and smaller amounts of → 4)-α-D-*N*-acetylglucosamine-6-sulfate (1→ 4)-β-D-glucuronic acid (1→ and → 4)-α-D-*N*-sulfoglucosamine (1→ 4)-β-D-glucuronic acid (1→ (*see* **Fig. 1B**). Heparin, which contains approx 2.7 sulfate groups per disaccharide unit, is more highly sulfated than HS, which contains less than one sulfate per disaccharide unit.

HS proteoglycans (PGs) are localized on the surface of many mammalian cells and in the extracellular matrix. HS proteoglycans are important for several different biological activities such as cell–cell and cell–protein interactions *(2)*. These biological activities are controlled mainly through the binding of a variety of proteins to the HS chains. Specific sequences in the HS chain are thought to be responsible for the binding of growth factors, protease inhibitors, and adhesion molecules. The use of HS-degrading enzymes can help in separating and identifying biological active oligosaccharides *(3)*.

HS can be degraded enzymatically by using heparin lyases from bacterial sources. The lyase enzymes degrade GAGs by endolytic cleavage *(4–6)*. The enzymes cut glucosamine–uronate linkage by elimination (*see* **Fig. 2**), leaving a C4–C5 unsaturated bond containing product that can be easily detected by ultraviolet (UV) absorbance. In contrast, mammalian heparanases cleave this linkage by hydrolysis.

Heparin lyases have been isolated from *Flavobacterium heparinum (7)*, *Bacteriodes* species *(8)*, *Bacteriodes heparinolyticus (9)*, and *Prevotella heparinolytica (10)*. Heparin lyases from *F. heparinum* have been purified to homogeneity, studied extensively *(11)*, and are available commercially from Sigma and Seikagaku.

From: *Methods in Molecular Biology, Vol. 171: Proteoglycan Protocols*
Edited by: R. V. Iozzo © Humana Press Inc., Totowa, NJ

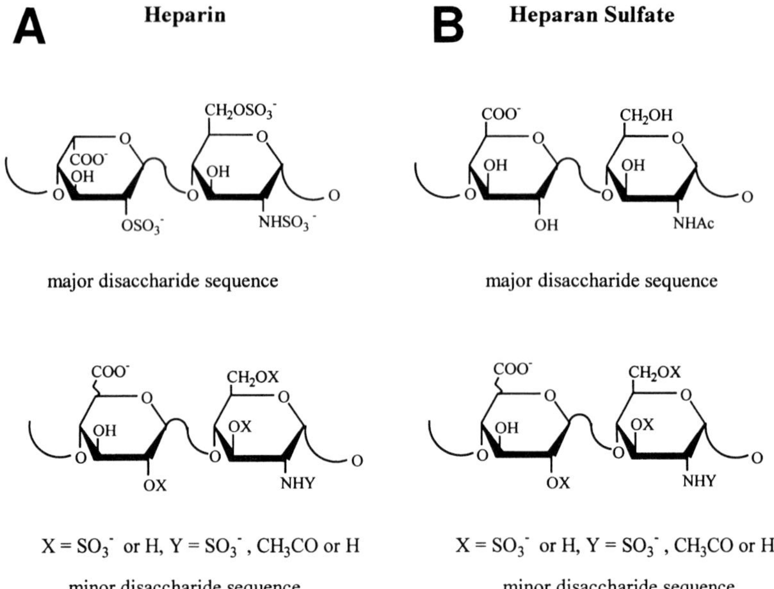


Fig. 1. **(A)** Structure of the major and minor disaccharide sequences of heparin. **(B)** Structure of the major and minor disaccharide sequences of heparan sulfate.

Three types of heparin lyases have been purified from *Flavobacterium*: heparin lyase I, heparin lyase II, and heparin lyase III (*see* **Fig. 2** and **Note 1**). Heparin lyase I acts primarily on heparin, heparin lyase II cleaves both heparin and HS, and heparin lyase III is active only on HS (*see* **Table 1**). The primary linkages cleaved by these enzymes and their relative activities toward heparin and HS are shown in **Table 1** and **Fig. 3**.

From both the DNA and amino acid sequences, there is only 15% alignment between heparin lyase I, II, and III *(12)*. There are certain conserved sequences such as the heparin-binding sites and the calcium-binding regions in heparin lyase I and III. Recently, chemical modification studies and site-directed mutagenisis have been used to help identify critical residues for enzyme activity *(13–17)*. Further studies and the crystal structures of the heparin lyases are needed to help us understand better the relationship between function and structure of these enzymes.

This chapter will explain how the heparin lyase enzymes can be used to degrade both heparin- and HS-containing samples and how to assay the activity of the enzyme. The heparin lyase enzymes can be used to identify the presence of HS/heparin in samples or to purify HS oligosaccharides for structural analysis.

## 2. Materials

### 2.1. Enzyme Preparation

1. Heparin Lyase I, II, or III. Enzymes can be ordered from Sigma (St. Louis, MO) and Siekagaku America (Falmouth, MA) (*see* **Notes 1** and **2**).

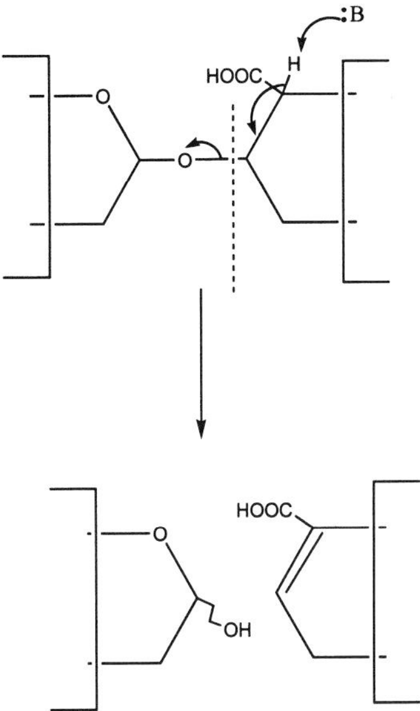


Fig. 2. Eliminative cleavage of GAGs by lyases.

2. Reagents required to make the appropriate buffers:
   a. Dibasic sodium phosphate (EM Science, Gibbstown, NJ).
   b. Phosphoric acid (Fisher Scientific, Fair Lawn, NJ).
   c. Sodium chloride (Fisher Scientific).
3. 500-µL polypropylene microcentrifuge tubes.

### 2.2. Enzyme Assay

1. HS (bovine kidney HS, sodium salt from Siekagaku America).
2. Heparin lyase solution (*see* **Subheading 3.1.**).
3. Buffer (*see* **Table 2**).
4. UV spectrophotometer.
5. 2 × 1-mL quartz cuvets.

### 2.3.Sample Digestion

1. HS or heparin samples (*see* **Note 3**) or samples containing radiolabeled HS or heparin.
2. Spectropor dialysis membrane (molecular-weight cutoff [MWCO] 1000) (Spectrum, Los Angeles, CA) or Centricon (YM3, MWCO 3000) centrifugal filter units (Millipore, Bedford, MA).
3. 500-µL polypropylene microcentrifuge tubes.
4. Heparin lyase solution.
5. Water baths at 30°C and 35°C for enzyme digestion and at 100°C to inactivate the enzyme reaction.

### 2.4. Product Analysis

High-performance liquid chromatography (HPLC), capillary electrophoresis (CE), gel-permeation chromatography, or polyacrylamide gel electrophoresis (PAGE) may be used to purify and analyze oligosaccharides prepared from HS/heparin.

**Table 1**
**Activity of Heparin Lyases**

| Activity and substrate conversion | Heparin lyase I | Heparin lyase II | Heparin lyase III |
|---|---|---|---|
| Heparin[a] | | | |
| Percent activity[b] | 100 | 60 | <1 |
| % Conversion[c] | 60 (80)[d] | 85 | 6 |
| Heparan sulfate[e] | | | |
| Percent activity | 10 | 100 | 100 |
| Percent conversion | 20 | 40 | 94 |

[a]Porcine mucosal heparin.

[b]Percent activity = [initial rate on the substrate examined/initial rate on substrate giving the highest activity] × (100).

[c]Percent conversion = [moles of linkages cleaved/total moles of hexosamine → uronic acid linkages] × (100).

[d]Bovine lung heparin.

[e]Bovine kidney heparan sulfate.

## 3. Methods

### 3.1. Preparation of Lyases for Use

1. Preparation of buffers: The following buffers can be stored at room temperature for over 1 mo (*see* **Note 4**).
   a. *For heparin lyase I*, prepare 50 m*M* sodium phosphate buffer containing 100 m*M* sodium chloride at pH 7.1. To prepare 1 L of buffer, dissolve 7.1 g of dibasic sodium phosphate and 5.8 g of sodium chloride into 900 mL of distilled water. Adjust the pH to 7.1 with phosphoric acid and bring the volume up to 1 L with distilled water.
   b. *For heparin lyase II and III*, prepare 50 m*M* sodium phosphate buffer by dissolving 7.1 g of dibasic sodium phosphate in 900 mL of distilled water. For heparin lyase II adjust the pH to 7.1 with concentrated phosphoric acid, and adjust to 7.6 for heparin lyase III. Adjust the final volume to 1 L with distilled water.
2. Aliquot samples:
   a. Dissolve 0.1 U of lyophilized enzyme in 100 µL of the appropriate buffer (*see* **Table 2**).
   b. Store the enzyme in 10-mU aliquots in 500-µL polypropylene tubes at –70°C (*see* **Note 5**).

### 3.2. Activity Assay for Lyases

1. Add 640 µL of the appropriate buffer (*see* **Table 2**) to a 1-mL quartz cuvet. Warm the cuvet to 30°C in a temperature-controlled UV spectrophotometer (*see* **Note 6**).
2. Thaw 10-µL aliquots of the appropriate enzyme solution (*see* **Table 1**) at room temperature.
3. Remove 90 µL of warm buffer out of the cuvet and transfer the solution into the tube containing the enzyme solution. Immediately transfer the entire 100 µL of buffer and enzyme back into the cuvet, which is incubating at 30°C.
4. Adjust the baseline of the spectrophotometer to zero at 232 nm.
5. Remove the cuvet from the spectrophotometer and add 50 µL of 20 mg/mL of the appropriate substrate HS/heparin (*see* **Table 1**) to the cuvet. Cover the cuvet with Parafilm and

Heparin lyase I

Heparin lyase II

Heparin lyase III

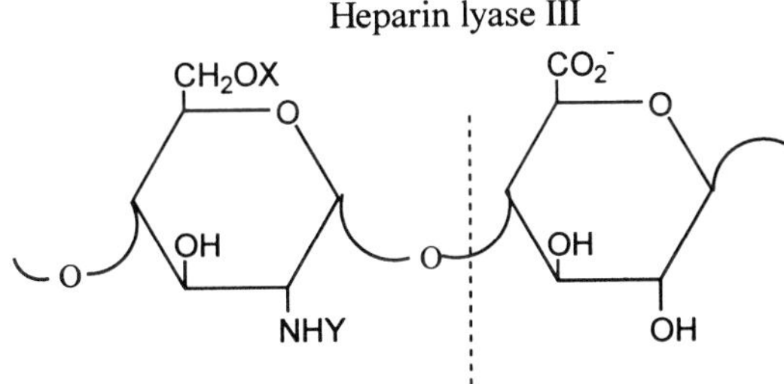


Fig. 3. Primary glycosidic linkages cut by hepain lyases. Abbreviations: X, H or $SO_3^-$; Y, $CH_3CO$ or $SO_3^-$. Heparin lyase II cleaves at either glucuronic or iduronic acid residues.

invert two times to mix. Remove the Parafilm and place the cuvet back into the spectrophotometer.

6. Within 30 s after the addition of substrate, begin to measure the absorbance continuously or at 30-s intervals for 2–10 min. Graph absorbance at 232 nm vs time. The initial rate is determined by measuring the slope of the linear portion of absorbance vs time.
7. Calculate the enzyme activity from the initial rate using the extinction coefficient ($\varepsilon = 3800\ M^{-1}$) for the reaction products (*see* **Note 7**). Each product formed has an unsaturated uronic acid residue at its nonreducing terminus that absorbs at 232 nm. The enzyme activity is calculated as

$$\text{Enzyme Activity} = \frac{(\Delta\text{Abs 232 nm/ min})\ (700\ \ \mu\text{L})}{(3800\ \text{M}^{-1})}$$

**Table 2**
**Properties of Heparin Lyases and Reaction Conditions**

| Enzyme | Substrate | MW (Da) | p*I* | $T_{opt}$[b] (°C) | Buffer system |
|---|---|---|---|---|---|
| Heparin lyase I [a](EC 4.2.2.7) | Heparin | 42,800 | 9.2 | 30 | 50 m*M* $NaPO_4$, 100 m*M* NaCl, pH 7.1 |
| Heparin lyase II | Heparin HS | 84,100 | 9.0 | 35 | 50 m*M* $NaPO_4$, pH 7.1 |
| Heparin lyase III (EC 4.2.2.8) | HS | 70,800 | 10 | 35 | 50 m*M* $NaPO_4$, pH 7.6 |

[a]EC is the Enzyme Commission number.
[b]$T_{opt}$, optimum temperature for the enzyme.

## 3.3. Sample Digestion

### 3.3.1. Complete Heparin Lyase-Catalyzed Depolymerization of a Sample Containing HS/Heparin

1. Dissolve samples, containing HS/heparin (*see* **Note 8**) in buffer and dialyze using a 1000-MWCO dialysis membrane or a Centricon (YM3, MWCO 3000) centrifugal filter unit (*see* **Note 9**).
2. Thaw 10 µL of the appropriate enzyme solution (*see* **Table 1**) at room temperature (assay if desired as described under **Subheading 3.2.**) and then add 40 µL of the appropriate buffer (*see* **Table 1**) to a 500-µL polypropylene microcentrifuge tube (*see* **Note 10**). Also add 50 µL of buffer to one tube as a blank control.
3. Add 50 µL of the HS/heparin-containing sample to each tube and mix by gently inverting.
4. Incubate for 8–12 h at the appropriate temperature as indicated in **Table 2** (*see* **Note 11**).
5. Terminate the reaction by heating the tubes at 100°C for 2–3 min (*see* **Note 12**).
6. Product formation can be determined by either UV detection, colormetric assay, or HPLC. Pure samples containing >10 µg of HS can be measured by absorbance. The C4–C5 unsaturated bond of the oligosaccharide product is a chromophore that can be measured at a $\lambda_{max}$ = 232 nm with a molar absorptivity of 5500 $M^{-1}$ in 30 m*M* hydrochloric acid (*see* **Note 13**) ***(18)***. If the sample contains a high concentration of protein, the Azure A metachromatic assay ***(19)*** should be used. For smaller samples or impure samples, gel-permeation chromatography using UV or colorimetric detection ***(20)*** can be used to measure the quantity of product (*see* **Note 14**).

### 3.3.2. Complete Heparin Lyase-Catalyzed Depolymerization of Radiolabeled HS

1. Dissolve GAG sample containing radiolabeled HS in 50 µL of sodium phosphate buffer. Dialyze sample against sodium phosphate buffer using 1000 MWCO dialysis membrane or a Centricon (YM3, MWCO 3000) centrifugal filter unit (*see* **Note 9**).
2. Thaw 10 µL of heparin lyase III solution at room temperature, immediately prior to use (*see* **Note 5**).

3. Add 30 µL of sodium phosphate buffer to the 500-µL polypropylene microcentrifuge tube containing the enzyme solution.
4. (*Optional*) (*see* **Note 15**). Add 1.7 µL each of 20-mg/mL chondroitin sulfate A, chondroitin sulfate C, and dermatan sulfate substrate solutions (34 µg of each GAG) to the enzyme in buffer.
5. Add 50 µL of radiolabeled heparan sulfate solution and incubate 8–12 h at 30°C.
6. Heat at 100°C for 2 min to inactivate the enzyme.
7. Analyzed depolymerized radioactive sample by gel-permeation chromatography using radioisotope detection methods (*see* **Note 14**).

### 3.4. Analysis of Product

The heparin and heparan sulfate oligosaccharides can be analyzed by gradient PAGE, capillary electrophoresis (CE), or strong-anion-exchange HPLC ***(21–24)***.

## 4. Notes

1. Heparin lyase I from *Flavobacterium heparinum* is sold as heparinase I by Sigma and as heparinase by Seikagaku. Heparin lyase II is sold as heparinase II by Sigma and as heparitinase II by Seikagaku. Heparin lyase III from *Flavobacterium heparinum* is sold as heparinase III by Sigma and as heparatinase or heparatinase I by Seikagaku.
2. Often the purchased lyophilized enzyme contains bovine serum albumin (BSA) as a stabilizer. For example, Sigma samples contain 25% BSA.
3. Samples consisting of tissue, biological fluids, PGs, and GAGs that contain microgram quantities of HS can often be analyzed directly using heparin lyases without the use of radioisotopes.
4. Since calcium is an activator for heparin lyase I and III, 20 m*M* sodium acetate buffer in the presence of 2 m*M* calcium acetate can be used with these enzymes. These enzymes are also compatible with a wide range of other biological buffers.
5. Storage: The lyophilized enzyme is stable at –20°C for at least 2 yr. The dissolved enzyme is stable when frozen at –20°C for 1 mo and for over a year at –70°C. The heparin lyases are sensitive to freeze-thawing, especially heparin lyase III. Once heparin lyase III samples are thawed, they should be used immediately.
6. If a temperature-controlled spectrophotometer is not available, activity can be measured at room temperature, or samples can be incubated in a water bath and the absorbance can be measured at fixed time points.
7. One unit is equal to 1 µmol product formed per minute.
8. If the sample is believed to contain HS, heparin lyase III should be used. Samples that contain heparin should be treated with heparin lyase I. In samples where the identity of the GAG is unknown or believed to be a mixture of HS and heparin use either heparin lyase II or an equal unit mixture of heparin lyase I, II, and III to ensure complete depolymerization.
9. The presence of metals, detergents, and denaturants can interfere with the activity of the lyases. Before digesting the samples, detergents should be removed by precipitation with potassium chloride or by using a detergent-removal column such as Biobeads (Bio-Rad). Urea and guanidine should be removed by exhaustive dialysis using controlled-pore dialysis membrane (MWCO 1000).
10. Additional enzyme (10- to 100-fold) may be required to break down small, resistant oligosaccharides (***25–26***).

11. If possible, gently shake the samples during digestion.
12. Following the use of a lyase, residual lyase activity can be destroyed by heating the reaction mixture to 100°C or by adding denaturants or detergents. Most lyases are cationic proteins and can be removed from anionic oligosaccharide products by passing the reaction mixture through a small cation-exchange column, such as SP-Sephadex (Sigma), adjusted to an acidic pH. The oligosaccharide products (void volume) are then recovered, readjusted to neutral pH, and analyzed. This method can also be used to remove BSA, an excipient found in many of the commercial enzymes, from the oligosaccharide products.
13. This assay is to be used with relatively pure GAGs. High concentrations of protein interfere with the measure of oligosaccharide production due to UV absorbance of the protein.
14. In gel-permeation chromotography of HS/heparin, following complete depolymerization using the appropriate heparin lyase, the products should elute close to the column's total volume, corresponding to an apparent molecular weight <1500 daltons (confirming the presence of heparin/HS), while the substrate (control without enzyme) should elute close to the column's void volume, corresponding to a molecular weight of >10,000 daltons.
15. When attempting to use heparin lyases to depolymerize radiolabeled samples that contain very small quantities of heparin or heparan sulfate, it is often useful to add cold substrate as a carrier so that the activity of heparin lyase can be distinguished from that of trace amounts of chondrotin lyases that may be present in heparin lyase preparations *(27)*.

## References


1. Linhardt, R. J., Ampofo, S. A., Fareed, J. Hoppensteadt, D., Mulliken, J. B., and Folkman, J. (1992) Characterization of a human heparin. *Biochemistry* **31**, 12,441–12,445.
2. Stringer, S. E. and Gallagher, J. T. (1997) Heparan sulphate. *Int. J. Biochem. Cell Biol.* **29**, 709–714.
3. Toida, T. and Linhardt, R. J. (1998) Structural analysis of heparan sulfate and heparan oligosacccharides. *Trends Glycosci. Glycotechnol.* **10**, 125–136.
4. Linhardt, R. J., Galliher, P. M., and Cooney, C. L. (1986) Polysaccharide lyases. *Appl. Biochem. Biotechnol.* **12**, 135–176.
5. Ernst, S., Langer, R., Cooney, C. L., and Sasisekharan, R. (1995) Enzymatic degradation of glycosaminoglycans. *Crit. Rev. Biochem. Mol. Biol.* **30**, 387–444.
6. Sutherland, I. W. (1995) Polysaccharide lyases. *FEMS Microbiol. Rev.* **16**, 323–347.
7. Payza, A. N. and Korn, E. D. (1956) Bacterial degradation of heparin. *Nature* **177**, 88–89.
8. Ahn, M. Y., Shin, K. H., Kim, D. H., Jung, E. A., Toida, T., Linhardt, R. J., and Kim, Y. S. (1998) Characterization of a *Bacteroides species* from human intestine that degrades glycosaminoglycans. *Can. J. of Microbiol.* **44**, 423–429.
9. Nakamura, T., Shibata, Y., and Fujimura S. (1988) Purification and properties of *Bacteroides heparinolyticus* heparinase (heparin lyase, EC 4.2.2.7). *J. Clin. Microbiol.* **26**, 1070–1071.
10. Watanabe, M., Tsuda, H., Yamada, S., Shibuta, Y., Nakamura, T., and Sugahara, K. (1998) Characterization of heparinase from an oral bacterium *Prevotella heparinolytica*. *J. Biochem.* **123**, 283–288.
11. Lohse, D. L. and Linhardt, R. J. (1992) Purification and characterization of heparin lyases from *Flavobacterium heparinum*. *J. Biol. Chem.* **267**, 24,347–24,355.
12. Godavarti, R. and Sasisekharan, R. (1996) A comparative analysis of the primary sequences and characteristics of heparinases I, II, and III from *Flavobacterium heparinum*. *Biochem. Biophys. Res. Commun.* **229**, 770–777.

13. Shriver, Z., Hu, Y., and Sasisekharan, R. (1998) Heparinase II from *Flavobacterium heparinum*. Role of histidine residues in enzymatic activity as probed by chemical modification and site-directed mutagenesis. *J. Biol. Chem.* **273**, 10,160–10,167.
14. Shriver, Z., Hu, Y., Pojasek, K., and Sasisekharan, R. (1998) Heparinase II from *Flavobacterium heparinum*. Role of cysteine in enzymatic activity as probed by chemical modification and site-directed mutagenesis. *J. Biol. Chem.* **273**, 22,904–22,912.
15. Godavarti, R. and Sasisekharan, R. (1998) Heparinase I from *Flavobacterium heparinum*. Role of positive charge in enzymatic activity. *J. Biol. Chem.* **273**, 248–255.
16. Sasisekharan, R., Leckband, D., Godavarti, R., Venkataraman, G., Cooney, C. L., and Langer, R. (1995) Heparinase I from *Flavobacterium heparinum*: the role of the cysteine residue in catalysis as probed by chemical modification and site-directed mutagenesis. *Biochemistry* **34**, 14,441–14,448.
17. Godavarti, R., Cooney, C. L., Langer, R., and Sasisekharan, R. (1996) Heparinase I from *Flavobacterium heparinum*. Identification of a critical histidine residue essential for catalysis as probed by chemical modification and site-directed mutagenesis. *Biochemistry* **35**, 6846–6852.
18. Linker, A. and Hovingh, P. (1972) Isolation and characterization of oligosaccharides obtained from heparin by the action of heparinase. *Biochemistry* **11**, 563–568.
19. Grant A. C., Linhardt, R. J., Fitzgerald, G. L., Park, J. J., and Langer R. (1984) Metachromatic activity of heparin and heparin fragments. *Anal. Biochem.* **137**, 25–32.
20. Bitter, T. and Muir, H. M. (1962) A modified uronic acid carbozole reaction. *Anal. Biochem.* **4**, 330–334.
21. Ampofo, S. A., Wang, H. M., and Linhardt, R. J. (1991) Disaccharide compositional analysis of heparin and heparan sulfate using capillary zone electrophoresis. *Anal. Biochem.* **199**, 249–255.
22. Linhardt, R. J. and Pervin, A. (1996) Separation of negatively charged carbohydrates by capillary electrophoresis. *J. Chromotogr. A.* **720**, 323–335.
23. Hileman R. E., Smith, A. E., Toida, T., and Linhardt, R. J. (1997) Preparation and structure of heparin lyase-derived heparan sulfate oligosaccharides. *Glycobiology* **7**, 231–239.
24. Sugahara, K. Tohno-oka, R., Yamada, S., Khoo, K.-H., Morris, H. R., and Dell, A. (1994) Structural studies on the oligosaccharides isolated from bovine kidney heparan sulfate and characterization of bacterial heparitinases used as substrates. *Glycobiology* **4**, 535–544.
25. Rice, K. G. and Linhardt R. J. (1989) Study of defined oligosaccharides substrates of heparin and heparan monosulfate lyases. *Carbohydr. Res.* **190**, 219–233.
26. Desai, U. R., Wang, H., and Linhardt, R. J. (1993) Substrate specificity of heparin lyases from *Flavobacterium heparinum*. *Arch. Biochem. Biophys.* **306**, 461–468.
27. Linhardt, R. J. (1994) Analysis of glycosaminoglycans with polysaccharide lyases, in *Current Protocols in Molecular Biology, Analysis of Glycoconjugates,* (Varki, A., ed.), Wiley-Interscience, Boston, MA, vol. 2, pp.p 17.13.17–17.13.32.

# 36

# Degradation of Chondroitin Sulfate and Dermatan Sulfate with Chondroitin Lyases

María José Hernáiz and Robert J. Linhardt

## 1. Introduction

Glycosaminoglycans (GAGs) are a family of complex linear polysaccharides characterized by a repeating core disaccharide structure typically comprised of an *N*-substituted hexosamine and an uronic acid residue. They can be categorized into four main structural groups: hyaluronate, chondroitin sulfate (CS)/dermatan sulfate (DS); heparan sulfate/heparin and keratan sulfate.

The biological roles of chondroitin and dermatan sulfate GAGs are poorly understood and their exact chemical structures have not been determined. Because enzymes are highly specific and act under mild conditions, enzymatic methods are often preferable over chemical methods for determining the structure of GAGs.

Enzymes that degrade GAGs have become increasingly important tools for understanding the biological roles of GAGs and the proteoglycans, including the regulation of various cellular process such as adhesion, differentiation, migration, and proliferation *(1)*. Utilizing these enzymes, design and preparation of GAG-based therapeutic agents might become possible *(2)*. Such drugs could have uses as antithrombotic agents, antiatherosclerotic agents, antiinflammatory agents, inhibitors of complement activation and regulators of cell growth, angiogenesis, and antiviral agents.

CS and DS are the most common type of GAGs in extracellular matrix proteoglycans *(1)*. CS is a hetereopolysaccharide made up largely of repeating disaccharide units, in which one sugar is *N*-acetyl-D-galactosamine and the other is D-glucuronic. These disaccharides can be sulfated at the 4- or 6-position of the *N*-acetylgalactosamine residue. The major classes are CS-A (chondroitin 4-sulfate), DS (CS-B), containing 4-sulfated *N*-acetylgalactosamine and iduronic acid, and CS-C (chondroitin 6-sulfate) (*see* **Fig. 1**).

From: *Methods in Molecular Biology, Vol. 171: Proteoglycan Protocols*
Edited by: R. V. Iozzo © Humana Press Inc., Totowa, NJ

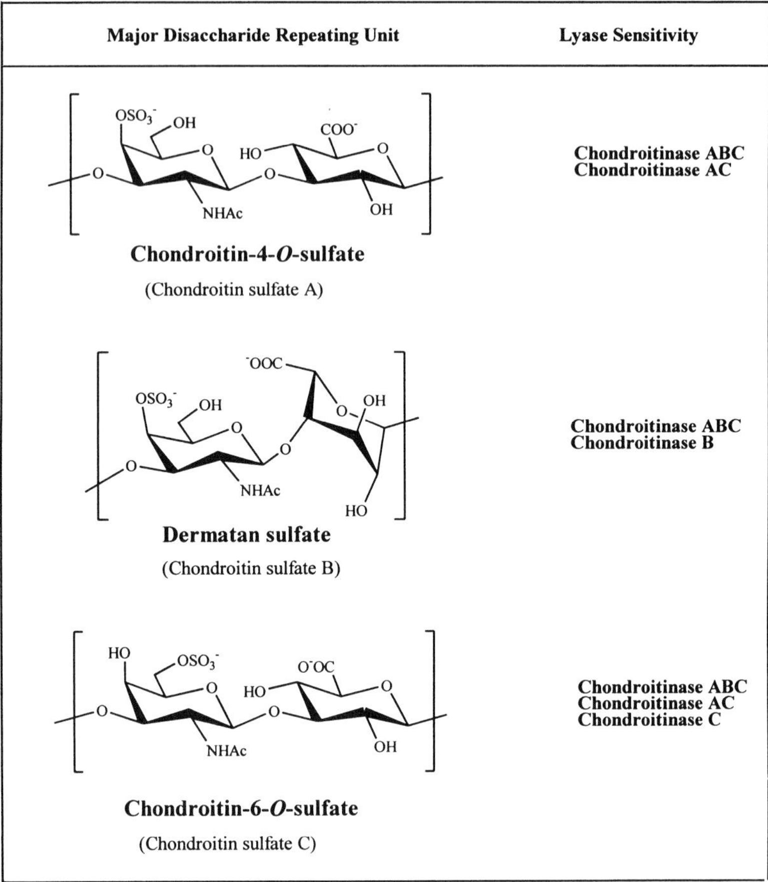


Fig. 1. Glycosidic linkages present in CS/DS and chondroitin lyases that act on these linkages.

Microorganisms are a major source of GAG-degrading enzymes *(3–5)*, particularly in the case of soil bacteria, which may depend on connective tissues in animal carcasses as a nutrient source. Based on their catalytic mechanism, GAG-degrading enzymes are divided into two distinct classes: prokaryotic enzymes, which are lyases that depolymerize GAGs by an elimination mechanism *(5)*, and eukaryotic enzymes, which act by hydrolysis *(6)* (*see* **Fig. 2**). The chondroitin lyases depolymerize the CS and DS, by an elimination mechanism, into oligosaccharides containing a $\Delta_{4,5}$-unsaturated uronic acid residue at the nonreducing end *(3–5)*. This residue exhibits an absorbance maximum at 232 nm, permitting the detection of the oligosaccharide products of the chondroitin lyases using ultraviolet (UV) spectroscopy.

Four classes of chondroitin lyases have been biochemically characterized: those that act on chondroitin, chondroitin-4-sulfate and chondroitin-6-sulfate (chondroitinase

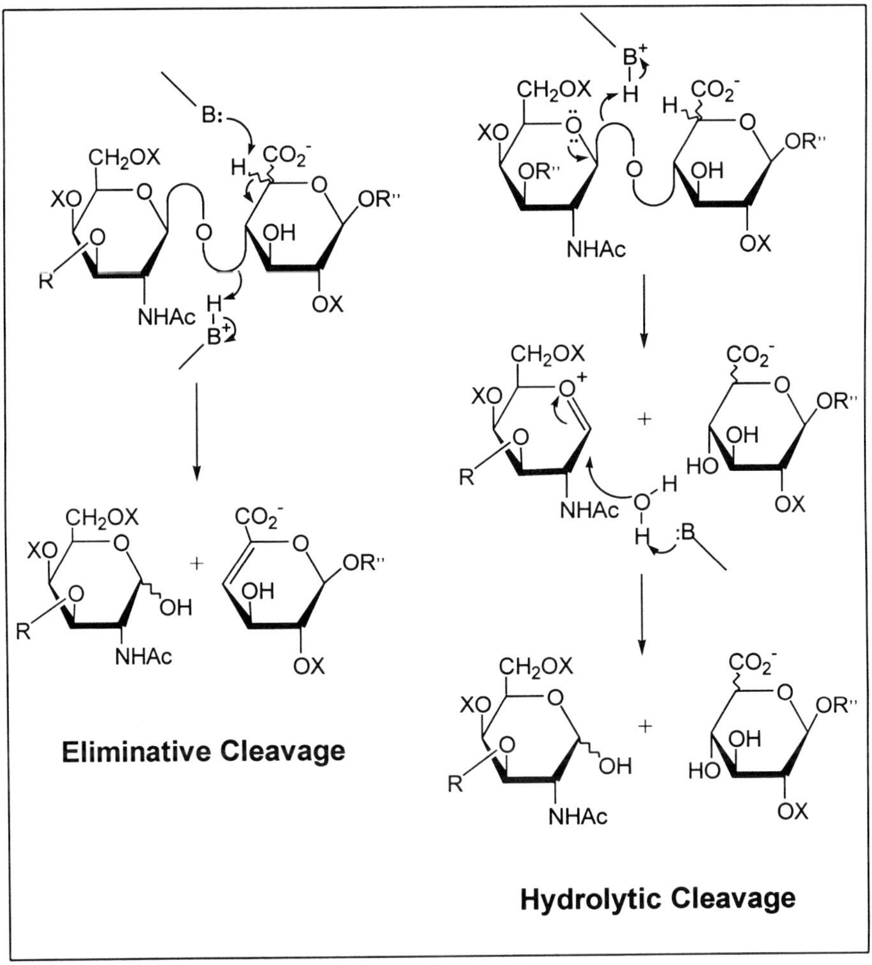

Fig. 2. Enzymatic mechanisms for chondroitin lyases (eliminative cleavage) and chondroitin hydrolases (hydrolytic cleavage), where B is a basic residue in the enzymes catalytic site, R is a monosaccharide (exolytic) or an oligosaccharide (endolytic), R' is H (exolytic) or an oligosaccharide (endolytic), and R'' is an oligosaccharide or polysaccharide.

AC or chondroitin AC lyase); dermatan sulfate (chondroitinase B or chondroitin B lyase); chondroitin-6-sulfate and hyaluronate (chondroitinase C or chondroitin C lyase); and an enzyme with broad substrate specificity that acts on both chondroitin and dermatan sulfate (chondroitinase ABC or chondroitin ABC lyase) (*see* **Fig. 1**). Most commercial preparations of chondroitin ABC lyase are a mixture of two enzymes with endo (ABC endolyase) and exo (ABC exolyase) activities *(7)*. The activity of these enzymes toward small oligosaccharide substrates differs substantially. Similarly, there are two chondroitin AC lyases, AC-I (endolyase) and AC-II (exolyase) ***(8–9)***.

**Table 1**
**Sources of Commonly Used Chondroitin Lyases**

| Chondroitinase | **Source** |
|---|---|
| Chondroitinase ABC (mixture of endolyase and exolyase) from *Proteus vulgaris* | Sigma<br>Seikagaku |
| Chondroitinase AC-I from *Flavobacterium heparinum* | Sigma<br>Seikagaku |
| Chondroitinase AC-II from *Arthrobacter aurescens* | Sigma<br>Seikagaku |
| Chondroitinase B from *Flavobacterium heparinum* | Sigma<br>Seikagaku |
| Chondroitinase C from *Flavobacterium heparinum* | Sigma |

Chondroitin lyases are most commonly obtained from *Proteus vulgaris*, *Arthrobacter aurescens*, *Bacteroides thetaiotaomicron, Bacteroides stercoris,* and *Flavobacterium heparinum*. Chondroitin lyases from *P. vulgaris*, *A. aurescens,* and *F. heparinum* have been purified to homogeneity and are commercially available (*see* **Table 1**). While little is know about the catalytic machinery of these enzymes, the recent publications of the three-dimensional structure of chondroitinase AC and B should shed light on their mechanism of action ***(10–12)***.

Determination of CS/DS oligosaccharide structure is a formidable analytical problem that has limited structure–activity relationship studies, and the development of improved methods is necessary for further progress. Current approaches involve the preparation of CS/DS oligosaccharides using chondroitin lyases followed by separation techniques including gel permeation chromatography (GPC) ***(13)***, strong anion exchange (SAX)-high-performance liquid chromatography (HPLC) ***(14)***, polyacrylamide gel electrophoresis (PAGE), ***(14)*** and capillary electrophoresis (CE) ***(13,15)***, that permit analysis of disaccharide composition. These provide important data on composition and domain structure but generally yield indirect and incomplete sequence information. Mass spectrometry (MS) has also been applied to the analysis of CS/DS oligosaccharides. Fast-atom bombardment (FAB-MS), electrospray ionization (ESI-MS), and matrix-assisted laser desorption/ionization (MALDI-MS) are capable of determining the molecular weight of oligosaccharides ***(13)***. Although muclear magnetic resonance (NMR) spectroscopy provides for the accurate determination of the chemical fine structure of small CS/DS oligosaccharides (containing 2–14 saccharide units), it requires a large amount of material ***(13,16–18)***.

What follows in this chapter are descriptions of the materials and methods required to use chondroitin lyase enzymes in the degradation of CS/DS-containing sample and how to assay the activity of these enzymes.

## 2. Materials

### *2.1. Enzyme Preparation and Storage*

1. Tris-HCl/sodium acetate buffer, 50 m*M* (*see* **Table 2** for pH).

**Table 2**
**Properties of the Chondroitinase Lyases**

| Chondroitin Lyase/Organism | MW (Da) | Buffer system | Opt. pH | Opt $T$ (°C) |
|---|---|---|---|---|
| Chondroitinase ABC from *Proteus vulgaris* | 150,000 | Tris-HCl/sodium acetate | 8.0 | 37 |
| Chondroitinase ACI from *Flavobacterium heparinum* | 76,000 | Tris-HCl/sodium acetate | 7.5 | 37 |
| Chondroitinase ACII from *Arthrobacter aurescens* | 76,000 | Tris-HCl/sodium acetate | 6.0 | 37 |
| Chondroitinase B from *Flavobacterium heparinum* | 55,000 | Tris-HCl/sodium acetate | 7.5 | 25 |
| Chondroitinase C from *Flavobacterium heparinum* | — | Tris-HCl/sodium acetate | 8.0 | 25 |

2. Chondroitin lyase. The decision of which lyase to use should be based on the specificity desired (*see* **Fig. 1** and **Table 1**).
3. 500-µL polypropylene microcentrifuge tubes.

### 2.2. Sample Preparation and Enzymatic Digestion

1. Tris-HCl/sodium acetate buffer, 50 m*M* (for pH *see* **Table 2**).
2. Chondroitin lyase solution.
3. CS- or DS-containing sample.
4. Spectropor dialysis tubing (1000-MWCO Spectrum or Centricon (YM3, & MWCO 3000, Millipore) centrifugal filter unit.
5. 500-µL polypropylene microcentrifuge tubes.

### 2.3. Assay Protocol

1. Tris-HCl/sodium acetate buffer, 50 m*M* (*see* **Table 2** for pH).
2. Chondroitin lyase solution.
3. CS- or DS-containing sample.
4. 500-mL polypropylene microcentrifuge tubes.
5. UV-spectrophotometer, temperature controlled.
6. 1-mL quartz cuvet.
7. Radiolabel-containing sample.
8. Dialysis membrane (MWCO 1000) or Centricon (YM3, MWCO 3000) centrifugal filter unit.
9. 500-µL polypropylene microcentrifuge tubes.
10. Water baths at 30° and 35°C for enzyme digestion and at 100°C for inactivation of the enzyme reaction.
11. Additional reagents and equipment for product analysis, such as, SAX-HPLC, GPC, PAGE, CE, MS, and NMR.

## 3. Methods

### 3.1. Enzyme Preparation and Storage

1. Dissolve the commercial enzyme (*see* **Note 1**) by adding buffer directly to each vial to afford a 4 mU/mL final concentration (*see* **Note 2**). Cap the vials tightly and gently agitate until the solids are completely dissolved.

2. Dispense 10-μL aliquots of the enzyme solution for storage.
3. Store tubes containing enzyme at –60 to –80°C (*see* **Note 3**).

## 3.2. Sample Preparation and Enzymatic Digestion

### 3.2.1. Complete Chondroitin Lyase-Catalyzed Depolymerization of a Sample

1. Dissolve sample, containing 1 μg to 1 mg CS or DS, in 1 mL of distilled water. Exhaustively dialyze sample against distilled water using 1000-MWCO dialysis membrane. Freeze-dry the nondializable retentate. Add 50 μL of Tris-HCl/sodium acetate buffer (*see* **Note 4**).
2. Thaw and assay activity of a frozen aliquot of enzyme (*see* **Subheading 3.3.**).
3. Add 40 μL of Tris-HCl/sodium acetate buffer containing CS/DS sample to 10 μL of chondroitin lyase solution in a 500-μL polypropylene microcentrifuge tube. Add 50 μL of Tris-HCl/sodium acetate buffer to another 500-μL polypropylene microcentrifuge tube to serve as a blank control.
4. Additional enzyme (10- to 100-fold) may be required to break down small, resistant oligosaccharides ***(19,20)***.
5. Incubate 50-μL sample for 8–12 h at 37°C (*see* **Notes 5** and **6**).
6. Heat 2 min at 100°C to terminate the reaction (*see* **Note 7**). Analyze the products by a method appropriate for its purity and concentration (*see* **Subheading 1.**).

### 3.2.2. Complete Chondroitin Lyase-Catalyzed Depolymerization of Radiolabeled GAGs

1. Dissolve GAGs sample containing radiolabeled ($^{35}$S, $^{14}$C, or $^{3}$H) CS or DS in 1 mL of distilled water. Exhaustively dialyze sample against water using 1000-MWCO dialysis membrane. Freeze-dry nondialyzable retentate. Add 50 μL of Tris-HCl/sodium acetate buffer. Alternatively, the radiolabeled sample can be buffer exchanged using a Centricon (YM3, 3000 MWCO) centrifugal filter unit.
2. Thaw 10 μL of chondroitin lyase solution at room temperature and use immediately.
3. Add 30 μL of samples containing radiolabeled CS or DS in Tris-HCl/sodium acetate buffer to the 500-μL polypropylene microcentrifuge tube containing enzyme.
4. Add 10 μL of unlabeled CS or DS (1 mg/mL in Tris-HCl/sodium acetate buffer) or 10 mL of Tris-HCL/sodium acetate (*see* **Note 8**).
5. GPC analysis of CS/DS following complete depolymerization by the appropriate chondroitin lyase (*see* **Note 9**) affords counts in fractions corresponding to a molecular weight < 1000 daltons SAX-HPLC or PAGE can also be used with radioisotope detection.

## 3.3. Assay Protocol

1. Add 640 μL of Tris-HCl/sodium acetate buffer to a 1-mL quartz cuvet. Warm the cuvet to 37°C in a temperature-controlled spectrophotometer (*see* **Note 10**).
2. Thaw a 10-μL aliquot of chondroitinase lyase solution at room temperature.
3. Take the cuvet out of the spectrophotometer, remove 90 μL of warm buffer and transfer it to enzyme solution. Immediately transfer entire 100 μL (buffer plus enzyme) back to the warm cuvet.
4. Place the cuvet in the spectrophotometer and set the absorbance at 232 nm ($A_{232}$) to zero.
5. Remove the cuvet from spectrophotometer and add 50 μL of 20-mg/mL CS or DS solution to initiate reaction. Seal the cuvet with Parafilm and invert once or twice to mix. Remove the Parafilm and return the cuvet to the spectrophotometer. To assay for chondroitin AC lyase activity, use CS A or C as substrate. To assay for chondroitin B lyase activity, use dermatan sulfate as substrate (*see* **Tables 1** and **2**).

6. Within 30 s after adding substrate begin to measure the absorbance continuously or at 30-s intervals for 2–10 min. Plot $A_{232}$ vs time.
7. Calculate the enzyme activity (1 U = 1 μmol product formed/min) from the initial rate (<5% reaction completion) using $\varepsilon = 3800\ M^{-1}$ for reaction products at pH 8.
8. Enzyme activity is calculated as: Enzyme activity = $(\Delta A_{232}/\text{min})$ (700 μL)/ 3800 $M^{-1}$. (Calculate the number of product molecules formed per substrate molecule from the $A_{232}$ measured at reaction completion.)

## 4. Notes

1. Protease contamination can also be present in the enzyme preparation. Commercial enzymes often contain bovine serum albumin (BSA) for stabilization during lyophilization, as this greatly reduces potential problems associated with proteolytic contamination.
2. Enzyme activity is defined differently by different suppliers. The definition of a milliunit used here is 1 nmol of unsaturated product formed/min (*see* **Subheading 3.3.** for assay protocol).
3. These enzymes can be stored in their lyophilized or reconstituted states at –20°C or –70°C for >1 yr. Once an enzyme is reconstituted, it should be aliquoted and frozen immediately. Single aliquots can be thawed to assay the enzyme and for use in an experiment. Chondroitinase ABC is very stable, but chondroitinase AC, B, and C are most susceptible to thermal inactivation ***(21,22)***. Lyase storage stability is enhanced by high (> 2 mg/mL) protein concentrations. This is often accomplished by addition of BSA.
4. Chondroitin lyases are compatible with a wide range of buffers, including succinate, acetate, ethylenediamine acetate, Tris-HCl, Bis-Trispropane-HCl, sodium phosphate, MOPS, TES, and HEPES ***(21)***.
5. Samples from of tissues, biological fluids, proteoglycans, and GAGs that contain microgram or greater quantities of CS/DS, which are not metabolically labeled, can be analyzed following treatment with chondroitin lyase.
6. Due to batch variations in enzymes, or the age of laboratory stocks, it is always advisable to test enzyme activities on a standard substrate before using them on valuable samples. This can easily be performed by incubating chondroitin lyase with 1.5 mg/mL of CS and monitoring the time course of the digest by the increase in absorbance at 232 nm. Once the digest appears to have ceased, a second addition of enzyme is useful to confirm that a true end point has been reached, rather than the enzyme having become prematurely inactivated. The quantity of enzyme and/or the incubation time can then be adjusted accordingly to guarantee maximal digestion of samples. The disaccharide yield using the appropriate chondroitin lyase should be > 95 %. Occasionally, CS/DS samples only partially digest, or even fail to digest at all. If the enzyme is selected correctly and is active, then the problem is in the sample quality, i.e., the presence of excess salts and/or buffer ions, certain divalent metals, denaturants, detergents, or other enzyme-inhibitory substances. Further sample clean-up is therefore necessary. Detergents should be removed by precipitation with potassium chloride or by using a detergent-removal column such as Biobeads (Bio-Rad). Urea and guanidine, salts, and metals should be removed by exhaustive dialysis using controlled-pore dialysis membrane (MWCO 1000).
7. Following the use of a lyase, residual lyase activity can be destroyed by heating the reaction mixture to 100°C or by adding denaturants or detergents. Most lyases are cationic proteins and can be removed from anionic oligosaccharide products by passing the reaction mixture through a small cation-exchange column, such as SP-Sephadex, adjusted to an acidic pH. The oligosaccharide products are then readjusted to neutral pH and analyzed. This method can also be used to remove BSA, an excipient found in many of the

commercial enzymes, from the oligosaccharide products. Also, chondroitin lyases can be immobilized using CNBr Sepharose, removed by filtration after the reaction, and reused *(23)*.

8. When attempting to use chondroitin lyase to depolymerize radiolabeled samples that contain very small quantities of chondroitin, it is often useful to add cold substrate as a carrier to prevent sample loss.
9. Chondroitin ABC lyase can be used to completely digest a mixture of CS. If hyaluronate and chondroitin sulfate are both present, it is advisable to use an equal-unit mixture of chondroitin ABC and AC lyases. For complete degradation of all GAGs *(24,25)*. Oversulfated chondroitin sulfates including chondroitin sulfate D, E, and trisulfated chondroitin sulfate are sensitive to chondroitin ABC lyase *(24,25)*.
10. If a temperature-controlled spectrophotometer is not available, activity can be measured at room temperature or the sample can be incubated in a water bath and the absorbance measured at fixed time points.

## References

1. Vogel, K. G. (1994) Glycosaminoglycans and proteoglycans, in *Extracellular Matrix Assembly and Structure*, (Yurchenco, P. D., Brik, D. E.. and Mechan, R. P. eds.), Academic Press, San Diego, CA.
2. Linhardt, R. J. and Toida, T. (1997) Heparin oligosaccharides: new analogues development and applications. in *Carbohydrates in Drug Design*, (Witczak, E. J. and Nieforth, K. A. eds.), Marcel Dekker, New York, NY.
3. Sutherland, I. W. (1995) Polysaccharide lyases. *FEMS Microbiol. Rev.* **16**, 323–347.
4. Ernst, S., Langer, R., Cooney, C. L., and Sasiskharan, R. (1995) Enzymatic degradation of glycosaminoglycans. *Crit. Rev. Biochem. Mol. Biol.* **30**, 387–444.
5. Linhardt, R. J., Galliher, P. M., and Cooney, C. L. (1986) Polysaccharide lyases. *Appl. Biochem. Biotechnol.* **12**, 135–176.
6. Medeiros, M. G. L., Ferreira, T. M. P. C., Leite, E. L., Toma, L., Dietrich, C. P., and Nader, H. B. (1998) New pathway of heparan sulfate degradation. Iduronate sulfatase and *N*-sulfoglucosamine 6-sulfatase act on the polymer chain prior to depolymerisation by a *N*-sulfo-glucosaminidase and glycuronidase in the mollusc *Tagelus gibbus*. *Compar. Biochem. Physiol. B—Biochem. Mol. Biol.* **119**, 539–547.
7. Hamai, A., Hashimoto, N., Mochizuki, H., Kato, F., Makiguchi, Y., Horie, K., and Suzuki, S. (1997) Two distinct chondroitin sulfate ABC lyases; an endoeliminase yielding tetrasaccharides and an exoeliminase preferentially acting oligosaccharides. *J. Biol. Chem.*, **272**, 9123–9130.
8. Jandik, K. A., Gu, K., and Linhardt, R. J. (1994) Action pattern of polysaccharide lyases on glycosaminoglycans. *Glycobiology* **4**, 289–296.
9. Gu, K., Liu, J., Pervin, A., and Linhardt, R.J. (1993) Comparison of the activity of two chondroitin AC lyases on dermatan sulfate. *Carbohydr. Res.* **244**, 369–377.
10. Li, Y., Matte, A., Su, H., and Cygler, M. (1999) Crystallization and preliminary X-ray analysis of chondroitinase B from *Flavobacterium heparinum*. *Acta Crystallogr. Biol. Crystallogr.* **55**, 1055–1057.
11. Huang, W., Matte, A., Li, Y., Kim, Y. S., Linhardt, R. J., and Cygler, M. (1999). Crystal structure of chondroitinase B from *Flavobacterium heparinum* and its complex with a disaccharide product at 1.7 A resolution. *Mol. Biol.* **294**, 1257–1269.
12. Fethiere, J., Shilton, B. H., Li, Y., Allaire, M., Laliberte, M., Eggimann, B., and Cygler, M. (1998) Crystallization and preliminary analysis of chondroitinase AC from *Flavobacterium heparinum*. *Acta Crystallogr. D—Biol. Crystallogr.* **54**, 279–280.

13. Yang, H. O., Gunay, N. S., Toida, T., Kubean, B., Yu, G., Kim, Y. S., and Linhardt, R. J. (2000) Preparation and structural determination of dermatan sulfate derived oligosaccharides. *Glycobiology* **10**, 1033–1040.
14. Linhardt, R. J., Desai, U. R., Liu, J., Pervin, A. Hoppensteadt, D., and Fareed, J. (1994) Low molecular weight dermatan sulfate as an antithrombotic agent: structure-activity relationship studies. *Biochem. Pharm.* **47**, 1241–1252.
15. Pervin, A., Al-Hakim, A., and Linhardt, R. J. (1994) Separation of glycosaminoglycan-derived oligosaccharides by capillary electrophoresis using reverse polarity. *Anal. Biochem.* **221**, 182–188.
16. Mourão, P. A. S., Pavão, M. S. G., Mulloy, B., and Wait, R. (1997) Chondroitin ABC lyase digestion of an ascidian dermatan sulfate. *Carbohydr. Res.* **300**, 315–321.
17. Lamb, D. J., Wang, H. M., Mallis, L. M., and Linhardt, R. J. (1992) Negative-ion fast-atom bombardment tandem mass spectrometry to determine sulfate and linkage position in glycosaminoglycan-derived disaccharides. *J. Am. Soc. Mass Spectrom.* **3**, 797–803.
18. Kitagawa, H., Tanaka, Y., Yamada, S., Seno, N., Haslam, S. M., Morris, H. R., Dell, A., and Sugahara, K. (1997) A novel pentasaccharide sequence GlcA(3-sulfate)(β1-3) GalNac(4-sulfate)(β1-4)(Fucα1-3)GlcA(β1-3)GalNAc(4-sulfate) in the oligosaccharides isolated from king crab cartilage chondroitin sulfate K and its differential susceptibility to chondroitinases and hyaluronidase. *Biochemistry* **36,** 3998–4008.
19. Rice, K. G. and Linhardt, R. J. (1989) Study of defined oligosaccharide substrates of heparin and heparan monosulfate lyases. *Carbohydr. Res.* **190**, 219–233.
20. Desai, U. R., Wang, H. M., and Linhardt, R. J. (1993) Specificity studies on the heparin lyases from *Flavobacterium heparinum. Biochemistry* **32**, 8140–8145.
21. Gu K., Linhardt, R. J., Laliberte, M., Gu, K., and Zimmerman, J. (1995) Purification, characterization and specificity of chondroitin lyases and glycuronidase from *Flavobacterium heparinum. Biochem. J.* **312**, 569–577.
22. Tsuda, H., Yamada, S., Miyazono, K., Yoshida, K., Goto, F., Tamura, J. I., Neumann, K. W., Ogawa, T., and Sugahara, K. (1999) Substrate specificity studies of *Flavobacterium* chondroitinase C and heparitinases towards the glycosaminoglycan-protein linkage region. Use of a sensitive analytical method developed by chromophore labeling of linkage glycoserines using dimethylaminoazobenzenesulfonyl chloride. *Eur. J. Biochem.* **262,** 127–33.
23. Park, Y., Yu, G., Gunay, N. S., and Linhardt, R. J. (1999) Purification and characterization of heparan sulphate proteoglycan from bovine brain. *Biochem. J.* **344**, 723–730.
24. Sugahara, K., Sadanaka, S., Takeda, K., and Kojima, T. (1996) Structural analysis of unsaturated hexasaccharides isolated form shark cartilage chondroitin sulfate D that are substrates for the exolytic action of chondroitin ABC lyase. *Eur. J. Biochem.* **239**, 871–880.
25. Linhardt, R. J. (1994) Analysis of glycosaminoglycans with polysaccharide lyases, in *Current Protocols in Molecular Biology, Analysis of Glycoconjugates*, (Varki, A. ed.), Wiley Interscience, Boston, MA, vol. 2., pp. 17.13.17–17.13.32.

# 37

# In Vitro Assays for Hyaluronan Synthase

**Andrew P. Spicer**

## 1. Introduction

Hyaluronan (HA) is synthesized at the plasma membrane as a free linear polymer of composition $[\beta 1 \rightarrow 4GlcA\beta 1 \rightarrow 3GlcNAc]^n$ *(1–3)*. All models suggest that polymerization occurs at the inner face of the plasma membrane while the polymer is coordinately translocated or extruded across the membrane to the extracellular face of the cell [for review, *see* *(2–5)*]. In mammals *(6)*, and all vertebrates (Spicer, unpublished data), HA is synthesized by any one of three HA synthases (HAS). The three HAS proteins are encoded by three related yet distinct genes *(6,7)*. All HAS proteins are predicted integral plasma membrane proteins with N-terminal and C-terminal transmembrane domains separating a large cytoplasmic domain [for review, *see* *(4)*]. Any one HAS protein is capable of catalyzing the *de novo* synthesis of HA *(8)*, suggesting that each is capable of specifically binding both UDP-sugar substrates and creating the alternate $\beta 1 \rightarrow 3$ and the $\beta 1 \rightarrow 4$ glycosidic bonds. Indeed, the related prokaryotic HA synthase, spHAS, from the Gram-positive bacterium, *Streptococcus pyogenes*, has been purified to apparent homogeneity and is capable of synthesizing high molecular mass HA chains in vitro, when provided with a source of UDP-GlcA, UDP-GlcNAc, and $Mg^{2+}$ ions *(9)*.

The vertebrate HAS proteins have not yet been successfully purified in a soluble and active form. Thus, all current approaches to the detection of HA synthase activity are performed on intact cells or on membrane preparations derived from cells transfected with a particular HA synthase expression vector or from cells that express endogenous HA synthase activity (*see* **Note 1**). In this chapter, three relatively simple procedures for the detection of HA synthase activity will be described.

From: *Methods in Molecular Biology, Vol. 171: Proteoglycan Protocols*
Edited by: R. V. Iozzo © Humana Press Inc., Totowa, NJ

## 2. Materials

### 2.1. General

All three protocols described here use cultured mammalian cell lines. These methods could be equally well applied to the investigation of cell lines derived from other vertebrates and potentially invertebrates. All cells are cultured on tissue culture plastic at 37°C in a humidified atmosphere of 5% $CO_2$. Most cell lines are cultured in medium containing 10% fetal bovine serum plus 5 m*M* L-glutamine and antibiotics (penicillin and streptomycin). Phospate-buffered saline (PBS), calcium- and magnesium-free, is used throughout for washing cell monolayers.

### 2.2. Particle Exclusion Assay

1. Fixed horse or sheep erythrocytes (Sigma), reconstituted to a final cell density of $5 \times 10^8$ cells/mL in PBS, supplemented with 1 mg/mL bovine serum albumin (BSA). Erythrocytes should be washed and pelleted at least twice in this buffer before use, to remove traces of sodium azide. These erythrocyte suspensions can be stored for long periods at 4°C, and need simply to be mixed well before each use to avoid clumps. Pipetting several times with a plugged Pasteur pipet is most effective.
2. *Streptomyces* hyaluronate lyase ***(10)*** (Sigma or Calbiochem) reconstituted to 400 turbidity-reducing units (TRU) per mL in 20 m*M* sodium acetate, pH 5.0, and stored as frozen aliquots (20 TRU/ 50-µL aliquot) at –80°C.
3. Inverted phase-contrast microscope with digital camera or other recording device.

### 2.3. In Vitro HA Synthase Assay

1. Sterile (autoclaved) hypotonic lysis buffer (LB): 10 m*M* KCl, 1.5 m*M* $MgCl_2$, and 10 m*M* Tris-HCl, pH 7.4, stored at 4°C.
2. Protease inhibitors: aprotinin, leupeptin, and phenyl methyl sulfonyl fluoride (PMSF) (Calbiochem); dissolved and used at the manufacturer's recommended concentrations.
3. Prechilled disposable cell scrapers or rubber policemen and 1-mL pipets.
4. Sterile screw-capped 2-mL microcentrifuge tubes.
5. Small (2-mL reservoir) Dounce homogenizer with type B pestle.
6. Refrigerated microcentrifuge, centrifuge, or microcentrifuge/centrifuge located in cold room.
7. Micro-BCA assay kit (Pierce) and flat-bottomed enzyme-linked immunosorbent assay (ELISA) plates.
8. 20% (w/vol) SDS solution.
9. UDP-[$^{14}$C]-glucuronic acid (>180 mCi/mmol; usually 250–350 mCi/mmol), NEN-Dupont, ICN, or Amersham. Can also use UDP-[$^{3}$H]-GlcNAc (NEN-Dupont, 20–45 Ci/mmol). Aliquot UDP-[$^{14}$C]-GlcA into 0.25-µCi aliquots (sufficient for one reaction) and store at –20°C.
10. 10× HA synthase buffer consisting of 50 m*M* dithiothreitol, 150 m*M* $MgCl_2$, and 250 m*M* HEPES, pH 7.1 (sterile-filtered and stored at 4°C).
11. UDP-GlcNAc (stock solution of 100 m*M* aliquotted and stored at –20°C) and UDP-GlcA (stock solution of 5 m*M* aliquotted and stored at –20°C). Dissolved in hypotonic lysis buffer.
12. *Streptomyces* hyaluronate lyase (reconstituted as described under **Subheading 2.2.2.**).
13. 20 m*M* sodium acetate, pH 5.0.
14. Whatman 3MM chromatography paper (big sheets), cut widthwise into strips 2 cm wide. Before cutting into strips, draw lines across the entire length of the sheet as follows: solid

line 2.5 cm from the top; dashed line 6 cm from the top; solid line 8 cm from the top; solid line 10 cm from the top. Cut into individual strips, then fold strips forward at first line, then backwards at second line. A large number of strips can be made in advance and stored in a suitable dry location.
15. Chromatography solvent consisting of a 130/70 mix of absolute ethanol/ 1 *M* ammonium acetate, pH 5.5. It is better to make the solvent fresh before each use by simply mixing ethanol with 1 *M* ammonium acetate, pH 5.5.
16. Chromatography chamber.
17. Scintillation vials, liquid scintillation fluid, and liquid scintillation counter. Alternatively, an imaging device capable of detecting $^{14}C$ could be used to scan each strip and determine the number of counts at the origin.

### 2.4. Metabolic Labeling with [$^3$H]-Acetate or [$^3$H]-Glucosamine

1. Cell line growing in appropriate medium at appropriate density.
2. [$^3$H]-glucosamine (20–45 Ci/mmol, NEN-Dupont or Amersham), or [$^3$H]-acetate (2–5 Ci/mmol, NEN-Dupont or Amersham).
3. 0.25% trypsin, 1 m*M* EDTA solution (Life Technologies).
4. 1.3% (w/vol) potassium acetate/95% ethanol solution.
5. Protease Type XIV (Sigma), resuspended in 100 m*M* Tris-HCl, pH 8.0.
6. 20 m*M* sodium acetate, pH 6.0.
7. *Streptomyces* hyaluronate lyase, reconstituted and aliquotted as described under **Subheading 2.2.2.**
8. 12% (w/vol) cetyl pyridinium chloride (CEPC) (Sigma) solution (made up in deionized water).
9. 0.05% (w/vol) CEPC, 50 m*M* NaCl solution.
10. Methanol.
11. Liquid scintillation counter, liquid scintillant, and scintillation vials.

## 3. Methods

### 3.1. General

Any of the three methods described here can be used to detect HA synthesis. The first, the particle exclusion assay, can be used as a rapid assessment of putative HA synthase activity in, e.g., transfected cells, or as a rapid assessment of possible endogenous HA synthase activity in established or primary cell cultures (*see* **Note 1**). The second in vitro assay is more quantitative in nature, providing a direct measure of the HA biosynthetic capacity of a given cell population (*see* **Note 2**). The third assay, metabolic labeling of HA, provides a measure of the HA being synthesized within a cell culture over a given period (*see* **Note 3**).

### 3.2. Particle Exclusion Assay

This very simple and now classical assay permits the visualization of the HA-dependent pericellular matrix *(11)* (*see* **Note 4**).

1. Cells are plated on tissue culture plastic at a suitable cell density to achieve 30–40% confluence the next day. Duplicate wells are established for each experiment.
2. Cells can either be assayed directly (if an endogenous HA synthase activity is being studied) or transfected with an expression vector carrying a putative HA synthase.

3. 48–72 h post-transfection or at least 24 h postplating (endogenous HA synthases), fixed red blood cells are added to cell cultures. Amounts added are as follows: 24-well plate, $1 \times 10^7$/well; 6-well plate, $5 \times 10^7$/well. Swirl dishes gently to distribute the cell suspension evenly.
4. Culture dishes are returned to the incubator for an additional 15 min to allow erythrocytes to settle on the monolayer.
5. Dishes are removed from the incubator (plates may need to settle on the microscope stage for a few minutes prior to viewing) and examined under phase contrast using the 10× and 20× objectives to scan individual wells for evidence of cells with pericellular coats. Positive cells are viewed using the 40× objective (*see* **Note 5**) and the image is captured for computer analysis (*see* **Note 6**) or displayed on a screen such that the outline of the cell and pericellular matrix can be traced. Tracing onto transparency paper works well for this approach.
6. For quantitation of relative pericellular matrix size, at least 20 cells are imaged per well. The relative pericellular coat size is expressed as the area of the pericellular matrix plus the area of the cell, divided by the area of the cell. Hence, a cell without a pericellular matrix will have a relative value of 1.0. In contrast, cells that are actively synthesizing HA may have ratios of 2.0 or more.
7. After imaging or tracing, 10 TRU of *Streptomyces* hyaluronate lyase (for 6-well plate) are added directly to one well of each pair (*see* **Note 7**). The second well receives an equivalent volume (25 µL) of 20 m*M* sodium acetate pH 5.0. Dishes are returned to the incubator and incubated for at least 1 h more at 37°C.
8. After 1 h, dishes are swirled to redistribute the erythrocyte layer, and **steps 4–6** are repeated.

### 3.3. In Vitro HA Synthase Assay

1. Cell cultures are established on 15-cm tissue culture plates. Generally, between 2 and 4 plates are necessary to obtain sufficient material for in vitro assays (*see* **Notes 8** and **9**).
2. Individual 15-mL conical tubes are filled (10–15 mL) with either PBS (2 per plate) or hypotonic lysis buffer (LB) (3 per plate) and placed on ice. These prechilled solutions will be used for the washes outlined below. Tubes can be reused each time. It is not necessary to cap tubes.
3. Prelabel 2-mL screw-capped microcentrifuge tubes and place on ice. Three tubes should be labeled for each plate. Label as follows: "nuclei," "membranes," and "lysate."
4. Cultures of subconfluent, proliferating cells (for endogenous HA synthase measurement) or transfected cells (72 h posttransfection is best) are removed from the tissue culture incubator and placed on ice in a large tray (*see* **Note 10**). Six plates can be processed at one time.
5. Culture medium is aspirated from each plate and plates are washed twice with cold, prealiquotted PBS. For each wash, buffer is poured gently onto each plate, the plate is swirled gently, then buffer is removed by aspiration.
6. Each plate is washed twice with cold hypotonic lysis buffer. After the second wash, LB is added to each plate and the plates are incubated on ice for 10 min to allow cells to swell.
7. After the 10 min incubation, each plate is treated separately in turn. Buffer is aspirated from the plate and the plate is tilted such that any remaining buffer drains to one side and can be removed by aspiration.
8. 1 mL of LB, supplemented with the protease inhibitors aprotinin and leupeptin (LB+), and prepared just prior to use, is added to the plate. Using a prechilled sterile cell scraper, the monolayer is scraped from the plate into the 1 mL of LB+ (*see* **Note 11**).

**Table 1**
**In Vitro HA Synthase Assay**

| *Component* | *Volume (or amount)* |
|---|---|
| Cell membranes (add last) | 25–100 μg (not to exceed 73.5–75.5 μL volume) |
| LB buffer | to a final volume of 100 μL[a] |
| 10X HA synthase buffer | 10 μL |
| UDP-GlcNAc (100 mM) | 1 μL (or 0 μL for specificity control reaction) |
| UDP-GlcA (5 mM) | 1 μL |
| Aprotinin (0.2 mg/mL in PBS) | 1 μL |
| Leupeptin (5 mg/mL in water) | 1 μL |
| UDP-[$^{14}$C]GlcA | 0.25 μCi (10 or 12.5 μL, depending upon supplier) |

[a]Calculate amount of LB required to bring final volume of each reaction to 100 μL before assembling reactions. Assemble each reaction according to the order in the table, adding the cell membranes last.

9. The suspension is transferred into the mortar of a prechilled, 2-mL Dounce homogenizer. 10 μL of PMSF is added to the suspension. Cells are disrupted by 5–10 twisting strokes using a B-type pestle and the homogenate is transferred to the prechilled tube labeled, "nuclei" (*see* **Note 12**). Leave tube on ice until all plates have been processed.
10. Pellet nuclei and organelles by spinning all tubes at 4000*g* for 5 min in a refrigerated microcentrifuge.
11. Transfer supernatants to prechilled, labeled tubes marked "membranes," and spin at high speed (approximately 20,000*g*) for 15 min in a refrigerated microcentrifuge to pellet cell membranes.
12. Transfer supernatants to prechilled, labeled tubes marked "lysate."
13. Carefully resuspend individual membrane pellets in 50 μL of LB+. Pool resuspended membranes from equivalent samples.
14. Remove 5–10 μL of each sample (after pooling) to a separate tube for determination of protein content using, for instance, a Micro-BCA assay. If the remainder of the sample will not be immediately used, store at –80°C until required (*see* **Note 13**).
15. Determine protein content using a Micro-BCA assay. Five microliters of each sample should be brought up to 205 μL by addition of 200 μL of LB plus 1% SDS (w/v) (*see* **Note 14**). The individual samples can then be split into duplicate wells of an ELISA plate (100 μL/well).
16. Assemble in vitro synthase assays as outlined in **Table 1**. Typically, between 25 and 100 μg of total membrane protein are used per assay (*see* **Notes 15–17**).
17. Incubate assays for 1–2 h (1 h is standard) at 37°C in a heat block or water bath. Establish duplicate or triplicate reactions wherever possible.
18. After incubation, stop the reactions by heating at 100°C for 3–5 min. Pulse-spin to collect condensate.
19. Thaw *Streptomyces* hyaluronate lyase (Hase) aliquots (one per reaction) at room temperature and label tubes to correspond to each reaction under study. Add 50 μL for each reaction to appropriately labeled HA lyase tubes. These tubes will represent the +Hase tubes. To the remaining 50 μL of each reaction, add 50 μL of 20 m*M* sodium acetate, pH 5.0. These tubes will serve as the mock treated group (-Hase). Incubate all tubes for a minimum of 3 h at 60°C.
20. Add 20% sodium dodecyl sulfate (SDS) to each reaction to a final concentration of 1% (w/v) and heat reactions for 5 min at 100°C.

21. Spot each reaction onto the origin (bounded by the lines at 8 and 10 cm from the top) of individual prelabeled paper chromatography strips, taking care to avoid spreading beyond the origin. This can most easily be achieved by applying each reaction as two separate additions, allowing drying between each application.
22. Assemble paper chromatography chamber and elute overnight with 130/70 (absolute ethanol/1 *M* ammonium acetate, pH 5.5) by descending paper chromatography (*see* **Notes 18** and **19**).
23. Cut out origins with scissors and transfer to individual scintillation vials containing 2 mL of deionized water. Ensure that paper is immersed in the water by mixing well.
24. Add liquid scintillation cocktail to each tube, mix well, and count using liquid scintillation counting (LSC) (*see* **Note 20**).
25. Assuming that dpm are approximately equivalent to cpm, the HA synthase activity of each sample can be determined from the LSC results and expressed as pmol/mg membrane protein/h. Calculations should take into account the specific activity of the labeled sugar and the relative molar amounts of unlabeled and labeled UDP-sugar. Compare results from +Hase (generally between 60% and 100% of the HA-dependent counts should be removed by Hase treatment) and –Hase, and reactions performed in the presence (+UDP-GlcNAc) and absence (–UDP-GlcNAc) of UDP-GlcNAc.

### 3.4. Metabolic Labeling with [$^3$H]-Glucosamine or [$^3$H]-Acetate

1. Aspirate medium and wash cell cultures briefly with PBS.
2. Replace cell culture medium with medium containing [$^3$H]-glucosamine (20 µCi/mL of medium) or [$^3$H]-acetate (100 µCi/mL of medium). Return cultures to the incubator and incubate for the desired amount of time for the study in question (*see* **Note 21**).
3. Collect medium, containing free radiolabeled glucosamine or acetate and radiolabeled macromolecules (including HA), and transfer to a 15-mL conical tube (*see* **Note 22**). Wash twice with PBS (2 mL for a single well of a 6-well plate) and add these wash solutions to the previously collected medium. This final solution represents the released or "cell-free HA."
4. Cell-surface HA can also be collected by trypsinization. This removes most of the remaining cell-surface localized HA. Trypsinize cells per the requirements for the cell line under study. Neutralize with complete medium containing serum, and pellet cells by centrifugation. Remove the supernatant to a fresh tube. This will represent the cell surface HA. If desired, this solution can be pooled with the cell-free HA, if total extracellular HA is being monitored. Alternatively, this solution can be treated separately as, "cell-surface HA."
5. Cell number should be determined using, for instance, a hemocytometer or Coulter counter.
6. Three volumes of 1.3% potassium acetate/95% ethanol are added to each solution and the samples are mixed and placed at –20°C overnight.
7. Total macromolecules are collected by centrifugation at 2500*g* for 15 min at 4°C.
8. Precipitates are diluted in 500 µL of 0.5% (w/v) protease XIV (Sigma) solution (in 100 m*M* Tris-HCl, pH 8.0) and incubated on a rocking platform overnight at 37°C (*see* **Note 23**).
9. Three volumes of 1.3% potassium acetate/95% ethanol are added to each sample. Samples are mixed well and placed at –20°C overnight.
10. Macromolecules are collected by centrifugation at 2500*g* for 15 min at 4°C.
11. Precipitates are resuspended in 1 mL of 20 m*M* sodium acetate, pH 6.0.
12. Five TRU (12.5 µL) of *Streptomyces* hyaluronate lyase, or 12.5 µL of 20 m*M* sodium acetate, pH 5.0, are added to duplicate 200-µL aliquots of each sample, and tubes are incubated overnight at 60°C (*see* **Note 24**).

13. 50-μL (0.25 volume) of 12% CEPC solution are added to each tube. Tubes are mixed well by vortexing and incubated at 37°C for 1 h.
14. Samples are centrifuged for 15 min at 10,000*g* at 4°C.
15. Pellets are washed with 500 μL of 0.05% CEPC, 50 m*M* NaCl, and spun again as described in **step 13**.
16. Final pellets can be resuspended in 150 μL of methanol and added directly to scintillation vials containing liquid scintillant, and analyzed by liquid scintillation counting (*see* **Note 25**). Comparison of Hase-treated and mock-treated will give a measure of the total HA-dependent counts in each sample. HA-dependent counts should be related to the cell density in the respective cultures.
17. Alternatively, if an indication of the relative length of labeled HA chains is desired, final pellets can be resuspended in a buffer suitable for size-exclusion chromatography.

## 4. Notes

1. Many established and primary cell lines synthesize significant amounts of HA through endogenous HA synthase activities. These include most embryonic fibroblasts, such as 3T3 and 3T6 cells, the mouse oligodendroglioma cell line (G26-24), human lung fibroblasts (WI38), primary keratinocytes and dermal fibroblasts, and various chondrocyte cultures. Most of these cell lines (for instance, 3T6, G26-24, WI38, and dermal fibroblasts) express predominantly Has2 (Spicer, unpublished data). There are several important cell lines however, that synthesize reduced levels or no detectable HA. These include the SV40-transformed African green monkey kidney cell lines, COS-1 and COS-7 (no detectable activity), human embryonic kidney (HEK293) cell line (no detectable activity), and various Chinese hamster ovary lines (low levels). The latter group of cell lines has proven extremely useful for functional screening of putative HA synthases through transfection experiments ***(6,8,16)***.
2. Endogenous HA synthase activities can be markedly affected by cell density ***(12)***. In general, HA synthase activity is highest in subconfluent, proliferating cultures and lowest in confluent, growth-arrested cultures ***(12)***. This may in part reflect regulation of HA synthase messenger RNA levels ***(13)***. Thus, the cell density is an important consideration in the overall experimental design. Any variations in cell density from one experiment to the next should, therefore, be noted.
3. Measurements of metabolically labeled HA can also be affected by receptor-mediated uptake and degradation (through hyaluronidase action). Hence, measurement of metabolically labeled HA may reflect the steady-state HA levels resulting from both synthesis and degradation. In cell cultures that are not actively endocytosing and degrading HA, extracellular (in the cell culture medium) HA levels may climb steadily and reach a steady-state level at confluence. In contrast, in those cell cultures that are actively endocytosing and degrading HA, extracellular HA levels may start to decline at confluence as the balance between synthesis and degradation is likely to shift in favor of uptake and degradation.
4. Some cell lines can assemble a HA-dependent pericellular matrix from exogenously supplied HA. This is dependent on the presence of proteoglycans and upon the expression of a cell surface HA receptor such as CD44 ***(14)***.
5. When viewing HA-dependent pericellular matrices in cell cultures grown in small wells, the cell density can vary across the well. In particular, cell densities may be much higher at the periphery of the well. Thus, it is important to consider local fluctuations in cell density when scanning a well.

6. Computer programs such as NIH-Image (http://rsb.info.nih.gov/nih-image/index.htmL) can be used very effectively to determine the relative areas of the HA-dependent pericellular matrix and the cell, either from directly captured images, or from scanned images of traced cells. Using NIH-Image, for instance, the area corresponding to the cell only is filled and the area calculated. Next, the pericellular matrix plus the cell are filled and the total area calculated. Data can be imported into a spreadsheet program such as Microsoft Excel for rapid calculation of ratios.
7. It is not necessary to remove the erythrocytes from each well during treatment with *Streptomyces* hyaluronate lyase.
8. The number of plates that will be necessary for a given experiment will be determined in part by the characteristics of the cell or cells under study. Transfected COS cells can be processed at 72 h posttransfection, at which time most cultures will be confluent. Under these conditions, 3 plates will usually yield between 100 and 200 μg of crude cell membranes.
9. The method detailed here describes the relatively small-scale preparation of membranes for in vitro HA synthase assay. If larger amounts of membranes are required, a larger Dounce homogenizer may be used along with 15-mL conical tubes and centrifuge tubes suitable for use in a larger centrifuge, rather than screw-capped microcentrifuge tubes. In these instances, cell suspensions from multiple 15-cm plates, or from roller bottles, can be homogenized at the same time.
10. It is vitally important to ensure that plates sit flat on the ice, such that there are no "dry" spots on the plate. Do not leave the plates empty for any length of time between washes.
11. When scraping the monolayer into the LB+, do not scrape back and forth. Instead, work systematically across the plate, from left to right.
12. It may be necessary to determine the optimal number of strokes required to break open cells leaving nuclei intact. This can be achieved through transferring a drop of the homogenate to a microscope slide, cover slipping, and observing under low (4× objective) and high (20× objective) magnification. Accept a lower efficiency of cell breakage rather than risk damage to nuclei. Chromatin released into the homogenate can induce aggregation of membrane vesicles and dramatically reduce yields.
13. Resuspended membrane preparations retain significant enzyme activity even if freeze-thawed up to 3–4 times. However, for maximum enzyme activity and reproducibility, membrane preparations should be aliquotted into 25- to 50-μL aliquots (depending on protein content) and stored at –80°C.
14. Addition of SDS is important in order to solubilize the cell membranes prior to the BCA assay. Likewise, it is important to serially dilute any standards (such as BSA), which may be used for the generation of a standard curve, in the same buffer (LB plus 1% SDS).
15. The synthase buffer described here contains $Mg^{2+}$ as the cation. The vertebrate HA synthases and the Streptococcal enzyme prefer $Mg^{2+}$. However, if this procedure is used to assay putative invertebrate or prokaryotic HAS enzymes, alternate cations, such as $Mn^{2+}$, might be considered before a membrane preparation or cell lysate is considered negative for HA synthase activity. Similarly, alternative reaction conditions, such as pH, temperature and substrate concentration, should be considered.
16. It is easiest to assemble reactions in 2-mL screw-capped tubes, as several rounds of heating and boiling are required for the whole procedure.
17. Assemble at least one specificity-control reaction in which UDP-GlcNAc (or the unlabeled sugar) is omitted from the reaction. HA synthesis requires both UDP-GlcA and

UDP-GlcNAc. Hence, any labeled product generated in the absence of UDP-GlcNAc is, by definition, not HA.

18. Place chromatography strips into the chromatography chamber as follows: the first fold (2.5 cm from the top of the strip) is placed under the glass rod; the second fold (6 cm from the top of the strip) takes the strip over the rail and allows it hang in a vertical manner. Once all strips have been placed into the chamber, solvent is added to the top of the reservoir.
19. If multiple chromatography strips are eluted at the same time, ensure that adjacent strips do not touch at their origins.
20. Set up the scintillation counter such that counts achieve 95% confidence, or proceed for 10 min, whichever comes first.
21. Generally, cells grown in individual wells of a 6-well tissue-culture plate will yield sufficient labeled HA over a 24- to 48-h period.
22. Approximately 70% or more of the radiolabeled HA synthesized by a given cell line will be found in the medium. Most of the remaining HA will be associated with the cell surface. This may occur through several mechanisms, including continued interaction with the HA synthase, interaction with specific cell-surface HA receptors, or through incorporation into the pericellular matrix through homophilic interaction with other HA chains, or through heterophilic interaction with specific HA-binding proteins. If desired, the intracellular HA fraction, a small percentage of the total HA, can be collected through lysis of the cell pellet using, for instance, treatment with a PBS solution supplemented with 0.5–1% (w/v) Triton X-100 or NP-40. The recent report demonstrating intracellular HA and its dynamic distribution during the cell cycle ***(15)*** suggests that intracellular HA may play previously unimagined roles in cell biology.
23. Protease XIV treatment will digest any protein molecules that may carry covalently linked, radiolabeled carbohydrate chains.
24. As molecules other than HA will be metabolically labeled by culture in the presence of [$^3$H]-glucosamine or [$^3$H]-acetate, it is important to determine the amount of radiolabeled HA in each sample. This can be determined by treatment of part of each sample with *Streptomyces* hyaluronate lyase.
25. Liquid scintillation counting will provide an accurate measure of the amount of HA produced by a given cell line. Size-exclusion chromatography, for instance as described ***(16)***, can provide a measure of the amount of HA and the relative molecular mass of the chains.

## References


1. Weissman, B. and Meyer, K. (1954) The structure of hyalobiuronic acid and of hyaluronic acid from umbilical cord. *J. Am. Chem. Soc.* **76**, 1753–1757.
2. Philipson, L. H. and Schwartz, N. B. (1984) Subcellular localization of hyaluronan synthetase in oligodendroglioma cells. *J. Biol. Chem.* **259,** 5017–1023.
3. Prehm, P. (1984) Hyaluronate is synthesized at plasma membranes. *Biochem. J.* **220,** 597–600.
4. Weigel, P. H., Hascall, V. C., and Tammi, M. (1997) Hyaluronan synthases. *J. Biol. Chem.* **272**, 13,997–14,000.
5. Weigel, P. H. (1998) Bacterial hyaluronan synthases. Hyaluronan Today Internet Glycoforum, http://www.glycoforum.gr.jp/science/hyaluronan/HA06/HA06E.html.
6. Spicer, A. P., and McDonald, J. A. (1998) Characterization and molecular evolution of a vertebrate hyaluronan synthase (HAS) gene family. *J. Biol. Chem.* **273,** 1923–1932.
7. Spicer, A. P., Seldin, M. F., Olsen, A. S., Brown, N., Wells, D. E., Doggett, N. A., Itano, N., Kimata, K., Inazawa, J., and McDonald, J. A. (1997) Chromosomal localization of the human and mouse hyaluronan synthase (HAS) genes. *Genomics* **41**, 493–497.

8. Itano, N., Sawai, T., Yoshida, M., Lenas, P., Yamada, Y., Imagawa, M., Shinomura, T., Hamaguchi, M., Yoshida, Y., Ohnuki, Y., Miyauchi, S., Spicer, A. P., McDonald, J. A., and Kimata, K. (1999) Three isoforms of mammalian hyaluronan synthases have distinct enzymatic properties. *J. Biol. Chem.* **274**, 25,085–25,092.
9. DeAngelis, P. L. and Weigel, P. H. (1994) Immunochemical confirmation of the primary structure of streptococcal hyaluronan synthase and synthesis of high molecular weight product by the recombinant enzyme. *Biochemistry* **33**, 9033–9039.
10. Ohya, T. and Kaneko, Y. (1970) Novel hyaluronidase from *Streptomyces*. *Biochim. Biophys. Acta* **198,** 607–609.
11. Clarris, B. J. and Fraser, J. R. (1968) On the pericellular zone of some mammalian cells in vitro. *Exp. Cell Res.* **49**, 181–193.
12. Matuoka, K., Namba, M., and Mitsui, Y. (1987) Hyaluronate synthetase inhibition by normal and transformed human fibroblasts during growth reduction. *J. Cell Biol.* **1104**, 1105–1115.
13. Watanabe, K. and Yamaguchi Y. (1996) Molecular identification of a putative human hyaluronan synthase. *J. Biol. Chem.* **271**, 22,945–22,948.
14. Knudson, W., Bartnik, E., and Knudson, C. B. (1993) Assembly of pericellular matrices by COS-7 cells transfected with CD44 lymphocyte-homing receptor genes. *Proc. Natl. Acad. Sci. (USA)* **90**, 4003–4007.
15. Evanko, S. P. and Wight, T. N. (1999) Intracellular localization of hyaluronan in proliferating cells. *J. Histochem. Cytochem.* **47**, 1331–1342.
16. Brinck, J. and Heldin, P. (1999) Expression of recombinant hyaluronan synthase (HAS) isoforms in CHO cells reduces cell migration and cell surface CD44. *Exp. Cell Res.* **252**, 342–351.

38

# Hyaluronidase Activity and Hyaluronidase Inhibitors

## *Assay Using a Microtiter-Based System*

**Susan Stair Nawy, Antonei B. Csóka, Kazuhiro Mio, and Robert Stern**

## 1. Introduction

Hyaluronidase is a term applied to a group of very dissimilar enzymes *(1–3)* that degrade hyaluronan (HA, hyaluronic acid), a high-molecular-weight glycosaminoglycan of the extracellular matrix. Some of these enzymes have the ability to degrade additional glycosaminoglycans, albeit at a slower rate. Most of the hyaluronidases from eukaryotes have both hydrolytic and transglycosidase activity, while those from bacteria operate by β-elimination. HA is prominent whenever rapid cell proliferation and movement occur, particularly during embryogenesis, wound healing, repair and regeneration, and in tumorigenesis *(4–8)*. Hyaluronidases regulate temporal and spatial distribution patterns of HA, critical during such processes. Hyaluronidases, often present at exceedingly low concentrations, are imbued with high but unstable specific activities. They can be difficult to detect, and their quantitation requires specialized techniques.

Previous methods for quantitation of hyaluronidase either lacked sensitivity, were slow and cumbersome, or required highly specialized reagents not available in most laboratories *(9–12)*. This accounts in part for the relative neglect, until recently, of this important group of enzymes. An improved ELISA-like assay was developed in which hyaluronidase activity could easily be detected in most biological samples *(13)*. The general technique is described here, together with notes regarding more specialized usage. The free carboxy groups of HA are biotinylated in a one-step reaction using biotin hydrazide. Standard and unknown samples of enzyme are subsequently allowed to react with the HA substrate covalently bound to the wells of 96-well microtiter plates. Residual substrate is detected with an avidin-peroxidase color reaction that can be read using a standard ELISA plate reader. A standard curve of hyaluronidase activity is run with each plate using serial dilutions of any hyaluronidase standard, and the

From: *Methods in Molecular Biology, Vol. 171: Proteoglycan Protocols*
Edited by: R. V. Iozzo © Humana Press Inc., Totowa, NJ

activity can then be expressed in absolute units. This rapid and sensitive technique, 1000 times more sensitive than the commonly used assays, facilitates evaluation of hyaluronidase in most biological samples including the conditioned media of cultured cells (*see* **Note 1**). The present assay facilitated the isolation and sequence analysis of the first vertebrate somatic hyaluronidase, human plasma hyaluronidase, Hyal-1 ***(14)***. This permitted detection of six paralogous hyaluronidase sequences in the human genome (***15***, *see* **Note 2**).

By using a standard amount of a defined hyaluronidase preparation, the procedure is easily converted to a rapid assay for hyaluronidase inhibitors. The latter, though first described in 1946, are a class of molecule about which very little is known ***(16–18)***. They are obviously important in the intricate modulation of HA deposition.

## 2. Materials

1. Human umbilical cord HA (ICN, Costa Mesa, CA), morpholineethane sulfate (MES), biotin, sulfo-NHS, DMSO stock solution, stock solution of 100 m*M* 1-ethyl-3-(3-dimethylaminopropyl) carbidodiimide (EDC), and 4 *M* guanidine-HCl.
2. Bovine testicular hyaluronidase (Wydase⟩, Wyeth-Ayerst Co., Philadelphia, PA). Covalink-NH plates, and PBS containing 2 *M* NaCl and 50 m*M* $MgSO_4$ (washing buffer).
3. COVALINK-NH microtiter plates (NUNC, Placerville, NJ), buffer: 0.1 *M* formate, pH 3.7, 0.1 *M* NaCl, 1% Triton X-100, 5 m*M* saccharolactone for lysosomal acid-active hyaluronidases, and 10.5 mL of PBS containing 0.1% Tween 20.
4. 96-well vinyl assay plates (Costar, Cambridge, MA).
5. N-hydroxysulfosuccinimide (Sulfo-NHS) and biotin hydrazide (Pierce, Rockford, IL).
6. Dimethyl sulfoxide (DMSO) and guanidine hydrochloride (Fisher Scientific, Pittsburg, PA).
7. O-Phenylenediamine (OPD, Calbiochem, La Jolla, CA).
8. Avidin-biotin complex (ABC kit, Vector Labs, Burlingame, CA).
9. All other reagents are from Sigma Chemical Company (St. Louis, MO).

## 3. Methods

### *3.1. Preparation of Biotinylated HA (bHA)*

1. Dissolve 100 mg of human umbilical cord HA in 0.1 *M* morpholineethane sulfate (Mes), pH 5.0, to a final concentration of 1 mg/mL, and allow to dissolve for at least 24 h at 4°C prior to the coupling with biotin.
2. Add sulfo-NHS to the HA-Mes solution to a final concentration of 0.184 mg/mL.
3. Dissolve biotin hydrazide in DMSO as a stock solution of 100 m*M* and add to the HA solution to a final concentration of 1 m*M*.
4. Prepare a stock solution of 100 m*M* 1-ethyl-3-(3-dimethylaminopropyl) carbidodiimide (EDC) in $dH_2O$ and add to the HA-biotin solution to a final concentration of 30 μ*M*. Stir overnight at 4°C.
5. Remove the unlinked biotin and EDC by the addition of 4 *M* guanidine-HCl and dialysing against 1000× volumes of $dH_2O$ with at least three changes. The dialyzed bHA can be aliquoted and stored at –20°C for up to several months.

### *3.2. Immobilization of bHA onto Wells of Microtiter Plates*

1. Dilute sulfo-NHS to 0.184 mg/mL in $dH_2O$ with the bHA at a concentration of 0.2 mg/mL and pipet into 96-well Covalink-NH plates at 50 μL per well.

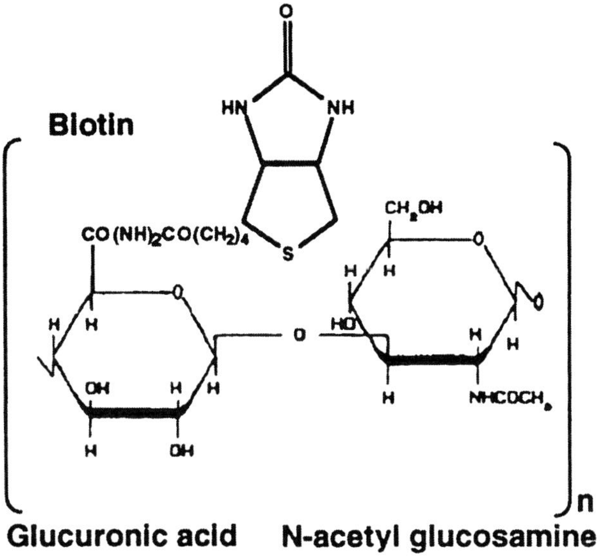

Fig. 1. Structure of the biotinylated disaccharide repeating unit of HA resulting from a reaction between HA, biotin-hydrazide, and EDAC (Reprinted by courtesy of Academic Press from **ref.** ***14***).

2. Dilute EDC to 0.123 mg/mL in $dH_2O$ and pipet into the Covalink plates with the HA solution yielding a final concentration of 10 µg/well HA and 6.15 µg/well EDC (*see* **Fig.1**).
3. Incubate the plates either overnight at 4°C or for 2 h at 23°C. Remove coupling solution by shaking and washing the plates three times in PBS containing 2 *M* NaCl and 50 m*M* $MgSO_4$ (washing buffer). The plates can be stored at 4°C for up to 1 wk in this buffer.

### 3.3. Assay for Hyaluronidase Activity

1. Equilibrate the plates with 100 µL/well assay buffer: 0.1 *M* formate, pH 3.7, 0.1 *M* NaCl, 1% Triton X-100, 5 m*M* saccharolactone for lysosomal acid-active hyaluronidases (*see* **Note 3**). Substitute 0.1 *M* formate, pH 4.5, for the PH-20 enzyme, and a neutral buffer for neutral-active enzymes.
2. Prepare a set of standards for the calibration of enzymatic activity against relative turbidity reducing units (rTRUs) by serial dilutions of Wydase or other hyaluronidase with a known activity, in the appropriate buffer, from 1.0 to $1 \times 10^{-6}$ rTRU and assay 100 µL/well, the standard enzyme reaction volume, in triplicate (*see* **Note 4**).
3. Pipet unknown samples into plate, 100 µL/well, in triplicate. Preliminary experiments must be performed to determine the proper dilution of the enzyme preparation. In order to establish a valid unit of enzyme activity, the range in which activity is proportional to levels of enzyme must be established (*see* **Figs. 2** and **3**). Samples from sources possessing very low activity may be enhanced by a preliminary immunoaffinity-purification step.
4. Incubate samples in plates for 30–60 min at 37°C.
5. Include positive and negative control wells (no enzyme or no avidin-biotin complex), in triplicate.

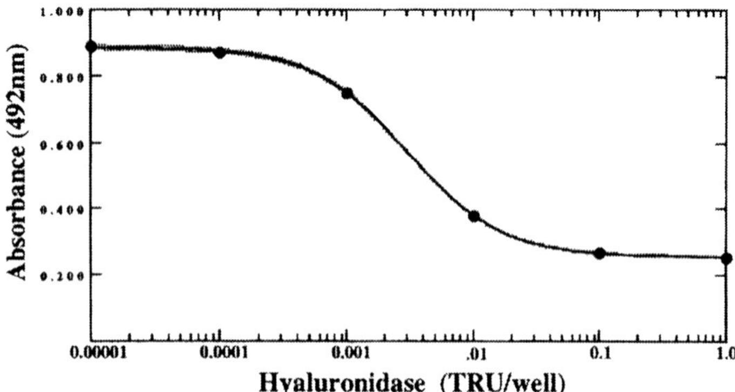


Fig.2. A four-parameter curve-fit of bovine testicular hyaluronidase assays as standard reactions performed at pH 3.7, diluted from 1.0 to $1 \times 10^{-6}$ rTRU/well. (Reprinted by courtesey of Academic Press from **ref. *14***.)

6. Terminate the reaction by addition of 100 μL/well of 6 *M* guanidine-HCl, followed by three washes using the washing buffer.
7. Prepare ABC in 10.5 mL of PBS containing 0.1% Tween 20 (preincubated for 30 min at room temperature during the hyaluronidase incubation).
8. Add 100 μL/well ABC solution, and incubate for 30 min at room temperature. Do not add the ABC solution to the negative control wells.
9. Wash the plate five times with the washing buffer, and add 100 μL/well of an OPD substrate by dissolving one 10-mg tablet of OPD in 10.5 mL of 0.1 *M* citrate-phosphate buffer, pH 5.3, plus 7 μL of 30% $H_2O_2$.
10. Incubate the plate in the dark for 5–10 min and read using a 492-nm filter in an ELISA plate reader (Titertek Multiskan PLUS, ICN) and monitor by computer using the Delta Soft II plate reader software from Biometallics (Princeton, NJ).
11. A standard curve is generated by a four-parameter curve-fit of the serial dilutions of the hyaluronidase standard and unknowns.

### 3.4. General Considerations

1. Since the activity of hyaluronidases can be unstable, particularly in the absence of detergents, samples should be kept on ice. Stability on freeze-thaw of any unknown enzyme preparation should be determined ahead of time, before large-scale experiments are undertaken and frozen.
2. Similarly, a variety of buffers should be compared, to establish the optimal buffer for a particular system. Acetate buffer may inhibit activity, compared to a formate buffer at the same pH, as is observed in embryonic chick brain extracts ***(19)***. Inclusion of low concentrations of reducing agents (1–5 m*M* dithiothreitol) should also be tested for their effect on activity. This is particularly important at low protein concentrations, less that 0.05 mg/mL.
3. Hyaluronidase inhibitors appear to be ubiquitous, and may interfere with detection of hyaluronidase activities, particularly in preparations that are crude extracts. Occasionally, extracts must be diluted out. The inhibitors appear to occur at more limiting concentrations than enzyme. Paradoxically, greater levels of apparent hyaluronidase activity are observed in plasma samples that are more diluted.

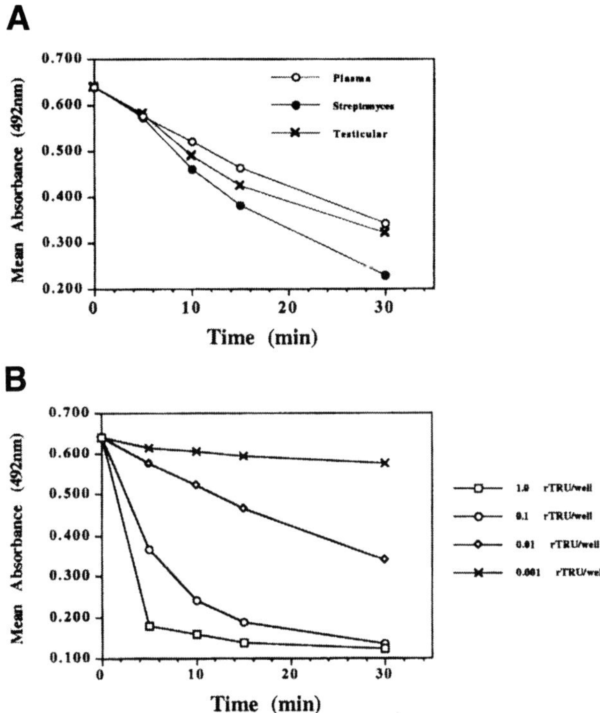


Fig.3. **(A)** Linearity of the enzyme reation over a 30-min incubation period, comparing three different hyaluronidases. 0.01 rTRU/well was utilized at each time and assayed in triplicate. **(B)** Kinetic analyis of log dilutions of immunoaffinity-purified recombinant human plasma hyaluronidase (Hyal-1). The enzyme, from 1.0 to 0.001 rTRU/well, was assayed from 0 to 30 min.

4. Some hyaluronidase inhibitors are $Mg^{2+}$-dependent *(20)*. For this reason, prior optimization of assay conditions is necessary for each new system. The effect of EDTA (1–10 m*M*, neutral pH) should be examined ahead of time. This may also explain why occasionally differences can be observed, in assays of hyaluronidase as well as of its inhibitors, between PBS (phosphate-buffered saline) and PBS-CMF (PBS calcium- and magnesium-free), or between serum and plasma that has been collected with EDTA or citrate.
5. For detecting low levels of activity, such as in cell culture media, sensitivity of the assay can be enhanced by increasing the incubation time from 1 h up to 12 h. This is useful for detecting low levels of activity. However, to maintain a valid unit of enzyme activity, time dependency must be demonstrated.
6. Any enzyme assay is time-dependent. The use of multichannel pipettes facilitates rapid addition of reagents and enzyme to the microtiter plates, ensuring reproducibility and standardizes the reaction time. The assay procedure can be further optimized by initially placing both the enzyme standards and the unknowns into a vinyl assay plate (Costar) in the same amounts and arrangement intended for the substrate-coated reaction plate (Covalink-NH). Once samples are loaded into the vinyl template plate, they can be transferred into the substrate-coated reaction plates with great speed and efficiency.

7. Kinetic analysis of data can be attempted. However, the HA substrate in this assay is bound to the microtiter plate, and solid–liquid interface interactions differ significantly from the true solution chemistry assumed by Michaelis-Menten analysis.

## 4. Notes

1. An inherited disorder involving absence of serum hyaluronidase activity was described recently ***(21,22)***. The present assay can easily be established as a routine clinical laboratory procedure and, requiring only 1-µL samples of serum for each determination, can be used for genetic screening.
2. A paralogous family of hyaluronidase-like sequences have been identified ***(15)***, three each at chromsomes 3p21.3 (HYAL1, 2, and 3) and 7q31.3 (HYAL4, SPAM1, the gene for the sperm-specific activity, PH-20, and HYALP1, a pseudogene). The product of HYAL2, Hyal-2, is an unusual hyaluronidase activity, degrading high-molecular-weight HA only to an intermediate 20-kDa-sized product. The present assay does not detect Hyal-2 activity, since the 20-kDa HA oligosaccharide would remain bound to the mirotiter plate. For the assay of Hyal-2 activity, the colorimetric assay of Reissig et al. is recommended ***(23)***. That assay is a measure of the new reducing N-acetylglucosamine terminus generated by each cleavage reaction. On the other hand, leech hyaluronidase, which is an endo-β-glucuronidase rather than an endo-β-N-acetylglucosaminidase, is not detected by the Reissig assay, but can be assayed using the present procedure.
3. The saccharolactone (D-saccharic acid-1,4-lactone, Sigma) is an inhibitor of β-glucuronidase activity ***(24)***. The enzyme catalyzes a reaction that produces false positives in the Reissig assay, generating reducing terminal N-acetylglucosamine residues. The importance of this in the microtiter assay has not been established. It may depend on the activity of β-glucuronidase in the particular extract being examined.
4. The definition of one relative turbidity reducing unit (rTRU) is defined as the amount of enzyme that reduces the turbidity producing capacity of 0.2 mg HA to 0.1 mg in 30 min at 37°C ***(25)***.

## Acknowledgments

This work was supported by National Institutes of Health (USA) Grant 1P50 DE/CA11912, to R. S., and by Lion Corporation, Kanagawa, Japan, to K. M.

## References

1. Kreil, G. (1995) Hyaluronidases—a group of neglected enzymes. *Protein Sci.* **4**, 1666–1669.
2. Frost, G. I., Csóka, T. and Stern, R. (1996) The Hyaluronidases: a chemical, biological and clinical overview. *Trends Glycosci. Glycotechnol.* **8**, 419–434.
3. Csóka, T. B., Frost, G. I., and Stern, R. (1997) Hyaluronidases in tissue invasion. *Invasion Metastasis*, **17**, 297–311.
4. Laurent, T. C. and Fraser, J. R. (1992) Hyaluronan. *Faseb J.* **6**, 2397–2404.
5. Laurent, T. C. (ed.), *The Chemistry, Biology and Medical Applications of Hyaluronan and its Derivatives*, Portland Press, London, UK, 1998.
6. Toole, B. P. (1991) Proteoglycans and hyaluronan in morphogenesis and differentiation, in *Cell Biology of the Extracellular Matrix* (Hay, E. D., ed.), Plenum Press, New York, NY, pp. 259–294.

7. Knudson, W., Biswas, C., Li, X. Q., Nemec, R. E., and Toole, B. P. (1989) The role and regulation of tumor-associtated hyaluronan, in *The Biology of Hyaluronan* (Evered, D., and Whelan, J., eds.), Wiley, Chichester, UK, pp. 150–169.
8. Weigel, P. H., Fuller, G. M., and LeBoeuf, R. D. (1986) A model for the role of hyaluronic acid and fibrin in the early events during the inflammatory response and wound healing. *J. Theor. Biol.* **119**, 219–234.
9. Dorfman, A., Ott, M. L., and Whitney, R. (1948) The hyaluronidase inhibitor of human blood. *J. Biol. Chem.* **174**, 621–629.
10. Benchetrit, L. C., Pahuja, S. L., Gray, E. D., and Edstrom, R. D. (1977) A sensitive method of the assay of hyaluronidase activity. *Anal. Biochem.* **79**, 431–437.
11. De Salegui, M., Plonska, H., and Pigman, W. (1967) A comparison of serum and testicular hyaluronidase. *Arch. Biochem. Biophys.* **121**, 548–554.
12. Stern, M. and Stern, R. (1992) An ELISA-like assay for hyaluronidase and hyaluronidase-inhibitors. *Matrix* **12**, 397–403.
13. Frost, G. I. and Stern, R. (1997) A microtiter-based assay for hyaluronidase activity not requiring specialized reagents. *Anal. Biochem.* **251**, 263–269.
14. Frost, G. I., Csóka, T. B., Wong, T., and Stern, R. (1997) Purification, cloning, and expression of human plasma hyaluronidase. *Biochem. Biophys. Res. Commun.* **236**,10–15.
15. Csóka, A. B., Scherer, S. W., and Stern, R. (1999) Expression analysis of six paralogous human hyaluronidase genes clustered on chromosomes 3p21 and 7q31. *Genomics* **60**, 356–361.
16. Haas, E. (1946) On the mechanism of invasion. I. Antinvasin I, An enzyme in plasma. *J. Biol. Chem.* **163**, 63–88.
17. Dorfman, A., Ott, M. L., and Whitney, R. (1948) The hyaluronidase inhibitor of human blood. *J. Biol. Chem.*, **223**, 621–629.
18. Moore, D. H. and Harris, T. N. (1949) Occurrence of hyaluronidase inhibitors in fractions of electrophoretically separated serum. *J. Biol. Chem.* **179**, 377–381.
19. Polansky, J. R., Toole, B. P., and Gross, J. (1974) Brain hyaluronidase: changes in activity during chick development. *Science* **183**, 862–864.
20. Mathews, M. B., Moses, F. E., Hart, W., and Dorfman, A. (1952) Effects of metals on the hyaluronidase inhibitor of human serum. *Arch. Biochem. Biophys.* **35**, 93–100.
21. Natowicz, M. R., Short, M. P., Wang, Y., Dickersin, G. R., Gebhardt, M.C., Rosenthal, D. I., Sims, K. B., and Rosenberg, A. E.(1996) Clinical and biochemical manifestations of hyaluronidase deficiency. *N. Engl. J. Med.* **335**, 1029–1033.
22. Triggs-Raine, B., Salo, T. J., Zhang, H., Wicklow, B. A., and Natowicz, M. R. (1999) Mutations in HYAL1, a member of a tandemly distributed multigene family encoding disparate hyaluronidase activities, cause a newly described lysozomal disorder, mucopolysaccharidosis IX. *Proc. Natl. Acad. Sci. (USA)* **96**, 6296–6300.
23. Reissig, J. L., Strominger, J. L., and Leloir, L. R. (1955) *J. Biol. Chem.* **217**, 298–305.
24. Orkin, R.W. and Toole, B. P. (1980) Isolation and characterization of hyaluronidase from cultures of chick embryo skin- and muscle-derived fibroblasts. *J. Biol Chem.* **255**, 1036–1042.
25. Tolksdorf, S., McCready, M. H., McCullagh, D. R., and Schwenk, E. (1949) The turbidometric assay of hyaluronidase. *J. Lab. Clin. Med.* **34**, 74–79.

# 39

# Detecting Hyaluronidase and Hyaluronidase Inhibitors

## *Hyaluronan-Substrate Gel and -Inverse Substrate Gel Techniques*

**Kazuhiro Mio, Antonei B. Csóka, Susan Stair Nawy, and Robert Stern**

## 1. Introduction

Hyaluronidases are a group of enzymes that degrade the glycosaminoglycan hyaluronan (HA, hyaluronic acid). Many types of hyaluronidases are reported, from prokaryotes to eukaryotes *(1,2)*. These enzymes have a wide variety of properties, including substrate specificity, inhibitor sensitivity, and a range of pH optima. *Streptomyces* hyaluronidase, and the venom hyaluronidases from bee, snake, and scorpion are active at neutral pH. Hyal-1, the best-characterized somatic hyaluronidase *(3–5)*, product of one of the six hyaluronidase-like sequences in the human genome *(6)*, is an acid-active enzyme with an optimum at pH 3.7. The sperm-specific PH-20 *(7)* has apparently two pH optima, pH 4.5 and 7.5, resulting possibly from two forms of the enzyme, membrane-bound and soluble *(8,9)*.

When analysis is being attempted for the first time of the hyaluronidase activity in a particular biological source, several problems are encountered. The pH optimum of the activity is not known. In addition to the very low levels of hyaluronidase in most tissues, there are hyaluronidase inhibitors. These appear to be ubiquitous, and are most apt to contaminate hyaluronidase preparations in the crude state, particularly in the early steps of an enzyme purification procedure. Little is known about these inhibitors, a class of molecule first described in 1946 *(10–12)*. However, there has been little progress in the isolation and characterization of this potentially important class of molecules.

Two types of gel procedures are described here that manage to overcome some of these problems. The HA-substrate gel procedure was developed specifically for the detection of hyaluronidase activity in crude extracts and in other complex mixtures *(13,14)*. This procedure can be run using a wide range of pH values and in various

From: *Methods in Molecular Biology, Vol. 171: Proteoglycan Protocols*
Edited by: R. V. Iozzo © Humana Press Inc., Totowa, NJ

buffers. The procedure also separates enzyme from inhibitor and permits detection of hyaluronidase activity in the presence of inhibitors. The inverse HA-substrate gel procedure was developed to examine precisely these hyaluronidase inhibitors ***(15)***.

The HA-substrate gel procedure requires addition of HA into the gel mixture before polymerization. The gel is stained with Alcian blue. Hyaluronidase activity appears as a white clearing on a light blue background. Double staining with Coomassie blue enhances the sensitivity of the procedure. Proteins that remain in the gel, particularly HA-binding proteins, appear as dark blue bands. These are eliminated by interposing a pronase digestion step. Some proteins, when present at a particularly high concentration, can prevent penetration of dye into the gel, thus causing a false positive. However, these are also eliminated by the pronase digeston step.

The two techniques can provide much useful information on the nature of hyaluronidase activities and their inhibitors: pH optima, ion and co-factor requirements, and their relative molecular sizes, even in the crudest extracts.

## 2. Materials

### 2.1. HA-Containing Sodium Dodecyl Sulfate (SDS) Gels

1. Human umbilical cord HA (ICN Biomedicals, Costa Mesa, CA). For the stock solution, dissolve 100 mg of HA in 100 mL of $H_2O$ (1 mg/mL). Continue mixing for at least 24 h at 4°C. Dilute to 0.4 mg/mL for the working solution.
2. Protein molecular-weight standards (Bio-Rad, Richmond, CA).
3. Acrylamide stock solution (30% acrylamide/0.8% bisacrylamide).
4. 4× Separating gel buffer: 0–1% 1.5 *M* Tris-HCl/0.1% SDS, pH 8.8. Dissolve 91 g of Tris base and 2 g of SDS in 300 mL of $H_2O$. Adjust to pH 8.8 with 1 *N* HCl. Add $H_2O$ to 500 mL total volume.
5. 20% ammonium persulfate.
6. TEMED (*N, N, N′,N′*-tetrametylethylenediamine).
7. Isobutyl alcohol.
8. 4× Stacking gel buffer: 0.5 *M* Tris-HCl/SDS, pH 6.8. Dissolve 6.05 g of Tris base and 0.4 g of SDS in 40 mL of $H_2O$. Adjust to pH 6.8 with 1 *N* HCl. Add $H_2O$ to 100 mL total volume.
9. 2× SDS/sample buffer: Dissolve 1.52 g of Tris base, 20 mL of glycerol, 2.0 g of SDS, 2.0 mL of 2-mercaptoethanol, and 1 mg of Bromphenol blue in 40 mL of $H_2O$. Adjust to pH 6.8 with 1 *N* HCl. Adjust to 100 mL with $H_2O$.
10. 5× SDS/electrophoresis buffer: Dissolve 15.1 g of Tris base, 72.0 g of glycine, and 5.0 g of SDS in $H_2O$, adjust to 1000 mL. Do not adjust the pH. It will be pH 8.3 when it becomes diluted. Dilute 1/5 with $H_2O$ just before use (*see* **Note 1**).

### 2.2. HA-Substrate Gels

1. 3% Triton X-100 solution: 3% Triton X-100 in 50 m*M* HEPES, pH 7.4.
2. pH 3.7 hyaluronidase assay buffer: 0.15 *M* NaCl in 0.1 *M* formate buffer, pH 3.7.
3. pH 7.4 hyaluronidase assay buffer: 0.15 *M* NaCl in 50 m*M* HEPES, pH 7.4.
4. Alcian blue staining solution. Dissolve 0.5 g of Alcian blue in 100 mL of 3% acetic acid solution.
5. Destaining solution for Alcian blue staining: 7% acetic acid.

6. Coomassie blue solution: Dissolve 50 mg of Coomassie brilliant blue R (*see* **Note 2**) in 50 mL of methanol. Add 10 mL of acetic acid and 40 mL of $H_2O$, mixing well.
7. Destaining solution for Coomassie blue staining: 50% methanol and 10% acetic acid.

### 2.3. Inverse HA-Substrate Gels

1. 3% Triton X-100 solution: 3% Triton X-100 in 50 m*M* HEPES, pH 7.4.
2. pH 7.4 hyaluronidase inhibitor assay buffer: Prepare 0.5 rTRU/mL bovine testicular hyaluronidase (Sigma): 0.15 *M* NaCl, 1 m*M* $MgCl_2$ in 50 m*M* HEPES, pH 7.4 (*see* **Note 3**).
3. 0.1 mg pronase/mL in phosphate-buffered saline (PBS), pH 7.4.
4. Alcian blue solution. Dissolve 0.5 g of Alcian blue in 3% acetic acid solution.
5. Destaining solution for Alcian blue staining: 7% acetic acid.
6. Destaining solution for Coomassie blue staining: 50% methanol and 10% acetic acid.
7. PBS.

### 2.4. Equipment

1. Gel casting cassette.
2. Electrophoresis cassette.
3. 37°C incubator.
4. Rotating shaker platform.
5. Transilluminator.
6. Constant-voltage power supply.
7. Pipet.
8. Pasteur pipet or 10-mL syringes with size 22 gage needles.
9. Glass Petri dish, 15 cm diameter.
10. Razor blades.

## 3. Methods

### 3.1. Electrophoresis in HA Gel

1. Gel electrophoresis is performed using a modification of the method of Laemmli ***(16)***. Set up the gel cassette.
2. Prepare the separating gel mixture by adding 5.0 mL of acrylamide stock solution, 3.75 mL of HA solution (0.4 mg/mL), 3.75 mL of separating gel buffer, 2.5 mL of water, and 100 μL of 20% ammonium persulfate.
3. Add 10 μL of TEMED. Gently swirl the flask to ensure even mixing. The addition of TEMED will initiate the polymerization reaction, so it is advisable to work fairly quickly at this stage.
4. Using a Pasteur pipet or syringe with a needle, transfer the separating gel mixture to the gel cassette carefully until it reaches 1 cm below the bottom of the sample loading comb.
5. Add isobutyl alcohol slowly to the top of the gel (~1 cm) to smooth out the gel surface.
6. Allow the gel to polymerize for 20 min.
7. While the separating gel is setting, prepare the stacking gel solution. Mix 0.65 mL of acrylamide stock solution, 1.25 mL of stacking gel buffer, 3.05 mL of water, and 50 μL of 20% ammonium persulfate.
8. When the separating gel has polymerized, pour off the overlaying isobutyl alcohol, and then wash twice with $H_2O$. Add 5 μL of TEMED to the stacking gel solution. Add the stacking gel solution to the gel cassette until the solution reaches the cutaway edge of the gel plate. Place the well-forming comb into this solution and leave to set. This will

take about 15 min. It is useful at this stage to mark the position of the bottom of the wells on the glass plates with a marking pen.

9. Carefully remove the comb from the stacking gel. Assemble the cassette in the electrophoresis tank. Fill the top reservoir with electrophoresis buffer, ensuring that the buffer fully fills the sample loading wells. Fill the bottom tank with electrophoresis buffer.
10. Mix samples with the same volume of 2× sample buffer and apply onto the gel (5–20 mL). Protein molecular-weight standards should be also applied to the end lane of the gel.
11. Connect the electrophoresis cassette to the power supply using a current of 15 mA/gel. Electrophoresis is stopped when the bromphenol blue reaches the bottom of the gel.

### 3.2. HA-Substrate Gel for the Detection of Hyaluronidase Activity

1. After electrophoresis, incubate the gel in 3% Triton X-100 for 1 h with agitation to remove SDS from the gel.
2. Transfer gels into the hyaluronidase assay buffer using a suitable pH. Rinse the gel twice with assay buffer.
3. Incubate on the rotating shaker for 16 h at 37°C.
4. Rinse the gel twice with distilled water.
5. Stain the gel in the Alcian blue solution for 1 h (*see* **Note 4**).
6. Place the gel in 7% acetic acid for destaining and fixation. Change solutions once every hour until bands of hyaluronidase can be observed clearly. This step requires 2–3 h.
7. Rinse the gel twice with distilled water, followed by two washes of 50% methanol/10% acetic acid.
8. Transfer the gel to the Coomassie blue solution. Incubate for 30 min.
9. Destain the gel with 50% methanol/10% acetic acid. Change solutions once every hour until bands appear on the gel. This step requires 2–3 h.
10. Analyze the bands of hyaluronidase activity by comparing with the lane containing protein molecular-weight standards. Hyaluronidase activity will appear as a white clearing on a pale blue background. Protein bands will stain blue (*see* **Note 6**).

### 3.3. Inverse HA-Substrate Gel for the Detection of Hyaluronidase Inhibitors

1. After electrophoresis, carefully separate the end lane containing the protein standards using a razor blade. Protein standards are subjected to Coomassie blue staining. Perform the following steps for the other portion of the gel.
2. Incubate the gel in 3% Triton X-100 in 50 m*M* HEPES, pH 7.4, with agitation to remove SDS from the gel.
3. Rinse the gel twice with 50 m*M* HEPES, pH 7.4.
4. Gels are transferred to the appropriate hyaluronidase-containing solution. For the detection of inhibitors of testicular hyaluronidase, e.g., 0.5 rTRU/mL of bovine testicular hyaluronidase, pH 7.4 is used. Incubate in the reaction solution for 16 h at 37°C on the rotating platform (*see* **Note 5**).
5. Rinse the gel twice with PBS.
6. Transfer the gel to a solution of 0.1 mg/mL pronase in PBS and incubate on the rotator for 4 h at 37°C. This step digests proteins that may produce false positives (*see* **Note 6**).
7. Rinse the gel twice with distilled water.
8. Transfer the gel to the Alcian blue solution, and stain the gel for 16 h.
9. Place the gel in 7% acetic acid for the destaining procedure. Change solutions once every hour until hyaluronidase inhibitor bands can be observed. This step usually requires 2–3 h.

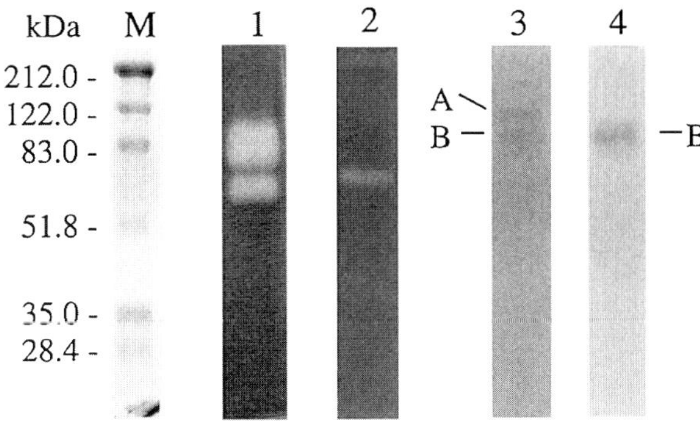


Fig. 1 Detection of hyaluronidase and hyaluronidase inhibitors using HA-substrate gel and inverse HA-substrate gel procedures. M), molecular-weight markers. *Lanes 1 and 2*: HA-substrate gel for the detection of hyaluronidase activities. Examination of bovine testicular hyaluronidase, PH-20 (lane 1), and human plasma hyaluronidase, Hyal-1 (lane 2), were performed at pH 7.4 and pH 3.7, respectively. *Lanes 3 and 4*: Inverse HA-substrate gel for the examination of hyaluronidase inhibitor in mouse serum. The HA-containing gel, to which mouse serum had been applied, was digested with 0.5 rTRU/mL of bovine testicular hyaluronidase at pH 7.4 (lane 3). The position of the hyaluronidase inhibitors corresponds to a band in which the HA remained undigested (lane 3, A). A false positive corresponds to a band of endogenous plasma glycoprotein and possible HA-binding protein. This band could be identified by running a corresponding gel that does not contain HA (lane 4, B). Lane 1, 0.5 rTRU bovine testicular hyaluronidase; lane 2, 0.5 μL human plasma; lanes 3 and 4; 4 μL human plasma.

10. Rinse the gel twice with distilled water.
11. Establish the molecular sizes of hyaluronidase inhibitors by comparison with protein molecular-weight standards.

## 4. Notes

1. This protocol is designed for two Bio-Rad minigels (8 × 10 cm.). For other sizes or thicknesses, volumes of stacking and separating gels, and operating current, must be adjusted accordingly.
2. Not all Coomassie blue preparations work equally well. We have found the product of BDH Chemicals, (Poole, UK) optimal. There are others that do not work at all.
3. Analysis of bovine testicular hyaluronidase on an HA-substrate gel demonstrates two forms of the enzyme (*see* **Fig.1**, lane 1) corresponding to the soluble and membrane-bound forms of the enzyme ***(8,9)***. Analysis of human serum on the HA-substrate gel performed at pH 3.7 reveals a band at 57 kDa, which is Hyal-1 (*see* **Fig. 1**, lane 2). On an inverse HA-substrate gel performed at pH 7.4, two bands, at 120 and 83 kDa, are identified in mouse serum (*see* **Fig. 1**, lane 3). The data can be compared using the pattern obtained from a conventional HA-free gel (*see* **Fig. 1**, lane 4). The 83-kDa band persists, demonstrating that this is an artifact, corresponding to an endogenous plasma glycoprotein. The 120-kDa band corresponds to one of the hyaluronidase inhibitors in mouse serum, an inhibitor of neutral-active PH-20 enzyme.
4. Alcian blue is commonly used for staining HA ***(17–21)***. Acidification of the Alcian blue is recommended. This enhances staining and prevents precipitation of dye. For optimal

results, staining should be performed for 16 h in a solution of 0.5% Alcian blue in 3% acetic acid. When sequential staining with Alcian blue and Coomassie blue is performed, staining for 1 h with Alcian blue is sufficient. For the inverse substrate gel procedure, a single staining step with Alcian blue is recommended.

5. The hyaluronidase digestion step in the inverse HA-substrate gel procedure is obviously critical but can be difficult, as the gel is very fragile at this stage. Enzyme activity can also vary with minor changes of pH and temperature. Optimization of the concentration of hyaluronidase in the gel digestion step may be required with each experiment. We routinely utilize three different levels of hyaluronidase with each experiment (0.25, 0.50, and 2.00 rTRU/mL) to avoid over- and underdigestion.
6. High levels of a protein can prevent dye penetration into the gels ***(9,10)*** and can generate false positive bands. Albumin introduces such an artifact when plasma and serum samples are examined. Albumin is also a HA-binding protein ***(22–24)***, which may explain its persistence in these HA-containing gels. Pronase treatment of the gels eliminates such false positives. Proteins present in lesser amounts will appear as blue bands following Coomassie blue staining. These should disappear if a pronase digestion step is interposed.

## Acknowledgements

This work was supported by Lion Corporation, Kanagawa, Japan, to K. M., and by National Institutes of Health (USA) Grant 1P50 DE/CA11912, to R. S.

## References

1. Kreil, G. (1995) Hyaluronidases-a group of neglected enzymes. *Protein Sci.* **4,** 1666–1669.
2. Csóka, T. B., Frost, G. I., and Stern, R. (1997) Hyaluronidases in tissue invasion. *Invasion Metastasis* **17**, 297–311.
3. Frost, G. I., Csóka, T. B., Wong, T., and Stern, R. (1997) Purification, cloning, and expression of human plasma hyaluronidase. *Biochem. Biophys. Res. Commun.* **236**, 10–15.
4. Csóka T.B., Frost G.I., Wong T., and Stern R. (1997) Purification and microsequencing of hyaluronidase isozymes from human urine. *FEBS Lett.* **417**, 307–310.
5. Csóka, T. B., Frost, G. I., Heng, H. H. Q., Scherer, S. W., Mohapatra G., and Stern R. (1998) The hyaluronidase gene HYAL1 maps to chromosome 3p21.2-3p21.3 in human and 9F1-F2 in mouse, a conserved candidate tumor suppressor locus. *Genomics* **48**, 63–70.
6. Csóka, A. B., Scherer, S. W., and Stern, R. (1999) Expression analysis of six paralogous human hyaluronidase genes clustered on chromosomes 3p21 and 7q31. *Genomics* **60**, 356–361.
7. Gmachl, M., Sagan, S., Ketter, S., and Kreil, G. (1993) The human sperm protein PH-20 has hyaluronidase activity. *FEBS Lett.* **336**, 545–548.
8. Cherr, G. N., Meyers, S. A., Yudin, A. I., VandeVoort, C. A., Myles, D. G., Primakoff, P., and Overstreet, J. W. (1996) The PH-20 protein in Cynomolgus macaque spermatozoa: identification of two different forms exhibiting hyaluronidase activity. *Develop. Biol.* **175,** 142–153.
9. Meyer, M. F., Kreil, G., and Aschbauer, H. (1997) The soluble hyaluronidase from bull testes is a fragment of the membrane-bound PH-20 enzyme. *FEBS Lett.* **413**, 385–388.
10. Haas, E. (1946) On the mechanism of invasion. I. Antinvasin I, An enzyme in plasma, *J. Biol. Chem.* **163**, 63–88.
11. Dorfman, A., Ott, M. L., and Whitney, R. (1948) The hyaluronidase inhibitor of human blood. *J. Biol. Chem.,* **223**, 621–629.

12. Moore, D. H. and Harris, T. N. (1949) Occurrence of hyaluronidase inhibitors in fractions of electrophoretically separated serum. *J. Biol. Chem.* **179**, 377–381.
13. Guntenhoener, M. W., Pogrel, M. A., and Stern, R. (1992) A substrate-gel assay for hyaluronidase activity. *Matrix* **12**, 388–396.
14. Miura, R. O., Yamagata, S., Miura, Y., Harada, T., and Yamagata, T. (1995) Analysis of glycosaminoglycan-degrading enzymes by substrate gel electrophoresis (zymography). *Anal Biochem.* **225**, 333–340.
15. Mio, K., Carette, O., Maibach, H. I., and Stern, R. (2000) A serum inhibitor of hyaluronidase. *J. Biol. Chem.,* July 24.
16. Laemmli, UK. (1970) Cleavage of structural proteins during the assembly of the head of bacteriophage T4. *Nature* **227**, 680–685.
17. Wardi, A. H. and Michos, G. A. (1972) Alcian blue staining of glycoproteins in acrylamide disc electrophoresis. *Anal. Biochem.* **49,** 607–609.
18. Cowman, M. K., Slahetka, M. F., Hittner, D. M., Kim, J., Forino, M., and Gadelrab, G. (1984) Polyacrylamide-gel electrophoresis and Alcian blue staining of sulphated glycosaminoglycan oligosaccharides. *Biochem. J.* **221**, 707–716.
19. Turner, R. E. and Cowman, M. K. (1985) Cationic dye binding by hyaluronate fragments: dependence on hyaluronate chain length. *Arch. Biochem. Biophys.* **237**, 253–260.
20. Wall, R. S. and Gyi, T. J. (1988) Alcian blue staining of proteoglycans in polyacrylamide gels using the "critical electrolyte concentration" approach. *Anal. Biochem.* **175**, 298–299.
21. Ghiggeri, G. M., Candiano, G., Ginevri, F., Mutti, A., Bergamaschi, E., Alinovi, R., and Righetti, P. G. (1988) Hydrophobic interaction of Alcian blue with soluble and erythrocyte membrane proteins. *J. Chromatogr.* **452**, 347–357.
22. Johnston, J. P. (1955) The sedimentation behavior of mixtures of hyaluronic acid and albumin in the ultracentrifuge. *Biochem. J.* **59**, 620–627.
23. Davies, M., Nichol, L. W., and Ogston, A. G. (1963) Frictional effects in the migration of mixtures of hyaluronic acid and serum albumin. *Biochim. Biophys. Acta* **75**, 436–438.
24. Gramling, E., Niedermeier, W., Holley, H. L., and Pigman, W. (1963) Some factors affecting the interaction of hyaluronic acid with bovine plasma albumin. *Biochim. Biophys. Acta* **69**, 552–558.

# III

# Interactions

# 40

## Affinity Coelectrophoresis of Proteoglycan–Protein Complexes

**James D. San Antonio and Arthur D. Lander**

### 1. Introduction

Affinity coelectrophoresis (ACE) was developed as a tool to measure the strengths of interaction between proteoglycans (PGs) or glycosaminoglycans (GAGs) and proteins, and to assess the specificity of the interaction (i.e., to detect and fractionate GAG or PG sample constituents that differentially bind to protein) *(1)*. In ACE, trace concentrations of radiolabeled GAG or PG are subjected to electrophoresis through agarose lanes containing protein at various concentrations. The electrophoretic pattern of the radiolabeled GAG or PG is then visualized by autoradiography, or using a phosphorimager, and the apparent dissociation constant ($K_d$) is calculated as the protein concentration at which the GAG or PG is half-shifted from being fully mobile at very low protein concentrations (or between protein-containing lanes) to being maximally retarded at saturating protein concentrations (*see* **Figs. 1–3**).

ACE holds many advantages over other means of studying GAG or PG–protein interactions since it: (1) uses only trace quantities of the interacting molecules (typically a microgram or less of GAGs, and a milligram or less of protein); (2) studies behaviors of native proteins and of radiolabeled PGs or GAGs that can be essentially unmodified (e.g., through metabolic radiolabeling) or minimally modified (e.g., by radioiodination); (3) can measure strengths of binding even for relatively weak interactions characteristic of GAG or PG-protein interactions (e.g., 100 n*M* or weaker $K_d$); (4) can detect protein-binding heterogeneity in a GAG or PG population and can even be used to isolate the differentially binding subpopulations for further analysis (*see* **Fig. 4**); and (5) is a low cost, simple, and rapid method, and is amenable to quantitative analysis.

The theory of ACE has been described in detail elsewhere *(1,2)* and will not be repeated here. Rather, this chapter will serve as a description of ACE methods, and will also include a protocol for the preparation of radioiodinated heparin samples, which are often useful in various ACE applications. However, before attempting ACE, there are several important questions which need to be addressed:

From: *Methods in Molecular Biology, Vol. 171: Proteoglycan Protocols*
Edited by: R. V. Iozzo © Humana Press Inc., Totowa, NJ

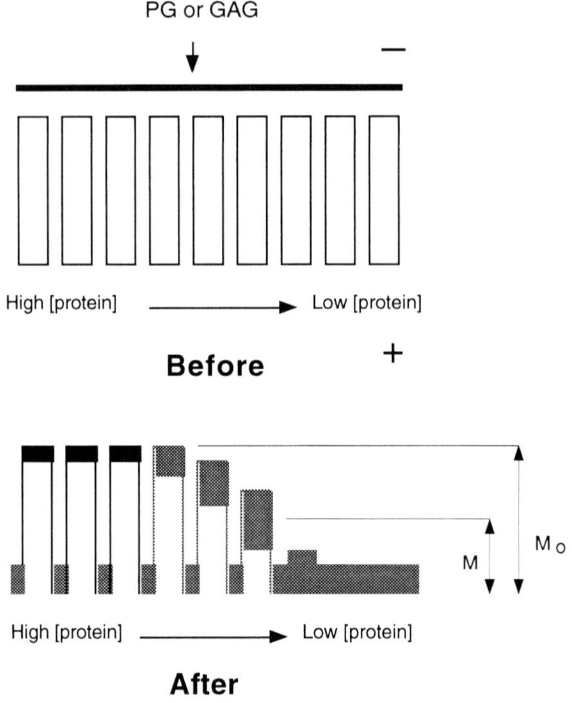


Fig. 1. Analytical ACE schematic. Top panel: ACE gels poured using a casting stand and Teflon combs and strips as shown in **Fig. 5** are used to create nine parallel rectangular wells, which are filled with protein–agarose mixtures, each at a different protein concentration. Radiolabeled GAG or PG is loaded into the slot above the protein-containing wells (shown as a dark line), and after electrophoresis of the GAG or PG through the protein-containing lanes, its migration as a function of protein concentration is visualized by autoradiography or phosphorimaging (shown here as a pattern of peaks and valleys). The degree of GAG/PG retardation at the various protein concentrations is used to calculate the apparent $K_d$ of GAG– or PG–protein binding (*see* text for details). Artwork by Shawn M. Sweeney.

---

Fig. 2. *(opposite page)* ACE analysis can reveal the affinity of interactions between PGs or GAGs and various proteins. For these experiments, syndecan-1 was electrophoresed through types I–VI collagens in ACE gels. **(A)** Images of PG migration patterns were obtained using a phosphorimager. The electrophoretograms indicate that some collagens bind strongly to syndecan-1 (e.g., type V), and others bind weakly (e.g., type II). Protein concentrations in n*M* are shown beneath gels. **(B)** Calculation of affinities of syndecan-1 for various human collagens. From each electrophoretogram in panel **(A)**, retardation coefficients *(R)* for syndecan-1 were determined (see text) and are plotted against protein concentration. Smooth curves represent nonlinear least-squares fits to the equation $R = R_\infty\ (1 + (K_d/[\text{protein}])^2)$. Data are adapted from ***(5)***.

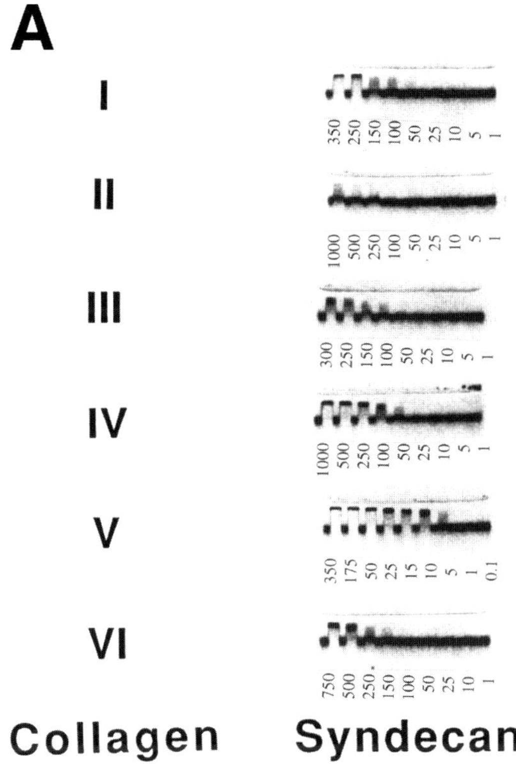
A
I
350 250 150 100 50 25 10 5 1
II
1000 500 250 100 50 25 10 5 1
III
300 250 150 100 50 25 10 5 1
IV
1000 500 250 100 50 25 10 5 1
V
350 175 50 25 15 10 5 1 0.1
VI
750 500 250 150 100 50 25 10 1
Collagen
Syndecan

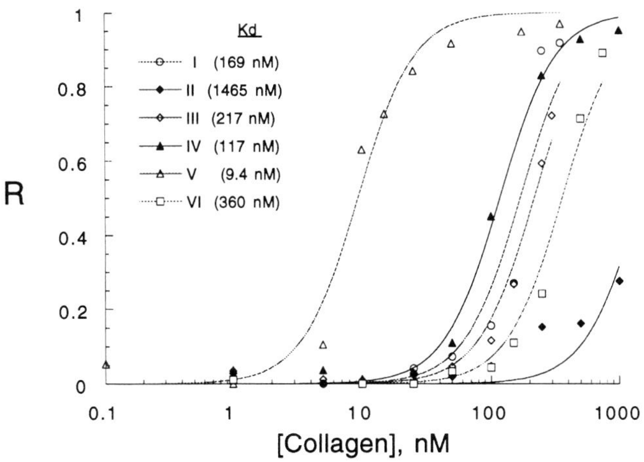
B
Kd
I (169 nM)
II (1465 nM)
III (217 nM)
IV (117 nM)
V (9.4 nM)
VI (360 nM)
R
0
0.2
0.4
0.6
0.8
1
0.1
1
10
100
1000
[Collagen], nM

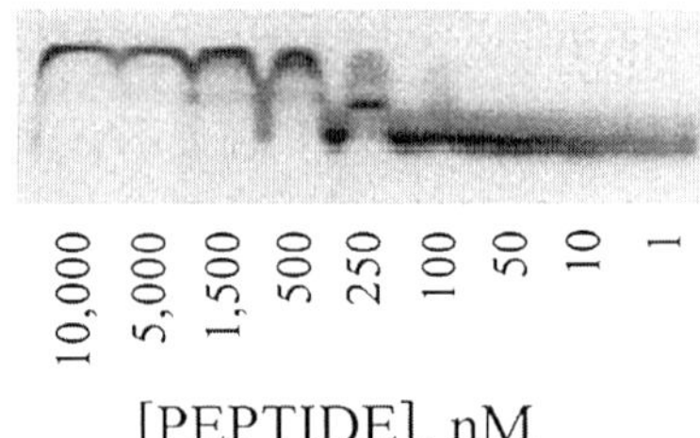


Fig. 3. ACE analysis can reveal selectivity in PG– or GAG–protein interactions. Example of ACE analysis of the interactions between a basic peptide and $^{35}$S-sulfate metabolically labeled PGs/GAGs secreted by endothelial cells in vitro. ACE gel image was obtained using a phosphorimager. At least two populations of PG/GAG, seen as two bands of radiolabeled material migrating with different mobilities at low protein concentrations (< 50 n*M*), indicates heterogeneity in size and/or charge density within the PG/GAG mixture. Potential heterogeneity in PG/GAG–peptide interactions is also obvious at a peptide concentration of 250 n*M*, in which a fractionation of the PG species through the peptide-containing lane is evident as a broad smear throughout the lane, and as a sharp band that migrates approximately halfway down the lane. Thus, components of the PG/GAG sample are binding strongly to the peptide (i.e., are retained closer to the top of the peptide-containing lane), and others are binding more weakly to the peptide (i.e., are not significantly retained and migrate further within the peptide-containing lane). In such cases preparative ACE can be used to recover differentially binding PG/GAG populations for further characterization. Data are adapted from ***(11)***.

### 1.1. Do I Have Enough Protein?

There is no way of knowing *a priori* how much of a protein sample one needs for an ACE gel, since it depends on a yet to be determined value, i.e., the affinity the protein will exhibit for GAGs or PGs. However, some general idea about the amounts of protein required can be derived from the following example. For type I collagen, a protein of $M_r \simeq$ 300,000 Da that exhibits a heparin-binding $K_d$ in the range of 100–200 n*M*, one needs 150 µg of protein per ACE gel (using the ACE gel dimensions specified under **Subheading 2.**). This amount allows for the creation of nine protein–agarose samples of 250-µL each, at concentrations of 1000, 500, 250, 100, 50, 25, 10, 5, and 1 n*M*. Since ACE gels should be repeated at least three times to derive a reasonable estimate of the $K_d$, then one would need a minimum of 450 µg of type I collagen for three experiments. Other proteins such as growth factors are much smaller than collagens, and often exhibit much higher affinities for GAGs and PGs, and thus can require considerably less protein for three experiments, i.e., on average < 50 µg total, or in some cases even much less—e.g., for basic fibroblast growth factor, < 1 µg is required.

### 1.2. Will the Protein Remain Native?

The native state of the protein, its solubility, and propensity to aggregate as a function of its concentration or solvent are key considerations, and must be determined for each protein used in ACE. For example, in the case of laminin, which tends to aggregate in solution, ethylendiamminetetraacetic acid (EDTA) can be supplemented to the protein samples and the ACE buffers to inhibit aggregation ***(3)***. In the case of the collagens,

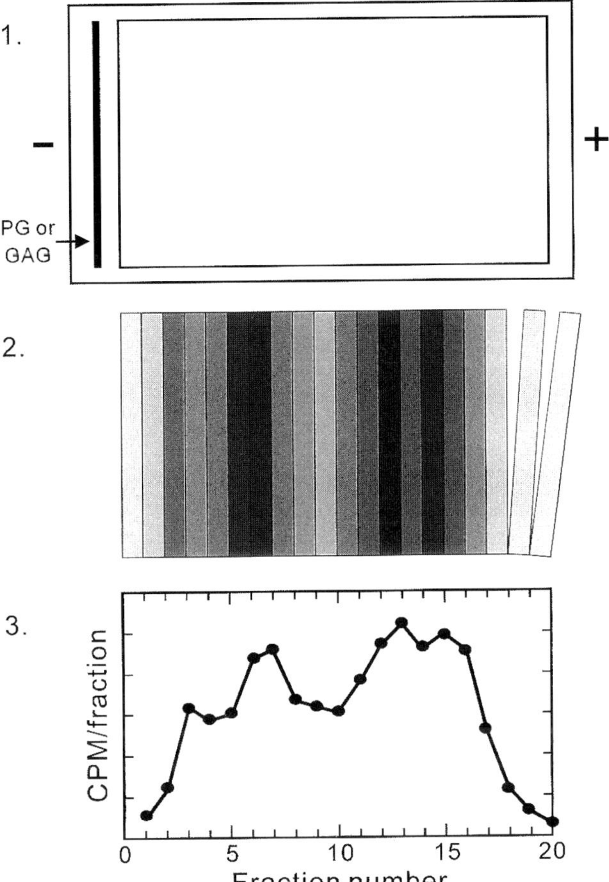


Fig. 4. Preparative ACE schematic. **Top panel**: A preparative ACE gel is poured using a casting stand as shown in **Fig. 5**, except instead of using protein well-forming Teflon combs, a single Plexiglas block is used to create one large rectangular well to be filled with a single protein–agarose mixture. Radiolabeled GAG or PG is loaded into the slot above the protein-containing wells (shown as a dark line to the left in the gel schematic), and electrophoresed through the protein-containing zone. **Middle panel**: The agarose gel surrounding the protein-containing zone is trimmed away, and the remaining protein-agarose block is sectioned into 2-mm-thick segments. The amount of radiolabeled GAG or PG in each segment is then determined. **Bottom panel**: Actual plot of CPM/fraction of heparin octasaccharide mixture electrophoresed through 1000 n*M* type I collagen, showing the partial resolution of four differentially binding populations. (From San Antonio and Lander, unpublished data). Artwork by Drew Likens.

which undergo fibrillogenesis within about 20 min after being brought from an acidic to a neutral solution, such fibrils are insoluble and are impossible to subject to a serial dilution, as is required in ACE. Thus, to avoid this problem one must bring collagen solutions from the acid soluble to the neutralized state, and then mixed into agarose and pipetted into ACE gels before fibrillogenesis occurs *(4,5)*.

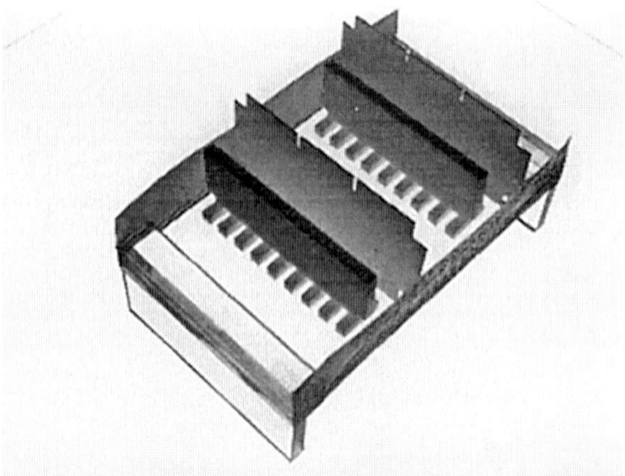

Fig. 5. Oblique view of apparatus for pouring two ACE gels, each with protein-containing lanes 15 mm in length. Plexiglas casting stand contains a clear piece of gel bond (not visible in this photograph), on which are placed two Teflon combs that are each used to create 9 agarose–protein-containing lanes, and two Teflon strips that are each used to create a GAG/PG loading slot. The stand is bordered on two sides by masking tape, which retains the agarose and Teflon strips in place. After filling the stand with agarose, upon solidification the combs and strips are removed, forming two ACE gel templates to be run as described in the text and shown diagramatically in **Fig. 1**.

### 1.3. Are the GAGs or PGs of Interest Suitable?

One of the requirements of ACE is that the GAG or PG concentration is much less than the *Kd* of GAG– or PG–protein binding *(1)*. Thus, radiolabeled GAGs and PGs are used at trace concentrations in ACE gels, i.e., generally far less than 1 µg of GAG or PG/gel will suffice. However, the GAG/PG must be radiolabeled or labeled otherwise, and present in great enough quantities to be detected (generally for radiolabeled samples at least 10,000 cpm/gel). Therefore, the GAGs or PGs must be metabolically radiolabeled with $^{35}$S-sulfate or $^{14}$C-D-glucosamine in culture and subsequently purified, or purified in their native forms but derivatized with, for example, Bolton-Hunter reagent *(6)*, fluoresceinamine *(1)*, or tyramine *(3)*, followed by radioiodination. Here we have presented a method for the tyramine endlabeling of heparin for use in ACE. Another factor that must be considered in the analysis of data from ACE gels is the potential multivalency of GAGs or PGs in terms of their interactions with proteins; this issue is addressed elsewhere *(1)*.

## 2. Materials

1. ACE casting apparatus (*see* **Fig. 5**): casting stage(s), protein well-forming comb(s), PG/GAG lane-forming comb(s), and tape (autoclave or equivalent). Combs are precision tooled from Teflon blocks (for protein well-forming combs) or sheets (for PG/GAG lane-forming combs). The ACE casting apparatus we most commonly use includes a casting stage made of Plexiglas with a gel platform of 100 long × 75 wide × 6 mm deep; protein well-forming combs consisting of nine parallel rectangular blocks spaced 3 mm apart, each 15 long × 4 × 4 mm; and PG/GAG lane-forming combs cut from a 25 × 75 mm

rectangle of 1-mm-thick Teflon. Rectangles of 4.5 × 10 mm are removed from two corners to produce a comb with one 66-mm edge, which is stood on its short edge and held upright in the casting apparatus by pressing the tape used to seal the apparatus against the overhanging tabs of the comb.

2. At least 1.2 L of running buffer (RB). To make 2 L, add 22.53 g of sodium 3-(*N*-morpholino)-2hydroxypropanesulfaonate (MOPSO) to 1.8 L of distilled water with stirring. Add 20.51 g of sodium acetate, anhydrous, or 34.02 g of sodium acetate, trihydrate. Use 5 *M* NaOH to bring the pH to 7.0. Bring to 2 L with distilled water. Store at 4°C, will last no longer than several months. This buffer can also be prepared as a 5× concentrated stock and diluted before use.
3. GelBond, 85 × 100 mm sheets (FMC, #53734).
4. 1.053% agarose: 1 g of low-melting-point (LMP) (Sea Plaque Agarose; FMC) in 95 mL of RB.
5. 2.22% agarose: 1 g of LMP agarose (Sea Plaque Agarose; FMC) in 45 mL of RB.
6. 10% CHAPS in distilled water.
7. Leveling device.
8. ACE gel running box (we use the Hoefer Super-Sub apparatus).
9. Low voltage electrophoresis power supply, capable of delivering 75 V.
10. Waterbath set to 37°C.
11. Boiling-water bath or microwave oven.
12. Circulating water chiller (unnecessary if cold-water tap is available at lab bench where the gel will be run).
13. Small space heater (1500 W).

## 3. Methods

### 3.1. Analytical ACE

This protocol is used to estimate the $K_d$ of GAG or PG binding to a protein, or to visualize heterogeneity in binding between a PG or GAG mixture and a protein.

1. Place the casting stand on a level benchtop. On a piece of GelBond, determine which is the bonding side by placing a drop of distilled water on one of the sides. If the drop beads up, it is the nonbonding side and is placed down on the casting apparatus; if the bead spreads out, it is the bonding side and is placed face up. Using scissors, cut the GelBond to fit the casting apparatus, then use 1–2 drops of distilled water to hold the nonbonding side in place. Press excess water out from between the GelBond and the casting stage and, using Kimwipes, make sure that the sides of the stage are dry.
2. Use autoclave tape to seal the edges of the casting stage, making sure to leave about 1 in (2.5 cm) excess on both ends so it can be folded against itself to make a tab to facilitate its removal later. Place the sample-forming comb(s) in the casting stand using the tape to hold it in place, then place the protein-lane forming comb by centering it and leaving a space of about 2 mm between it and the sample comb. Using the 15-mm-long protein well-forming combs and the gel casting stand specified here, either one or two ACE gels can be poured per stand.
3. When agarose is made fresh, to promote its rapid dissolution allow at least 20 min for it to soak in room temperature RB before the mixture is boiled. Place a glass bottle containing the 1.053% agarose into the water bath, making sure it cannot tip and that the cap is very loose, so that upon heating it will not explode. The bottle should be left in the boiling-water bath or microwaved until the agarose has come to a boil and is

completely melted. Check that the agarose is completely mixed by swirling the bottle to see that the liquid appears homogeneous.

4. To a polypropylene tube, add 1.0 mL of 10% CHAPS and bring to volume with 20 mL with 1.053% boiling hot agarose, place the lid on tightly and gently invert a few times; take care not to create bubbles. Pour the agarose quickly into the casting apparatus, making sure that the agarose flows evenly between the combs. If any of the combs shifted in position during the pouring, readjust them into proper position while the agarose is still hot. Allow the agarose to solidify completely before removing the tape (usually at least 30 min). If any agarose leaks out of the casting stand, or there are significant numbers of bubbles remaining in the agarose after it has solidified, the gel must be discarded.
5. While the gel is cooling, perform a serial dilution of your test protein, into nine concentrations including several that should exceed the $K_d$, several that are close to it, and several that should be considerably lower (*see* **Note 1**). Number nine polypropylene tubes and add the correct amount of diluent, usually RB, to each. Usually, each sample is diluted in RB to twice its desired final concentration, since later they will be mixed 1/1 with 2.0% agarose before being loaded into ACE gels (*see* **step 8**). An exception is for collagen samples, which are dissolved in 0.5 *N* acetic acid at eight times their final concentration, and are quickly mixed 1/1 with 0.5 *N* NaOH, and then 1/1 with 2× RB, before being mixed 1/1 with 2.0% agarose.
6. After the gel has solidified, remove the well-forming combs first, by slowly and gently sliding the comb backwards until it separates from the agarose at the front end, then sliding it forward until it separates from the agarose at the rear end. Then, while holding the casting stand firmly against the bench surface, tilt the comb and gently pull it away from the agarose. Be careful not to rip the gel or deform the lanes. Small rips can be repaired by reinserting the comb and pipetting hot 1.0% agarose (prepared as before) into the gel. Should small clumps of agarose remain in sample lanes, these can be removed using a pipet tip affixed to a vacuum line. The PG/GAG lane-forming comb is next removed by holding the comb in place while slowly peeling the tape away from each of its sides, then the comb is lifted straight up, out of the agarose. If rips occur in the agarose between the sample-forming slot and the gel edge, these can be repaired by reinserting the comb and pipetting new 1.0% agarose (prepared as before) onto the damaged regions of the gel. If the PG/GAG lane-forming comb has been removed before the gel is fully solidified, the sample well may collapse upon itself and the gel must be discarded. Remove the gel (which should be firmly attached to the gel bond) from the casting stand. Using a waterproof marker, mark on the gel bond the position of the PG/GAG loading lane, and any necessary notes about the samples to be added to the protein-containing wells.
7. To a polypropylene tube, add 10% of the total volume needed of 10% CHAPS and bring to volume with boiling-hot 2% agarose, place the lid on tightly and gently invert a few times to mix. Using the ACE gel dimensions specified under **Subheading 2.**, each ACE gel will require 1.125 mL of CHAPS/2% agarose, although extra volume should be prepared, as some is lost on the side of the sample tube and on the outside of pipet tips. Place the tube in the 37°C bath to equilibrate for later use (*see* **Note 2**).
8. To load the test proteins into the gel wells, make sure the gel is level and located near the 37°C water bath. From the 2% agarose tube, withdraw an amount equal to the total volume within each of the nine sample tubes, which would be 125 μL. Add this volume to the ninth sample tube (which contains the most dilute protein sample) and mix it by

agitating the pipet tip back and forth while rapidly drawing the solution in and out of the tip 10 times, taking care not to generate bubbles in the mixture. Then add the mixture (now totaling 250 μL) to lane 9, which is the rightmost of the nine lanes of the gel, taking care to overfill the lane slightly. Any excess agarose that spills over into the zone between sample lanes will quickly solidify and will not interfere with the running of the gel, whereas underfilling the protein lanes results in anomalous GAG/PG migration through the gel. Move on to the remaining samples, filling the lanes in the following order: 7, 5, 3, 1, 8, 6, 4, and 2. The key to success during this step is to mix the samples thoroughly but to make sure to work quickly enough that the agarose–protein mixture does not gel in the tube. If need be, the mixing can be done while the sample tube remains partially submerged in the water bath.

9. Prepare the PG or GAG sample by mixing the labeled material in sufficient quantity and activity that it can be detected (usually at least 10,000–20,000 cpm of $^{35}$S-radiolabeled material is sufficient for a good 3-d exposure in a phosphorimager cassette); tracking dye(s) (we use 0.05% each of bromophenol blue and xylene cyanol); sucrose so that the sample will sink through the RB during gel loading (5% w/v), and enough RB to bring the volume to approximately 100 μL/gel (*see* **Note 3**). Once the labeled sample is mixed, it should be vortexed and any insolubles pelleted at 13,000*g* for 2 min. When adding the sample to the gel, be careful not to disturb any pellet that may be at the bottom of the tube.
10. The electrophoresis apparatus is next prepared. Make sure the apparatus is resting properly on a level stir plate. Attach the cooling hoses to a water source, either *cold* tap water or a circulating water chiller set to 10°C, and turn on the water. Fill the apparatus with cold running buffer to roughly 3 mm above the platform, then place the ACE gels with the PG/GAG lane end of the gel oriented closest to the cathode, and use glass microscope slides to hold them in place. Turn on the stir plate to purge air trapped in the buffer recirculator, then turn the stir plate off. Add more buffer if needed, to be sure that the gels are covered by about 3 mm of buffer. Using a pipetter with a 200-μL gel-loading pipet tip, load the labeled sample into the lane, drawing it across the gel while discharging the sample, so that it fills evenly. Take care not to nick the gel with the pipet tip during sample loading—this leads to an uneven sample front and/or rapid loss of the sample into the running buffer during electrophoresis (*see* **Note 4**).
11. Attach the cover to the gel box, plug the leads into the power supply, and turn it on to deliver 60–80 V. After the tracking dye has entered the gel (in approximately 5 min), turn on the stir plate to an intermediate setting but take care that the resulting agitation of the buffer does not cause the gels to float away from the platform.
12. Run the gel for the appropriate time. For example, for gels containing protein-containing lanes 15-mm-long, heparin samples should generally run for about 1.0 h, by this time the heparin should have migrated most if not all of the way through the lane. PGs and other GAGs may migrate at different rates depending on their relative sizes, charges, etc. The position of the tracking dyes can be used as a guide to assess when the gels are finished—for example, the dyes that we use (*see* **step 9**) typically move at approximately one-third the rate of heparin in the electrophoretic field. After electrophoresis, turn off the power supply and remove the gel box cover. Remove the gels and place them on an elevated surface that will not impede the flow of warm air around them. We use 8-cm-high plastic test tube racks placed about 30 cm in front of a warm air source supplied by a small personal space heater, such as the Holmes (HFH 195, 1500 W). Allow the gels to dry at the high heat setting for at least 8.0 h; they are dry when the agarose has flattened to a thin clear sheet that is not sticky to the touch.

### 3.2. Data Analysis

ACE gel electrophoretograms can be visualized by autoradiography or phosphorimaging, and the approximate $K_d$ of GAG– or PG–protein binding can often be estimated by visual inspection. For example, from the phosphorimages of ACE gels shown in **Fig. 2A**, it can be seen that the affinity of syndecan type I collagen binding can be estimated as 50 n$M \leq K_d \leq$ 250 n$M$, since within these concentration ranges it is evident that the PG is half-shifted from being fully retarded at high protein concentrations to being fully mobile at very low protein concentrations. However, such gels can also be analyzed quantitatively, by first measuring GAG or PG mobility using a Phosphorimager by scanning protein-containing lanes and determining relative radioactivity per 88-µm pixel along the length of each lane (*see* **Fig. 2B**) *(3)*. GAG or PG mobility is taken as the pixel position that divides the curves representing the distribution of GAG or PG into halves of equal areas. The retardation coefficient $R$ is next calculated for each lane as the GAG or PG migration position in that lane divided by its mobility in a protein-free lane ($R = (M_o – M)/M_o$, where $M_o$ is the mobility of free GAG or PG, and $M$ is its mobility through protein; *see* **Fig. 1**). Under appropriate experimental conditions as described previously, $R$ is proportional to the fractional saturation of GAG or PG by protein, so that values of the equilibrium binding constant may be determined from the relationship between $R$ and protein concentration *(1)*. Data are analyzed graphically, by curve-fitting to the equation $R = R_\infty/(1 + K_d/[\text{protein}]^n$; where $R_\infty$ represents the value of $R$ at full saturation (i.e., at an arbitrarily high protein concentration), and $n$ is a coefficient that reflects cooperativity of binding *(1)*. Nonlinear least-squares fits are calculated using a graphics program, e.g., the Kaleidagraph program (Synergy Software, Reading, PA). The above analysis provides apparent $K_d$ values for GAG– or PG–protein interactions, but does not indicate whether the sample of GAG or PG exhibits heterogeneity in protein binding. Such heterogeneity is usually apparent by visual inspection of the ACE gel electrophoretogram, as evidenced by a broad smearing of the GAG or PG migration front throughout the length of protein-containing lanes, or by the presence of multiple bands of the sample at concentrations near the binding $K_d$ (*see* **Fig. 3**). When binding heterogeneity is evident, subpopulations of GAGs or PGs that bind strongly or weakly to protein can be isolated using preparative ACE, and subjected to further analysis.

### 3.3. Preparative ACE

This method is used to fractionate a heterogeneous population of GAGs or PGs that bind differentially to a protein. In this technique, radiolabeled GAGs or PGs are subjected to electrophoresis through agarose containing a single concentration of protein *(1,3)* as shown schematically in **Fig 4**. After individual species from the GAG or PG mixture are isolated, their affinity for protein can be analyzed by standard ACE methods, or they can be subjected to other analyses.

1. A 1% low-melting agarose/CHAPs solution in ACE electrophoresis buffer is prepared as detailed previously and is poured hot onto a piece of GelBond fitted within a Plexiglas gel casting tray where a 4 × 7 cm Plexiglas block and a Teflon PG/GAG lane-forming comb are positioned. After the agarose solidifies, removal of the block and strip leaves a 4 × 7 cm well with a 66 × 1 mm slot 2 mm away from and parallel to one of the short edges of the 4 × 7 cm well (*see* **Fig. 4**, top).

2. A protein is prepared at a concentration that will achieve maximal separation of GAG or PG species of interest as determined by analytical ACE (e.g., for the protein in **Fig. 3**, a concentration of 250 n*M* may be suitable). The sample is then mixed with agarose as described in **step 8** of the analytical ACE methods to a final agarose concentration of 1.0%, and is loaded into the 4 × 7 cm well and allowed to solidify.
3. Gels are submerged under electrophoresis buffer, radiolabeled GAG or PG is loaded into the 66 × 1 mm slot, and electrophoresis is carried out, all as detailed in **steps 9–12** of the analytical ACE methods.
4. After electrophoresis, gels are removed and are not dried, but rather are affixed to a surface marked with a millimeter grid, with the electrophoretic origin at the top. For a surface we use a sheet of 1-mm lined graph paper under a clear plastic sheet, taped onto a flat Styrofoam block.
5. Regions of the gel to the left and right of the 4 × 7 cm block as well as the 3 mm of the block itself that are immediately adjacent to the left and right edges are cut away and discarded.
6. The resultant 3.4 cm × 7 cm gel is sectioned into 2-mm segments perpendicular to the direction of electrophoresis, using a piece of surgical suture thread drawn tightly between the hands.
7. Each agarose segment is lifted with a flat-headed metal spatula and placed in a tube, and the amount of radiolabeled GAG or PG in each segment is measured with a gamma counter, or by liquid scintillation counting of a melted gel aliquot. To melt gel segments, tubes are placed in a ≥70°C water bath. They may be pooled or divided into aliquots while they are liquid, and then stored frozen. Alternatively, they may be melted at 70°C and then brought to 6 *M* in urea by addition of solid urea (urea blocks gelation of the agarose). In this case they will remain liquid for days, even at 4°C. Such samples may then be mixed with running dyes and loaded directly into analytical ACE gels (*see* **Note 5**), or subjected to other analyses.

### 3.4. Iodination and Molecular-Weight Fractionation of Heparin

Heparin is often used as a model compound in studies of GAG– or PG–protein interactions because it potentially contains a large number of different protein-interactive domains, is structurally homologous to heparan sulfates that are common to many cell surface and extracellular matrix PGs, and is inexpensive and readily available from commercial sources. Commercial heparin is polydisperse in $M_r$, often averaging about 15 kDa. However, in protein-binding studies the use of low-molecular-weight heparin is advantageous, since it minimizes factors that complicate binding analysis, such as multivalency *(1)*. Thus, here we have included methods we use to tyramine-derivatize, radioiodinate, and $M_r$-fractionate commercial heparin to be used in ACE analysis. We have found that derivatization of heparin with tyramine is more suitable than with fluoresceinamine, as the latter may artifactually enhance heparin-binding affinity for protein *(3)*.

### 3.5. Tyramine Labeling of Heparin

1. Dissolve 9 mg of heparin in 750 µl of 5% (w/v) solution of tyramine in formamide, and place in a sealed tube. We use porcine intestinal mucosa heparin (grade 1A; Sigma) in our experiments.
2. Heat to 80°C for 1 h and cool to room temperature (the solution may turn yellow).
3. Add 1 mg of Na cyanoborohydride, seal, and incubate overnight at room temperature.
4. Dilute with 9 volumes of distilled water.
5. Dialyze against distilled water using 1-kDa *Mr* cutoff dialysis tubing.
6. Lyophilize the sample.

7. Resuspend in a small volume of water and measure heparin concentration [e.g., using the Dische assay *(7)*].
8. Measure tyramine content by $OD_{278}$. As heparin absorbs to a small extent at 278 nm, tyramine concentrations must be corrected by subtracting this background. Tyramine content may be estimated from corrected OD values by the formula: tyramine (mg/mL) = $0.0824 \times OD_{278}$.

### 3.6. Iodination of Tyramine-Heparin

1. Dissolve 400 µg of Iodogen/mL (Pierce) of dichloromethane. Add 50 µL to the bottom of 5-mL glass test tubes; a glass Pasteur pipet connected to a rubber hose affixed to a nitrogen tank can be used to direct a slow stream of nitrogen gas over the sample in the bottom of the rotating tube. Tyramine-coated glass tubes can be covered with Parafilm and stored under vacuum at room temperature for years.
2. Dilute 2.5 µg of tyramine-labeled heparin in 50 µL of 0.25 *M* Tris-HCl, pH 7.5.
3. Rinse an Iodogen-coated tube gently with 500 µL of 0.25 *M* Tris-HCl, pH 7.5 to wash off any unbound Iodogen. Visually inspect tube to ensure that the Iodogen coat remains intact.
4. React 50 µL of heparin solution and 5 mCi of $^{125}I$ at room temperature for 6 min with intermittent agitation, then add an additional 50 µL of buffer for 6 min more with agitation. The addition of extra buffer helps prevent the Iodogen from increasing the pH, which inhibits iodination.

### 3.7. Desalting of $^{125}I$-Heparin on G-25

After iodination of the heparin sample, it must be desalted over a G-25 column (prepared in a 2.0-mL disposable tissue culture pipet) to remove unbound $^{125}I$.

1. Equilibrate and elute the column with 0.5 × RB.
2. Draw off buffer from column bed.
3. Load the $^{125}I$-heparin sample onto the column and position a test tube rack with numbered Eppendorf tubes underneath.
4. Collect the column eluate in fractions of 2 drops/tube.
5. Make sure to replenish RB as the column runs.
6. Monitor the passage of $^{125}I$-heparin through the column using a Geiger counter.
7. As fractions are collected, monitor their relative radioactivity using a Geiger counter at a fixed distance from tubes.
8. Record activities and plot elution profile to determine where bound (the smaller of the two peaks, which elutes first) and free (the larger of the two peaks, which elutes second) isotope are eluting. Generally, 10–15 fractions are collected.
9. Pool the fractions containing the bound isotope.
10. Discard the column and the fractions containing the unincorporated isotope.

### 3.8. Molecular–Weight Fractionation of $^{125}I$-Heparin on G-100

Pooled fractions containing $^{125}I$-heparin from the desalting column are prepared for G–100 chromatography and $M_r$-fractionated as follows; our typical column dimesions are 300 × 10 mm. A typical elution profile of $^{125}I$-heparin from a G-100 column is shown in **Fig. 6**.

1. Prepare the sample for G-100 chromatography by mixing $^{125}I$-heparin (generally < 0.5 mL) to 5% sucrose (w/v) plus 10 µL of a saturated phenol red solution. Bring to 2.0 mL with running buffer, clarify at 13,000*g* for 2 min, and load on column.
2. Collect fractions of about 8–10 drops (i.e., from 0.5–1.0 mL).

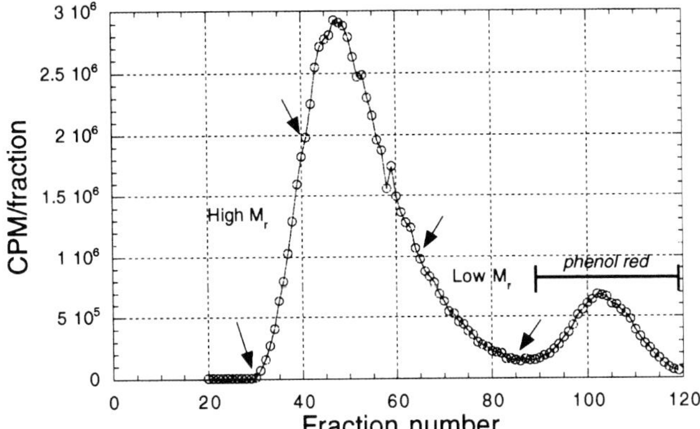


Fig. 6. Elution profile of $^{125}$I-tyramine-heparin from G-100 column. The larger of the two peaks, eluting first, represents the radiolabeled heparin sample; the smaller peak, eluting second, is unincorporated $^{125}$I-iodine. The last 12% (marked with arrows) of sample to elute represents the low-$M_r$ heparin chains of about ≤ 6 kDa.

3. Remove and count several microliters of each fraction using a gamma counter.
4. Plot column profile indicating elution position of phenol red, which should largely overlap with free iodine peak. *See* **Fig. 6** for an example of a typical profile.
5. The first 12% of radioactive material to elute is the high-molecular-weight fraction.
6. The following 76% to elute is the medium-molecular-weight fraction.
7. The remaining 12% to elute is the low-molecular-weight fraction of $M_r \approx 6$ kDa ***(8–10)***.

### 3.8.1. Storage of $^{125}$I-Heparin Samples

1. Pool fractions within each category and cryoprotect with bovine serum albumin (BSA) to 0.1 mg/mL (a 2.0-mg/mL stock of BSA can be used).
2. The low-$M_r$ fraction, which is the fraction most commonly used in binding experiments, should be divided into 50- to 100-µL aliquots.
3. The remaining fractions should be divided into samples of about 1.0 mL.
4. Store samples at –80°C. These can be used for approximately 3–6 mo.

## 4. Notes

1. The accuracy of the ACE technique relies on knowing the exact concentration of the test protein.
2. If the 2% agarose mixture is not equilibrated to 37°C when it is mixed with test proteins (*see* **step 8** under **Subheading 3.1.**), the proteins may become denatured.
3. Tubes containing 5× stock mixtures of dyes/sucrose should be premixed, filtered, and stored frozen.
4. Loading of the GAG/PG sample is the most technically difficult step of this procedure; thus, one should practice loading mock samples into ACE gels before working with radioactive GAG or PG samples.
5. Since urea does not migrate in the electrophoretic field, its removal from samples before electrophoresis is not required; due to the density of urea, it is not necessary to add sucrose.

## References

1. Lee, M. K., and Lander, A. D. (1991). Analysis of affinity and structural selectivity in the binding of proteins to glycosaminoglycans: development of a sensitive electrophoretic approach. *Proc. Natl. Acad. Sci. (USA)* **88,** 2768–2772.
2. Lim, W. A., Sauer, R. T., and Lander, A. D. (1991). Analysis of DNA-protein interactions by affinity coelectrophoresis. *Meth. Enzymol.* **208,** 196–210.
3. San Antonio, J. D., Slover, J. Lawler, J., Karnovsky, M. J., and Lander, A. D. (1993). Specificity in the interactions of extracellular matrix proteins with subpopulations of the glycosaminoglycan heparin. *Biochemistry* **32,** 4746–4755.
4. San Antonio, J. D., Lander, A. D., Wright, T. C., and Karnovsky, M. J. (1992). Heparin inhibits the attachment and growth of Balb c/3T3 fibroblasts on collagen substrata. *J. Cell. Physiol.* **150,** 8–16.
5. San Antonio, J. D., Karnovsky, M. J., Gay, S., Sanderson, R. D., and Lander, A. D. (1994). Interactions of syndecan-1 and heparin with human collagens. *Glycobiology* **4,** 327–332.
6. LeBaron, R. G., Hook, A., Esko, J. D., Gay, S. and Hook, M. (1989). Binding of heparan sulfate to type V collagen. *J. Biol. Chem.* **264,** 7950–7956.
7. Dische, Z. (1947). A new specific color reaction of hexuronic acids. *J. Biol. Chem.* **167,** 189–192.
8. Yamada, K. M., Kennedy, D. W., Kimata, K., and Pratt, R. M. (1980). Characterization of fibronectin interactions with glycosaminoglycans and identification of active proteolytic fragments. *J. Biol. Chem.* **255,** 6055–6063.
9. Laurent, T. C., Tengblad, A., Thunberg, L., Hook, M., and Lindhal, U. (1978). The molecular-weight dependence of the anti-coagulant activity of heparin. *Biochem. J.* **175,** 691–701.
10. Jordon, R., Beeler, D., and Rosenberg, R. D. (1979). Fractionation of low molecular weight heparin species and their interactions with antithrombin. *J. Biol. Chem.* **254,** 2902–2913.
11. Verrecchio, A., Germann, M. W., Schick, B. P., Kung, B., Twardowski, T., and San Antonio, J. D. (2000). Design of peptides with high affinities for heparin and endothelial cell proteoglycans *J. Biol. Chem.* **275,** 7701–7707.

# 41

# Binding Constant Measurements for Inhibitors of Growth Factor Binding to Heparan Sulfate Proteoglycans

**Kimberly E. Forsten and Matthew A. Nugent**

## 1. Introduction

A large number of proteins (>100) have been demonstrated to bind to heparan sulfate proteoglycans, and in many instances these interactions have important biological consequences *(1)*. The best-studied example is the fibroblast growth factor (FGF) family of proteins. The FGFs regulate a wide range of cellular functions, including proliferation, differentiation, and migration. The best-characterized FGF family member, basic fibroblast growth factor (bFGF) or FGF-2, has been shown to require interaction with heparan sulfate on cell surfaces in order to induce maximal activity *(2,3)*. However, interaction of bFGF with heparan sulfate within the extracellular matrix can limit bFGF diffusion and access to cell surfaces *(4,5)*. Thus the interaction of bFGF with heparan sulfate has been targeted as a site for regulation of both endogenous and exogenously administered bFGF. For example, bFGF and many other heparin-binding proteins have been demonstrated to play pivotal roles in the growth of the new blood vessels (angiogenesis), a process that is essential for both efficient wound healing and the development of malignant tumors. Indeed, inhibition of endogenous bFGF activity has been suggested as a possible treatment to prevent tumor growth and metastasis, whereas enhancement of angiogenesis by pharmacological bFGF has been proposed to stimulate repair of damaged tissue (i.e., ischemic heart muscle) *(6)*. In both instances, effective treatments might make use of small compounds, which inhibit bFGF binding to heparan sulfate proteoglycans. These compounds might have applications as inhibitors of bFGF binding and activity at cell surfaces, or in turn they might enhance the transport of added bFGF through connective tissue and allow cell stimulation distant from the administration site. Compounds that bind either to the growth factor or to heparan sulfate could block bFGF binding to heparan sulfate, yet might have very different effects biologically. To screen various potential inhibitors of growth factor/heparan sulfate binding effectively, a simple, rapid, and semi-quantitative assay is required.

From: *Methods in Molecular Biology, Vol. 171: Proteoglycan Protocols*
Edited by: R. V. Iozzo © Humana Press Inc., Totowa, NJ

In this chapter we report a simple assay to determine equilibrium binding constants for inhibitors of bFGF/heparan sulfate proteoglycan binding that is based on the semiquantitative retention of sulfated proteoglycans on cationic nylon filters *(7)*. Using a modification of previously published conditions, samples containing proteoglycan and growth factor are subject to filtration through cationic nylon. While the large negatively charged proteoglycans are retained on the filter, only a small fraction of growth factor is retained unless bound to the proteoglycan (*see* **Fig. 1**). Thus, using standard preparations of bFGF and heparan sulfate proteoglycan (isolated from bovine aortic endothelial cell conditioned media [mostly perlecan]), baseline measurements were made to determine the bFGF/heparan sulfate proteoglycan binding affinity. A range of concentrations of potential inhibitors are then included in the reaction, and decreased bFGF retention on the cationic membrane is measured. These data are then analyzed by fitting to a generalized equation for binding using a commercially available software package (Mathematica, version 3.0, Wolfram Research) to determine the binding constant for each inhibitor (binding to either bFGF or heparan sulfate). This assay has been established with bFGF and has been shown to be effective with inhibitors that bind either bFGF or heparan sulfate proteoglycans. Minor modifications of the base assay should allow it to be transferable for the analysis of a large number of heparin-binding proteins.

## 2. Materials

1. Human recombinant $^{125}$I-bFGF is prepared by a modification of the Bolton-Hunter procedure *(8)* and frozen in single-use aliquots to avoid freeze–thaw cycles.
2. Proteoglycans (E-HS) are purified from bovine aortic endothelial cell conditioned medium as described previously *(4)* (*see* **Note 1**). Briefly, confluent bovine endothelial cells (isolated from freah calf aorta or purchased from Coriell Cell Repositories) are established in Dulbecco's Modified Eagle's Medium (DMEM) with 10% calf serum. The medium is then replaced by DMEM without any additives and the cells incubated for 1 h at 37°C as a wash, followed by a 24 h incubation in fresh DMEM. Conditioned medium is collected and centrifuged at 3000*g* for 30 min at 4°C to remove cell debris. The conditioned media is equilibrated with urea (1 *M* final) for at least 30 min prior to being applied to an anion-exchange column (Q-sepharose, Pharmacia-LKB) in Tris buffer (50 m*M* Tris-HCl, pH 8.0, 0.15 *M* NaCl, 1 *M* urea). The column is washed extensively with 50 m*M* Tris-HCl, pH 8.0, 0.3 *M* NaCl, 1 *M* urea, and the proteoglycan eluted in the same buffer containing 1.5 *M* NaCl. Proteoglycan fractions are identified using the dimethylmethylene blue (DMMB) dye binding assay *(9)* and dialyzed extensively in 50 m*M* Tris-HCl, pH 8.0, 0.15 *M* NaCl. Proteoglycan concentration is based on the mass of glycosaminoglycan using the DMMB assay *(9)* with a bovine kidney heparan sulfate standard (Sigma).
3. Incubation buffer: 50 m*M* Tris-HCl, pH 8.0, 0.15 *M* NaCl, 2 mg/mL bovine serum albumin. BSA, biotech grade (Fisher), worked well.
4. Cationic membrane (Zeta-Probe membrane) and Dot-blot apparatus (Bio-Dot Microfiltration Apparatus) (Bio-Rad)
5. Test compound, we have used glycosaminoglycans (Sigma), protamine sulfate (TCI America), and sucrose octasulfate.

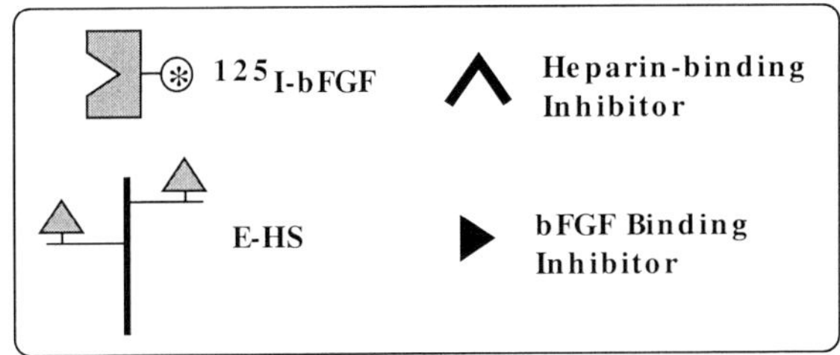


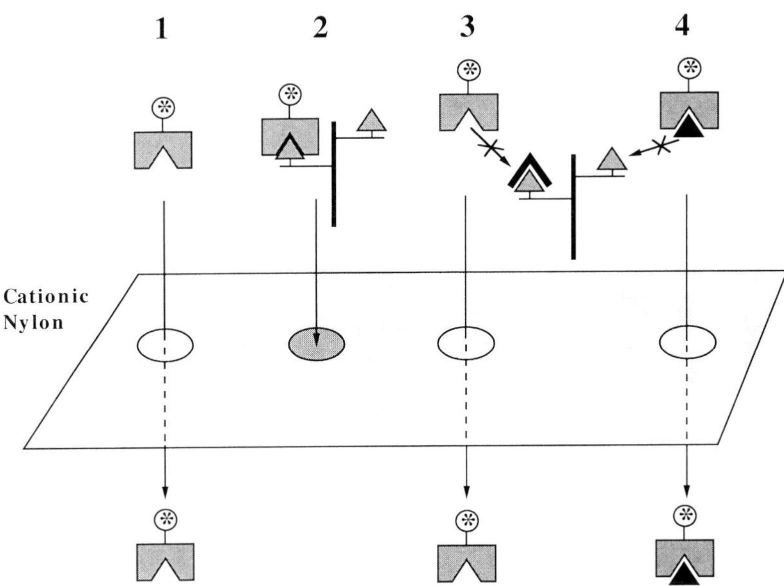


Fig. 1. Schematic of cationic nylon filter binding assay for growth factor/heparan sulfate proteoglycan interactions. Filtration of various mixtures of growth factor, heparan sulfate proteoglycan (E-HS), and compounds that block either the heparin-binding domain on the growth factor or the protein-binding domain on the heparan sulfate chains, through cationic nylon filters can be used to measure the various binding affinities quantitatively. (1) Cationic nylon filters do not specifically retain small proteins such as FGF-2. (2) FGF-2 bound to heparan sulfate proteoglycans will be retained on the filter, providing a direct measure of FGF-2/heparan sulfate proteoglycan binding. (3) Compounds that bind to E-HS and block the growth factor-binding domain can reduce FGF-2 retention on the filter in relation to the binding affinity of the inhibitor compound for E-HS. (4) Compounds that bind to FGF-2 and block its interaction with E-HS can reduced FGF-2 retention in relation to the binding affinity of the inhibitor compound for FGF-2. As an example we have provided a description of how this assay can be used to determine the FGF-2–binding affinity for E-HS and sucrose octasulfate, and the binding affinity of E-HS for protamine sulfate.

## 3. Method

### 3.1. Experimental Protocol

1. Label and arrange 0.5-mL microcentrifuge tubes—1 per sample, 3 per condition.
2. Prepare incubation buffer—Buffer may be made in advance and filter-sterilized for later use. No degassing is needed.
3. Add the needed incubation buffer to each vial—the volume should be 0.2 mL minus the volume of $^{125}$I-bFGF, inhibitor, and E-HS needed. Controls should be conducted with vials containing only $^{125}$I-bFGF, and $^{125}$I-bFGF and the inhibitors, to determine the non-proteoglycan–mediated adsorption of $^{125}$I-bFGF to the filter. These values should be substrated from the experimental points with E-HS to determine the level of $^{125}$I-bFGF bound to E-HS.
4. Add the E-HS, inhibitor, and $^{125}$I-bFGF in that order. As a standard assay we have used 40 ng of E-HS and 0.2 ng of $^{125}$I-bFGF in a final volume of 0.2 mL. Under these conditions greater than 80% of the E-HS is retained with and without bFGF addition, and only ~5% of the bFGF will adsorb to the filter nonspecifically. Initial experiments should be conducted with a range of $^{125}$I-bFGF concentrations to determine the binding affinity and capacity of each E-HS preparation before inhibitor competition data can be analyzed. After each addition of $^{125}$I-bFGF, vortex gently to mix and place in an incubation rack. If the assay is done at a temperature other than room temperature (25°C), be sure that the buffers are preequilibrated at the desired temperature and that vials are placed immediately in a temperature equilibrated rack.
5. Incubate for desired time (generally 60 min to reach a steady state in bound versus unbound bFGF).
6. If incubation is shorter then 30 min, the cationic membrane should be prepared prior to incubation (**steps 3–5**). The membrane is cut to fit the apparatus (cover all holes) and then incubated on a rocking platform for at least 20 min in incubation buffer. Longer incubations (up to 1 h) do not adversely affect the assay.
7. The membrane is placed on the dot-blot apparatus and the top plate is tightened using a criss-cross pattern. The apparatus is attached to a vacuum pump (GAST, model ROA-P131-AA, Manufacturing Corp., Benton Harbor, MI) and the apparatus tightened again under vacuum. Experiments should be done under a 25 mmHg vacuum.
8. Close off the apparatus to the vacuum and, using a multipipeter, add 0.2 mL of incubation buffer to each well.
9. When the incubation is within 30 s of completion, turn on the vacuum and pull the wash buffer through the membrane. Add samples (0.2 mL/well) in sets of three. Wash each well of the triplicate set once with 0.2 mL of incubation buffer before adding the next set of samples to the membrane. Allow all sample fluid to filter before adding the incubation buffer. Repeat for each set of triplicate samples. Air bubbles may arise and interfere with the filtration. Gently pipet the liquid near the surface of the membrane to mix, being careful not to puncture the membrane.
10. After all samples have been added, do two additional washes with 0.2 mL of incubation buffer per well.
11. Loosen the screws holding the unit in place and remove the top plate. Turn off the vacuum and carefully remove the membrane.
12. Allow the membrane to dry sample side up on a paper towel and then cut out the individual membrane dots corresponding to each individual well. Place each sample dot in a vial and count the radioactive bFGF in a gamma counter. Sample data are shown in **Fig. 2**.

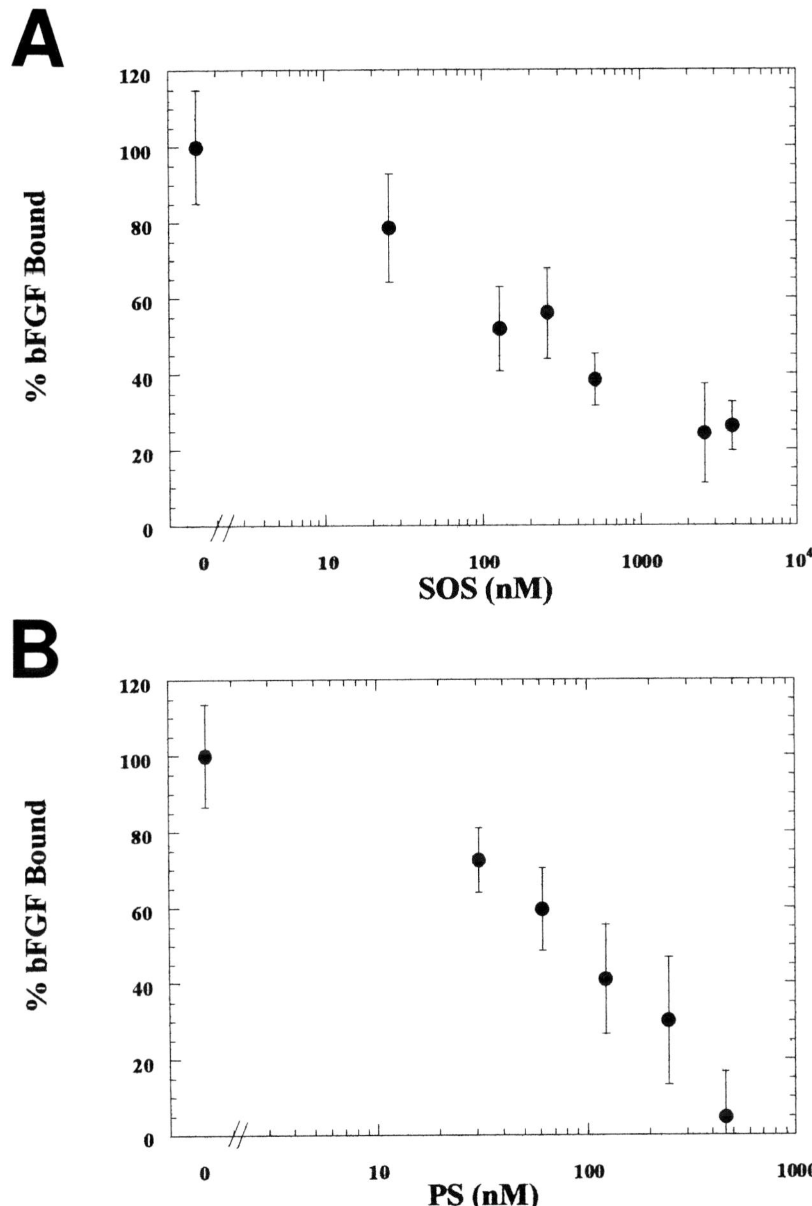


Fig. 2. Inhibitors of bFGF-E-HS Binding. E-HS (200 ng/mL), bFGF (0.056 n*M*), and inhibitor ([A] sucrose octasulfate or SOS [B] protamine sulfate or PS) were incubated for 1 h and then filtered across the cationic membrane. bFGF retention following incubation with SOS or PS in the absence of E-HS was subtracted from total binding as nonspecific binding. 100% bFGF bound corresponds to specific bFGF retention following incubation with E-HS in the absence of inhibitor. The average ± standard error of at least triplicate samples is shown.

## 3.2 Analysis

### 3.2.1. Ligand–Proteoglycan Binding

1. The output data will represent the amount of bFGF retained on the filter for a given added concentration. The amount retained in the absence of E-HS should be subtracted from that with E-HS to generate data that represent specific E–HS bound bFGF.
2. Varying the concentrations of bFGF while holding constant the concentration of E-HS (with no inhibitor), this data set should be analyzed to determine the binding capacity (amount of bFGF bound per unit of E-HS) and the binding affinity ($K_{D1}$), assuming a simple monovalent binding reaction,

$$P + L \overset{K_{D1}}{\longleftrightarrow} C_1$$

which, at steady state assuming negligible ligand depletion, is described by

$$C_1 \quad \frac{R_T L_0}{K_{D1} + L_0}$$

where $R_T$ is the binding capacity, and $L_0$ is the concentration of FGF added. The programming to determine $K_{D1}$ and $R_T$ is outlined below.
3. Mathematica software (Version 3.0, Wolfram Research) is launched, and a new notebook page is initialized. Commands as outlined below are entered into the notebook page, and the "shift" and "enter" keys are pressed simultaneously at the end of each instructional set (symbolized in the procedure by " ♦ ")
4. The statistical package must be loaded by typing (*see* **Note 2**):

<<Statistics`NonlinearFit` ♦

5. Enter the volume of test solution (in liters) and the molecular weight of the growth factor (bFGF) being investigated:

volume = 0.0002; ♦

mwF = 18000; ♦

6. Enter the amount of bFGF added per test well ($L_0$) ⟨ng⟩

Fadd = {0.05, 0.05, 0.05, 0.05, 0.05, 0.05, 0.10, 0.10, 0.10, 0.10, 0.10, 0.10, 0.20, 0.20, 0.20, 0.20, 0.20, 0.20, 0.20, 0.20, 0.20, 0.20, 0.20, 0.20, 0.50, 0.50, 0.50, 0.50, 0.50, 0.50, 0.50, 0.50, 0.50, 0.75, 0.75, 0.75, 1.0, 1.0, 1.0, 1.0, 1.0, 1.0, 1.0, 1.0, 1.0, 2.0, 2.0, 2.0, 2.0, 2.0, 2.0, 2.0, 2.0, 2.0, 2.5, 2.5, 2.5, 4.0, 4.0, 4.0, 5.0, 5.0, 5.0, 5.0, 5.0, 5.0, 6.0, 6.0, 6.0, 10.0, 10.0}; ♦

7. Enter the measured values of bFGF retained ($C_1$) ⟨ng⟩

Fbound = {.0025, .0035, .0034, .0013, .0055, .0033, .0067, .0069, .0076, .0028, .0033, .0035, .017, .020, .018, .0063, .0076, .0063, .017, .016, .018, .016, .017, .017, .079, .034, .043, .029, .034, .032, .021, .022, .020, .047, .056, .057, .080, .13, .099, .13, .071, .078, .037, .039, .086, .088, .066, .070, .21, .24, .19, .17, .16, .16, .25, .20, .22, .32, .32, .30, .17, .37, .27, .22, .20, .17, .51, .48, .45, .44, .39, .42}; ♦

8. Convert these values ($L_0$ and $C_1$) from ng to n*M* and store as data sets 1 and 2:

dat1 = Fadd/(volume*mwF) ; ♦

dat2 = Fbound/(volume*mwF); ♦

## Mathematica Fitting Program for Ligand – Proteoglycan Binding

```
<< Statistics`NonlinearFit`

volume = 0.0002;
mwF = 18000;
Fadd = {.05, .05, .05, .05, .05, .05, 0.10, 0.10, 0.10, 0.10, 0.10, 0.10, 0.20, 0.20,
  0.20, 0.20, 0.20, 0.20, 0.20, 0.20, 0.20, 0.20, 0.20, 0.20, 0.50, 0.50, 0.50, 0.50,
  0.50, 0.50, 0.50, 0.50, 0.50, 0.75, 0.75, 0.75, 1.0, 1.0, 1.0, 1.0, 1.0, 1.0, 1.0,
  1.0, 1.0, 2.0, 2.0, 2.0, 2.0, 2.0, 2.0, 2.0, 2.0, 2.0, 2.5, 2.5, 2.5, 4.0,
  4.0, 4.0, 5.0, 5.0, 5.0, 5.0, 5.0, 5.0, 6.0, 6.0, 6.0, 10.0, 10.0, 10.0};
Fbound = {.0025, .0035, .0034, .0013, .0055, .0033, .0067, .0069, .0076,
  .0028, .0033, .0035, .017, .020, .018, .0063, .0076, .0063, .017, .016,
  .018, .016, .017, .017, .079, .034, .043, .029, .034, .032, .021, .022,
  .020, .047, .056, .057, .080, .13, .099, .13, .071, .078, .037, .039,
  .086, .088, .066, .070, .21, .24, .19, .17, .16, .16, .25, .20, .22,
  .32, .32, .30, .17, .37, .27, .22, .20, .17, .51, .48, .45, .44, .39, .42};


dat1 = Fadd / (volume * mwF);
dat2 = Fbound / (volume * mwF);
data = {dat1, dat2};
binding = Transpose[data];


ParameterCITable /.
 NonlinearRegress[binding, ligand * rt / (kd + ligand), {ligand}, {rt, kd}]
```

| | Estimate | Asymptotic SE | CI |
|---|---|---|---|
| rt | 0.228852 | 0.0385813 | {0.151904, 0.3058} |
| kd | 2.34021 | 0.624793 | {1.0941, 3.58632} |

Fig. 3. Mathematica program to solve for ligand–proteoglycan affinity and binding sites. The code and data are input as shown. The ; following each line suppresses output and may be eliminated if the user prefers. The output is displayed at the bottom and includes the asymptotic standard error and the asymptotic confidence intervals.

9. Group the data ($L_0$, $C_1$):

   data = {dat1, dat2}; ♦

   binding = Transpose[data]; ♦

10. Perform the calculation for the nonlinear regression—ParameterCITable will output the $K_{D1}$ and $R_T$ values at the end with the standard error and the confidence interval. The units for each will be in n*M*:

    ParameterCITable /. NonlinearRegress[binding, ligand*Rt/(Kd1+ligand), {ligand}, {Rt,Kd1}] ♦

11. $K_{D1}$ is 2.3 ± 0.6 n*M*. $R_T$ is 0.23 ± 0.04 n*M*. The complete Mathematica program is shown in **Fig. 3**. To convert $R_T$ to sites/ng E-HS (*see* **Note 3**):

$$\text{sites} = R_T \frac{\text{volume}}{\text{PG}} \cdot N_{\text{AV}}$$

where PG is the amount of E-HS added ⟨ng⟩, $N_{AV}$ is Avogadro's number <6.02 * $10^{23}$ sites mol>. $R_T$ for this system is $7.0 \pm 1.2 \times 10^8$ sites/ng.

Although similar, the specifics for the ligand-binding inhibitor and the proteoglycan binding inhibitor do differ slightly and will each be outlined below. As example inhibitors, we include data for sucrose octasulfate (ligand-binding inhibitor) and protamine sulfate (proteoglycan-binding inhibitor)

### 3.2.2. Ligand-Binding Inhibitors (Sodium Sucrose Octasulfate)

An analysis of the inhibitor effectiveness is conducted based on steady-state data acquisition with two independent first-order reactions occurring in solution.

$$P + L \overset{K_{D1}}{\longleftrightarrow} C_1 \tag{1}$$

$$S + L \overset{K_{D2}}{\longleftrightarrow} C_2 \tag{2}$$

where $P$ is E-HS, $L$ is the growth factor (bFGF), $C_1$ is the binding complex of E-HS and bFGF, $S$ is a growth factor binding inhibitor (SOS), and $C_2$ is the binding complex of SOS and bFGF.

Assuming conservation of mass, the following equation at the steady state can be solved for the $K_D$ value ($K_{D2}$):

$$(P_0 - C_1)(L_0 - C_1) - (P_0 - C_1)C_2 = K_{D1}C_1 \tag{4}$$

where $$C_2 = \frac{1}{2}\left[(L_0 - C_1 + S_0 + K_{D2}) - \sqrt{(L_0 - C_1 + S_0 + K_{D2})^2 - 4S_0(L_0 - C_1)}\right]$$

Knowing $P_0$ and $K_{D1}$ from steady-state analysis in the absence of inhibitor (*see* **Subheading 3.2.**), experiments are run at a constant level of E-HS and bFGF ($L_0$) and various levels of inhibitor ($S_0$). Retention is a measurement of $C_1$. The programming to determine $K_{D2}$ is outlined below.

1. Mathematica software (Version 3.0, Wolfram Research) is launched, and a new notebook page is initialized. Commands as outlined below are entered into the notebook page, and the "shift" and "enter" key are pressed simultaneously at the end of each instructional set (symbolized in the procedure by " ♦ ")
2. The statistical package must be loaded by typing (*see* **Note 2**)

```
<<Statistics`NonlinearFit`    ♦
```

3. The known parameters—$P_0$ and $K_{D1}$—with both in units of n*M*:

```
P0 = 0.23;    ♦
KD1 = 2.3;    ♦
```

4. Enter the ligand concentration, $L_0$ ⟨n*M*⟩, the reaction volume, volume ⟨L⟩, the molecular weight of the inhibitor, mwI ⟨g/mol⟩, and the molecular weight of the growth factor, mwF ⟨g/mol⟩:

```
L0 = 0.056;    ♦
volume = 0.0002;    ♦
mwI = 1159;    ♦
mwF = 18000;    ♦
```

5. Enter the quantity of inhibitor ($S_0$) added ⟨ng⟩:

s = {0, 0, 0, 0, 5, 5, 5, 25, 25, 25, 50, 50, 50, 100, 100, 100, 500, 500, 500, 750, 750, 750}; ♦

6. Enter the measured values of bFGF retained ($C_1$) ⟨ng⟩:

c = {.018, .015, .022, .021, .02, .016, .013, .011, .0061, .011, .0082, .0061, .013, .0094, .010, .0076, .0072, .0015, .0052, .0073, .0040, .0039}; ♦

7. Convert these values ($S_0$ and $C_1$) to n$M$ from ng and store as data sets A and B:

dat1 = s/(volume*mwI) ; ♦

dat2 = c/(volume*mwF); ♦

8. Enter $K_{D1}C_1$ as a set of data:

dat3 = Kd1*dat2; ♦

9. Group the data ($S_0$, $C_1$, $K_{D1}C_1$):

data = {dat1, dat2, dat3}; ♦

SOS = Transpose[data]; ♦

10. Perform the calculation for the nonlinear regression—ParameterCITable will output the $K_{D2}$ value at the end, with the standard error and the confidence interval. Entering an estimate or initial guess for $K_{D2}$ will aid in the fitting process (the initial guess in our example was 300; *see* **Note 4**). It should be noted that the $C_2$ value corresponds to one of the roots of the quadratic equation (note the minus sign shown bold below)—the alternate root yielded a negative value for $K_{D2}$ for this case.

ParameterCITable /. NonlinearRegress[SOS, (Po-C1)*(Lo-C1)-(Po-C1)*((Lo-C1+So+Kd2) **–** ((Lo-C1+So+Kd2)^2-4*So*(Lo-C1))^0.5)*0.5, {So,C1}, {Kd2,300}] ♦

11. $K_{D2}$ for SOS was 0.25 ± 0.08 μ$M$.

### 3.2.3. Proteoglycan Binding Inhibitors (Protamine Sulfate)

Inhibitor effectiveness analysis is based on steady-state data acquisition with two independent first-order reactions occurring in solution.

$$P + L \overset{K_{D1}}{\longleftrightarrow} C_1 \tag{1}$$

$$\mathrm{PS} + L \overset{K_{D3}}{\longleftrightarrow} C_3 \tag{2}$$

where $P$ is E-HS, $L$ is the growth factor (bFGF), $C_1$ is the binding complex of E-HS and bFGF, PS is a proteoglycan binding inhibitor (protamine sulfate), and $C_3$ is the binding complex of PS and bFGF.

Assuming conservation of mass, the following equation at the steady state can be solved for the $K_D$ value ($K_{D3}$):

$$(L_0 - C_1)(P_0 - C_1) - (L_0 - C_1)\, C_3 = K_{D1}C_1 \tag{3}$$

where $C_3 = \frac{1}{2}\left[(P_0 - C_1 + \mathrm{PS}_0 + K_{D3}) - \sqrt{(P_0 - C_1 + \mathrm{PS}_0 + K_{D3})^2 - 4\mathrm{PS}_0(P_0 - C_1)}\right]$

Knowing $P_0$ and $K_{D1}$ from steady-state analysis in the absence of inhibitor (*see* **Subheading 3.2.1.**), experiments are run at a constant level of E-HS and bFGF ($L_0$) and various levels of inhibitor ($PS_0$). Retention is a measurement of $C_1$.

1. Mathematica software (Version 3.0, Wolfram Research) is launched, and a new notebook page is initialized. Commands as outlined below are entered into the notebook page, and the "shift" and "enter" keys are pressed simultaneously at the end of each instructional set (symbolized in the procedure by " ♦ ")
2. The statistical package must be loaded by typing (*see* **Note 2**):

<<Statistics`NonlinearFit` ♦

3. The known parameters are then initialized—$P_o$ and $K_{D1}$—with both in units of n*M*:

$P_0 = 0.23;$ ♦

$K_{D1} = 2.3;$ ♦

4. Enter the ligand concentration - $L_0$ ⟨n*M*⟩, the reaction volume, volume ⟨L⟩, the molecular weight of the inhibitor - mwI ⟨g/mol⟩, and the molecular weight of the growth factor - mwF ⟨g/mol⟩:

$L_0 = 0.056;$ ♦

volume = 0.0002; ♦

mwI = 6500; ♦

mwF = 18000; ♦

5. Enter the quantity of inhibitor ($S_0$) added ⟨ng⟩:

s = {0, 0, 0, 40, 40, 40, 80, 80, 80, 160, 160, 160, 320, 320, 320, 600, 600, 600}; ♦

6. Enter the measured values of bFGF retained ($C_1$) ⟨ng⟩:

c = {.018, .022, .019, .014, .017, .014, .009, .013, .014, .01, .012, .003, .001, .006, .011, .003, .002, -.003}; ♦

7. Convert these values ($PS_0$ and $C_1$) from ng to n*M* and store as data sets A and B:

dat1 = s/(volume*mwI) ; ♦

dat2 = c/(volume*mwF); ♦

8. Enter $K_{D1}C_1$ as a set of data:

dat3 = Kd1*dat2; ♦

9. Group the data ($PS_0$, $C_1$, $K_{D1}C_1$):

data = {dat1, dat2, dat3}; ♦

Protamine = Transpose[data]; ♦

10. Perform the calculation for the nonlinear regression—ParameterCITable will output the $K_{D2}$ value at the end or with the standard error and the confidence interval. Entering an estimate or initial guess for $K_{D3}$ will aid in the fitting process (the initial guess in our example was 75) (*see* **Note 4**). It should be noted that the $C_3$ value corresponds to one of

the roots of the quadratic equation (note the minus sign shown bold below)—the alternate root yielded a negative value for $K_{D3}$ for this case.

ParameterCITable /. NonlinearRegress[Protamine, $(L_0 - C_1)*(P_0 - C_1) - (L_0 - C_1) * ((P_0 - C_1 + PS + K_{D3}) \mathbf{-} ((P_0 - C_1 + PS + K_{D3}) \wedge 2 - 4 * PS *(P_0 - C_1)) \wedge 0.5) * 0.5$, $\{PS,C_1\}$, $\{K_{D3},75\}$] ♦

10. $K_{D3}$ for protamine sulfate was $0.1 \pm 0.02$ μ*M*.

### 3.3. Summary

Proteoglycan binding is a physiologically important means of regulating growth factor activity and availability. This chapter details a fast and simple assay for quantifying inhibitors of growth factor–proteoglycan binding. Although data and the corresponding analysis focused on inhibitors of bFGF interactions, the method should be easily transferable to other proteoglycan-binding growth factors and molecules.

## 4. Notes

1. Many protocols for isolation of proteoglycans involve the use of denaturing agents such as urea. These agents can interfere with protein–proteoglycan interactions and should be removed from proteoglycan preparation prior to binding experiments.
2. When loading the Statistics packages, it is important to use ` (grave accent) as opposed to ‘ (open quote) in the statement: <<Statistics`NonlinearFit`.
3. In calculating the growth-factor binding sites per nanogram of proteoglycan, it is important to enter the $R_T$ value in moles per liter, volume in liters, and proteoglycan in nanogram to generate binding sites per nanogram (*see* **Subheading 3.2.1, step 11**).
4. When fitting parameters using NonlinearRegress, a warning may be issued:

   **NonlinearFit::lmpnocon : Warning: The sum of squares has achieved a minimum, but at least one parameter estimate fails to satisfy either an accuracy goal of 1 digit (S) or a precision goal of 1 digit (s). These goals are less strict than those for the sum of squares, specified by AccuracyGoal ->6 and PrecisionGoal ->**

   If this occurs, it is best to rerun your regression fit with the output $K_D$ used as your new initial guess. This may be easily done by highlighting the right sidebar and pressing ♦. Repeat this procedure until the warning is not issued.

## References


1. Conrad, E. (1998) *Heparin Binding Proteins*, Academic Press, San Diego, CA.
2. Rapraeger, A. C., Krufka, A., and Olwin, B. B. (1991) Requirement of heparan sulfate for bFGF-mediated fibroblast growth and myoblast differentiation. *Science* **252,** 1705–1708.
3. Fannon, M. and Nugent, M. A. (1996) Basic fibroblast growth factor binds its receptors, is internalized, and stimulates DNA sythesis in Balb/c3T3 cells in the absence of heparan sulfate. *J. Biol. Chem.* **271,** 17949–17956.
4. Forsten, K. E., Courant, N. A., and Nugent, M. A. (1997) Endothelial proteoglycans inhibit bFGF binding and mitogenesis. *J. Cell. Physiol.* **172,** 209–220.
5. Dowd, C. J., Cooney, C. L., and Nugent, M. A. (1999) Heparan sulfate mediates bFGF transport through basement membrane by diffusion with rapid reversible binding. *J. Biol. Chem.* **274,** 5236–5244.

6. Isner, J. M. (1996) The role of angiogenic cytokines in cardiovascular disease. *Clin. Immunol. Immunopathol.* **80,** S82–S91.
7. Rapraeger, A. and Yeaman, C. (1989) A quantitative solid-phase assay for identifying radiolabeled glycosaminoglycans in crude cell extracts. *Anal. Biochem.* **179,** 361–365.
8. Nugent, M. A. and Edelman, E. R. (1992) Kinetics of basic fibroblast growth factor binding to its receptor and heparan sulfate proteoglycan: a mechanism for cooperativity. *Biochemistry* **31,** 8876–8883.
9. Farndale, R. W., Buttle, D. J., and Barrett, A. J. (1986) Improved quantitation and discrimination of sulphated glycosaminoglycans by use of dimethylmethylene blue. *Biochim. Biophys. Acta* **883,** 173–177.

# 42

## Interaction of Proteoglycans with Receptor Tyrosine Kinases

**David K. Moscatello and Renato V. Iozzo**

### 1. Introduction

The control of cell proliferation depends on the interactions between growth factors and their specific receptor-activated signaling pathways. It is well accepted that the local extracellular matrix can modulate cellular responses to a given signal in several ways, such as by modulating the affinity of the ligand for its cognate receptor ***(1)***, by binding and limiting availability of a growth factor ***(1–3)***, or by influencing proteolytic processing and internalization ***(3)***. However, it has only recently been shown that "structural" components of the extracellular matrix can interact directly with, and activate, receptor tyrosine kinases (RTKs). This was first shown by Vogel et al. and Shrivastava et al., who demonstrated that the "orphan" receptor tyrosine kinases DDR1 and DDR2 in fact bind fibrillar collagen ***(4,5)***. This binding required the native triple-helical structure of collagen and showed much slower kinetics than observed with other ligand–receptor interactions ***(5)***. Decorin ***(6–8)***, a member of a family of small leucine-rich proteoglycans ***(3)***, binds to fibrillar collagen and is an important regulator of matrix assembly ***(9–11)***. Decorin content is elevated in the tumor stroma of colon cancer ***(11)***, and ectopic expression of decorin inhibits cell growth ***(11–13)***. The growth-suppressive properties of decorin are independent of p53 or retinoblastoma proteins but require functional p21 protein (Waf1/Cip1/Sdi1) ***(13–16)***.

We recently reported that decorin activated the epidermal growth factor receptor (EGFR) in A431 squamous carcinoma cells and other transformed cell lines. This signaling was mediated by the protein core of decorin and induced MAP kinase activation and a protracted upregulation of endogenous p21, thereby leading to growth suppression ***(17)***. This activation occurred as a result of a direct interaction of decorin with the EGF receptor ***(17,18)***. However, even "small" proteoglycans such as decorin (~100 kDa, with a ~42-kDa core protein) are quite large in comparison with most ligands of receptor tyrosine kinases, which are typically 6–25 kDa, a fact that complicates analysis of crosslinking studies. Thus, in combination with their relatively

From: *Methods in Molecular Biology, Vol. 171: Proteoglycan Protocols*
Edited by: R. V. Iozzo © Humana Press Inc., Totowa, NJ

promiscuous binding properties, establishing a direct interaction of proteoglycans with receptor tyrosine kinases requires a number of methodological approaches. Examples of a few such techniques are presented here, but the reader is encouraged to consider other chapters in this volume as well, since other approaches may be useful in the context of different ligand–receptor systems. While the sample protocols presented all involve the EGFR and decorin, the same approaches should prove useful for other receptor–proteoglycan interactions.

## 2. Materials

### 2.1. Cells, Culture Media, and Growth Factors

1. A431, HT-1080, and a wide variety of other human tumor cell lines can be obtained from the American Type Culture Collection (Rockville, MD), as can CHO and NIH-3T3 cells. The latter two cell lines are widely used for transfection and overexpression of proteins of interest.
2. Fetal bovine serum (FBS) is used at 10% (v/v) to culture A431 and most human tumor cell lines. Ten percent calf serum (CS) is used to grow NIH-3T3 cells and transfectants. Both can be obtained from Hyclone (Logan, UT) or Life Technologies (Rockville, MD). A431 and 3T3 cells are grown in Dulbecco's modified Eagle medium (Life Technologies or Mediatech) (Herndon, VA). RPMI 1640, D-PBS, epidermal growth factor, and trypsin (0.25% for A431, 0.05% for 3T3) are all obtained from Life Technologies.
3. Conditioned media for soluble receptor ectodomain or recombinant proteoglycan production are readily concentrated using Amicon Ultrafree-15 centrifugal filter units with Biomax-50 membranes (50,000-dalton nominal molecular weight limit) and filter-sterilized with 0.45-µm pore vacuum filter units (Millipore, Bedford, MA).

### 2.2 Antibodies and Biochemical Reagents

1. Anti-EGFR and anti-phosphotyrosine antibodies can be obtained from Promega (Madison, WI) and Upstate Biotechnology (Lake Placid, NY), respectively. Antibodies to a variety of EGFR epitopes are also available from Calbiochem-Novabiochem Corp. (San Diego, CA).
2. Bovine serum albumin (BSA, Cohn fraction V) and all other reagents not specified are from Sigma Chemical (St. Louis, MO).
   a. For blocking cells, a 0.2% BSA (w/v) solution in RPMI 1640 is filter sterilized and stored at 4°C.
   b. Membranes are blocked with 2% BSA in TBST (TBS [100 m*M* Tris-HCl, pH 7.5, 150 m*M* NaCl] plus 0.1% Tween 20) or 5% nonfat dry milk in TBST.
   c. Kits for labeling by the Iodo-Gen method are available from Pierce Chemical (Rockford, IL). [$^{125}$I]NaI is from Amersham (Arlington Heights, IL).
   d. Active, affinity-purified EGFR was purchased from Sigma Chemical (cat. no. E2645). [$\gamma$-$^{32}$P] ATP (6000 Ci/mmol), enhanced chemiluminescence (ECL) reagents and Hybond ECL membranes were from Amersham. Nitrocellulose membranes from Schleicher & Schuell (Keene, NH) also work well.
   e. Immulon 4HXB wells are from Dynex Technologies (Chantilly, VA).
   f. The membrane-impermeable crosslinker bis[sulfosuccinimidyl]-suberate (BS$^3$) is from Pierce. Nickel-nitrilo-triacetic acid (Ni-NTA) columns for purification of His$_6$-tagged recombinant proteins are from Qiagen (Valencia, CA).

3. PBS/TDS: 10 m$M$ $Na_2HPO_4$, 150 m$M$ NaCl, 1% Triton X-100, 0.5% sodium deoxycholate, 0.1% sodium dodecyl sulfate, 0.02% sodium azide, 0.004% sodium fluoride, and 1 m$M$ sodium orthovanadate, pH 7.25. The stock is stored at 4°C, and phenylmethylsulfonyl fluoride (made as a 100× stock in isopropanol and stored at –20°C) is added to 100 µg/mL immediately before use.
4. In vitro reactions with EGFR kinase are performed in 20 m$M$ HEPES, pH 7.4, 2 m$M$ $MnCl_2$, 10 m$M$ $p$-nitrophenyl phosphate, 40 m$M$ $Na_3VO_4$, 0.01% BSA with 15 m$M$ ATP (kinase buffer). For blocking wells in radioligand binding assays, 0.1% BSA in TBS with 2 m$M$ $CaCl_2$, 2 m$M$ $MgCl_2$, and 0.02% $NaN_3$ is used.

## 3. Methods

### 3.1. Binding Competition with Known Receptor Ligands

1. Growth of cells. Cells are seeded into 24-well (16-mm) plates and grown for 2 d or until just confluent. Sufficient wells must be plated for triplicate determinations of all conditions. A431 cells are commonly used to analyze binding to the EGFR.
2. Recombinant human decorin (*see* **Note 1**) is labeled to high specific activity (~8 × $10^6$ cpm/mg) with [$^{125}$I]NaI (Amersham) by the Iodo-Gen method (Pierce Chemical).
3. The cells are washed twice with D-PBS, then blocked with 1 mL of RPMI-1640 containing 0.2% BSA at 4°C or on ice for 1 h (*see* **Note 2**).
4. The blocking medium is removed, and the cells are then incubated with 0.5 mL of 0.2% BSA/ RPMI-1640 containing *ca.* 200 ng (1.5 × $10^6$ cpm) $^{125}$I-decorin with or without the appropriate concentrations of specific ligand. In this example, EGF is added at concentrations from 10 to 300 ng/mL. A 100-fold excess of unlabeled test ligand is added to triplicate wells to correct for background binding.
5. The binding media are removed, and the wells are washed three times with cold PBS. The cells are then dissolved in 1 mL of 1 $M$ NaOH and the radioactivity measured in a gamma counter. The means ± SD of the triplicate determinations are then calculated.

### 3.2. Chemical Crosslinking to Receptors

1. A431 cells (or other cells of interest) are plated in 6-well (35-mm) plates and grown until nearly confluent. The cells are incubated in serum-free medium (e.g., DMEM only) for 24–36 h.
2. The cells are washed twice with D-PBS, then blocked with 1 mL of RPMI-1640 containing 0.2% BSA at 4°C or on ice for 1 h.
3. The cells are incubated with decorin (DCN), decorin protein core (ΔDCN)(100 µg/mL), or EGF (16 n$M$), or combinations thereof, at 4°C or on ice for 1 h.
4. The cells are washed three times with 2 mL of cold PBS per well and incubated for 10 min at room temperature in PBS with 15 m$M$ crosslinker $BS^3$. If the lysates are to be subjected to immunoprecipitation, the cells are scraped into 0.5 mL/well ice-cold lysis buffer, centrifuged at 12,000$g$ in microcentrifuge tubes for 10 min at 4°C, and the supernatants are transferred to clean tubes. If the lysates are only to be analyzed by sodium dodecyl sulfate-polyacrylamide gel electrophoresis (SDS-PAGE), the cells may be lysed directly in SDS sample buffer. The cell lysates are separated by SDS-PAGE in thin (0.75-mm), 3–15% acrylamide gels, transferred to nitrocellulose membranes, and subjected to Western immunoblotting with anti-EGFR antibody *(18)*.
5. After exposing the blot to film, the autoradiogram is analyzed for the presence of receptor monomers and dimers; the latter should be present only in lanes with crosslinker. In the case of studies involving the EGFR, the monomeric receptor migrates at an $M_r$ of 170 kDa,

and the dimers at an $M_r$ of 340 kDa. Slower-migrating species may also be observed if the ligand is also crosslinked; in the present instance, a single decorin molecule bound to a EGFR dimer would be expected to migrate at an $M_r$ of ~440 kDa, and two decorin molecules per dimer would be expected to migrate at an $M_r$ of ~540 kDa (*see* **Note 3**).

### 3.3. Interaction with Soluble Receptor Ectodomain

1. A431 cells synthesize a secreted form of EGFR of ~105 kDa lacking the transmembrane and intracytoplasmic domains ***(19)***. A431 cells are grown to near confluence in standard medium, washed twice with D-PBS, and incubated with serum-free medium for 24–48 h (*see* **Note 4**).
2. Medium conditioned by confluent A431 cells is concentrated by centrifugation in Ultrafree-15 (50,000-nmwl) units at 4°C according to the manufacturer's protocol. The conditioned medium is then filter-sterilized (0.45-μm-pore filters) or centrifuged at high speed to remove insoluble material. $NaN_3$ (0.02% final concentration) may be added to inhibit microbial growth. Such preparations are ideally used fresh, but may be stored for a few days at 4°C, or mixed with an equal volume of glycerol and stored at –20°C.
3. Serial dilutions of BSA, decorin or its protein core are slot- or dot-blotted onto nitrocellulose membranes using a vacuum manifold and blocked overnight at 4°C (or for 1 h at room temperature) with 5% FBS and 5% nonfat milk (*see* **Note 5**).
4. The blots are washed several times in TBST and incubated with the serum-free medium conditioned by the A431 cells. The blots are again washed three times, then incubated with gentle agitation using an antibody against the ectodomain of the human EGFR (e.g., anti-LEEKK of the human EGFR N-terminal sequence, affinity purified on peptide linked to Sepharose) at 1 μg/mL in BSA/TBST for 1–2 h at room temperature. The antibody used should not interfere with the binding, in this case because it recognizes the N-terminal end of the EGF receptor, distant from the known ligand-binding site. It is desirable, however, to use an antibody that does recognize the ligand-binding domain as an additional control (e.g., monoclonal antibody 225 raised against the EGF-binding domain of the EGFR, **ref. 20**).
5. The membranes are washed three times with TBST and incubated with the appropriate secondary antibody (against either rabbit polyclonal or mouse monoclonal antibody) for 1 h at room temperature. Alternatively, radiolabeled secondary antibodies may be used.
6. Finally, the membranes are incubated with anti-rabbit or anti-mouse IgG coupled to horseradish peroxidase (1/5000) in TBS-T, washed three times, and incubated with enhanced chemiluminescence reagent. Multiple exposures are performed to guarantee linearity, and the intensities of the bands are quantitated by laser scanning densitometry.

### 3.4. In Vitro Phosphorylation of the Receptor Tyrosine Kinase

1. Constant amounts (~300 ng) of immunopurified EGFR are incubated with kinase buffer alone or containing various concentrations (5–20 μg) of decorin, decorin protein core, or collagen type I. A control tube should include 100 ng of EGF.
2. After 15 min of incubation in kinase buffer, 1 μCi [$\gamma$-$^{32}$P]ATP and 0.2% NP40 are added in a final volume of 60 μL. The mixtures are incubated for an additional 10 min, and the reactions are terminated by boiling in SDS sample buffer.
3. The samples are separated by SDS-PAGE, and the gel is either dried, or the proteins are transferred to nitrocellulose. The latter technique enables one to use immunoblotting to confirm the identities of the labeled proteins. Phosphorylated proteins are visualized by autoradiography. Control samples omitting either [$\gamma$-$^{32}$P]ATP or EGFR should show no activity (*see* **Fig. 1**).

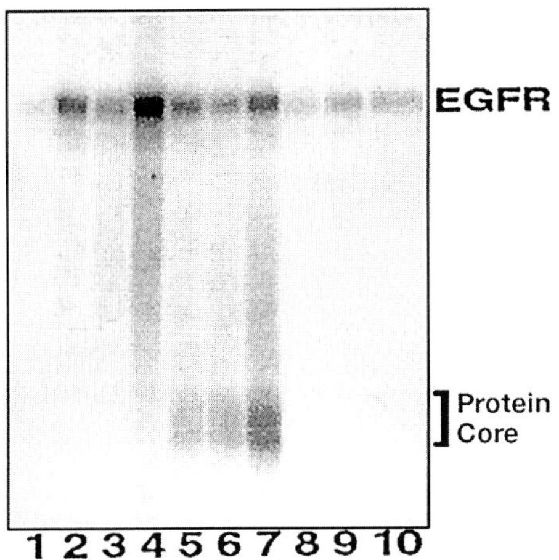


Fig. 1. In vitro phosphorylation assay using immunopurified EGFR and decorin, its protein core, or collagen. Constant amounts (~300 ng) of immunopurified EGFR were preincubated with buffer alone (*lane 1*) or containing decorin (*lanes 2–4*; at 5, 10, and 20 µg, respectively), decorin core protein (*lanes 5–7*; at 5, 10, and 20 µg, respectively), or collagen (*lanes 8–10*; at 5, 10, and 20 µg, respectively). After 15 min in kinase buffer, 1 µCi of [$\gamma$-$^{32}$P]ATP and 0.2% Nonidet P-40 were added. The mixtures were incubated for an additional 10 min, stopped by boiling in SDS buffer, and analyzed by SDS-PAGE and autoradiography. Note the strong EGFR autophosphorylation induced by decorin and decorin core protein relative to the collagen control.

### 3.5. Interaction of Proteoglycan with Purified Receptor Kinase

1. Column protocol: Recombinant decorin with an N-terminal tag of six histidine residues (6 His) allows for a rapid and efficient purification via Ni-NTA affinity chromatography ***(21)***.
2. Purified EGFR is phosphorylated to a high specific activity using the EGF control reaction in vitro phosphorylation procedure described in **Subheading 3.4., steps 1** and **2**, except that the reactions are not boiled in sample buffer and labeled receptor is separated from unincorporated label by Sephadex G-50 chromatography (Roche Quick-spin columns).
3. Constant amounts of $^{32}$P-labeled EGFR are incubated with increasing concentrations of decorin, decorin core protein, or control proteins for 30 min at 4°C under gentle agitation. The spin columns are equilibrated with three column volumes of binding buffer (300 m*M* $NaH_2PO_4$, pH 8.0, 300 m*M* NaCl), and then the samples are applied and spun at 750 g for 5 min.
4. Following two consecutive washes, the bound decorin/EGFR complexes are eluted with buffer containing 250 m*M* imidazole. Aliquots of the fractions are counted in a scintillation counter and the remainder are analyzed by SDS-PAGE and autoradiography.
5. Solid-phase binding protocol: Immulon 4HXB wells are coated with 1 µg (~22 pmol) of recombinant decorin in 100 µL of PBS for 2 h at 37°C or overnight at 4°C. The wells are washed three times with PBS and blocked with 0.1% BSA in TBS supplemented with 2m*M* $CaCl_2$, 2 m*M* $MgCl_2$, and 0.02% $NaN_3$ (blocking buffer) for 1 h.
6. Purified EGFR is labeled as described in **step 2**, 0.1–10 pmol $^{32}$P-EGFR is added in blocking buffer, and the wells are incubated under gentle shaking (60 rpm) for 4–14 h at 4°C.

Background binding is corrected for by incubation with at least 100-fold molar excess unlabeled EGFR. All samples should be done in triplicate. After incubation the wells are washed three times with ice-cold TBST and measured using a scintillation counter. Data are analyzed by Scatchard analysis, e.g., using the Ligand program ***(18)***.

## 4. Notes

1. It is imperative that the proteoglycans in question be purified by techniques that permit retention of a "native" conformation ***(21)***. Many commercially available extracellular matrix proteins and proteoglycans, particularly those from tissues such as skin or tendon, are prepared by harsh extraction methods that result in denatured products. Not surprisingly, such preparations have greatly reduced or no biological activity ***(18,22)***, and of course are unsuitable for use in studies of signaling interactions. Recombinant or endogenous materials isolated from cell cultures are generally suitable, as they can be purified by relatively gentle procedures (*see* Chaps. 1, 4, 20, 21; and **ref. 21** for relevant protocols). Wherever possible, it is also desirable to use a closely related proteoglycan as a control for nonspecific binding. For example, we use biglycan ***(17,21)*** as a control for decorin in experiments with EGFR; biglycan does not bind or activate the EGFR ***(17,22)***. We also use a deglycosylated proteoglycan, or mutated recombinant protein lacking the glycosaminoglycan chain ***(17,21,23)***, to ascertain the importance of the glycosaminoglycan chain in the interaction.
2. Only purified proteins such as bovine serum albumin should be used as blocking agents in binding experiments. Gelatin cannot be used, as it is composed of collagen, with which many proteoglycans can interact. Other commonly used blocking agents such as nonfat milk or sera contain growth factors or matrix components that may interact with either the receptor or proteoglycan of interest, respectively.
3. Very large complexes (>400 kDa) do not penetrate standard polyacrylamide gels, and are not efficiently transferred to membranes for blotting. A possible alternative approach would be the use of cleavable crosslinkers such as 3,3'-dithiobis[sulfosuccinimidyl] propionate (Pierce). Crosslinked receptor–proteoglycan complexes are immunoprecipitated from lysates, the cleaving reagent is added, the samples are boiled in SDS sample buffer, separated by SDS-PAGE, and analyzed by immunoblotting against both the receptor and the proteoglycan (or labeled ligand may be used).
4. Alternatively, constructs encoding ectodomains of other receptors of interest may be transfected into NIH-3T3 cells by standard methods. Conditioned media would be produced and collected in the same fashion as for EGFR ectodomain from A431, although the kinetics of expression would have to be determined empirically. Expression of the soluble receptor would be confirmed by immunoblotting, and an epitope tag such as 6 ↔ His would provide a convenient "handle" for purification and isolation of receptor–ligand complexes.
5. Alternatively, the reciprocal experiment may be performed; that is, the soluble receptor preparation may be slot-blotted and incubated with proteoglycan solution. The blots would then be probed with antibody to the proteoglycan, or labeled proteoglycan could be used.

## References


1. Iozzo, R. V. and Murdoch, A. D. (1996) Proteoglycans of the extracellular environment: clues from the gene and protein side offer novel perspectives in molecular diversity and function. *FASEB J.* **10,** 598–614.
2. Somasundaram, R. and Schuppan, D. (1996) Type I, II, III, IV, V, and VI collagens serve as extracelular ligands for the isoforms of platelet-derived growth factor (AA, BB, and AB). *J. Biol. Chem.* **271***(43)***,** 26884–26891.

3. Iozzo, R. V. (1997) The family of the small leucine-rich proteoglycans: key regulators of matrix assembly and cellular growth. *Crit. Rev. Biochem. Mol. Biol.* **32,** 141–174.
4. Vogel, W., Gish, G. D., Alves, F., and Pawson T. (1997) The discoidin domain receptor tyrosine kinases are activated by collagen. *Mol. Cell.* **1,**13–23, 1997.
5. Shrivastava, A., Radziejewski, C., Campbell, E., Kovac, L., McGlynn, M., Ryan, T. E., Davis, S., Goldfarb, M. P., Glass, D. J., Lemke, G., and Yancopoulos G. D. (1997) An orphan receptor tyrosine kinase family whose members serve as nonintegrin collagen receptors. *Mol. Cell.* **1,** 25–34.
6. Krusius, T. and Ruoslahti, E. (1986) Primary structure of an extracellular matrix proteoglycan core protein deduced from cloned cDNA. *Proc. Natl. Acad. Sci. (USA).* **83,** 7683–7687.
7. Day, A. A., McQuillan, C. I., Termine, J. D., and Young, M. R. (1987) Molecular cloning and sequence analysis of the cDNA for small proteoglycan II of bovine bone. *Biochem. J.* **248,** 801–805.
8. Fisher, L. W., Termine, J. D., and Young, M. F. (1989) Deduced protein sequence of bone small proteoglycan I (biglycan) shows homology with proteoglycan II (decorin) and several nonconnective tissue proteins in a variety of species. *J. Biol. Chem.* **264,** 4571–4576.
9. Toole, B. P. and Lowther, D. A. (1968) The effect of chondroitin sulphate-protein in the formation of collagen fibrils *in vitro. Biochem. J.* **109,** 857–866.
10. Vogel, K. G., Paulsson, M., and Heinegård, D. (1984) Specific inhibition of type I and type II collagen fibrillogenesis by the small proteoglycan of tendon. *Biochem. J.* **223,** 587–597.
11. Adany, R., Heimer, R., Caterson, B., Sorrell, J. M., and Iozzo. R. V. (1990) Altered expression of chondroitin sulfate proteoglycan in the stroma of human colon carcinoma. Hypomethylation of PG-40 gene correlates with increased PG-40 content and mRNA levels. *J. Biol. Chem.* **265,** 11389–11396.
12. Santra, M., Skorski, T., Calabretta, B., Lattime, E. C., and Iozzo, R. V. (1995) *De novo* decorin gene expression suppresses the malignant phenotype in human colon cancer cells. *Proc. Natl. Acad. Sci. (USA)* **92,** 7016–7020.
13. Santra, M., Mann, D. M., Mercer, E. W., Skorski, T., Calabretta, B., and Iozzo, R. V. (1997) Ectopic expression of decorin protein core causes a generalized growth suppression in neoplastic cells of various histogenetic origin and requires endogenous p21, an inhibitor of cyclin-dependent kinases. *J. Clin. Invest.* **100,** 149–157.
14. De Luca, A., Santra, M., Baldi, A., Giordano, A., and Iozzo, R. V. (1996) Decorin-induced growth suppression is associated with upregulation of p21, an inhibitor of cyclin-dependent kinases. *J. Biol. Chem.* **271,** 18961–18965.
15. Harper, J. W., Adami, G. R., Wei, N., Keyomarsi, K., and Elledge, S. J. (1993) The p21 Cdk-interacting protein Cip1 is a potent inhibitor of G1 cyclin-dependent kinases. *Cell.* **75,** 805–816.
16. El-Deiry, W. S., Tokino, T., Velculescu, V. E., Levy, D. B., Parsons, R., Trent, J. M., Lin, D., Mercer, W. E., Kinzler, K. W., and Vogelstein, B. (1993) WAF1, a potential mediator of p53 tumor suppression. *Cell.* **75,** 817–825.
17. Moscatello, D. K., Santra, M., Mann, D. M., McQuillan, D. J., Wong, A. J., and Iozzo, R. V. (1998) Decorin suppresses tumor cell growth by activating the epidermal growth factor receptor. *J. Clin. Invest.* **101,** 406–412.
18. Iozzo, R. V., Moscatello, D. K., McQuillan, D. J., and Eichstetter, I. (1999) Decorin is a biological ligand for the epidermal growth factor receptor. *J. Biol. Chem.* **274,** 4489–4492.

19. Weber, W., Gill, G. G., and Spiess, J. (1984) Production of an epidermal growth factor receptor-related protein. *Science* **224,** 294–297.
20. Fan, Z., Lu, Y., Wu, X., and Mendelsohn, J. (1994) Antibody-induced epidermal growth factor receptor dimerization mediates inhibition of autocrine proliferation of A431 squamous carcinoma cells *J. Biol. Chem.* **269,** 27595–27602.
21. Ramamurthy, P., Hocking, A. M., and McQuillan, D. J. (1996) Recombinant decorin glycoforms. Purification and structure. *J. Biol. Chem.* **271,** 19578–19584.
22. Hocking, A. M., Strugnell, R. A., Ramamurthy, P., and McQuillan, D. J. (1996) Eukaryotic expression of recombinant biglycan. Post-translational processing and the importance of secondary structure for biological activity. *J. Biol. Chem.* **271,** 19571–19577.
23. Mann, D. M., Yamaguchi, Y., Bourdon, M. A., and Ruoslahti, E. (1990) Analysis of glycosaminoglycan substitution in decorin by site-directed mutagenesis. *J. Biol. Chem.* **265,** 5317–5323.

# 43

# Measuring Single-Cell Cytosolic $Ca^{2+}$ Concentration in Response to Proteoglycans

**Sandip Patel and Renato V. Iozzo**

## 1. Introduction

$Ca^{2+}$ is a crucial biological messenger involved in a host of diverse cellular processes *(1)*. Many hormones, growth factors, and neurotransmitters raise cytosolic $Ca^{2+}$ levels through activation of phospholipase C, which catalyzes the production of the second messenger, inositol 1,4,5-trisphosphate ($IP_3$) *(2)*. G-protein-linked receptors and receptor tyrosine kinases couple to distinct isoforms of phospholipase C *(3)*. Once generated, $IP_3$ raises cytosolic $Ca^{2+}$ levels by activating $Ca^{2+}$ channels located on the membranes of intracellular $Ca^{2+}$ stores *(4,5)*.

Extracellular matrix constituents are also known to induce $Ca^{2+}$ signals through activation of integrins *(6)*. More recently, we have demonstrated that decorin, a member of the small leucine-rich proteoglycan family *(7)*, mediates cytosolic $Ca^{2+}$ increases through a novel action on the epidermal growth factor (EGF) receptor (**Fig. 1**) *(8)*. Indeed, interaction of the extracellular matrix with cell surface receptor tyrosine kinases may be more widespread than previously thought ***(9–11)***. Changes in cytosolic [$Ca^{2+}$] are therefore likely to be important in mediating the effects of the extracellular matrix on cell function. In this chapter, the principles of measuring single cell cytosolic [$Ca^{2+}$] are discussed.

Measuring cytosolic [$Ca^{2+}$] at the single cell level using imaging approaches has several advantages over monitoring [$Ca^{2+}$] in populations of cells. In many cells, submaximal hormone stimulation induces complex changes in cytosolic [$Ca^{2+}$], including $Ca^{2+}$ oscillations *(12)*, which in population studies are "averaged" out and remain undetected. Such temporal averaging also distorts the kinetics of even relatively simple changes in cytosolic [$Ca^{2+}$]. For example, if all cells do not respond under a particular condition or do so with differing latencies, the average population response will underestimate the peak and rate of rise in cytosolic [$Ca^{2+}$] of the responsive cells. In addition, population studies provide no spatial information concerning the $Ca^{2+}$

From: *Methods in Molecular Biology, Vol. 171: Proteoglycan Protocols*
Edited by: R. V. Iozzo © Humana Press Inc., Totowa, NJ

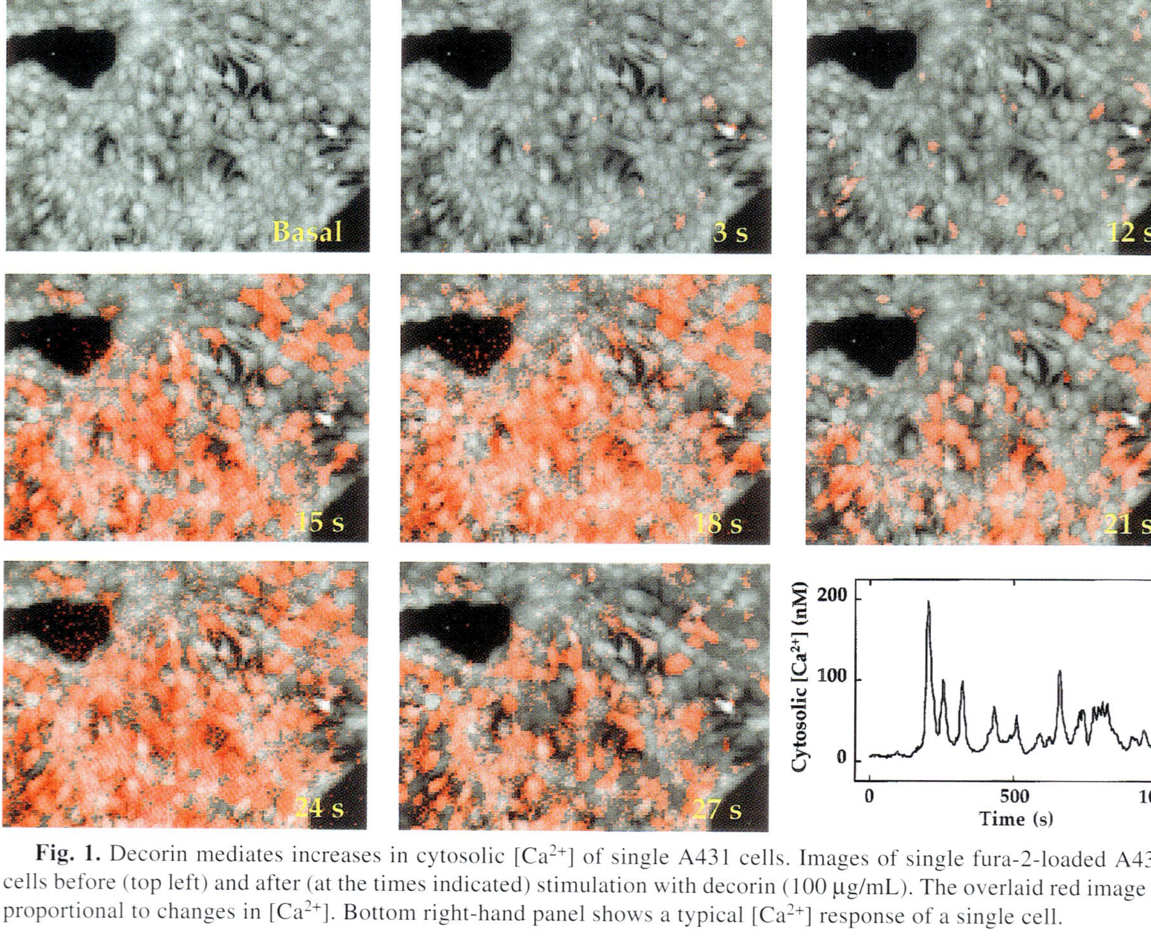


**Fig. 1.** Decorin mediates increases in cytosolic [$Ca^{2+}$] of single A431 cells. Images of single fura-2-loaded A431 cells before (top left) and after (at the times indicated) stimulation with decorin (100 µg/mL). The overlaid red image is proportional to changes in [$Ca^{2+}$]. Bottom right-hand panel shows a typical [$Ca^{2+}$] response of a single cell.

increase within the cell. Indeed, $Ca^{2+}$ signals often originate from a specific cellular locus and travel throughout the cytosol as planar or even spiral "waves" ***(12)***. Finally, population measurements do not distinguish signals from damaged or contaminating cells.

### 1.1. Fluorescent $Ca^{2+}$-Sensitive Indicators

The use of fluorescent $Ca^{2+}$-sensitive indicators is currently the most popular method for monitoring cytosolic [$Ca^{2+}$]. Fluorescence is the process whereby a molecule in an excited state dissipates some of its energy prior to relaxation, upon absorbing a photon of light, such that the emitted light is of lower energy and thus longer wavelength than that of the exciting light. Fluorescent $Ca^{2+}$ indicators are molecules that change their fluorescence properties upon binding $Ca^{2+}$, most of which are based on the $Ca^{2+}$ chelators, ethylene-bis(oxyethylenenitrilo)tetraacetic acid (EGTA) and 1,2-bis(2-Aminophenoxy)ethane-N,N,N'N'-tetraacetic acid (BAPTA). Many $Ca^{2+}$ indicators are commercially available for measuring [$Ca^{2+}$] [*see* **ref. *(13)***]. Factors to be considered when choosing a dye are its molar extinction coefficient and quantum yield—that is, the efficiency of light absorption and the amount of light emitted relative to what was absorbed. In order to maximize fluorescence signals in response to $Ca^{2+}$, a dye with an affinity appropriate to the range of anticipated [$Ca^{2+}$] changes should be chosen. Upon stimulation, cytosolic [$Ca^{2+}$] can be elevated several fold from resting levels of < 200 n*M*.

$Ca^{2+}$-sensitive indicators can be classified according to the nature of the fluorescence change that $Ca^{2+}$ binding induces. Ratiometric (dual-wavelength) indicators undergo a shift in their fluorescence spectra (excitation or emission) upon binding $Ca^{2+}$, whereas intensiometric (single wavelength) dyes do not. Fura-2 ***(14)*** is a ratiometric indicator that is currently the most commonly used $Ca^{2+}$-sensitive dye in imaging studies (**Fig. 2A**). Important properties of this dye include the following:

1. In the absence of $Ca^{2+}$, the excitation and emission peaks are 360 and 510 nm, respectively.
2. In the presence of saturating $Ca^{2+}$ levels, the excitation peak is shifted to 340 nm with no appreciable change in the emission spectra.
3. The isosbestic ($Ca^{2+}$-independent) wavelength is 360 nm.
4. Reciprocal changes in the fluorescence of the dye occur upon $Ca^{2+}$ binding when excited either side of the isosbestic wavelength.
5. At 340-nm excitation, fluorescence intensity increase as the [$Ca^{2+}$] increases.
6. At 380-nm excitation, fluorescence intensity decreases as the [$Ca^{2+}$] increases.
7. The ratio of the two intensities at 340 and 380 nm is proportional to [$Ca^{2+}$].

Fluo-3 is an example of an intensiometric indicator. This visible wavelength dye has excitation and emission peaks of ~500 and ~525 nm, respectively. This dye is particular useful in that in it undergoes a large increase in fluorescence upon binding $Ca^{2+}$. Additionally, the absolute florescence of the $Ca^{2+}$-free dye is extremely low, thus further improving the signal-to-noise ratio. Ratiometric dyes, however, are advantageous over intensiometric dyes in several respects. For intensiometric indicators, fluorescence is proportional to the concentration of both $Ca^{2+}$ and the dye. Since changes in cellular dye concentration can occur during experimentation, such

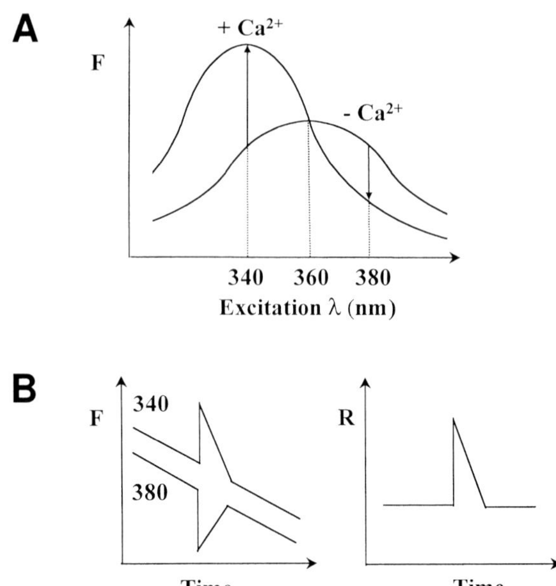


**Fig. 2. (A)** Excitation spectra of fura-2 in the presence of zero and saturating $Ca^{2+}$. The graph shows fluorescence *(F)* of the dye measured at 510 nm at a range of excitation wavelengths. Binding of $Ca^{2+}$ reduces the excitation peak from 360 to 340 nm. Fluorescence of the dye therefore increases at 340 nm and decreases at 380 nm in response to $Ca^{2+}$. No change in fluorescence occurs at 360 nm. **(B)** Ratio measurements cancel out nonreciprocal changes in fura-2 fluorescence. A schematic fura-2 timecourse recording showing a change in [$Ca^{2+}$] manifest as a transient increase and decrease in fluorescence at excitation wavelengths of 340 and 380 nm, respectively (left). The response is superimposed on a progressive slow decrease in fluorescence at both wavelengths. These non-reciprocal ($Ca^{2+}$-independent) changes in fluorescence are eliminated after calculation of the fluorescence ratio (340/380, right).

changes are difficult to distinguish from actual changes in [$Ca^{2+}$] when using intensiometric indicators. For example, in many cells, dye can be actively extruded from the cytosol or undergo photobleaching (*see* **Note 1**), which results in a decrease in fluorescence intensity. This decrease in fluorescence would appear as a decrease in [$Ca^{2+}$] when measured at a single wavelength. For ratiometric dyes such as fura-2, however, nonreciprocal changes in the fluorescence intensities are cancelled out after calculation of the fluorescence ratio (**Fig. 2B**). Ratiometric recording is also less prone to artefacts arising from cell movement, uneven dye distribution, or inhomogeneities in cell thickness. Fura-2, then is the dye of choice for single cell $Ca^{2+}$ imaging.

### 1.2. Loading Fluorescent $Ca^{2+}$ Indicators Into Cells

Most $Ca^{2+}$ indicators are available as acetoxymethyl (AM) esters, rendering them cell permeable *(15)*. Once inside the cell, endogenous esterases cleave the ester bonds, releasing the negatively charged dye and essentially trapping it within the cell. Thus, simple incubation of the cells in a physiological medium supplemented with the AM

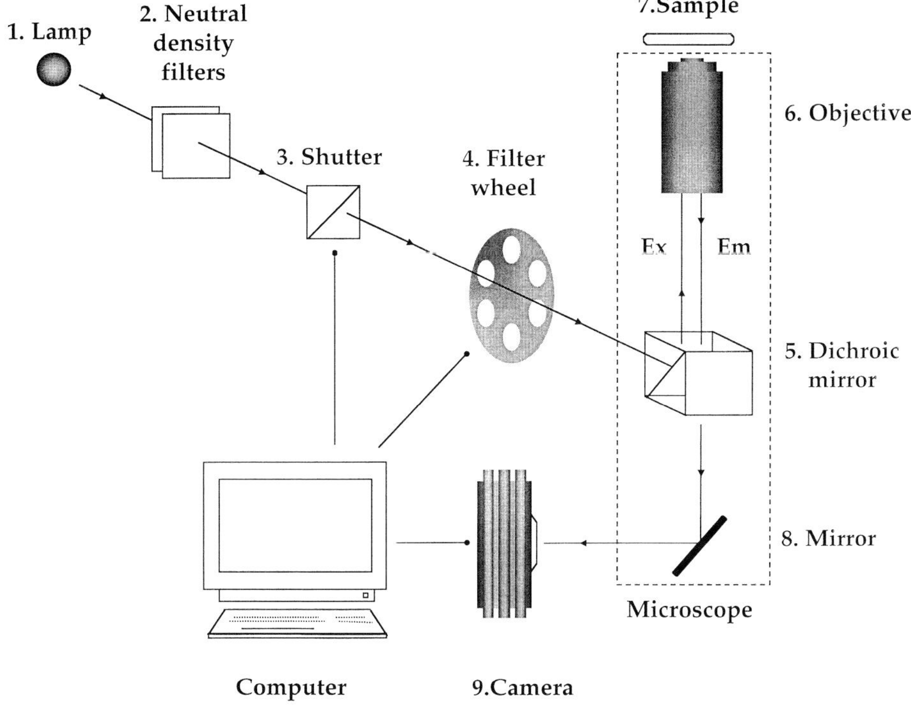


Fig. 3. Schematic representation of a typical imaging system. Light from an arc lamp (1) is attenuated by a neutral density filter (2) and passed through a shutter (3) to a filter wheel (4) that selects the appropriate excitation wavelength. The excitation light (Ex) then enters the microscope and is reflected by a dichroic mirror (5) up into the objective (6) to the sample (7) loaded with the fluorescent dye. The emitted fluorescence (Em) is then collected by the objective and directed back to the dichroic mirror, where it is transmitted to a second mirror (8) that reflects the light to the camera (9). The shutter, filter wheel, and camera are computer-controlled.

ester for a defined period of time results in significant accumulation of the dye within the cell (*see* **Subheading 3.1.**). The ease by which these dyes can be introduced into cells is a major advantage of using these dyes for monitoring [$Ca^{2+}$].

### 1.3. Instrumentation

Central to an imaging system (**Fig. 3**) is the epifluorescence microscope that houses appropriate optics to allow delivery of the exciting light and isolation of the emitted fluorescence. Briefly, cells grown on a glass cover slip are loaded with a fluorescent indicator and placed into a holder on the stage of the microscope. Excitation light of the appropriate wavelength for the indicator is selected (usually with a filter) and directed to the cells via the objective lens through which the cells are imaged. Part of the resulting emitted fluorescence is collected by the objective and discriminated from the excitation light by a dichroic mirror. The light is then focused onto a camera, digitized, and stored to a computer. By periodically illuminating the cells through the use of a computer-controlled shutter, a series of images is collected. Regions of interest, corresponding to individual cells (or subcellular regions) are selected by the user, and data are extracted from the image at each time point. Fluorescence intensities/ratios can then be calibrated to [$Ca^{2+}$] for individual cells in the entire field. A brief description of the principle components of an imaging system is given in **Table 1**.

Many imaging systems are now available commercially. Normally, an inverted as opposed to an upright microscope configuration is adopted, since access to the sample (for perfusion or microinjection) is unhindered. The choice of camera is one of the most important considerations in low-light-level studies. Cameras can be broadly classified according to their output, which may in the form of a standard video signal (e.g., silicon-intensified target cameras) that is later digitized or direct digital readout (e.g., cooled charged-coupled-device cameras). The latter provide better signal-to-noise ratios but are inherently slower. High-sensitivity cameras allow illumination intensity to be reduced without compromising signal, thus reducing photobleaching of the dyett. Imaging formats such as 1024 by 1024 pixels are typical. The range of digitisation (8–16 bit) determines the dynamic range (256–65536 intensity levels per pixel) (*see* ***ref. 16*** for further details).

### 1.4 Calibration of Fluorescence Data

Once fluorescence data are acquired, it is relatively straightforward to convert them to [$Ca^{2+}$]. Data should first be corrected for background fluorescence that is, cell-independent ("instrument") noise. For intensiometric indicators, fluorescence intensity *(F)* is related to [$Ca^{2+}$] by the following equation:

$$[Ca^{2+}] = \frac{(F - F_{min}) \cdot K_d}{(F_{max} - F)}$$

where $F_{min}$ and $F_{max}$ are the fluorescence intensities of the $Ca^{2+}$-free and $Ca^{2+}$-saturated dye, respectively, and $K_d$ is the dissociation constant of the dye for $Ca^{2+}$. With these indicators, an *in situ* calibration is necessary since measured fluorescence is dependent on dye concentration (*see* **Subheading 3.4.1.**). This involves exposing cells to a $Ca^{2+}$ ionophore such as ionomycin in order to equilibrate $Ca^{2+}$ across the plasma membrane, and then setting the extracellular [$Ca^{2+}$] to zero and saturating levels to derive $F_{min}$ and $F_{max}$, respectively. The parameters are thus determined under similar dye loading levels as those during experimentation.

For ratiometric indicators, the fluorescence ratio, R ($Ca^{2+}$-bound fluorescence/$Ca^{2+}$-free fluorescence) is related to [$Ca^{2+}$] by the following equation:

$$[Ca^{2+}] = \frac{(R - R_{min})}{(R_{max} - R)} \cdot \frac{K_d \cdot S_{f2}}{S_{b2}}$$

where $R_{min}$ and $R_{max}$ are the fluorescence ratios of the $Ca^{2+}$-free and $Ca^{2+}$-saturated dye, respectively, $K_d$ is the dissociation constant of the dye for $Ca^{2+}$, and $S_{f2}/S_{b2}$ is the ratio of fluorescence values for the $Ca^{2+}$-free and $Ca^{2+}$ saturated dye at the wavelength used to monitor the $Ca^{2+}$-free indicator. For fura-2, *R* is the fluorescence of the dye at 340 nm excitation/fluorescence of the dye at 380 nm excitation and $S_{f2}$ /$S_{b2}$ is determined at 380 nm. At 37°C, the $K_d$ of fura-2 is 224 n*M*.

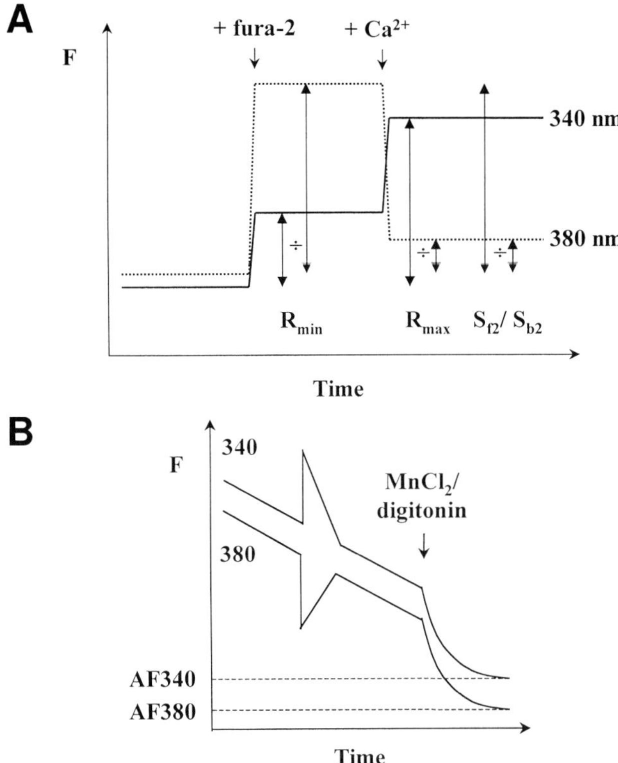


Fig. 4. In vitro calibration of ratiometric fluorescence data. **(A)** A cytosol-like calibration medium (initially $Ca^{2+}$-free) is used to determine calibration parameters (*see* **Table 2**). For fura-2, fluorescence of the medium at 340-nm (solid line) and 380-nm (dotted line) excitation is acquired first in the absence of dye to determine background fluorescence. Fura-2 free acid (1 μ*M*) and then $CaCl_2$ (500 μ*M*) are added to determine $R_{min}$, $R_{max}$, and $S_{f2}/S_{b2}$ as shown. These parameters are used to convert acquired ratio values to [$Ca^{2+}$] using **Eq. (2)**. **(B)** Autofluorescence (AF) is measured with either $MnCl_2$ or digitonin (see text) and subtracted from all experimental data prior to calculation of the fluorescence ratio.

$R_{min}$ and $R_{max}$ (from which $S_{f2}/S_{b2}$ is calculated) can be determined *in situ* with ionophore as with intensiometric indicators. However, as ratios are independent of dye concentration, calibration parameters can also be obtained (for the particular instrument configuration employed) in vitro that is, in the absence of cells (*see* **Fig. 4A**; **Subheading 3.3.2.**). This is achieved using an intracellular-like solution supplemented with the free (de-esterified) form of the dye in the presence of BAPTA and excess $Ca^{2+}$ to determine the fluorescence of the $Ca^{2+}$-free and $Ca^{2+}$-bound dye, respectively. Unlike *in situ* calibrations, which need to be performed at the end of each experimental run, in vitro calibrations need only be performed once on the day of experimentation. Furthermore, it is often difficult to completely equilibrate $Ca^{2+}$ across the plasma membrane in *in situ* calibrations. However, a major assumption in in vitro calibrations is that the conditions mimic the environment of the dye within the cell, which may not necessarily be the case.

**Table 1**
**Components of a Single-Cell Imaging System**

| Component | Function | Notes |
|---|---|---|
| Arc Lamp | Provide illumination | Xenon arc lamps are preferred over mercury lamps, because they provide a more even illumination over the range of wavelengths required for excitation commonly used $Ca^{2+}$ indicators. |
| Neutral-density Filter | Attenuate light source | Attenuates light evenly at all wavelengths. |
| Computer-controlled shutter | Provide illumination only | Prevents premature burning out of filters shutter during acquisition and unnecessary photobleaching. |
| Excitation selector | Provide appropriate wavelength of light for optimal excitation of the fluorophore | Normally achieved with a barrier filter. For dual-excitation studies (as with fura-2), it is necessary to have a device for the rapid changing of the excitation wavelength. This can be achieved with a computer-controlled rotating filter wheel, which houses the appropriate excitation filters ***(18,19)***. Alternatively, a monochromator system can be employed, which is more flexible in that the number of available wavelengths is essentially unlimited. |
| Dichroic mirror | Reflect excitation light to, and isolate emitted fluorescence from sample | Designed to reflect a specific range of wavelengths while allowing light of other wavelengths to pass through. Chosen such that the cutoff for passing light is greater than the wavelength of exciting light. Excitation light is therefore reflected up to the sample, whereas the resulting emitted light (of longer wavelength) passes through. For fura-2, a dichroic mirror with a cutoff of 400 nm is appropriate to separate excitation light (340/380 nm) and fluorescence emission (peak 510 nm). |
| Objective | To deliver excitation light and collect emitted fluorescence | 16 × or 20 × magnification is sufficient to accurately resolve most single cells. Higher magnification (40 × /63 ×) is required for subcellular recording. Must efficiently transmit light at wavelengths to be used. Fura-2, for example requires quartz or Fluor objectives in order to pass light of 340 nm. Lenses with high numerical apertures (NA), an index of light-collecting capacity, are preferred. |
| Sample chamber | Chamber that fits onto microscope stage and houses the cover slip containing adherent cells | In an inverted microscope configuration, the cover slip itself forms the base of the chamber. It is usually thermostatted to physiological temperatures. |

| | | |
|---|---|---|
| Emission selector | Select appropriate emission wavelengths | A filter designed to pass a broad range of wavelengths (long-pass) is normally used, in order to maximize the fluorescence signal, but sufficiently removed from the cutoff wavelength of the dichroic mirror in order to filter out stray excitation light. For Fura-2, a 420- to 600-nm filter can be used with a 400-nm cutoff dichroic mirror.<br>For dual-emission dyes such as indo-1, a filter system similar in design to that used to select alternate excitation wavelengths can be fitted to the exit port of the microscope. |
| Camera | Capture emitted fluorescence | Silicon-intensified target (SIT) cameras or cooled charged-coupled-device (CCD) cameras are commonly used. |

Before calculation of fluorescence ratios in dye-loaded cells, it is important to subtract any fluorescence not emanating from $Ca^{2+}$-sensitive dye from the measured signal. Such autofluorescence can derive from many sources, including flavoproteins and reduced adenine nucleotides. Autofluorescence is determined by first eliminating the fluorescence due to the dye and then subtracting the residual fluorescence from the data. For fura-2, this can be achieved by introducing $Mn^{2+}$ into the cell (using ionophore), which quenches dye fluorescence (*see* **Fig. 4B**, **Subheading 3.4.1.**). Note that with intensiometric indicators, autofluorescence is cancelled out in both the numerator and denominator of Eq. (1), and thus does not need to be determined separately.

In some cells, AM loading can result in accumulation of the dye into noncytosolic compartments such as the endoplasmic reticulum. Since $Ca^{2+}$ levels in the endoplasmic reticulum are much higher than in the cytosol, compartmentalized dye will, under normal conditions, remain saturated, thereby contributing to background fluorescence. In this case, the $Mn^{2+}$ quench method for determining autofluorescence is unsuitable, because ionomycin will equilibrate $Mn^{2+}$ into most cellular compartments. An alternative method to determine autofluorescence in cells where there is significant compartmentalization of dye (or when using dyes, such as fluo-3, that are not completely quenched by $Mn^{2+}$) is to permeabilize selectively using, the plasma membrane with detergents such as digitonin (*see* **Fig. 4B**, **Subheading 3.4.2.**). This method effects release of just the cytosolic dye from the cell. More problematic is compartmentalization of dye into organelles such as mitochondria, where indicators used for measuring cytosolic $Ca^{2+}$ will also report mitochondrial $Ca^{2+}$ changes. In such cases "mixed" signals will result, since cytosolic and mitochondrial $Ca^{2+}$ changes may occur asynchronously (*see* **Note 2**).

## 2. Materials

1. Loading of the cells with $Ca^{2+}$ indicator and data acquisition are performed in an extracellular-like, imaging medium (*see* **Table 2** for composition), which should be prepared fresh on the day of experimentation. Alternatively, the medium can be prepared in bulk, sterile filtered, aliquoted, and stored at 4°C.
2. Autofluorescence and calibration media (*see* **Table 2**) are made in bulk and stored at 4°C.
3. AM esters of $Ca^{2+}$ indicators (1 m*M*) should be reconstituted in dimethylsulfoxide (DMSO), aliquoted into single-use vials, and stored at –20°C.
4. Free acids of $Ca^{2+}$ indicators (1 m*M*) should be reconstituted in $H_2O$, aliquoted into single-use vials, and stored at –20°C.
5. Ionomycin (2 m*M*) should be prepared in DMSO, aliquoted into single-use vials, and stored at –20°C.
6. Digitonin (10 mg/mL) and $MnCl_2$ (1 *M*) stock solutions should be prepared in $H_2O$ and stored at room temperature.

## 3. Methods

### 3.1. Dye Loading

1. Cells should be plated onto glass (no. 1 thickness) cover slips and cultured until 70–80% confluent in the appropriate tissue culture medium.
2. Remove tissue culture medium and rinse cover slips twice with 2–3 mL of prewarmed imaging medium.

**Table 2**
**Solutions for Single Cell Ca$^{2+}$ Imaging**

| | Imaging medium | Autofluorescence medium | Calibration medium |
|---|---|---|---|
| NaCl | 121 m*M* | 10 m*M* | 10 m*M* |
| KCl | 4.7 m*M* | 120 m*M* | 120 m*M* |
| $MgCl_2$ | 1.2 m*M* | 2 m*M* | 2 m*M* |
| $CaCl_2$ | 2 m*M* | 150–300 μ*M* | — |
| BAPTA | — | 500 μ*M* | 500 μ*M* |
| $KH_2PO_4$ | 1.2 m*M* | — | — |
| $NaHCO_3$ | 5 m*M* | — | — |
| Glucose | 10 m*M* | — | — |
| BSA | 0.25% | — | — |
| HEPES | 20 m*M* | 20 m*M* | 20 m*M* |
| | pH 7.4 @ 37°C | pH 7.2 @ 37°C | pH 7.2 @ 37°C |

3. Incubate cover slip with fresh imaging medium supplemented with the appropriate concentration of Ca$^{2+}$ indicator and incubate with gentle agitation (*see* **Note 3**).
4. Remove medium and rinse cover slips twice with 2–3 mL of imaging medium.
5. Transfer cover slip to incubation chamber.
6. Leave for sufficient time (e.g., 10 min) to effect complete de-esterification of the dye.

## 3.2. Data Acquisition

1. Experiments can be performed either in a static chamber or by continual perfusion of the cells. The former approach is recommended, because much smaller quantities of (usually precious) materials are required. Indeed, experiments can be performed in volumes as small as 0.3 mL.
2. Capture images for at least 60 s prior to stimulation of the cells in order to obtain an accurate measure of the basal [Ca$^{2+}$] and to characterize any possible spontaneous changes in [Ca$^{2+}$] (*see* **Note 4**).
3. For static chambers additions can be made by complete removal (by pipet) of the medium and addition of fresh medium supplemented with the test agent. Alternatively, a small aliquot of the medium can be removed, mixed with the appropriate volume of a concentrated stock solution of the test agent, and the entire volume added back. This method is less prone to addition artefacts than complete exchange of the medium.

## 3.3. Calibration of Fluorescence Data

### 3.3.1. In Situ Calibration

1. This method can be used to calibrate fluorescence data from both intensiometric and ratiometric indicators. For ratiometric indicators, autofluorescence (*see* below) must be subtracted prior to calculation of the ratio.
2. At the end of the experimental run, rinse cells into imaging medium (without added Ca$^{2+}$) supplemented with 2 m*M* BAPTA and add 2 μ*M* ionomycin to equilibrate Ca$^{2+}$ across the cell membrane.
3. Reinitiate data acquisition and monitor until fluorescence intensity reaches a stable plateau to obtain $F_{min}$ (for intensiometric indicators) or $R_{min}$ (for ratiometric indicators).

4. Add 10 m$M$ $CaCl_2$ to saturate the dye with $Ca^{2+}$.
5. Reinitiate data acquisition and monitor until fluorescence intensity reaches a stable plateau to obtain $F_{max}$ (for intensiometric indicators) or $R_{max}$ (for ratiometric indicators).
6. Use **eq. (1)** (for intensiometric) or **eq. (2)** (for ratiometric indicators) to calculate [$Ca^{2+}$].

#### 3.3.2. In Vitro Calibration

1. This method is suitable for ratiometric indicators for a given instrument configuration.
2. Place 2 mL of calibration buffer (*see* **Table 2**) in the incubation chamber and acquire 5–10 images. This is the background fluorescence, which should be subtracted from subsequent data (below) at both wavelengths before calculation of ratios.
3. Add 1 µ$M$ of the free acid form of the indicator and acquire 5–10 images. Calibration medium is initially $Ca^{2+}$-free, thus the fluorescence ratio gives $R_{min}$.
4. Add 500 µ$M$ $CaCl_2$ to saturate the dye and acquire 5–10 images. This fluorescence ratio is $R_{max}$.
5. Use **eq. (2)** to convert the acquired experimental ratio values (corrected for autofluorescence) to [$Ca^{2+}$] at each time point.

### 3.4. Determination of Autofluorescence

#### 3.4.1. $Mn^{2+}$ Quench

1. At the end of the experimental run, remove medium and rinse twice with 2 mL of imaging buffer.
2. Reinitiate data acquisition and add $MnCl_2$ (2–4 m$M$) and ionomycin (2 µ$M$).
3. Monitor until fluorescence intensity falls to a stable plateau.
4. Subtract final image from all acquired images.

#### 3.4.2. Digitonin Permeabilization

1. At the end of the experimental run, remove medium and rinse cells twice with 2 mL of autofluorescence buffer.
2. Reinitiate data acquisition and add digitonin (20–50 µg/mL).
3. Monitor until fluorescence intensity falls to a stable plateau.
4. Subtract final image from all acquired images.

## 4. Notes

1. Photobleaching and dye extrusion can be distinguished by briefly interrupting data acquisition and comparing the fluorescence intensity before and after resuming acquisition. If the fluorescent intensities are comparable, then the loss of signal is due to photobleaching, in which case illumination intensity should be reduced by decreasing exposure time and/or increasing attenuation of the excitation source. Dye extrusion, however, should not be affected by stopping acquisition, thus the fluorescence signal should continue to fall. Loss of dye can be slowed by reducing the working temperature and/or including organic anion-transport inhibitors such as bromosulphophthalein (100 µ$M$) in the imaging medium to inhibit active extrusion pathways (also *see* **Note 2**).
2. Compartmentalized dye can be recognized by a punctate cellular dye distribution and/or a dim nucleus and can be reduced by reducing the loading temperature and/or loading time. Alternatively, the free acid (de-esterified) form of the dye can be introduced directly into the cytosol by micropipet or dialysis through a patch pipet. This method, although more disruptive, has the added advantage of allowing delivery of dyes conjugated to inert dextrans, thereby preventing dye extrusion (*see* **ref. *17*** for further discussion).

3. The conditions for dye loading are highly dependent on cell type and should be determined empirically. Cells are typically loaded for up to 1 hour with 1–10 μ*M* of the AM ester. Dye loading can be increased by premixing the dye with dispersants such as Pluronic F-127 (0.02%) prior to dilution into imaging medium (to increase dye solubilization) and/or by decreasing cell density. Care should also be taken not to overload cells with dye, since at high concentrations, the indicator may buffer $Ca^{2+}$ increases, resulting in sluggish or blunted responses.
4. As a rule, minimize the time that cells are exposed to the excitation light to prevent photobleaching (*see* **Table 1**). This is achieved by adjusting shutter exposure time and attenuation (neutral-density filter). Also, the delay between capturing successive images (acquisition delay) will determine total exposure time. A common problem encountered during acquisition is uneven fluorescence throughout the imaging field. This is likely to be due to misalignment of the excitation source. Also, excitation bulbs have a finite life span. Bulbs should be replaced if high frequency noise becomes a problem during experimentation.

## Acknowledgements

We would like to thank Dr. David McQuillan for providing decorin, and G. C. Churchill and A. Galione for helpful discussions. This work was supported by a Wellcome Prize Travelling Research Fellowship.

## References

1. Berridge, M. J., Bootman, M. D., and Lipp, P. (1998) Calcium—a life and death signal. *Nature* **395,** 645–647.
2. Berridge, M. J. (1993) Inositol trisphosphate and calcium signalling. *Nature* **361,** 315–325.
3. Rhee, S. G. and Bae, Y. S. (1997) Regulation of Phosphoinositide-specific Phospholiapse C Isoenzymes. *J. Biol. Chem.* **272,** 15045–15048.
4. Joseph, S. K. (1996) The inositol triphosphate receptor family. *Cell Signal.* **8,** 1–7.
5. Patel, S., Joseph, S. K., and Thomas, A. P. (1999) Molecular properties of inositol 1,4,5-trisphosphate receptors. *Cell Calcium.* **25,** 247–264.
6. Sjaastad, M. D. and Nelson, W. J. (1997) Integrin-mediated calcium signaling and regulation of cell adhesion by intracellular calcium. *BioEssays* **19,** 47–55.
7. Iozzo, R. V. (1999) The biology of the small leucine-rich proteoglycans. *J. Biol. Chem.* **274,** 18843–18846.
8. Patel, S., Santra, M., McQuillan, D. J., Iozzo, R. V., and Thomas, A. P. (1998) Decorin activates the epidermal growth factor receptor and elevates cytosolic $Ca^{2+}$ in A431 carcinoma cells. *J. Biol. Chem.* **273,** 3121–3124.
9. Vogel, W., Gish, G. D., Alves, F., and Pawson, T. (1997) The Discoidin Domain Receptor Tyrosine Kinases Are Activated by Collagen. *Mol. Cell* **1,** 13–23.
10. Shrivastava, A., Radziejewski, C., Campbell, E., Kovac, L., McGlynn, M., Ryan, T. E., Davis, S., Goldfarb, M. P., Glass, D. J., Lemke, G., and Yancopoulos, G. D. (1997) An Orphan Receptor Tyrosine Kinase Family Whose Members Serve as Nonintegrin Collagen Receptors. *Mol. Cell* **1,** 25–34.
11. Schlessinger, J. (1997) Direct Binding and Activation of Receptor Tyrosine Kinases by Collagen. *Cell* **91,** 869–872.
12. Thomas, A. P., Bird, G. St. J., Hajnóczky, G., Robb-Gaspers, L. D., and Putney, J. W.,Jr. (1996) Spatial and temporal aspects of cellular calcium signaling. *FASEB J.* **10,** 1505–1517.
13. Haughland, R. P. (1996) *Handbook of Fluorescent Indicators.* Molecular Probes, Eugene, OR.
14. Grynkiewicz, G., Poenie, M., and Tsien, R. Y. (1985) A new generation of $Ca^{2+}$ indicators with greatly improved fluorescence properties. *J. Biol. Chem.* **260,** 3440–3450.

15. Tsien, R. Y. (1981) A non-disruptive technique for loading calcium buffers and indicators into cells. *Nature* **290,** 527–528.
16. Thomas, A. P. and Delaville, F. (1991) The use of fluorescent indicators for measurement of cytosolic-free calcium concentration in cell populations and single cells, in *Cellular Calcium. A Practical Approach* (McCormack, J. G. and Cobbold, P. H., eds.). Oxford University Press, Oxford, UK, pp. 1–54.
17. Morgan, A. J. and Thomas, A. P. (1999) Single cell and subcellular measurement of intracellular $Ca^{2+}$ concentration ($[Ca^{2+}]_i$). *Meth. Mol. Biol.* **114,** 93–123.
18. Tsien, R. Y. and Harootunian, A. T. (1990) Practical design criteria for a dynamic ratio imaging system. *Cell Calcium.* **11,** 93–109.
19. Dunn, K. and Maxfield, F. R. (1998) Ratio Imaging Instrumentation. *Meth. Cell Biol.* **56,** 217–236.

# 44

# Serum Amyloid A Peptide Interactions with Glycosaminoglycans

## Evaluation by Affinity Chromatography

**John B. Ancsin and Robert Kisilevsky**

## 1. Introduction

Amyloid is a pathological deposit of protein and glycosaminoglycans (GAGs) that can lead to the destruction of tissue architecture and function *(1,2)*. To date, 18 unrelated proteins are known precursors of amyloid deposits *(3)*. Each one is associated with a specific disease, such as Alzheimer's disease, adult–onset diabetes, inflammatory disorders, and some cancers, but regardless of the type of precursor protein involved, all amyloids have common tinctorial and structural characteristics. Fibrils are composed of two or more filaments 3 nm in diameter twisted around each other forming nonbranching fibrils 7–10 nm in diameter with a crossed β–pleated sheet conformation. Congo red stains these deposits, and when viewed under polarizing light they exhibit a red/green birefringence characteristic™ for amyloid.

Available evidence suggests that the deposition of amyloid involves a nidus or protofilament around which amyloid fibrillogenesis occurs, and heparan sulfate (HS) facilitates this pathological process *(2,4,5)*. To understand better the mechanism by which this HS–dependent transformation takes place, a number of studies have been undertaken to characterize potential amyloid precursor–HS–binding activities. Heparin–HS–binding activity has been reported for five amyloid precursors, Aβ and βPP *(6,7)*, tau *(8,9)*, prions *(10,11)*, amylin *(12,13)* and serum amyloid A *(14)*. The latter is an acute–phase apoprotein of HDL for which we have recently defined its GAG–binding site.

Heparin/heparan sulfate–binding activity for serum amyloid A (apoSAA) was localized to its carboxyl–terminal end, residues 77–103, by affinity chromatography

From: *Methods in Molecular Biology, Vol. 171: Proteoglycan Protocols*
Edited by: R. V. Iozzo © Humana Press Inc., Totowa, NJ

of apoSAA2 1.1 CNBr fragments. The relative importance of the six basic residues within this region was demonstrated by employing different mutant synthetic peptides in which one or more of these basic residues were either deleted or replaced with alanine. The GAG–binding potential of protein/peptides is generally tested only on heparin, but we were interested to see if apoSAA had an affinity for any other GAGs. Chondroitin sulfate, dermatan sulfate, hyaluronan, and heparan sulfate charged columns were generated with Sepharose 4B and Affi–Gel and their binding affinity tested. In addition, the influence on binding activity of two popular coupling reactions used to link GAGs to column matrices were compared and found to have an influence on apoSAA binding.

## 2. Materials

### 2.1. Glycosaminoglycan Charged Columns

1. Heparin, heparan sulfate, chondroitin sulfate, dermatan sulfate, and hyaluronan were all purchased from Sigma (St. Louis, MO). Sepharose 4B was purchased from Pharmacia Biotech (Baie d'Urfé, Quebec), 1–ethyl–3–3–dimethyl–aminopropyl carbodiimide (EDAC), Heparin–Affigel and Affigel–102 were purchased from Bio–Rad (Hercules, CA).
2. Glycosaminoglycan assay reagents: 0.2% NaCl; GAG standards (1 mg/mL in 0.2% NaCl); 0.005% toluidine blue (5 mg/100 mL 0.01 *N* HCl, 0.2% NaCl); ethanol (absolute) and hexane.
3. Other reagents: 50 m*M* acetate pH 6.0, 20 m*M* Tris–HCl, pH 7.5, 1 *M* ethanolamine, pH 9.0,0.1 *M* sodium acetate, pH 4.0, 0.1 *M* $NaHCO_3$, pH 8.3, cyanogen bromide (Sigma) stock solution 1 g/mL in *N, N*–dimethylformamide, 70% formic acid, and $N_{2\,(gas)}$.
4. Waters HPLC system (Waters Chromatography, Milford, MA) with a model 680 automated gradient controller, model 501 pump units, series 440 absorbance detector connected to a Waters 740 data module integrator.

## 3. Methods

### 3.1. GAG–Sepharose Columns

Cyanogen bromide (CNBr) activation is a common method used in the preparation of affinity resins. It is a relatively simple reaction that works well with agarose–based column matrices and has minimal secondary effects on ligand chemistry. Briefly, at alkaline pH the hydroxyl groups on the agarose resin react with CNBr. A high molar excess of CNBr is required, because most of it (1) reacts with water, generating inert cyanate ions, or (2) forms cyanate esters, which become hydrolyzed, or (3) reacts with matrix hydroxyls to form imidocarbonates. The remaining active cyanate esters react with amino groups on lysines of proteins, or hydroxyls on carbohydrate chains.

$$\text{heparin-OH} + \text{N}{\equiv}\text{C-O-}\bigcirc \Rightarrow \text{heparin-O-}\overset{\text{HN}}{\overset{\|}{\text{C}}}\text{-O-}\bigcirc$$

1. GAGs were linked to Sepharose 4B based on the method of Smith et al., ***(15)***. Sepharose 4B was washed with 20 bed volumes of water, resuspended in 1 volume of water, and transferred to a beaker with a stir bar on ice. GAGs were dissolved in water at 2 mg/mL and also placed on ice.

2. The two solutions were mixed and, while stirring, the pH was adjusted to pH 11 with NaOH (5 *N*). Cyanogen bromide, 1 g/mL in *N,N*–dimethylformamide, was added dropwise to a final concentration of 31.3 mg/mL.
3. The pH was maintained at about 11 by adding NaOH for 15 min, then left stirring for 18 h at room temperature. The pH gradually returns to approximately pH 8.
4. The gel was then washed with 20 bed volumes of water followed by 1 *M* ethanolamine, pH 9.0, to block excess reactive groups. After further washing with 10 bed volumes of (1) water, (2) 0.1 *M* sodium acetate pH 4.0, and (3) 0.1 *M* $NaHCO_3$, pH 8.3, the column gel was equilibrated in 20 m*M* Tris–HCl, pH 7.5.

### 3.2. GAG–Affigel Columns

GAGs can be covalently linked to agarose matrices by activation of the uronic acid carboxyls with EDAC coupling the GAG chain to an amino–functionalized matrix such as Affi–Gel–102 (Bio–Rad) (*see* **Notes 3** and **4**).

$$\text{heparin-COO}^- + H_2N\text{-}\bigcirc + \text{carbodiimide} \Rightarrow \text{heparin-}\overset{\overset{\text{O}}{//}}{\text{C}}\text{-NH-}\bigcirc$$

1. Affigel–102 (4 mL) was washed with 20 bed volumes of 50 m*M* acetate, pH 6.0, and then 8 mg of GAG in 4 mL of the same buffer was mixed with the gel.
2. The coupling reaction was initiated by the addition of 32 mg of EDAC, adjusting the pH to 5 with 1 N HCl and allowing the reaction to proceed for 3 h at room temperature.

### 3.3. Assay of GAG Coupling Efficiency to Matrix (Toluidine Blue Method)

1. Add 0–35 µL of GAG to tubes containing 0.75 mL of 0.2% NaCl—this is your standard. Also dilute the GAG column matrix 1/4 with 0.2% NaCl (0.75 mL total). Blank tubes should contain 0.2% NaCl or Sepharose 4B in 0.2% NaCl.
2. To all tubes, add 0.75 mL of toluidine blue solution and vortex for 30 s. Next add 1 mL of hexane to each tube and vortex for another 30 s. GAG–dye complexes should precipitate at the water–hexane interface. Centrifuge at 735*g* (3000 rpm) for 5 min. Aspirate the hexane and the top half of the water layers and discard. Dilute a sample of the water layer 1/10 with absolute ethanol, mix, and read at 631 nm within 30 min. Care must be taken not to get any of the GAG–dye complex precipitate into the sample to be assayed.
3. The amount of GAG linked to columns ranged from 0.5 – to 0.75–mg/mL of gel.

### 3.4. Affinity Chromatography on GAG–columns

#### 3.4.1. Low–Pressure Liquid Chromatography

1. Peptides were dissolved in 20 m*M* Tris–HCl, pH 7.5, and loaded onto a 6–mL Heparin Affigel column preequilibrated in the same buffer (*see* **Fig. 1**).
2. The column was then washed at 30 mL/h with 4 bed volumes of buffer followed by a 0–1 *M* NaCl linear concentration gradient. Fractions were collected (0.6 mL) and their absorbency determined at 214 nm (*see* **Notes 5** and **6**).

#### 3.4.2. High–Pressure Liquid Chromatography (HPLC)

1. Different GAG–charged matrices were also packed into a 3–mL stainless steel column and equilibrated with 20 m*M* Tris–HCl, pH 7.5, at 0.5 mL/min using a Waters HPLC system (*see* **Figs. 2** and **3**).

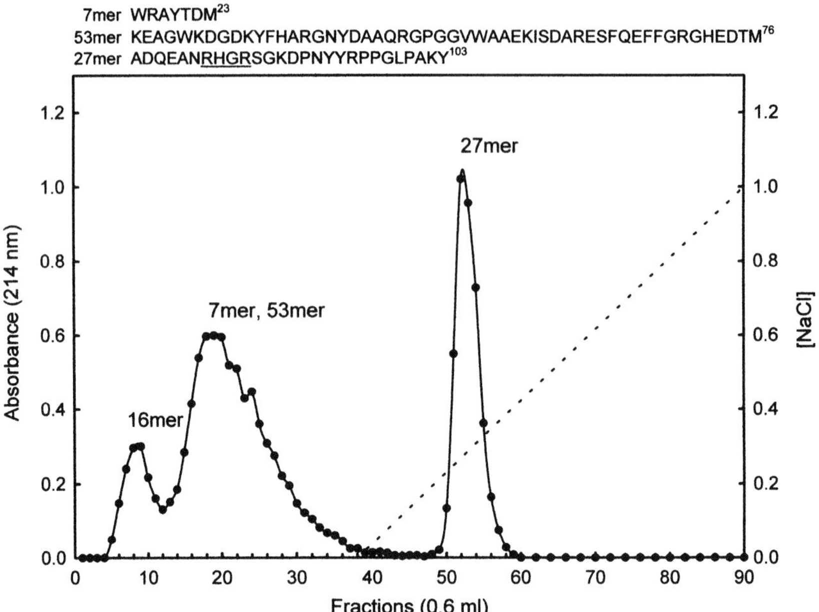


Fig. 1. Elution profile of the apoSAA1.1 CNBr peptides on a heparin–Affigel column: testing apoSAA1.1 CNBr fragments for heparin–binding activity. ApoSAA1.1 CNBr cleavage fragments are shown above the graph; 16mer (residues 1–16), 7mer (residues 17–23), 53mer (residues 24–76), and 27mer (residues 77–103). A heparin–binding consensus sequence, (XBBXBX, X = nonbasic, B = basic) is underlined in m27mer.

2. Samples (20–80 μg in 150–200 μL) were injected onto the column, washed with 3 bed volumes (18 min) of the same buffer, and then developed with a 0–1.0 *M* NaCl linear gradient for 10 bed volumes (60 min). The eluate was monitored continuously at 214 nm and the absorbance plotted against the retention time (RT).
3. Generally, unbound peptides/proteins eluted 6.5–7.0 min after loading and, based on the RTs for the bound peptides/proteins, the NaCl concentration at which desorption took place was calculated; desorption [NaCl] = (RT –6.5 min –18 min) / 60 min.

### 3.5. ApoSAA2 Peptides

#### 3.5.1. CNBr Peptides

1. ApoSAA1.1 was cleaved at Met–X peptide bonds with cyanogen bromide (CNBr). Protein was dissolved in 70% formic acid at 1 mg/mL, to which 5.5 mg/mL CNBr was added (250 molar excess over Met) plus freeL–Trp (5 molar excess over Met) to protect Trp residues.
2. The reaction was carried out under nitrogen, at room temperature overnight, and then the solvent was evaporated by vacuum centrifugation.
3. The reaction was evaluated and peptides purified, by reverse–phase (RP)–HPLC. A semipreparative C–18 Vydac column was equilibrated with 10% acetonitrile, 0.1%

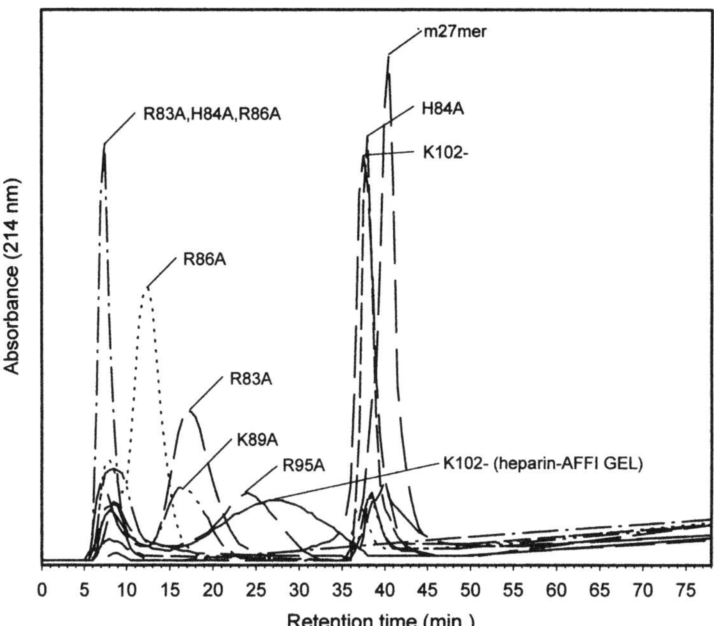


Fig. 2. Identification of the basic residues of m27mer that are important for heparin binding. Chromatography was performed on heparin–Sepharose 4B and the elution profiles were similar to that seen with heparin–Affigel (not shown) with the exception of K102–. Carbodiimide treatment of HS interferred with binding (HS–Sepharose + carbodiimide). The sequence of m27mer is shown above the graph.

TFA. Peptides were dissolved in 40% formic acid, filtered through a 0.2–µm filter, or centrifuged at 10,000*g*, and loaded onto the column, washed for 5 min at 3 mL/min, then developed with a 1.5%/min acetonitrile linear concentration gradient for 30 min, followed by 4.3%/min acetonitrile for 10 min, bringing the elution buffer to 98% acetonitrile by 45 min.

4. The separated peptides were identified by amino–terminal sequencing and molecular–weight determination by mass spectroscopy carried out at the Alberta Peptide Institute (Edmonton, Alberta, Canada).

### 3.5.2. Synthetic Peptides

1. A series of eight apoSAA1.1 peptides corresponding to wild–type mouse apoSAA1.1, residues 77–103, (m27mer) six peptides in which one or more of the basic residues were replaced with A, and one 20mer residues 77–96 (missing K102) were synthesized by Multiple Peptide Systems (San Diego, California, U.S.A.). (*see* **Notes 1** and **2**).

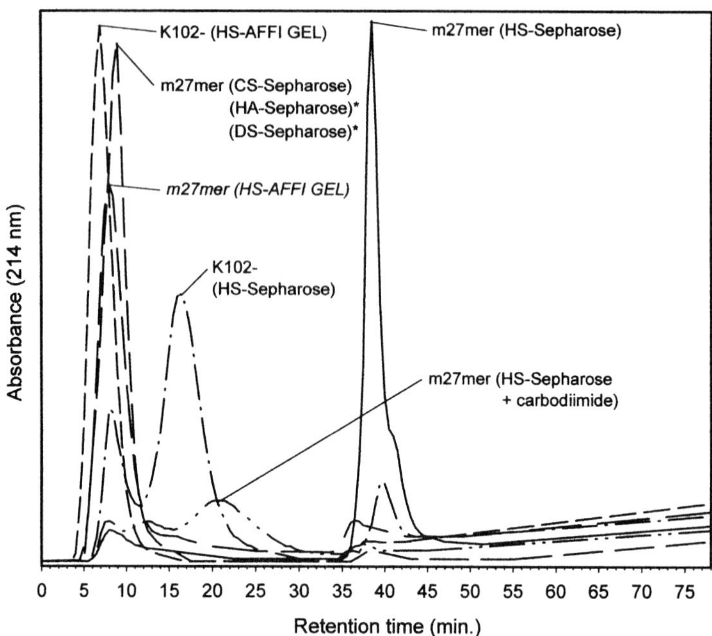


Fig. 3. Determination of the GAG–binding specificity for m27mer. Heparan sulfate (HS), chondroitin sulfate (CS), dermatan sulfate (DS), and hyaluronan (HA) were covalently linked to either Sepharose 4B of Affigel 102. The elution profiles on DS–Sepharose and HA–Sepharose (asterisks) were very similar to that of CS–Sepharose.

m27mer (mouse apoSAA1.1$_{77-103}$) ADQEAN**RH**G**R**SG**K**DPNYY**R**PPGLPA**K**Y
R83A ADQEANA**H**G**R**SG**K**DPNYY**R**PPGLPA**K**Y
H84A ADQEAN**R**AG**R**SG**K**DPNYY**R**PPGLPA**K**Y
R86A ADQEAN**RH**GASG**K**DPNYY**R**PPGLPA**K**Y
R83A,H84A,R86A ADQEANAAGASG**K**DPNYY**R**PPGLPA**K**Y
K89A ADQEAN**RH**G**R**SGADPNYY**R**PPGLPA**K**Y
R95A ADQEAN**RH**G**R**SG**K**DPNYYAPPGLPA**K**Y
K102– (mouse apoSAA1.1$_{77-96}$) ADQEAN**RH**G**R**SG**K**DPNYY**R**P

2. Purity of the peptides was analyzed by RP–HPLC and, where required, the peptide was re–purified.

## 4. Notes

1. To localize GAG–binding sites to specific peptide sequences, limited chemical or enzymatic cleavage is required of the native protein. Additional flanking sequences to maintain appropriate secondary structure and cleavage near basic residues (K, H, R) should be avoided. Chemical cleavage at M–X with CNBr, or at W–X with BNPA–skatole is a good starting point, since both residues occur with relatively low frequency in proteins (M 2.4% and W 1.3%) ***(16)***.

2. Whenever possible, conservative residue replacements should be performed. The strategy for deciding on replacement of residues in GAG–binding sites should take into account potential changes in peptide solubility, secondary structure (i.e., conformational preferences α–helix, β–strand, β–bend turn), and the volume of the amino acid chosen. The basic residues most often shown to be crucial for GAG binding (K, H, R) are α–helix formers ***(16)***, and A can be a satisfactory replacement. However, if solubility of the peptide is adversely affected then Q may be a useful alternative. Also, Qs van der Waals volume is more similar to K, H, and R than A; Q = 114 $Å^3$ vs A = 69 $Å^3$, compared to K = 135 $Å^3$, H = 118 $Å^3$, and R = 148 $Å^3$.
3. All GAGs, CCS, DS, HA, HS, and heparin can be covalently coupled to column matrices, and determination of the GAG–binding specificity of any peptide using these columns can be easily carried out. Heparin is quite popular and inexpensive, and is often used as a cheap substitute for HS. While heparin and HS are synthesized in a similar fashion, sulfation and uronic acid epimer (glucuronic vs iduronic acid) content and distribution are quite different. Also HS is most often the physiological ligand for most proteins and peptides that can bind heparin. We have observed that while heparin–binding activity was useful in localizing the GAG–binding site on apoSAA1.1, the minimal peptide sequence required for binding to heparin was at least 7 residues shorter then that required for HS binding. In addition, deletion of K102 destroyed binding for HS–Sepharose, but not for heparin–Sepharose implying that the K102–peptide was binding to different oligosaccharide sequences.
4. Type of support matrix. Sepharose vs Affigel helps define the type of chemical linkage group available to react with one or more of the GAG's reactive side groups. This is an important factor, since the linkage of GAG to matrix can influence binding activity. We found that the basic residue requirements were found to be different for heparin, depending on whether it was linked via its COO– or HO– groups. We observed that K102–bound heparin–Sepharose but not heparin–Affigel. A similar discrepancy was seen with the m27mer peptide, which bound HS only when linked to Sepharose, not Affigel. Furthermore, treatment of HS–Sepharose with carbodiimide indicated that blocking of the carboxyls on HS was likely responsible for the loss of binding activity.
5. Preliminary screening of protein/peptide for GAG–binding activity can be carried out relatively quickly with 1– to 10–mL columns. Samples can be loaded in 0–0.15 *M* NaCl, followed by an increasing linear NaCl concentration gradient to 1 or 2 *M* NaCl totaling 10–20 bed volumes to develop the elution profile. Distinguishing between different peptides with similar affinities for the GAG column was more readily achieved with the columns connected to HPLC. The HPLC allowed for more precise solvent delivery and gradient formation, highly reproducible retention times, and accurate determination of NaCl concentrations required for protein/peptide desorption.
6. FPLC– and HPLC–grade heparin columns are also available from a number of suppliers.

## References

1. Sipe, J. D. (1994) Amyloidosis. *Crit. Rev. Clin. Lab. Sci.* **31,** 325-354.
2. Magnus, J. H. and Stenstad, T. (1997) Proteoglycans and the extracellular matrix in amyloidosis. *Amyloid: Int. J. Exp. Clin. Invest.* **4,** 121–134.
3. Kazatchkine, M. D., Husby, G., Araki, S., Benditt, E. P., Benson, M. D., Cohen, A. S., Frangione, B., Glenner, G. G., Natvig, J. B., and Westermark, P. (1993) Nomenclature of amyloid and amyloidosis. *Bull. WHO* **71,** 105–108.

4. Kisilevsky, R. and Fraser, P. (1996) Proteoglycans and amyloid fibrillogenesis, in *The Nature and Origin of Amyloid Fibrils* (Bock, G. R. and Goode, J.A., eds.), Wiley, Chichester, UK, pp. 58–67.
5. Kisilevsky, R. and Fraser, P. (1997) Aβ amyloidogenesis: unique, or variation on a systemic theme? *Crit. Rev. Biochem. Mol. Biol.* **32,** 361–404.
6. Narindrasorasak, S., Lowery, D., Gonzalez–DeWhitt, P., Poorman, R. A., Greenberg, B. D., and Kisilevsky, R. (1991) High affinity interactions between the Alzheimer's beta–amyloid precursor proteins and the basement membrane form of heparan sulfate proteoglycan. *J. Biol. Chem.* **266,** 12,878–12,883.
7. Leveugle, B., Scanameo, A., Ding, W., and Fillit, H. (1994) Binding of haparan sulfate glycosaminoglycan to beta–amyloid peptide: inhibition by potentially therapeutic polysulfated compounds. *Neuroreport* **5,** 1389–1392.
8. Goedert, M., Jakes, R., Spillantini, M. G., Hasegawa, M., Smith, M. J., and Crowther, R. A. (1996) Assembly of microtubule–associated protein tau into Alzheimer–like filaments induced by sulphated glycosaminoglycans. *Nature* **383,** 550–552.
9. Hasegawa, M., Crowther, R. A., Jakes, R., and Goedert, M. (1997) Alzheimer–like changes in microtubule–associated protein tau induced by sulfated glycosaminoglycans—inhibition of microtubule binding, stimulation of phosphorylation, and filament assembly depend on degree of sulfation. *J. Biol. Chem.* **272,** 33,118–33,124.
10. Gabizon, R., Meiner, Z., Halimi, M., and Ben–Sasson, S. A. (1993) Heparin–like molecules bind differentially to prion–protein and change their intracellular metabolic fate. *J. Cell Physiol.* **157,** 319–325.
11. Caughey, B., Brown, K., Raymond, G. J., Katzenstein, G. E., and Threscher, W. (1994) Binding of the protease sensitive form of PrP (prion protein) to sulfated glycosaminoglycans and Congo red. *J. Virol.* **68,** 2135–2141.
12. Watson, D. J., Lander, A. D., and Selkoe, D. J. (1997) Heparin–binding properties of the amyloidogenic peptides Aβ and amylin. Dependence on aggregation state and inhibition by Congo red. *J. Biol. Chem.* **272,** 31617–31624.
13. Castillo, G. M., Cummings, J. A., Yang, W. H., Judge, M. E., Sheardown, M. J., Rimvall, K., Hansen, J. B., and Snow, A. D. (1998) Sulfate content and specific glycosaminoglycan backbone of perlecan are critical for perlecan's enhancement of islet amyloid polypeptide (amylin) fibril formation. *Diabetes* **47,** 612–620.
14. Ancsin, J. B. and Kisilevsky, R. (1999) The heparin/heparan sulfate–binding site on apo–Serum Amyloid A: implication for the therapeutic intervention of amyloidosis. *J. Biol. Chem.* **274,** 7172–7181.
15. Smith, P. K., Mallia, A. K., and Hermanson, G. T. (1980) Colorimetric method for the assay of heparin content in immobilized heparin preparations. *Anal. Biochem.* **109,** 466–473.
16. Creighton, T. E. (1993) *Proteins Structure and Molecular Properties*. Freeman, New York.

# 45

# Interactions of Lipoproteins with Proteoglycans

**Kevin Jon Williams**

## 1. Introduction

### *1.1. General*

Significant physiological and pathophysiological processes involve interactions of specific lipoproteins with specific proteoglycans. So far, these interactions fall into two large groups. The first is interactions with heparan sulfate proteoglycans (HSPGs), primarily on the cell surface that lead to cellular uptake of the lipoproteins. These pathways are of interest because they involve endocytic machinery, intracellular itineraries, and regulation that are generally distinct from classic, coated pit-mediated endocytosis *(1)*. Moreover, many important nonlipoprotein ligands, such as infectious agents, growth factors, platelet secretory products, and proteins implicated in Alzheimer's disease, bind to the same HSPGs, and lipoproteins are a convenient model ligand to study HSPG-mediated catabolism. The second group of lipoprotein-proteoglycan interactions involves chondroitin sulfate proteoglycans (CSPGs), primarily in the extracellular matrix, leading to retention of cholesterol-rich lipoproteins. Many lines of evidence now support the concept that retention of lipoproteins within the arterial wall is the key initial step in provoking atherosclerosis, the major killer in Western countries *(2,3t)*.

The following sections contain a brief primer on general lipoprotein biology and methods (*see* **Subheading 1.2.**); then essential background for methods used to study lipoprotein-HSPG interactions, with a focus on endocytosis (*see* **Subheading 1.3.**); background relevant to molecular methods to study HSPG-mediated endocytosis, with a focus on specific domains within the syndecan-1 core protein (*see* **Subheading 1.4.**); and then five detailed, basic protocols (*see* **Subheadings 2–4**). Because of space limitations, lipoprotein–CSPG interactions are not covered in detail here, although key methods can be found in the published literature *(4–13)*.

From: *Methods in Molecular Biology, Vol. 171: Proteoglycan Protocols*
Edited by: R. V. Iozzo © Humana Press Inc., Totowa, NJ

### 1.2. Primer on General Lipoprotein Biology and Methods

Lipoproteins are noncovalent complexes of lipid and protein that allow the body to transport many hydrophobic substances through the aqueous environment of blood. All normally occurring mammalian lipoproteins have the same basic structure, known as the oil-drop model *(14)*: a central core of hydrophobic lipid, chiefly triacylglycerols and esterified cholesterol, surrounded by a layer of amphipathic molecules, chiefly phospholipids, unesterified cholesterol, and proteins known as apolipoproteins or apoproteins (**Fig. 1**). The most widely used nomenclature defines mammalian lipoproteins by their densities: high-density lipoprotein (HDL, 1.063 g/mL < $d$ < 1.21 g/mL), low-density lipoprotein (LDL, 1.019 < $d$ < 1.063 g/mL), and very low-density lipoprotein (VLDL, $d$ < 1.006 g/mL). Sometimes, intermediate-density lipoproteins are referred to as a separate class (IDL, 1.019< $d$ < 1.019 g/mL). In addition, there are two lipoproteins specifically associated with meals: the chylomicron ($d$ < 0.96 g/mL), which appears in plasma in the post-prandial state and transports lipids, chiefly triacylglycerols, that have been ingested and absorbed; and the chylomicron remnant, which can appear in the VLDL or IDL density ranges and is the particle that remains after peripheral tissues have extracted most of the triglycerides from circulating chylomicrons. An abnormal particle, β-VLDL, which appears in individuals with certain genetic abnormalities in apolipoprotein (apo) E or after prolonged administration of high-cholesterol diets to experimental animals, is commonly used as an experimental substitute for chylomicron remnants. Importantly, the different classes of lipoproteins have distinct lipid compositions and apoprotein constituents (**Table 1**) and hence distinct metabolic roles.

LDL and β-VLDL are the lipoproteins most commonly studied with proteoglycans. They are abundant, easy to isolate, easy to label either radioactively or fluorescently, and they exhibit important interactions with proteoglycans in vivo. Both contain apoB, a large amphipathic protein that cannot move between lipoproteins and hence serves as an excellent anchor for tags to track entire particles. In contrast, all other known apoproteins readily exchange between lipoprotein particles, and surface and core lipids are moved between lipoproteins by lipid transfer proteins in human plasma. Both LDL and β-VLDL bind cell-surface LDL receptors in vitro, but are also known to undergo substantial catabolism in vitro *(1,15)* and in vivo *(16,17)* independent of LDL receptors. Both of these particles have been implicated in atherosclerotic vascular disease.

Isolation of these particles from plasma requires sequential ultracentrifugation at accelerations sufficient to overcome Brownian motion. Several companies in the U.S. sell human LDL, such as Sigma Chemical Company (St. Louis, MO), CalBiochem (La Jolla, CA), Molecular Probes (Eugene, OR), and Academy Bio-Medical (Houston, TX), but as with most materials, it can be less expensive to isolate it yourself, particularly for long-term needs [ultracentrifugal isolation is described in detail in *(18,19)*]. Radioactive labeling is best performed by the iodine monochloride method *(19–22)*, which covalently attaches $^{125}$I to tyrosyl residues in apoproteins, with minimal disruption of double bonds in the fatty acyl side chains of lipoprotein lipids. Fluorescent labeling usually employs DiI or DiO, which are generally nontoxic lipophilic compounds that insert into cell membranes or lipoprotein surfaces, but then are poorly

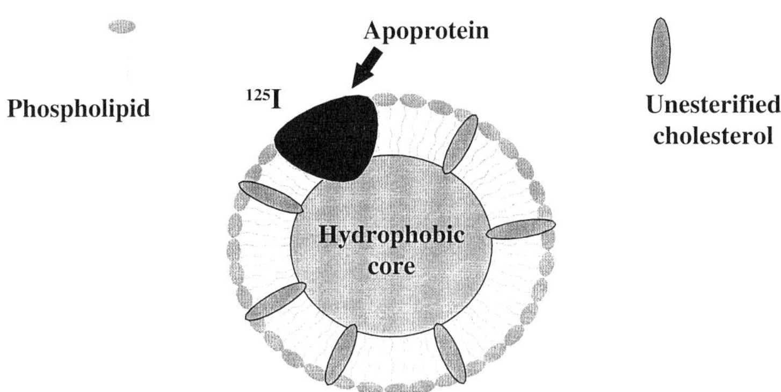


Fig. 1. General schematic of normal lipoprotein structure. All normal human plasma lipoproteins exhibit the same basic structure, known as the oil-drop model, which consists of a single layer of amphipathic molecules (phospholipids, apoproteins, and unesterified cholesterol) surrounding a hydrophobic core (cholesteryl ester or triacylglycerides). The fatty acyl side chains of each phospholipid molecule point inward toward the hydrophobic core, while the polar head group is exposed on the particle surface. Apoproteins are similarly oriented, with a hydrophobic face associating with lipid, and the hydrophilic face directed outward. Owing to its alcohol group, unesterified cholesterol is associated mainly with the phospholipid molecules of the particle surface, although some partitions into the core. This overall arrangement keeps hydrophobic molecules and domains shielded from the surrounding aqueous environment of blood and interstitial fluid. Only one schematic apoprotein has been drawn, although many species of lipoproteins have several protein molecules per particle. Apoproteins are often convenient sites to place radioactive tags, shown here as an $^{125}$I label (*see* **Subheading 1.2.** for more details).

**Table1**
**Physical Characteristics of the Major Plasma Lipoproteins in Humans**

| Particle | Diameter (nm) | Core lipid | Principal Apoproteins |
|---|---|---|---|
| Chylos | 80–1000 | Tg (diet) | $B_{48}$, AI,AII,CII,E |
| VLDL | 30–80 | Tg (liver) | $B_{100}$, CII,E |
| LDL | 18–28 | ChE | $B_{100}$ |
| HDL | 5–12 | ChE | AI, AII, occasionally E |

Abbreviations: Chylos, chylomicrons and remnants; VLDL, very low-density lipoprotein; LDL, low-density lipoprotein; HDL, high-density lipoprotein; diet, originating from dietary intake and secreted by the intestines; liver, synthesized and secreted from the liver; Tg, triacylglycerol; ChE, cholesteryl ester; $B_{100}$, the full-length apolipoprotein-B, which is originates primarily from the liver; $B_{48}$, a truncated version of apolipoprotein-B that contains the N-terminal 48% of the full molecule and originates primarily from the intestine.

exchangeable to other particles or surfaces. These moieties emit at 571 nm (orange) and 501 nm (green), respectively, and can therefore be used with standard optical filters for rhodamine and fluorescein during multicolor imaging. Fluorescently labeled lipoproteins are available commercially (Molecular Probes, Eugene, OR, and Leiden, The Netherlands), and so protocols from the literature for preparing these particles ***(22,23)*** are not reproduced here. Quantitative methods using fluorescently labeled lipoproteins have been developed ***(24)***, but $^{125}$I-labeled lipoproteins remain the standard in catabolic studies and will be the focus of the rest of this chapter.

The parameters most commonly measured after incubation of $^{125}$I-lipoproteins with cells are surface binding, intracellular accumulation, and lysosomal degradation of the particles. Typically, monolayers of essentially confluent cells are used, usually fibroblasts, macrophages, or hepatocytes, that have been preincubated overnight in cholesterol-poor medium to stimulate expression of LDL receptors ***(19)***. Cells are then incubated at 37°C for 2–6 h in the continuous presence of $^{125}$I-labeled LDL or β-VLDL (usually 1–10 µg lipoprotein protein per millilliter of medium, although wider concentration ranges are used for formal assessment of the $K_d$ and $V_{max}$). During this time, particles bind to the cell surface, become internalized, and are then delivered to lysosomes. In the lysosomes, $^{125}$I-labeled protein is degraded to individual amino acids, but [$^{125}$I]monoiodotyrosine, the radiolabeled form of tyrosine, does not charge the tyrosine-specific tRNA ***(25)***, and so it simply leaks out of the cells back into the media. At the end of the incubation, cells are chilled to 4°C to stop further metabolism. The culture media are harvested for isolation of [$^{125}$I]monoiodotyrosine as an indication of lysosomal degradation (*see* **Subheading 3.1.**) ***(19,26)***. The cell monolayers are rinsed twice for 10 min with chilled phosphate- or Tris-buffered saline, pH 7.4, supplemented with 2 mg of fatty acid-free bovine serum albumin per milliliter, then twice rapidly with chilled buffered saline without albumin. Surface-bound particles are released from LDL receptors and other sites by a 30-min incubation at 4°C in buffered saline with 10 mg of heparin/mL. The cell monolayers are rinsed rapidly once more in buffered saline at 4°C, this rinse is pooled with the heparin wash, and the radioactivity in an aliquot is determined by gamma counting. The cell monolayers are then dissolved in 0.1 *M* NaOH. One aliquot of dissolved cells is used for gamma counting to assess intracellular accumulation of labeled material, and another aliquot is used to assess total cellular protein content. Radioactivity results are converted to mass of lipoprotein protein using the known specific activity of the particular radiolabeled preparation, and results are expressed as nanograms of catabolized lipoprotein protein per milligram of total cell protein. To verify that particle degradation involves lysosomes, cells are treated with chloroquine (150 µ*M*), an inhibitor of lysosomal proteases ***(19)***, beginning 0–45 min before the incubation with $^{125}$I-labeled lipoproteins, and continuing until the end of the experiment. Typically, ligand degradation is inhibited by 85–90%, and there is a corresponding increase in the accumulation of intracellular radioactive material.

### 1.3. Background for Methods Used to Study Lipoprotein-HSPG Interactions

Several modifications in the general methods outlined above are necessary for the study of lipoprotein–HSPG interactions. First, there are several distinct genetic families of cell-surface HSPGs, prinicpally syndecans, perlecan, and glypicans, and it is most informative to use cellular preparations that have one predominant cell-surface HSPG, or else a limited and known combination of these HSPGs. For the study of syndecans (**Fig. 2A**, left schema), we have used mainly Chinese hamster ovary (CHO) cells that we transfected with an expression construct for the human syndecan-1 core protein *(27)*. For perlecan, we have used the WiDr colon carcinoma line, which expresses perlecan but no other proteoglycans (American Type Culture Collection [ATCC], Manassas, VA, cat. no. CCL 218, also known as HT-29) *(28–32)*. For glypicans, there are reports of transfected mammalian cells in the literature *(33)*.

Second, the binding of lipoproteins to HSPGs in vivo is facilitated by bridging molecules, such as lipoprotein lipase (LpL), apoE, defensins, and hepatic lipase, each of which has a hydrophobic face that adheres to the lipoprotein surface and a cationic face that binds HS. Thus, studies in vitro typically involve examination of cellular catabolism of lipoproteins in the absence and in the presence of one of these molecules, and the arithmetic increase in each of the catabolic parameters is calculated *(27,32,34)*. We have preferred to use LpL because, unlike apoE, it does not bind LDL receptors, and unlike hepatic lipase, it is relatively easy to isolate in large quantities [*see* **Subheading 3.2.**, adapted from references *(35,36)*]. Typically, we add 5 μg of $^{125}$I-labeled lipoprotein protein per milliliter of medium, without or with 5 μg LpL/mL. Alternatively, cells that naturally secrete these bridging molecules *(37)* or cells transfected to express them [e.g., *(38,39)*] can be used. The role of HSPGs in the increased catabolism upon addition of LpL is verified by heparitinase digestion of the cells, which typically abolishes ~90% of LpL-dependent catabolism *(27,34,37,40,41)*; by the use of HS-negative CHO mutants *(34,40,42,42a)*; by addition of very low concentrations of heparin (<100 μg/mL) that are insufficient to interfere with LDL receptor binding but are able to displace surface-bound LpL and other bridging molecules *(34,42a)*; or by pre-incubation of cells in chlorate to block sulfation of glycosaminoglycan side chains *(43–45)*. Heparitinase digestion usually involves preincubation of cells in serum-free medium at 37°C for several hours to allow the cells to clear surface-bound serum-derived molecules, then an initial digestion with heparitinase for 60–90 min at 37°C before addition of ligand, and finally an incubation of cells with ligand, but in the continuous presence of heparitinase, to avoid rapid regeneration of side chains *(34)*. Apoproteins normally found on LDL, VLDL, and β-VLDL include apoB and apoE, both of which bind HS, although lipoprotein concentrations around 100 μg/mL are usually required before this binding becomes a significant contributor to total cellular catabolism in the absence of added bridging molecules *(15,45)*. These higher concentrations may be physiological, and LDL receptor-independent clearance of lipoproteins is substantial in vivo *(16,17)*. Interestingly, two common, naturally occurring polymorphisms of apoE, one of which has been associated with Alzheimer's diease *(46)*, show substantial differences in their catabolism by cells through a pathway mediated by HSPGs *(47)*, particularly syndecan HSPGs *(48)*.

Third, careful attention must be paid to catabolic contributions by members of the LDL receptor family. These contributions take the form of direct internalization, which appears simply as background in measurements of lipoprotein catabolism in the absence of LpL, and synergistic interactions, in which cell-surface HSPGs and LDL

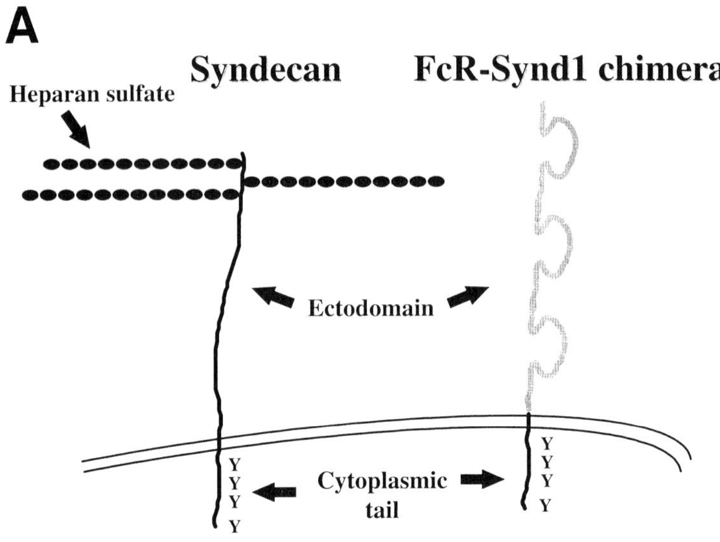


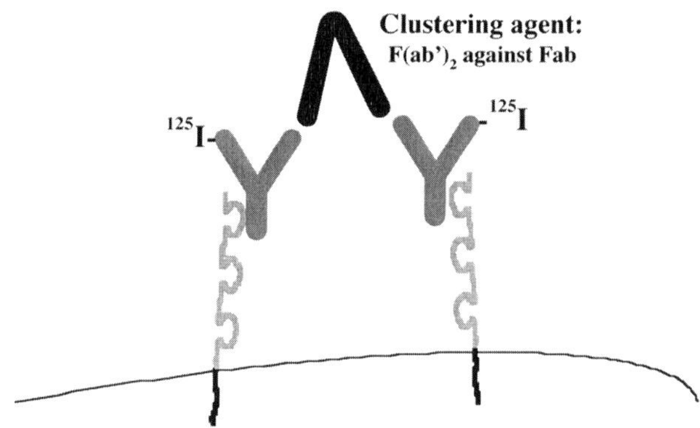


Fig. 2. The FcR-Synd1 chimera, a molecular tool to study the syndecan transmembrane and cytoplasmic domains. **(A)** Comparison of the syndecan-1 HSPG (left schema) and the FcR-Synd1 chimera (right schema). As indicated, the chimera contains the ectodomain of the IgG Fc receptor-Ia linked to the highly conserved transmembrane and cytoplasmic domains of the human syndecan-1 core protein *(27)*. **(B)** Clustering of FcR-Synd chimeras. Two chimeric molecules are shown, each bound to an $^{125}$I-labeled nonimmune human IgG. Clustering is accomplished by adding the clustering agent, that is, goat F(ab')$_2$ fragments against the Fab domain of human IgG. This clustering agent cannot interfere with the binding of whole IgG to the chimeras, because it has no Fc domain to bind to the chimeras, nor does it interact with the Fc domain of the IgG ligands.

receptor family members co-operate in ligand catabolism. Direct internalization via LDL receptor family members is easier to deal with. LDL can be 35% reductively methylated (mLDL; *see* **Subheading 3.3.**), which abolishes its ability to bind LDL receptors ***(49)*** while having no effect on the binding to cell-surface HSPGs in the presence of LpL ***(27,32)***. In an experimental tour de force, a series of transgenic mice were created that express different site-directed mutants of human apoB, including one variant with defective LDL receptor binding but unaltered affinity for proteoglycans, and another that binds LDL receptors but not proteoglycans ***(9,10,10a)***. Monoclonal antibodies against specific domains of apoB and apoE have been reported and used in studies of lipoprotein–proteoglycan interactions ***(50,51)*** (many of these antibodies are available for purchase from the University of Ottawa Heart Institute, Ottawa, Ontario, Canada). Cellular LDL receptors can be blocked with a polyclonal antibody ***(52)*** or with a monoclonal antibody produced by a clone available from the ATCC (#CRL 1691), but this monoclonal antibody must already be bound to cells before the lipoproteins are added ***(53,54)***. Cellular LDL receptors are readily suppressed in most cell types by an overnight incubation in medium supplemented with 1–2 μg of 25-hydroxycholesterol, which potently downregulates LDL receptors, plus 20-40 μg of cholesterol per milliliter to protect the cells from the toxicity of 25-hydroxycholesterol ***(19)***, although cholesterol-induced alterations in glycosaminoglycan synthesis have been reported ***(55)***. Suppression of LDL receptors must be verified by a substantial reduction in $^{125}$I-LDL binding: some macrophage-like cell lines ***(56)*** and cells expressing cholesterol 7α-hyroxylase ***(57)***, such as hepatic parenchymal cells, maintain significant LDL receptor expression despite the presence of abundant sterol. Unlike proteoglycans, LDL receptor family members depend on calcium ions for binding ***(19)***, which can be chelated by EDTA ***(45,51)***. Finally, LDL receptor-negative human fibroblasts (the ATCC has several lines) ***(19)*** and CHO cells ***(58)***, as well as an LDL receptor knockout mouse (Jackson Laboratory, Bar Harbor, ME, cat. no. 002207) ***(59)***, are available. Owing to their simplicity, our methods of choice for controlling ligand binding to the LDL receptor in vitro have been LDL methylation or cellular supplementation with sterols.

The LDL receptor-related protein (LRP), a cell-surface LDL receptor family member that has been reported to bind LpL, apoE, and hepatic lipase, is readily blocked by the 39-kDa receptor-associated protein (RAP), which is available as a recombinant protein ***(60)***. Unfortunately, RAP is the human homolog of mouse heparin-binding protein 44 ***(61)***, readily binds heparin ***(61)***, and therefore might also compete for binding to HS side chains, particularly ones that are rich in heparin-like domains. In the experience of our laboratory ***(27,32)***, RAP has had small or no effects on the binding of LpL-enriched $^{125}$I-labeled mLDL to the HSPGs of CHO or WiDr cells, although the binding of RAP to cell-surface HSPGs remains controversial ***(62,63)***. To our knowledge, no studies have systematically examined the binding of RAP to hepatic parenchymal cells, which synthesize HS with abundant heparin-like domains ***(64)***. LRP-deficient cell lines ***(65–67)*** and liver-specific knockout mice ***(68)*** have been reported, although the status of their proteoglycans is unknown. $^{125}$I-labeled RAP is prepared using Iodobeads (Pierce Chemical Company, Rockford, IL) instead of the iodine monochloride method, owing to the lack of unsaturated lipid in RAP, and this radiolabeled molecule can serve as a convenient ligand for coated pit-mediated internalization, provided it is known to bind poorly to the particular HSPGs that are expressed by the cell type of interest *(27)*. $^{125}$I-RAP can be released from cell-surface LRP by a protease cocktail, to distinguish surface-bound from internalized material *(27,69)*.

Synergistic interactions between cell-surface HSPGs and LDL receptor family members are thought to involve either the transfer of ligand from HSPGs to LDL receptor family members or the formation of ternary complexes of HSPGs, HSPG-bound ligands, and LDL receptor family members *(40,41,70,71)*. No direct evidence to date has distinguished these possibilities. Quantification of synergy between cell-surface HSPGs and LDL receptors requires measurement of the catabolism of $^{125}$I-labeled native LDL and $^{125}$I-mLDL, each in the absence and presence of LpL. Four catabolic components can then be computed *(32)*: (1) the LDL receptor-independent, LpL-independent component, which is usually referred to as nonspecific uptake or assay background and is measured by the catabolism of $^{125}$I-mLDL in the absence of LpL; (2) the LDL receptor-dependent, LpL-independent component, which is the classical LDL receptor pathway and is computed by the difference between the catabolism of $^{125}$I-LDL vs $^{125}$I-mLDL; (3) the LDL receptor-independent, LpL-dependent component, which involves cell-surface HSPGs and is computed by the difference between the catabolism of $^{125}$I-mLDL in the presence vs the absence of LpL; and (4) a synergistic component, which requires cooperation between LDL receptors and LpL and is computed as the increase in $^{125}$I-LDL catabolism upon addition of LpL minus the increase in $^{125}$I-mLDL catabolism upon addition of LpL. Total catabolism of $^{125}$I-labeled native LDL in the presence of LpL equals the sum of these four components, reflecting the ability of LpL-enriched $^{125}$I-LDL to participate in these four potential uptake mechanisms *(32)*. Similar information can be obtained using cellular sterol enrichment instead of LDL methylation to manipulate binding to the LDL receptor, or heparitinase digestion instead of omission of LpL to manipulate binding to cell-surface HSPGs. These alternative methods tend to be more cumbersome and somewhat less effective and, as noted above, cellular sterol enrichment has been reported to affect glycosaminoglycan synthesis *(55)*. In WiDr cells under our experimental conditions, we have found that the four components respectively account for approximately 4%, 15%, 57%, and 23% of total LpL/$^{125}$I-LDL catabolism. In other words, most LpL/$^{125}$I-LDL enters these cells via perlecan directly (component 3) without any assistance from LDL receptors *(31,32)*.

Fourth, careful attention must be paid to the unusual characteristics of HSPG-mediated ligand internalization. To assess the efficiency of endocytosis, we calculate ligand internalization as the sum of intracellular accumulation plus degradation *(27)*. Internalization calculated in this fashion takes into account ligand still within the cells, as well as ligand that had been taken up by cells but then degraded into amino acids, which the cells release to the culture medium. To examine in detail the kinetics of ligand internalization and degradation, we incubate labeled ligands with cells in serum-free medium for 1 h at 4°C, to allow surface binding without further catabolism, and then the cells are washed at 4°C to remove unbound material. Fresh media at 37°C with no ligands are added, and incubations are continued at 37°C for various times, usually 15 min to 24 h *(27,32)*. Assays for surface-bound, intracellular, and degraded ligand are then performed, and ligand internalization is calculated. In some experiments, TCA-precipitable radioactivity in the media is also quantified (**Subheading 3.1.**), as an indication of retroendocytosis or desorption from the cell surface during the incubation at 37°C. In addition, because the principle method under **Subheading 3.1.** for assessing lysosomal degradation requires complete breakdown of labeled apoproteins into individual amino acids, supplementary tests for partial breakdown products can also be informative (also given under **Subheading 3.1.**). From these studies, we have found that syndecan-mediated endocytosis of LpL/$^{125}$I-mLDL proceeds with a $t_{1/2}$ of ~1 h *(27)*, and that perlecan-mediated internalization exhibits a $t_{1/2}$ of ~5 h *(31,32)*. By contrast, the $t_{1/2}$ for coated pit-mediated internalization is ~10 min *(27,32,72,73)*.

HSPG-mediated pathways for ligand internalization can also be distinguished through the use of metabolic inhibitors, particularly genistein (0–400 μ*M*), a tyrosine kinase inhibitor *(74)*, and cytochalasin D (0–2 μ*M*), which disrupts the actin cytoskeleton *(75)*. Typically, ligands are bound to the cell surface at 4°C, unbound ligand is washed away, and then specific inhibitors are added simultaneously with fresh medium at 37°C. Cells are incubated at 37°C until ~30–50% of surface-bound ligand has been internalized in the absence of inhibitors, that is, 45 min for syndecan and 2 h for perlecan (incubations longer than 2 h can encounter fading of the genistein effect) *(27,76)*. To allow comparison with coated pit-mediated internalization, we employ a 30-min pre-incubation at 37°C in the presence of these inhibitors then a 10- to 15–min incubation in the presence of inhibitors plus $^{125}$I-RAP *(27)* or surface-bound $^{125}$I-LDL *(32)*. Coated pit-mediated endocytosis is insensitive to genistein, whereas both syndecan- and perlecan-mediated internalization are inhibited *(27,32)*. Coated pit endocytosis exhibits limited sensitivity to cytochalasin *(27,32,77)*, whereas syndecan-mediated endocytosis is readily inhibited *(27)*, and perlecan-dependent internalization is slightly enhanced by this agent *(32)*. Thus, the three pathways exhibit distinctive kinetics of intenalization and different dependence on tyrosine kinases and the actin cytoskeleton.

### *1.4. Background for Molecular Methods to Study HSPG-Mediated Endocytosis*

The underlying model is that each cell-surface HSPG, like other receptors, is organized into domains that mediate specific functions. Portions of the HS side chains serve as the ligand-binding domains, while regions within the core protein are responsible for recruiting the specific cellular machinery used for internalization of that particular species of HSPG [*see* *(1,33,78)*]. In this light, the distinct pathways for cellular uptake of ligands bound to syndecan HSPGs versus the perlecan HSPG can be understood as a consequence of the differences between the two core proteins, which lack any significant homology to each other.

To examine this model, we constructed the FcR-Synd1 chimera, which consists of the ectodomain of the IgG Fc receptor-Ia linked to the highly conserved transmembrane and cytoplasmic domains of the human syndecan-1 core protein (**Fig. 2A**, right schema) *(27)*. Thus, this chimera contains the portions of the syndecan-1 core protein that contact the plasma membrane and cellular interior. The ligand for the chimera is $^{125}$I-labeled nonimmune human IgG, which we prepare using Iodobeads. We have made this chimera and cells that express it available to other investigators, so the protocols for construction and transfection are not reproduced here.

The FcR-Synd1 chimera offers many advantages, primarily by providing a simple and easily controlled experimental system. First, contributions to ligand catabolism from other cell-surface molecules, such as LDL receptor family members and HSPGs, are completely eliminated if the chimera is expressed in a cell type with no endogenous Fc receptors. Thus, we stably express FcR-Synd1 in CHO cells, which we have already used to study syndecan-mediated endocytosis *(27)*. Second, ligand clustering, which appears to trigger efficient syndecan-mediated endocytosis, can be induced at will through the use of a clustering agent (*see* **Fig. 2B** and **Subheading 3.4.**) *(27)*. Third, the ligand, $^{125}$I-IgG, does not deliver any lipids to the cells, nor is it broken apart by detergents. Thus, the role of cold Triton-insoluble, cholesterol-rich membrane rafts in this endocytic pathway can be more easily examined than when relying on lipoproteins as ligands (*see* **Subheading 3.5.**) ***(79,80)***. Fourth, in terms of kinetics and the dose-responses to different metabolic inhibitors, the cellular uptake of surface-bound, clustered $^{125}$I-IgG is identical to the uptake of LpL/$^{125}$I-mLDL, which supports the model described above *(27)*.

Most important, the syndecan domains within the FcR-Synd1 chimera are readily deleted or mutated by standard molecular techniques, thereby allowing identification of molecular determinants for each stage of this endocytic pathway ***(81)***.

## 2. Materials

### 2.1. Purification of [$^{125}$I]Monoiodotyrosine from Culture Media, as an Assay of Degradation of $^{125}$I-Proteins in Lysosomes

1. Two sets of borosilicate glass tubes, 12 × 75 mm, with one tube in each set for each sample of medium.
2. One set of 12 × 75-mm plastic tubes (e.g., Sarstedt, Newton, NC, cat. no. 55.476) with push-in stoppers (e. g., Sarstedt cat. no. 65.809).
3. 50% (w/v) solution of trichloroacetic acid, kept at 4°C, 0.25 mL for each sample of medium.
4. Freshly made solution of 40% (w/v) of KI, 5 µL for each sample of medium.
5. 30% $H_2O_2$, 20 µL for each sample of medium.
6. Chloroform, analytic grade, 1–2 mL for each sample of media.
7. Table-top refrigerated centrifuge (e.g., a Beckman GS-6R).
8. Fume hood.
9. Mixing platform for test tubes, such as a Vortex Genie.
10. Gamma counter.
11. Additional materials for **Subheading 3.1.**, **step 8** (optional):
    a. 20% (w:v) solution of trichloroacetic acid, kept at 4°C.
    b. 0.1 N NaOH solution.
12. Additional materials for **Subheading 3.1, step 9** (optional):
    a. Buffered saline with 10 mg heparin/mL, kept at 4°C.
    b. 1% solution of sodium dodecyl sulfate (SDS).
    c. SDS-polyacrylamide gel and electrophoresis apparatus.
    d. Materials for autoradiography, preferable a PhosphorImager (Molecular Dynamics, Sunnyvale, CA).

### 2.2. Isolation of Bovine Milk LpL

1. 3.6 L of raw, fresh whole cows' milk (it is best to milk the cows on the morning the isolation begins, and the milk must be promptly chilled and kept cold).
2. 160 mL of heparin-agarose beads (Bio-Rad cat. no.153-6173 is suitable and inexpensive).
3. One 4-L vacuum Erlenmyer flask with a large scintered glass funnel.
4. One 4-L beaker.
5. One 250-mL beaker.
6. A refrigerated table-top centrifuge with centrifuge bottles, to accommodate about 4 L of liquid (Four Beckman 750-mL centrifuge bottles in a GS-6R centrifuge work well).
7. Cheesecloth or any other sort of fine mesh that is lint-free.
8. Seven 700-mL plastic culture flasks with canted necks.
9. A magnetic stir plate, kept at 4°C.
10. A rocking platform, kept at 4°C.
11. A fraction collector, kept at 4°C.
12. A glass chromatography column, 2.5 cm × 30 cm (e.g., Bio-Rad cat. no. 737-2532), kept at 4°C, and connected to the fraction collector.
13. A small funnel for the glass chromatography column (e.g., Bio-Rad cat. no. 731-0003).
14. A spectrophotometer, to measure absorbance at 280 nm.
15. Distilled water, chilled to 4°C.
16. Clean, powder-free gloves or a large, clean spatula.
17. Buffers, chilled to 4°C:
    a. 2 L of 0.4 *M* NaCl, 10 m*M* Tris/HCl, pH 7.4
    b. 2 L of 0.75 *M* NaCl, 10 m*M* Tris/HCl, pH 7.4
    c. 1 L of 1.5 *M* NaCl, 10 m*M* Tris/HCl, pH 7.4
    d. 3 L of phosphate-buffered saline, pH 7.4
    e. 200 mL of phosphate-buffered saline, pH 7.4, with 0.02% sodium azide
18. Air-tight 2-mL freezer vials: these must have screw caps that close on an O-ring, to avoid sublimation of $H_2O$ during storage in a freezer (e.g., Sarstedt cat. no. 72692005).
19. 8 *M* urea, 1.5 *M* NaCl, in phosphate buffer, pH 7.4, at room temperature (optional).

### 2.3. Reductive Methylation of LDL

1. 1.5 L of lipoprotein buffer (150 m*M* NaCl, 0.3 m*M* $Na_2EDTA$, pH 7.4) at 4°C.
2. A preparation of LDL or $^{125}I$-LDL of known protein concentration, in lipoprotein buffer at 4°C.
3. 0.1 *M* $NaCNBH_3$ (sodium cyanoborohydride; Aldrich) in saline at 4°C.
4. 133.2 m*M* HCHO in distilled water (mix 10 µL of Fisher F79-500 ACS-grade formaldehyde, which is 37% w/w, i.e., 13.32 *M*, with 990 µL of distilled water) at 4°C.
5. Dialysis membranes, such as 30-kDa cutoff dialysis tubes (e.g., Spectra/Por, carried by Fisher Scientific, Pittsburgh, PA) or 10-kDa membrane cassettes (e.g., Slide-A-Lyzer, Pierce), which are very convenient.
6. 0.45-µm syringe filter, such as from Millipore (Bedford, MA, USA), and a syringe.

### 2.4. Special Methods Allowed by the Use of the FcR-Synd1 Chimera: Control of Ligand Clustering

1. CHO cells expressing the FcR-Synd1 chimera, grown in Ham's F-12 medium with 10% fetal calf serum to a confluent monolayer.
2. Ice tray or cold room.
3. Phosphate- or Tris-buffered saline, pH 7.4, at 4°C.

4. Serum-free cell-culture medium at 4°C (Ham's F-12 medium supplemented with 0.2% fatty acid-free bovine serum albumin, Sigma Chemical, St. Louis, MO, cat. no. A-8806).
5. Serum-free cell-culture medium at 37°C (optional).
6. $^{125}$I-labeled nonimmune human IgG in lipoprotein buffer: 150 m*M* NaCl, 0.3 m*M* $Na_2$EDTA, pH 7.4 at 4°C. Nonimmune human IgG is commercially available (e.g., from Rockland Immunochemicals, Inc., Gilbertsville, PA) and can be radioiodinated with Iodobeads (Pierce Chemical, Rockford, IL).
7. Clustering agent: Unlabeled goat F(ab')$_2$ raised against the Fab domain of human IgG (e.g., from Rockland Immunochemicals).

### 2.5. Special Methods Allowed by the Use of the FcR-Synd1 Chimera: Assessment of Cold Triton Insolubility

1. CHO cells expressing the FcR-Synd1 chimera, incubated as described under **Subheading 3.4.** (i.e., $^{125}$I-IgG bound to the cell surface, then incubated at 4°C or 37°C, in the presence or absence of the clustering agent).
2. Phosphate- or Tris-buffered saline, pH 7.4, at 4°C.
3. 1% Triton X-100 solution: 150 m*M* NaCl, 10 m*M* Tris, 50 m*M* sodium acetate, 5 m*M* EDTA, 1% (v/v) Triton X-100, pH 7.5, at 4°C.
4. Optional reagents:
   $^{125}$I-labeled RAP.
   1% Triton X-100 solution, at 37°C.
   1% solution of octylglucoside, at 4°C.
   Serum-free culture medium with 2 μ*M* cytochalasin D. The cytochalasin D is added as a 2 m*M* stock solution in DMSO.
   Serum-free culture medium with 50 m*M* cyclodextrins (Sigma Chemical Company, 2-hydroxypropyl-β-cyclodextrin, cell-culture tested, cat. no. C 0926).

## 3. Methods

### 3.1. Purification of [$^{125}$I]Monoiodotyrosine from Culture Media, as an Assay of Degradation of $^{125}$I-Proteins in Lysosomes (see Note 1)

1. After incubation of cells at 37°C with $^{125}$I-labeled ligands, the cells are placed on ice, and the media are harvested. As a control, culture wells without cells are preincubated with serum-containing media, rinsed, and incubated at 37°C with medium and $^{125}$I-labeled ligands. Media are harvested from these cell-free wells and processed in parallel with the samples from the experimental wells.
2. Place 1.0 mL of each medium sample into the first set of borosilicate glass tubes, into which has been added 0.25 mL of cold 50% trichloroacetic acid. This will precipitate undegraded $^{125}$I-labeled proteins, as well as other proteins in the medium, such as albumin. Vortex and place at 4°C for 30 min.
3. Centrifuge these tubes in a standard table-top centrifuge (e.g., 500–1300*g*, 1700–2800 rpm in a Beckman GS-6R, $r_{avg}$ ~15 cm) at 4°C for 30 min.
4. Harvest 500 μL of supernatant and place into the second set of borosilicate glass tubes. Add 5 μL of fresh 40% KI as a carrier, mix, then under a fume hood, add 20 μL of 30% $H_2O_2$ and mix again. The $H_2O_2$ converts radioactive iodide ions into [$^{125}$I]iodine.
5. Add 1.5–2.0 mL of chloroform to each tube, to extract free [$^{125}$I]iodide. Mix, then let stand for 15 min at room temperature. The lower, chloroform phase should become purple, and the upper, aqueous phase should lose color, until only a slightly yellowish tint remains. If substantial transfer of color does not occur, remix.

6. Remove 200 μL of the upper (aqueous) phase and place into the plastic tubes with snap-on caps. Quantify the $^{125}$I-radioactivity in a gamma counter.
7. To correct for dilution during these extractions, all results are multiplied by 6.5625, which will give the counts per minute of [$^{125}$I]monoiodotyrosine in the original 1.0 mL from each sample of medium. Results from culture wells without cells are subtracted as background, and should represent < 0.02% of the initially added radioactivity. (*See* **Note 2**).
8. (Optional) In some experiments, the $^{125}$I-labeled ligands are not present continuously during the incubation at 37°C, but instead are bound to the cell surface ahead of time, followed by a rinse to remove unbound ligand (*see* **Subheading 1.3.,** "Fouth, careful attention ..."). In this case, the trichloroacetic acid precipitate of medium (*see* **step 2**) will contain material that has desorbed from the cell surface or has undergone retroendocytosis. To quantitate radioactivity in this precipitate, remove the supernatant, then rinse the pellet twice in 1 mL of cold 20% trichloroacetic acid solution. The rinsed pellet can be counted directly, or dissolved in 0.1 N NaOH. Most of these experiments are performed in serum-free medium with 0.2% bovine serum albumin; hence, a protein assay of the re-suspended pellet can serve as a standard for recovery.
9. [Optional; adapted from *(32)*] In some experiments, it may be desirable to look for partial degradation products within the cells. After incubation at 37°C with $^{125}$I-labeled ligands, the cells are placed on ice, and residual surface-bound material, which is presumably undegraded, is removed by a 30-min incubation at 4°C in buffered saline with 10 mg of heparin/mL. The cells are then solubilized in SDS and subjected to polyacrylamide gel electrophoresis under reducing or nonreducing conditions, followed by autoradiography. A small amount of the original $^{125}$I-labeled ligand is electrophoresed on the same gel, to indicate undegraded material. Radiolabel on lower-molecular-weight fragments within the cells indicates the generation of partial degradation products, and the presence of full-sized labeled material indicates persistent, intact ligand protein within the cells.

### 3.2. Isolation of Bovine Milk LpL (see Note 3)

1. Place the 3.6 L of cold, raw, fresh milk into a 4-L beaker on a stir plate at 4°C. Add 84.1 g of solid NaCl, to bring the final concentration of added NaCl to 0.4 *M*. Stirring should be fast enough for the salt to dissolve within about 1 h, but not fast enough to cause foaming.
2. While the salted milk is stirring, place 140 mL of heparin-agarose beads into the sintered glass funnel attached to the 4-L vacuum flask. Wash the beads with 3L of chilled distilled water, then 2 L of chilled phosphate-buffered saline. At the end of the washes, leave a small amount of phosphate-buffered saline with the beads to keep them wet, then swirl the material and pour it into a 250-mL beaker and keep at 4°C until use.
3. Place the salted milk into centrifuge bottles and spin at ~1,000*g* (e.g., 2600 rpm in a Beckman GS-6R centrifuge) for 45 min at 4°C. The lipid in the milk will collect in a thick layer on the top. Remove this layer by skimming with the spatula or by scooping it out by hand while wearing powder-free gloves. LpL in unhomogenized cows' milk does not associate with the cream, and so we discard the lipid layer. Filter the infranatant (i.e., the skimmed milk) through cheesecloth or lint-free mesh, to remove any residual clumps of cream.
4. Distribute the skimmed milk into seven 700-mL canted-neck culture flasks. Carefully add to each flask an equal amount of the washed heparin-agarose slurry, which must be swirled occasionally to keep the heparin beads evenly distributed.
5. Close the culture flasks and place them on the rocker at 4°C overnight. Rocking should be fast enough that the heparin beads mix with the skimmed milk without settling, but not fast enough to form bubbles.

6. The next morning, pour the contents of each culture flask, one at a time, into the sintered glass funnel attached to the 4-L vacuum flask. Set the vacuum strength to allow only a slow, trickling flow of liquid through the funnel, and never allow any heparin beads to become completely dry. Use a small amount of the 0.4 *M* NaCl buffer to recover any beads that remain in the flasks. Wash the heparin beads now in the funnel with the remainder of the 2 L of 0.4 *M* NaCl buffer, again without letting the beads become dry.
7. Wash the beads with 1.5 L of 0.75 *M* NaCl buffer. This wash should be performed at a slightly slower rate than with the 0.4 *M* NaCl buffer, and enough buffer should be left with the beads at the end of the wash to allow a slurry to form upon swirling.
8. Swirl the contents of the sintered funnel and pour into the 2.5 × 30 cm glass column, using the small fitted funnel to avoid spillage. The glass column should be connected to a fraction collector, in a cold room or a chromatography refrigerator at 4°C. Rinse the column with the remaining 0.5 L of 0.75 *M* NaCl buffer, and discard the flowthrough. Fluid will flow through the column of heparin-agarose beads under gravity only, with no need for a pump.
9. Elute the LpL from the heparin-agarose beads using the 1.5 *M* NaCl buffer, while collecting 5-mL fractions. LpL should elute around fraction #20. Pool fractions with an $A_{280} > 0.200$. Aliquot this material into air-tight freezer vials, usually 0.1–0.5 mL/vial, and store at –80°C until use.
10. Set one aliquot aside for SDS-PAGE/Coomassie stain, to verify the presence of a single band at 55 kDa, and protein assay, which should be around 200–250 µg/mL. Because the effect of LpL on lipoprotein catabolism via HSPGs is structural, not enzymatic, an assay to verify enzymatic activity in each LpL preparation used for that purpose is not necessary (the enzymatic assay can be found in ***(82)*** and the citations therein).
11. The heparin-agarose beads can be placed directly into buffered saline with azide for storage then reused several times. Alternatively, some laboratories prefer to wash the beads in the 8 *M* urea buffer before reuse. Note that a small volume of beads is lost with each use.

### 3.3. Reductive Methylation of LDL (see Note 4)

1. To each 1 mg of apoB in a preparation of LDL or $^{125}$I-LDL, add 60 µL of the 0.1 *M* $NaCNBH_3$ solution and mix (final concentration of $NaCNBH_3$ in the mixture should be around 10–20 m*M*).
2. 1.29 µmole of HCHO is then added per milligram of apoB (~2/1 molar ratio of formaldehyde to lysine residues), i.e., 9.68 µL of the 133.3 m*M* HCHO solution per milligram of apoB. Mix, then incubate the mixture for 18 h at 4°C. (*See* **Note 5.**)
3. The methylation reaction is stopped by dialysis at 4°C against at least three changes of lipoprotein buffer. The methylated LDL is then sterilized by passage through a 0.45-µm syringe filter and placed into a sterile tube for storage at 4°C. As before, protein concentration is assessed by the SDS-Lowry assay ***(83)*** (Sigma cat. no. P-5656).

### 3.4. Special Methods Allowed by the use of the FcR-Synd1 Chimera: Control of Ligand Clustering

1. (**Fig. 2B**; *see* **Note 6**) Place the CHO-FcRSynd1 cells on ice and rinse three times with cold buffered saline.
2. Add $^{125}$I-IgG in cold serum-free medium (3–5 µg of IgG/mL final concentration) to the cells and incubate at 4°C for 1 h, to allow cell-surface binding.
3. Rinse the cells again at 4°C with buffered saline to remove unbound ligand.
4. Add fresh serum-free medium, with or without the clustering agent (4 µg/mL final concentration). This medium can be added at either 4°C or 37°C, and then the cells are incubated at whichever of these two temperatures was chosen. Certain steps in this endocytic pathway, such as the development of cold Triton insolubility, will develop even if the cells are kept

at 4°C during ligand clustering, indicating that this process can be independent from active cellular metabolism ***(79)***. Ligand internalization via this pathway, however, does require active cellular metabolism and will not proceed at 4°C. (*See* **Note 7.**)

5. Cells and media are harvested at serial time points, to quantitate surface-bound ligand, intracellular accumulation, lysosomal degradation, and cold Triton insolubility.

   Surface-bound particles are released from the FcR-Synd1 chimera by a 2-min incubation of cells in acidified phosphate-buffered saline (pH 2.5) at 4°C ***(27,84)***, and the radioactivity in an aliquot is determined by gamma counting. The remaining cell monolayers are rinsed rapidly once more in buffered saline, then dissolved in 0.1 *M* NaOH. One aliquot is used for gamma counting to assess intracellular accumulation of labeled material, and another aliquot is used to assess total cellular protein content. Lysosomal degradation is assessed according to **Subheading 3.1.**, and Cold Triton Insolubility according to **Subheading 3.5.**

### *3.5. Special Methods Allowed by the use of the FcR-Synd1 Chimera: Assessment of Cold Triton Insolubility (see Note 8)*

1. Place the CHO-FcRSynd1 cells on ice and rinse three times with cold buffered saline. If the cells had been incubated at 4°C, they will remain at this temperature. If the cells had been incubated at 37°C, this step will now cool them to 4°C.
2. Cold Triton solubility is assayed by incubation of cells for 5 min at 4°C in a 1% solution of chilled Triton X-100. Solubilized material is assayed directly by $^{125}$I-radioactivity released into the solution. Cold Triton-insoluble material is dissolved in 0.1 *N* NaOH and quantified by gamma counting. Data are expressed as the percentage of total cell-associated ligand (cold Triton soluble plus insoluble) that is cold Triton insoluble. $^{125}$I-labeleled RAP is an attractive control, because it contains no lipid and hence is not disrupted by Triton, it becomes rapidly internalized, and yet it does not exhibit cold Triton insolubility ***(79,80)***.
3. To distinguish cold Triton insolubility that arises from membrane effects vs cytoskeletal binding, we have two approaches:
   a. The use of cytochalasin D to disrupt the cytoskeleton without affecting cholesterol-rich membrane patches. Cytochalasin D (0–2 μ*M*) is added to cells in serum-free medium for a 30-min incubation at 37°C before the addition of $^{125}$I-IgG (i.e., before **step 5** 1 and 2 in **Subheading 3.4.**), and this inhibitor is kept in all incubation solutions, except saline rinses, until the addition of the 1% Triton solution. Alternatively, cytochalasin D can be added at the same time as the clustering agent.
   b. The use of octylglucoside, warm Triton, or pretreatment with cyclodextrins to disrupt cholesterol-rich membrane patches but without affecting the cytoskeleton. Solubility in octylglucoside (1%) or warm Triton (37°C) is assessed similarly to **step 2**. Pretreatment with cyclodextrins, which remove unesterified cholesterol from the cell membrane, involves a 1-h incubation at 37°C before the addition of $^{125}$I-IgG (i.e., before **step 1** in **Subheading 3.4.**), and this compound is kept in subsequent incubation solutions if the cells are warmed to 37°C upon addition of the clustering agent (**step 4** in **Subheading 3.4.**). If the cells are kept at 4°C, no additional exposure to cyclodextrin is needed. Notice that the addition of $^{125}$I-IgG or the clustering agent will not replenish cell-membrane cholesterol, whereas the addition of LpL-enriched $^{125}$I-LDL or $^{125}$I-methylated LDL will.

## 4. Notes

1. This method was adapted from references ***(19,26)***.
2. The method is very easy to perform, generally gives tight replicates, and can be used with a variety of $^{125}$I-labeled ligands, such as lipoproteins, RAP, LpL, and others. The method

has two limitations. First, it assesses only the complete degradation of an $^{125}$I-labeled protein into individual amino acids. To detect partial proteolytic fragments, cellular extraction then SDS-PAGE and autoradiography can be performed (*see* **Subheading 3.1., step 9**, optional). Second, the oxidation-extraction step (*see* **Subheading 3.1., step 5**) removes free $^{125}$I and substantially lowers the background for the assay, but it also eliminates any evidence of deiodination by the cells. In other words, the results of this assay will be a slight underestimate of the total degradation of $^{125}$I-labeled protein by cells ***(19,32)***.

3. This method was adapted from **refs.** ***35*** and ***36***.
4. This method was adapted from **refs.** ***49*** and ***85***.
5. This is a very easy, reliable method to modify LDL to abolish its ability to bind LDL receptors, but without changing particle charge and without introducing modifications that allow binding to other receptors. The method attaches methyl groups to lysine residues in apoB, generating the dimethylamino derivative. Each molecule of $apoB_{100}$, which is essentially the only protein of human LDL, contains 357 lysine residues and has a protein molecular weight of 512,723 dalton. Thus, each milligram of $apoB_{100}$ detected in a protein assay contains roughly $7 \times 10^{-7}$ mol of lysine residues. The degree of methylation can be varied from 0% to 63% by altering the concentration of $NaCNBH_3$ and the ratio of HCHO per lysyl residue in apoB. The method given here achieves 35% methylation, to assure elimination of LDL receptor binding ***(49)***. To generate $^{125}$I-labeled methylated LDL, radioiodination is performed first ***(19–21)***, then reductive methylation.
6. This method was adapted from reference ***(27)***.
7. Efficient endocytosis via syndecan HSPGs is a multi-step process that is triggered by clustering of the syndecan transmembrane and cytoplasmic domains ***(27)***. Clustering is followed by lateral movement into cold Triton-insoluble membrane rafts and then recruitment of tyrosine kinases and the actin cytoskeleton to bring the ligands into the cellular interior ***(27,79,81)***. The FcR-Synd1 chimeric construct allows for control of ligand clustering and easy assessment of cold Triton insolubility.
8. This method was adapted from ***(79–81,86–88)***.

## References


1. Williams, K. J. and Fuki, I. V. (1997) Cell-surface heparan sulfate proteoglycans: dynamic molecules mediating ligand catabolism. *Curr. Opin. Lipidol.* **8,** 252–261.
2. Williams, K. J. and Tabas, I. (1995) The response-to-retention hypothesis of early atherogenesis. *Arterioscler. Thromb. Vasc. Biol.* **15,** 551–561.
3. Williams, K. J. and Tabas, I. (1998) The response-to-retention hypothesis of atherogenesis reinforced. *Curr. Opin. Lipidol.* **9,** 471–474.
4. Tabas, I., Li, Y., Brocia, R. W., Xu, S. W., Swenson, T. L., and Williams, K. J. (1993) Lipoprotein lipase and sphingomyelinase synergistically enhance the association of atherogenic lipoproteins with smooth muscle cells and extracellular matrix. A possible mechanism for low density lipoprotein and lipoprotein(a) retention and macrophage foam cell formation. *J. Biol. Chem.* **268,** 20419–20432.
5. Hurt-Camejo, E., Olsson, U., Wiklund, O., Bondjers, G., and Camejo, G. (1997) Cellular consequences of the association of apoB lipoproteins with proteoglycans. Potential contribution to atherogenesis. *Arterioscler. Thromb. Vasc. Biol.* **17,** 1011–1017.
6. Pentikäinen, M. O., Lehtonen, E. M., and Kovanen, P. T. (1996) Aggregation and fusion of modified low density lipoprotein. *J. Lipid Res.* **37,** 2638–2649.

7. Pentikäinen, M. O., Öörni, K., Lassila, R., and Kovanen, P. T. (1997) The proteoglycan decorin links low density lipoproteins with collagen type I. *J. Biol. Chem.* **272,** 7633–7638.
8. O'Brien, K.D., Olin, K.L., Alpers, C.E., Chiu, W., Ferguson, M., Hudkins, K., Wight, T. N., and Chait, A. (1998) Comparison of apolipoprotein and proteoglycan deposits in human coronary atherosclerotic plaques: colocalization of biglycan with apolipoproteins. *Circulation* **98,** 519–527.
9. Borén, J., Olin, K. L., Lee, I., Chait, A., Wight, T., and Innerarity, T. L. (1998) Identification of the principal proteoglycan-binding site in LDL. A single point mutation in apo-B100 severely affects proteoglycan interaction without affecting LDL receptor binding. *J. Clin. Invest.* **101,** 2658–2664.
10. Borén, J., Olin, K., O'Brien, K. D., Arnold, K. S., Ludwig, E. H., Wight, T. N., Chait, A., and Innerarity, T. L. (1998) Engineering non-atherogenic low density lipoproteins: direct evidence for the "Response-to-Retention" hypothesis of atherosclerosis. *Circulation* **98, (suppl.)**, I–314 (abstract).

10a. Borén, J., Gustafsson, M., Skålén, K., Flood, C., Innerarity, T.L. (2000) Role of extracellular retention of low density lipoproteins in atherosclerosis. *Cur. Opin. Lipidol.* **11**: 451–456.

11. Mazany, K. D., Peng, T., Watson, C. E., Tabas, I., and Williams, K. J. (1998) Human chondroitin 6-sulfotransferase: cloning, gene structure, and chromosomal localization. *Biochim. Biophys. Acta* **1407,** 92–97.
12. Fukuta, M., Kobayashi, Y., Uchimura, K., Kimata, K., and Habuchi, O. (1998) Molecular cloning and expression of human chondroitin 6- sulfotransferase. *Biochim. Biophys. Acta* **1399,** 57–61.
13. San Antonio, J. D. and Lander, A. D. (2001). Affinity coelectrophoresis of proteoglycan–protein complexes, in *Proteoglycan Protocols* (R.V. Iozzo, ed.), Humana Press, Totowa, NJ, pp. 401–414.
14. Shen, B. W., Scanu, A. M., and Kezdy, F. J. (1977) Structure of human serum lipoproteins inferred from compositional analysis. *Proc. Natl. Acad. Sci. (USA)* **74,** 837–841.
15. Al-Haideri, M., Goldberg, I. J., Galeano, N.F., Gleeson, A., Vogel, T., Gorecki, M., Sturley, S. L., and Deckelbaum, R. J. (1997) Heparan sulfate proteoglycan-mediated uptake of apolipoprotein E- triglyceride-rich lipoprotein particles: a major pathway at physiological particle concentrations. *Biochemistry* **36,** 12766–12772.
16. Kesaniemi, Y. A., Witztum, J. L., and Steinbrecher, U. P. (1983) Receptor-mediated catabolism of low density lipoprotein in man. Quantitation using glucosylated low density lipoprotein. *J. Clin. Invest.* **71,** 950–959.
17. Goldstein, J. L. and Brown, M. S. (1977) Atherosclerosis: the low-density lipoprotein receptor hypothesis. *Metabolism* **26,** 1257–1275.
18. Havel, R. J., Eder, H. A., and Bragdon, J. H. (1955) The distribution and chemical composition of ultracentrifually separated lipoproteins in human serum. *J. Clin. Invest.* **34,** 1345–1353.
19. Goldstein, J. L., Basu, S. K., and Brown, M. S. (1983) Receptor-mediated endocytosis of low-density lipoprotein in cultured cells. *Meth. Enzymol.* **98,** 241–260.
20. McFarlane, A. S. (1958) Efficient trace-labelling of proteins with iodine. *Nature* **182,** 53.
21. Bilheimer, D. W., Eisenberg, S., and Levy, R. I. (1972) The metabolism of very low density lipoprotein proteins. I. Preliminary in vitro and in vivo observations. *Biochim. Biophys. Acta* **260,** 212–221.
22. Innerarity, T. L., Pitas, R. E., and Mahley, R. W. (1986) Lipoprotein-receptor interactions. *Meth. Enzymol.* **129,** 542–565.
23. Pitas, R.E., Innerarity, T. L., Weinstein, J. N., and Mahley, R. W. (1981) Acetoacetylated lipoproteins used to distinguish fibroblasts from macrophages in vitro by fluorescence microscopy. *Arteriosclerosis* **1**, 177–185.
24. Teupser, D., Thiery, J., Walli, A. K., and Seidel, D. (1996) Determination of LDL- and scavenger-receptor activity in adherent and non-adherent cultured cells with a new single-step fluorometric assay. *Biochim. Biophys. Acta* **1303,** 193–198.

25. Scherberg, N. H., Seo, H., and Hynes, R. (1978) Incorporation of radioiodotyroisines into proteins formed during cell-free translation. *J. Biol. Chem.* **253,** 1773–1779.
26. Bierman, E. L., Stein, O., and Stein, Y. (1974) Lipoprotein uptake and metabolism by rat aortic smooth muscle cells in tissue culture. *Circ. Res.* **35,**136–150.
27. Fuki, I. V., Kuhn, K. M., Lomazov, I. R., Rothman, V. L., Tuszynski, G. P., Iozzo, R. V., Swenson, T. L., Fisher, E. A., and Williams, K. J. (1997) The syndecan family of proteoglycans: novel receptors mediating internalization of atherogenic lipoproteins *in vitro*. *J. Clin. Invest.* **100,** 1611–1622.
28. Chen, T. R., Drabkowski, D., Hay, R. J., Macy, M., and Peterson, W., Jr. (1987) WiDr is a derivative of another colon adenocarcinoma cell line, HT-29. *Cancer Genet. Cytogenet.* **27,** 125–134.
29. Iozzo, R. V. (1984) Biosynthesis of heparan sulfate proteoglycan by human colon carcinoma cells and its localization at the cell surface. *J. Cell Biol.* **99,** 403–417.
30. Iozzo, R. V. (1987) Turnover of heparan sulfate proteoglycan in human colon carcinoma cells. A quantitative biochemical and autoradiographic study. *J. Biol. Chem.* **262,** 1888–1900.
31. Fuki, I. V., Iozzo, R. V., and Williams, K. J. (1996) Perlecan heparan sulfate proteoglycan (HSPG): a novel, distinct receptor for atherogenic lipoproteins. *Circulation* **94(suppl. I)**, 698–699 (abstract).
32. Fuki, I. V., Iozzo, R. V., and Williams, K. J. (2000). Perlecan heparan sulfate proteoglycan: a novel receptor that mediates a distinct pathway for ligand catabolism. *J. Biol. Chem.* **275,** 25742–25750.
33. Liu, W., Litwack, E. D., Stanley, M. J., Langford, J. K., Lander, A. D., and Sanderson, R. D. (1998) Heparan sulfate proteoglycans as adhesive and anti-invasive molecules. Syndecans and glypican have distinct functions. *J. Biol. Chem.* **273,** 22825–22832.
34. Williams, K. J., Fless, G. M., Petrie, K. A., Snyder, M. L., Brocia, R. W., and Swenson, T. L. (1992) Mechanisms by which lipoprotein lipase alters cellular metabolism of lipoprotein(a), low density lipoprotein, and nascent lipoproteins. Roles for low density lipoprotein receptors and heparan sulfate proteoglycans. *J. Biol. Chem.* **267,** 13284–13292.
35. Socorro, L., Green, C. C., and Jackson, R. L. (1985) Preparation of a homogeneous and stable form of bovine milk lipoprotein lipase. *Prep. Biochem.* **15**, 133–143.
36. Saxena, U., Witte, L. D., and Goldberg, I. J. (1989) Release of endothelial cell lipoprotein lipase by plasma lipoproteins and free fatty acids. *J. Biol. Chem.* **264,** 4349–4355.
37. Rumsey, S. C., Obunike, J. C., Arad, Y., Deckelbaum, R. J., and Goldberg, I. J. (1992) Lipoprotein lipase-mediated uptake and degradation of low density lipoproteins by fibroblasts and macrophages. *J. Clin. Invest.* **90,** 1504–1512.
38. Ji, Z. S., Fazio, S., Lee, Y. L., and Mahley, R. W. (1994) Secretion-capture role for apolipoprotein E in remnant lipoprotein metabolism involving cell surface heparan sulfate proteoglycans. *J. Biol. Chem.* **269,** 2764–2772.
39. Ji, Z. S., Lauer, S. J., Fazio, S., Bensadoun, A., Taylor, J. M., and Mahley, R. W. (1994) Enhanced binding and uptake of remnant lipoproteins by hepatic lipase-secreting hepatoma cells in culture. *J. Biol. Chem.* **269,** 13429–13436.
40. Eisenberg, S., Sehayek, E., Olivecrona, T., and Vlodavsky, I. (1992) Lipoprotein lipase enhances binding of lipoproteins to heparan sulfate on cell surfaces and extracellular matrix. *J. Clin. Invest.* **90,** 2013–2021.
41. Mulder, M., Lombardi, P., Jansen, H., van Berkel, T. J., Frants, R. R., and Havekes, L. M. (1993) Low density lipoprotein receptor internalizes low density and very low density lipoproteins that are bound to heparan sulfate proteoglycans via lipoprotein lipase. *J. Biol. Chem.* **268,** 9369–9375.

42. Esko, J. D. (1991) Genetic analysis of proteoglycan structure, function and metabolism. *Curr. Opin. Cell Biol.* **3,** 805–816.
42a. Higazi, A. A-R., Nassar, T., Ganz, T., Rader, D.J. Udassin, R., Bdeir, K., et al. (2000) The α-defensins stimulate proteoglycan-dependent catabolism of low-density lipoprotein by vascular cells: a new class of inflammatory apolipoprotein and a possible contributor to atherogenesis. *Blood* **96**:1393–1398.
43. Humphries, D. E. and Silbert, J. E. (1988) Chlorate: a reversible inhibitor of proteoglycan sulfation. *Biochem. Biophys. Res. Commun.* **154,** 365–371.
44. Sehayek, E., Olivecrona, T., Bengtsson-Olivecrona, G., Vlodavsky, I., Levkovitz, H, Avner, R, and Eisenberg, S. (1995) Binding to heparan sulfate is a major event during catabolism of lipoprotein lipase by HepG2 and other cell cultures. *Atherosclerosis* **114,** 1–8.
45. Seo, T. and St.Clair, R. W. (1997) Heparan sulfate proteoglycans mediate internalization and degradation of β-VLDL and promote cholesterol accumulation by pigeon macrophages. *J. Lipid Res.* **38,** 765–779.
46. Corder, E. H., Saunders, A. M., Strittmatter, W. J., Schmechel, D. E., Gaskell, P. C., Small, G. W., Roses, A. D., Haines, J. L., and Pericak-Vance, M. A. (1993) Gene dose of apolipoprotein E type 4 allele and the risk of Alzheimer's disease in late onset families. *Science* **261,** 921–923.
47. Ji, Z. S., Pitas, R. E., and Mahley, R. W. (1998) Differential cellular accumulation/retention of apolipoprotein E mediated by cell surface heparan sulfate proteoglycans. Apolipoproteins E3 and E2 greater than E4. *J. Biol. Chem.* **273,** 13,452–13,460.
48. Pitas, R. E., Ji, Z.-S., Fuki, I. V., Williams, K. J., and Mahley, R. W. (1998) ApoE isoforms and neurite outgrowth: possible role of heparan sulfate proteoglycans. *Neurobiol. Aging* **19(4S),** S222 (abstract).
49. Lund-Katz, S., Ibdah, J. A., Letizia, J. Y., Thomas, M. T., and Phillips, M. C. (1988) A $^{13}$C NMR characterization of lysine residues in apolipoprotein B and their role in binding to the low density lipoprotein receptor. *J. Biol. Chem.* **263,** 13,831–13,838.
50. Goldberg, I. J., Wagner, W. D., Pang, L., Paka, L., Curtiss, L. K., DeLozier, J. A., Shelness, G. S., Young, C. S. H., and Pillarisetti, S. (1998) The $NH_2$-terminal region of apolipoprotein B is sufficient for lipoprotein association with glycosaminoglycans. *J. Biol. Chem.* **273,** 35,355–35,361.
51. Burgess, J. W., Gould, D. R., and Marcel, Y. L. (1998) The HepG2 extracellular matrix contains separate heparinase- and lipid-releasable pools of ApoE. Implications for hepatic lipoprotein metabolism. *J. Biol. Chem.* **273,** 5645–5654.
52. Koo, C., Wernette-Hammond, M. E., and Innerarity, T. L. (1986) Uptake of canine β-very low density lipoproteins by mouse peritoneal macrophages in mediated by a low density lipoprotein receptor. *J. Biol. Chem.* **261,** 11194–11201.
53. Beisiegel, U., Schneider, W. J., Goldstein, J. L., Anderson, R. G. W., and Brown, M. S. (1981) Monoclonal antibodies to the low density lipoprotein receptor as probes for study of receptor-mediated endocytosis and the genetics of familial hypercholesterolemia. *J. Biol. Chem.* **256,** 11,923–11,931.
54. van Driel, I. R., Brown, M. S., and Goldstein, J. L. (1989) Stoichiometric binding of low density lipoprotein (LDL) and monoclonal antibodies to LDL receptors in a solid phase assay. *J. Biol. Chem.* **264**, 9533–9538.
55. Vijayagopal, P., Figueroa, J. E., Guo, Q., Fontenot, J. D., and Tao, Z. (1996) Marked alteration of proteoglycan metabolism in cholesterol-enriched human arterial smooth muscle cells. *Biochem. J.* **315**, 995–1000.

56. Tabas, I., Weiland, D. A., and Tall, A. R. (1986) Inhibition of acyl coenzyme A:cholesterol acyl transferase in J774 macrophages enhances down-regulation of the low density lipoprotein receptor and 3-hydroxy-3-methylglutaryl-coenzyme A reductase and prevents low density lipoprotein-induced cholesterol accumulation. *J. Biol. Chem.* **261,** 3147–3155.
57. Dueland, S., Trawick, J. D., Nenseter, M. S., MacPhee, A. A., and Davis, R. A. (1992) Expression of 7 $\alpha$-hydroxylase in non-hepatic cells results in liver phenotypic resistance of the low density lipoprotein receptor to cholesterol repression. *J. Biol. Chem.* **267,** 22,695–22,698.
58. Sege, R. D., Kozarsky, K. F., and Krieger, M. (1986) Characterization of a family of gamma-ray-induced CHO mutants demonstrates that the ldlA locus is diploid and encodes the low- density lipoprotein receptor. *Mol. Cell Biol.* **6,** 3268–3277.
59. Ishibashi, S., Brown, M. S., Goldstein, J. L., Gerard, R. D., Hammer, R. E., and Herz, J. (1993) Hypercholesterolemia in low density lipoprotein receptor knockout mice and its reversal by adenovirus-mediated gene delivery. *J. Clin. Invest.* **92,** 883–893.
60. Williams, S. E., Ashcom, J. D., Argraves, W. S., and Strickland, D. K. (1992) A novel mechanism for controlling the activity of $\alpha_2$ macroglobulin receptor/low density lipoprotein receptor-related protein. Multiple regulatory sites for 39-kDa receptor-associated protein. *J. Biol. Chem.* **267,** 9035–9040.
61. Strickland, D. K., Ashcom, J. D., Williams, S., Battey, F., Behre, E., McTigue, K., Battey, J. F., and Argraves, W. S. (1991) Primary structure of $\alpha_2$ macroglobulin receptor-associated protein. Human homologue of a Heymann nephritis antigen. *J. Biol. Chem.* **266,** 13364–13369.
62. Ji, Z-S. and Mahley, R. W. (1994) Lactoferrin binding to heparan sulfate proteoglycans and the LDL receptor-related protein. Further evidence supporting the importance of direct binding of remnant lipoproteins to HSPG. *Arterioscler. Thromb.* **14,** 2025–2032.
63. Vassiliou, G. and Stanley, K. K. (1994) Exogenous receptor-associated protein binds to two distinct sites on human fibroblasts but does not bind to the glycosaminoglycan residues of heparan sulfate proteoglycans. *J. Biol. Chem.* **269,** 15,172–15,178.
64. Lyon, M., Deakin, J. A., and Gallagher, J. T. (1994) Liver heparan sulfate structure. A novel molecular design. *J. Biol. Chem.* **269,** 11,208–11,215.
65. Willnow, T. E. and Herz, J. (1994) Genetic deficiency in low density lipoprotein receptor-related protein confers cellular resistance to Pseudomonas exotoxin A. Evidence that this protein is required for uptake and degradation of multiple ligands. *J. Cell Sci.* **107,** 719–726.
66. FitzGerald, D. J., Fryling, C. M., Zdanovsky, A., Saelinger, C. B., Kounnas, M., Winkles, J. A., Strickland, D., and Leppla, S. (1995) Pseudomonas exotoxin-mediated selection yields cells with altered expression of low-density lipoprotein receptor-related protein. *J. Cell Biol.* **129,** 1533–1541.
67. Sehayek, E., Wang, X. X., Vlodavsky, I., Avner, R., Levkovitz, H., Olivecrona, T., Olivecrona, G., Willnow, T. E., Herz, J., and Eisenberg, S. (1996) Heparan sulfate-dependent and low density lipoprotein receptor-related protein-dependent catabolic pathways for lipoprotein lipase in mouse embryonic fibroblasts. *Isr. J. Med. Sci.* **32,** 449–454.
68. Rohlmann, A., Gotthardt, M., Hammer, R. E., and Herz, J. (1998) Inducible inactivation of hepatic LRP gene by cre-mediated recombination confirms role of LRP in clearance of chylomicron remnants. *J. Clin. Invest.* **101,** 689–695.
69. Chappell, D. A., Fry, G. L., Waknitz, M. A., Iverius, P-H., Williams, S. E., and Strickland, D. K. (1992) The low density lipoprotein receptor-related protein/$\alpha_2$ macroglobulin receptor binds and mediates catabolism of bovine lipoprotein lipase. *J. Biol. Chem.* **267,** 25,764–25,767.
70. Mahley, R. W., Ji, Z. S., Brecht, W. J., Miranda, R. D., and He, D. (1994) Role of heparan sulfate proteoglycans and the LDL receptor-related protein in remnant lipoprotein metabolism. *Ann. N.Y. Acad. Sci.* **737,** 39-52.

71. Beisiegel, U. (1995) Receptors for triglyceride-rich lipoproteins and their role in lipoprotein metabolism. *Curr. Opin. Lipidol.* **6,** 117–122.
72. Brown, M. S. and Goldstein, J. L. (1986) A receptor-mediated pathway for cholesterol homeostasis. *Science* **232,** 34–47.
73. Weigel, P. H. and Oka, J. A. (1982) Endocytosis and degradation mediated by the asialoglycoprotein receptor in isolated rat hepatocytes. *J. Biol. Chem.* **257,** 1201–1207.
74. Akiyama, T., Ishida, J., Nakagawa, S., Ogawara, H., Watanabe, S., Itoh, N., Shibuya, M., and Fukami, Y. (1987) Genistein, a specific inhibitor of tyrosine-specific protein kinases. *J. Biol. Chem.* **262,** 5592–5595.
75. Sampath, P. and Pollard, T.D. (1991) Effects of cytochalasin, phalloidin, and pH on the elongation of actin filaments. *Biochemistry* **19(30),** 1973–1980.
76. Luton, F., Buferne, M., Davoust, J., Schmitt-Verhulst, A.M., and Boyer, C. (1994) Evidence for protein tyrosine kinase involvement in ligand- induced TCR/CD3 internalization and surface redistribution. *J. Immunol.* **153,** 63–72.
77. Khoo, J. C., Miller, E., McLoughlin, P., and Steinberg, D. (1988) Enhanced macrophage uptake of low density lipoprotein after self-aggregation. *Arteriosclerosis* **8,** 348–358.
78. Yanagishita, M. (1992) Glycosylphosphatidylinositol-anchored and core protein- intercalated heparan sulfate proteoglycans in rat ovarian granulosa cells have distinct secretory, endocytotic, and intracellular degradative pathways. *J. Biol. Chem.* **267,** 9505–9511.
79. Fuki, I. V. and Williams, K. J. (1998) Direct internalization of atherogenic lipoproteins by syndecan heparan sulfate proteoglycans is a multi-step process independent of coated pits. *Circulation* **98 (suppl.),** I–109 (abstract).
80. Fuki, I. V., Meyer, M. E., and Williams, K. J. (1999). Transmembrane and cytoplasmic domains of syndecan mediate a multi-step endocytic pathway involving detergent-insoluble membrane rafts. *Biochem. J.* **351,** 607–612.
81. Fuki, I. V., Mazany, K. D., and Williams, K. J. (1999) Molecular determinants of syndecan-mediated endocytosis, a novel, multi-step pathway independent of coated pits. *Circulation* **100 (suppl.),** I–706 (abstract).
82. Williams, K. J., Petrie, K. A., Brocia, R. W., and Swenson, T. L. (1991) Lipoprotein lipase modulates net secretory output of apolipoprotein B in vitro. A possible pathophysiologic explanation for familial combined hyperlipidemia. *J. Clin. Invest.* **88,** 1300–1306.
83. Markwell, M. A. K., Haas, S. M., Bieber, L. L., and Tolbert, N. E. (1978) A modification of the Lowry procedure to simplify protein determination in membrane and lipoprotein samples. *Anal. Biochem.* **87,** 206–210.
84. Davis, W., Harrison, P. T., Hutchinson, M. J., and Allen, J. M. (1995) Two distinct regions of Fcγ RI initiate separate signalling pathways involved in endocytosis and phagocytosis. *EMBO J.* **14,** 432–441.
85. Jentoft, N. and Dearborn, D. G. (1983) Protein labeling by reductive alkylation. *Methods Enzymol.* **91,** 570–579.
86. Rapraeger, A., Jalkanen, M., and Bernfield, M. (1986) Cell surface proteoglycan associates with the cytoskeleton at the basolateral cell surface of mouse mammary epithelial cells. *J. Cell Biol.* **103,** 2683–2696.
87. Miettinen, H. M. and Jalkanen, M. (1994) The cytoplasmic domain of syndecan-1 is not required for association with Triton X-100-insoluble material. *J. Cell Sci.* **107,** 1571–1581.
88. Carey, D. J., Bendt, K. M., and Stahl, R. C. (1996) The cytoplasmic domain of syndecan-1 is required for cytoskeleton association but not detergent insolubility. Identification of essential cytoplasmic domain residues. *J. Biol. Chem.* **271,** 15,253–15,260.

# 46

# Hyaluronan and Hyaluronan-Binding Proteins

## *Probes for Specific Detection*

**Bryan P. Toole, Qin Yu, and Charles B. Underhill**

## 1. Introduction

Hyaluronan is a uniformly repetitive, linear glycosaminoglycan (GAG) composed of disaccharides of glucuronic acid and N-acetylglucosamine: [-β(1,4)-GlcUA-β(*1,3*)-GlcNAc-]$_n$. The polymer usually consists of 2,000–25,000 disaccharides, giving rise to molecular weights ranging from $10^6$ to $10^7$ dalton. Hyaluronan has unusual physical and biochemical properties, and fulfills several distinct physiological functions that contribute both to structural properties of tissues and to cell behavior during formation or remodeling of tissues ***(1–4)***. First, hyaluronan contributes directly to tissue homeostasis and biomechanics due to its unique charge characteristics and biophysical properties. Second, interactions of hyaluronan with link proteins and proteoglycans are of fundamental importance to the structural integrity of extracellular and pericellular matrices. Third, interactions with cell-surface hyaluronan receptors mediate significant influences on cell behavior during morphogenesis, tissue remodeling, inflammation, and diseases such as cancer and atherosclerosis ***(1–4)***.

The potential importance of hyaluronan in development, tissue remodeling and disease has led to the need for specific detection methods. However, until the mid-1980s the only method available for detection of hyaluronan was indirect, that is, by using specific hyaluronidases together with nonspecific staining methods. Subsequently, specific hyaluronan-binding proteins were adapted for use as morphological probes ***(5–7)***. The most commonly employed probe has been labeled hyaluronan-binding polypeptides prepared from the cartilage proteoglycan complex of aggrecan, link protein, and hyaluronan ***(5,8–10)***. This method has also been applied to electron microscopy ***(11)***. However, the most widespread use of this preparation has been for localizing hyaluronan in tissue sections by light microscopy, as described herein.

From: *Methods in Molecular Biology, Vol. 171: Proteoglycan Protocols*
Edited by: R. V. Iozzo © Humana Press Inc., Totowa, NJ

Many functions of hyaluronan are mediated by hyaluronan-binding proteins ***(2–4)***. The identification and molecular characterization of hyaluronan-binding proteins has been facilitated by the use of biotinylated hyaluronan (bHA) as a probe in transfer blots. For example this method has been used to demonstrate the hyaluronan-binding ability and specificity, and for dissecting the binding domains, of several hyaluronan-binding proteins ***(12–15)***. Although this method can also be adapted to morphological localization ***(16)***, its major use has been biochemical; this latter application is described herein.

Binding of hyaluronan to the surface of cells is mediated by hyaluronan receptors, the major receptor being CD44 ***(2–4)***. However the presence of CD44 on the cell surface does not necessarily indicate that a given cell will bind hyaluronan, since many factors contribute to binding—for example, glycosylation, alternative splicing, and clustering within the cell membrane ***(17,18)***. Consequently, convenient and reliable means of measuring hyaluronan binding to cells have been developed, the most common of which involves the use of fluorescein-tagged hyaluronan followed by analysis by flow cytometry; this application is also described herein.

## 2. Materials

### 2.1. Preparation of Biotinylated Hyaluronan-Binding Protein (bHABP)

1. Bovine nasal cartilage (Pel-Freez, Rogers, AR).
2. Surform pocket plane (Stanley Tools); cheesecloth (prewashed); Whatman no. 1 filter paper (Whatman, Clinton, NJ).
3. 4.0 *M* guanidium HCl, 0.5 *M* Na acetate, pH 5.8.
4. HEPES buffer: 0.1 *M* HEPES containing 0.1 *M* Na acetate, pH 7.3.
5. Trypsin (type III; Sigma, St. Loius, MO).
6. Soybean trypsin inhibitor (Type I-S; Sigma).
7. Sulfo-NHS-LC-Biotin (EZ-Link; Pierce, Rockford, IL).
8. Hyaluronan-Sepharose ***(19,20)***: hyaluronan (100 mg; Seikagaku, Falmouth, MA) is partially digested with testicular hyaluronidase (2 mg [type VIII, Sigma], in 30 mL of 0.05 *M* Na acetate/0.15 *M* NaCl, pH 5.0, at room temperature for 3 h), boiled for 10 min, and then mixed with 10 mL of EAH-Sepharose 4B (Pharmacia Biotech, Uppsala, Sweden) and 250 mg of 1-ethyl-3-(3-dimethylaminopropyl)-carbodiimide HCl (Sigma). The mixture is adjusted to pH 5 with 0.1 *M* HCl and incubated at room temperature for 24 h, after which 2 mL of acetic acid is added and the mixture is reincubated for 6 h. The gel is then washed with 1 *M* NaCl, 0.05 *M* formate buffer (pH 3), and distilled water.

### 2.2. Localization of Hyaluronan in Tissue Sections

1. 4.0% formaldehyde, dissolved in PBS-A.
2. PBS-A: calcium and magnesium-free phosphate-buffered saline.
3. 3% hydrogen peroxide, dissolved in methanol.
3. 2 μg/mL bHABP, dissolved in 10% calf serum in PBS-A (filtered through 0.45-μm filter before use).
5. Vectastain Elite ABC kit (Vector Laboratories, Burlingame, CA).
6. 5 U/mL *Streptomyces* (Calbiochem, San Diego, CA), dissolved in 10% calf serum in PBS-A.
7. 200 μg/mL hyaluronan (Seikagaku) in 10% calf serum in PBS-A; stock solution: 5 mg/mL in distilled water; store at –20°C.

8. 1 mg/mL hyaluronan oligomers in 10% calf serum in PBS-A. For preparation of hyaluronan oligomers ***(20)***, hyaluronan is first digested with testicular hyaluronidase (2 mg [type VIII, Sigma], in 30 mL of 0.05 *M* Na acetate/0.15 *M* NaCl, pH 5.0, at 37°C for 3–24 h). The digest is boiled for 10 min, then passed over a column of Sephadex G-50 (Pharmacia; 1.5 × 250 cm) in ammonium acetate buffer, pH 5.0. Fractions are assayed for total uronic acid ***(21)*** and terminal hexosamine ***(22)***, and those fractions with uronic acid to hexosamine ratios of 5–10 are pooled.

### 2.3. Preparation of Biotinylated Hyaluronan (bHA)

1. 5 mg/mL hyaluronan (Healon, Pharmacia, or purest preparation from Seikagaku) in PBS-A.
2. MES buffer: 0.1 *M* 2-N-morpholino ethanesulfonic acid (Sigma), pH 5.5.
3. 50 m*M* ImmunoPure biotin-LC-hydrazide (Pierce), freshly dissolved in dimethyl sulfoxide.
4. EDC buffer: 100 mg/mL 1,ethyl-3-[3-dimethylaminopropyl] carbodiimide hydrochloride (Pierce) in 0.1 *M* MES buffer, pH 5.5.

### 2.4. Detection of Hyaluronan-Binding Proteins in Blots

1. Routine supplies for sodium dodecyl sulfate polyacrylamide gel electrophoresis (SDS-PAGE).
2. Hybond-ECL nitrocellulose membranes (Amersham, Arlington Heights, IL).
3. Transfer buffer: 0.025 m*M* Tris, 0.19 *M* glycine, 10% methanol, 0.005% SDS, pH 8.3.
4. TTBS: 0.2% Tween-20, 0.05 *M* Tris-HCl, 0.15 *M* NaCl pH 8.0.
5. Blocking buffer: 20% calf serum (or 1% BSA, 1% nonfat milk) in TTBS.
6. TBS: 0.05 *M* Tris-HCl, 0.15 *M* NaCl, pH 8.0.
7. 10% hydrogen peroxide in TBS.
8. 10 μg/mL bHA in TTBS.
9. 200 μg/mL hyaluronan (Seikagaku) in TTBS; stock solution: 5 mg/mL in distilled water; store at –20°C.
10. 1 mg/mL hyaluronan oligomers in TTBS (*see* **Subheading 2.2.**).
11. 1/1000 dilution of Streptavidin-horseradish peroxidase (Amersham) in TTBS.
12. ECL Western blot detection reagents and Hyperfilm-ECL (Amersham).

### 2.5. Preparation of Fluorescein-Labeled Hyaluronan (Fl-HA)

1. 5 mg/mL hyaluronan (Healon, Pharmacia; or purest preparation from Seikagaku) in water.
2. 3.8 mg/mL fluoresceinamine (Aldrich, Milwaukee, WI) in methanol.
3. 46 m*M* cyclohexyl isocyanide (Aldrich) in methanol.
4. 250 m*M* acetaldehyde (Aldrich) in water.

## 3. Methods

### 3.1. Specific Localization of Hyaluronan in Tissue Sections (8)

#### 3.1.1. Preparation of Biotinylated Hyaluronan-Binding Protein (bHABP) (see Note 1)

1. Shred bovine nasal cartilage with Surform pocket plane (or grind under liquid nitrogen in Wiley mill); mix cartilage with 10 volumes of 4.0 *M* guanidium HCl, 0.5 *M* Na acetate, pH 5.8, and shake overnight at 4°C.
2. Pour mixture through several layers of cheesecloth, then centrifuge at 10,000*g*, 45 min, 4°C; filter supernatant through filter paper on Buchner funnel; dialyze supernatant thoroughly against running tap water followed by distilled water; lyophilize for storage.
3. Dissolve 2.0 g of lyophilized cartilage by stirring in 50 mL of 0.1 *M* HEPES buffer at 4°C overnight (or add appropriate amount of solid HEPES and Na acetate to solution if

lyophilization is omitted); add 1.0 mg of trypsin and incubate for 2 h at 37°C; add 1.0 mg of soybean trypsin inhibitor.

4. Adjust pH to 8.0 with NaOH; add 0.1 mg of sulfo-NHS-LC-biotin per milligram protein and incubate for 1–2 h at room temperature.
5. Dialyze preparation against 4.0 *M* guanidium HCl, 0.5 *M* Na acetate, pH 5.8; mix with 100 mL of hyaluronan-Sepharose (*see* **Subheading 2.1.**) equilibrated with the same buffer; dialyze mixture against 9 volumes of distilled water at 4°C overnight on a rotary shaker table.
6. Wash hyaluronan-Sepharose thoroughly with 3.0 *M* NaCl, then elute bHABP with 4.0 *M* guanidine HCl, 0.5 *M* Na acetate, pH 5.8; dialyze preparation against 0.15 *M* NaCl.
7. Aliquot and store at -20°C in 50/50 glycerol and 0.15 *M* NaCl. (*See* **Note 2**).

### 3.1.2. Localization of Hyaluronan in Tissue Sections

1. Fix tissue in 4.0% formaldehyde in PBS-A for 2 h at room temperature; wash twice with PBS-A; dehydrate through 30%, 70%, 95%, and 100% ethanol, then xylene; embed in paraffin (*see* **Note 3**).
2. Cut sections at 8–10 µm; incubate sections with 3% hydrogen peroxide in methanol for 1 h to inactivate endogenous peroxidases; rinse twice with distilled water; equilibrate in PBS-A for 5 min.
3. Incubate sections for 1 h with 2 µg/mL bHABP dissolved in PBS-A containing 10% calf serum. Wash sections 5 times for 1 min each in PBS-A.
4. Use the Vectastain Elite ABC kit for detection of the bHABP.
5. As controls for specificity, pretreat sections with 5 U/mL *Streptomyces* hyaluronidase in PBS-A at 37°C for 30 min; or pretreat with 200 µg/mL hyaluronan or 1 mg/mL hyaluronan oligomers (*see* **Note 4**) for 30 min at room temperature, in PBS-A containing 10% calf serum, then incubate with 4 µg/mL bHABP, together with competing agent, as above.

## 3.2. Specific Detection of Hyaluronan-Binding Proteins in Transfer Blots (16)

### 3.2.1. Preparation of Biotinylated Hyaluronan (bHA)

1. Dissolve hyaluronan in PBS-A at 5 mg/mL, and dialyze against 0.1 *M* MES buffer overnight at 4°C (*see* **Note 5**).
2. Add 50 m*M* biotin-LC-hydrazide, freshly dissolved in dimethyl sulfoxide (DMSO), to give a final concentration of 1 m*M* (*see* **Note 6**).
3. Add freshly prepared EDC buffer to give a final concentration of 10 m*M* EDC and stir overnight at room temperature.
4. Dialyze against PBS-A at 4°C for 2 days.
5. Remove any precipitate that forms during the reaction by centrifugation. Aliquot and store at –20°C for no more than 6 months (it is preferable to make the bHA preparations in small batches to avoid extended storage).

### 3.2.2. Detection of Hyaluronan-Binding Proteins in Blots

1. Separate proteins (10–20 µg per lane for crude extracts; 0.1–0.5 µg per lane for purified proteins) by 10% SDS-PAGE under reducing (0.1 *M* DTT) or nonreducing conditions.
2. Transfer proteins to Hybond-ECL nitrocellulose membranes in transfer buffer at 300 mA (constant current) for 1.5 h at 4°C. After transfer, wash the membranes with TTBS, then block with 20% calf serum (or 1% BSA, 1% nonfat milk) in TTBS at 37°C for 1 h, and wash 3 more times with TTBS. If interference from endogenous peroxidase activity occurs, treat the membranes with 10% hydrogen peroxide in TBS at room temperature for 15 min and then wash twice with TBS, before blocking with 20% calf serum in TTBS.

3. Incubate the membranes with 10 μg/mL bHA in TTBS at room temperature for 2 h. To establish specificity, preincubate parallel membranes (or strips of membrane) with 200 μg/mL unlabeled hyaluronan or 1 mg/mL hyaluronan oligomers (*see* **Note 4**) in TTBS at room temperature for 1 h. The bHA is then added, together with unlabeled competing reagent, and the mixture incubated for 2 h more.
4. Wash the membranes 3–4 times with TTBS for 5–10 min each, and treat with Streptavidin-peroxidase conjugate in TTBS (1/1000 dilution) for 30 min at room temperature. Wash 3 times with TTBS for 5 min each, incubate the membranes with ECL Western blot detection reagents for 1 min, and expose using Hyperfilm-ECL (*see* **Note 7**).

## 3.3. Measurement of Hyaluronan Binding to Cell Surface Receptors by Flow Cytometry (23)

### 3.3.1. Preparation of Fluorescein-Labeled Hyaluronan

1. Dissolve 3.8 mg of fluoresceinamine in 1 mL of methanol; add dropwise to hyaluronan solution (~50 mg in 10 mL of water) with constant stirring (*see* **Note 5**).
2. Add 1.5 mL of 46 m*M* cyclohexyl isocyanide in methanol and 0.5 mL of 250 m*M* acetaldehyde and stir for 1 h.
3. Add 3 volumes denatured alcohol, centrifuge, and wash precipitate several times with alcohol; dry under vacuum (*see* **Note 8**).

### 3.3.2. Measurement of Hyaluronan Binding by Flow Cytometry

1. Detach cells from tissue culture plates with EDTA, wash with media and then with PBS-A.
2. Add 5–50 μg/mL fluorescein-labeled hyaluronan; incubate on ice for 2 h.
3. Wash 3 times with PBS-A; resuspend in PBS-A; analyze in FACScan (*see* **ref. 24**).
4. Controls for specific binding include pre-incubation with antibody to CD44, unlabeled hyaluronan or hyaluronan oligomers (*see* **Note 4**).

## 4. Notes

1. The bHABP is prepared from the proteoglycan complex of cartilage, a ternary complex containing proteoglycan (aggrecan), link protein, and hyaluronan. Aggrecan and link protein contain specific hyaluronan-binding domains that bind the three components together. In the method described, the complex is first extracted in 4 *M* guanidinium HCl, conditions that dissociate the components of the complex. The complex is then reassociated by dialysis against water and treated with trypsin. Trypsin treatment is performed in the associated state, so as to protect the hyaluronan-binding sites in link protein and aggrecan. The resulting complex, that is, link protein plus the hyaluronan-binding domain of aggrecan bound to hyaluronan, is biotinylated while the hyaluronan-binding sites are still protected. The complex is dissociated again in 4 *M* guanidinium HCl and then biotinylated link protein plus hyaluronan-binding domain of aggrecan are purified by affinity chromatography on hyaluronan-Sepharose. Purified bHABP is now available from Seikagaku.
2. SDS-PAGE analysis of the final product shows two bands, one at ~70–80 kDa (the hyaluronan-binding domain of aggrecan) and one at ~43 kDa (link protein).
3. Other methods of processing and sectioning, (cryostat, plastic, etc.), may be used.
4. A higher concentration of oligomer than polymer is used for competition, since the former has a lower affinity for most hyaluronan-binding proteins *(25)*. Competition with chondroitin sulfate and heparin can also be used for comparison.
5. It is very important that pure hyaluronan be used for preparation of biotinylated and fluorescein-labeled hyaluronan.

6. We calculated the amount of biotin-LC-hydrazide to be used from the approximate molar concentration of carboxyl groups in the HA preparation, such that a maximum of 1 out of 10–20 carboxyl groups in the HA would be labeled. Thus, 80–90% of the carboxyl groups would remain unaltered. We have found that this degree of conjugation yields sufficient labeling for sensitive detection while preserving full reactivity of the bHA.
7. If high background occurs, perform each step in TTBS containing 10% calf serum. Sometimes biotin or streptavidin also causes nonspecific binding in the procedure. In such cases the sample should be preabsorbed, before reaction with bHA, by mixing the protein extract with biotin-agarose beads (Sigma; 3/1, vol/vol) for 20 min at room temperature, followed by removal of the beads by centrifugation. The supernatant is then mixed with Streptavidin-agarose beads (Sigma; 3/1, vol/vol) for 20 min, followed by removal of these beads.
8. Approximately 7% of the carboxyl groups of hyaluronan are labeled with fluorescein under these conditions.

## References


1. Hascall, V. C. and Laurent, T. C. (1999) Hyaluronan: structure and physical properties. Glycoforum: Science of hyaluronan. http://www.glycoforum.gr.jp
2. Knudson, W. and Knudson, C. B. (1999) The hyaluronan receptor, CD44. Glycoforum: Science of hyaluronan. http://www.glycoforum.gr.jp
3. Sherman, L., Sleeman, J., Herrlich, P., and Ponta, H. (1994) Hyaluronate receptors: key players in growth, differentiation, migration and tumor progression. *Curr. Opin. Cell Biol.* **6,** 726–733.
4. Toole, B. P. (2000) Hyaluronan, in *Proteoglycans: Structure, Biology, and Molecular Interactions* (Iozzo, R., ed.), Marcel Dekker, New York, NY, pp. 61–92.
5. Knudson, C. B. and Toole, B. P. (1985) Fluorescent morphological probe for hyaluronate. *J. Cell Biol.* **100,** 1753–1758.
6. Delpech, B., Bertrand, P., Maingonnat, C., Girard, N., and Chauzy, C. (1995) Enzyme-linked hyaluronectin: a unique reagent for hyaluronan assay and tissue location and for hyaluronidase activity detection. *Anal. Biochem.* **229,** 35–41.
7. Fenderson, B. A., Stamenkovic, I. and Aruffo, A. (1993) Localization of hyaluronan in mouse embryos during implantation, gastrulation and organogenesis. *Differentiation* **54,** 85–98.
8. Green, S. J., Tarone, G., and Underhill, C. B. (1988) Distribution of hyaluronate and hyaluronate receptors in the adult lung. *J. Cell Sci.* **90,** 145–156.
9. Alho, A. M. and Underhill, C. B. (1989) The hyaluronate receptor is preferentially expressed on proliferating epithelial cells. *J. Cell Biol.* **108,** 1557–1565.
10. Yu, Q. and Toole, B. P. (1997) Common pattern of CD44 isoforms is expressed in morphogenetically active epithelia. *Dev. Dyn.* **208,** 1–10.
11. Ripellino, J. A., Klinger, M. M., Margolis, R.U., and Margolis, R.K. (1985) The hyaluronic acid binding region as a specific probe for the localization of hyaluronic acid in tissue sections. *J. Histochem. Cytochem.* **33,** 1060–1066.
12. Yang, B., Yang, B. L., Savani, R. C., and Turley, E. A. (1994) Identification of a common hyaluronan binding motif in the hyaluronan binding proteins RHAMM, CD44 and link protein. *EMBO J.* **13,** 286–296.
13. Grammatikakis, N., Grammatikakis, A., Yoneda, M., Yu, Q., Banerjee, S. D., and Toole, B. P. (1995) A novel glycosaminoglycan-binding protein is the vertebrate homologue of the cell cycle control protein, Cdc37. *J. Biol. Chem.* **270,** 16198–16205.

14. Yang, B., Yang, B. L., and Goetinck, P. F. (1995) Biotinylated hyaluronic acid as a probe for identifying hyaluronic acid-binding proteins. *Anal. Biochem.* **228,** 299–306.
15. Melrose, J., Little, C. B., and Ghosh, P. (1998) Detection of aggregatable proteoglycan populations by affinity blotting using biotinylated hyaluronan. *Anal. Biochem.* **256,** 149–157.
16. Yu, Q. and Toole, B. P. (1995) Biotinylated hyaluronan as a probe for detection of binding proteins in cells and tissues. *BioTechniques* **19,** 122–129.
17. English, N. M., Lesley, J. F., and Hyman, R. (1998) Site-specific de-N-glycosylation of CD44 can activate hyaluronan binding, and CD44 activation states show distinct threshold densities for hyaluronan binding. *Cancer Res.* **58,** 3736–3742.
18. Kincade, P. W., Zheng, Z., Katoh, S., and Hanson, L. (1997) The importance of cellular environment to function of the CD44 matrix receptor. *Curr. Opin. Cell Biol.* **9,** 635–642.
19. Tengblad, A. (1979) Affinity chromatography on immobilized hyaluronate and its application to the isolation of hyaluronate binding proteins from cartilage. *Biochim. Biophys. Acta* **578,** 281–289.
20. Banerjee, S. D. and Toole, B. P. (1991) Monoclonal antibody to chick embryo hyaluronan-binding protein: changes in distribution of binding protein during early brain development. *Dev. Biol.* **146,** 186–197.
21. Bitter, T. and Muir, H. (1962) A modified uronic acid carbazole reaction. *Anal. Biochem.* **4,** 330–334.
22. Reissig, J. L., Strominger, J. L., and Leloir, L. R. (1955) A modified colorimetric method for the estimation of N-acetylamino sugars. *J. Biol. Chem.* **217,** 959–966.
23. Zeng, C., Toole, B. P., Kiney, S. D., Kuo, J. W., and Stamenkovic, I. (1998) Inhibition of tumor growth in vivo by hyaluronan oligomers. *Int. J. Cancer* **77,** 396–401.
24. Lesley, J., Kincade, P. W. and Hyman, R. (1993) Antibody-induced activation of the hyaluronan receptor function of CD44 requires multivalent binding by antibody. *Eur. J. Immunol.* **23,** 1902–1909.
25. Underhill, C. B., Chi-Rosso, G., and Toole, B. P. (1983). Effects of detergent solubilization on the hyaluronate-binding protein from membranes of simian virus 40-transformed 3T3 cells. *J. Biol. Chem.* **258,** 8086–8091.

# 47

# Novel Confocal-FRAP Analysis of Carbohydrate–Protein Interactions Within the Extracellular Matrix

**Philip Gribbon, Boon C. Heng, and Timothy E. Hardingham**

## 1. Introduction

Extracellular matrices (ECMs) contain a mixture of fibrillar and nonfibrillar macromolecular components, which interact through a range of covalent and noncovalent associations to form a composite structure *(1–3)*. It is the ECM that defines the architecture, the form, and the biomechanical properties of many tissues. In order to perform their functional roles, many of the important noncollagenous components of extracellular matrices are required to be immobilized or focally located within tissues *(4,5)*. Positioning of macromolecules within tissues occurs as a consequence of specific protein–protein and protein–carbohydrate interactions. Aggrecan, the large aggregating proteoglycan, is noncovalently associated with hyaluronan via its N-terminal domain and this association is further stabilised by link protein *(6)*. A large immobilized hydrated aggregate structure is formed, with up to 50 aggrecan molecules per hyaluronan chain, which gives articular cartilage the ability to resist the high compressive loads generated during joint articulation. Other examples of specific intermolecular associations include the binding of cell surface integrins to fibronectins and collagenous proteins (*7*) and growth factors to heparan sulphate (*8*). A variety of methods have been developed to investigate the affinity and specificity of these associations and their sensitivity to environmental factors such as pH and metal ion concentration. Many techniques for analyzing binding equilibria or kinetics require either the ligand or substrate to be linked to a solid support. As a consequence of derivitization and high local concentrations, binding equilibria can be significantly perturbed.

Confocal fluorescence recovery after photobleaching (confocal FRAP) offers a method for investigating the interactions of molecules at high concentrations, the formation of molecular networks, and the permeability of such networks to "reporter"

From: *Methods in Molecular Biology, Vol. 171: Proteoglycan Protocols*
Edited by: R. V. Iozzo © Humana Press Inc., Totowa, NJ

molecules of different sizes. Additionally, confocal FRAP can be used to quantify protein–protein binding and protein–carbohydrate binding under true equilibrium conditions with a minimum of chemical modifications ***(9–13)***.

Experimentally, confocal FRAP involves viewing a solution of a fluorescently labeled molecule with a confocal microscope, locally creating a bleached area in the solution, followed by observing the subsequent redistribution of fluorescence (*see* **Notes 1** and **2**). The bleaching can be achieved rapidly by brief high-power laser illumination. In simple solutions of monodisperse macromolecules, the measurement of the recovery of fluorescence yields a long time translational lateral self-diffusion coefficient. This describes the movement of the macromolecule through a matrix of similar macromolecules and can be used to investigate self-association and molecule entanglement at high concentration ***(11)***. The lateral tracer diffusion coefficient describes the movement of a macromolecule within a matrix composed of a high concentration of another macromolecule. Self and tracer diffusion coefficients are a function of the hydrodynamic friction caused by the solvent and the hindrances, steric or otherwise, offered by other components in the solution. If a macromolecule associates specifically with one or more components in a matrix, and its diffusion coefficient is sufficiently changed in the process, then confocal FRAP can be used to determine the affinity and specificity of the interaction ***(13)***.

The application of confocal FRAP to characterize the specific binding of a low-molecular-weight protein to a high-molecular-weight ligand will be discussed in detail here. The movement of a fluorescently labeled macromolecule A, diffusing within a matrix of an unlabeled macromolecule B, will be considered. The translational diffusion coefficients of A, B, and the complex AB are $D_A$, $D_B$ and $D_{AB}$, respectively. The binding equilibrium between A and B is defined as

$$A + B \Leftrightarrow AB \tag{1}$$

and at equilibrium, the dissociation constant ($K_d$) is given by:

$$K_d = \frac{[A][B]}{[AB]} \tag{2}$$

where [ ] represents concentration. In a confocal FRAP experiment only the fluorescence recovery of component A is measured. At equilibrium, A exchanges rapidly between bound and free states that have characteristic lifetimes of $\tau_{bound}$ and $\tau_{free}$ respectively. If the fluorescence recovery of A is observed over time $t$, where $t >> (\tau_{bound} + \tau_{free})$, then the redistribution of A results from diffusion during multiple free and bound phases. The measured diffusion coefficient ($D_m$) can then be defined as

$$D_m = [D_A \times \langle T_{free} \rangle] + \{D_{AB}\ [1 - \langle T_{free} \rangle]\} \tag{3}$$

where $\langle T_{free} \rangle$ is the fraction of time t that macromolecule A was free during time t. Since $t >> (t_{bound} + t_{free})$, then $\langle T_{free} \rangle \simeq A_{free}$, the fraction of free A at equilibrium. Similarly, if $D_A >> D_B$, then $D_B \simeq D_{AB}$ and **Eq. 3** can be re-expressed as

$$A_{free} = \frac{D_m - D_B}{D_A - D_B} \tag{4}$$

The expression for the inverse Scatchard plot is then

$$\frac{D_A - D_B}{D_m - D_B} = \frac{1}{n} + \frac{K_d}{n[B]_{bs}} \tag{5}$$

where $[B]_{bs}$ is the concentration of binding sites on B, and $n$ is the number of molecules of A bound per molecule of B. Values for $K_d$ are determined by measuring $D_m$ as a function of the concentration of substrate B. Magnitudes of $D_A$ and $D_B$ are determined independently.

Confocal FRAP can be used to characterise competitive binding. Consider an equilibrium between A and B, in the presence of a competitor, C:

$$A + B + C \Leftrightarrow AB + AC + BC \tag{6}$$

This equilibrium can be analyzed by confocal FRAP if (1) the translational diffusion coefficient of C, ($D_c$), is very different from that of B, and (2) binding between B and C is minimal, that is, [BC] = 0.

Confocal FRAP has recently been used to analyze the association between the G1 domain of aggrecan and hyaluronan. This interaction is ideally suited to analysis by this technique because (1) the interaction is noncovalent, (2) the diffusion coefficients of G1 and HA differ considerably, and (3) the diffusion coefficient of the G1-HA complex is similar to that of hyaluronan (HA) ***(14)***. The G1 domain of aggrecan was labeled with fluorescein isothiocyanate (FITC) at 2.5 mol FITC per mole of G1 (*see* **Subheading 3.1.1.**) and its diffusion coefficient in increasing concentrations of hyaluronan (800 kDa) at pH 7.4 was determined (*see* **Fig. 1**). The diffusion coefficient of the FITC-G1 was reduced with increasing hyaluronan concentration. Tracer diffusion studies on nonbinding FITC-G1, which had been reduced and alkylated, showed the reduction in FITC-G1 mobility was not due to the steric effect of the hyaluronan ***(14)***. The self-diffusion of fluoresceinamine (FA)-labeled hyaluronan (800 kDa) was then determined independently by conventional confocal FRAP. An inverse Scatchard plot was constructed from the FITC-G1-binding isotherm (*see* **Fig. 2**). This gave $K_d$ as $4 \times 10^{-8}$ *M*, and $n = 1$, (*see* **Fig. 2**) which agrees well with dissociation constants determined by solid-phase-type assays ***(15,16)***.

The confocal FRAP technique was used to determine the minimum length of hyaluronan oligosaccharide required for binding to a single G1 domain. The low molecular weight of the competitor oligosaccharides compared to the FITC-G1 and the high-molecular-weight hyaluronan makes this an ideal system for analysis by confocal FRAP. A mixture of FITC-G1 (5.5 µ*M*) and 800-kDa hyaluronan (12.5 m*M* of $HA_{10}$ binding sites) was prepared and >99% of the FITC-G1 was found to be bound to the hyaluronan. Hyaluronan oligosaccharides (21 m*M*) of length 4–14 monosaccharides were added ***(17)***, and the diffusion coefficient of the FITC-G1 was determined (*see* **Fig. 3**). It can be seen that hyaluronan oligosaccharides compete effectively with full-length hyaluronan and cause an increase in the diffusion coefficient of FITC-G1. The method shows that a minimum length of 10 hyaluronan monosaccharides is needed for maximal binding to FITC-G1 ***(15)***.

The confocal FRAP technique is a novel and direct method for investigating the many protein–protein and protein–carbohydrate interactions that contribute greatly to the organization and function of extracellular matrices. It can be applied over a

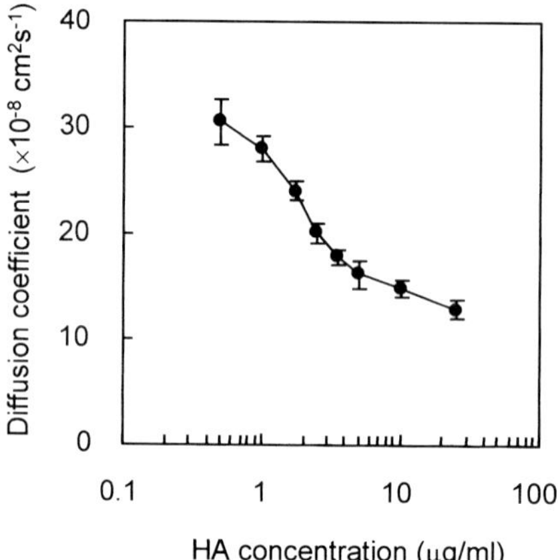


Fig. 1. Translational diffusion coefficient of FITC-G1 (80 μg/mL) as a function of increasing concentration of hyaluronan (800 kDa). The self-diffusion of G1 is $3.3 \times 10^{-8}$ cm$^2$/s. All measurements in PBS (pH 7.4) at 25°C

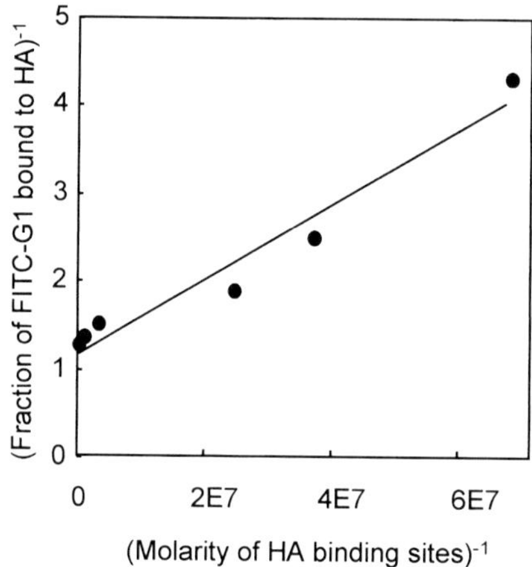


Fig. 2. Inverse Scatchard type plot of data from **Fig. 1.** Analysis of binding between FITC-G1 (80 μg/mL) and hyaluronan (800 kDa). The dissociation constant was calculated as $4 \times 10^{-8}$ *M* and $n = 1$ (*see* **Eq. 6**).

wide range of conditions, including high concentrations, and in the presence of multiple interacting components. Confocal FRAP also has the major advantage that interactions are analyzed under near-equilibrium conditions, which allows the analysis of the weak interactions that are potentially important in defining matrix properties.

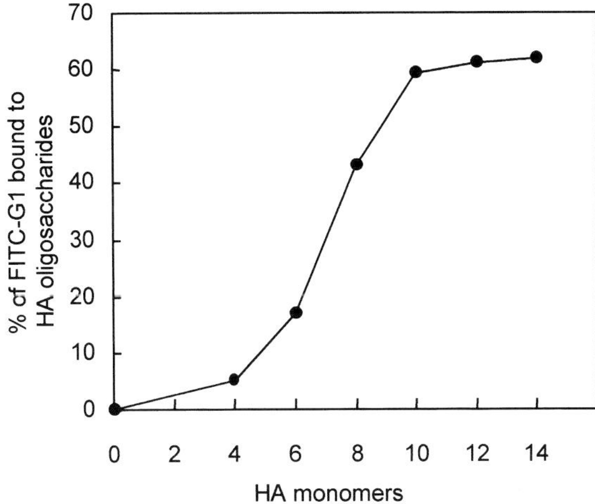


Fig. 3. Competition between high-molecular-weight hyaluronan (800 kDa, (25 µg/mL) and hyaluronan oligosaccharides (4mer to 14mer, 21 µ*M*) for binding to FITC-G1 (88 µg/mL). All measurements in PBS (pH 7.4) at 25°C.

## 2. Materials

### 2.1. Protein Labeling

1. Protein labeling buffer, 0.2 *M*: $NaHCO_3$ (1.48 g /100 mL), $Na_2CO_3$ (0.24 g/100 mL), pH 9.0. Filter 0.2 µm before use. Should be prepared fresh, keeps at 4°C for 2 wk.
2. PBS: 8 g of NaCl, 0.2 g of KCl, 1.44 g of $Na_2HPO_4$∞• $2H_2O$, 0.2 g of $KH_2PO_4$, made up to 1 L, pH 7.4. Filter 0.2 µm before use, keep at 4°C.

### 2.2. Photobleaching Experiments

1. Upright confocal laser scanner and fluorescence microscope: The authors use a 100-mW argon-ion laser with a MRC-1000 scanner (Bio-Rad, Hemel Hempstead, UK). Macros written in a simple programming language (MPL) control data acquisition and analysis. The majority of confocal microscopes include similar facilities for user application development.
2. Microscope objectives of numerical aperture < 0.5 (magnification: ↔10). A low numerical aperture gives a more uniform and parallel bleach volume.
3. Cavity microscope slides, 12-mm cavity diameter, 30 µL volume (Scientific Lab Supplies, Nottingham, UK).

## 3. Methods

### 3.1. Sample Preparation and Fluorescence Labeling

For studies of interactions involving carbohydrate components of the extra cellular matrix (ECM), glycosaminoglycans such as chondroitin sulfate and hyaluronan are commercially available (e.g., Sigma, Seikagaku) from several tissue sources (e.g., bacteria, cock's comb, umbilical cord). Commercially available FITC-labeled proteins are sometimes unsuitable for binding studies, as they can be excessively substituted

(>10 mol FITC per mole protein) and their binding and diffusion properties may differ significantly from those of the native protein. Many important matrix proteins, such as the G1 domain of aggrecan, are not commercially available and will have to be extracted from tissues using standard protocols ***(14)***. Functional binding properties of proteins can normally be preserved if they are labeled with a low degree of FITC substitution, as outlined next.

#### *3.1.2. Protein Labeling*

1. Dissolve protein at 0.5 mg/mL in carbonate-labeling buffer (pH 9.0) by mixing with gentle rotation for 24 h at 4°C.
2. Add 0.1 mg of FITC to 1 mL of labeling buffer. Immediately, add the cold FITC solution dropwise until the final molar ratio of FITC to protein is 20/1 and incubate overnight at 4°C with gentle rotation.
3. Remove unconjugated FITC by exhaustive dialysis against de-ionized water at 4°C and freeze-dry to constant mass.

#### *3.1.3. Preparation of Samples for Confocal FRAP Analysis*

FITC concentration in final solutions for analysis should be kept below 1 m*M* to avoid fluorescence self-quenching effects.

1. Make up solutions of labeled protein and matrix macromolecules by weight and equilibrate at 4°C for 48 h. In competition studies, add competitor species and equilibrate for a further 24 h at 4°C.
2. Pipet 30 uL of sample onto a cavity slide and seal under a 20-mm-diameter circular cover slip using nail polish; allow 10 min to dry. Place slide on microscope stage and allow to reach thermal equilibrium.

### *3.2. Microscope Setup and Data Acquisition*

In a confocal FRAP experiment a prebleach image is acquired, the sample is bleached, and a recovery image series is collected and analyzed. Scanning confocal microscopes excite fluorescence with a laser and detect emitted light with a photomultiplier (PMT). The essential elements of a protocol, based on a square bleach and a moments recovery analysis will be given (*10,11*). The analysis is not dependent on the shape of the bleach as long as it is symmetrical about the bleach centre.

1. Determine instrument settings for bleach and prebleach/recovery and save as separate setup configurations to be accessed by the bleaching software. Typically, for prebleach/recovery, scan time = 2 s, laser intensity = 1% of maximum, PMT gain = 1000 V, iris size = 75% of maximum. Typical bleach settings are scan time = 2 s, laser intensity = 100%, PMT gain = 0, iris size = minimum, (*see* **Notes 1** and **2**).
2. At the start of the bleaching and data acquisition program, a square region for bleaching is defined at the center of the field of view (*see* **Note 3**).
3. The microscope is set for recovery and one or more prebleach images are scanned, centered on the proposed bleach. Scan dimensions are 4↔ the bleach width, optical zoom = 1. Several prebleach images may be averaged to reduce noise.
4. The microscope switches to bleach configuration and sample bleaching is performed at 100% laser power.

5. The microscope resets to the prebleach setup, and a series of recovery images (up to 50) is collected. For slowly moving species or where there is strong binding, two or more images should be taken consecutively and averaged to reduce noise (*see* **Note 4**).
6. Final images should be taken at two long time points and compared to determine whether recovery approaches completeness. This is important, as incomplete recovery will indicate the presence of irreversible binding.
7. The image series is processed and analyzed. This operation is ideally carried out on a separate computer (*see* **Notes 1–4**).
8. Less than 0.01% of the total sample volume is typically included in the bleach. This allows a number of repeat experiments to be made on a single sample. The authors normally perform 6 measurements per sample.

### 3.3. Data Analysis

The data analysis strategy for confocal FRAP is based on the variance method of Kubitscheck et al. for determining the diffusion properties of monodisperse populations of macromolecules ***(10,11)*** (*see* **Note 5**). This method can be applied to binding studies, provided the validity of **Eq. 3** is ensured (*see* **Notes 6** and **7**). The redistribution of fluorescence is measured over the whole imaged area, which includes the area inside and immediately outside the bleached region (*see* **Notes 3** and **4**). The straightforward computing routine developed by the authors combines the image analysis tools in the MRC-1000 Microscope Programming Language (MPL) and the data analysis capabilities found in Excel.

## 4. Notes

1. Each recovery series is checked to ensure that no convective motion of the solution has occurred.
2. The effects of nonuniformity in illumination and detector response are corrected for by dividing each recovery image by the prebleach image.
3. Pixels contained within three bleach widths of the center are quantified and a normalized radial intensity distribution is determined for each image, with the bleach center set as the origin.
4. Intensity profile data is passed from MPL to Excel. The radial distribution of bleached fluorophores is determined by subtracting radial intensity values from the image background intensity.
5. The variance of the radial distribution of bleached fluorophores is calculated and plotted as a function of time (*see* **ref. 11** for background details of how to calculate the variance values). The lateral translational diffusion coefficient is calculated from the gradient determined from a linear least-squares fit to the data.
6. Independent noninteracting multiple diffusing species with greatly differing diffusion coefficients show characteristic multicomponent recovery of total intensity in the bleached area. Typically there is fast initial recovery as the smaller components redistribute, followed by a slower recovery phase as the mobility of larger species dominates. This also results in plots of variance against time being nonlinear.
7. In dynamic binding equilibria, fluorescent molecules redistribute between bound and unbound forms and fast and slow components are not distinguishable in the recovery curve. Plots of variance vs time are essentially linear, and the equilibrium mixture behaves similarly to a monodisperse species having an intermediate diffusion coefficient, providing the experiment is designed so that **Eq. 3** is valid.

## References


1. Winlove, C. P. and Parker, K. H. P. (1995) The physiological functions of extracellular matrix macromolecules, in *Interstitium, Connective Tissue and Lymphatics* (Reed, R. K., McHale, N. G., Bert, J. L., Winlove, C. P., and Laine, G. A., eds.), Portland Press, London, UK, pp. 137–165.
2. Comper, W. D. and Laurent, T. C. (1978) Physiological functions of connective tissue polysaccharides. *Physiol. Rev.* **58,** 255–316.
3. Comper, W. D., ed. (1996) *Extracellular Matrix.* Harwood, Amsterdam, The Netherlands.
4. Maroudas, A. (1976) Balance between swelling pressure and collagen tension in normal and degenerate cartilage. *Nature* **260,** 808–809.
5. Grodzinsky, A. J. (1983) Electromechanical and physicochemical properties of connective tissue. *CRC Crit. Rev. Bioeng.* **14,** 133–199.
6. Hardingham, T. E. and Fosang, A. (1992) Proteoglycans: many forms and many functions. *FASEB. J.* **6,** 861–870.
7. Mould, A. P., Garratt, A. N., PuzonMcLaughlin, W., Takada. Y., and Humphries, M. J. (1998) Regulation of integrin function: evidence that bivalent-cation-induced conformational changes lead to the unmasking of ligand- binding sites within integrin alpha 5 beta 1. *Biochem.J.* **33,** 821–828.
8. Rahmoune, H., Rudland, P. S., Gallagher, J. T., and Fernig, D. G. (1998) Hepatocyte growth factor scatter factor has distinct classes of binding site in heparan sulfate from mammary cells. *Biochemistry* **37,** 6003–6008.
9. Axelrod, D., Koppel, D. E., Schlessinger, J., Elson, E., and Webb, W. W. (1976) Mobility measurements by analysis of fluorescence photobleaching recovery kinetics. *Biophys. J.* **16,** 1055–1069.
10. Kubitscheck, H., Wedekind, P., and Peters R. (1994) Lateral diffusion measurements at high spatial resolution by scanning microphotolysis in a confocal microscope. *Biophys. J.* **67,** 946–965.
11. Gribbon, P., and Hardingham, T. E. (1998) Macromolecular diffusion of biological polymers measured by confocal fluorescence recovery after photobleaching. *Biophys. J.* **75,** 1032–1039.
12. Blonk, J. C. G., Don, A., Van Aalst, H., and Birmingham, J. J. (1993). Fluorescence photobleaching in the confocal scanning light microscope. *J. Microsc.* **169,** 363–374.
13. Kaufmann, E. N., and Jain, R. K., (1991) Measurement of mass transport and reaction parameters in bulk solutions using photobleaching—reaction limited binding regime. *Biophys. J.* **60,** 596–610.
14. Hardingham, T. E., Ewins, R. J. F., and Muir, H. (1976) Cartilage proteoglycans: structure and heterogeneity of the protein core and the effects of specific protein modifications on the binding to hyaluronate. *Biochem. J.* **157,** 127–143.
15. Heinegard, D. and Hascall, V. C. (1974) Aggregation of cartilage proteoglycans. 3. Characterisation of proteins isolated from trypsin digests of aggregates. *J. Biol. Chem.* **249,** 4250–4256.
16. Nieduszynski, I. A., Sheehan, J. K., Phelps, C. F., Hardingham, T. E., and Muir, H. (1980) Equilibrium binding studies of pig laryngeal aggrecan with hyaluronate fractions. *Biochem. J.* **185,** 927–934.
17. Cowman, M. K., Balazs, E. A., Bergmann, C. W., and Meyer, K. (1981) Preparation and circular-dichroism analysis of sodium hyaluronate oligosaccharides and chondroitin. *Biochem.* **20,** 1379–1385

# 48

## Regulatory Roles of Syndecans in Cell Adhesion and Invasion

**J. Kevin Langford and Ralph D. Sanderson**

### 1. Introduction

Evidence accumulated over the last decade demonstrates that proteoglycans profoundly influence many cell behaviors such as growth factor binding, apoptosis, cell adhesion, and cell invasion. The major family of transmembrane proteoglycans is the syndecans, all of which have heparan sulfate chains that interact with various soluble and insoluble molecules within the extracellular matrix *(1)*. Our lab has developed two experimental techniques to examine the biological role of the syndecans. One assay examines syndecan-1–mediated cell–cell adhesion (i.e., aggregation assay). The other examines the syndecan-1–mediated inhibition of cell invasion.

The aggregation assay has the advantage of simplicity. Cells suspended in buffer are allowed to interact by gravity and then counted on a hemocytometer for the number of cells in aggregates *(2,3)*. This assay is rapid, yields highly reproducible results, and is easily manipulated by the inclusion of promoters or inhibitors of cell aggregation.

To investigate the invasive potential of myeloma cells, we employ a hydrated native type I collagen gel. First, the distance, or depth of invasion is measured by phase microscopy. Next, protease digestions facilitate the harvest of noninvasive cells from the top of the gel, followed by subsequent digestion of the gel to remove the invasive cells. After counting cells on a hemocytometer or Coulter counter, a simple calculation determines the percentage of invasive cells. The major advantage of this model is that it allows quantification of both the distance and extent of cell invasion. For an extensive review of in vitro and in vivo techniques used to examine cell migration or cell invasion, see Mareel et al. *(4)*.

Both of the above assays have proven to be powerful tools to probe the biology of the syndecans as they relate to tumor cell adhesion and invasion. We have used these assays to demonstrate that, following transfection with the cDNA for syndecan-1, ARH-77 cells gain the ability to form cellular aggregates and are rendered noninvasive in collagen gels *(2,5)*. These assays have been further used to examine the behavior of

From: *Methods in Molecular Biology, Vol. 171: Proteoglycan Protocols*
Edited by: R. V. Iozzo © Humana Press Inc., Totowa, NJ

ARH-77 cells expressing mutant or chimeric proteoglycans in order to investigate the molecular mechanisms of syndecan-1 function.

## 2. Materials

### 2.1. Cell Aggregation Assay

1. Aggregation buffer: For 500 mL of aggregation buffer add: 0.2 g of KCl ; 0.03 g of $KH_2PO_4$; 4.0 g of NaCl; 0.046 g of $Na_2HPO_4 \cdot 7H_2O$; 5 mL of a 1.0 *M* HEPES solution; 0.175 g of $NaHCO_3$; 5.0 g of bovine serum albumin (BSA). Next, add water to 500 mL and adjust pH to 7.0. Then, sterile filter and store at 4°C.
2. Heparin solution (1 mg/mL porcine heparin (Sigma; St. Louis, MO) in sterile deionized water).
3. Hemocytometer or Coulter counter.
4. Disposable transfer pipets (5-mL capacity; Fisher, Pittsburgh, PA).

### 2.2. Invasion Assay

1. Rat tail type I collagen stock (Collaborative Biochemical Products; Bedford, MA).
2. Sodium bicarbonate solution, 7.5% (w/v) (Gibco BRL; Grand Island, NY).
3. Complete medium: Final concentration of each component is as follows: 1 × RPMI solution; 2m*M* L-glutamine; and 1 × antibiotic-antimycotic solution (Mediatech, Herndon, VA); 5% fetal calf serum (Atlanta Biologicals, Norcross, GA).
4. 10 × RPMI solution (Mediatech, Herndon, VA).
5. Sterile distilled deionized water (Mediatech Inc., Herndon, VA).
6. 24-well cell culture plate.
7. Trypsin (0.25%) EDTA (0.1%) solution (Mediatech, Herndon, VA).
8. Collagenase solution (0.5 mg/mL of collagenase type 2 (Worthington Biochemical, Freehold, NJ) in 1 × RPMI).
9. Phosphate-buffered saline (PBS) + EDTA (0.5 m*M*) solution.
10. Hemocytometer or Coulter counter.

## 3. Methods

### 3.1. Cell Aggregation Assay (see Fig. 1)

1. Prewarm aggregation buffer to 37°C (*see* **Note 1**).
2. Collect $5.0 \times 10^6$ cells in an eppendorf tube.
3. Centrifuge the cells at 300*g* for 5 min and remove the supernatant.
4. Wash the cell pellet in 1 mL of aggregation buffer and centrifuge again.
5. Using a transfer pipet, vigorously resuspend the cells in 1 mL of aggregation buffer. During initial experiments, visually inspect the cells using a phase microscope to ensure that the cells are in a single-cell suspension.
6. Split the sample into two equal volumes (i.e., 500 µL).
7. To one tube, add 65 µL of a heparin solution (*see* **Note 2**).
8. To the other tube, add 65 µL of sterile deionized water.
9. Incubate the cells undisturbed at 37°C for 1 h.
10. Resuspend the cells by gently pipetting the solution 5 times. This is a critical step in that pipetting must be uniformly gentle or aggregates may be disrupted (*see* **Notes 3** and **4**).
11. Cells are counted in both the heparin-containing and heparin-free samples using a hemocytometer. A single cell or aggregates containing more than 3 cells are counted as 1. However, in aggregates consisting of 2 or 3 cells, each cell is counted (*see* **Note 5**).

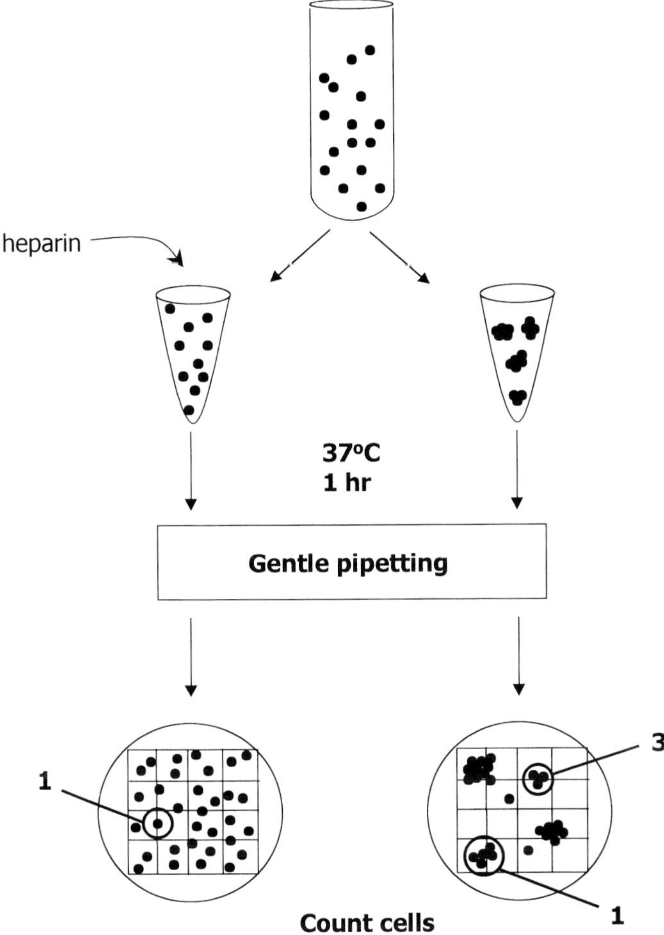


Fig. 1. Diagram of the aggregation assay. Cells from each sample are separated into two tubes containing equal numbers of cells. Heparin is added to one tube (to block aggregation), while the other cells are allowed to aggregate. Following gentle pipetting, the number of cells in each tube is determined. Single cells and aggregates containing more than three cells are counted as 1. Cells within groups of 2 or 3 are counted individually.

12. The percentage of cells in aggregates can now be determined as follows:

$$\frac{\text{total cells} - (\text{single cells} + \text{aggregates})}{\text{total cells}} \times 100 = \%\ \text{cells in aggregates}$$

total cells = number of single cells in heparin-containing sample

single cells + aggregates = number of single cells plus number of aggregates in the heparin-free sample

## 3.2. Invasion Assay

### 3.2.1. Preparing and Seeding Collagen Gels

1. Determine the total volume of collagen gel solution required. Use 1 mL of collagen gel solution for each well of the 24-well plate (example: 10 wells will require 10 mL of collagen gel solution).
2. Prepare the collagen gel solution: Add cell-culture grade deionized water to approximately one-half of the total collagen gel solution required to an appropriately sized culture tube (*see* **Note 6**).
   Next, add the collagen stock solution to the final working (*see* **Notes 7–9**). Mix by inverting the tube several times and place on ice. Then add 1/10 volume of 10 × RPMI dropwise while gently shaking the tube. After addition of 10 × RPMI, mix by inverting the tube several times and place on ice. Now add sodium bicarbonate solution dropwise while gently shaking the tube to achieve a pH of approximately 7.2–7.4 (*see* **Note 10**). Again, mix by inverting the tube several times and place on ice.
3. Degas the solution by allowing it to sit undisturbed for approximately 20 min on ice (*see* **Note 11**).
4. Gently, so as not to introduce bubbles, pipet 1 mL of the collagen gel solution into each well of the 24-well plate.
5. Place the plate containing the collagen gel solution into a tissue culture incubator at 37°C and 5% $CO_2$ for at least 1 h. Polymerization can be assessed by tilting the plate 45°C. If the solution remains liquid, discard.
6. To equilibrate the gel, carefully pipet 1 mL of complete medium to the surface of each collagen gel, and place back into the incubator for at least 1 h (*see* **Note 12**). For convenience, the gels may be equilibrated overnight.
7. To each gel, carefully add cells diluted in 1 mL of complete medium and culture at 37°C in 5% $CO_2$ for 48 h (*see* **Notes 13–15**).

### 3.2.2. Quantification of Maximum Invasive Depth

The depth of cell invasion is determined by measuring the distance from the top of the gel to the leading front of migrating cells using the calibrated micrometer present on the fine-focus dial of an inverted phase microscope (*see* **Fig. 2**). The leading front is defined as the point at which two of the leading cells within a given field are in the same focal plane under 200× magnification.

1. Measurements are taken in five fields within each well; using the center of the well as a landmark and as the first point (center point), the four additional measurements are taken in fields selected by moving the stage to points north, south, east, and west of center. The distance to each field is defined as half the distance from the center point to the edge of the well.
2. Starting near the bottom of the gel, turn the fine-focus dial so the focal plane moves toward the surface of the gel. When two cells appear in the same focal plane, begin counting revolutions until you reach the surface of the gel.
3. To calculate the distance from the gel surface to cells at the leading front, multiply the number of revolutions of the fine-focus dial by the number of μm/revolution. For the Nikon Diaphot used in our lab, 1 revolution moves the focal plane 100 μm.

### 3.2.3. Quantification of the Percentage of Invasive Cells

The percentage of invading cells is determined by using a limited trypsin and collagenase treatment to remove the noninvasive cells (i.e., cells attached at or near the gel

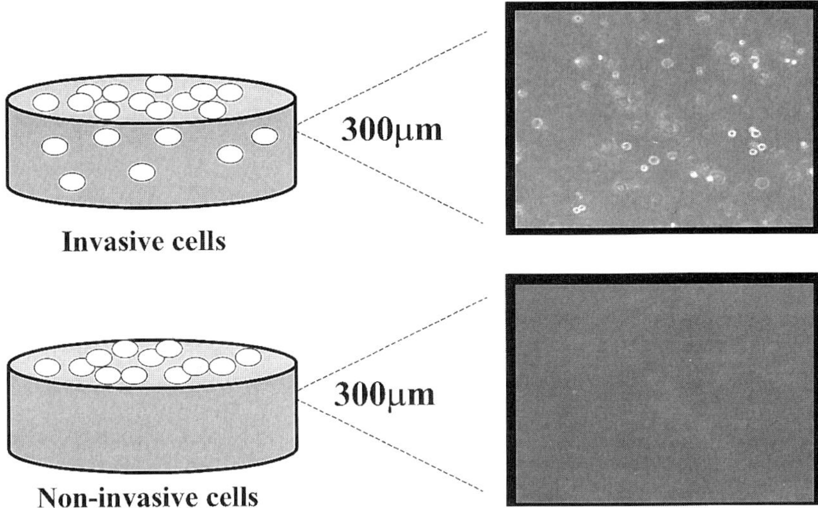


Fig. 2. Measuring depth of cell invasion. Following the incubation of the cells on the collagen gels, the distance from the surface of the gel to the cells of the leading front is determined using the calibrated micrometer (on the fine-focus dial) of an inverted phase microscope. Illustrations of collagen gels with an invasive or noninvasive cell line are drawn on the left side of the figure. Photomicrographs taken of cells within collagen gels are shown on the right. In this example, for the invasive cell culture, invading cells are visible 300 μm below the surface of the gel. However, for noninvasive cells, the field at 300 μm below the surface of the gel is devoid of cells.

surface). This is followed by complete digestion of the gel with collagenase and recovery of the invasive cells (*see* **Fig. 3**). Once the numbers of invading and no+ninvading cells are determined, the percentage of invading cells can be calculated.

1. Collection of the noninvasive cell population:
   a. Aspirate the medium from the surface of each gel and discard.
   b. Carefully wash the gel with 0.5 mL of PBS + 5 m*M* EDTA solution and collect by aspiration into the appropriate tube (*see* **Note 12**). Repeat this wash/collection step two additional times. Place all solutions (i.e., washes, trypsin, and collagenase solutions) collected from each well into the same 15-mL culture tube.
   c. Incubate the collagen with 0.5 mL of trypsin/EDTA solution at room temperature for 30 min and collect the medium and released cells by aspiration. Although this digestion does not release many cells, it does prepare the collagen for the collagenase treatment.
   d. Wash the gels twice and collect the solution two times as in **step b**.
   e. Incubate the gels with 0.5 mL of collagenase solution (0.5 mg/mL of collagenase in 1 × RPMI, prewarmed to 37°C) for approximately 10 min at 37°C (*see* **Notes 16–17**).
   f. Quickly but carefully remove the collagenase solution and wash the gels 3 times as in **step b**.

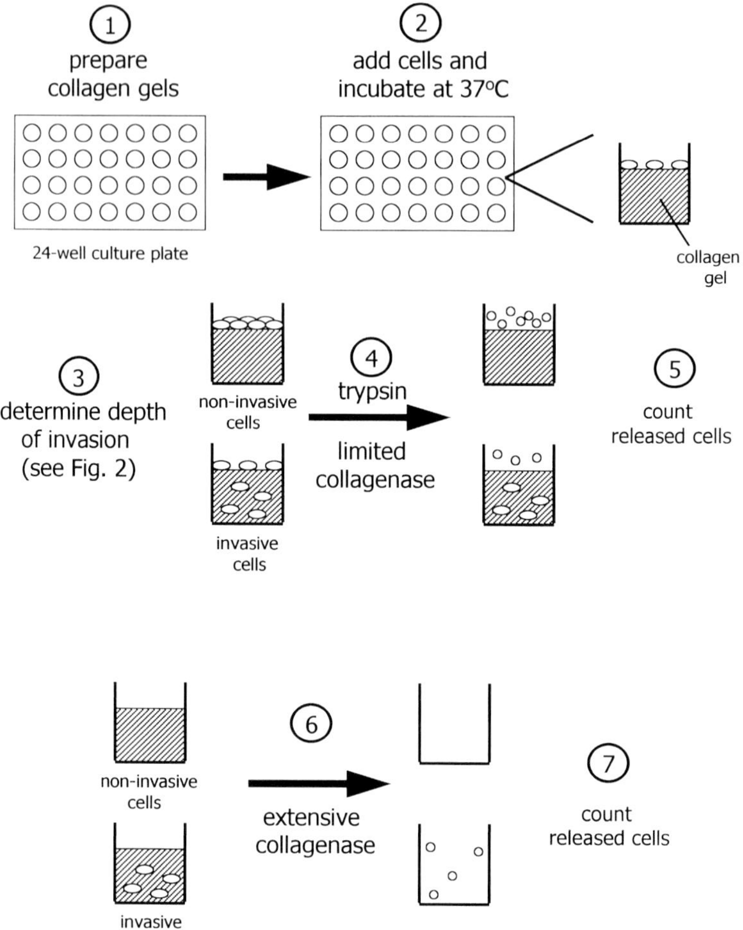


Fig. 3. Diagram of cell invasion assay. Collagen gels are prepared and seeded with cells. After the appropriate incubation period, first the depth of cell invasion is determined, followed by determination of the percent of cell invasion. Noninvading cells are recovered from the surface of the gels by trypsin and limited collagenase digestions and counted. Next, following complete digestion of the remaining collagen with collagenase, the invading cells are collected and counted. The percentage of cell invasion for each gel is calculated by dividing the number of invading cells by the total number of cells recovered from the collagen gel (noninvading + invading cells) and multiplying by 100.

g. Centrifuge the tubes containing the noninvasive cell population at 300*g* for 10 min and resuspend the cell pellet in an appropriate amount of buffer for cell counting (i.e., hemocytometer or Coulter counter).

2. Collection of the invasive cell population:
   a. Add 0.5 mL of collagenase solution to each well and incubate at 37°C until the gels are completely digested (approximately 1.5 h).

   b. Collect the solution by aspiration into individual tubes. Place all solutions collected from each well into the same 15-mL culture tube.
   c. Wash the wells and collect the solution by aspiration as in **step b**.
   d. Centrifuge the cells and count as in g above.
3. Determine percentage of invasive cells:

$$\frac{\text{invasive cells}}{\text{noninvasive cells} + \text{invasive cells}} \times 100 = \%\ \text{invasive cells}$$

## 4. Notes

1. Because the buffer used for this assay is highly conducive to bacterial growth, care must be taken to use fresh and uncontaminated solutions. The buffer should be sterile filtered, stored at 4°C, and handled using aseptic techniques in order to reduce contamination.
2. The heparin solution is used to disrupt cell aggregation and thus obtain single cells that are counted and used in the calculation of the percentage of cell aggregation. This is possible because aggregation, in our cell lines, is mediated via the heparan sulfate chains of syndecan-1. It may be necessary to use another disrupting agent for other cell types.
3. The force used to resuspend the cells prior to counting is the only subjective step in the entire assay. Considering this, it has been helpful to gain the assistance of a colleague to code the tubes containing the cells and thus "blind" the study.
4. A disposable transfer pipet (5-mL capacity) works well for resuspending the cells prior to counting. Holding the opening of the pipet at the 0.5-mL line of a 1.5-mL Eppendorf tube and gently applying pressure to the bulb has yielded consistent results when resuspending cells. Also, this pipet has a large enough opening so cell aggregates are not mechanically dispersed as they would if forced through a narrow aperture.
5. When counting cells using a hemocytometer, for each sample, count the cells within the four 1-mm$^2$ corners and divide by 4 to obtain an average cell number. This is done for cells in buffer containing heparin and cells in heparin-free aggregation buffer. Cells in groups of 2 and 3 cells may not represent aggregated cells but may simply be cells passively associated. Therefore, cells in these groups are counted individually.
6. Because of the duration of the invasion assay, care must be taken to ensure the sterility of the solutions and collagen gels during gel preparation.
7. A major consideration in the preparation of the gels is the collagen concentration best suited to yield cell invasion. The concentration of the collagen dramatically affects the rate of cell invasion. At high collagen concentrations, invasion may be inhibited all together; whereas at low concentration, with little to no resistance, all cells may simply fall through the large spaces of the gel. Therefore, a concentration of collagen that allows suitable invasion in an appropriate period of time and allows experimental manipulation of invasion (i.e., inhibition) must be determined empirically for each cell line. For example, although we use a collagen concentration of 0.5 mg/mL and an assay time of 48 h for myeloma cells *(5)* for breast cancer cells, the time was extended to 72 h *(6)*.
8. Although a collagen stock may be obtained from several sources, we have found that the collagen produced by Collaborative Biomedical Products yields consistent results and reproducible gels.
9. Prepare the collagen gel solution on ice, making certain that the solution remains cold during setup, to prevent premature polymerization of the collagen.

10. The pH of the collagen gel is a variable that influences the integrity of the gel and thus the invasive extent of the cells. The pH is adjusted by varying the amount of sodium bicarbonate added to the collagen gel solution. At slightly more acidic pH (i.e., 6.8–7.0) the gels do not polymerize as well, and are unstable. While this allows cells to invade faster, the gels become fragile, making it difficult to add media or wash solutions to the gels without damaging the surface. Conversely, a more basic pH (i.e., 7.6–8.0) creates a gel through which cells have difficulty invading. With experience, one can determine the optimal pH by carefully examining the color of the solution. The appropriate pH results in a "salmon pink" color. For initial experiments, a small amount of collagen gel solution can be applied to pH test strips for accurate pH determination. If the pH of the collagen gel solution becomes too basic, discard the solution.
11. If the collagen gel solution is not properly degassed, trapped air bubbles will compromise the integrity of the gel and allow cells to invade at an artificially rapid rate.
12. While washing the gels and changing the media, care must be taken not to damage the gels. If a hole appears during the quantification process, the gel must be discarded. When pipetting, slowly allow each drop of liquid to spread over the surface of the gel or liquid surface rather than drop from a distance. The impact of a large droplet of liquid may damage the gel.
13. The density of cells growing in culture before to the start of the invasion assay may also affect the results. For myeloma cells, the most consistent results are obtained from cells growing between 50% and 75% confluency. Myeloma cells at low density in fresh media, or overconfluent cells in exhausted media, do not invade well.
14. Cells placed on the collagen gels should not have extensive cell–cell contact. This applies at the beginning and end of the experiment. Thus, the size and mitotic rate of the cells to be used dictate the density at which they are added to each collagen well. For myeloma cells, we use $5.0 \times 10^4$ cells per collagen gel.
15. One of the most necessary requirements for obtaining interpretable results is the inclusion of positive and negative controls (i.e., invasive and noninvasive cells respectively). This ensures that the collagen gel is permissive for invasive cells yet still able to prevent noninvasive cells from passively falling through the spaces between collagen fibers. By including these two controls, data can be reported either as raw data *(5)* or as the percent invasion relative to the controls *(3)*. The latter method compensates for interassay variability.
16. The most critical step in the quantification process is the collagenase treatment used to remove the noninvasive cell population from the gel surface. Overdigestion of the gel will result in removal of many of the invasive cells, while underdigestion will fail to remove the cells at the surface. Both situations produce inaccurate data.
17. Because the majority of cells in the noninvading controls are attached at the surface of the gel, they form a "landmark" that can be used to determine the time required to remove the noninvasive cells from the surface of all gels. After 5 min at 37°C of the first collagenase digestion, the plate can be removed from the incubator and the cells on the surface examined under an inverted phase microscope. If the majority of cells in the noninvading control are in suspension, proceed to the next step. If many cells remain attached at the gel surface, return the plate to the incubator for 1-min intervals and examine until the noninvading cells are released from the gel surface.

## References

1. Bernfield, M., Gotte, M., Park, P. W., Reizes, O., Fitzgerald, M. L., Lincecum, J., and Zako, M. (1999) Functions of cell surface heparan sulfate proteoglycans. *Annu. Rev. Biochem.* **68,** 729–777.

2. Stanley, M. J., Liebersbach, B. F., Liu, W., Anhalt, D. J., and Sanderson, R. D. (1995) Heparan sulfate-mediated cell aggregation. Syndecans-1 and -4 mediate intercellular adhesion following their transfection into human B lymphoid cells. *J. Biol. Chem.* **270,** 5077–5083.
3. Langford J. K., Stanley M. J., Cao D., and Sanderson R. D. (1998) Multiple heparan sulfate chains are required for optimal syndecan-1 function. *J. Biol. Chem.* **273,** 29,965–29,971.
4. Mareel, M. M., Baetselier, P. D., and Van Roy, F. M. (1991) *Mechanisms of Invasion and Metastasis*, CRC Press, Boca Raton, FL.
5. Liebersbach, B. F. and Sanderson, R. D. (1994) Expression of syndecan-1 inhibits cell invasion into type I collagen. *J. Biol. Chem.* **269,** 20013–20019.
6. Kelly, T., Yan, Y., Osborne, R. L., Athota, A. B., Rozypal, T. L., Colclasure, J. C., and Chu, W. S. (1998) Proteolysis of extracellular matrix by invadopodia facilitates human breast cancer cell invasion and is mediated by matrix metalloproteinases. *Clin. Exp. Metastasis.* **16,** 501–512.

# 49

# Optical Biosensor Techniques to Analyze Protein-Polysaccharide Interactions

**David G. Fernig**

## 1. Introduction

Networks of interacting molecules, operating from the outside of the cell to the cell nucleus, regulate cell behavior. Optical biosensors provide a means of analyzing these interactions and possess key advantages over other methods: posttranslationally modified proteins, secondary gene products such as polysaccharides, chemically synthesized molecules, and nucleic acids are all equally susceptible to analysis; a quantitative description of an interaction is obtained; the structural rules and the kinetics governing the formation of multimolecular assemblies can be probed.

The principles of measurement underlying optical biosensors have been reviewed in detail elsewhere *(1)*, and only a basic description will be given here. Optical biosensors consist of a sensor surface, on one side of which reside the optics that enable measurements to be made and on the other side of which resides the liquid phase (*see* **Fig. 1**). The optical system generates an exponentially-decaying evanescent wave that penetrates into the liquid phase. The greatest sensitivity occurs where the evanescent wave is strongest, that is, closest to the surface. In general, useful measurements can be made within about 200 nm of the surface. The optics essentially measure changes in refractive index, so the signal obtained depends on the refractive index within this 200 nm.

The essence of experiments using optical biosensors is that one partner of a molecular interaction is immobilized on the surface and the interaction of the other partner(s) is then followed in real time by adding them to the liquid phase. It is usual to refer to the *immobilized molecule* as the *immobilized ligand* or *ligand* and the soluble partner(s) as the *soluble ligate(s)* or *ligate(s)*. The signal obtained from proteins and nucleic acids depends solely on the amount of material at the surface. Hence with these macromolecules, the optical biosensor acts as a very sensitive mass sensor. However, glycosaminoglycans have a low refractive index, which may vary with the degree of sulfation (Fernig, D. G., unpublished). Therefore, compared to proteins, glycosami-

From: *Methods in Molecular Biology, Vol. 171: Proteoglycan Protocols*
Edited by: R. V. Iozzo © Humana Press Inc., Totowa, NJ

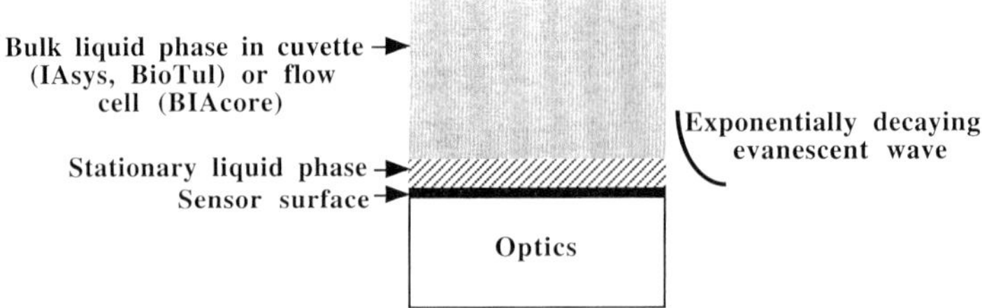


Fig. 1. Schematic of an optical biosensor.

noglycans give a small signal on a mass basis, and this signal cannot be related directly to the amount of material, since it may depend on composition. For these reasons, experiments should always be designed such that the proteoglycan, its glycosaminoglycan chains or oligosaccharides derived from the latter are immobilized on the surface and the protein partner(s) are used as soluble ligate. Unfortunately, proteoglycans often contain more than one glycosaminoglycan chain, and each chain may contain more than one binding site for the protein partner of interest. Multivalent ligands usually have high avidity. Molecules with high avidity exacerbate the major artifacts encountered in optical biosensors (*see* **Subheading 3.5.5.**).

There are currently four commercial instruments on the market, BIAcore (Pharmacia Biosensor, Uppsala, Sweden), BIOS-1 (Artificial Sensing Instruments, Zurich, Switzerland), BioTul (BioTul, Munich, Germany), and IAsys (Affinity Sensors, Cambridge, UK). Two of these instruments, the BIAcore and the BioTul, use surface plasmon resonance to produce the evanescent wave in order to probe the liquid phase and thus have a gold film at the sensor surface. The ability of thiol groups to bond gold is used to modify the surface so as to make it amenable to the attachment of biological macromolecules. Available surfaces include carboxymethyl dextran for BIAcore and BioTul, and for BIAcore only, carboxymethyl dextran derivatized with streptavidin and nitrilotriacetic acid, short carboxymethyl dextran, dextran with a low degree of carboxylation, planar hydrophobic, carboxylated, and gold. In contrast, the BIOS-1 and IAsys use an evanescent wave generated by waveguiding to probe the liquid phase. The IAsys has a layer of metal oxide at the sensor surface, which enables molecular deposition of a variety of groups, and, in addition to carboxymethyl dextran, hydrophobic, amino, carboxyl, and biotin surfaces are available. There are two issues to bear in mind regarding the surfaces. First, they will possess holes at the atomic level and thus some gold or metal oxide may be exposed to the experiment. Second, carboxymethyl dextran gels possess hydrophobic pockets, which may either cause nonspecific binding artifacts or be a useful functionality.

The liquid phase, in which experiments take place (*see* **Fig. 1**), consists of two components, a homogenous bulk phase that is mixed by a vibrational stirrer (IAsys) or by flow (BioTul and BIAcore), and a thin stationary phase next to the sensor surface, whose constituents exchange by diffusion with the bulk phase. The ligand immobilized on the sensor surface is within the stationary phase. The depth of the stationary phase depends on the efficiency of mixing, which in turn determines the likelihood of diffusion artifacts (*see* **Subheading 3.5.5.**). The vibrational stirrer in IAsys should always be set to maximum. Since it induces chaotic mixing at the interface between the bulk phase and the stationary phase, the latter is reduced to a minimum and exchange between the two phases is efficient. In flow systems, mixing depends on the rate of flow (BioTul and BIAcore), and this should always be set to the maximum. The issue of mixing is compounded with laminar flow systems (BIAcore), since, as the bulk phase approaches the stationary phase, the rate of flow decreases, thus reducing the efficiency of mixing at the bulk phase–stationary phase interface.

## 2. Materials

### 2.1. Biotinylation

1. N-hydroxysuccinimide (NHS) amino caproate (LC) biotin (Pierce) or hydrazide-LC-biotin (Pierce), 50 m*M*, dissolved in dimethyl sulfoxide. It is essential to use biotin with the aminocaproate spacer arm. Stored in 10-μL aliquots at –70°C, the NHS-LC-biotin and hydrazide-LC-biotin are stable for at least 4 mo.
2. Proteoglycan or peptidoglycan chains, normally desalted and freeze-dried.
3. 3 *M* Tris-HCl, pH 7.2.
4. Sephadex G-25 desalting column (1 × 25 cm).

### 2.2. Biotinylation of Amino Groups

1. Add 10 μL of a 50 m*M* solution of NHS-LC-biotin in dimethyl sulfoxide to 100 μg of proteoglycan in 100 μL of distilled water, mix, and allow the reaction to proceed at room temperature for 24 h.
2. Two further additions of 10 μL of NHS-LC-biotin may be made, either over the initial 24 h or over a subsequent 48 h. Blocking of unreacted NHS groups on biotin is achieved by the addition of 10 μL of 3 *M* Tris-HCl, pH 7.2, followed by a 10 min incubation at room temperature.

### 2.3. Biotinylation of Reducing Ends

1. Add 10 μL of a 50 m*M* solution of hydrazide-LC-biotin in dimethyl sulfoxide to 100 μg of oligosaccharide in 100 μL of distilled water, mix, and allow the reaction to proceed at room temperature for 24 h. Two further additions of 10 μL of hydrazide-LC-biotin may be made either over the initial 24 h or over a subsequent 48 h.

### 2.4. Removal of Free Biotin

1. Free biotin is removed by fractionation on a Sephadex G-25 column (1 × 25 cm, flow rate 0.5 mL/min), equilibrated in distilled water, and calibrated with high-molecular-weight blue dextran and potassium dichromate. The biotinylated proteoglycans or peptidoglycans elute in the void volume and are then lyophilized.

### 2.5. Capture of Biotinylated Ligands, Surface Regeneration and Storage

1. PBST (phosphate-buffered saline, pH 7.2 with 0.02 % [v/v] Tween-20).
2. Biotinylated ligand between 1 μg/mL and 1 mg/mL dissolved in PBST (*see* **Subheading 3.4.**).
3. Instrument stirrer setting: 100 (maximum).

### 2.6. Regeneration

1. 2 *M* NaCl, 10 m*M* $NaH_2PO_4$, pH 7.2.
2. 20 m*M* HCl.
3. 2 *M* guanidine-HCl, pH 7.0, freshly made and the highest grade available, for example, Aristar from BDH, (Poole, UK).

## 3. Methods

### 3.1. Background Binding and Choice of Surface

Optical biosensors are extremely sensitive and precise instruments. Since all instruments offer parallel/multiple channels, the temptation is to subtract any nonspecific binding by using a control surface. There are two drawbacks to this approach. When nonspecific binding is considerable, the signal is a fraction of the noise, which is always a poor basis for any type of measurement. In addition, a true control surface is almost impossible to generate, since, by definition, only the experimental surface has immobilized ligand. If an unrelated ligand is used on the control surface, the potential effects of immobilization of this ligand on the ligate should be explored. For example, commonly used "blocking" proteins such as bovine serum albumin *(2)* bind weakly a large number of protein ligates and are often not suitable. The search for a suitable control surface for nonspecific binding rapidly becomes recursive. It is far more profitable to spend time identifying a surface that maximizes the signal-to-noise ratio. In the author's experience, protocols that have no background binding can be identified with a little effort (*see* **Fig. 2**). Ideally, two different surfaces, at least one of which should be planar (*see* **Subheading 3.5.5.**) should be identified.

If a capture system, such as streptavidin for biotinylated ligands, is to be used, following the identification of suitable surfaces, the tests for nonspecific binding should be repeated on a surface with the capture system immobilized. If nonspecific binding occurs, then, in addition to the procedures outlined in Fig. 2, alternative capture systems should be explored. In the case of streptavidin, substitution with avidin or neutravidin often reduces background binding to zero.

### 3.2. Immobilization of Proteoglycans, Glycosaminoclycan Chains and Oligosaccharides

The immobilization of ligands is accomplished by chemically activating a functional group on the surface, e.g., carboxyl on carboxymethyl dextran or amino on aminosilane. Since excellent protocols for ligand immobilization are supplied by the manufacturers, only points specific to proteoglycans will be considered here. There are major difficulties in the efficient direct coupling of proteoglycans, glycosaminoglycan chains, and oligosaccharides to surfaces, due to the highly anionic character of these molecules, which is a consequence of the carboxyl and sulfate groups present on the saccharides. Thus, electrostatic uptake strategies, which are used to concentrate ligand on carboxymethyl dextran surfaces to increase the efficiency of immobilization, will not work, since the isoelectric point of the polysaccharide chains is lower than that of the carboxyl groups on the matrix. Moreover, due to the high negative

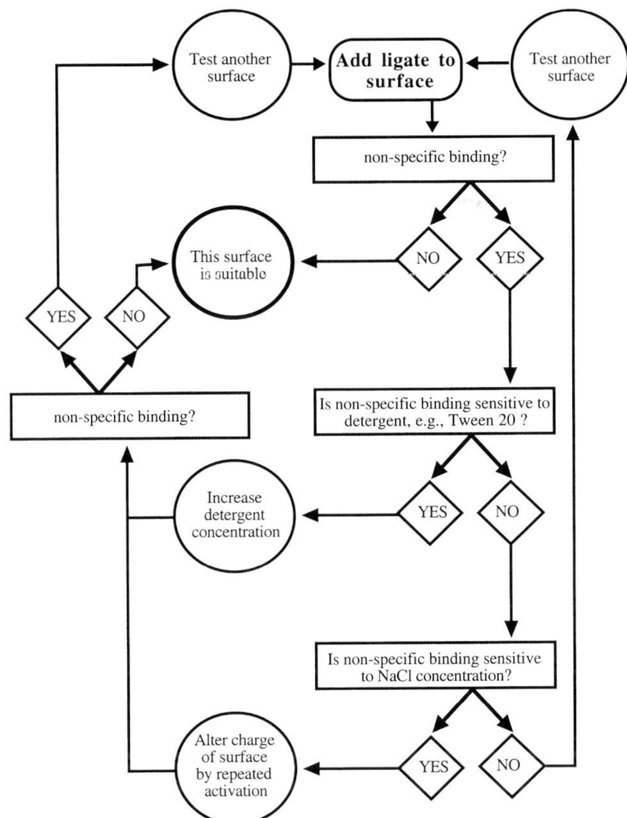


Fig. 2. Identification of zero background surfaces. Ligate is added in the buffer chosen for binding assays at the highest concentration likely to be used in these assays. The most commonly used detergent is Tween-20. In 0.5% (v/v) Tween-20 no background binding is observed with crude lysates of tissues or cells on streptavidin-derivatized IAsys aminosilane surfaces (Wainwright, G. and Fernig, D.G., unpublished observations). Changing the charge of the surface at physiological pH, e.g., negative charge of carboxymethyl dextran, is accomplished by repeated activation of the charged groups of the surface. An elegant example is provided in **ref.** ***(8)***.

charge of the glycosaminoglycan chains, proteoglycans and their saccharide components will not penetrate readily into the carboxymethyl dextran gel in the absence of a counterion. To obtain reasonable levels of immobilized ligand in the absence of electrostatic uptake requires a high concentration of ligand (>1 mg/mL) during the coupling reaction. Samples of proteoglycan, glycosaminoglycan chains, and oligosaccharides are usually fairly precious. A capture system such as biotin–streptavidin avoids the problems inherent to the direct coupling of proteoglycans and derived polysaccharides, and so is the preferred method for the immobilization of these ligands. In addition, the capture of biotinylated proteoglycans and polysaccharides allows the oriented immobilization of the ligand, which optimizes ligate binding (*see* **Subheading 3.5.5.**).

Proteoglycans and glycosaminoglycan chains isolated as peptidoglycans can be conveniently biotinylated on amino groups. Free amino groups may also occur along the polysaccharide chain. If required, the relative level of biotinylation of peptide and polysaccharide can be estimated, with respect to a specific ligate, e.g., basic fibroblast growth factor (bFGF) *(3)*. Oligosaccharides produced by enzymatic cleavage of glycosaminoglycan chains, e.g., heparinase, will contain a reducing end. Again this provides a unique functionality that can readily be biotinylated.

### 3.3. Capture of Biotinylated Ligands, Surface Regeneration and Storage

#### 3.3.1. Capture

Streptavidin, avidin, or neutravidin should be immobilized on the appropriate surface according to the instructions of the instrument manufacturer. The following protocol is suitable for the IAsys instrument. It will have to be adapted for the other instruments. In particular, high flow rates (100 µL/min) should be maintained throughout the capture reaction in BIAcore and BioTul to ensure efficient mixing and hence an even capture of biotinylated ligand over the sensor surface.

The cuvet is equilibrated in 50 µL of PBST. Replace the PBST with 20 µL biotinylated proteoglycan, peptidoglycan chains, or oligosaccharides, usually at concentrations between 1 µg/mL and 1 mg/mL in PBST. The contact time of the biotinylated ligand is generally 30 min.

If required, the unbound biotinylated material may be recovered by pausing the experiment, withdrawing the cuvet and removing the bulk phase with a pipet.

Wash the cuvette 5 × 50 µL PBST.

Before using the cuvet for measurements, perform two cycles of regeneration, followed by one binding reaction and one cycle of regeneration.

Before using a stored cuvet, perform one cycle of regeneration, followed by one binding reaction and one cycle of regeneration.

#### 3.3.2. Regeneration

Regeneration of the surface serves to remove all bound ligate, thus returning the ligand to its original state. The importance of efficient regeneration cannot be overstated, since for optical biosensors to be useful the same surface, (*see* e.g., **Subheading 3.5.**), must be used for multiple, comparable measurements. It is essential that regeneration protocols do not remove immobilized ligand or chemically alter the structure of the immobilized ligand. Luckily, glycosaminoglycans are robust ligands and the biotin–streptavidin bond is extremely strong, resistant to 2 *M* NaCl, 20 m*M* HCl, and 2 *M* guanidine. Most protein-glycosaminoglycan interactions depend heavily on ionic bonding between the sulfate groups of the polysaccharide and the amino groups of the protein. Therefore 2 *M* NaCl is usually sufficient to regenerate the surface. When 2 *M* NaCl fails to remove all the bound the ligate, additional regeneration steps with 20 m*M* HCl and, in extreme cases, with 2 *M* guanidine-HCl, may be required.

#### 3.3.3. Storage

Cuvettes (BioTul, IAsys) and sensor chips (BIAcore) can be stored wet or dry at 4°C. For wet storage (<1 wk) of cuvets, put 70 µL PBST with 0.02 % (w/v) $NaN_3$ in the cuvet and wrap with Parafilm to prevent evaporation. For dry storage, cuvets should be washed 5 times with water, emptied, and stored upside down. Sensor chips (BIAcore) can be stored in a similar fashion in a screw-top tube. Immobilized proteoglycans, peptidoglycan chains, and oligosaccharides are very stable and will retain their full ligate-binding capability after months of storage. The major cause of deterioration of sensor surfaces is the effect of incomplete regeneration, which results in a gradual loss of binding sites.

### 3.4. Control of the Amount of Immobilized Ligand

In some applications it is important to control the amount of immobilized ligand, and in general, it is advisable to repeat experiments with different levels of immobilized ligand. In addition, the interaction of proteins with the glycosaminoglycan chains of proteoglycans is dependent on the presence of specific sequences of saccharides. Therefore equal amounts of proteoglycan isolated from different sources will not necessarily contain the same number of binding sites for a given ligate. The difficulty of reliably detecting the amount of immobilized proteoglycan means that this quantity must be determined empirically by calculating the total number of binding sites on the surface, $B_{\max}$ (*see* **Subheading 3.5.5.**). To vary the amount of immobilized ligand, vary the concentration of biotinylated ligand during ligand capture (*see* **Subheading 3.3.**).

### 3.5. Binding Assays

A single binding assay is illustrated in **Fig. 3**, and the schematic (*see* **Fig. 4**) shows the changes occurring at the surface in the bulk phase.

#### 3.5.1. Footprinting and Multimolecular Complexes

In all footprinting experiments, it is essential that the ligates are used at concentrations > $K_d$ for the ligand, when the ligate will occupy >50% of the binding sites on the ligand. The main measurement made in footprinting experiments is the extent of ligate bound, which requires the binding reaction to be at or near equilibrium. Consequently, these experiments take a relatively long time. The extent of binding is calculated in two ways, the results of which should be identical. First, the extent of ligate binding can be determined directly from the binding curve, where it is equal to the response at equilibrium minus the response before the addition of ligate, minus the bulk shift. Second, the nonlinear curve-fitting software supplied by the manufacturer is used to calculate the extent of binding from the association curve.

The protocols described in **Figs 5** and **6** can be adapted for the other instruments, though the high flow rates required for efficient mixing will require quite extensive use of ligates.

#### 3.5.2. Footprinting with a Competitor of the Ligate

These experiments test the hypothesis that two ligates bind to sites on the ligand that are the same, overlap, or interfere with each other. In the example shown (*see* **Fig. 5**), the two ligates are bFGF and a synthetic peptide bFGF(127–140). The experiment illustrates how the dissociation rate constant of a ligate determines the experi-

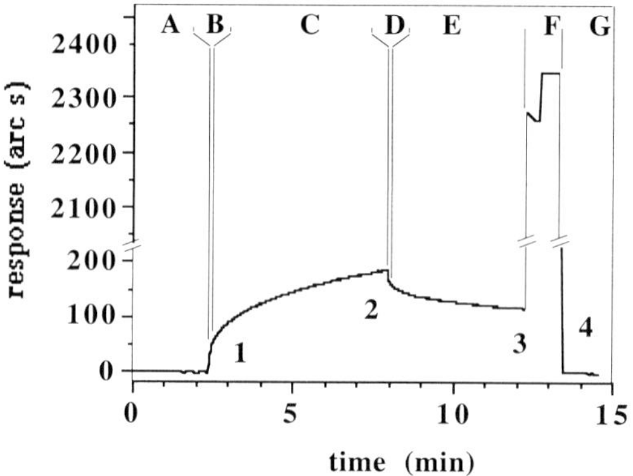


Fig. 3. A binding assay. Binding of bFGF (final concentration 3 µg/mL or 167 n*M*) to biotinylated heparan sulfate from Rama 27 cell culture medium, immobilized on a streptavidin derivatized carboxymethyl dextran cuvet. A response of 200 arc seconds is equal to 1 ng/mm$^2$ of protein in the carboxymethyl dextran gel. The binding curve is described by a series of events (arabic numerals) and regions (letters). **Event 1:** ***Addition of ligate*.** Binding is initiated by the addition of 1 µL of 90 µg/mL bFGF to the 29 µL of PBST in the cuvet. **Event 2:** *PBST washes.* At the end of the association reaction, the cuvet is washed 3 times with 50 µL of PBST to initiate the dissociation reaction. **Event 3:** *Regeneration.* To regenerate the surface, the cuvet is washed 3 times with 2 *M* NaCl, 10 m*M* $NaHPO_4$, pH 7.2, and left in this solution for 1 min. **Event 4:** *Return to starting conditions.* The cuvet is washed 3 times with 50 µL of PBST and then once with 29 µL of PBST. **Region A** is the baseline, 29 µL of PBST in the cuvet. **Region B** is the 3–5 s region immediately after the addition of ligate (event 1), where mixing and bulk shifts occur—the latter result from the difference in refractive indices of PBST and PBST containing 3 µg/mL bFGF. **Region C** is the association phase from which the $k_{on}$ and the extent of binding are calculated. **Region D** is the 3–5 s region immediately following the PBST washes (event 2), where mixing and bulk shifts occur. **Region E** is the dissociation phase from which $k_{diss}$ is calculated. **Region F** is regeneration, which is initiated by the addition of 2 *M* NaCl (event 3). **Region G** is the new baseline with 29 µL of PBST in the cuvet (event 4).

mental protocol. The bFGF dissociates slowly from heparin, whereas the peptide bFGF(127-140) dissociates more rapidly *(4)*.

In the first footprinting experiment (**Fig. 5**, 7.9–104 min), once bFGF binding has reached a maximum, dissociation is initiated (**Fig. 5**, 29.7 min). At 43.8 minutes, the dissociation of bound bFGF from the immobilized heparin is negligible compared to the association of the peptide bFGF(127–140), so the peptide is added. In the second experiment (116.9–206.1 min) the order of addition of the ligates is reversed. This experiment has to contend with the relatively fast dissociation rate of the peptide from heparin. After the binding of the peptide has reached a maximum, 1 µL of the bulk phase is replaced with 1 µL of an identical solution (PBST containing 100 µg/mL bFGF (127–140) and 90 µg/mL bFGF. In this way the equilibrium between the soluble and bound peptide is only perturbed by the bFGF binding to the immobilized heparin.

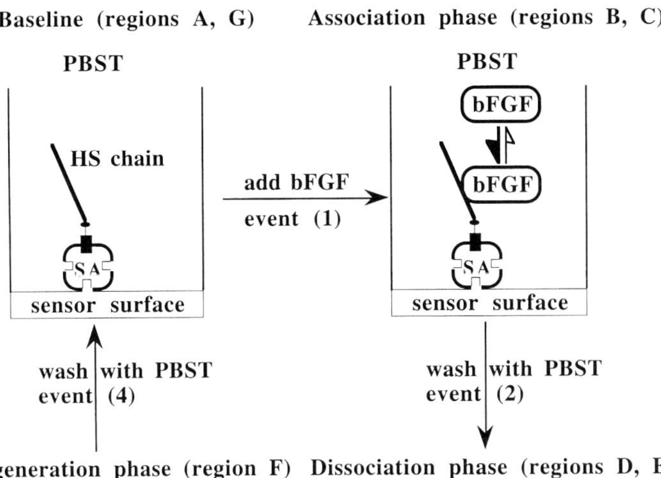


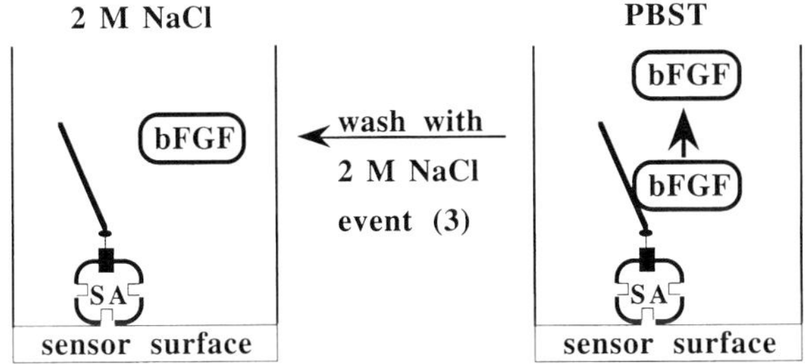


Fig. 4. Schematic of molecular changes occurring at the sensor surface in the course of the experiment in **Fig. 3.** SA, streptavidin.

The concentration of the ligates and their molecular weight have a considerable effect on this type of footprinting experiment. The concentrations of peptide and bFGF relative to their $K_d$ for heparin *(4)* were chosen such that the peptide in the first experiment (43.8 min) could displace bFGF, but in the second experiment bFGF was unlikely to displace the peptide. The signal produced by proteins and nucleic acids in optical biosensors is directly proportional to molecular weight. In this example the molecular weights of bFGF and bFGF(127–140), 18 and 1.6 kDa, respectively, differ by over 10-fold. Thus displacement of bFGF in the first experiment (43.8 min) by the peptide would cause a large decrease in signal, whereas bFGF is unlikely to displace the peptide in the second experiment and cause an increase in signal. Other experiments, however, may be set up differently due to the type of ligates, and the results not be so clearcut. In these cases, if ligate dissociation is slow, antibodies can be used at the end of the experiment to quantify the amount of each ligate that is bound. Thus an antibody to bFGF could be added at 94.1 min, and once this has bound maximally, antibody to the peptide would be added.

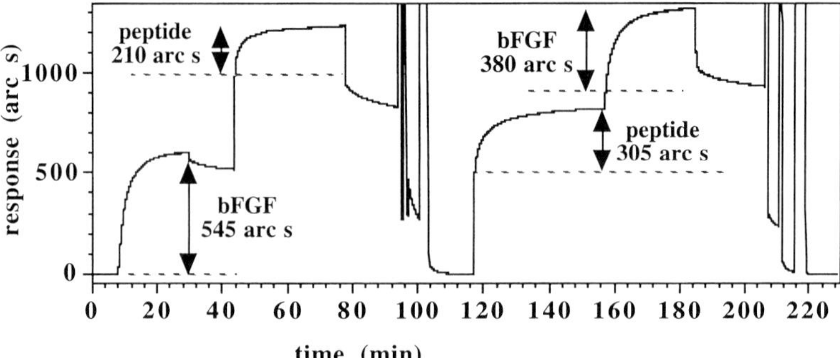


Fig. 5. Footprinting with a competitor of the ligate in an IAsys optical biosensor. In the first experiment, the extent of binding of bFGF to biotinylated porcine mucosal heparin, captured on streptavidin immobilized on a carboxymethyl dextran surface, is measured. Then the synthetic peptide bFGF(127–140) is added to determine the extent of binding of the peptide to porcine mucosal heparin in which over half the binding sites or bFGF are occupied. In the second experiment the order of addition of the ligates is reversed, so the synthetic peptide is added first, followed by bFGF. This experiment measures the extent of binding of the synthetic peptide to porcine mucosal heparin and the extent of binding of bFGF to porcine mucosal heparin in which over half the peptide binding sites are occupied. The table below is a detailed description of experiment.

Calculation of extent of binding: In the first experiment the bulk shifts do not have to be explicitly subtracted, since the two binding reactions start and finish in PBST. The extent of binding of bFGF and the bFGF(127-140) peptide is denoted by the vertical two headed arrows. In the second experiment, the situation is a little more complex, since the cuvette is not returned to PBST after the bFGF(127-140) peptide has bound, due to the relatively fast rate of dissociation of the peptide-heparin complex. In this case the bulk shift is taken as the first 5 s of the binding reaction and an offset baseline (horizontal dotted lines) is used to calculate the extent of binding (two headed arrow).

| Time (min) | Event |
|---|---|
| 0 | Establish a baseline with 29 μL of PBST. |
| 7.9 | Add 1 μL of 90 μg/mL bFGF to the cuvet. |
| 29.7 | Maximal binding of bFGF, wash 3 × with 50 μL of PBST; leave the cuvet in 29 μL of PBST. |
| 43.8 | Add 1 μL of 3 μg/mL bFGF(127–140) peptide. |
| 77.7 | Maximal binding of bFGF(127–140). |
| 77.7 | Wash 3 × with 50 μL of PBST; leave the cuvet in 29 μL of PBST. |
| 94.1 | Regenerate with multiple washes of 2 *M* NaCl. |
| 104 | Establish a baseline in 29 μL of PBST. |
| 116.9 | Add 1 μL of 3 mg/mL bFGF(127–140) peptide. |
| 156.8 | Maximal binding of bFGF(127–140); remove 1 μL from the cuvet. |
| 156.8 | Add 1 μL of 90 μg/mL bFGF in PBST containing 100-μg/mL bFGF(127–140). |
| 184.6 | Wash 5 × times with PBST and allow dissociation to proceed. |
| 206.1 | Regenerate with multiple washes of 2 *M* NaCl to recover the baseline. |

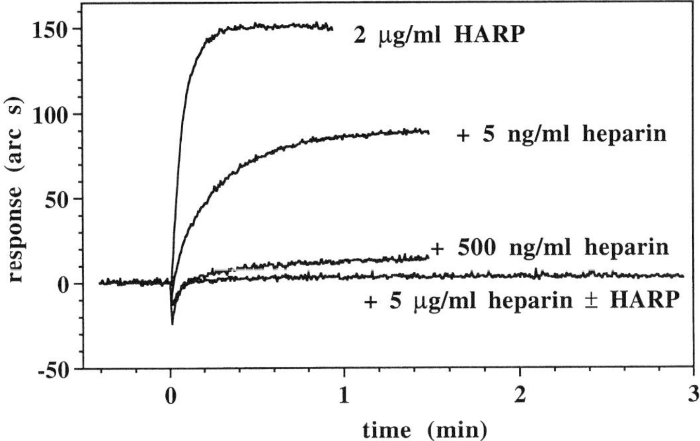


Fig. 6. Footprinting with a competitor of the ligand. In the control binding reaction, 3 μL HARP (20 μg/mL in PBST) is added to a cuvet containing 27 μL of PBST. In the subsequent competitions, 3 μL of PBST containing 20 μg/mL HARP and 50-ng/mL, 5 μg/mL, or 50 μg/mL heparin is added. The signal from the last experiment is indistinguishable from the background. The fourth binding curve, which is again indistinguishable from background, controls for any signal that might be generated by the competitor alone; in this case 3 μL of PBST containing no HARP and 50 μg/mL of heparin is added.

### 3.5.3. Footprinting with a Competitor of the Ligand

These experiments test the hypothesis that the region of the ligate that recognizes the ligand can also recognize the soluble competitor. In the example shown (*see* **Fig. 6**), the effect of heparin on the binding of heparin affin regulatory peptide (HARP) ***(5,6)*** to immobilized heparin is determined.

### 3.5.4. Multimolecular Complexes

The analysis of multimolecular complexes, for example where ligate (1) binds to heparan sulfate and ligate (2) binds to ligate (1), is a simple extension of the above footprinting experiments. The only constraint is that the dissociation rate of ligate (1) from the ligand is sufficiently slow (as in the case of bFGF in **Fig. 5**) to allow the cuvette to be washed and returned to binding buffer prior to the addition of ligate (2).

### 3.5.5. Kinetics

One of the key uses of optical biosensors is the rapid determination of the kinetics of a molecular interaction. The instrument manufacturers provide ample information on the theoretical considerations behind the measurements of binding kinetics. This section will therefore deal only with artefacts.

The major artefact associated with the determination of kinetics in optical biosensors is the generation of second phase binding kinetics by diffusion limitations (the so-called mass-transport artefact) and steric hindrance ***(1,7)***. Diffusion limitations arise when the rate of diffusion of the soluble ligate (dependent on its diffusion coefficient, $D$) from the bulk, stirred solution, through the boundary layer of immobile solution, which exists next to the surface of the sensor, is equivalent or slower than the apparent on rate, $k_{on}$. In this case, after the initial rapid depletion of soluble ligate from the solution near the surface, the observed association kinetics reflect diffusion rates rather than association rates. The diffusion artefact also affects the measurement of the dissociation rate constant, $k_{diss}$. $k_{diss}$ should be independent of ligate concentration. If $k_{diss}$ is found to increase with increasing ligate concentration, it is likely that dissociated ligate is rebinding to unoccupied ligand faster than it is diffusing into the bulk phase. Multivalent ligates, which possess a high avidity, represent a special case of this effect. The steric hindrance of binding sites arises when the immobilized ligand is at a high density and/or randomly oriented, and is most prominent on 3-dimensional carboxymethyldextran surfaces *(7)*.

There are two ways to investigate whether diffusion is indeed rate-limiting:

1. Decreasing the rate of flow or of stirring to determine the rate at which diffusion becomes limiting. A drawback is that this is rather insensitive and in stirred systems the relationship between diffusion and the rate of stirring may not be linear.
2. Increasing the viscosity of the binding buffer (diffusion is inversely proportional to viscosity) by the addition of glycerol. This is the more sensitive method.

To avoid these possible artefacts, experimental design should always include the following:

1. The minimum amount of ligand required to give a useful signal should be immobilized. This is determined empirically.
2. Oriented immobilization of ligand rather than random reduces steric hindrance between the surface and the binding site on the immobilized ligand.
3. Since $k_{on}$ depends directly on concentration, keeping the concentrations of ligate as low as possible reduces the possibility of the rate of diffusion controlling the binding reaction.
4. To avoid rebinding during dissociation, $k_{diss}$ should be determined at ligate concentrations where the majority of the binding sites are occupied. If rebinding is still a problem, soluble ligand can be added as a competitor to the dissociation buffer [e.g., *(8)*].
5. Diffusion into the bulk phase from the stationary phase is faster with planar than three-dimensional, e.g., carboxymethyl dextran, surfaces, so some experiments should always use a planar surface *(7)*.

### 3.6.6. Microaffinity Chromatography

Optical biosensors provide the opportunity to carry out microaffinity chromatography. In these experiments, the immobilized ligand is used to fish for a specific target in a mixture. Once the target is bound, the regeneration step is used to elute the target into the bulk phase. Recovery of the target is simplest in cuvet-based instruments. IAsys cuvets come in analytical and preparative formats. Analytical surfaces have a mask that covers all but 4 $mm^2$ of the surface, so the total amount of ligand is low, thus preventing depletion of ligate from the bulk phase during the binding reaction. Preparative surfaces ("Select") do not have the mask and the area of the surface is 16 $mm^2$, which allows the immobilization of fourfold more ligand and the capture of a correspondingly larger amount of ligate. Microaffinity chromatography can be used either as a means to identify the steps required to prepare an efficient conventional

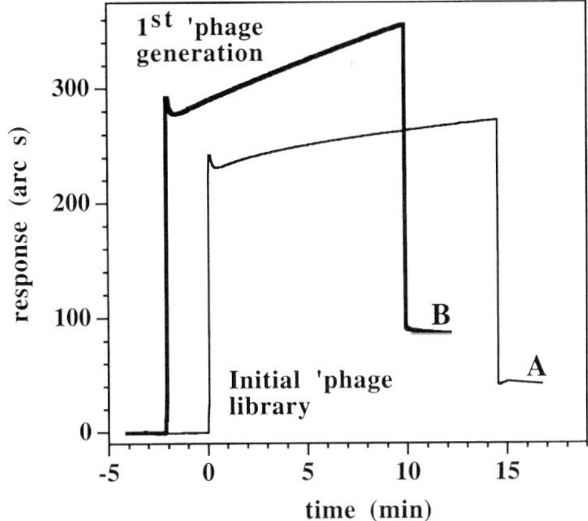


Fig. 7. Selection of phage that bind heparin. Biotinylated porcine mucosal heparin was immobilized on a biotin select surface. Wild-type phage do not bind to this surface at NaCl concentrations as low as 10 m*M* (Rahmoune, H., and Fernig, D. G. unpublished). The cuvet is cleaned with phenol prior to the addition of phage, to eliminate contaminating wild-type phage from the environment and all solutions are autoclaved prior to use. Phage from the initial library (70 µL at $10^{11}$ pfu/mL) are added to the cuvette (0 min, thin line). After a large bulk shift due to the high concentration of phage, a low but significant amount of binding is observed (about 20 arc s). After washing the surface with PBST, bound phage (**A**, the first generation) are eluted (equivalent to surface regeneration) with either 2 *M* NaCl or 1 mg/mL heparin in PBST, both of which were equally effective. These first-generation phage are grown up and added (70 µL at $10^{11}$ pfu/mL) to the same surface (–2 min, thick line). Clearly the first generation is enriched in phage that are able to bind to the heparin compared to the initial library, since the extent of binding is double and the $k_{on}$ is considerably faster. Bound phage (**B**, the second generation) were collected as before and grown up.

chromatography column or as the preparative chromatography step itself. The only constraint on the use of optical biosensors as microchromatography systems is that nonspecific binding is undetectable.

The example shown is an experiment that used the optical biosensor to isolate heparin-binding phage from a library that displayed a constrained 7-amino acid peptide (*see* **Fig. 7** and Chapter 50). Twenty clones were isolated from the second-generation phage and, compared to the initial starting library, the DNA sequences encoding the peptide library in second-generation phage always contained codons for basic amino acids, illustrating the success of the method. The clear advantage of using the optical biosensor as a microaffinity chromatography system is that the instrument provides a readout in real time of the binding events. It is therefore possible to troubleshoot protocols in real time and thus save considerable amounts of time and precious materials. In addition, since nonspecific binding is negligible, the fold-purification achieved is dramatic. The major inconvenience is that only small amounts of material are recovered, which require either a simple amplification step (as in the case of the phage) or suitable downstream microanalytical facilities.

## Acknowledgements

The author would like to thank the Association for International Cancer Research, the Biotechnology and Biological Sciences Research Council, the Cancer and Polio Research Fund, the Leverhulme Trust, the Medical Research Council, the Mizutani Foundation for Glycoscience, the North West Cancer Research Fund, and The Royal Society for financial support, and his colleagues in Liverpool for help and advice.

## References

1. Schuck, P. (1997) Use of surface plasmon resonance to probe the equilibrium and dynamic aspects of interactions between biological macromolecules. *Annu. Rev. Biophys. Biomol. Struct.* **26,** 541–566.
2. Peters, T. (1996) *All About Albumin : Biochemistry, Genetics, and Medical Applications.* Academic Press, San Diego, CA.
3. Rahmoune, H. J., Gallagher, T., Rudland, P. S., and Fernig, D. G. (1998) Interaction of heparan sulphate from mammary cells with extracellular regulatory proteins. Acidic and fibroblast growth factor: regulation of the activity of bFGF by high and low affinity binding sites in heparan sulphate. *J. Biol. Chem.* **273,** 7303–7310.
4. Kinsella, L., Chen, H.-L., Smith, J. A., Rudland, P. S., and Fernig, D. G. (1998) Interactions of putative heparin-binding domains of basic fibroblast growth factor and its receptor, FGFR-1, with heparin using synthetic peptides. *Glycoconjugate J.* **15,** 419–422.
5. Courty, J. M., Dauchel, C., Caruelle, D., Perderiset, M., and Barritault, D. (1991). Mitogenic properties of a new endothelial-cell growth-factor related to pleiotrophin. *Biochem. Biophys. Res. Commun.* **180,** 145–151.
6. Vacherot, F., Delbe, J., Heroult, M., Barritault, D., Fernig, D. G. and Courty, J. (1999) Glycosaminoglycans differentially bind HARP and modulate its biological activity. *J. Biol. Chem.* **274,** 7741–7747.
7. Edwards, P. R., Gill, A., Pollardknight, D. V., Hoare, M., Buckle, P. E., Lowe, P. A., and Leatherbarrow, R. J. (1995) Kinetics of protein-protein interactions at the surface of an optical biosensor. *Anal. Biochem.* **231,** 210–217.
8. Sadir, R., Forest, E., and LortatJacob, H. (1998) The heparan sulfate binding sequence of interferon-gamma increased the on rate of the interferon-gamma-interferon-gamma receptor complex formation. *J. Biol. Chem.* **273,** 10919–10925.

# 50

# Phage Display Technology to Obtain Antiheparan Sulfate Antibodies

**T. H. van Kuppevelt, G. J. Jenniskens, J. H. Veerkamp, G. B. ten Dam, and M. A. B. A. Dennissen**

## 1. Introduction

Antibodies have proven to be valuable tools in research on proteoglycans. They have been used extensively to study the tissue expression patterns of proteoglycans at the light as well as at the electron microscopical level. In addition, they have been frequently applied as (immuno)precipitating agents, in immunoaffinity chromatography, and—in some cases—as blocking agents. Most antibodies available are directed against the core protein of proteoglycans. Only a few are reactive with the glycosaminoglycan moiety. This is due to the largely nonimmunogenic character of glycosaminoglycans. Those antibodies that have been raised to glycosaminoglycans were obtained using proteoglycans as antigen, rather than the glycosaminoglycan chains as such. Here, we describe the use of phage display technology to obtain antibodies to glycosaminoglycans, as exemplified by heparan sulfate. Phage display allows the generation of antibodies to "self" antigens. The antibody is "displayed" at the surface of the phage by fusion to a coat protein ***(1,2)***. In the protocol described here, a semisynthetic antibody phage display library ["synthetic scFv library # 1", ***(3)***] was used, consisting of $>10^8$ different clones, each expressing one unique antibody. In principle, any antibody phage display library can be used. The synthetic library #1 contains 50 different $V_H$ genes with synthetic complementarity-determining region 3 segments (CDR3), which contain a random sequence, encoding 4–12 amino acid residues. Only one light-chain gene is present. In the library, only the variable parts of the heavy and light chains are expressed, joined to each other by a linker sequence to form so-called single-chain variable fragments (scFv). All antibodies contain a cMyc tag for identification with anti-cMyc antibodies.

From: *Methods in Molecular Biology, Vol. 171: Proteoglycan Protocols*
Edited by: R. V. Iozzo © Humana Press Inc., Totowa, NJ

A major advantage of the phage display system is that once a phage expressing the antibody has been selected, the DNA encoding the antibody is available. This opens the realm of molecular-biological techniques (e.g., large-scale production in bacteria, easy purification using His-tags, fusion of the antibodies to other proteins). Phage display-derived antibodies may be of value in characterizing the structural heterogeneity of heparan sulfate and other glycosaminoglycans.

This chapter describes the selection and characterization of anti-heparan sulfate antibodies and their coding genes using antibody phage display technology.

## 2. Materials

### 2.1. Selection of Phages Displaying Antibodies Reactive with Heparan Sulfate Using Biopanning

1. To avoid any carryover of phages during selections, several precautions need to be taken. The use of sterile disposable plastic ware and devoted pipetes is highly recommended. Nondisposable plastic ware should be soaked for 1 h in 2% (v/v) hypochlorite, followed by thorough washing and autoclaving. Glassware should be baked at 200ºC for at least 4 h. Use aerosol-resistant pipet tips (Molecular Bio-Products) when working with bacteria or phages. It is recommended to work in a laminar-flow cabinet or in a fume cabinet. Clean the workplace (benchtops, etc.) with 10% (v/v) hypochlorite before and after each working day. Clean pipetes etc., daily by wiping the outside with 0.1 *M* NaOH.
2. Bacterial strain: *Escherichia coli* TG1 ***(3)*** suppressor strain (K12, Δ(*lac-pro*), *sup*E, *thi*, *hsd*Δ(5/F'*tra*D36, *pro*AB, *lac*I$^q$, *LacZ*Δ(M15) (*see* **Note 1**).
3. VCS-M13 helper phages (Stratagene) (*see* **Note 2**) used at a titer of $1 \times 10^{12}$ CFu/mL. Alternatively, M13 KO7 helper phages (Pharmacia) can be used.
4. Glycerol stock of the (semi)-synthetic scFv Library #1 [Dr. G. Winter, Cambridge University, Cambridge, UK ***(3)***], stored at –80ºC.
5. 2XTY: 1.6% (w/v) Bacto-Trypton, 1.0% (w/v) Bacto-Yeast extract (Gibco BRL), and 0.5% (w/v) NaCl.
6. 40% (w/v) glucose (Sigma) in $H_2O$ sterilized by filtering using a 0.2-µm filter (Schleicher & Schuell).
7. Ampicillin (Sigma) and kanamycin (Gibco BRL).
8. 2XTY containing 100 µg of ampicillin/mL and 1% (w/v) glucose.
9. 2XTY containing 100 µg of ampicillin/mL and 25 µg kanamycin/mL.
10. Minimal medium: Autoclave 450 mL of 2.2% (w/v) Bacto-Agar (Gibco BRL) in $H_2O$. Cool the solution down to 60°C and add, after sterile filtering with a 0.2-µm filter (Schleicher & Schuell), 50 mL of 10XM9, 0.5 mL of 20% (w/v) $MgSO_4$, 2.5 mL of 40% (w/v) glucose, and 0.25 mL of 1% (w/v) thioamine.
    a. 10 × M9 medium: 0.60 *M* $K_2HPO_4$, 0.33 *M* $KH_2PO_4$, 76 m*M* $(NH_4)_2SO_4$, 17 m*M* trisodium citrate · $2H_2O$, pH 7.4 (adjust with phosphate component).
11. Polyethylene glycol (PEG)/NaCl: 20% (w/v) PEG 6000 (Serva) containing 2.5 *M* NaCl.
12. Phosphate-buffered saline (PBS): 0.14 *M* NaCl, 8.1 m*M* $Na_2HPO_4$ and 1.5 m*M* $NaH_2PO_4 \cdot 2H_2O$, 2.7 m*M* KCl, pH 7.4 (adjust with phosphate component).
13. Microlon immunotubes, 12/55 mm, 4 mL (Greiner).
14. Heparan sulfate from bovine kidney (Seikagaku).
15. Marvel: dried skimmed milk (Premier Beverages, Stafford, UK).
16. PBS containing 2% (w/v) Marvel.
17. PBS containing 4% (w/v) Marvel.

18. Parafilm (American National Can).
19. Polyoxyethylenesorbitan monolaurate (Tween-20, Sigma).
20. PBS containing 0.1% (v/v) Tween-20.
21. 100 m*M* triethylamine (Merck): add 700 µL (7.18 M) to 50 mL of $H_2O$ (prepare on the day of use).
22. 1 *M* Tris-HCl, pH 7. 4.
23. Sterile 50-mL tubes (Greiner).
24. TYE: 1.5% (w/v) Bacto-Agar, 0.8% (w/v) NaCl, 1.0% (w/v) Pepton, and 0.5% (w/v) Bacto-Yeast extract.
25. TYE containing 100 µg of ampicillin/mL and 1% (w/v) glucose.
26. Nunclon TC dish 245 × 245 × 25 mm (Nunc); 94/15 Petri dish (Greiner).
27. Glycerol (Sigma).
28. 2XTY containing 15% (v/v) glycerol.

### 2.2. Screening for Bacterial Clones Expressing Heparan Sulfate-Binding Antibodies Using ELISA

1. 96-well flat-bottom and 96-well round-bottom sterile Cellstar plates (Greiner).
2. Sterile toothpicks.
3. 2XTY containing 100 µg of ampicillin/mL and 1% (w/v) glucose.
4. 2XTY containing 100 µg of ampicillin/mL and 0.1% (w/v) glucose.
5. Isopropylthio-β-D-galactoside (IPTG, Gibco BRL).
6. 2XTY containing 100 µg of ampicillin/mL and 9 m*M* IPTG.
7. Glycerol (Gibco BRL).
8. 96-well Microlon ELISA plates, nonsterile (Greiner).
9. PBS.
10. PBS containing 2% (w/v) Marvel (*see* **Note 3**).
11. PBS containing 4% (w/v) Marvel.
12. PBS containing 0.1% (v/v) Tween-20.
13. Anti-c-Myc antibody; hybridoma culture supernatant (clone 9E10, mouse IgG; *see* **Note 4**).
14. PBS containing 2% (w/v) Marvel and 0.2% (v/v) Tween-20.
15. Alkaline phosphatase-conjugated rabbit anti-mouse IgG antibodies (DAKO).
16. PBS containing 1% (w/v) Marvel and 0.1 % (v/v) Tween-20.
17. 0. 9% (w/v) NaCl.
18. 1 *M* diethanolamine (Fluka) containing 0.5 m*M* $MgCl_2$, pH 9.8.
19. 4-Nitrophenyl phosphate disodium salt (hexahydrate) (P-NPP, Merck).

### 2.3. Detection of Antibodies Expressed by Bacterial Clones Using an Immunoblot Assay

1. 0.45-µm nitrocellulose filter (Schleicher & Schuell).
2. Whattman 3 MM paper.
3. PBS.
4. PBS containing 3% (w/v) Marvel and 1% (v/v) Tween-20.
5. PBS containing 2% (w/v) Marvel and 0.2% (v/v) Tween-20.
6. Anti-cMyc antibody; hybridoma culture supernatant (clone 9E10, mouse IgG; *see* **Note 4**).
7. PBS containing 0.1% (v/v) Tween-20.
8. Alkaline phosphatase-conjugated rabbit anti-mouse IgG antibodies (DAKO).
9. PBS containing 1% (w/v) Marvel and 0.1% (v/v) Tween-20.
10. 1 *M* diethanolamine containing 0.5 m*M* $MgCl_2$, pH 9.8.

11. *p*-Nitro blue tetrazolium chloride (NBT, Merck), and 5-bromo-4-chloro-3-indolyl sulfate *p*-toluidine salt (BCIP, Research Organics). Add to 10 mL of 1 *M* diethanolamine containing 0.5 m*M* $MgCl_2$ (pH 9.8): 45 μL of NBT (stock: 75 mg/mL 70% [v/v] dimethylformamide) and 35 μL of BCIP (stock 50 mg/mL dimethylformamide).

## 2.4. Screening for "Full-Length" Inserts Using Polymerase Chain Reaction (PCR) and for $V_H$ Gene Diversity Using DNA Fingerprinting

1. TYE containing 100 μg of ampicillin/mL and 1% (w/v) glucose.
2. 94/15 Petri dishes.
3. Sterile toothpicks.
4. 5 m*M* dNTP, Taq DNA polymerase (5 U/μL), and 10 × PCR buffer containing 15 m*M* $MgCl_2$ (Promega).
5. Primers: LMB3 (5'-CAGGAAACAGCTATGAC-3'), fd-SEQ1 (5'-GAATTTTCTGTATGAGG-3').
6. Mineral oil (Sigma).
7. Restriction enzyme *Bst*NI (10 U/μL), NEBuffer 2 (New England Biolabs).
8. DNA marker: ϕX174/*Hae*III Molecular Weight Marker 4 (Eurogentec).
9. SeaKem agarose, NuSieve 3:1 agarose (FMC Bioproducts).
10. 10 mg ethidium bromide (Sigma)/mL $H_2O$.
11. 10× TBE buffer: 12.1% (w/v) Tris, 5.1% (w/v) boric acid, and 3.7% (w/v) EDTA.
12. 5× DNA sample buffer: 0.25% (w/v) bromophenol blue, 0.25% (w/v) xylene cyanol FF, and 40% (w/v) sucrose.
13. QIAprep Spin Miniprep Kit (Qiagen), ABI PRISM™ Big Dye Terminator Cycle Sequencing Ready Reaction Kit (Perkin Elmer).
14. Sequencing primer: (5'- GCCACCTCCGCCTGAACC- 3'), annealing temperature: 65°C.
15. 2XTY containing 100 μg of ampicillin/mL and 1% (w/v) glucose.
16. 2XTY containing 15% (v/v) glycerol.
17. Sterile cryovials (Greiner).

## 2.5. Production of Culture Supernatant Containing Antibodies

1. Glycerol stock of a bacteria producing an anti-heparan sulfate scFv antibody, stored at –80°C.
2. TYE containing 100 μg of ampicillin/mL and 1% (w/v) glucose.
3. 94/15 petri dish.
4. 2XTY containing 100 μg of ampicillin/mL and 1% (w/v) glucose.
5. 2XTY containing 100 μg of ampicilin/mL and 0.1% (w/v) glucose.
6. IPTG (Gibco BRL).
7. 10× protease inhibitor mix: 0.1 *M* EDTA, 250 m*M* iodacetamid, 1 *M* *N*-ethylmaleimide, 1% (w/v) $NaN_3$, 1.5 mTIU aprotinin/mL, 0.1% (w/v) pepstatin A, and 1 m*M* phenylmethylsulfofluoride in $H_2O$.

## 2.6. Production of Periplasmic Fraction Containing Antibodies

1. *See* **Subheadings 2.5.1. – 2.5.7.**
2. Periplasmic fraction buffer (PPF): adjust 0.5 *M* boric acid (Gibco BRL) to pH 8.0 with 0.5 *M* sodium borate (Sigma). Take 20 mL of the adjusted solution and add 1.6 mL of 5 *M* NaCl, 0.25 mL of 0.2 *M* EDTA, and adjust the volume to 50 mL with $H_2O$.
3. 0.2 μm disposable filter holder.
4. Dialyzing membrane (Spectra/Por, cutoff value 10 kDa).
5. PBS.

### 2.7. Evaluation of Specificity of Antibodies Using Immunofluorescence Analysis

1. Tissue specimens, snap-frozen in liquid isopentane cooled with liquid nitrogen, and stored at –70°C.
2. 5-µm tissue cryosections (stored at –20 or –80°C).
3. Heparinase III (0.006 units/mL, Sigma) in incubation buffer (50 m*M* NaAc and 50 m*M* $Ca(Ac)_2$, pH 7.0).
4. Chondroitinase ABC (1 unit/mL, Seikagaku) in incubation buffer (25 m*M* Tris-HCl, pH 8.0).
5. Slides and cover slips.
6. PBS.
7. Blocking solution: PBS containing 1% (w/v) BSA .
8. Washing solution: PBS.
9. Primary antibody solution: add 1 volume of periplasmic fraction of an anti-heparan sulfate antibody to 1 volume of blocking solution.
10. Anti-cMyc (9E10) antibody (hybridoma culture supernatant) diluted 1/1 with blocking solution (*see* **Note 4**).
11. Fluorophore (Alexa-488)-conjugated anti-mouse IgG antibody (Molecular Probes) solution: Make a dilution (1/200–1/500) in PBS containing 0.5% (w/v) BSA.
12. Mowiol (Hoechst) embedding solution.

### 2.8. Evaluation of Specificity of Antibodies by ELISA

1. ELISA plates (Microlon, Greiner).
2. PBS.
3. Washing solution: PBS containing 0.1% (v/v) Tween-20.
4. Blocking solution: PBS containing 1% (w/v) BSA.
5. Anti-heparan sulfate (primary) antibody solution: add 1 volume of a bacterial culture supernatant containing an anti-heparan sulfate antibody to 1 volume of blocking solution, or use diluted periplasmic fraction.
6. Anti-cMyc (9E10) antibody (hybridoma culture supernatant) diluted 1/1 with blocking solution (*see* **Note 4**).
7. Alkaline phosphatase-conjugated rabbit anti-mouse IgG antibodies diluted 1/1000 in PBS containing 0.5% (w/v) BSA.
8. Substrate solution: 1 mg of *p*-nitrophenyl phosphate/mL diethanolamine solution (1 *M* diethanolamine, 0.5 m*M* $MgCl_2$, pH 9.8).
9. A number of molecules necessary for evaluation of crossreactivity of the antibodies. These may include heparan sulfate from various sources, heparin, dermatan sulfate, chondroitin 4 and 6-sulfate, keratan sulfate, dextran sulfate, hyaluronate, and DNA.

## 3. Methods

### 3.1. Selection of Phages Displaying Antibodies Reactive with Heparan Sulfate Using Biopanning

For a schematic outline of the selection procedure, *see* **Fig. 1**.

#### 3.1.1. Growth of Antibody Phage Display Library

1. Inoculate 50 mL of 2XTY containing 100 µg of ampicillin/mL and 1% (w/v) glucose with about $5 \times 10^8$ bacteria from a glycerol stock of the (semi)-synthetic scFv Library #1 (*3*).

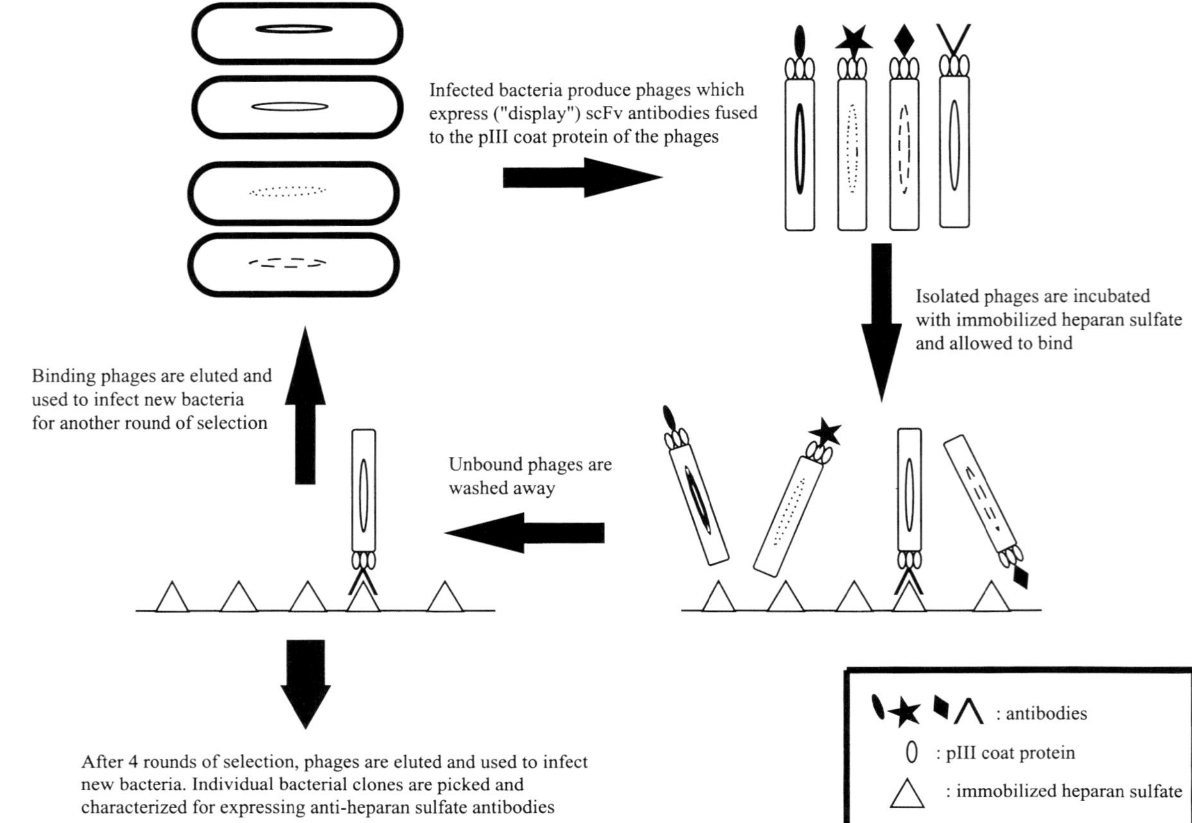


Fig. 1. Schematic representation of the selection of phage display-derived anti-heparan sulfate antibodies by biopanning.

2. Grow the culture, while shaking, at 37°C until an absorbance at 600 nm of 0.5 is reached (*see* **Note 5**).
3. Pipet 10 mL of the culture into a sterile 50-mL tube and add about $4 \times 10^{10}$ VCS-M13 helper phages (*see* **Note 6**).
4. Incubate in a water bath, without shaking, at 37°C for 30 min.
5. Spin the infected culture at 3000*g* for 10 min at room temperature. Decant the supernatant and resuspend the pellet in 30 mL of 2XTY containing 100 µg of ampicillin and 25 µg of kanamycin/mL.
6. Add the 30 mL of bacterial suspension to 270 mL of prewarmed (30°C) 2XTY containing 100 µg of ampicillin and 25 µg of kanamycin/mL (*No glucose*! *see* **Note 7**). Incubate, while shaking, at 30°C for 16–20 h to allow for large-scale (antibody-displaying) phage production.
7. Inoculate 5 mL of 2XTY with a single *E. coli* TG1 colony from a minimal medium plate and grow, while shaking, at 37°C for 16–20 h. This culture will be used in **Subheading 3.1.3, step 16**.

### 3.1.2. Isolation of Phages

1. Spin the culture from **step 6** under **Subheading 3.1.1** at 10,000*g* for 10 min at 4°C to remove bacteria. Decant the supernatant containing the phages into another sterile bucket.
2. Add 60 mL of ice-cold PEG/NaCl to the supernatant, mix well by inverting the bucket at least 30 times, and leave the bucket on ice for 1 h. In this step (and **step 3**) phages are precipitated (*see* **Note 8**).
3. Spin the phages at 10,000*g* for 30 min at 4°C. Resuspend the pellet in 40 mL of ice-cold, sterile milli-Q water. Transfer the suspension to a 50-mL tube and add 8 mL of ice-cold PEG/NaCl. Mix well (as in **step 2**) and leave for 30 min on ice.
4. Spin the mixture at 3000*g* for 30 min at 4°C. Decant the supernatant and remove the remains with a pipet. Respin briefly and remove residual PEG/NaCl. Invert the tube on a piece of paper tissue, and leave for 30 min to drain any remaining PEG/NaCl.
5. Resuspend the pellet in sterile PBS and spin at 3000*g* for 10 min at 4°C to remove any remaining bacterial debris. Decant the supernatant containing the phages into a sterile tube and store at 4°C until use.

### 3.1.3. Selection of Phages Binding to Heparan Sulfate

1. Add 2 mL of a 20-µg heparan sulfate/mL solution to an immunotube for 16 h, 4°C. Wash the immunotube 3 times with PBS and block the tube with PBS containing 2% (w/v) Marvel. Fill the tube to the brim, cover it with Parafilm, and incubate for at least 2 h at room temperature to avoid nonspecific binding of phages to the surface of the tube. This step should be performed early in the day, so the immunotube will be ready when the phages used for biopanning (**step 5, Subheading 3.1.2.**) are obtained.
2. Wash the blocked tube 3 times with PBS. Add 2 mL of PBS containing 4% (w/v) Marvel and 2 mL of phage supernatant (**step 5, Subheading 3.1.2.**) to the tube, cover with Parafilm, and incubate for 30 min on an under-and-over turntable (room temperature), followed by standing for 90 min (room temperature).
3. Discard the phage suspension and wash the tube 20 times with PBS containing 0.1% (v/v) Tween-20, followed by 20 washes with PBS.
4. Remove the last remains of PBS and elute the bound phages with 1 mL of 100 m*M* triethylamine. Cover the tube with Parafilm and rotate for 10 min on an under-and-over turntable at room temperature.

5. Add the 1 mL of eluted phages to a 50-mL tube containing 0.5 mL of 1 *M* Tris-HCl (pH 7.4) for pH neutralization. Also add 200 µL of 1 *M* Tris-HCl (pH 7.4) to the remaining phages in the immunotube. At this point phages can be stored at 4°C for a short period of time (up to 2 d), or used the same day to infect *E. coli* TG1 cells. The latter is recommended.
6. Add 1 mL of the eluted phages from the 50-mL tube (**Subheading 3.1.3., step 5**) to 9 mL of exponentially growing *E. coli* TG1 cells in a 50-mL tube. Add 4 mL of the TG1 culture to the remaining phages in the immunotube (**step 5**). Incubate both cultures for 30 min at 37°C in a water bath, without shaking, to allow for infection. Exponentially growing TG1 culture is prepared as follows: Inoculate 50 mL of 2XTY with 0.5 mL of overnight culture from **Subheading 3.1.1., step 7**. Grow the culture, while shaking, at 37°C until an absorbance at 600 nm of 0.4–0.5 is reached. The culture obtained is ready for infection (*see* **Note 5**).
7. Pool both infected TG1 cultures. Take 100 µL of the pooled culture and make 4 serial dilutions ($1/10^2$, $1/10^4$, $1/10^6$, and $1/10^8$) in 2XTY containing 100 µg of ampicillin/mL and 1%(w/v) glucose. Plate 100 µl of these dilutions on 94/15 TYE plates containing 100 µg of ampicillin/mL and 1% (w/v) glucose, and grow for 16–20 h at 37°C. Calculate the titer from these dilutions (*see* **Note 9**).
8. Spin the rest of the pooled culture at 3000*g* for 10 min at room temperature. Decant the supernatant and resuspend the pellet in 1 mL of 2XTY. Spread the cell suspension on a Nunclon TC plate with TYE containing 100 µg of ampicillin/mL and 1% (w/v) glucose and grow for 16–20 h at 37°C.
9. Add 5 mL of ice-cold 2XTY containing 15% (v/v) glycerol to the Nunclon TC dish and scrape the bacterial cells from the plate with a glass spreader. Take 50 µL of the bacterial suspension and use it for inoculation of 50 mL 2XTY containing 100 µg of ampicillin/mL and 1% (w/v) glucose as in **Subheading 3.1.1., step 1**. Store the remaining bacteria at –70°C (*see* Note 10). Repeat the selection procedure for another 3–4 rounds (all steps **Subheading 3.1.1.–3.1.3.**).

### 3.2. Screening for Bacterial Clones Expressing Heparan Sulfate-Binding Antibodies Using ELISA

1. Pick individual bacterial clones, using sterile toothpicks, from the serial dilution plates of the last 2 selection rounds (*see* **Subheading 3.1.3., step 6**) and inoculate 100 µL of 2XTY containing 100 µg of ampicillin/mL and 1% (w/v) glucose (one clone per well!) in sterile 96-well flat-bottom tissue culture plates. Secure the lid with tape and grow for 16–20 h at 37°C while gentle shaking. These plates will be the master plates.
2. Transfer 2 µL of bacterial culture (**step 1**) to the corresponding wells of sterile 96-well round-bottom tissue culture plates with 200 µL 2XTY containing 100 µg of ampicillin/mL and 0.1%(w/v) glucose (*see* **Note 11**). Secure the lid with tape and grow at 37°C, while shaking, until an absorbance at 600 nm of 0.9 is reached (*see* **Note 12**). Add to each well 25 µL of 2XTY containing 100 µg of ampicillin/mL and 9 m*M* IPTG. Incubate the plates, while shaking gently, at 30°C for 16–20 h. Add glycerol to the master plates to a final concentration of 15% (v/v), mix well, and store the plates at –70°C until further use.
3. Incubate wells from ELISA plates with 100 µl of a 10-µg heparan sulfate/mL solution for 16 h at 4°C. Wash 3 times with PBS and block with PBS containing 2% (w/v) Marvel for 1 h at 37°C. Wash the plates 3 times with PBS containing 0.1% (v/v) Tween-20.
4. Spin the 96-well round-bottom plates (**step 2**) at 1800*g* for 10 min at room temperature.
5. Add 60 µL of the culture supernatant (**step 4**) to 60 µl of PBS containing 4% Marvel, transfer 100 µl to wells of ELISA plates, and incubate for 1 h at room temperature. Use 50 µL of the culture supernatant in an immunoblot assay used for evaluation of antibody production (*see* **Subheading 3.3.**). The immunoblot assay can be performed simultaneously with the ELISA.

6. Discard the culture supernatant and wash the ELISA plates 6 times with PBS containing 0.1% (v/v) Tween-20.
7. Add 100 µL of 9E10 hybridoma supernatant diluted 1/1 with PBS containing 2% (w/v) Marvel and 0.2% (v/v) Tween-20 to the wells and incubate for 1 h at room temperature.
8. Discard the 9E10 solution and wash the ELISA plates 6 times with PBS containing 0.1% (w/v) Tween-20.
9. Add 100 µL of alkaline phosphatase-conjugated rabbit-anti-mouse IgG antibodies, diluted 1/1000 diluted in PBS containing 1% (w/v) Marvel and 0.1% (v/v) Tween-20 to the wells and incubate for 1 h at room temperature.
10. Discard the fluid and wash the ELISA plates 5 times with PBS containing 0.1% (v/v) Tween-20 followed by 1 wash with 0.9% (w/v) NaCl.
11. Add 100 µL 1 *M* diethanolamine containing 1-mg/mL P-NPP and 0.5 m*M* $MgCl_2$ (pH 9.8) to the wells and incubate, in the dark, at room temperature until color development is optimal.
12. Read the absorbance at 405 nm (*see* **Note 13**).

### 3.3. Detection of Antibody Expression by Bacterial Clones Using an Immunoblot Assay

1. Cut two Whattman 3 MM papers and one 0.45-µm nitrocellulose filter to the size required for use in a 96-well dot-blot apparatus.
2. Soak the papers and the filter in PBS for 10 min. Apply the papers and filter to the dot-blot apparatus (nitrocellulose filter on top). Make sure to remove all air bubbles.
3. Transfer 50 µL of culture supernatant from **step 4** under **Subheading 3.2.** to the dot-blot apparatus and pull the fluid through the filter by vacuum suction.
4. Remove the nitrocellulose filter from the apparatus and air-dry the filter for 20 min at room temperature.
5. Block the filter in a container with PBS containing 3% (w/v) Marvel and 1% (v/v) Tween-20 for 1 h at room temperature, while shaking. The volume of fluids used in this step, as well as in the following steps, depends on the size of the container used. Make sure the filter is sufficiently covered with fluid.
6. Discard the blocking solution and add 9E10 hybridoma supernatant diluted 1/1 with PBS containing 2% (w/v) Marvel and 0.2% (v/v) Tween-20. Incubate, while shaking, for 1 h at room temperature.
7. Discard the solution and wash 3 times for 10 min, while shaking, with PBS containing 0.1% (v/v) Tween-20.
8. Add alkaline phosphatase-conjugated rabbit anti-mouse IgG antibodies diluted 1/1000 in PBS containing 1% (w/v) Marvel and 0.1% (v/v) Tween-20. Incubate, while shaking, for 1 h at room temperature.
9. Discard the antibody solution and wash, while shaking, 2 times with PBS containing 0.1% (v/v) Tween-20 for 10 min, 1 time with PBS for 5 min, and 1 time with 1 *M* diethanolamine containing 0.5 m*M* $MgCl_2$ (pH 9.8) for 5 min.
10. Add NBT/BCIP substrate solution and incubate, while shaking, at room temperature until the color develops. Discard the substrate solution and wash the filter a couple of times with water (*see* **Note 14**).

### 3.4. Screening for "Full-Length" Inserts Using PCR and for $V_H$ Gene Diversity Using DNA Fingerprinting

In this procedure, the positive clones will be analyzed for the presence of DNA (about 1 kbp) encoding the scFv antibody. In addition, as an initial screening for the diversity of the clones, restriction enzyme analysis will be performed. (Also *see* **Fig. 2**).

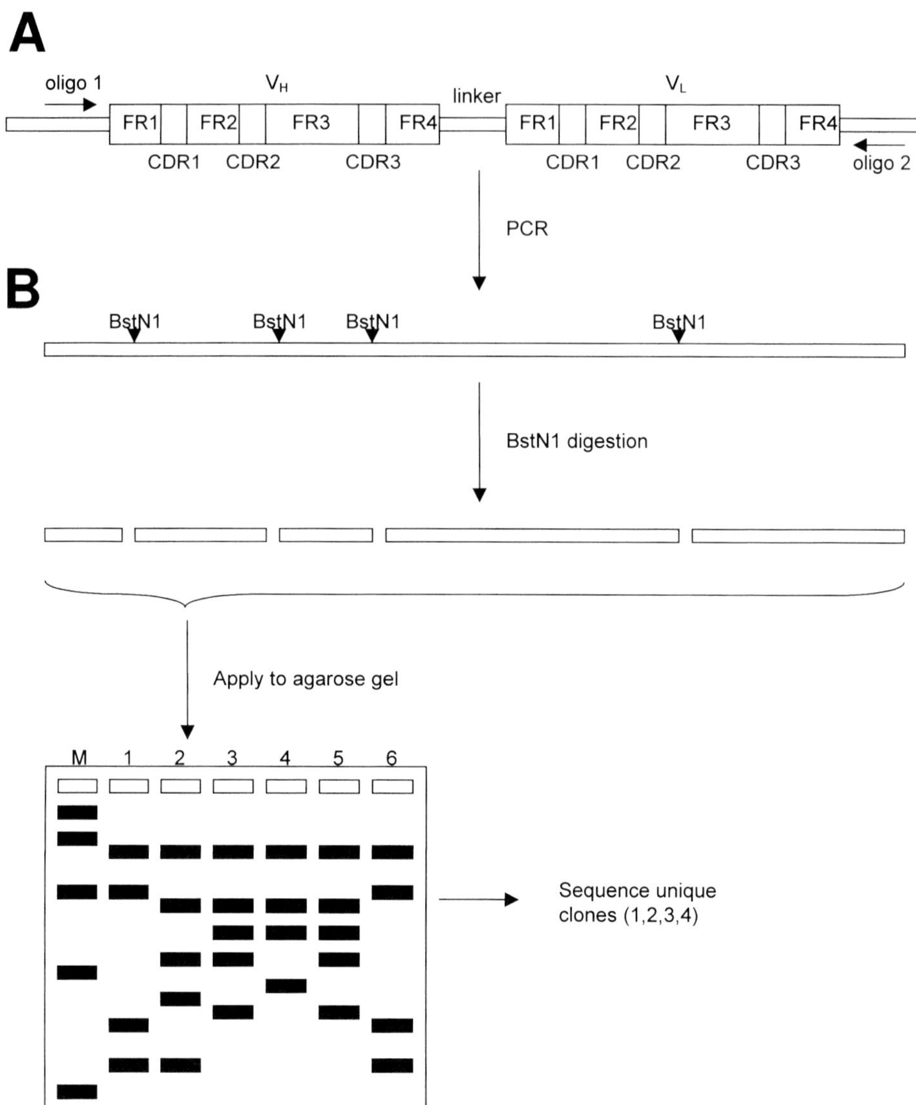


Fig. 2. Restriction fragment analysis of clones expressing anti-heparan sulfate antibodies. Polymerase chain reaction (PCR) is preformed to amplify the region encoding the scFv antibody **(A)**, using a set of primers flanking the $V_H$ and $V_L$ segments. The resulting (full-length) PCR fragments **(B)** harbor a unique pattern of BstN1 cleavage sites ($CC{*}^{A}/_{T}GG$), serving as a fingerprint. Following digestion with restriction enzyme BstN1 **(C)**, the fragments are separated on an agarose gel and the restriction patterns of each clone are compared **(D)**. Clones with unique restriction patterns are selected and sequenced to establish the $V_H$ family, germ line segment (DP number, *see* **ref. *4***) and $V_H$ -CDR3 sequence (randomized in the library used; *see* **ref. *3***).FR, framework region; CDR, complementary determining regions; M, DNA marker.

1. Select heparan sulfate-positive clones (as detected by ELISA) from the corresponding master plate (*see* **Subheading 3.2.**). Plate bacteria on 94/15 dishes with TYE containing 100 µg of ampicillin/mL and 1% (w/v) glucose to obtain single colonies.
2. Pick a single colony of each clone (mark the colony on the back of the Petri dish) with a sterile toothpick and transfer the cells into the following PCR-mixture: 34.5 µL $H_2O$, 5.0 µL 10X PCR buffer, 2.5 µL 20X dNTP (5 m*M* each), 2.5 µL LMB3 primer (10 pmol/µL), 2.5 µL fd-SEQ1 primer (10 pmol/µL), 0.5 µL T*aq* polymerase (5 U/µL). Overlay the PCR mixture with a droplet of mineral oil and use the following PCR program: 10 min at 94°C, 30 cycles of 1 min at 94°C, 1 min at 60°C, 2 min at 72°C, 10 min at 72°C, cool to 4°C.
3. Take 4 µL of the PCR-mixture and add 1 µL of 5X DNA sample buffer. Run the samples on a 1% (w/v) Seakem agarose gel (add 3.0 µL ethidium bromide [10 mg/mL] to 75 mL of agarose solution). Include DNA marker (250 ng) in one of the lanes. Run the gel at 50 V in an appropriate volume of 1X TBE buffer. Analyze the gel on a UV transilluminator. PCR products of about 1000 base pairs indicate full-length clones (*see* **Note 15**).
4. For DNA fingerprinting take 20 µL of the PCR-mixture and add 20 µL of the following restriction enzyme mix: 17.8 µL $H_2O$, 2.0 µL 10X NEbuffer 2, 0.2 µL *Bst*NI (10 U/µL). Overlay with mineral oil and incubate at 60°C for 3 h. Take 8 µL and add 2 µL 5X DNA sample buffer. Run the restriction mixtures on a 4% (w/v) 3/1 NuSieve agarose gel as described in **step 3**, **Subheading 3.4.**, Differences in banding pattern indicate different clones (*see* **Note 16** and **Fig. 2**).
5. Take with a sterile toothpick unique clones from the plate with marked colonies (**step 2**, **Subheading 3.4.**) and add to 10 mL 2XTY containing 100 µg ampicillin/mL and 1% (w/v) glucose. Grow for 16–20 h, while shaking, at 37°C.
6. Take 1.5 mL of the bacterial culture (**step 5**, **Subheading 3.4.**) for preparation of phagemid DNA, used for sequencing and for long-term storage (*see* **Note 17**). For phagemid isolation and sequencing we use materials described in **steps 13** and **14**, **Subheading 2.4.**
7. Spin the rest of the culture at 3000*g* for 10 min at 4°C. Decant the supernatant and resuspend the pellet in 1 mL of ice-cold 2XTY containing 15% (v/v) glycerol. Aliquot the bacterial suspension into several sterile cryovials and store at –70°C.

### 3.5. Production of Culture Supernatant Containing Antibodies

1. Inoculate 5 mL of 2XTY containing 100 µg ampicillin/mL and 1% (w/v) glucose with a single colony from a 94/15 dish with TYE containing 100 µg ampicillin/mL and 1% (w/v) glucose, and derived from a glycerol stock. Grow for 16–20 h, while shaking, at 37°C.
2. Inoculate 500 mL of 2XTY containing 100 µg ampicillin/mL and 0.1% (w/v) glucose with 5 mL of culture (**step 1**, **Subheading 3.5.**) and grow, while shaking, at 37°C until an absorbance at 600 nm of 0.5 –0.8 is reached.
3. Add IPTG to a final concentration of 1 m*M* and grow the culture, while shaking, at 30°C for 16–20 h.
4. Put the culture on ice for 20 min.
5. Spin the culture at 3000*g* for 10 min at 4°C. Add 0.1 volume of 10X protease inhibitor mix (*see* **step 7**, **Subheading 2.5.**) to the supernatant and store in aliquots at 4°C, if used directly, or at –70°C (*see* **Note 18**).

### 3.6. Production of Periplasmic Fraction Containing Antibodies

1. *See* **step 1, Subheading 3.5.**
2. *See* **step 2, Subheading 3.5.**
3. Add IPTG to a final concentration of 1 m*M* and grow the culture, while shaking, at 30°C for 3 h.

4. Put the culture on ice for 20 min.
5. Spin the culture at 3000*g* for 10 min at 4°C. Decant the supernatant and resuspend the bacterial pellet in 5 mL of ice-cold "periplasmic fraction" (PPF) buffer (*see* **step 2, Subheading 2.6.**).
6. Vortex the bacterial suspension vigorously for 10 s and spin at 48,000*g* for 30 min at 4°C.
7. Filter the supernatant (periplasmatic fraction) through a 0.2-μm disposable filter.
8. Dialyze the periplasmatic fraction for 16–20 h against 5 L of PBS at 4°C.
9. Add 0.1 volume of 10X protease inhibitors (*see* **step 7, Subheading 2.5.**). Store the periplasmatic fraction at 4°C, if used directly, or at –70°C (*see* **Note 18**).

### 3.7. Evaluation of Specificity of Antibodies Using Immunofluorescence Analysis

1. All incubation steps are carried out in a humid atmosphere at room temperature.
2. Incubate cryosections with heparinase III or chondroitinase ABC overnight at 37°C. As a control for enzyme reactivity, incubation buffer without enzyme is used (*see* **Notes 19** and **20**).
3. Rinse 3 times in PBS and block in PBS containing 1% (w/v) BSA for 30 min.
4. Incubate cryosections with primary antibody solution (anti-heparan sulfate antibodies) for 60 min.
4. Remove primary antibody solution carefully and rinse once and wash 3 times (10 min) with PBS.
5. Incubate cryosections with mouse anti-cMyc antibody solution for 60 min.
6. Remove antibody solution carefully and rinse once and wash 3 times (10 min) with PBS.
7. Incubate cryosections with fluorophore-conjugated anti-mouse IgG antibody solution for 60 min.
8. Remove antibody solution carefully and rinse once and wash 3 times (10 min) with PBS.
9. Incubate cryosections in 100% methanol for 10 s for dehydration.
10. Air-dry sections and use mowiol for embedding.
11. Analyze staining patterns by fluorescence microscopy (*see* **Note 21**). **Figures 3** and **4** are examples of staining patterns.

### 3.8. Evaluation of Specificity of Antibodies Using ELISA

The reactivity of the anti-heparan sulfate antibodies with other molecules can be analyzed by ELISA in two ways: (1) by application of antibodies to wells of microtiter plates coated with the test molecule, or (2) by an inhibition assay in which the antibodies are incubated with the test molecule.

#### *Method 1*

1. All incubation steps are carried out at room temperature.
2. Coat wells with test molecules (*see* **step 9, Subheading 2.8.**) by incubation with 10 μg of test molecules/mL solution for 16 h at 4°C.
3. Wash the wells 6 times with PBS containing 0.1% (v/v) Tween-20.
4. Block the free binding sites with 200 μL of blocking solution for 1 h.
5. Wash the wells 6 times with PBS containing 0.1% (v/v) Tween-20.
6. Incubate the wells with 100 μL of anti-heparan sulfate (primary) antibody solution for 60 min.
7. Wash the wells 6 times with PBS containing 0.1% (v/v) Tween-20.
8. Incubate the wells with 100 μL of mouse anti-cMyc antibody solution for 60 min.

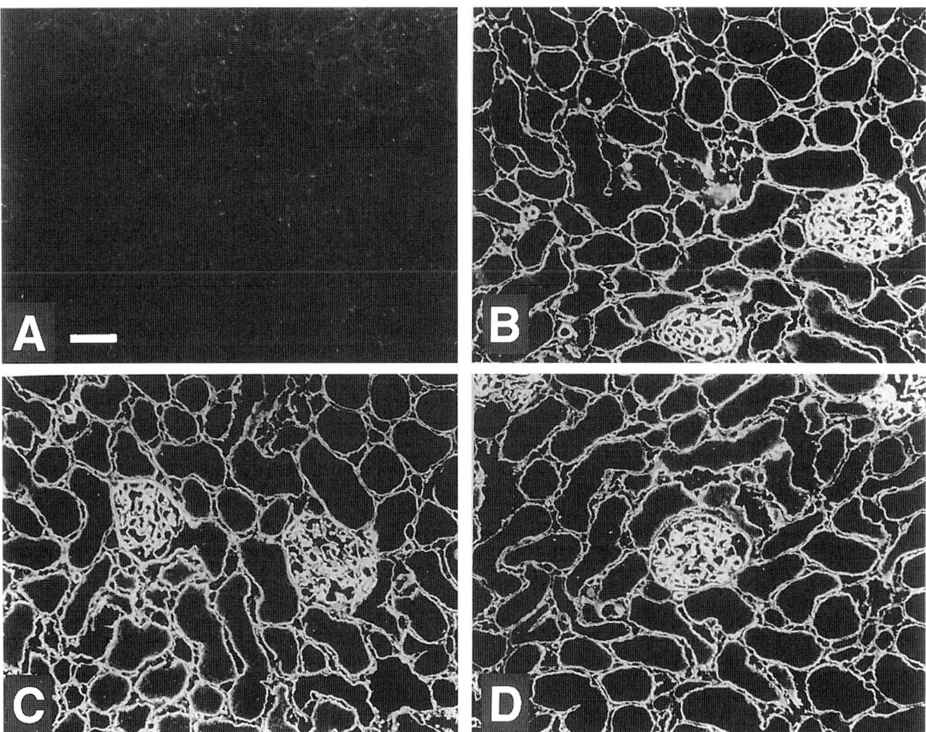


Fig. 3. Specificity of anti-heparan sulfate antibody HS3G8. Rat kidney cryosections were treated with heparinase III (***A***), heparinase III incubation buffer (***B***), chondroitinase ABC (***C***), and chondroitinase ABC incubation buffer (***D***). Next, sections were incubated with periplasmic fraction containing the antibody. Bound antibodies were visualized using mouse anti-cMyc IgG followed by Alexa 488-conjugated goat anti-mouse IgG. Bar= 50 μm. *Source:* from **ref. 5.**

9. Wash the wells 6 times with PBS containing 0.1% (v/v) Tween-20.
10. Incubate the wells with 100 μL of alkaline phosphatase-conjugated rabbit anti-mouse IgG antibody solution for 60 min.
11. Wash the wells 4 times with PBS containing 0.1% (v/v) Tween-20.
12. Wash the wells 2 times with 0.9% (v/v) NaCl.
13. Incubate the wells with 100 μL of substrate solution (in the dark) until color development is optimal.
14. Read absorbance at 405 nm.

### Method 2

1. Add 6 μg of test molecule to 150 μl antibody solution and incubate overnight at room temperature.
2. Transfer 100 μl of this solution to a well coated with heparan sulfate and incubate for 60 min.
3. Proceed as under **Method 1**, starting with **step 7**.

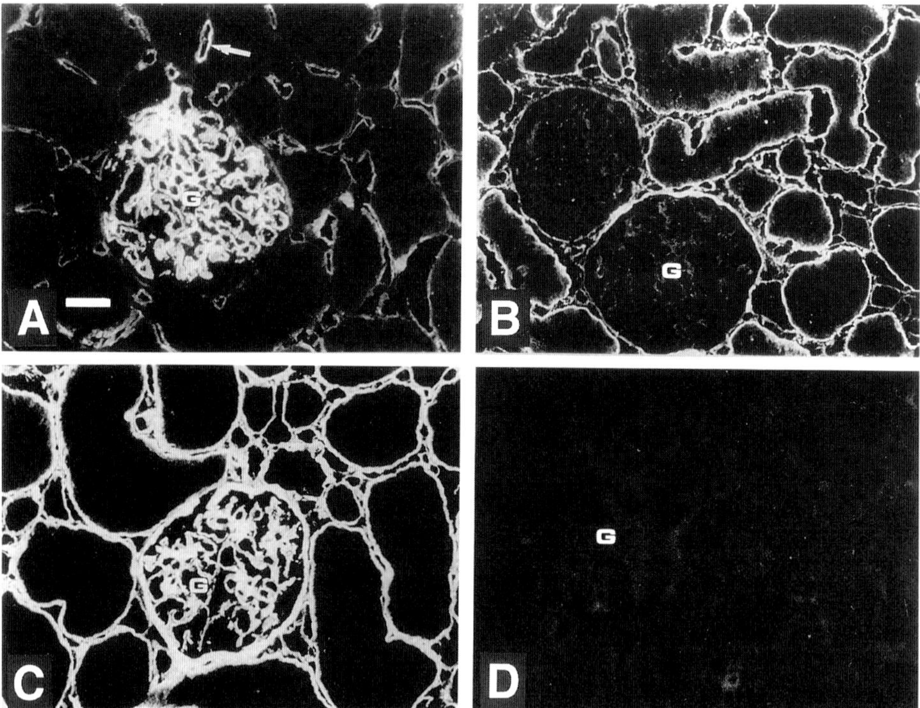


Fig. 4. Immunostaining of rat kidney with three different anti-heparan sulfate scFv antibodies. Cryosections were incubated with periplasmatic fractions of bacteria expressing antibody HS4C3 (**A**), HS4D10 (**B**), HS3G8 (**C**), and anti-filaggrin (**D**)(control; filaggrin is not present in the kidney). Bound antibodies were visualized using mouse anti-cMyc IgG followed by Alexa 488-conjugated goat anti-mouse IgG. Bar=25 µm. All three anti-heparan sulfate antibodies stain differently, indicating reactivity with different heparan sulfate species. G, glomerulus. Arrow in **a**: peritubular capillary. *Source* from **ref. 5**.

## 4. Notes

1. *E. coli* TG1 is a T–phage resistant strain that harbors a mutated tRNA gene. The mutated tRNA will suppress the UAG amber (stop) codon. A glutamine will be substituted for the amber codon allowing the expression of scFv-pIII fusion protein on the phage tip.
2. VCS-M13 helper phages provide phage coat proteins and enzymes necessary for phage rescue.
3. Except for the scFv antibody selection procedure, Marvel can be substituted with bovine serum abumin (BSA) in the same concentrations.
4. The 9E10 hybridoma cell line is available from the ATCC (American Type Culture Collection). Alternatively, a polyclonal rabbit anti c-Myc antibody (A14, Santa Cruz Biotechnology) can be used.
5. M13 phages infect $F^+$ *E. coli* via sex pili. For production of sex pili, *E. coli* needs to be grown into the log phase (absorbance at 600 nm of 0.4–0.5) at 37°C. When grown to a higher density, sex pili are lost very rapidly. A log phase culture can be kept on ice for no

longer than 30 min. Have the eluted and neutralized phages (**step 5, Subheading 3.1.3.**) ready for immediate infection.

6. Take the remaining 40 mL of the culture, spin it down and resuspend the pellet in 1 mL 2XTY. Spread it on a 245 ↔ 245 ↔ 25 Nunclon TC dish containing TYE , 100 μg ampicillin/mL and 1% (w/v) glucose, and grow overnight at 37°C. Harvest the cells in 1–2 mL of ice-cold 2XTY containing 15% (v/v) glycerol and store this stock in 50-μL aliquots (about $10^8$ clones) at –70°C for other selections.
7. Glucose represses transcription of the scFv-pIII fusion protein through the *lac* operon in the phagemid.
8. Besides concentrating the phages, this step is also necessary for removing any soluble antibodies, since the TG1 suppression of the amber codon is never complete.
9. After each selection round an increase in titer is expected, indicating enrichment of binding clones.
10. The glycerol stocks are used as a backup. When a subsequent selection round fails, use 50 μl of stock for a new round of selection.
11. In this step, 0.1% (w/v) glucose is added to suppress the expression of scFv antibodies until a sufficient number of cells is produced for large-scale antibody production. The total amount of glucose will be metabolized at an absorbance at 600 nm of 0.9.
12. In case no suitable apparatus is available for measering absorbance, grow for 3 h, while shaking, at 37°C before adding IPTG. Bacterial growth should be clearly visible.
13. Include negative controls (e.g., supernatant without scFv antibodies, omission of culture supernatant).
14. Clones that are weakly positive in both the immunoblot assay and in the ELISA may still be clones of interest. They may react weakly in ELISA because of poor scFv antibody production.
15. Non–full-length clones should be ignored, since they are notoriously unspecific binders.
16. Fingerprinting is a quick method for looking for clone diversity. This method is very useful when a large number of positive clones is to be examined.
17. Next to storage of bacteria, it is recommended to store phagemid DNA of a selected clone in 70% (v/v) ethanol at –70°C. You can analyze the DNA sequence for the $V_H$- gene number using the Sanger center's germline query (http://www.sanger.ac.uk/DataSearch/gq_search.shtml). With the gene number the $V_H$ family can be identified using a paper by Tomlinson et al. ***(4)***.
18. The stability of scFv antibodies is variable. Some antibodies can be stored at 4°C for weeks to months, whereas others will stay immunoreactive only for a couple of days. Most antibodies can be stored at –70°C for months up to years. Bacterial supernatants containing scFv antibodies are suitable for ELISA, but are generally not suitable for immunohistochemistry. Periplasmic fractions, in which the antibodies are more concentrated, are suitable for both. The use of the HB2151 nonsuppressor strain of *E. coli* generally gives a higher yield of (soluble) antibodies.
19. To analyze the HS specificity of the antibody, tissue sections are pretreated with heparinase III, which digests all heparan sulfates. As a control, pretreatment with chondroitinase ABC, which digests chondroitin sulfates and dermatan sulfates, is performed.
20. As a control to verify effectiveness of heparinase III treatment, an antibody (3G10, Seikagaku) directed against heparan sulfate stubs, generated by the enzyme, may be used.
21. The stained tissue sections can be kept for up to 6 mo. The fluorescent tag (Alexa 488) is very stable, and fading of the signal hardly occurs. Store frozen or at –20°C. Background

staining can often be eliminated by additional blocking steps with 1–2% (w/v) BSA or with serum (1–5% v/v) from the same species in which the tertiary antibody is raised. Extra washing steps can also lower the background signals.

## References


1. Borrebaeck, C. A. K., ed. (1995) *Antibody engineering*. Oxford University Press, New York, NY.
2. Kay, K. B., Winter, J., and McCafferty, J. (1996) *Phage Display of Peptides and Proteins*. Academic, San Diego, CA.
3. Nissim, A., Hoogenboom, H. R., Tomlinson, I. M., Flynn, G., Midgley, C., Lane, D., and Winter, G. (1994) Antibody fragments from a 'single' pot, phage display library as immunochemical reagents. *EMBO J.* **13,** 692–698.
4. Tomlinson, I. M., Walter, G., Marks, J. D., Llewelyn, M. B., and Winter, G. (1992) The repertoire of human germline $V_H$ sequences reveals about fifty groups of $V_H$ segments with different hypervariable loops. *J. Mol. Biol.* **227,** 776–798.
5. Van Kuppevelt, T. H., Dennissen, M. A. B. A., Van Venrooij, W. J., Hoet, R. M. A., and Veerkamp, J. H. (1998) Generation and application of type-specific anti-heparan sulfate antibodies using phage display technology. *J. Biol. Chem.* **273,** 12,960–12,966.

# 51

# Tissue-Specific Binding by FGF and FGF Receptors to Endogenous Heparan Sulfates

**Andreas Friedl, Mark Filla, and Alan C. Rapraeger**

## 1. Introduction

Heparan sulfate proteoglycans (HSPGs) bind via their heparan sulfate (HS) glycosaminoglycan chains to a large variety of extracellular ligands. These ligands include components of the extracellular matrix, other cell surface receptors, viruses, proteases and their inhibitors, and growth factors. The interaction of growth factors with HS has been proposed to affect growth factor function, if by no means other than limiting growth factor diffusion. For certain growth factors, however, work over the past decade has shown that HS binding has a more direct role in signaling. It is proposed that the HS chain participates directly in the assembly of these growth factors with their signaling receptor and may even act as a regulator of the signaling *(1,2)*. This role for HSPGs has been shown for epidermal growth factor family members, most notably heparin-binding EGF and the heregulins *(3)*, hepatocyte growth factor/scatter factor *(4)*, and for the members of the fibroblast growth factor (FGF) family *(5,6)*.

The role of HSPGs as regulators of growth factor action is best characterized for the FGF family, and this family will be the focus of this chapter. HSPGs regulate FGF signaling by participating in the formation of a ternary signaling complex comprised of the FGF ligand, FGF receptor tyrosine kinase (RTK), and HS *(7,8)*. HS-binding sites on both the FGF ligand *(9)* and the RTK *(10)* facilitate assembly of this signaling complex. Because the HS structure is known to have considerable variation, particularly in its sulfation pattern, it becomes an important question as to whether this variation is specific and serves to regulate the formation of these signaling complexes. The 19 FGF family members currently known signal through RTKs that are encoded by four genes (FGFR1–4). In addition, all FGFRs with the exception of FGFR4 are subject to RNA splice variation with profound effects on ligand specificity [reviewed in *(11)*]. HSPGs have been shown to be both promoters and inhibitors of FGF signaling

From: *Methods in Molecular Biology, Vol. 171: Proteoglycan Protocols*
Edited by: R. V. Iozzo © Humana Press Inc., Totowa, NJ

*(12,13)*. This dual role can be explained by differential binding of domains within the HS chain to FGF ligand and its RTK. For example, a HS that contains a specific sulfation sequence for both the FGF and the RTK is predicted to behave as a stimulator of FGF signal transduction, whereas a sulfation sequence that binds the FGF ligand but fails to recognize the RTK is postulated to be a competitive inhibitor. Specificity in the binding interactions between HS and the other signaling partners is made possible by the remarkable diversity of HS polysaccharide chains. In fact, the information density present in HS chains exceeds that of nucleic acids or polypeptides *(14)*.

Experimental evidence is also accumulating in support of the notion that *specific* HSPG core proteins bear specific FGF-modulating HS chains *(15,16)*. It is unclear whether this is a uniform behavior of a specific core protein, since literature reports have attributed FGF stimulatory and inhibitory activities to basically all existing HSPG classes, or whether this depends on the cell type that expresses the HSPG. Divergent reports on the activity of specific HSPGs are likely to reflect differences in the cell type of origin and metabolic state of the experimental model. In vivo, HSPGs are widely distributed, but show remarkable tissue- and cell type-specific patterns of expression. It is also becoming increasingly clear that HSPGs are not passive FGF co-receptors, but are themselves dynamically regulated with dramatic effects on FGF signal transduction *(17)*.

An important question regarding HSPG regulation of FGF signaling, therefore, is whether the HS chains are expressed with specific regulatory properties and whether this specific expression occurs within a specific tissue or cell type. Much of the information on HS regulation of FGF signaling has been acquired using heparan sulfate isolated in bulk from tissues. These extracts would of course represent a mixture of HSPGs stemming from multiple cellular and extracellular sources, and any location-specific information would inescapably be lost. Although the question can also be addressed by extracting HSPGs from cells in culture and examining the effects of the isolated HSPG preparations on FGF signaling, the clear disadvantage of this approach is that there is no reason to believe that cell lines that have escaped normal growth control would still be equipped with HSPGs equivalent to their counterparts in vivo. Therefore, we have developed different assays that allow us to localize HS chains *in situ* and to explore their abilities to assemble a ternary complex with FGF and RTK, thereby providing direct information of their potential role in FGF signaling *(18,19)*.

The strategy that will be described is to reconstitute the signaling complex *in situ* in a stepwise fashion by adding each of its components as a binding probe (illustrated schematically in **Fig. 1**). As a first step, the HS chains in the tissue need to be localized. This can be achieved by using an antibody directed against the "stubs" remaining on HSPGs after heparitinase digestion (anti-delta-HS antibody 3G10) *(20)*. This localization is important for establishing whether all of the HS participates in formation of a signaling complex, or whether this is limited to only a specific HS population. Since monoclonal antibody 3G10 reacts with HS sugar moieties rather than the protein, it detects all HSPGs regardless of the core protein identity. Next, HS capable of binding the FGF under investigation needs to be identified. This is done by incubating the tissue sections with the FGF and detecting bound growth factor (*see* **Fig. 1A**). Due to

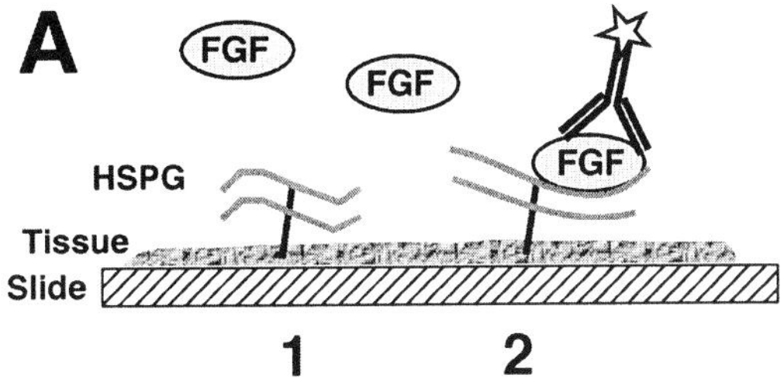


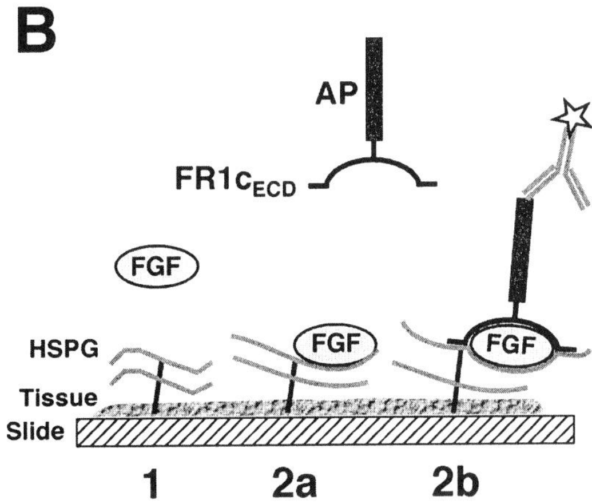


Fig. 1. Schematic representation of the *in situ* binding assays. **(A)** Assay to detect growth factor binding to tissue HSPGs. Tissue sections are incubated with growth factor and bound growth factor is detected with antibody directed against the growth factor protein or a biotin tag. This assay distinguishes the following classes of HSPGs: (1) HSPGs that do not bind FGF and are expected not to play a role in signaling; (2) HSPGs that bind FGF and may be positive or negative regulators of signaling. **(B)** Assay to distinguish positive and negative regulators of growth factor signaling. This assay allows the differentiation of the following classes of HSPGs: (1) HSPGs that do not bind FGF; (2a) HSPGs that bind FGF, but the HSPG/growth factor complex does not bind soluble receptor (These HSPGs are expected to sequester FGF and therefore not to promote signaling.); (2b) HSPGs that not only bind FGF, but assemble the complete signaling complex (These HSPGs are expected to promote signaling.). AP, alkaline phosphatase; $FR1c_{ECD}$, FGFR-1c extracellular domain.

the abundant expression of HS is tissues, this can be performed relatively easily using either frozen and paraffin-embedded tissue sections. Finally, the ability of the FGF/HS complexes to bind FGF RTK is tested by adding a soluble tagged receptor construct (*see* **Fig. 1B**). The binding pattern of this construct to the FGF provides a detailed picture of which HS populations are able to bind the FGF and promote its binding to RTK.

## 2. Materials

### 1.1. Preparation of Biotinylated FGF

1. Growth factors: FGF is commercially available from a number of companies. These are usually expressed in bacteria.
2. Heparin-agarose beads: commercially available from Sigma (St. Louis, MO).
3. PBS: 140 m*M* NaCl, 2.7 m*M* KCl, 10 m*M* $Na_2HPO_4$, 1.8 m*M* $KH_2PO_4$, pH 7.4.
4. Reaction buffer: 0.2 *M* sodium bicarbonate in double distilled water (pH 8.1).
5. Bead wash buffer: 20 m*M* HEPES, 200 m*M* NaCl, pH 7.4.
6. Elution buffer: PBS, 2.5 *M* NaCl (sodium chloride concentration may vary depending on affinity of FGF for heparin).
7. Biotinylation reagent: Sulfo-NHS-biotin (Pierce, Rockford, IL, cat. no. 21217ZZ) dissolved in double-distilled H2O as stock at 100 mg/mL (225 mM).

### 2.2. Purification of FR1c-AP–Binding Probe

1. COS-7 selection medium: DME culture medium (500 mL) containing 10% calf serum to which is added 10 mL of freshly thawed stocks of 4 m*M* L-glutamine, 5 mL 1% penicillin/streptomycin, and 500 μg/mL Geneticin (G418 sulfate).
2. COS-7 cell culture medium: Same as selection medium, without added Geneticin.
3. Filter: 0.45 μm filter in a Büchner funnel for filtration of conditioned culture medium.
4. Anti-AP affinity column: 1 mL of commercial preparation of monoclonal anti-AP conjugated to agarose beads (Sigma, cat. no. A-2080).
5. AP column elution buffer: 0.1 *M* glycine (pH 2.5).
6. Neutralization buffer: 1 *M* Tris (pH 8.0).
7. Human placental alkaline phosphatase: Dilute in PBS to use as a standard for comparison with purified FR1c-AP (Sigma, cat. no. P1391).
8. 2× AP assay substrate: 2 *M* diethanolamine, 1 m*M* $MgCl_2$ (dissolved at 10 m*M* concentration in water prior to addition), 20 m*M* L-homoarginine, 12 m*M* *p*-nitrophenylphosphate (Sigma 104; Sigma cat. no. 1040-1G).

### 2.3. Preparation of Tissue Sections and General Histology Material

1. Frozen tissue embedding medium, for example, Tissue Tek OCT compound (Sakura, Torrence, CA, cat. no. 4583 ).
2. Charged histology slides, for example, Fisher "Plus" slides (Fisher Scientific, Pittsburgh, PA, cat. no. 12-550-15).
3. Cover slips, e.g. Fisher "Finest" (Fisher Scientific, cat. no. 12-548-5P).
4. Slide-staining racks (optional), for example, Shandon Sequenza racks and cover plates (Shandon, Pittsburgh, PA).
5. Humidified chamber (as an alternative to staining racks).
6. Fixation of frozen sections: 70 vol% ethanol in double-distilled $H_2O$: 4% paraformaldehyde (e.g., EMS, Ft. Washington, PA, cat. no. 15710) in PBS.
7. Autofluorescence reduction reagents: 0.5mg/mL sodium borohydride, prepared in 4°C double-distilled $H_2O$; 0.1 *M* glycine, in PBS.
8. Blocking buffer: Bovine serum albumin (2%) in TBS, or 10% serum (e.g., normal swine serum, Dako, Carpinteria, CA, X0901)
9. Washing buffer TBS: 150 m*M* NaCl, 10 m*M* Tris-HCl, pH 7.4.
10. Fluorescently labeled secondary antibody: Alexa-conjugated secondary antibody (e.g., Alexa-546-goat-anti-mouse, Molecular Probes, Eugene, OR, cat. no. A-11003).

11. Detection system for brightfield microscopy, for example, Dako EnVision+ system, HRP (DAB), cat. no. K4006, or Dako LSAB+ peroxidase, Dako, cat. no. K0679.
12. Nuclear counterstain, optional (applied after staining or binding reaction), for example, Hoechst 33258 for fluorescence microscopy, Meyer's hematoxylin (e.g., Dako, cat. no. S3309) for bright-field microscopy.

### 2.4. Localization of Heparan Sulfate

1. Anti-heparan sulfate antibody: Mouse monoclonal antibody, clone 3G10, Seikagaku, cat. no. 370260 (Seikagaku-USA [now Cape Cod], Ijamsville, MA).
2. Heparitinase enzyme: Heparitinase (mixture of heparitinase I and II, e.g., Seikagaku, cat. no. 100703).
3. Heparitinase buffer: 50 m*M* HEPES, 50 m*M* NaOAc, 150 m*M* NaCl, 9 m*M* $CaCl_2$, 0.1% BSA, pH 6.5.

### 2.5. Binding of FGF and FR1c-AP to Tissue Sections

1. FGF: Biotinylated (*see* **Subheading 2.1.**) or native growth factor.
2. Anti-biotin antibody, for example, mouse monoclonal (Jackson, Cat. # 200-002-096, Jackson ImmunoResearch, West Grove, PA) or goat polyclonal (Vector, cat. no. SP-3000, Vector Laboratories, Burlingame, CA).
3. Anti-placental alkaline phosphatase antibody, for example, mouse monoclonal (Sigma, clone 8B6, cat. no. A2951) or rabbit polyclonal (Biomeda, cat. no. A67, Foster City, CA).
4. Heparitinase enzyme: Heparitinase (mixture of heparitinase I and II, e.g. Seikagaku, cat.no. 100703), and heparitinase II, ICN, cat. no. 190102, Costa Mesa, CA.
5. Heparitinase buffer: 50 m*M* HEPES, 50 m*M* NaOAc, 150 m*M* NaCl, 9 m*M* $CaCl_2$, 0.1% BSA, pH 6.5.

## Methods

### 3.1. Preparation of Biotinylated FGF

The *in situ* assays require growth factor ligand and soluble receptor-binding probes. Our experiments have been performed using human recombinant FGFs expressed in yeast or bacteria. We have used both biotinylated and native FGFs in binding experiments. Biotinylated FGFs offer the advantage that their biotin tag allows detection without potential background signaling from endogenous FGFs. However, using a chemically modified FGF introduces potential variability in labeling efficiency and growth factor inactivation during the biotinylation reaction. In our experience, high labeling efficiency leads to some loss of activity, which may be of concern. An alternative is to use native FGF and observe its localization using high-quality antibodies. The obvious caveat to this approach is that endogenous FGFs may potentially also be detected. In practice, this possibility has not been a problem, perhaps because tissue fixation destroys immunoreactivity of endogenous FGFs sufficiently to prevent their facile detection. In our routine work, sections consistently lack any endogenous signal if the incubation step with FGF ligand is omitted before the detection step. In some embryonic tissues, endogenous FGF can be detected, but its staining is often minor and can be easily accounted for. For some FGFs and other heparan sulfate-binding growth factors, high-quality antibodies allowing *in situ* localization may not be available. In this case, biotinylation of ligand may be necessary. Alternatively, the growth factor could be expressed with a short epitope tag peptide sequence, presumably without affecting its activity (*see* **Note 1**).

1. Lyophilized growth factor is dissolved in PBS at 1 mg/mL concentration, aliquoted into 50-μL aliquots, and snap-frozen in liquid nitrogen. It is important to store and handle the growth factor in polypropylene containers, as it binds avidly to glass.
2. One FGF aliquot (50 μg) is thawed and diluted to 0.5 mL in reaction buffer.
3. Heparin-agaose beads (e.g. 200 μL slurry equaling 100 μL bead volume) are prewashed 1 time with 1 mL elution buffer and 3 times with 1 mL bead wash buffer by centrifugation using 1.5-mL polypropylene centrifuge tubes. The beads are then equilibrated in 1 mL of reaction buffer for 10 min.
4. The prewashed beads are combined with the FGF solution and incubated at room temperature for 10 min, resuspending the beads several times during the binding reaction. The FGF binds quantitatively to the beads.
5. Sulfo-NHS-biotin stock solution is added to a final concentration of 20 mM. The beads are incubated in biotinylation reagent at room temperature for 5 min.
6. The reaction is terminated by washing the beads 6 times with bead washing buffer.
7. Biotinylated growth factor is batch-eluted in 200 μL of elution buffer containing NaCl at a concentration appropriate for the FGF used (e.g., 2.5 *M* for FGF-2, 1.0 *M* for FGF-7). The 2-min incubation is followed by centrifugation, and the process is repeated.
8. The concentration of the biotinylated FGF pooled from the two elution washes is determined by antibody binding (Westerns blots, ELISA) and comparison with native FGF standards. The activity of the FGF is compared with native FGF standards in mitogenesis assays.

### 3.2. Expression and Purification of RTK-Binding Probe

Soluble FGF RTK is expressed as fusion protein containing the extracellular (ligand-binding) portion of the RTK at the amino terminus and carboxy-terminal human placental alkaline phosphatase, which serves as a tag (e.g., FR-AP). Fusion constructs containing the extracellular portion of the FGFR1c and most other FGF receptors and their splice variants have been designed and generated by D. Ornitz (Washington University, St. Louis, MO) ***(21)***. FR1c-AP has been extensively characterized in binding studies in vitro and will be used as the example here. This soluble receptor fusion protein shows HS-dependent binding interactions similar to native full-length transmembrane receptor. The alkaline phosphatase tag offers a means for facile purification and the advantage that the protein is easily traceable with a simple enzyme activity assay. Because human placental alkaline phosphatase is a dimer, the fusion construct also forms a dimer under physiological conditions. Whether this dimer formation is critical for its binding activity is uncertain.

1. FR1c-AP is expressed in mammalian COS-7 cells. This cell line is easily transducible and stably transfected clones can be quickly generated. Using the CMV promoter in the expression vector (e.g., pCDNA3.1, Invitrogen, Carlsbad, CA) assures high-level expression. Expression in yeast or insect cells would also be possible, but bacterial expression would likely not produce satisfactory results because RTK binding requires glycosylation and disulfide bridges.

2. Transfected COS-7 cells are grown in selection medium in T175 flasks. Once confluent, the cells are suspended by trypsinization, and plated in 850-$cm^2$ roller bottles (1 flask per roller bottle) in 100 mL of culture medium. The bottles are gassed with 5% $CO_2$/air mixture, capped tightly, and placed in a 37°C roller bottle apparatus. Once the roller bottle is 50–60% confluent, the medium is replaced daily. The conditioned medium is saved, snap-frozen in liquid nitrogen, and stored at –70°C.
3. The anti-AP antibody affinity column is stored in the cold room in 0.02% $NaN_3$ as a preservative. Before using the column, wash with 5–10 bed volumes of PBS.
4. Conditioned medium from the transfected COS cells is filtered to remove cell debris and 250 mL are applied to a 1-mL anti-AP antibody affinity column. The conditioned medium typically contains 1–10 µg/mL FR1c-AP.
5. After washing the column with 5 column volumes of PBS, FR1c-AP is eluted from the column with 3.6 mL of AP elution buffer. Fractions (0.3 mL) of the eluate are collected directly into an equal volume of neutralization buffer. Immediate neutralization is important to preserve the activity of AP and probably also ligand-binding activity. Following elution, the eluate is dialyzed against PBS.
6. The concentration of purified FR1c-AP is determined is several ways. It is quantified in an AP assay using *p*-nitrophenylphosphate as a substrate and comparing the activity with human placental alkaline phosphatase as standard. Aliquots containing FR1-AP are mixed with an equal volume of 2× AP assay substrate and incubated at 37°C for 10, 20, and 30 min before reading absorbance at $A_{405}$. Second, its purity is assessed using SDS-PAGE gels stained with Coomassie blue and concentration is estimated using human PLAP as a standard. Finally, once the preparations are shown to be pure, protein is determined using a BCA protein assay (Pierce).

### 3.3. Preparation of Tissue Sections

The binding and localization procedure can be performed using either frozen or paraffin-embedded tissue sections. When fresh-frozen tissues are being examined, the tissue should be embedded in OCT or equivalent compound (*see* **Note 2**).

1. *(Frozen sections)* 4-µm sections are thaw-mounted on charged slides (e.g., Fisher Plus slides) and immediately dipped in 70% ethanol and then fixed in 4% paraformaldehyde in PBS at room temperature. The fixation procedure is crucial for reproducible FR1-AP binding of high intensity.
   *(Paraffin-embedded sections)* Formalin-fixed and paraffin-embedded tissues are also cut at 4 µm and mounted on charged slides. Paraffin sections are rehydrated according to standard histology technique by melting and exposure to xylene and a series of graded alcohols.
2. When bound protein probes are to be detected by fluorescence microscopy, autofluorescence is reduced by treating the tissues with sodium borohydride (0.5 mg/mL; prepared in 4°C double-distilled $H_2O$, and then warmed to room temperature). Following a 10 min incubation, the slides are washed in PBS and incubated with glycine (0.1 M) in PBS overnight at 4°C.
3. To reduce nonspecific binding, the sections are blocked before addition of antibody (for 3G10 staining) or before growth factor incubation with either blocking solution.
4. The staining and *in situ* binding reactions can be carried out in a humidified chamber. We have also found vertical slide staining racks (Sequenza System) to be very useful, especially when staining large numbers of slides. The cover plates used in the Sequenza racks create a capillary space above the tissue section, allowing for even and consistent antibody/growth factor exposure. The staining intensities are somewhat reduced when using this system, compared to slide incubation in the humid chamber.

### 3.4. Localization of Heparan Sulfate

The detection of heparan sulfate relies on cleavage of the glycosaminoglycan by a highly specific heparitinase. Preparations of this enzyme are commercially available, and different forms recognize different sulfation or sugar sites in the heparan sulfate chain. Because these enzymes are lyases rather than hydrolases, they generate a novel unsaturated glucuronate residue at the nonreducing terminus of the chain fragment that remains attached to the core protein. This unsaturated glucuronate is antigenic, and antibodies (e.g., mAb 3G10) for its detection are commercially available. Thus, detection of the heparan sulfate in the tissue requires only a single clip by the heparitinase somewhere in the length of the chain. Antibody 3G10 and the immunohistochemistry method have been developed by David and co-workers ***(20)***:

1. The tissue sections (frozen sections or rehydrated paraffin sections) are treated with heparitinase. We use generic heparitinase at 10 mIU/mL in heparitinase buffer for 4 h at 37°C, replenishing the enzyme after 2 h.
2. After blocking for 30 min, the sections are incubated with anti-delta-HS antibody 3G10 at 2 μg/mL for 60 min. After washing, the bound antibody is detected by a fluorescent secondary antibody (frozen sections) or an HRP-conjugated secondary antibody (paraffin sections).
3. An important control is to omit the heparitinase digestion step. Sections stained without this prior step lack the unsaturated glucuronate antigen, and antibody staining is typically seen to be absent.

### 3.5. Binding of FGF and FR1c-AP to Tissue Sections

1. After blocking, the tissue sections are incubated with FGF (e.g., FGF-2 at 3–10 nM) in blocking solution for 60 min.
2. Unbound FGF is removed by washing 5 times in TBS. After washing, continue to **step 3** to detect bound FGF by antibody staining, or to **step 5** to carry out FR1c-AP binding to the FGF/HS complexes.
3. After washing in TBS to remove unbound growth factor, the sections are incubated with anti-FGF-2 antibody, or with anti-biotin antibody for 60 min.
4. Bound FGF is detected with fluorescent (frozen) or HRP-conjugated secondary antibody or avidin (for biotinylated FGF).
5. Alternatively, rather than detect the bound FGF ligand with antibody, sections containing bound, exogenous FGF are incubated with FR1c-AP at 10–30 n*M* for 60 min at room temperature.
6. Unbound FR1c-AP is removed by 5 washes in TBS.
7. Bound FR1c-AP is detected with anti-placental alkaline phosphatase antibody. Antibody is detected as described below.

### 3.6. Important Binding Controls

The purpose of the binding assays is to detect (1) HSPGs capable of binding FGF ligand and (2) HSPGs capable of forming the complete FGF signaling complex. Therefore, the controls have to prove that the FGF-binding sites are HSPGs and that soluble receptor binding is FGF-dependent. The HS identity of the FGF-binding sites can be demonstrated by treating the tissue sections with heparitinase before the FGF binding step. Bacterial heparitinase has a high degree of substrate specificity and will not degrade other glycosaminoglycans such as chondroitin sulfate or hyaluronate. Best results are obtained by treating the sections for 4 h with a mixture of generic heparitinase and heparitinase II, both at 10 mIU/mL in heparitinase buffer, for 4 h at 37°C. The enzyme is replenished after 2 h. Using the combination of the two enzymes with different heparan sulfate-specific scission sites results in more efficient signal reduction. FGF dependence of soluble receptor binding is shown by omitting the FGF incubation step.

### 3.7. Detection Systems

For antibody detection by fluorescence microscopy, we recommend secondary antibodies conjugated with Alexa dyes by Molecular Probes (Eugene, OR). The Alexa fluorochromes are available for excitation and emission wavelength compatible with commonly used filters, deliver a signal of high intensity, and are very resistant to fading. For viewing with a regular bright-field microscope and for paraffin sections, the DAKO detection kits have been used by us with good success. The polymer-based Envision-Plus® or the biotin-avidin-based LSAB-Plus detection systems combine high sensitivity with low background.

## 4. Notes

1. The combined application of these *in situ* assays has allowed us to distinguish HSPGs that bind FGF-2 from those that bind FGF-7 ***(18)***, and to classify HSPGs as promoters and inhibitors of FGF-2 signaling complex formation ***(19)***. In human skin, an organ governed by signaling of different members of the FGF family, distinct differences are detected between keratinocyte surface HSPGs (likely members of the syndecan and glypican families of HSPGs) and epidermal basement membrane HSPGs (predominantly perlecan). Specifically, we find that keratinocyte surface HSPGs bind both FGF-2 and FGF-7, while epidermal basement membrane HSPGs bind FGF-2 but not FGF-7 (*see* **Fig. 2B,C**). Similarly, distinct differences between cell surface and basement membrane HSPGs are detected with regard to soluble FGF RTK binding. Epidermal surface HSPGs immobilize FR1c-AP, while epidermal basement membrane HSPGs failed to so (*see* **Fig. 2E**), despite the fact that HSPGs in this extracellular location bind ample amounts of FGF-2 (*see* **Fig. 2C**). These *in situ* techniques enable one to examine HSPG location and function in model systems where HSPGs have been found to be dynamically regulated and where FGFs play important roles, such as during development and in pathological conditions. The applicability of the assays to paraffin-embedded archival tissues opens the door to retrospective analysis of pathological specimens.
2. A few limitations of the *in situ* binding assays have to be emphasized. The assays do not allow the clear assignment of a certain HSPG species to FGF binding or signaling; they localize HS-binding activities and, when combined with immunohistochemical detection of HSPG core proteins, the HS may be shown to co-localize with a specific core protein. Also, it cannot be ruled out that endogenously produced HS-binding ligands may mask some of the binding sites on tissue HSPGs. In pilot experiments, we have rinsed tissue sections with 2 *M* NaCl before fixation to remove potential blocking ligands. No difference in binding patterns is detected, but this possibility cannot entirely be excluded. Despite these limitations, the ligand and soluble receptor binding assays provide important information on HSPG functions in vivo. These are tools to formulate hypotheses that can subsequently be vigorously tested with in vitro models.

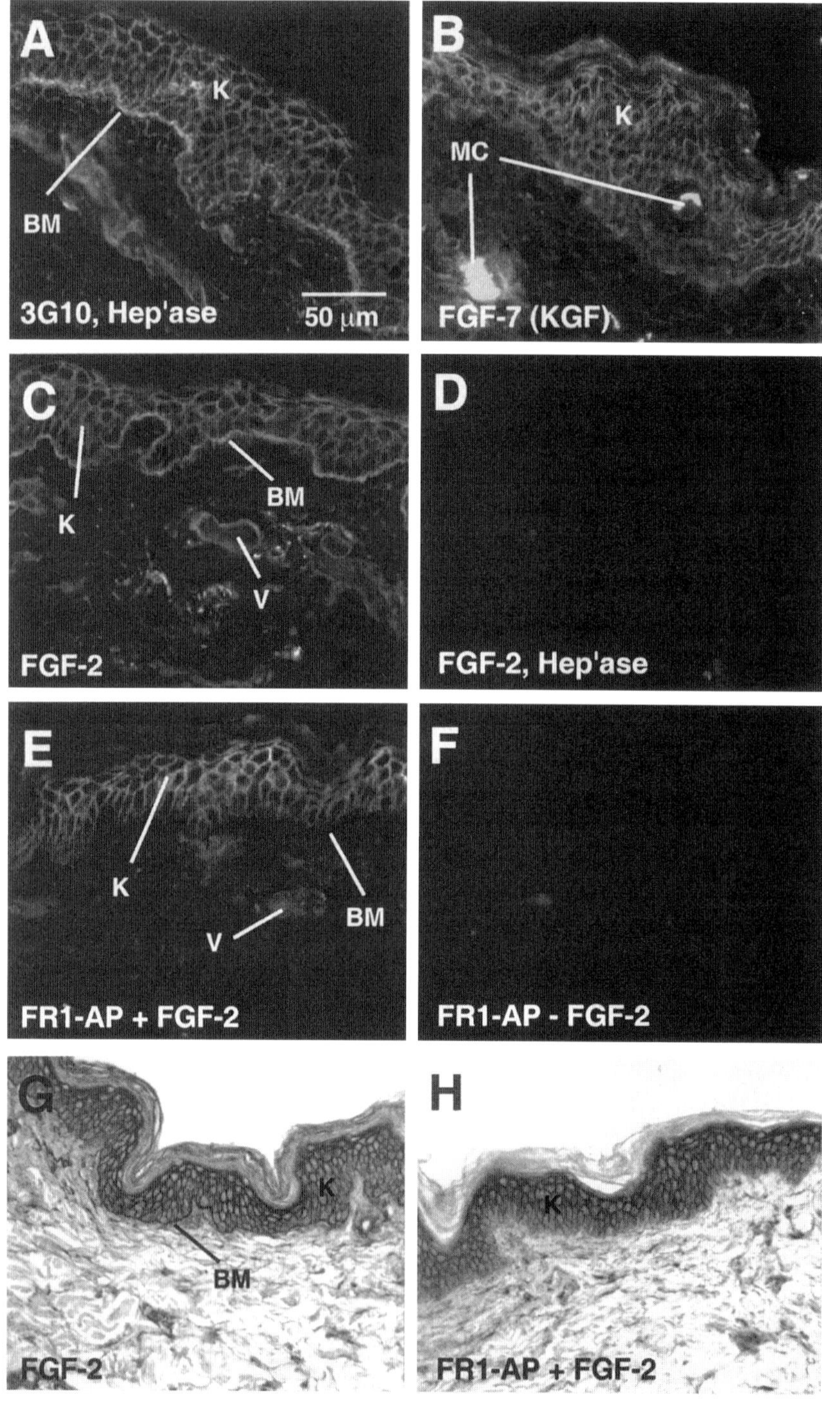
A
K
BM
3G10, Hep'ase
50 μm
B
K
MC
FGF-7 (KGF)
C
K
BM
V
FGF-2
D
FGF-2, Hep'ase
E
K
BM
V
FR1-AP + FGF-2
F
FR1-AP - FGF-2
G
K
BM
FGF-2
H
K
FR1-AP + FGF-2

## References

1. McKeehan, W. L., Wang, F.,and Kan, M. (1998) The heparan sulfate-fibroblast growth factor family: diversity of structure and function. *Prog. Nucleic Acid Res. Mol. Biol.* **59,** 135–176.
2. Rapraeger, A. C. (1995) In the clutches of proteoglycans: how does heparan sulfate regulate FGF binding? *Chem. & Biol.* **2,** 645–649.
3. Aviezer, D. and Yayon, A. (1994) Heparin-dependent binding and autophosphorylation of epidermal growth factor (EGF) receptor by heparin-binding EGF-like growth factor but not by FGF. *Proc. Natl. Acad. Sci. (USA)* **91,** 12,173–12,177.
4. Naka, D., Ishii, T., Shimomura, T., Hishida, T., and Hara, H. (1993) Heparin modulates the receptor-binding and mitogenic activity of hepatocyte growth factor on hepatocytes. *Exp. Cell Res.* **209,** 317–324.
5. Rapraeger, A. C., Krufka, A., and Olwin, B. B. (1991) Requirement of heparan sulfate for bFGF-mediated fibroblast growth and myoblast differentiation. *Science*, **252**: 1705–1708.
6. Yayon, A., Klagsbrun, M., Esko, J. D., Leder, P., and Ornitz, D. M. (1991) Cell surface, heparin-like molecules are required for binding of basic fibroblast growth factor to its high affinity receptor. *Cell*, **64**: 841–848.
7. Pantoliano, M. W., Horlick, R. A., Springer, B. A., Van Dyk, D. E., Tobery, T., Wetmore, D. R., Lear, J. D., Nahapetian, A. T., Bradley, J. D., and Sisk, W. P. (1994) Multivalent ligand-receptor binding interactions in the fibroblast growth factor system produce a cooperative growth factor and heparin mechanism for receptor dimerization. *Biochemistry* **33,** 10,229–10,248.
8. Plotnikov, A. N., Schlessinger, J., Hubbard, S. R., and Mohammadi, M. (1999) Structural basis for FGF receptor dimerization and activation. *Cell* **98,** 641–650.
9. Ornitz, D. M., Herr, A. B., Nilsson, M., Westman, J., Svahn, C. M., and Waksman, G. (1995) FGF binding and FGF receptor activation by synthetic heparan-derived di- and trisaccharides. *Science* **268,** 432–436.
10. Kan, M., Wang, F., Xu, J., Crabb, J. W., Hou, J., and McKeehan, W. L. (1993) An essential heparin-binding domain in the fibroblast growth factor receptor kinase. *Science* **259,** 1918–1921.
11. Johnson, D. E. and Williams, L. T. (1993) Structural and functional diversity in the FGF receptor multigene family. *Adv. in Cancer Res.* **60,** 1–41.

---

Fig. 2. *(opposite page)* Binding of FGF-2 and FR1c-AP soluble receptor alkaline phosphatase fusion protein to sections of human skin. Tissue HS is identified in frozen sections **(A–F)** or paraffin sections **(G, H)** of human skin. Signal is detected by immunofluorescence **(A–F)**, or immunoperoxidase **(G, H)**. **(A)** Detection of all HSPGs with antibody 3G10 after heparitinase digestion. **(B)** Binding of biotinylated FGF-7 (200 nM). **(C)** Binding of FGF-2 (10 nM). **(D)** Binding of FGF-2, but digestion of tissue section with heparitinase before growth factor incubation step. **(E)** Binding of soluble receptor FR1c-AP (30 nM) to tissue section preincubated with FGF-2 (10 nM). **(F)** Binding of soluble receptor FR1c-AP (30 nM) to tissue section omitting the FGF-2 incubation step. **(G)** Binding of FGF-2 (10 nM) to paraffin section of skin. **(H)** Binding of soluble receptor FR1c-AP (10 nM) to paraffin section of skin preincubated with FGF-2 (10 nM). Magnification: 400× **(A-F)**, see also scale bar; 200× **(G, H)**. Abbreviations: BM, basement membrane; K, keratinocytes; MC, mast cell; V, blood vessels). **(A,C,E)** reproduced with permission from Chang, Z., et al. (2000) *FASEB J* **14,** 137–144. **(B)** reproduced with permission from Friedl, A., et al. (1997) Am. J. Pathal. **150,** 1443–1455.

12. Guimond, S., Maccarana, M., Olwin, B. B., Lindahl, U., and Rapraeger, A. C. (1993) Activating and inhibitory heparin sequences for FGF-2 (basic FGF). Distinct requirements for FGF-1, FGF-2, and FGF-4. *J. Biol. Chem.* **268,** 23,906–23,914.
13. Reich-Slotky, R., Bonneh-Barkay, D., Shaoul, E., Bluma, B., Svahn, C. M., and Ron, D. (1994) Differential effect of cell-associated heparan sulfates on the binding of keratinocyte growth factor (KGF) and acidic fibroblast growth factor to the KGF receptor. *J. Biol. Chem.* **269,** 32,279–32,285.
14. Venkataraman, G., Shriver, Z., Raman, R., and Sasisekharan, R. (1999) Sequencing complex polysaccharides. *Science* **286,** 537–542.
15. Joseph, S. J., Ford, M. D., Barth, C., Portbury, S., Bartlett, P. F., Nurcombe, V., and Greferath, U. (1996) A proteoglycan that activates fibroblast growth factors during early neuronal development is a perlecan variant. *Development* **122,** 3443–3452.
16. Bonnehbarkay, D., Shlissel, M., Berman, B., Shaoul, E., Admon, A., Vlodavsky, I., Carey, D. J., Asundi, V. K., Reichslotky, R., and Ron, D. (1997) Identification of glypican as a dual modulator of the biological activity of fibroblast growth factors. *J. Biol. Chem.* **272,** 12,415–12,421.
17. Nurcombe, V., Ford, M. D., Wildschut, J. A., and Bartlett, P. F. (1993) Developmental regulation of neural response to FGF-1 and FGF-2 by heparan sulfate proteoglycan. *Science* **260,** 103–106.
18. Friedl, A., Chang, Z., Tierney, A., and Rapraeger, A. C. (1997) Differential binding of FGF-2 and FGF-7 to basement membrane heparan sulfate: Comparison of normal and abnormal human tissues. *Am. J. Path.* **150,** 1443–1445.
19. Chang, Z., Meyer, K., Rapraeger, A. C., and Friedl, A. (2000) Differential ability of heparan sulfate proteoglycans to assemble the fibroblast growth factor receptor complex in situ. *FASEB J.* **14,** 137–144.
20. David, G., Bai, X. M., Van der Schueren, B., Cassiman, J. J., and Van den Berghe, H. (1992) Developmental changes in heparan sulfate expression: in situ detection with mAbs. *J. Cell Biol.* **119,** 961–975.
21. Ornitz, D. M., Yayon, A., Flanagan, J. G., Svahn, C. M., Levi, E., and Leder, P. (1992) Heparin is required for cell-free binding of basic fibroblast growth factor to a soluble receptor and for mitogenesis in whole cells. *Mol. Cell Biol.* **12,** 240–247.

# Index

**A**

A431 squamous carcinoma cells, 427, 436
  fura-2-loaded, 436
Acid,
  acetic, 16
  anthranilic acid (2-aminobenzoic acid, 2AA), 131
  low-pH nitrous, 132
Acrylic monomer resin (LR white), 277
ADAMTS,
  family of metalloproteinases, 336
Adenosine 3'-phosphate 5'-phosphosulfate (PAPS), 92, 97, 105, 112
Adenosine-5'-phosphosulfate kinase, 325
Affinity blotting, 62,
  hyaluronan,
    biotinylated, 53–66
Affinity coelectrophoresis (ACE), 401–414
  analytical, 402
  casting apparatus, 406
  gel,
    electrophoretograms, 410
    running box, 407
  preparative gel, 405, 410
Affinity probe, 54
Aggrecan, 53, 149, 487
  cartilage,
    purification of, 55
  G1, 305
    -containing, 56
    -proteoglycan species, 58
  MMP-3, 58
  species,
    Western analysis, 341
Aglycones, 317
Alcian blue, 11, 16, 150, 159
  biological fluids, 159–174
  precipitation, 164
  preparation for electrophoresis, 163
  quantitation of proteoglycans, 159–174
  removal of, 164
  staining, 176, 177
  tetravalent cationic dye, 160
Alkaline borohydride,
  elimination, 30
Alkaline phosphatase-conjugated goat anti-mouse IgG, 55
Alkaline phosphatase-conjugated goat anti-mouse IgM, 55
Alkaline phosphatase-conjugated goat anti-rabbit IgG, 55
Alkylation, 72
Alpha-D-xylosides, 321
α–N-Acetylglucosaminidase, 137
Alzheimer's disease, 449
AMAC (Reductive amination with 2-aminoacridone), 119
Amino acid sequence analysis,
  internal, 67-77
  N-terminal, 6777
Amino sugars,
  α–linked, 350
  β-linked, 350
Anion-exchange,
  chromatography, 3, 5, 20, 38
  HPLC, 141–148
    analytical and preparative, 141–148
  strong (SAX) HPLC, 141
Anthranilate synthase (*trpE*), 221

From: *Methods in Molecular Biology, Vol. 171, Proteoglycan Protocols*
Edited by: R. V. Iozzo © Humana Press Inc., Totowa, NJ

Antisense,
expression, 251–260
constitutive, 251–260
inducible, 251–260, 256
mRNA, 251
Apolipoprotein (apo),458, 459
Artefact,
mass-transport, 515
Assays,
aggregation, 495
cationic nylon filter binding, 417
cell aggregation, 496
dot-blot reflectance, 167
immunoblot, 521
in vitro HA synthase, 374
luciferase, 256
modified carbazole, 24
particle exclusion, 374
tube assorbance, 162
ATP sulfurylase, 325
Autofluorescence,
calibration media, 444
determining, 444

**B**

β-elimination, 88
β-glucuronidase, 106
Bacterial clones,
antibodies expressed, 521
expressing heparan sulfate-binding antibodies, 521
immunoblot assay, 521
Basic fibroblast growth factor (FGF-2), 311, 415
Benzamindine, 10,
Beta-D-xylosides, 317
BIAcore, 506
Biglycan, 152
Biopanning, 520
BIOS-1, 506
BioTul, 506
Blot analysis,
Northern, 257
Southern, 248, 256
Western analysis of aggrecan species, 339–341
Bolton-Hunter,
procedure, 416
reagent, 406
Brevican, 36, 336

**C**

$Ca^{2+}$ ionophore, 440
Calibration,
autofluorescence media, 444
of fluorescence data, 445
Canine adult menisci, 153
CAPAGE (Composite agarose polyacrylamide gel electrophoresis), 54, 63,
Cartilage,
adult, 124
aggrecan,
purification of, 55
chick embryo epiphyseal, 105
fetal, 124
proteoglycan, 53–66
Catabolic fragments, 335–346
*C. elegans,* 41
CellFECTIN reagent, 43
Cell invasion,
maximum invasive death, 498
Cell proliferation,
control of, 427
Cellulose acetate strips, 175
Centrifugation, 37
zonal rate, 152
CHAPS, 8, 38, 43, 83, 343
Chase ABC digestion, 342
Chemiluminescence signal, 343
Chinese hamster ovary cells (CHO), 309
Chlorate,
inhibition of sulfation, 326
CHO (Chinese hamster ovary cells), 309
Cholesterol,
unesterified, 459
Chondroitin,
ABC lyase, 365
AC lyases, 365
lyase-catalyzed depolymerization, 368
of radiolabeled GAGs, 368
Chondroitinase, 330, 333
ABC, 10, 16, 69, 118, 152, 294, 365
ACII (*Arthrobacter aurescens),* 124
-resistant macromolecules, 190
Chondroitin/dermatan sulfate,
composition of, 117–128
Chondroitin-6-sulfate, 160, 169

Chondroitin sulfate,
degradation of, 363–372
in *Drosophila,* 44
Chondrosarcomas,
cell cultures, 342
swarm rat, 105
Chromatography,
affinity, 59, 235
metal chelating, 215
anion-exchange, 3, 5, 20, 38
anti-perlecan immunoaffinity, 29
DEAE, 29
-sepharose, 222
gel, 5, 6
hydrophobic interaction (HIC), 222, 225
micellar electrokinetic capillary, 193
microaffinity, 516
paper, 113
size-exclusion, 20
Chymotrypsin, 76
Cleavage,
endolytic, 353
nitrous acid, 347
N-sulfated glycosaminoglycans, 349
reduction of products with $NaBH_4$, 349
with pH 4 nitrous acid, 349
Cloning,
ring, 258
vector, 246
Cloth,
polyester, 311
CMV promoter, 252
Collagen,
rat tail type I, 496
Colloidal gold-conjugated secondary antibody, 285
Column,
anti-perlecan immunoaffinity, 28
CarboPac PA1,
polymer-based, 146
DEAE, 38, 41
ion-exchange, 76
ProPac PA1, 143, 146
polymer-based, 146
sepharose column, 28
Composite agarose polyacrylamide gel electrophoresis (CAPAGE),54, 63
Confocal-Frap (confocal fluorescence recovery after photobleaching), 487
analysis, 487–494
of carbohydrate-protein interactions, 487–494
Congo red, 449
Coomassie blue, 16
Cornea,
adult, 124
Coverslips,
gelatinized carbon-coated, 306
Crosslinker bis[sulfosuccinimidyl]-suberate ($BS^3$), 428
CsCI density gradient centrifugation, 55
Culture supernatant,
containing antibodies, 522
production of, 529
Curve fitting, 169
Cyanogen bromide (CNBr), 450
Cytochalasin D, 465
Cytochrome c., 271
Cytoplasmic expression, 234
Cytosolic $Ca^{2+}$, 435
proteoglycans, 435–448
single-cell, 435–448

**D**

DE52 resin, 14
DEAE
cellulose, 10, 12
chromatography, 29
column, 38, 41
buffers, 83
isolation of,
sulfaltated proteoglycans, 47
preparative microchromatography, 81
-sepharose
chromatography, 222
column, 28
Fast Flow, 43, 51
Decorin, 16, 152,
human,
purification of recombinant, 221–229
Δ-4,5-Glycuronage-2-sulfatase, 137
Δ–4,5-Glycuronidase, 137
Δ-4,5-Unsaturated uronate, 135
Densitometry, 169
Density gradient, 112
ultracentrifugation, 3

Dermatan sulfate (DS),
  degradation of, 363
  structural characterization, 187
Dialysis,
  tubing, 10
Diaminobenzidine, 276, 282
Digitonin permeabilization, 446
Disaccharides, 51
  composition of heparin/HS
  Δ–, 182
    analysis of, 183
    laser-induced fluorescence, 183
    sulfated, 189–190,
    ultraviolet detection, 183
  $^{3}$H-labeled, 143
  unsaturated, 142
    HS/heparin, 142
DNA,
  genomic isolation, 248
  transfection by electroporation, 247
Doxycycline, 253
*Drosophila,*
  chondroitin sulfate,
    microdetermination of, 44
  glycosaminoglycans, 43
    analysis of, 41–52
    extractions of, 44
  heparan sulfate,
    microdetermination of, 44
  larvae, 50
  *melanogaster,* 41
  proteoglycans, 43
    analysis of, 41–52
  pupae, 50
  purification of
    proteoglycans, 42
  S-2 cells, 47
Dulbecco's Minimal Essential Medium, 267
Dye,
  alcian blue,
    tetravalent cationic dye, 160
  Alexa, 543
  extrusion, 446
  intensiometric (single wavelength) dyes, 437
  loading, 444
  thiazine dye, 150

**E**

Ecdysone, 51
Edman,
  chemistry, 67
  degradation, 67
EDTA, 83
EGFR (epidermal growth factor receptor), 427
  receptor ectodomain, 430
EHS (Englebreth-Holm-Swarm), 105
Electron microscopy,
  transmission, 274
Electrophoresis,
  alcian blue, 163
  capillary (CE), 181, 193–198,
  capillary gel (CGE), 193,
    glycosaminoglycans, 194
    hyaluronan, 194
  capillary zone (CZE), 181–192, 181,
    glycosaminoglycans, 194
  cellulose acetate, 175–180
  composite gel, 149–158
  FACE gel, 120
  fluorophore-assisted carbohydrate, 117–128
  in HA Gel, 393
  in HCl, 178
  in zinc acetate, 178
  monodimensional, 175
  of glycosaminoglycans, 176
  zone, 175
Electroporation, 255
Electrospray ionization (ESI-MS), 366
Elimination mechanism, 364
ELISA, 521
*En bloc stain,* 281
Endobetagalactosidase, 336
Endogenous acceptor substrates, 108
Endolyase,
  ABC, 365
  AC-I, 365
  AC-II, 365
Endonuclease digestions, 257
Endothelial cells,
  human, 28
  purification of perlecan from, 27–34
Englebreth-Holm-Swarm (EHS), 105
Enhanced chemiluminescence, 258

Enzymes,
polymerizing, 105
preparations, 108
Epidermal growth factor receptor (EGFR), 427
anti-EGFR, 428
Ethanol precipitation of GAG, 85
Exoglycosidases, 129
Exolyase,
ABC, 365
Exosulphatases, 129
Extracellular matrices, 35
Extract concentration and desalting, 20
Extraction, 4

**F**

4 *M* Guanidine HCl, 4
Factor Xa cleavage site, 206
Fast-atom bombardment (FAB-MS), 366
FcR-Synd1 chimera, 462, 466
FGF (fibroblast growth factor),
-2, 32
-SAP, 314
binding to tissue sections, 542
biotinylated,
preparation of, 539
tissue-specific binding, 535–546
Fibroblast growth factor (FGF),
-2, 32
biotinylated,
preparation of, 539
-SAP, 314
tissue-specific binding, 535–546
Flow cytometry, 483
Fluoresceinamine, 406
Fluorescein isothiocyanate (FITC), 489
Footprinting,
and multimolecular complexes, 511
with competitor of the ligate, 511, 514
FR1c-AP,
-binding probe, 538
purification of, 538
binding to tissue sections, 542

**G**

G418, 254
Galactose (Gal), 103
Gel,
electrophoresis,
agarose/acrylamide, 151
composite, 149–151
of proteoglycans, 10, 15
filtration chromatography, 5, 6,
high-density polyacrylamide, 130
imaging, 121
Genapol X-100, 8
Gene knockout,
of proteoglycan core protein, 240
targeting knockout, 240
targeting strategies, 240
Genistein, 465
Glucosamine,
N-sulfated, 146
-6-sulfatase, 137
-3-sulphatase, 137
Glucuronate-2-sulfatase, 137
Glucuronic acid (GlcA), 103
Glucuronidase, 137
Glycosaminoglycan(s), 3
analysis of, 41–52
antibodies, 329–334
assay, 451
assembly, 320
autoradiography, 176
binding, 454
biosynthesis,
xyloside priming of, 317–324
chains,
immobilization of, 508
columns,
affigel, 451
charged, 450
-sepharose, 450
-containing peptides
-deficient mutants, 309–316
*Drosophila,* 43
extraction from, 44
from, 41–52
electrophoresis, 176
cellulose acetate, 175–180
inhibition, 319, 320
intact, 193–198
interactions with,
serum amyloid A peptide, 449–456
keratan sulfate,
biosynthesis of, 111

Glycosaminoglycan(s) (cont.),
  lyase (s), 329–334
    digestion, 118
  N-acetylated,
    hydrazinolysis of, 349
  N-sulfated, 349
  oligomeric, 193–198,
  priming, 319
  quantitation of, 9
  radiolabeled,
    chondroitin lyase-catalyzed depolymerization, 368
  semiquantitative analysis of, 175
  sulfated, 326
  sulfation, 326
  -synthesizing enzyme systems, 103
Glycosidases, 69, 137
Glycosidic linkages, 364
  cut by heparin lyases, 357
Glycosylphosphatidylinositol, 35
Glycosyltransferases, 112
Golgi,
  fractions, 110
  membranes, 105
Gradient,
  potassium dihydrogen phosphate, 143
  sodium chloride, 143
Granulocyte colony stimulating factor (G-CSF), 262
Guanidium thiocyanate protocols, 257

**H**

HA
  -binding proteins, 392
  biotinylated (bHA),
    preparation of, 384
  -containing sodium dodecyl slufate (SDS) gels, 392
  -substrate gels, 393, 394
HA-sepharose 4B, 63
HABPs (*see* Hyaluronan, binding proteins), 54
HeLa,
  207
Heparan sulfate, 176
  -binding antibodies, 521
  biosynthesis of, 79
  degradation of, 347-352,
    by nitrous acid, 347–352
    with heparin lyases, 353–362
Heparan sulfate *(cont.)*,
  endogenous, 535–546,
  in *Drosophila,* 44
  inhibition of, 325–328
    by chlorate, 325–328
  integral glycan sequencing, 129–139
  localization of, 542
  phage displaying, 520
  proteoglycans,
    binding, 416
    growth factor binding to, 415–426
    -mediated endocytosis, 465
      molecular methods to study, 465
  $^{35}$S-labeled, 79-90
Heparatinase, 69
Heparin, 176
  $^{125}$I-Heparin,
    desalting, 412
    molecular weight fractionation, 412
    storage of, 413
  desulfated, 107
  integral glycan sequencing, 129–139
  iodination, 411
    of tyramine-heparin, 412
  molecular-weight fractionation, 411
  tyramine labeling of, 411
Heparinase, 142
Heparin lyase(s), 353–362,
  I, 358, from *Flavobacterium heparinum,* 359
  II, 358
  III, 358
  -catalyzed depolymerization, 358
    of radiolabeled HS, 358
  cutting of glycosidic linkages, 357
  reaction conditions, 358
Heparitinase, 330
  I, 142
  II, 142
  III, 333
  digestion, 461
Herpes simplex virus, 310
Hexa-histidine sequence, 206
Histidine-6 (His6) tag, 291
Homogenization, 37
HPLC, 49, 74
  analysis, 44, 50
  strong anion-exchange, 141
    analytical and preparative, 141–148

HSP 70,
promoters for, 51
HT-1080$^{c}$, 207
Human intervertebral disk, 153
Hyaluronan, 176, 479–486,
binding,
to cell-surface receptors, 483
-binding domain, 149
-binding proteins (HABPs), 54, 479–486,
detection of, 481
biotinylated (bHA), 54, 480
affinity blotting, 53–66
binding protein, 273
preparation of, 480
preparation of, 63,
disaccharide composition of, 117–128
flow cytometry, 483
histochemistry of, 273
intracellular localization, 274
localization of, 480
specific in tissue sections, 481
synthase(s), (HAS), 373
in vitro assays for, 373–382
ultrastructure localization, 274
Hyaluronidase, 391–397, 479
activity, 383–390, 385, 394
inhibitor, 383–390, 391–397, 394
*Streptomyces,* 281, 336, 342
Hydrolysis, 364
Hygromycin, 241, 254
Hyperphase solution, 272

**I**

IAsys, 506
$^{125}$I-cytochrome c, 150
Iduronate-2-sulfatase, 137
Iduronidase, 137
Image, 341
Imidazole, 294
Immunoblotting, 149–158
Immunolabeling, 283
Inclusion body solubilization, 236
Indicators,
fluo-3 intensiometric, 437
fluorescent $Ca^{2+}$,
loading into cells, 438
-sensitive, 437
ratiometric (dual-wavelength), 437
fluorescence data, 441
Integral glycan sequencing (IGS), 130
of heparan sulfate, 129–139
of heparin saccharides, 129–138
Intensiometric (single wavelength) dyes, 437
Internal amino acid sequence analysis, 67–77
*In vitro,*
transcription, 208
translation, 208
Ion-exchange column, 76
Ionomycin, 440

**K**

Kaleidagraph program, 410
Karnovsky's fixative, 274
Keratanase, 69, 336
Keratan sulphate, 169
Kinetics,
second phase binding, 515

**L**

Labeling,
metabolic,
with [$^{3}$H]-acetate, 375
with [$^{3}$H]-glucosamine, 375
methionine, 299
sulfate, 299
Laser-induced fluorescence (LIF), 182
Ligand(s),
affinity-conversion approach, 91–101
binding competition with, 429
-binding inhibitors, 422
-binding sites
cell-free synthesis, 93
biotinylated, 507
clustering,
control of, 467
immobilized,
control of amount, 511
-proteoglycan binding, 420
Lipofectamine, 293
Lipoproteins,
biology, 458
high-density (HDL), 458
interactions with proteoglycans, 457–478
intermediate-density (IDL), 458

Lipoproteins *(cont.)*,
  lipase (LpL), 461
    isolation of bovine milk, 467
  low-density (LDL), 458
    receptor related protein (LRP)
    reductive methylation of, 467
  major plasma,
    physical characteristics, 459
  methods, 458
  structure, 459
  very low-density (VLDL), 458
LR white (acrylic monomer resin), 277
Lyase(s)
  chondroitin, 363–372,
    ABC lyase, 365
    AC lyases, 365
    -catalyzed depolymerization, 368
      of radiolabeled GAGs, 368
  digestion,
    fluorescent derivitization of, 120
    glycosaminoglycan, 118,
  glycosaminoglycan, 329–334
    digestion, 118
  heparin lyase(s), 353–362,
    I, 358
      from *Flavobacterium heparinum,* 359
    II, 358
    III, 358
    -catalyzed depolymerization, 358
      of radiolabeled HS, 358
    cutting of glycosidic linkages, 357
    reaction conditions, 358
Lysosomes,
  degradation of $^{125}$I-proteins, 466

**M**

MALDI mass spectrometry, 130
Mathematica software, 420
Matrix-assisted laser desorption/ionization (MALDI-MS), 366
Metallothionein,
  promoters for, 51
Methotrexate, 44
Microsomal fractions, 105
Molecular probes, 543
Molecular spreads, 271, 278
Monoclonal,
  Ab 2-B-6, 342
  anti proteoglycand Δ-di-OS
    -4S, 332
    -6S, 332
  Δ-deparan sulfate, 332
[$^{125}$]Monoiodotyrosine,
  culture media, 466
  purification of, 466
Mutants,
  glycosaminoglycan-deficient mutants, 309–316
  selection of, 312

**N**

NaCl,
  gradient elution, 45
N-acetyl-D-galactosamine (GalNAc), 347
N-acetyl-D-glucosamine (GlcNAc), 347
N-acetylgalactosamine (GalNAc), 103
  deacetylation of, 106
N-acetylglucosaminidase, 134
N-acetylglucosamine (GlcNAc), 103, 479,
  deacetylation of, 106
N-Deacetylase activity, 111
N-ethylmaleimide, 10
N-sulfated glucosamine ($GlcNSO_3$), 347
N-terminal
  amino acid sequence analysis, 67–77
  analysis,
    preparation for, 71
  sequencing of,
    proteoglycan core proteins, 337
N-unsubstituted glucosamine residue (GlcN), 347
Neo-epitope antiserium (anti-KEEE), 340
Neomycin
  -resistant clones, 255
    generation of stable, 255
Ni NTA agarose, 294
Ni resin,
  equilibration, 297
  loading, 297
Nitrous acid, 146
  cleavage, 347
  degradation, 142
    of heparan sulfate, 347–352
  pH 1.5 reagent, 348
  pH 4.0 reagent, 348

Nonlinear regression, 424
Northern blot analysis, 257
NP-40

**O**

Oligosaccharides,
immobilization, 508
N-linked, 205
Optical biosensor techniques, 505–518
Orcinol, 10
reagent, 14

**P**

Papain, 10,
PAPS (adenosine 3'-phosphate 5'-phosphosulfate), 92, 97, 105, 112, 325
Paraformaldehyde-lysine-periodate fixative, 302
PCR,
screening recombinants, 212
PEG, 23
Periplasmic,
expression, 235
fraction, 522
production of, 529
Perlecan, 27,
anti-
immunoaffinity chromatography, 29
immunoaffinity column, 28
detection of, 31
purification of 27–34
quantitation of, 31
Phage display,
antibodies, 520
technology, 519–534
obtain antiheparin sulfate antibodies, 519–534
Phenol extraction, 85
Phenylmethylsulfonyl fluoride, 10,
Phospholipid, 459
Phosphotyrosine,
anti-phosphotyrosine antibodies, 428
Photobleaching, 446
experiments, 491
Plaque amplification
6-well plates, 212
Plasmid pTRE-LUC, 256
Plastic removal, 284
Plating,
replica, 312
selection, 312
PLUS reagent (Life Technology), 293
pMAL-c2, 232
-vectors, 232
Polyanions, 198
Polymerization, 109
Polyvinylidene fluoride (PVDF), 154
Ponceau, S., 76
Positive-negative selection (PNS), 240
Post-staining, 283
Posttranslational modifications, 205
Poxvirus family, 203
Product quantitation, 121
Promotorless,
strategy, 240
targeting vectors, 241
Protamine sulfate (PS), 419, 423
Protease,
antipain, 297
aprotinin, 297, 377
benzamidine, 297
-free ABC (*Proteus vulgaris),* 124
inhibitors, 10, 42, 297
aprotinin, 374
leupeptin, 374
phenyl methyl sulfonyl fluoride (PMSF), 374
leupeptin, 297, 377
Protein(s),
core,
deglycosylated, 339
heat-shock, 292
-heparan sulfate complexes
quantitation of, 93
by solid-phase capture, 93
hybrid,
anthranilate synthase, 223
hylauronan-binding (HABPs), 54, 479–486,
detection of, 481
LDL receptor-related, 463
link, 53
maltose-binding (MBP), 231, 232
-polysaccharide interactions, 505–518
receptor-associated (RAP), 463
ribosome-inactivating, 311

Proteoglycans, 3
  alcian blue, 159–174
  analysis of, 41–52
  assembly, 319
  binding inhibitors, 423
  cartilage, 53–66
  cell-surfaced, 80
  core protein, 240, 335–346
    detection of, 329–334
    sequencing of, 337
  cytosolic $Ca^{2+}$,
    single-cell, 435–448
  engineered,
    characterization of, 281–300
    expression of, 281–300
    intracellular localization, 301–308
  expressed,
    immunofluorescence localization, 301
  from *Drosophila,* 41–52
  G1,
    -containing aggrecan, 56
    species, 58
  gel electrophoresis of, 10, 15
  gene,
    cell-mediated transfer, 261–270
    knockout, 240
    targeting, 239–250
  heparan sulfate,
    growth factor binding, 415–426
  histochemistry of, 272,
  immobilization of, 508
  immunocytochemistry, 275
  interaction, 427–434
    with lipoproteins, 457–478
    with purified receptor kinase, 431
  isolation of, 3–8, 9–18,
    -enriched fractions, 38
  ligand-proteoglycan binding, 420
  membrane-bound, 109
  morphological evaluation, 271–290
  nervous tissue,
    characterization of, 35–40
    isolation of, 35–40
  prokaryotic expression, 231–238
    purification,
      from *Drosophila,* 42
      from mineralized tissues, 19–26
  quanitation of, 159–174

Proteoglycans *(cont.)*,
  recombinant expression,
    in mammalian cells, 201–219
  small leucine-rich (SLRP) 427
  smith-degraded, 112
  somatic cells, 239–250
  sulfated,
    DEAE isolation, 47
Proteolysis, 335
Purification,
  procedures, 3–8,
Puromycin antibiotic, 241


**Q, R**


Q-sepharose column, 6
Radioactivity,
  TCA-precipitable, 465
Radiography,
  processing of, 298
Reagent,
  Bolton-Hunter, 406,
  CellFECTIN, 43
  hydrazine, 349
  nitrous acid,
    pH 1.5 reagent, 348, 349
    pH 4.0 reagent, 348, 349
  orcinol,14
  PLUS (Life Technology), 293
  sodium borohydride, 349
Receptor,
  -associated protein (RAP), 463
  chemical crosslinking, 429
  LDL-related protein (LRP), 463
  tyrosine kinase, 427–434
    in vitro phosphorylation, 430
Recombinant(s),
  amplification, 212
  proMMP-3, 55
  purification,
    human decorin, 221–229
  screening recombinants,
    by PCR, 212
Reduction, 72
Reductive amination,
  procedure, 125
  with 2-aminoacridone (AMAC), 119
Residues,
  andydromannose, 146
  critical for GAG binding, 454

Retroviral systems,
replication-defective, 261
transfection of target cells, 262
Reynolds lead citrate, 277
Ruthenium red (RR), 275

**S**

Saccharide
derivatization of, 131
with fluorophore 2AA, 131
fluorotagging, 119
Safranin O, 24
Scanning, 169
SDS PAGE, 54
gel electrophoresis of proteoglycans, 10, 15
preparative, 58
processing of, 298
Seeding,
peri-adventitial
on decellular carotids, 263
retrovirally transfected rat smooth muscle cells, 263
Sephadex G-50,
Serum amyloid A (apoSAA), 449
Signalling pathways,
receptor-activated, 427
SLRP (small leucine-rich proteoglycans), 427
Sodium sucrose octasulfate, 422
Solvent exchange, 4, 5
Southern blot analysis, 248, 256
Spectrometry,
mass, 68
Stable transfected colonies,
expansion, 247
recovery, 247
Stem cells,
embryonic, 239
Steric hindrance, 515
Sucrose-containing cacodylate buffer, 302
Sucrose octasulfate (SOS), 419
Sulfatases, 137
Sulfotransferase(s), 112
activity, 97
isoforms, 91–101
soluble, 110
Syndecans,
cell adhesions, 495–503
invasion, 495–503

**T**

T7
bacteriophage,
hybrid expression system, 201–219
promoter, 203
TEMED, 13, 23
Tendons,
adult flexor, 12
tensile swells, 16
Tet-off system, 252
Tet-on system, 252
Tetracycline repressor (TetR), 252
Tetracycline responsive element (TRE), 252
Texas Red-conjugated goat IgG anti-rabbit IgG, 304
Thiazine dye, 150
3-*O*-sultotransferase (3OST), 91
activity, 98
Tissues,
embryonic,
rat brain, 35
Toluidine blue, 150
method, 451
Transfection procedure, 296
Transilluminators, 131
Trifluoroacetic acid, 74
Triton,
assessment of, 468
cold insolubility, 468
X-100, 8, 38, 39, 109
solution, 320
Trypinization, 98
2-cyanoacetamide, 41
Tyramine, 406
Tyrosine kinase,
inhibitor, 465
receptor, 427-434

**U**

UDP
-[$^{14}$C]GlcA, 377
-GalNAc, 104
-GlcA, 103, 377
-GlcNAc, 104, 377
-Xyl, 103
Ultraviolet (UV),
detection, 198
spectroscopy, 364

UMR-106, 207

**V**

V8 protease, 76
Vaccinia virus,
  expression vector, 202
  hybrid expression system, 201–219
Vascular endothelial growth factor, 32
Vector,
  cloning, 246
  linearization, 246
  plasmid,
    construction of antisense, 254
  promotorless targeting, 241
VEGF (Vascular endothelial growth factor), 32
Versican, 53, 154
Virus,
  simplex,
    VP16 activation domain, 252
  stock,
    amplification of, 214
    titration, 213

**W, Z**

Western blotting, 62, 150, 156
  analysis of aggrecan species, 340, 341
Zero background surfaces,
  identification of, 509